World – Political

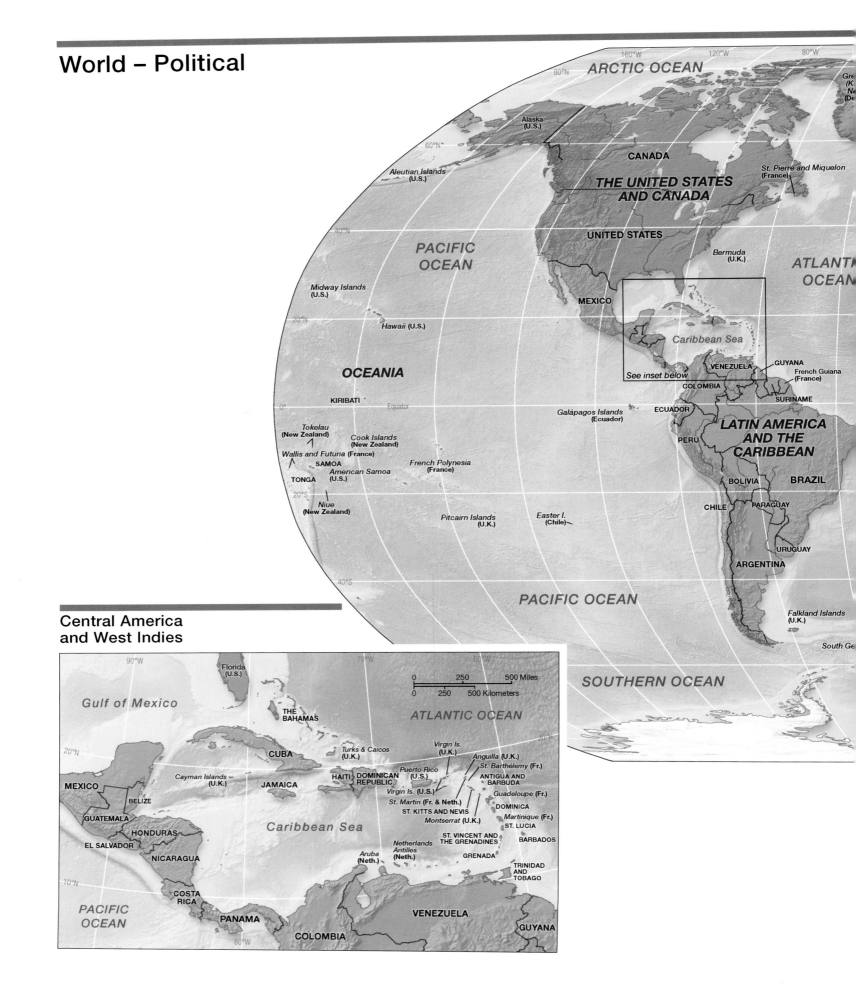

ARCTIC OCEAN

Alaska (U.S.)

Aleutian Islands (U.S.)

CANADA

THE UNITED STATES AND CANADA

St. Pierre and Miquelon (France)

UNITED STATES

PACIFIC OCEAN

Bermuda (U.K.)

ATLANTIC OCEAN

Midway Islands (U.S.)

MEXICO

Caribbean Sea

See inset below

VENEZUELA

GUYANA

French Guiana (France)

Hawaii (U.S.)

COLOMBIA

SURINAME

OCEANIA

KIRIBATI

Equator

Galápagos Islands (Ecuador)

ECUADOR

LATIN AMERICA AND THE CARIBBEAN

Tokelau (New Zealand)

Cook Islands (New Zealand)

PERU

Wallis and Futuna (France)

SAMOA

American Samoa (U.S.)

French Polynesia (France)

TONGA

BOLIVIA

BRAZIL

Niue (New Zealand)

CHILE

PARAGUAY

Pitcairn Islands (U.K.)

Easter I. (Chile)

URUGUAY

ARGENTINA

PACIFIC OCEAN

Falkland Islands (U.K.)

South Ge

SOUTHERN OCEAN

Central America and West Indies

Florida (U.S.)

Gulf of Mexico

THE BAHAMAS

ATLANTIC OCEAN

250   500 Miles

250   500 Kilometers

CUBA

Turks & Caicos (U.K.)

Virgin Is. (U.K.)

Anguilla (U.K.)

St. Barthélemy (Fr.)

MEXICO

Cayman Islands (U.K.)

JAMAICA

HAITI

DOMINICAN REPUBLIC

Puerto Rico (U.S.)

ANTIGUA AND BARBUDA

BELIZE

Virgin Is. (U.S.)

St. Martin (Fr. & Neth.)

Guadeloupe (Fr.)

GUATEMALA

ST. KITTS AND NEVIS

DOMINICA

HONDURAS

Montserrat (U.K.)

Martinique (Fr.)

ST. LUCIA

EL SALVADOR

Caribbean Sea

NICARAGUA

ST. VINCENT AND THE GRENADINES

BARBADOS

Aruba (Neth.)

Netherlands Antilles (Neth.)

GRENADA

COSTA RICA

TRINIDAD AND TOBAGO

PACIFIC OCEAN

PANAMA

VENEZUELA

GUYANA

COLOMBIA

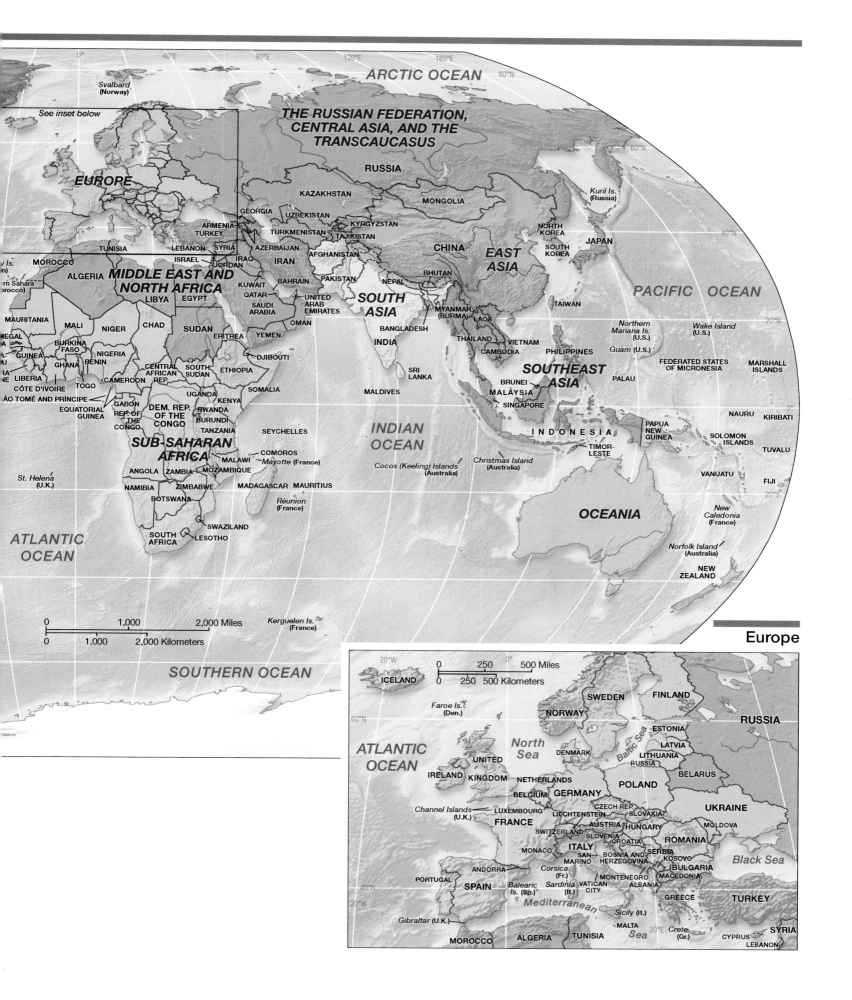

A Conceptual Exploration of World Regions & the Issues Critical to Geography Today

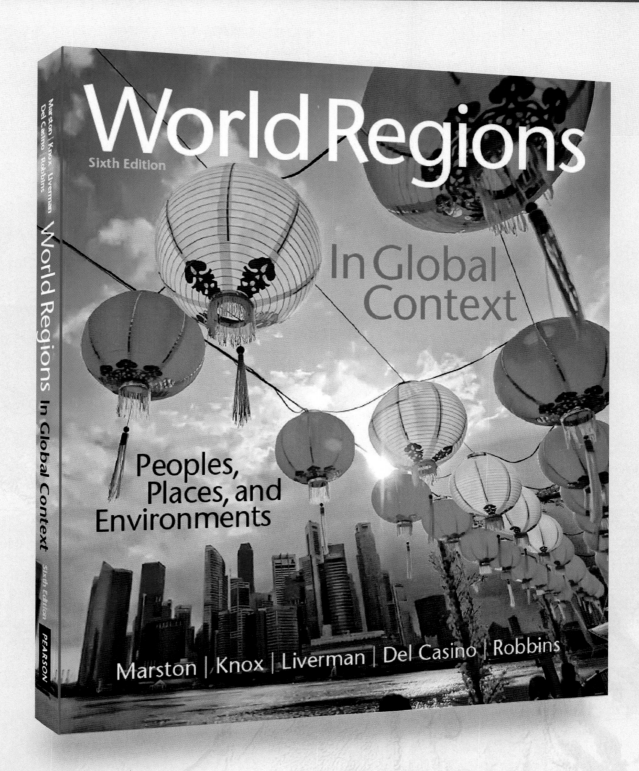

World Regions
Sixth Edition

In Global Context

Peoples, Places, and Environments

Marston | Knox | Liverman | Del Casino | Robbins

PEARSON

The Important Issues & People
Shaping World Regions

NEW! Sustainability in the Anthropocene features explore efforts to develop more sustainable lifestyles, cities, or food systems in each region by highlighting a specific project or place where people are implementing solutions that are socially, economically, and ecologically sustainable.

Water Scarcity and Quality

The future of water scarcity and water's availability and quality are serious issues in the Middle East and North Africa. As the region's population grows and the regional economy demands more water for the manufacturing sector as well as for the production of food, the scarcity of water resources will become more acute.

The Causes of Water Scarcity A number of factors contribute to the region's future map of water scarcity (FIGURE 4.1.1). In some places, economic conditions make water access difficult, while in other places, changes in global climate and overuse of freshwater resources have pushed many countries in the region to the brink. For example, excessive extraction of water from oasis wells has been occurring for such a long period that oases are dying. The countries of the Persian Gulf region have some of the highest per capita water use rates in the world, nearly double that of Europe. This fact further exacerbates the region's problem of water security.

The Future of Water Quality The expansion of the agricultural economy of the region appears, on the surface, as a positive economic indicator. But, chemical fertilizers for agriculture have a direct impact on the quality of the water available for drinking. Conflict in the region has also had a direct impact on water quality in the region, as fighting directly impacts water quality assurance systems as well as basic access to clean water (FIGURE 4.1.2).

> "As the region's population grows and the regional economy demands more water for the manufacturing sector as well as for the production of food, the scarcity of water resources will become more acute."

▲ FIGURE 4.1.1 **Map of Water Scarcity** The reasons for water scarcity are tied to the physical use and depletion of water resources, a result of climate change, and economic. What this suggests is that water scarcity is both an ecological and social issue.

1. What are some of the factors that are worsening water scarcity in the region?

2. What is degrading water quality in the region? What are the implications for the region's growing population if water quality continues to deteriorate?

▶ FIGURE 4.1.2 **Child in Syria Seeking Water** A Syrian boy, who fled with his family from the violence in their village in 2012, carries a plastic container as he walks to fill it with water at a refugee camp in the Syrian village of Atma near the Turkish border with Syria.

138

Refugees Flee the Violence of the Syrian Civil War

Imagine waking up one morning and remembering that you are no longer sleeping in your home country but are now a refugee in a foreign land. For the millions of people displaced by conflict in Syria a few have found their way to countries as distant and different as the United Kingdom. A woman from Syria has managed just this, first by making her way to the Mediterranean Coast and then across the sea to Europe. Another woman, named Nor, was fortunate to find herself in a refugee camp when she was given further refuge in the United Kingdom (FIGURE 4.4.1).

For many of those who remain in the refugee camps, however, life is difficult. The camps remain under-resourced, families are crowded together in temporary shelters, and weather conditions make everyday life difficult, particularly in the winter months. For many, particularly the approximately 1.6 million children who have been displaced by the war, the trauma of war can be felt each day in the camp. Many women, children, and men have been victims of torture, rape, or other sorts of violence. Others have seen their family members killed. The camps, while providing some relief from the violence of the war in Syria, do not end the trauma from the war (FIGURE 4.4.2). Many want to flee to Europe or other destinations where they hope their lives will be better.

> "Many women, children, and men have been victims of torture, rape, or other sorts of violence. Others have seen their family members killed."

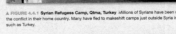

1. Where do we find many of Syria's refugees today?

2. What are the living conditions of those who remain in the refugee camps?

Syrian Refugees in U.K.
https://goo.gl/t.gys.kr

▲ FIGURE 4.4.1 **Syrian Refugees Camp, Qtma, Turkey** Millions of Syrians have been displaced by the conflict in their home country. Many have fled to makeshift camps just outside Syria in countries, such as Turkey.

▶ FIGURE 4.4.2 **Syrians Awaiting Asylum** Syrians occupied parts of France's Calais Port in 2013 in an attempt to pressure the French and British government to grant them the right of asylum.

169

NEW! Faces of the Region features explore the experience of different people within each world region, asking what is it like to be a young person in one place or what is it like to grow older in another place. It looks at the everyday, real-life experience of migration and asks how and in what ways changing demographics in each world region may be impacting how people come to know and understand their place in the world.

Current & Compelling Stories from the Regions

UPDATED! Geographies of Indulgence, Desire, and Addiction features link people in one world region to people throughout the world through a discussion of the local production and global consumption of regional commodities, helping students appreciate the links between producers and consumers around the world as well as between people and the natural world. New topics include luxury cars, beer and wine, and trekking.

▲ **C06** Marchers in 2014 in Washington, DC protest the choking death of Eric Garner by an NYC police officer.

▲ **FIGURE 1.32 An Election in India** India is the world's most populous democracy, and people turn out to vote in local, regional, and national elections.

2.1 Geographies of Indulgence, Desire, and Addiction

Beer and Wine

Today, Britain is famous for many varieties of ale and beer. But before the cold snap of the Little Ice Age, Britons also produced and enjoyed their own wines. Until the Late Medieval period, viticulture—the cultivation of grape vines for winemaking—was common throughout Europe, extending as far north as Britain and Scandinavia during the Medieval Warm Period that lasted from about 950 to 1250 c.e. Greek civilization had established viticulture by the 6th century b.c.e. Under the Roman Empire, viticulture spread west along the north shores of the Mediterranean and along the valleys of rivers in France and Spain and north to the North Sea and the Baltic. By the 1st century c.e., wine had become a commodity of indulgence, desire, and—for some—addiction throughout Europe. But sharply reduced average tempera-

TABLE 2.1.1 Alcohol Consumption in Europe			
Wine Consumption, 2013		Beer Consumption, 2013	
Country	Amount in 1,000 hL*	Country	Amount in 1,000 hL*
France	28, 81	Germany	85,588
Italy	21,795	United Kingdom	42,422
Germany	20,300	Poland	37,388
United Kingdom	12,738	Spain	35,169
Spain	9,100	France	19,421
Portugal	4,551	Italy	17,504
Netherlands	3,585	Romania	16,300
Greece	3,300	Czech Republic	15,278
Belgium and Luxembourg	3,054	Netherlands	11,690
Switzerland	2,650	Turkey	9,047

SOURCE: Wine Annual Report and Statistics Wine Annual Organisation Internationale de la Vigne du Vin, May 2014. *hL = hectoliters; 1 hectoliter = 26.4 gallons

SOURCE: Brewers of Europe, http://www.brewersofeurope.org/uploads/myorms-files/documents/publications/2014/statistics_2014_web_2.pdf *hL = hectoliters; 1 hectoliter = 26.4 gallons

> "The beer–wine division between northern and southern Europe has persisted since the Little Ice Age."

▼ FIGURE 2.1.1 The grape harvest in Burgundy, France.

tures during the Little Ice Age meant that viticulture retreated to Mediterranean Europe, leaving northerners to satisfy their need for alcohol with grain-based beverages, namely beer and spirits. It was a division that has characterized patterns of alcohol consumption ever since (TABLE 2.1.1). Incidentally, since North America was populated mostly by northern Europeans, beer and spirits, rather than wine, became characteristic of alcohol consumption in the United States and Canada.

Nevertheless, these divisions are by no means absolute. In the 16th century, Spanish and Portuguese overseas expansion saw the introduction of viticulture to the New World—to Mexico in the 1520s, Peru in the 1530s, Chile in the 1550s, and Florida in the 1560s. The British introduced viticulture to Virginia in the 1600s, and the first vineyards in California were established by Franciscan missions in the 1770s. Meanwhile, in Europe, demographic growth and increasing prosperity rapidly expanded the market for wine.

Today, the exclusivity of Europe's best wines is protected by strict systems of regulation. In France, for example, the Appellation Contrôlée system was introduced to guarantee the authenticity of wines, district by district. Such regulations have been important in reinforcing the appeal of wine as a commodity of indulgence and desire. Fine wines such as those from Burgundy, France (FIGURE 2.1.1) denote affluence and distinction. Meanwhile, recent climate change has meant that the frontier of viticulture in Europe has begun to shift northward. At the same time, the consumption of wine in northern Europe has increased significantly, especially among young, middle-class households. Brewers and distillers have responded to the competition by producing specialized cask ales and flavored spirits.

1. List the environmental factors that created the geographic division between beer and wine in Europe.

2. Compare and contrast other examples of indulgent commodities. In your analysis, do you find them to be sourced locally or imported from other places?

▼ **FIGURE 1.3.2 Youth Demonstrate as Part of the Yo Soy 132 March in Mexico City, Mexico** Popular protests by 131 Mexican youth was augmented by a larger social movement, Yo So 132 "I am 132," which developed as a protest movement against presidential candidate, Enrique Peña Nieto, and the press, which the protesters suggested were not covering the election fairly.

NEW! The latest stories & data from the regions. Updates include: the European response to the Syrian refugee crisis; the Syrian civil war and the rise of Islamic State of Iraq and Syria (ISIS); the recent Russian annexation of Crimea; natural disasters in Southeast Asia and the threat of rising sea-levels due to climate change in Oceania; and the growing connections between China and Africa.

A Changing World Geography

3.1 Emerging Regions

The Arctic

For hundreds of years, explorers and sailors have sought a route through the **Northwest Passage**, an ice-choked waterway spanning the Arctic Sea between the Atlantic and Pacific Oceans, north of the Canadian and Russian mainlands. Recently, global warming has accomplished what generations of explorers and investors have failed to do—it has opened up the Northwest Passage to easier shipping. This opening significantly shortens the shipping distance between Shanghai and New York or Tokyo and London, making a globalizing world all the more tightly connected. In 2014, the Chinese-owned cargo ship "Nunavik" became the first-ever to make the journey without an icebreaker escort, carrying 23,000 tons of cargo to the port of Bayuquan, China. This journey is 40% shorter than going by way of the Panama Canal.

Ice-Free Arctic Summers? As temperatures rise, the extent of ice in the Arctic is decreasing rapidly, especially in the summer (**FIGURE 3.1.1**). Arctic sea ice levels are at their lowest since records have been kept. Projections based on current trends suggest that the Arctic Ocean will be free of summer ice sometime between 2030 and 2080. Far sooner, the route across the Arctic will be reliably open to global commerce, and for the first time, the seafloor will be accessible to extensive resource development involving drilling for oil and natural gas.

Far-Reaching Effects These historic changes will have devastating impacts on the wildlife of the region. Polar bears will effectively be deprived of their natural habitat and ultimately be found only in zoos. And the opening of the ice means the creation of an entirely new world region—a contested prize for many world powers, a novel area for tourism, a critical source of resources, and a connected path between the worlds of the Atlantic and Pacific. The geopolitical contest for the control of this area has already begun, with Russia, Norway, Denmark, Canada, and the United States marking territory and making legal claims on the region (**FIGURE 3.1.2**). In August 2007, the Russian government sent two tiny submarines to plant the Russian flag on the Arctic seafloor (**FIGURE 3.1.3**). In 2015, Denmark, which controls Greenland, hopes to follow suit by making a formal claim to the United

Nations for control over a large portion of the Arctic seabed.

Greenland's Future Although physiographically considered part of North America, Greenland has its own indigenous populations (Kalaallisut-speaking Inuits) and wildlife (polar bear, musk ox, narwhal, and walrus) and is politically an overseas territory of Denmark. The emergence of a new geostrategic region around the North Pole, and the recently established semi-independent status of Greenland, reinforces this ambiguity. While Greenland is clearly a "victim" of global climate change, through the loss of its ice sheets, wildlife habitat, and indigenous human livelihoods, its position also allows it to assert claims on minerals and oil and gas reserves. This gives Greenland considerable influence and economic opportunity, alongside Canada, Russia, and the United States, though they remain under the nominal control of a small European power, Denmark.

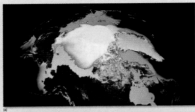
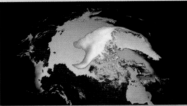

▲ FIGURE 3.1.1 The Melting Polar Ice Cap Shifting patterns of summer ice cover in the Arctic region between (a) 1979 and (b) 2014.

> "The geopolitical contest for the control of this area has already begun."

1. For whom does the transformation of the Arctic pose problems and for whom does it present opportunities?

2. What new connections does the opening of the Arctic create between other existing world regions?

Arctic Sea Ice News and Analysis

https://goo.gl/UtQfs9

▲ FIGURE 3.1.2 Claims on the New Arctic Frontier The national boundaries and the competing claims on this emerging region make this a dynamic place. Each of the five countries bordering the Arctic Ocean has claimed an Exclusive Economic Zone (EEZ), an area where they hold exclusive rights to drill, fish, or mine. Several claims, shown with hash marks, are claimed by nations but not recognized by the international community and depend on contested information about the shape and extension of the continental shelf.

▲ FIGURE 3.1.3 Conflicting Claims to the Arctic Seafloor (a) Russia planted its flag in 2007 on an undersea formation, called the Lomonosov Ridge, a gesture seen by other Arctic powers as a land grab. (b) In 2014, Denmark also claimed the ridge, arguing it is an extension of Greenland's continental shelf.

Future Geographies

The world is in transition, and the distribution of people, money, resources, opportunities, and crises is necessarily changing as a result. Each of the chapters in this book contains a "Future Geographies" section that speculates about the future of different regions, while "Emerging Regions" sections introduce new regions that are forming now.

Population Boom or Bust?

Many forces are lowering birthrates around the world, leading to leveling off or declining regional populations. In some places, populations may continue to rise, however. **FIGURE 1.49** shows projected population changes according to the UN Estimates up to and 2050.

Emerging Resource Regions

The expansion of the global economy and the globalization of industry will boost the overall demand for raw materials of every kind, and this will spur the development of some previously less exploited but resource-rich regions in Africa and Asia (**FIGURE 1.50**). The emergence of these new resource regions has enormous implications, as in Africa, where new geopolitical, cultural, and economic relationships are forming to rising powers like China and India.

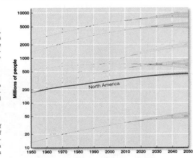

▲ FIGURE 1.49 Population Geography of the Future The range of projections is reflected in the widening bars further out in time.

▲ FIGURE 1.50 Global Resource Production of Important and Rare Elements As historically peripheral countries develop and use more of the world's resources, demands for scarce materials will increase. What are some of the applications of rare materials shown on the map (like Cobalt and Platinum)? What makes them important?

Dynamic Data Visualization & Critical Analysis

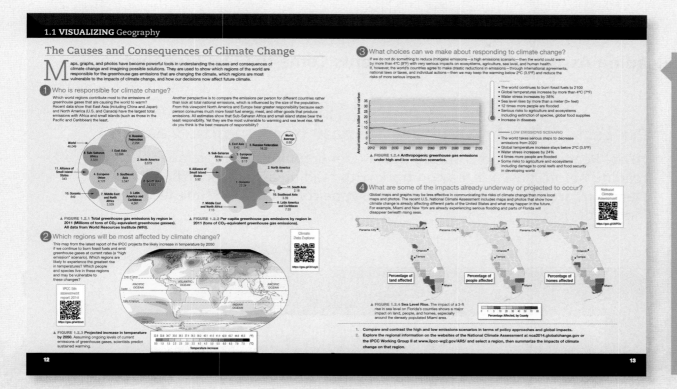

1.1 VISUALIZING Geography

The Causes and Consequences of Climate Change

Maps, graphs, and photos have become powerful tools in understanding the causes and consequences of climate change and imagining possible solutions. They are used to show which regions of the world are responsible for the greenhouse gas emissions that are changing the climate, which regions are most vulnerable to the impacts of climate change, and how our decisions now affect future climate.

1 Who is responsible for climate change?

Which world regions contribute most to the emissions of greenhouse gases that are causing the world to warm? Recent data show that East Asia (including China and Japan) and North America (U.S. and Canada) have the largest total emissions with Africa and small islands (such as those in the Pacific and Caribbean) the least.

Another perspective is to compare the emissions per person for different countries rather than look at total national emissions, which is influenced by the size of the population. From this viewpoint North America and Europe bear greater responsibility because each person consumes much more fossil fuel energy, meat, and other goods that produce emissions. All estimates show that Sub-Saharan Africa and small island states bear the least responsibility. Yet they are the most vulnerable to warming and sea level rise. What do you think is the best measure of responsibility?

▲ FIGURE 1.2.1 Total greenhouse gas emissions by region in 2011 (Millions of tons of CO₂-equivalent greenhouse gasses). All data from World Resources Institute (WRI).

▲ FIGURE 1.2.2 Per capita greenhouse gas emissions by region in 2011 (tons of CO₂-equivalent greenhouse gasses).

2 Which regions will be most affected by climate change?

This map from the latest report of the IPCC projects the likely increase in temperature by 2050 if we continue to burn fossil fuels and emit greenhouse gases at current rates (a "high emission" scenario). Which regions are likely to experience the greatest rise in temperatures? Which people and species live in these regions and may be vulnerable to these changes?

▲ FIGURE 1.2.3 Projected increase in temperature by 2050. Assuming ongoing levels of current emissions of greenhouse gases, scientists predict sustained warming.

3 What choices can we make about responding to climate change?

If we do not do something to reduce (mitigate) emissions—a high emissions scenario—then the world could warm by more than 4°C (9°F) with very serious impacts on ecosystems, agriculture, sea level, and human health. If, however, the world's countries agree to make drastic reductions in emissions—through international agreements, national laws or taxes, and individual actions—then we may keep the warming below 2°C (3.5°F) and reduce the risks of more serious impacts.

▲ FIGURE 1.2.4 Anthropogenic greenhouse gas emissions under high and low emission scenarios.

4 What are some of the impacts already underway or projected to occur?

Global maps and graphs may be less effective in communicating the risks of climate change than more local maps and photos. The recent U.S. National Climate Assessment includes maps and photos that show how climate change is already affecting different parts of the United States and what may happen in the future. For example, Miami and New York are already experiencing serious flooding and parts of Florida will disappear beneath rising seas.

▲ FIGURE 1.2.4 Sea Level Rise. The impact of a 3-ft rise in sea level on Florida's counties shows a major impact on land, people, and homes, especially around the densely populated Miami area.

1. Compare and contrast the high and low emissions scenarios in terms of policy approaches and global impacts.
2. Explore the regional information on the websites of the National Climate Assessment at nca2014.globalchange.gov or the IPCC Working Group II at www.ipcc-wg2.gov/AR5/ and select a region, then summarize the impacts of climate change on that region.

DATA Analysis

As discussed in this chapter, human activities in the environment directly influence climate change, which affects each region in different ways. Rising sea levels due to climate change directly impact Small Island Developing States (SIDS). Yet SIDS have contributed to less than 1% of the global emissions that cause temperatures and sea levels to rise. Tulun Atoll is one such Pacific island where the population is currently facing forced relocation. Take a deeper look at climate change and SIDS by first reading the 2014 *Guardian* article, "Island nations shouldn't be left to drown from climate change" at http://www.theguardian.com/commentisfree and respond to these following questions:

1. What are the factors causing sea levels to rise? In addition to sea-level rise, how will climate change affect Pacific islands and the surrounding ocean?

2. What is the "call to action" expressed by Tuilaepa Aiono Sailele Malielegaoi, the prime minister of Samoa to other world leaders?

3. How have New Zealand and Australia responded to climate migration of Pacific Islanders?

Going deeper, search Vimeo.com for "Chief Bernard Tunim in Copenhagen / COP15 (2009)" to see this address delivered by Chief Bernard Tunim, a leader and fisherman from Tulun Atoll at the UN Climate Change Conference in Copenhagen, 2009.

4. What does Chief Tunim say about what his people and island are experiencing with regard to food security, changing water levels, and the island's shoreline?

5. In Chief Tunim's view, what and who is destroying Tulun Atoll?

6. How does Chief Tunim rate the relative usefulness of the policies of international organizations in comparison with the everyday experience of his own people? Explain your answer.

7. What does Chief Tunim say about the government's relocation program for his people? Are his criticisms justified? Explain why or why not.

8. Do you agree with Chief Tunim that Europeans and North Americans should curb their consumption of fossil fuels to combat climate change? Explain your answer.

9. Reflecting on both the *Guardian* article and Chief Tunim's address, how much responsibility do you think the high-polluting nations have to the smaller nations who bear the greatest impacts? What percentage of this responsibility should be response, and what part should constitute preventative measures?

Sea-Level Rise

https://goo.gl/yADxc4

Continuous Learning
Before, During, and After Class

BEFORE CLASS

Mobile Media and Reading Assignments Ensure Students Come to Class Prepared.

Pre-Lecture Reading Quizzes are easy to customize & assign

NEW! Reading Questions ensure that students complete the assigned reading before class and stay on track with reading assignments. Reading Questions are 100% mobile ready and can be completed by students on mobile devices.

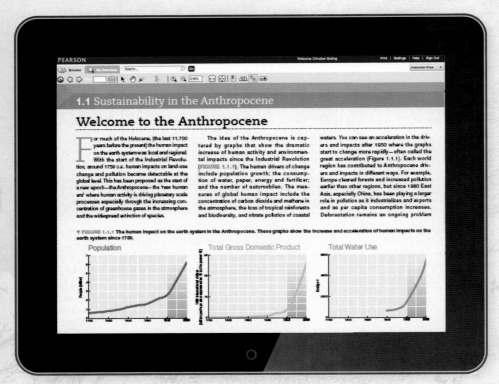

NEW! Dynamic Study Modules personalize each student's learning experience. Created to allow students to acquire knowledge on their own and be better prepared for class discussions and assessments, this mobile app is available for iOS and Android devices.

Pearson eText in MasteringGeography

gives students access to the text whenever and wherever they can access the internet. eText features include:

- Now available on smartphones and tablets.
- Seamlessly integrated videos and other rich media.
- Fully accessible (screen-reader ready).
- Configurable reading settings, including resizable type and night reading mode.
- Instructor and student note-taking, highlighting, bookmarking, and search.

with MasteringGeography

DURING CLASS

Learning Catalytics™ and Engaging Media

What has Professors and Students excited? Learning Catalytics, a 'bring your own device' student engagement, assessment, and classroom intelligence system, allows students to use their smartphone, tablet, or laptop to respond to questions in class. With Learning Catalytics, you can:

- Assess students in real-time using open ended question formats to uncover student misconceptions and adjust lecture accordingly.

- Automatically create groups for peer instruction based on student response patterns, to optimize discussion productivity.

"My students are so busy and engaged answering Learning Catalytics questions during lecture that they don't have time for Facebook."
Declan De Paor, *Old Dominion University*

Enrich Lecture with Dynamic Media

Teachers can incorporate dynamic media into lecture, such as Videos, MapMaster Interactive Maps and Geoscience Animations.

MasteringGeography™

MasteringGeography delivers engaging, dynamic learning opportunities—focusing on course objectives and responsive to each student's progress—that are proven to help students absorb world regional geography course material and understand challenging geography processes and concepts.

AFTER CLASS

Easy to Assign, Customizable, Media-Rich, and Automatically Graded Assignments

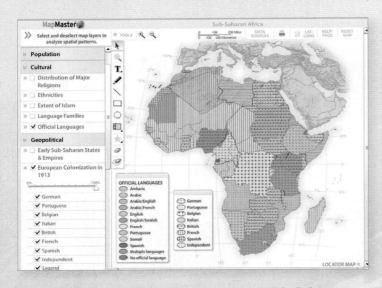

MapMaster Interactive Map Activities
are inspired by GIS, allowing students to layer various thematic maps to analyze spatial patterns and data at regional and global scales. This tool includes zoom and annotation functionality, with hundreds of map layers leveraging recent data from sources such as NOAA, NASA, USGS, United Nations, and the CIA.

NEW! Geography Videos from such sources as the BBC and *The Financial Times* are now included in addition to the videos from Television for the Environment's Life and Earth Report series in **MasteringGeography**. Approximately 200 video clips for over 25 hours of video are available to students and teachers and **MasteringGeography**.

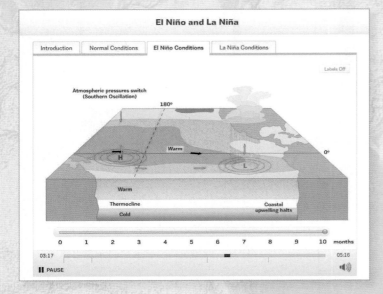

GeoScience Animations visualize complex physical geoscience concepts, and include audio narration.

www.MasteringGeography.com

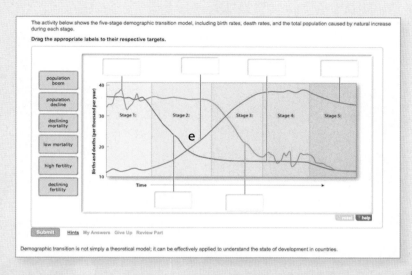

The activity below shows the five-stage demographic transition model, including birth rates, death rates, and the total population caused by natural increase during each stage.

Drag the appropriate labels to their respective targets.

population boom
population decline
declining mortality
low mortality
high fertility
declining fertility

Submit   Hints   My Answers   Give Up   Review Part

Demographic transition is not simply a theoretical model; it can be effectively applied to understand the state of development in countries.

NEW! GeoTutors. Highly visual coaching items with hints and specific wrong answer feedback help students master the toughest topics in geography.

UPDATED! Encounter (Google Earth) activities provide rich, interactive explorations of world regional geography concepts, allowing students to visualize spatial data and tour distant places on the virtual globe.

Data SIO, NOAA, U.S. Navy, NGA, GEBCO
Image Landsat
Image IBCAO

Google earth

Map Projections
Map Projection Properties: Spatial Relationships and Characteristics

Introduction | Earth's Graticule | Map Projection Properties | Map Projection Classes | Using Map Projections

Spatial Relationships and Characteristics | Distortion on Projections | Equal Area Projections | Conformal Projections | Compromise Projections   SHOW TEXT

Equal Area (Equivalent)
Preserves area.
Does not preserve shape or angles.

Conformal
Preserves shape and angles.
Does not preserve area.

00:58   01:11

REPLAY   PREVIOUS   PLAY   NEXT

Map Projections interactive tutorial media helps reinforce and remediate students on the basic yet challenging Chapter 1 map projection concepts.

World Regions

Sixth Edition

In Global Context

Peoples, Places, and Environments

Sallie A. Marston
University of Arizona

Paul L. Knox
Virginia Tech

Diana M. Liverman
University of Arizona

Vincent J. Del Casino, Jr.
University of Arizona

Paul F. Robbins
University of Wisconsin, Madison

PEARSON

Senior Geography Editor: Christian Botting
Project Manager: Sean Hale
Program Manager: Anton Yakovlev
Development Editor: Jonathan Cheney
Media Producer: Ziki Dekel
Editorial Assistant: Michelle Koski
Director of Development: Jennifer Hart
Program Management Team Lead: Kristen Flathman
Project Management Team Lead: David Zielonka
Production Management: Lindsay Bethoney, Lumina Datamatics, Inc.
Copyeditor, Compositor: Lumina Datamatics, Inc.

Design Manager: Mark Ong
Interior and Cover Designer: Richard Leeds
Rights & Permissions Project Manager, Management: Rachel Youdelman
Photo Researcher: Eric Schrader
Manufacturing Buyer: Maura Zaldivar-Garcia
Executive Product Marketing Manager: Neena Bali
Senior Field Marketing Manager: Mary Salzman
Marketing Assistant: Ami Sampat
Cover Photo Credit: WSBoon Images/Getty Images

Library of Congress Cataloging-in-Publication Data.
Names: Marston, Sallie A., author.
Title: World regions in global context : peoples, places, and environments /
    Sallie A. Marston, Paul L. Knox, Diana M. Liverman, Vincent J. Del Casino,
    Paul F. Robbins.
Description: Sixth Edition. | Boston : Pearson, [2019] | Earlier editions
    cataloged under title. | Includes bibliographical references and index.
Identifiers: LCCN 2015035836| ISBN 9780134183640 (alk. paper) | ISBN
    0134183649 (alk. paper)
Subjects: LCSH: Geography—Textbooks. | Globalization—Textbooks.
Classification: LCC G116 .W675 2019 | DDC 910--dc23
LC record available at http://lccn.loc.gov/2015035836

About Our Sustainability Initiatives

Pearson recognizes the environmental challenges facing this planet, as well as acknowledges our responsibility in making a difference. This book is carefully crafted to minimize environmental impact. The binding, cover, and paper come from facilities that minimize waste, energy consumption, and the use of harmful chemicals. Pearson closes the loop by recycling every out-of-date text returned to our warehouse.

Along with developing and exploring digital solutions to our market's needs, Pearson has a strong commitment to achieving carbon-neutrality. As of 2009, Pearson became the first carbon- and climate-neutral publishing company, having reduced our absolute carbon footprint by 22% since then. Pearson has protected over 1,000 hectares of land in Columbia, Costa Rica, the United States, the UK, and Canada. In 2015, Pearson formally adopted The Global Goals for Sustainable Development, sponsoring an event at the United Nations General Assembly and other ongoing initiatives. Pearson sources 100% of the electricity we use from green power and invests in renewable energy resources in multiple cities where we have operations, helping make them more sustainable and limiting our environmental impact for local communities.

The future holds great promise for reducing our impact on Earth's environment, and Pearson is proud to be leading the way. We strive to publish the best books with the most up-to-date and accurate content, and to do so in ways that minimize our impact on Earth. To learn more about our initiatives, please visit **https://www.pearson.com/social-impact/sustainability/environment.html**

www.pearsonhighered.com

ISBN 10: 0-134-18364-9; ISBN 13: 978-0-134-18364-0 (Student edition)
ISBN 10: 0-134-26301-4; ISBN 13: 978-0-13426301-4 (Instructor's Review Copy)

Brief Contents

Contents

1 World Regions in Global Context 2

2 Europe 48

3 The Russian Federation, Central Asia, and the Transcaucasus 90

4 Middle East and North Africa 130

5 Sub-Saharan Africa 176

6 The United States and Canada 222

7 Latin America and the Caribbean 258

10 Southeast Asia 382

11 Oceania 426

Preface

We live in a world of global interconnection and dynamic change. This means that if we want to understand the human condition or the changing environment, we have to look at both our local community and the wider world. We have to challenge our assumptions about what we think we know. We have to work together. *World Regions in Global Context* provides a framework for understanding the global connections that affect relationships within world regions, while also recognizing that the events that take place locally can have an impact on a global scale. Of course, no textbook can provide the answers to all the complex questions about the forces that fuel these global connections and local changes. That's why we have classes, students, teachers, travel, and other ways of understanding the world! But *World Regions in Global Context* can shed some light on the dynamic and complex relationships between people and the world they inhabit. This book gives students the basic geographical tools and concepts they need to understand the complexity of today's global geography and the world regions that make up that geography.

New to the 6th Edition

The 6th edition of *World Regions in Global Context* has been thoroughly revised by the authors and editorial team based on reviews from teachers and scholars in the field. Every line and graphic in the book has been reviewed and edited for maximum clarity and effectiveness. The text has been significantly edited to provide additional space for infographics, data-driven maps, and images. The new edition includes significant changes as well as a number of new features that make the revised text more accessible and engaging.

- Global change, especially climate change, is becoming an increasingly pressing issue as is responses to that change. The 6th edition takes up this concern by more overtly incorporating a discussion of environmental change in each chapter through the reorganized subsection titled **Environment, Society, and Sustainability**. The increasing emphasis on sustainable solutions to climate change and other environmental challenges is marked by the addition of other features in the text as well, including a new box feature.

- **Sustainability in the Anthropocene** This feature provides an example of efforts to develop more sustainable lifestyles, cities, or food systems in this era of the "Anthropocene"—a newly proposed geologic era of human influences. In each region we have highlighted a specific project or place where people are implementing solutions that are socially, economically, and ecologically sustainable.

- **Faces of the Region** explores the experience of different people within each world region. It takes up the challenge of asking what is it like to be a young person in one place, or what is it like to grow older in another place, or what it's like to grow up in a place that is experiencing dramatic change. It looks at the everyday, real-life experiences of migration and generational change and asks how and in what ways changing demographics in each world region may be impacting how people come to know and understand their place in the world.

- Geography is strongly invested in the use of maps and other visual data. The **Visualizing Geography** feature has been updated with a new emphasis on infographics and maps that encourage data and visual analyses. It builds on and extends that tradition with extensive use of visualizations and maps to focus on issues such as global sea-level rise, the consequences of conflict in the Middle East and North Africa, and the migration of Muslim populations into Europe.

- Every chapter review includes a new **Data Analysis** feature in which students apply chapter concepts and answer critical thinking questions based on data accessed via Quick Response (QR) links to Web sites of governments, nongovernmental organizations, and other important sources of data related to regional, economic, social, and political developments.

- Recognizing the importance of population dynamics as a factor in many regional challenges, the Culture and Populations section of each chapter contains a section, **Demographic Change**, with updated population statistics and trends as well as new population pyramids helping students to visualize the societal impacts of population change.

- The **maps, images, graphs, and tables** that make up the text's visual program have been revised. Readers will notice that many maps now include images that highlight key features. The photo program for this edition has also been substantially revised with newer and different photos. We have added questions that prompt students to look more carefully at some of the graphics and images.

- We have **updated the histories, stories, and current events** in each chapter. As readers know, the world has changed a lot since the previous edition of this book. To respond to these changes, we have included stories on the European response to the Syrian refugee crisis; the Syrian civil war and the rise of Islamic State of Iraq and Syria (ISIS); the recent Russian annexation of Crimea; natural disasters in Southeast Asia and the threat of rising sea-levels due to climate change in Oceania; and the growing connections between China and Africa, for example. New and updated information has been added to all the special feature material as well, including all the new **Geographies of Indulgence, Desire, and Addiction** features on luxury cars, beer and wine, and trekking.

- **Chapter 1** now includes a new section on how one can begin "Thinking Like a Geographer."

Objectives and Approach

World Regions in Global Context has two primary objectives. The first is to provide a body of knowledge about world regions and their distinctive political and economic practices, cultural and environmental landscapes, and sociocultural attributes. The second is to emphasize that although there is diversity among world regions, all world regions are connected through new and changing relationships. This approach informs the book's thematic structure, which is organized to engage readers in a discussion of environmental, social, historical, economic, and territorial change as well as cultural practices and demographic shifts.

Thematic Structure

This book is built on an opening chapter that describes how one thinks like a geographer. The 10 regional chapters follow, explore, and elaborate on the concepts laid out in **Chapter 1**. In each chapter, we balance discussions of global interconnections with local realities. To do this systematically, we divide each regional chapter into four major categories, each highlighting a set of themes that are central to understanding world regions.

Environment, Society, and Sustainability

We begin each chapter with a discussion of the physical and environmental context of the region; this includes a discussion of climate and climate change; geological resources, risks, and water; ecology, land, and environmental management; and sustainability. Our aim is to demonstrate how environment is shaped by, and shapes, the region's inhabitants over time.

History, Economy, and Territory

This section focuses on the historical geographic context for each world region and illustrates how the economies and territories that make up each world region have evolved over time. Included are discussions of historical landscapes and legacies; economy, accumulation, and the production of inequality; and territory and politics.

Culture and Populations

This section explores the cultures and populations of each world region. This section emphasizes the relationships between population change and settlement patterns, while exploring the importance of urbanization in each region. This section is broken down into three subsections focusing on culture, religion, and language; cultural practices, social differences, and identity; and demography and urbanization.

Future Geographies

In keeping with the theme of this textbook, which emphasizes ongoing change, each chapter concludes with a brief discussion of some of the key issues facing each world region, projecting how they are likely to develop in the coming years and decades.

Pedagogy and Content Enrichment

The book includes a number of important pedagogical devices to help readers understand the complex processes that connect our world and make it different.

Learning Outcomes and Learning Outcomes Revisited

On the opening pages of each regional chapter, we provide a list of *Learning Outcomes*. This list directs students to the key take-away points in the chapter. They are intentionally broad, drawing from a number of different discussions throughout each chapter. At the end of the chapter, we return to these learning outcomes and offer brief comments on them. The *Learning Outcomes Revisited* section helps readers grapple with some of the larger conceptual material and focuses student review and also includes key questions.

Apply Your Knowledge

Apply Your Knowledge questions ask readers to synthesize the information in the text and respond to applied questions that link back to the chapter's broad learning outcomes. Readers will find six to eight of these question in each chapter. Many have been updated with QR links to Web sites where students can access current data that deepens their understanding of regional issues.

Special Content Features

New and updated box features provide students with an opportunity for in-depth exploration of key chapter content. In addition to a new emphasis on data analysis, the 6th edition's box features include critical thinking questions to encourage students to self-assess and reflect on what they have learned.

■ **Visualizing Geography** In each chapter, we use cutting-edge cartography and data visualization techniques to introduce readers to a current geographic issue. Visual data provide a powerful way to convey information and analyze geographic processes in action, encouraging students to ask, "What types of geographic data can I use to answer the pressing questions of the day?"

■ **Emerging Regions** This feature emphasizes global and local change and underscores the importance that these new regions have now and may have in the future. Readers are encouraged to explore *Emerging Regions* with an eye toward asking how world regional geography changes over time and how it might look different in the future.

■ **Faces of the Region** This section explores the experience of different people within each world region. It takes up the challenge of asking what is it like to be a young person in one place or what is it like to grow older in another place. It looks at the everyday, real-life experience of migration and asks how and in what ways changing demographics in each world region may be impacting how people come to know and understand their place in the world.

■ **Geographies of Indulgence, Desire, and Addiction** This feature links people in one world region to people throughout the world through a discussion of the local production and global consumption of regional commodities, helping students appreciate the links between producers and consumers around the world as well as between people and the natural world.

■ **Sustainability in the Anthropocene** This feature provides an example of efforts to develop more sustainable lifestyles, cities, or food systems in each region by highlighting a specific project or place where people are implementing solutions that are socially, economically, and ecologically sustainable.

MasteringGeography™

MasteringGeography™ now features an expansive library of BBC video clips, a new next generation of Geographic Information System (GIS)–inspired MapMaster interactive maps, Dynamic Study Modules for World Regional Geography, a responsive-design eText 2.0 version of the book, and more.

Conclusion

This book is the product of conversations among the authors, colleagues, students, and the editorial team about how best to teach a course on world regional geography. In preparing the text, we have tried to help students make sense of the world by connecting conceptual materials to the most compelling current events. We have also been careful to represent the best ideas the discipline of geography has to offer by mixing cutting-edge and innovative theories and concepts with more classical and proven approaches and tools. Our aim has been to show how important a geographic approach is for understanding the world and its constituent places and regions.

Acknowledgments

We are indebted to many people for their assistance, advice, and constructive criticism in the course of preparing this book. Among those who provided comments on various drafts and editions are the following professors:

Donald Albert, *Sam Houston State University;* Martin Balinsky, *Tallahassee Community College;* Brad Baltensperger, *Michigan Technological University;* Karen Barton, *University of North Colorado;* Max Beavers, *University of Northern Colorado;* Richard Benfield, *Central Connecticut State University;* William H. Berentsen, *University of Connecticut;* Keshav Bhattarai, *Central Missouri State University;* Warren R. Bland, *California State University, Northridge;* Brian W. Blouet, *College of William and Mary;* Sarah Blue, *Northern Illinois University;* Pablo Bose, *University of Vermont;* Jean Ann Bowman, *Texas A&M University;* John Christopher Brown, *University of Kansas;* Stanley D. Brunn, *University of Kentucky;* Joe Bryan, *University of Colorado: Boulder;* Michelle Calvarese, *California State University, Fresno;* Craig Campbell, *Youngstown State University;* Xuwei Chen, *Northern Illinois University;* Jessie Clark, *University of Oregon;* David B. Cole, *University of Northern Colorado;* Joseph Corbin, *Southern New Hampshire University;* Jose A. da Cruz, *Ozarks Technical Community College;* Tina Delahunty, *Texas Tech University;*

Cary W. de Wit, *University of Alaska, Fairbanks;* Catherine Doenges, *University of Connecticut-Stamford;* Lorraine Dowler, *Pennsylvania State University;* Dawn Drake, *Missouri Western State University;* Brian Farmer, *Amarillo College;* Caitie Finlayson, *Florida State University;* Ronald Foresta, *University of Tennessee;* Gary Gaile, *University of Colorado;* Roberto Garza, *University of Houston;* Jay Gatrell, *Indiana State University;* Mark Giordano, *Oregon State University;* Dusty Girard, *Brookhaven College;* Qian Guo, *San Francisco State University;* Devon A. Hansen, *University of North Dakota;* Julie E. Harris, *Harding University;* Russell Ivy, *Florida Atlantic University;* Rebecca Johns, *University of Southern Florida;* Kris Jones, *Saddleback College;* Tim Keirn, *California State University, Long Beach;* Marti Klein, *Mira Costa College;* Lawrence M. Knopp, *University of Minnesota, Duluth;* Debbie Kreitzer, *Western Kentucky University;* Robert C. Larson, *Indiana State University;* Alan A. Lew, *Northern Arizona University;* John Liverman, *independent scholar;* Max Lu, *Kansas State University;* Donald Lyons, *University of North Texas;* Taylor Mack, *Mississippi State University;* Brian Marks, *Louisiana State University;* Chris Mayda, *Eastern Michigan University;* Eugene McCann, *Simon Fraser University;* Tom L. McKnight, *University of California, Los Angeles;* M. David Meyer, *Central Michigan University;* Sherry D. Moorea-Oakes, *University of Colorado, Denver;* Barry Donald Mowell, *Broward Community College;* Darla Munroe, *The Ohio State University;* Tim Oakes, *University of Colorado;* Nancy Obermeyer, *Indiana State University;* J. Henry Owusu, *University of Northern Iowa;* Rosann Poltrone, *Arapahoe Community College;* Jeffrey E. Popke, *East Carolina University;* Kevin Raleigh, *University of Cincinnati;* Henry O. Robertson, *Louisiana State University, Alexandria;* Robert Rundstrom, *University of Oklahoma;* Yda Schreuder, *University of Delaware;* Anna Secor, *University of Kentucky;* Daniel Selwa, *Coastal Carolina University;* Sangeeta Singh, *Metropolitan State University of Denver;* Christa Ann Smith, *Clemson University;* Richard Smith, *Harford Community College;* Barry D. Solomon, *Michigan Technical University;* Joseph Spinelli, *Bowling Green State University;* Kristin Sziarto, *University of Wisconsin-Milwaukee;* Liem Tran, *Florida Atlantic University;* Syed (Sammy) Uddin, *William Paterson University/St. John's University;* Samuel Wallace, *West Chester University;* Matthew Waller, *Kennesaw State University;* Gerald R. Webster, *University of Alabama;* Julie Weinert, *Southern Illinois University;* Mark Welford, *Georgia Southern University;* Clayton Whitesides, *Coastal Carolina University;* Sharon Wilcow, *University of Texas, Austin;* Keith Yearman, *College of Du Page;* and Anibal Yanez-Chavez, *California State University, San Marcos.*

Special thanks go to our editorial team at Pearson Education, Christian Botting, Sean Hale, and Anton Yakovlev; to our fantastic developmental editor, Jonathan Cheney, and our project manager, Lindsay Bethoney at Lumina Datamatics; to Eric Schrader for photo research; to Kevin Lear and International Mapping for their creative work with the art program; and to Rachel Youdelman for her work on permissions for text and line art. We would also like to thank our excellent research assistants, Jennifer McCormack and Fiona Gladstone.

Sallie A. Marston
Paul L. Knox
Diana M. Liverman
Vincent J. Del Casino Jr.
Paul F. Robbins

About the Authors

Sallie A. Marston

Sallie Marston received her PhD in geography from the University of Colorado, Boulder. She is currently a professor in the School of Geography and Development at the University of Arizona. Her research focuses on the political and cultural aspects of social life, with particular emphasis on sociospatial theory. She is the recipient of the College of Social and Behavioral Sciences' Outstanding Undergraduate Teaching Award as well as the University of Arizona's Graduate College Graduate and Professional Education Teaching and Mentoring Award. She teaches an undergraduate course on community engagement through school gardens and another on culture and political economy through the HBO television show, *The Wire*. She is the author of over 85 journal articles, book chapters, and books and serves on the editorial board of several scientific journals. She has coauthored, with Paul Knox, the introductory human geography textbook, *Human Geography: Places and Regions in Global Context*, also published by Pearson.

Paul L. Knox

Paul Knox received his PhD in geography from the University of Sheffield, England. After teaching in the United Kingdom for several years, he moved to the United States to take a position as professor of urban affairs and planning at Virginia Tech. His teaching centers on urban and regional development, with an emphasis on comparative study. He has written several books on aspects of economic geography, social geography, and urbanization and serves on the editorial board of several scientific journals. In 2008, he received the Association of American Geographers Distinguished Scholarship Award. He is currently a University Distinguished Professor at Virginia Tech, where he also serves as Senior Fellow for International Advancement.

Diana M. Liverman

Diana Liverman received her PhD in geography from the University of California, Los Angeles. Born in Accra, Ghana, she is the codirector of the Institute of the Environment and Regents Professor of Geography and Development at the University of Arizona. She has taught geography at Oxford University, Pennsylvania State University, and the University of Wisconsin–Madison. Her teaching and research focus on global environmental issues, environment and development, and Latin America. She has served on several national and international advisory committees dealing with environmental issues and climate change and has written about topics such as natural disasters, climate change, trade and environment, resource management, and environmental policy.

Vincent J. Del Casino Jr.

Vincent J. Del Casino Jr. received his PhD in geography from the University of Kentucky in 2000. He is currently vice provost for digital learning and student engagement, associate vice president for student affairs and enrollment management, and professor in the School of Geography and Development at the University of Arizona. He was previously professor and chair of Geography at California State University, Long Beach. His research interests include social and health geography, with a particular emphasis on human immunodeficiency virus (HIV) transmission, the care of people living with HIV and acquired immunodeficiency syndrome (AIDS), and homelessness in Southeast Asia as well as the United States. His teaching focuses on social geography, geographic thought, and geographic methodology. He also teaches a number of general education courses in geography, including world regional geography, which he first began teaching as a graduate student in 1995.

Paul F. Robbins

Paul Robbins received his PhD in geography from Clark University in 1996. He is currently the director of the Nelson Institute for Environmental Studies at the University of Wisconsin–Madison. Previously, he taught at the University of Arizona, Ohio State University, the University of Iowa, and Eastern Connecticut State University. His teaching and research focus on the relationships between individuals (e.g., homeowners, hunters, professional foresters), environmental actors (e.g., lawns, elk, mesquite trees), and the institutions that connect them. He and his students seek to explain human environmental practices and knowledge, the influence the environment has on human behavior and organization, and the implications this holds for ecosystem health, local community, and social justice. Robbins's past projects have examined chemical use in the suburban United States, elk management in Montana, forest product collection in New England, and wolf conservation in India.

Digital & Print Resources

For Teachers & Students

This edition provides a complete human geography program for students and teachers.

MasteringGeography™ with Pearson eText

The **Mastering** platform is the most widely used and effective on-line homework, tutorial, and assessment system for the sciences. It delivers self-paced coaching activities that provide individualized coaching, focus on the teacher's course objectives, and are responsive to each student's progress. The Mastering system helps teachers maximize class time with customizable, easy-to-assign, and automatically graded assessments that motivate students to learn outside of class and arrive prepared for lecture.

MasteringGeography™ offers the following:

- **Assignable activities** that include GIS-inspired MapMaster™ Interactive Map activities, Encounter World Regional Geography Google Earth™ Explorations, Geography Video activities, Geoscience Animation activities, Map Projection activities, coaching activities on the toughest topics in geography, end-of-chapter questions and exercises, reading quizzes, and Test Bank questions.
- **A student Study area** with GIS-inspired MapMaster™ Interactive Maps, Geography Videos, Geoscience Animations, "In the News" RSS Feeds, Web links, glossary flashcards, chapter quizzes, an optional Pearson eText that includes versions for iPad and Android devices and more.

Pearson eText gives students access to the text whenever and wherever they can access the Internet. The eText pages look exactly like the printed text and include powerful interactive and customization functions, including links to the multimedia.

Features of Pearson eText include the following:

- Now available on smartphones and tablets
- Seamlessly integrated videos and other rich media
- Fully accessible (screen-reader ready)
- Configurable reading settings, including resizable type and night reading mode
- Instructor and student note-taking, highlighting, bookmarking, and search

Teaching College Geography: A Practical Guide for Graduate Students and Early Career Faculty by the Association of American Geographers (0136054471) This two-part resource provides a starting point for becoming an effective geography teacher from the very first day of class. Part One addresses "nuts-and-bolts" teaching issues. Part Two explores being an effective teacher in the field, supporting critical thinking with GIS and mapping technologies, engaging learners in large geography classes, and promoting awareness of international perspectives and geographic issues.

Aspiring Academics: A Resource Book for Graduate Students and Early Career Faculty by the Association of American Geographers (0136048919) Drawing on several years of research, this set of essays is designed to help graduate students and early career faculty start their careers in geography and related social and environmental sciences. Aspiring Academics stresses the interdependence of teaching, research, and service—and the importance of achieving a healthy balance of professional and personal life—while doing faculty work. Each chapter provides accessible, forward-looking advice on topics that often cause the most stress in the first years of a college or university appointment.

Practicing Geography: Careers for Enhancing Society and the Environment by the Association of American Geographers (0321811151) This book examines career opportunities for geographers and geospatial professionals in business, government, nonprofit, and educational sectors. A diverse group of academic and industry professionals share insights on career planning, networking, transitioning between employment sectors, and balancing work and home life. The book illustrates the value of geographic expertise and technologies through engaging profiles and case studies of geographers at work.

Television for the Environment Earth Report Videos on DVD (0321662989) This three-DVD set helps students visualize how human decisions and behavior have affected the environment, and how individuals are taking steps toward recovery. With topics ranging from the poor land management promoting the devastation of river systems in Central America to the struggles for electricity in China and Africa, these 13 videos from Television for the Environment's global Earth Report series recognize the efforts of individuals around the world to unite and protect the planet.

Television for the Environment Life World Regional Geography Videos on DVD (013159348X) From the Television for the Environment's global Life series, this two-DVD set brings globalization and the developing world to the attention of any world regional geography course. These 10 full-length video programs highlight matters such as the growing number of homeless children in Russia, the lives of immigrants living in the United States trying to help family still living in their native countries, and the European conflict between commercial interests and environmental concerns.

Television for the Environment Life Human Geography Videos on DVD (0132416565) This three-DVD set is designed to enhance any human geography course. These DVDs include 14 full-length video programs from Television for the Environment's global Life series, covering a wide array of issues affecting people and places in the contemporary world, including the serious health risks of pregnant women in Bangladesh, the social inequalities of the "untouchables" in the Hindu caste system, and Ghana's struggle to compete in a global market.

Learning Catalytics

Learning Catalytics™ is a "bring your own device" student engagement, assessment, and classroom intelligence system. With Learning Catalytics, you can:

- assess students in real time, using open-ended tasks to probe student understanding.
- understand immediately where students are and adjust your lecture accordingly.
- improve your students' critical thinking skills.
- access rich analytics to understand student performance.
- add your own questions to make Learning Catalytics fit your course exactly.
- manage student interactions with intelligent grouping and timing. Learning Catalytics™ has grown out of 20 years of cutting-edge research, innovation, and implementation of interactive teaching and peer instruction. Available integrated with MasteringGeography™.

For Teachers

Instructor Resource Manual Download (0134142667) *The Instructor Resource Manual*, originally created by one of this book's coauthors, Vincent Del Casino Jr., follows the new organization of the main text. Strategies for Teaching Key Concepts provide teachers with a focused plan of action for every class session. Web Exercises tie in with associated Interactive Maps, and Additional Resources such as journals and Web sites are provided.

TestGen/Test Bank download (0134142640) TestGen is a computerized test generator that lets teachers view and edit *Test Bank* questions, transfer questions to tests, and print the test in a variety of customized formats. This *Test Bank* includes approximately 1,000 multiple-choice, true/false, and short answer/essay questions. Questions are correlated to the book's Learning Outcomes, the U.S. National Geography Standards, and Bloom's Taxonomy to help teachers to better map the assessments against both broad and specific teaching and learning objectives. The *Test Bank* is available in Microsoft Word® and also importable into Blackboard.

Instructor Resource DVD (0134142780) Everything teachers need, where they want it. The *Instructor Resource DVD (IRC)* helps make teachers more effective by saving them time and effort. All digital resources can be found in one, well-organized, easy-to-access place.

The IRC DVD includes the following:

- All textbook images as JPEGs, PDFs, and PowerPoint™ Presentations
- Pre-authored Lecture Outline PowerPoint™ Presentations, which outline the concepts of each chapter with embedded art and can be customized to fit teachers' lecture requirements
- CRS "Clicker" Questions in PowerPoint™ format, which correlate to the book's Learning Outcomes, the U.S. National Geography Standards, and Bloom's Taxonomy

- The TestGen software, *Test Bank* questions, and answers for both MACs and PCs
- Electronic files of the *Instructor Resource Manual* and *Test Bank*

This Instructor Resource content is also available completely online via the Instructor Resources section of **www.MasteringGeography.com** and **www.pearsonhighered.com/irc**.

For Students

Goode's World Atlas 23rd Edition (0133864642) *Goode's World Atlas* has been the world's premiere educational atlas since 1923. It features over 260 pages of maps, from definitive physical and political maps to important thematic maps that illustrate the spatial aspects of many important topics. The 23rd edition includes over 160 pages of digitally-produced reference maps, as well as new thematic maps on global climate change, sea level rise, CO_2 emissions, polar ice fluctuations, deforestation, extreme weather events, infectious diseases, water resources, and energy production, and more.

Pearson's Encounter Series

Pearson's **Encounter** series provides rich, interactive explorations of geoscience concepts through Google Earth™ activities, exploring a range of topics in regional, human, and physical geography. For those who do not use MasteringGeography™, all chapter explorations are available in print workbooks as well as in online quizzes at **www.mygeoscienceplace.com**, accommodating different classroom needs. Each Exploration consists of a worksheet, online quizzes, and a corresponding Google Earth™ KMZ file.

- *Encounter World Regional Geography* Workbook and Website by Jess C. Porter (0321681754)
- *Encounter Human Geography* Workbook and Web site by Jess C. Porter (0321682203)
- *Encounter Physical Geography* Workbook and Web site by Jess C. Porter and Stephen O'Connell (0321672526)

Dire Predictions: Understanding Global Warming 2nd edition by Michael Mann and Lee R. Kump (0133909778) For any science or social science course in need of a basic understanding of Intergovernmental Panel on Climate Change (IPCC) reports, periodic reports from the IPCC evaluate the risk of climate change brought on by humans. But the sheer volume of scientific data remains inscrutable to the general public, particularly to those who may still question the validity of climate change. In just over 200 pages, this practical text presents and expands upon the essential findings in a visually stunning and undeniably powerful way to the lay reader. Scientific findings that provide validity to the implications of climate change are presented in clear-cut graphic elements, striking images, and understandable analogies. The 2nd Edition covers the latest climate change data and scientific consensus from the ongoing Fifth Assessment Report and integrates links to media and active learning to capture learning opportunities for students. The text is also available in various eText formats, including an eText upgrade option in MasteringGeography.

People carry bags of coltan down a hill from the Mudere mine, near Rubaya, Democratic Republic of Congo (DRC). Miners dig 50 meters underground for the minerals before transporting them to a nearby river where they are separated before being sold to dealers. Mine accidents are common in DR Congo, where raw materials are mined for the manufacture of many commercial items, including electronics.

World Regions in Global Context

H ere is an experiment you shouldn't try. Grab your cell phone, throw it on the ground, stomp on it, and pick through the pieces. Amid the remnants, you can find the world. The screen was manufactured in Mexico. The microprocessor chip was assembled in a factory in China, owned by a company in South Korea, funded by investment from the United States. The software code that runs the phone was designed by a programmer in India. The electronics are made from materials found in copper mines in Chile and coltan mines in the Democratic Republic of the Congo (DRC), and the lead that soldered together the circuit board comes from Australia. Your cell phone cannot exist without the resources and knowledge of all these different world regions.

The objects we use in our daily lives are produced through international linkages and are central to the processes of globalization. Globalization reflects a world where places and people are increasingly connected. Thanks to these connections, resources and products as well as ideas, languages, culture, and music flow from place to place, making places seem more *similar*. And yet places remain strikingly *different* in spite of these similarities. Why?

If you visited all the places involved in the production of your phone, you would find well-educated, highly paid technicians living in Bangalore, India. In Mexico, the urban-based factory that produced the screen employs workers who migrated from rural areas. The Chilean copper mine is an enormous pit mine, three miles wide and a half-mile deep, drawing and polluting water from local communities. In Australia mines are located on lands where indigenous people struggle for their rights, and in the DRC the mining of coltan has fueled conflicts. In all these places, cell phones have become the way people connect to each other, but these places are *different* because of the economic, cultural, and environmental transformations that happen when they connect to global networks. This process is regionalization—a world where novel cultures, ideas, and products emerge from the mix of elements into new unique regions. The conclusion you can draw smashing your cell phone and considering its global origins is: *places are different because they are connected.*

Learning Outcomes

▶ *Compare* and contrast the concepts of globalization and regionalization.

▶ *Describe* the Anthropocene's global impacts on earth systems and analyze related environmental issues and sustainability choices.

▶ *Differentiate* between forms of economic activity and explain why these forms vary around the globe.

▶ *Explain* contemporary economic development trends and describe the main indicators of social and economic advancement.

▶ *Identify* the global, regional, and national actors that play a vital role in the world today.

▶ *Explain* the implications of globalization and regionalization for world regions and cultures.

▶ *Provide* examples of how the global distribution of languages and religions is changing.

▶ *Apply* the demographic transition model and use population pyramids to explain how and why regional population growth rates rise and fall.

Thinking Like a Geographer

Geography is the study of global relationships involving everything from how people earn a living to how they interact with the environment. Geographers seek to understand where things are, why they are there, and how they are connected. **Geography** comes from the Greek word *geographia,* which translates as "writing the world." Geographers map, travel, and measure the world to provide rich accounts of Earth's characteristics. Geographers investigate the physical features of Earth and its atmosphere, the spatial organization and distribution of human activities, and the complex interrelationships between people and the natural and built (meaning "human-made" or "human-altered") environments in which they live. Geographers—with their knowledge of the world and its connection to our communities, economy and environment—play important roles in business and government, education, health and environmental management and are well positioned to understand our rapidly changing world with its risks and opportunities.

Geographers do this through the study of **physical geography**, which is concerned with climate, weather patterns, landforms, soil formation, and plant and animal ecology and through **human geography**, which focuses on the spatial organization of human activity and how humans make Earth into a home. **Environmental geography** connects physical and human geography, as geographers also study the relationship between humans and the natural and built environments in which they live. The power of **world regional geography** lies in its ability to describe and examine global geographic processes, while at the same time explaining *why* and *how* certain patterns emerge on Earth. This book uses physical, human, and environmental geography to explore relationships within and among **world regions** (FIGURE 1.1).

Place and the Making of Regions

World regions can best be thought of as an aggregation of **places** and the connections that develop between those places over time. Places themselves are dynamic, with changing properties and fluid boundaries that are the product of a wide variety of environmental and human factors. Places exert a strong influence, for better or worse, on people's physical well-being, opportunities, and lifestyle choices. Places also contribute to people's collective memory and are powerful emotional and cultural symbols. The meanings given to place may be so strong that they become a central part of the shared identity of the people experiencing them. A **sense of place** refers to the feelings evoked among people as a result of the experiences and memories they associate with a place and to the symbolism they attach to that place. A sense of place develops out of the human capacity to reorganize the natural world into a built environment. Geographers think of the built environment as **landscape**, Earth's surface as transformed by human activity. As a product of human actions over time, landscape provides evidence about our character

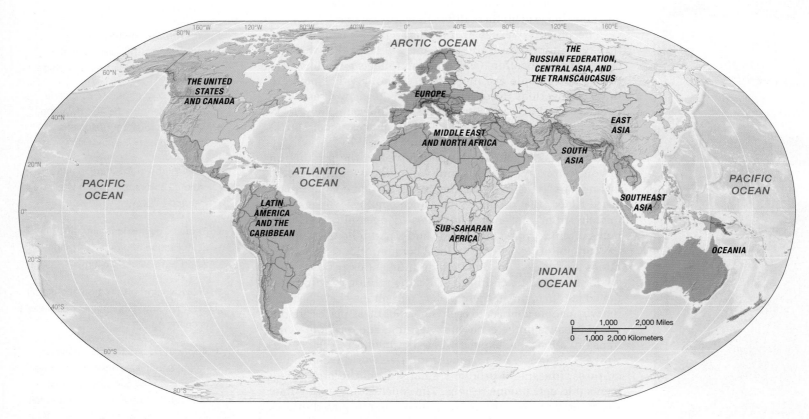

▲ **FIGURE 1.1 World Regions** This map highlights the expanse of each of the ten world regions discussed in this book.

and experience, our struggles and human triumphs. Through an analysis of landscape, geographers compare the meanings of the natural environment and built environment in the context of different places and regions.

Regions are best thought of as the connections that emerge between and among places over time. When this happens at a global scale—between different countries, for example—we identify these as *world regions*. At the same time, people's own conceptions of place, **region**, and identity may generate strong feelings of regionalism. **Regionalism** is a term used to describe the strong feeling of collective identity often shared by people who inhabit a region with distinctive characteristics. The feelings that one has toward places and regions also generate one's geographical imagination. A **geographical imagination** is how people think about the world around them—their own places and the places of others. Combined with critical thinking, a geographical imagination allows geographers to understand changing meanings of social identity and the relationships among people, places, and regions.

Maps and Mapping

Geographers use many tools to study the world, including maps as well as statistical and qualitative techniques. There is not one singular way that geographers ask and answer questions related to change over time and across space. Geographers do, however, rely on maps to illustrate the patterns and processes

of world regional geographies (see Appendix for more detail). A **map** is a visual representation and generalization of the world (**FIGURE 1.2**). Maps can locate places using a coordinate **system** of latitude and longitude. Maps also represent the names that people ascribe to places and the relationships that exist between places. Maps help geographers ask questions about the relationship between different sociocultural, political-economic, or environmental distributions, human activities and living experiences as well as uses of the natural environment. Maps are not neutral objects, as every single map is created through a series of choices about what should and what should not appear on it (**FIGURE 1.3**). A map set at the global **scale** tends to be more general than one at regional, national, or even local scale.

Mapping the world is complicated by the dynamic nature of the world itself, its changing features, and its transforming regions. On a constantly changing Earth, every map is only a snapshot. This basic reality about mapping reflects the larger challenge posed by this book, which is to explain how and why the map of world regions looks the way it does. Some regions that we take for granted now would have made no sense to people in the past. The Ancient Celts or Romans would never have recognized "Europe" as a coherent world region 2,000 years ago. How did Europe become what we recognize today? With this sort of question in mind, this chapter introduces the basic tools and fundamental concepts that geographers use to study the world and describes the conceptual framework that informs the subsequent chapters.

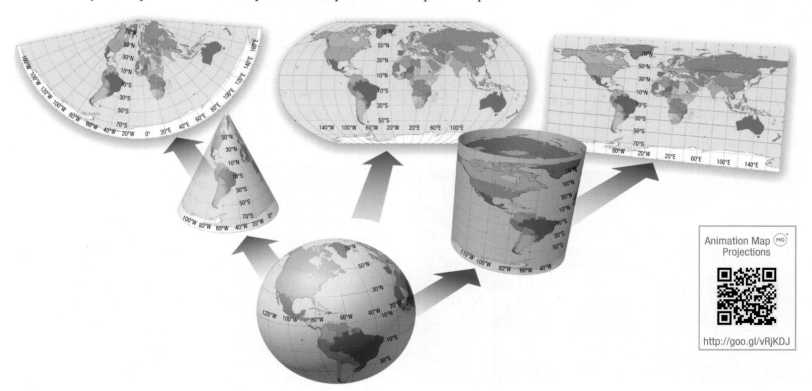

Animation Map (MG)
Projections

http://goo.gl/vRjKDJ

▲ **FIGURE 1.2 Maps and Mapping** All maps are partial representations of the world. The projection of the world from a spherical object to a flat map always produces certain distortions in distance, direction, area, or size. There are many different map projections that geographers use to measure, assess, and analyze global and regional patterns and processes. Understanding the reason for choosing one map projection or one approach to mapping data over another is one of the core critical thinking skills that all geographers must develop over time. **What makes a map a representation of reality and not reality itself? What are the choices that cartographers must make when making a map?**

◀ **FIGURE 1.3 Tabula Rogeriana** Muhammed al-Idrisi, an Islamic cartographer, had a strong impact on mapmaking worldwide. Tabula Rogeriana is a "map of the known world," which al-Idrisi produced in 1154 for King Roger of Sicily. It includes Europe, Asia, and North Africa. The Islamic tradition places the south at the top of the map, in contrast to many world maps today. The map became the basis of many other maps of the world by both Islamic and European cartographers.

"It is the process of making new global connections—through trade, migration, or environmental exchange—that allows or causes regions to change."

Globalization and Regionalization

The world has always been global. Since *Homo sapiens* walked out of East Africa and long after the moment when McDonald's began to appear in malls in Kenya (**FIGURE 1.4**), the environments, economies, and societies of the globe have been tied together. In today's world, these connections have intensified and become more widespread in a process geographers call

▲ **FIGURE 1.4 A Mall in Kenya** A shopping mall is more than just a place of consumption; it is an iconic marker of a certain form of development. The concept of the mall has been globalized over the last 50 years, and malls can now be found throughout the world. Most malls provide goods and services tied to global products as well as goods unique to the local market. Malls also play valuable roles as public spaces.

globalization. Globalization is a system of elements—political-economic, sociocultural, environmental—linked together so that changes in one element often result in changes in another. Some scholars predict that the most recent wave of globalization will result in unprecedented consolidation and homogenization of the world's ecologies, economies, and societies. They stress that globalization is a process that breaks down boundaries, makes places similar, and connects them by encouraging the flow of ideas, products, and practices.

And yet parts of the world retain their uniqueness and new world regions may emerge over time. We use the term **regionalization** to describe how and why new regions emerge. As we will see, it is the process of making new global connections that allows or causes world regions to change. These connections mean that world regions are:

- best studied by considering how they interact and develop as part of wider global political-economic, sociocultural, and environmental *systems*;

- best conceptualized as *interdependent,* as they affect, and are affected by, each other; and

- best understood as products of *change* over time.

These three themes are intertwined in the processes of globalization and regionalization, the twin forces that generate a world of regions that is *both* globally interconnected and locally differentiated. Globalization becomes an engine of regionalization and regional differences can contribute to globalization. Put another way, it is the process of making new global connections—through trade, migration, or environmental exchange—that allows or causes regions to change. These connections have far-reaching effects. They create global and regional trade networks, ethnic neighborhoods in cities, new consumer products and ways of shopping, and even new migrant communities (**FIGURE 1.5**). They may lead to the formation of new ecological communities or new agricultural systems based on imported crops and animals. By studying world regions, we can understand why and how differences emerge, even as global processes connect the world's regions in new and important ways. That *places are different because they are connected* is the single central lesson of this book.

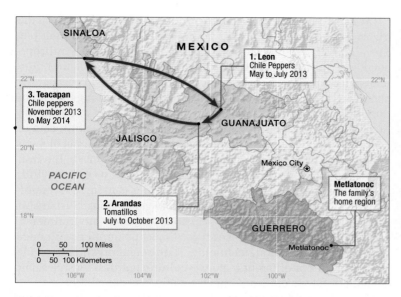

▲ **FIGURE 1.5 Migration Networks Among Mexico's Farmworkers** This map represents the movement of one migrant farm family in Mexico over the course of a year. They are certainly not alone, as thousands of people follow such patterns across Mexico and between Mexico and other countries, such as the United States. **Why do farmers migrate in Mexico? Can you think of other examples of economically-driven agricultural migration patterns in the world today?**

Apply Your Knowledge

1.1 Identify three examples of how globalization has affected your local community.

1.2 Using the examples you selected, list the ways in which your local community influences globalization. For example, a "big box" store is part of the global economy, but often sells products or services tailored for different regions, such as urban or rural essentials, Southwest or Northeast specialties.

Ecological Footprint

https://goo.gl/TcIL1T

A World of Regions

Exploring the interconnections among world regions not only helps explain the contemporary world, but it also allows us to think about where the world might go from here. The world we grew up in, and all the regions we know now, will not and cannot be the ones we will inhabit in the future. New regions and clusters are developing as places in the world connect in new ways. To highlight the changing nature of regions, consider that the regions and countries described in this edition of this book are already different from those in the previous edition published only three years ago. Regional changes in politics and government (as in the struggle over Crimea between Ukraine and Russia), the continuing emergence of economic power centers (such as Brazil and China) in what used to be called the underdeveloped world, and new regional opportunities and challenges (such as the Arctic melting as a result of global warming) demonstrate the ever-changing nature of world regions (**FIGURE 1.6**).

In an effort to address the emerging and future topics that affect each region, we introduce several *Emerging World Regions*

▲ **FIGURE 1.6 The Emergence of a New Country** The emergence of the Republic of South Sudan in 2011 was celebrated in many ways. In this photo a citizen of the new country waves a flag as part of the independence celebration.

throughout the text. An **emerging world region** is an area where loosely connected locations are developing shared characteristics that differentiate them from other world regions, past and present. These areas may become increasingly important to global relationships or systems. For example, the Arctic, which has often been viewed in fragments (as part of a number of different world regions, such as the United States and Canada, Europe, and Russia), is now linked closely together through human migration, international trade, and shared environmental problems. An emerging world region may also be noncontiguous—it might not share borders with other partners in the region. This is the case for new regions, such as BRICS—Brazil, Russia, India, China, and South Africa—which have strong regional connections even though they are spread widely across the planet.

Organizing and Exploring the World's Regions

The world region concept is a useful tool for organizing and understanding information about the world. Accordingly, the framework for the study of world regions in this chapter provides the structure for the 10 world regional chapters that follow. Each chapter is organized around a set of themes common to every world region, though unique in each.

- *Environment, Society, and Sustainability:* How environments change and are changed by people

- *History, Economy, and Territory:* How history, economics, and politics evolve over time

- *Culture and Populations:* How people and cultures all around the world interact and change

- *Future Geographies:* How contemporary regional differences and new global forces are likely to impact important real world issues in coming years

You will find that within each of these areas of analysis, global systems connect world regions and, as a result, produce differences between them. The remainder of this chapter explores the core concepts of this thematic framework.

Environment, Society, and Sustainability

The environment can be understood as everything that surrounds us—air, water, plants and animals, buildings, and even society and culture. In this book we use the term environment to describe the physical and ecological setting for human activities where the environment is critical to the study of world regions. Environmental characteristics that are studied by physical geographers and other Earth scientists include rainfall, temperature, vegetation, soils, wildlife, geology, and landforms. World regions are shaped as the environment influences many opportunities for societies, but also as people transform the environment. A physical environment with extreme cold, little water or frequent storms, and unstable geology can pose great challenges for human survival, yet humanity now occupies extreme and hazardous environments in places such as the Arctic region of Russia or drought and earthquake prone California in the United States (**FIGURE 1.7**).

Although some still call our physical and biological surroundings the 'natural' environment, almost all aspects of the Earth system have now been transformed by human action and there is very little untouched 'nature'. And humans, as one of many species occupying the planet, are part of nature as well.

Much of our evolution as a species took place during the Pleistocene epoch. Scientists have traditionally divided Earth's history into epochs lasting thousands of years during which geological conditions produce characteristic rock layers and fossils. The Pleistocene epoch lasted from about 2.5 million to 11,700 years ago and included major glaciations when much of North America and Europe were covered with ice, with ecosystems dominated by now extinct large mammals such as mammoths and with the emergence of modern humans. The Pleistocene ended when the ice retreated and warmer stable temperatures allowed for the development of agriculture and the expansion of human populations during the most recent epoch called the Holocene.

▼ **FIGURE 1.7 San Andreas Fault, California** Two tectonic plates sliding past each other cause frequent earthquakes along the San Andreas Fault, visible in this image.

We now live in the **Anthropocene**—the period of Earth's history where human activity dominates the earth system (see Sustainability in the Anthropocene: "Welcome to the Anthropocene" on pp. 18–19). In the last 200 years we have cleared more than half of the world's forest cover, polluted rivers and oceans with chemicals and plastic, warmed the climate by doubling the carbon dioxide content of the atmosphere, and contributed to the extinction of hundreds of species. In short, human activities now occur on such a vast scale that we are altering the air, water, and ecology in ways that risk the sustainability of many places on Earth. Environmental sustainability is a concept that challenges us to live within the constraints of the earth's system without causing irreversible damage to it or harming the lives of future generations. Sustainability requires evaluation of our decisions and their environmental impacts, including our choices regarding consumption, affluence, production, population, technology, and social organization. The characteristics of different regions—culture, politics, lifestyles, and economy—have significant impacts on their sustainability.

In each world region, we discuss how climate, geology, and ecology have influenced the development of the region. We also discuss the human use of the environment in the Anthropocene and describe what is being done in each region to confront the challenges of sustainability and environmental change.

Climate and Climate Change

Weather and climate are ever-present aspects of the environment that impact our lives. **Weather** is the current state of temperature and precipitation (it is a cold day or it is raining) at a particular time and place. **Climate** is the average weather or typical conditions of temperature, precipitation (e.g., rain, snow), and other weather variables (e.g., humidity, wind) at a location over the longer term (this is generally a cold place or a wet place or summers are hot). Climate—the saying goes—is what we expect, weather is what we get.

Our weather and climate are products of the **climate system**—the effects of the sun's energy with the interactions of air (atmosphere), water (hydrosphere), ice (cryosphere), landforms (lithosphere), and ecosystems (biosphere). The climate is not the same everywhere because places receive different amounts of sunlight and have different atmospheric compositions (e.g., because of dust or pollution); amounts of water, snow, and ice; and dissimilar landforms and ecosystems. But climate regions are also connected. As the sun heats one region and cools another, masses of air rise and fall and flow with winds and currents from one place to another, bringing moisture that can fall as rain or transporting pollution across the globe. If any of these components of the climate system change, the average temperature or precipitation may also change local and global conditions in a process called **climate change**.

Regional Climate The climates of world regions are influenced by a number of basic factors. These include the orientation to the sun at different times of the year and the associated variations in solar radiation; the configuration of land, sea, and mountains; the resulting **atmospheric circulation** of air and ocean currents that

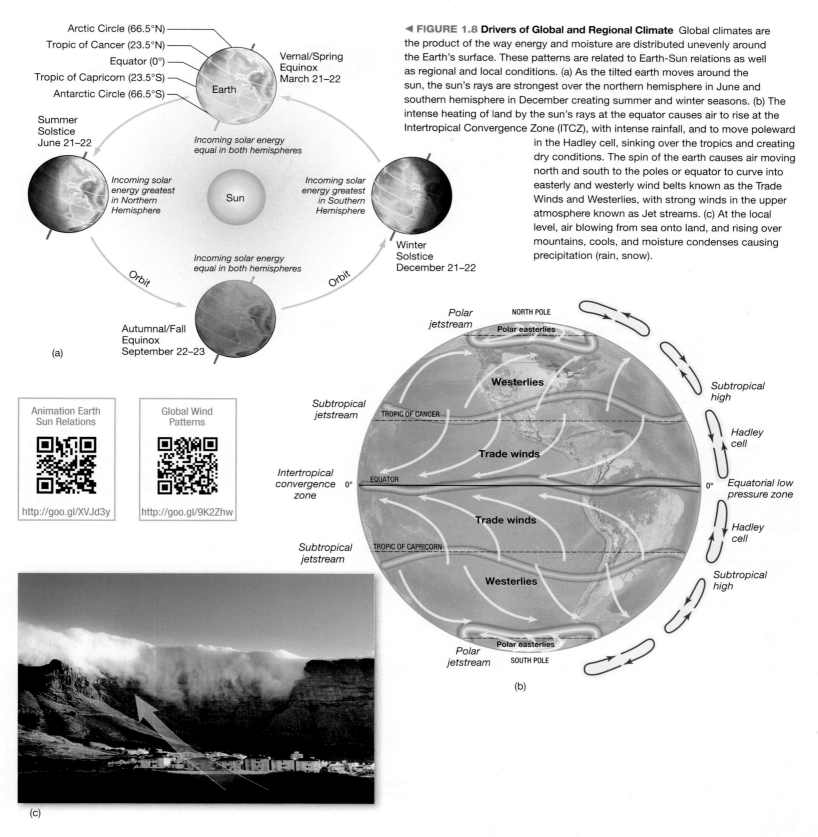

▲ **FIGURE 1.8 Drivers of Global and Regional Climate** Global climates are the product of the way energy and moisture are distributed unevenly around the Earth's surface. These patterns are related to Earth-Sun relations as well as regional and local conditions. (a) As the tilted earth moves around the sun, the sun's rays are strongest over the northern hemisphere in June and southern hemisphere in December creating summer and winter seasons. (b) The intense heating of land by the sun's rays at the equator causes air to rise at the Intertropical Convergence Zone (ITCZ), with intense rainfall, and to move poleward in the Hadley cell, sinking over the tropics and creating dry conditions. The spin of the earth causes air moving north and south to the poles or equator to curve into easterly and westerly wind belts known as the Trade Winds and Westerlies, with strong winds in the upper atmosphere known as Jet streams. (c) At the local level, air blowing from sea onto land, and rising over mountains, cools, and moisture condenses causing precipitation (rain, snow).

transport heat and moisture from one place to another; and precipitation processes (**FIGURE 1.8**).

These influences combine to create climatic patterns across the world that can be classified according to temperature and moisture characteristics. The most commonly used classification, shown in **FIGURE 1.9**, is based on that of Köppen, which has five major types of climate: tropical, dry, temperate, continental and highland. Subdivisions indicate whether seasons are wet or dry, warm or cool, and the presence of ice.

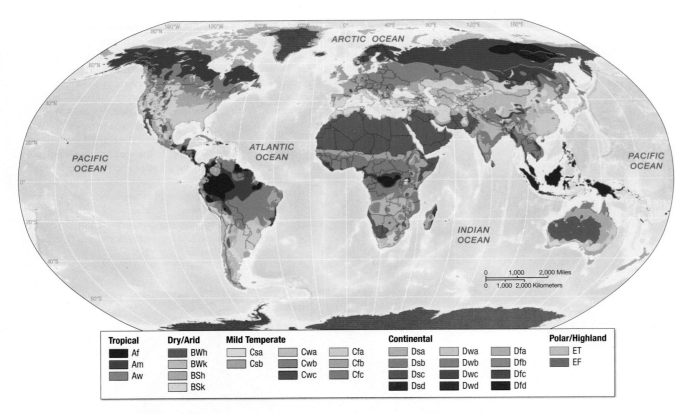

Major Köppen Climate Categories

The first level recognizes six major climatic types with each group being designated by a capital letter. These major climate categories have the following broad characteristics:

A - Tropical Moist Climates: These are very warm climates found in the tropics that experience high quantities of precipitation. The primary distinguishing characteristic of these climates is all months have average temperatures above 18°C (64°F).

B - Dry Climates: These are climates that experience little precipitation during most of the year. Further, potential losses of water from evaporation and transpiration greatly exceed atmospheric input.

C - Moist Mid-latitude Climates with Mild Winters: In these climates, summer temperatures are warm to hot and winters are mild. The primary distinguishing characteristic of these climates is the coldest month has an average temperature between 18°C (64°F) and -3°C (27°F).

D - Moist Mid-Latitude Climates with Cold Winters: In these climates, summer temperatures are warm and winters are cold. The primary distinguishing characteristic of these climates is the average temperature of warmest month exceeds 10°C (50°F), and average temperature of coldest is below -3°C (27°F).

E - Polar Climates: These climates have very cold winters and summers, with no real summer season. The primary distinguishing characteristic of these climates is the warmest month has an average temperature below 10°C (50°F).

H - Highland Climates: These are climates that are strongly influenced by the effects of altitude. As a result, the climate of such locations is rather different from places with low elevations at similar latitudes.

Secondary Köppen Climate Categories

At the secondary level, the major climate groups are further subdivided according to the seasonal distribution of precipitation, the degree of aridity, or the presence/absence of permanent ice.

Lowercase letters f, w, and s are used to distinguish precipitation patterns and are only applicable to A, C, and D climates. Thus, for example, within the major climate category, A - Tropical Moist Climates, are the subcategories:

Af – Tropical Wet
Aw – Tropical Wet and Dry
Am – Tropical Monsoon

Uppercase W and S identify desert (arid) or steppe (semiarid) climate subtypes for the Dry Climates (B) major category. Thus:

BW - Dry Arid (Desert)
BS - Dry Semiarid (Steppe)

For the Polar Climates (E), the secondary letters F and T distinguish whether the site is covered by permanent ice fields and glaciers or free of snow and ice during the summer season. Thus:

ET - Polar Tundra
EF - Polar Ice Cap

Major climate types B, C, and D can be further sorted according to a third category. This level is used to distinguish particular temperature characteristics found in these climates.

In B climates, the lowercase letter h identifies a subtropical location where average annual temperature is above 18°C (64°F). Cooler mid-latitude Dry Climates are distinguished with a lowercase k. Thus:

BWh – Dry Arid Low Latitudes
BWk – Dry Arid Mid-Latitudes

For C and D climates the tertiary level letters a, b, c, and d are used to distinguish different monthly temperature characteristics. Thus:

Cfa – Humid Subtropical
Cfb - Marine - Mild Winter
Cfc - Marine - Cool Winter

Climate Change and a Warming World

Although the climate can remain stable for centuries, global and regional climates have varied over time, especially as a result of slight changes in the tilt of Earth's axis and its orbit around the Sun that cooled Earth into ice ages. The landscapes of places such as western Canada, the U.S. Rocky Mountains, the Andes, Scotland, and Norway show the marks of former ice cover from periods when it was so cold that rivers of ice (called *glaciers*) or massive sheets of ice (*ice caps*) covered much of the world (refer to Chapter 11). The remnants of the ice still remain in Antarctica and the Arctic, and the highest mountain regions. Global and regional climates cool temporarily when the dust from volcanic eruptions blocks sunlight. Rainfall may also vary over the centuries as a result of slight shifts in atmospheric circulation.

Scientists agree that humanity is now changing the climate in a dramatic way. One of the most significant markers of the Anthropocene is the rapid increase in atmospheric **greenhouse gases**, which have caused the Earth to warm. We have altered the composition of the atmosphere through burning of coal, oil, and gas, using more electricity, cutting down forests, and increasing the global livestock herd. All these activities emit greenhouse gases—especially carbon dioxide and methane—that trap heat within

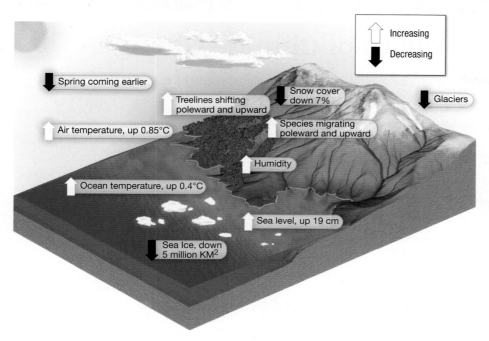

◀ **FIGURE 1.10 Indicators of a Warming World** Since 1950, scientists have observed many different types of evidence that the climate is already changing, including increase in land and sea temperatures, decrease in ice and snow cover, increase in sea level, and movement of plants and animal species toward cooler areas.

This book's regional chapters discuss "hotspots" for climate change and climate vulnerability, where it will get warmer and drier with serious impacts for poorer and disadvantaged populations. These hotspots include southern Africa, the U.S. Southwest, Australia, the Mediterranean, and the Arctic. While some regions become hotter and drier, others may experience more intense precipitation and flood risk (**FIGURE 1.11**).

There is a spirited debate about what to do about climate change. Should we reduce emissions (termed mitigation) as fast as possible and does this mean shifting from fossil fuels to renewable energy? Should we learn how to adapt to living in a warmer, more hazardous world? Which regions are most responsible for the emissions, which regions are most severely impacted and how, and who should respond and when? You can see some of these patterns in Visualizing Geography: "The Causes and Consequences of Climate Change" on pp. 12–13.

International actions to respond to climate change are framed by the 1992 **United Nations Framework Convention on Climate Change (UNFCCC)**, ratified by almost all countries with the goal of reducing the risks of dangerous anthropogenic climate change through reducing emissions (mitigation), coping with the effects (adaptation) and financial assistance to developing countries. In 1997, the Kyoto Protocol—a treaty in which countries promised to reduce greenhouse gas emissions by at least 5% between 2008 and 2012—was signed by a smaller group of developed countries. It included mechanisms for carbon trading.

the atmosphere, resulting in a **global warming** of the atmosphere and surface. This warming has resulted in higher average temperatures in most world regions, decreases in ice and snow cover, and changes in rainfall patterns and climate extremes such as heat waves, droughts, and severe storms (**FIGURE 1.10**).

The United Nations (UN) has established an expert group of hundreds of scientists—the Intergovernmental Panel on Climate Change (IPCC). The IPCC's 2014 report—Climate Change 2014—confirms that Earth is already warming, caused by increases in greenhouse gases that have reached the highest levels in more than 800,000 years. If current trends in emissions continue, Earth could warm more than 2°C (3.5°F) this century with even greater warming in polar and temperate regions. The IPCC also forecasts that droughts and floods will become more intense, sea level will rise by several feet, and the oceans will become more acidic from carbon dioxide. A warmer, more extreme climate will have significant implications for plants and wildlife, and many may need to move if their habitat becomes too warm or dry. The human impacts are also serious: Climate change, according to the IPCC, will affect the security of food and water in regions that become drier, increase the intensity of weather disasters and forest fires, and alter patterns of insect-borne and other diseases.

Social vulnerability to climate extremes and changes—**climate vulnerability**—is as important as the physical location and severity of high temperatures, droughts, or floods. High temperatures and droughts, for example, can have much more severe human impacts on health and food security where people do not have air conditioning or irrigation. The poor are often more vulnerable to climate extremes because they cannot afford to protect themselves—through locating homes in safer locations, equipping farms with irrigation, purchasing disaster insurance, or accessing government support. Geographers have played an important role in defining and mapping climate vulnerability—the characteristics of people or places that affect their capacity to anticipate, cope with, resist, and recover from the impact of a natural hazard. These characteristics can include income, gender, age, and ethnicity.

▼ **FIGURE 1.11 Climate Vulnerability** Women and children are often the most vulnerable to climate extremes such as floods.

The Causes and Consequences of Climate Change

Maps, graphs, and photos have become powerful tools in understanding the causes and consequences of climate change and imagining possible solutions. They are used to show which regions of the world are responsible for the greenhouse gas emissions that are changing the climate, which regions are most vulnerable to the impacts of climate change, and how our decisions now affect future climate.

1 Who is responsible for climate change?

Which world regions contribute most to the emissions of greenhouse gases that are causing the world to warm? Recent data show that East Asia (including China and Japan) and North America (U.S. and Canada) have the largest total emissions with Africa and small islands (such as those in the Pacific and Caribbean) the least.

Another perspective is to compare the emissions per person for different countries rather than look at total national emissions, which is influenced by the size of the population. From this viewpoint North America and Europe bear greater responsibility because each person consumes much more fossil fuel energy, meat, and other goods that produce emissions. All estimates show that Sub-Saharan Africa and small island states bear the least responsibility. Yet they are the most vulnerable to warming and sea level rise. What do you think is the best measure of responsibility?

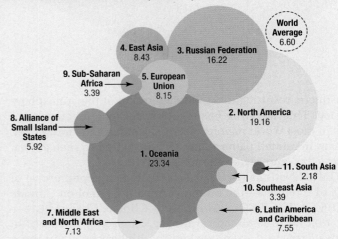

▲ **FIGURE 1.2.1 Total greenhouse gas emissions by region in 2011 (Millions of tons of CO$_2$-equivalent greenhouse gasses). All data from World Resources Institute (WRI).**

▲ **FIGURE 1.2.2 Per capita greenhouse gas emissions by region in 2011 (tons of CO$_2$-equivalent greenhouse gas emissions).**

2 Which regions will be most affected by climate change?

This map from the latest report of the IPCC projects the likely increase in temperature by 2050 if we continue to burn fossil fuels and emit greenhouse gases at current rates (a "high emission" scenario). Which regions are likely to experience the greatest rise in temperatures? Which people and species live in these regions and may be vulnerable to these changes?

IPCC 5th assessment report 2014

https://goo.gl/wl30aX

Climate Data Explorer

https://goo.gl/CK1ey9

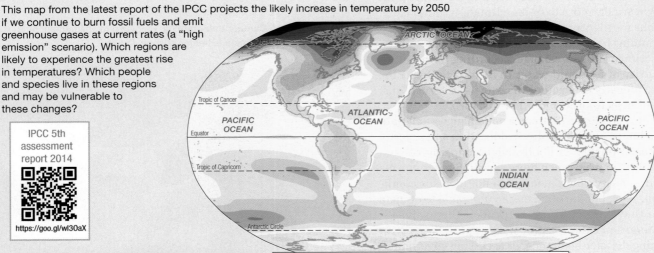

▲ **FIGURE 1.2.3 Projected increase in temperature by 2050.** Assuming ongoing levels of current emissions of greenhouse gases, scientists predict sustained warming.

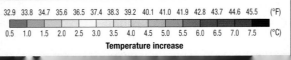

32.9	33.8	34.7	35.6	36.5	37.4	38.3	39.2	40.1	41.0	41.9	42.8	43.7	44.6	45.5	(°F)
0.5	1.0	1.5	2.0	2.5	3.0	3.5	4.0	4.5	5.0	5.5	6.0	6.5	7.0	7.5	(°C)

Temperature increase

③ What choices can we make about responding to climate change?

If we do not do something to reduce (mitigate) emissions—a high emissions scenario—then the world could warm by more than 4°C (9°F) with very serious impacts on ecosystems, agriculture, sea level, and human health. If, however, the world's countries agree to make drastic reductions in emissions—through international agreements, national laws or taxes, and individual actions—then we may keep the warming below 2°C (3.5°F) and reduce the risks of more serious impacts.

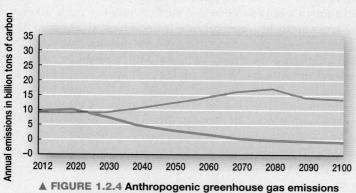

▲ **FIGURE 1.2.4 Anthropogenic greenhouse gas emissions under high and low emission scenarios.**

HIGH EMISSIONS SCENARIO

- The world continues to burn fossil fuels to 2100
- Global temperatures increase by more than 4°C (7°F)
- Water stress increases by 38%
- Sea level rises by more than a meter (3+ feet)
- 12 times more people are flooded
- Serious risks to agriculture and ecosystems including extinction of species, global food supplies
- Increase in diseases

LOW EMISSIONS SCENARIO

- The world takes serious steps to decrease emissions from 2020
- Global temperature increase stays below 2°C (3.5°F)
- Water stress increases by 24%
- 4 times more people are flooded
- Some risks to agriculture and ecosystems including damage to coral reefs and food security in developing world

④ What are some of the impacts already underway or projected to occur?

Global maps and graphs may be less effective in communicating the risks of climate change than more local maps and photos. The recent U.S. National Climate Assessment includes maps and photos that show how climate change is already affecting different parts of the United States and what may happen in the future. For example, Miami and New York are already experiencing serious flooding and parts of Florida will disappear beneath rising seas.

National Climate Assessment

https://goo.gl/UVFf3v

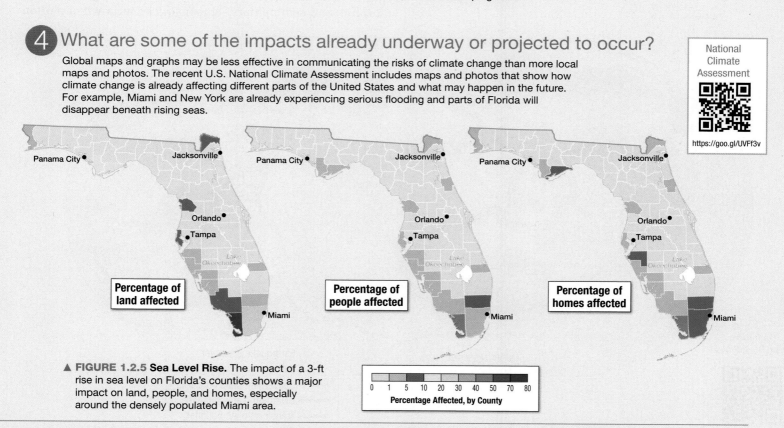

▲ **FIGURE 1.2.5 Sea Level Rise.** The impact of a 3-ft rise in sea level on Florida's counties shows a major impact on land, people, and homes, especially around the densely populated Miami area.

Percentage Affected, by County
0 1 5 10 20 30 40 50 70 80

---

1. **Compare and contrast the high and low emissions scenarios in terms of policy approaches and global impacts.**
2. **Explore the regional information on the websites of the National Climate Assessment at nca2014.globalchange.gov or the IPCC Working Group II at www.iipcc-wg2.gov/AR5/ and select a region, then summarize the impacts of climate change on that region.**

Increasing awareness of the risks of climate change, and that much deeper cuts were needed, produced the 2015 Paris Agreement, in which all countries committed to keep warming under 2°C. Countries promised much deeper cuts by 2020 and beyond—the U.S. up to 28%, the European Union 40%, and Costa Rica to become carbon neutral (net zero emissions) by using renewables and protecting its forests. Major emitters such as India and China also promised to significantly reduce their growth in emissions. At least $100 billion a year was set as a financial goal for helping least developed and other vulnerable countries reduce emissions and adapt to the impacts and losses of climate change. Even with these significant commitments however, scientists still suggest temperatures may rise as high as 3.5°C before starting to fall, unless further cuts are made soon.

Apply Your Knowledge

1.3 What factors affect climate? How are diverse regional climates connected in a system?

1.4 Research a recent news article discussing the severity of climate change. Who is quoted in the article and why are they considered an expert? What data are used to support the opinions expressed? Do you agree with the claims expressed in the article and why?

Geological Resources, Risks, and Water

Like the global climate system, the geologic system of Earth helps produce regions by uplifting mountains, forming water drainage and river systems, and producing diverse resources and hazards around the world. Earth's physical features greatly influence how people live in different landscapes. The study of the physical processes that create these features is known as **geomorphology**. Underlying these physical forces on the Earth's surface are processes at work deep within Earth's molten core and surface crust, which drive the slow movement of large "plates" of solid rock that constitute the landmasses called continents. As plates interact, they create geological features. Where plates collide or converge, they form mountains. Where plates pull apart or diverge, they form deep ocean canyons and valleys. These **plate tectonics** also create unstable geological conditions—transform boundaries—where blocks of Earth's crust move suddenly alongside or under each other causing earthquakes (**FIGURE 1.12**). Where molten rock is forced to the surface, we often find volcanoes (**FIGURE 1.13**).

Resources and Risks Earth's surface is composed of many types of rocks that geologists general classify as igneous, metamorphic, and sedimentary. Igneous rocks form when molten

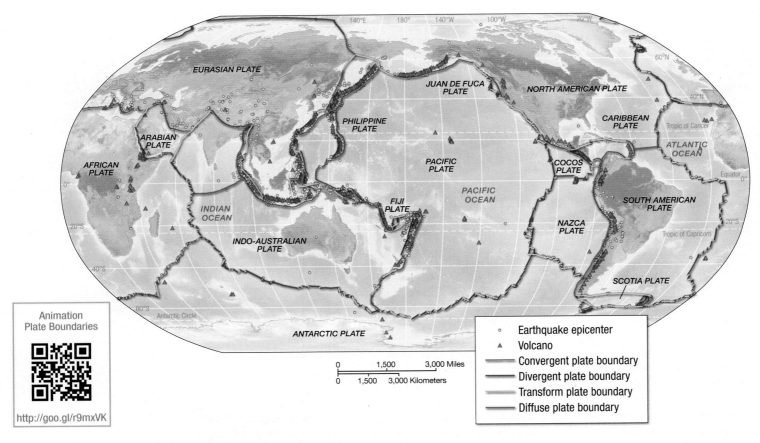

Animation
Plate Boundaries

http://goo.gl/r9mxVK

- ○ Earthquake epicenter
- ▲ Volcano
- — Convergent plate boundary
- — Divergent plate boundary
- — Transform plate boundary
- — Diffuse plate boundary

▲ **FIGURE 1.12 Active Earthquake and Volcanic Zones** Tectonic plates interact along their boundaries, producing earthquakes and volcanoes. The most tectonically active regions include the Ring of Fire around the Pacific. Some volcanoes occur at "hotspots" far from plate boundaries. **What areas of the planet are most seismically inactive? Why is that?**

▲ **FIGURE 1.13 Volcanic Activity in Hawaii** A lava flow from the Kilauea Volcano on the Big Island of Hawaii moves toward the town of Pahoa in 2014. Kilauea is one of the world's most active volcanoes and marks a hotspot on the Pacific Plate where molten lava emerges away from plate boundaries.

▲ **FIGURE 1.14 Japan and the 2011 Tsunami** In March 2011 a major earthquake off the coast of Japan triggered a tidal wave that killed more than 15,000 people, caused a leak at the Fukushima nuclear reactor, and destroyed coastal communities such as Kesennuma.

rock cools within the inner core of Earth or when it emerges as volcanic lava; they are often associated with valuable resources such as uranium, diamonds, and granite. Metamorphic rocks are also associated with plate tectonics because they form under intense heat and pressure and include useful resources for building and manufacturing, including marble and slate. The sedimentary rocks occur where sediment (sand, pebbles, and plants) is deposited by rivers, lakes, ice, or wind and solidifies over time into rocks, including coal and limestone, or compresses into other materials including oil and iron ore.

Geological conditions pose serious hazards and risks to humanity, especially in the tectonically active zone around the Pacific—called the **Ring of Fire**—where earthquakes and volcanoes threaten the lives and homes of residents of the western Americas from Alaska to Chile, Asia, including Japan and Indonesia, and Pacific islands such as Hawaii and New Zealand. For example, a major earthquake in Japan in 2011 damaged buildings and caused a tsunami that destabilized the Fukushima nuclear plant (**FIGURE 1.14**).

River Formation and Water Management When rain falls on the land, it can erode (cut into) the surface, especially on slopes, and will start to channel water into streams and rivers flowing toward the ocean or lakes in interior low lands. Flows are especially intense on mountain slopes, during heavy rainstorms, and when the snow melts in the spring. During ice ages and under current glaciers, ice flows, and also carves out deep valleys that usually contain rivers or lakes when ice melts. Rivers and lakes are critical sources of freshwater for ecosystems and for human activity. They are often locations for human settlement, including cities, industry, and agriculture, which benefit from the adjacent water resources,

including the rich sediments deposited by rivers that create fertile soil, fisheries, the use of water for drinking, transportation, and for electricity generation through hydropower. Most of the regions in this book have large concentrations of people in major river basins and around lakes—such as the Great Lakes in North America, the Rhine in Europe, and the Ganges in South Asia. But when heavy precipitation causes rivers to overflow, the resulting flooding can destroy property and farmland and place lives at risk. As with climate, some people are more socially and economically vulnerable than others to floods.

Apply Your Knowledge

1.5 How do geologic conditions pose risks for human societies? In what ways is water a key component in these conditions?

1.6 Use one of the listed links to identify three recent disasters associated with geological or extreme climate conditions and research their impacts on society: http://www.unisdr.org/we/inform/disaster-statistics http://www.ifrc.org/publications/ and http://www.usgs.gov/natural_hazards How did the combination of physical geography and social vulnerability contribute to loss of life and economic impacts for different locations and social groups? How might the impacts have been reduced?

Natural Hazards

https://goo.gl/1ZnIM5

Ecology, Land, and Environmental Management

The interactions of climate, geomorphology, and human activity are the major influences on the global distribution of living things. Over large regions of similar climate and physical conditions, we can identify major **biomes** (also called *ecoregions*), with vegetation closely correlated with temperature and rainfall: forests where rainfall is high, grasslands where precipitation is less and temperatures are moderate, and deserts where it is dry (**FIGURE 1.15**). The science of ecology studies the interactions between living organisms (e.g., plants and animals) and their physical surroundings and classifies them into different communities called **ecosystems**.

Ecosystems and Biodiversity Decline A few large ecosystems host the wealth of Earth's species and genetic variability. This **biodiversity**—the variety and differences in the types and numbers of species in different regions of the world—has been a boon for human beings for more than 10,000 years, first providing food and clothing from hunting, gathering, and fishing wild species, and then through the selective cultivation and breeding of plants and animals in the process of domestication and agriculture. Biodiversity and ecosystems provide a range of **ecosystem services** to society, including food, building materials, and medicine, as well as enjoyment, recreation, cultural values, and enhancement of water resources (**FIGURE 1.16**).

Biodiversity is lost as land is converted from diverse forests and grasslands to crops, ranches, and settlements, and as a growing population overharvests wildlife and fisheries. Although it is impossible to estimate the exact rate of loss today, the International Union for Conservation of Nature (IUCN) estimates that one in eight birds, one in four mammals, and one in three amphibians (e.g., frogs) are threatened with extinction. All of the world regions have suffered serious declines in biodiversity in all ecosystems. For example, Latin America and the Caribbean hosts more than half of our world's biodiversity. It has lost vast areas of tropical forests and some associated species in the Amazon basin, and unique ecosystems on the islands of the Caribbean, including corals and fish. Efforts to restore biodiversity include **rewilding** of ecosystems through the reintroduction of key species like wolves into the Yellowstone ecosystem in North America, the restoration of forests, and even genetic breeding

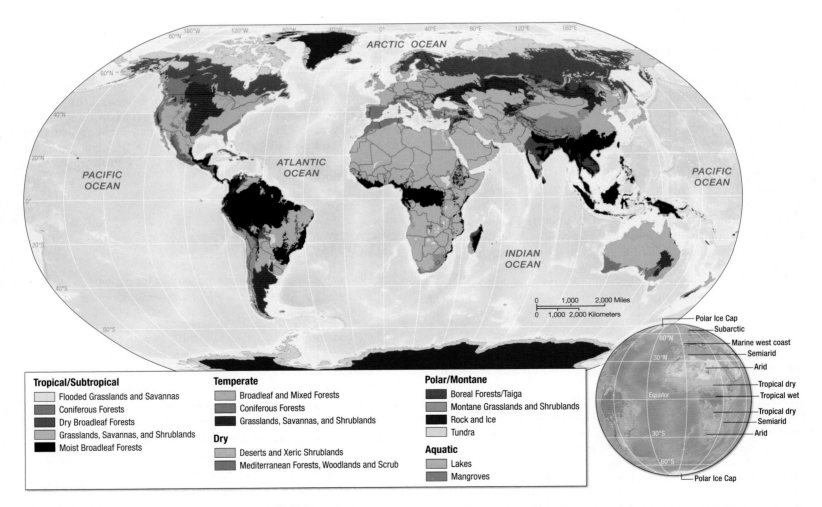

▲ **FIGURE 1.15 Global Biomes** The geography of vegetation can be classified into biomes with close links to the map of climate zones (Figure 1.9). The major biomes shown here include tropical, temperate, dry, and polar/montane, each subdivided according to vegetation types such as grasses (savannas, grasslands, montane, taiga, and tundra) and forests (coniferous, broadleaf, and boreal).

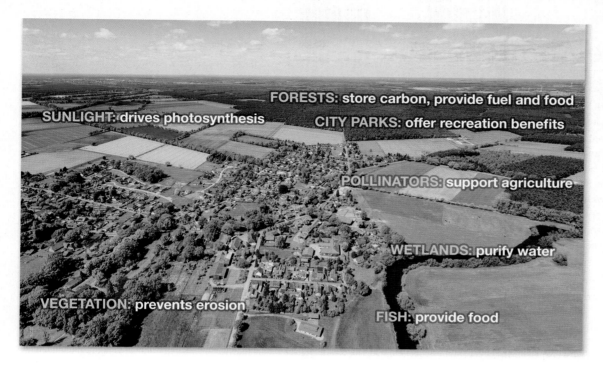

◀ **FIGURE 1.16 Ecosystem Services** Humanity receives many useful services from ecosystems including food, recreation, wood, and processes that help keep air and water clean and reduce flooding and heat waves.

◀ **FIGURE 1.17 Rewilding** Fauna that populated North America during the Pleistocene included mammoths, saber tooth cats, dire wolves, and a zebra like horse. Reintroducing major species or their equivalents is imagined to encourage the return of pre-human ecosystems.

Human-Influenced Ecologies Human activities are by no means always ecologically destructive. Wherever people travel or migrate, they bring other species with them. Sometimes this is intentional, as when humans introduce new agricultural species, and sometimes it is accidental, as in the case of many pests and predators. The selective breeding of crops and animals for food and fiber has evolved to include genetic modification (GM) and biological manipulation to increase production or pest resistance. Roughly one-tenth of cropland is now planted to GM varieties including corn, soybeans, and cotton. While some people are concerned about the implications of these crops for native agricultural diversity, human health, or local ecosystems, others see them as the creative solution to food security and climate change.

Humans may not be acting intentionally when species from one place "hitchhike" to new locations along with humans. Nonetheless, these introduced plants and animals can thrive at their destination, interacting with native species and habitats to create new ecological mixes, land covers, and ecosystems. **Invasive species** can overtake pastures, disrupt water systems, kill or drive other species to extinction, and wreak expensive and unanticipated havoc on important local plant or animal resources. The total costs of losses due to these invasive species in the United States exceed $138 billion annually.

of cattle in efforts to produce cattle with some of the characteristics of extinct wild cattle called auroch or the extinct animals of the Pleistocene period (**FIGURE 1.17**). Efforts to protect biodiversity range from endangered species protection and reserves to international treaties such as the Convention on Biodiversity.

Welcome to the Anthropocene

For much of the Holocene, (the last 11,700 years before the present) the human impact on the Earth system was local and regional. With the start of the Industrial Revolution, around 1750 C.E., human impacts on land-use change and pollution became detectable at the global level. This has been proposed as the start of a new epoch—the Anthropocene—the 'new human era' where human activity is driving planetary scale processes, especially through the increasing concentration of greenhouse gases in the atmosphere and the widespread extinction of species.

The idea of the Anthropocene is captured by graphs that show the dramatic increase of human activity and environmental impacts since the Industrial Revolution (**FIGURE 1.2.1**). The human drivers of change include population growth; the consumption of water, paper, energy and fertilizer; and the number of automobiles. The measures of global human impact include the concentration of carbon dioxide and methane in the atmosphere, the loss of tropical rainforests and biodiversity, and nitrate pollution of coastal

waters. You can see an acceleration in the drivers and impacts after 1950 where the graphs start to change more rapidly—often called the great acceleration. Each world region has contributed to Anthropocene drivers and impacts in different ways. For example, Europe cleared forests and increased pollution earlier than other regions, but since 1980 East Asia, especially China, has been playing a larger role in pollution as it industrializes and exports and as per capita consumption increases. **Deforestation** remains an ongoing problem across world regions

▼ **FIGURE 1.2.1 The human impact on the earth system in the Anthropocene** These graphs show the increase and acceleration of human impacts on the earth system since 1750.

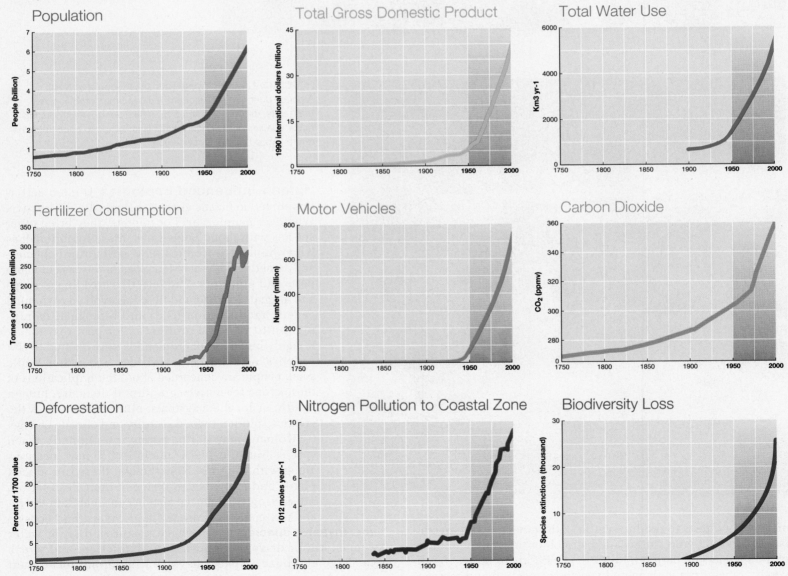

(FIGURE 1.2.2). One indication of current impacts is the **ecological footprint** of nations and regions—the area required to provide the renewable resources humanity is using and to absorb its waste.

The Anthropocene epoch raises important questions for humanity. If our activities are transforming our planet, do we need to think about managing these changes at a global level? Which of these changes are dangerous to our survival or that of other species? What does the onset of the Anthropocene mean for international cooperation and for our individual responsibilities for pollution and other environmental impacts? In all world regions the recognition of the Anthropocene has drawn attention to the urgency of sustainable development (FIGURE 1.2.3).

"The Anthropocene epoch raises important questions for humanity."

1. **Visit the Global Footprint Network website: http://www.footprintnetwork.org. Select "Footprint for Nations" from the "Footprint Basics" menu, and choose three countries. Compare the ecological footprints of the countries you selected. What makes one country's footprint different from another?**

2. **Analyze your individual eco-footprint, using the footprint calculator on the homepage of the Global Footprint Network. How does your lifestyle compare to your country's average and to those of the countries you selected?**

◄ FIGURE 1.2.2 **Deforestation and the Anthropocene** Humans have cut large areas of forest for agriculture, timber, and fuel in many regions, such as here in Pará, Brazil.

◄ FIGURE 1.2.3 **Anthropocene landscapes in Dubai** Humans have transformed this desert coastal environment by reengineering the coast, vegetating the landscape, and constructing soaring buildings.

Zebra mussels from the Black Sea are now thriving in the Great Lakes of the United States, while water hyacinth from the Amazon has spread throughout the tropics of Africa and South Asia, choking waterways and creating new ecosystems.

Apply Your Knowledge

1.7 What are some of the challenges to biodiversity around the world? Do you think rewilding is a viable solution to some of these issues? Why or why not?

1.8 The World Wildlife Fund has the goal of protecting ecosystems and biomes and has identified priority places based on their biodiversity and the potential for them to be damaged. Select one of these special places from the map at http://www. worldwildlife.org/places. What type of ecosystem did you select? What are the threats to this ecosystem?

Threats to Biodiversity

https://goo.gl/3iyhsv

▼ **FIGURE 1.18 Sustainability in Tucson, Arizona** Three of the authors teach and live in Tucson, where there are efforts to move toward more sustainable food, water, and energy systems. (a) Water is conserved through rainwater harvesting and the use of wastewater for irrigating parks. (b) Sustainable agriculture includes students growing local food varieties using water conservation and recycled compost. (c) Electricity generated by solar energy reduces greenhouse gas emissions.

(a)

(b)

(c)

Sustainability and the Future

There is a growing concern that the recent rate of change in climate and ecology in the Anthropocene has brought us close to thresholds that, if exceeded, could lead to rapid and irreversible changes and to serious risks for much of humanity and other species. We are reaching danger zones with regards to climate change, biodiversity loss, and pollution that present serious problems for the security of food, water, energy, and health, especially for vulnerable populations. There have been calls for humanity to reduce our impact on the planet through adopting principles of **sustainability**, which allow us to meet current and future human needs while simultaneously preserving our world's precious environmental resources. Sustainability is often seen to include goals of meeting economic, social, and environmental needs at the same time. In business this is known as the "triple bottom line" of serving profit, people, and the planet through making money, while also providing fair wages and community benefits and minimizing environmental impact. Some are cynical about business claims for sustainability, suggesting that the few actions they take to protect the environment are a form of "greenwashing" or are only implemented as a result of public pressure or government regulation.

Sustainability has become an important element of policy and actions in all world regions as a result of public pressures, leadership and international agreements, which promote more sustainable approaches to everyday living and the economy. For example, sustainable food systems usually promote the conservation of water and soils and the use of compost; reduce the use of polluting chemicals; and promote recycling, community participation, and animal welfare. Sustainable energy systems focus on conservation, efficiency, and the use of renewable sources such as solar energy. Sustainable cities are designed to reduce the use of the automobile, provide healthy environments and promote homes and buildings that are built to conserve energy and water (**FIGURE 1.18**). For example, in the South Asian world region, India has curbed air pollution by mandating the use of natural gas in urban transportation fleets, while within the European world region the European Union (EU) has a large number of laws to promote sustainability including regulations to support renewable energy, control greenhouse gas emissions, ensure water quality, and protect biodiversity.

Apply Your Knowledge

1.9 What are some of the factors that motivate business to develop sustainable practices? How can consumers motivate business sustainability?

1.10 Identify a company or city that has a published sustainability policy or report. List at least three actions the company or city is taking to promote sustainability.

History, Economy, and Territory

The historical, economic, and political geography of the world is a story of interconnection and isolation, movement and settlement, growth and retreat. Places in the world have responded differently over time to changing events. This section examines the geographic complexity of the past and the present. It does so by highlighting how the processes of globalization and regionalization foster global interconnection and regional economic and political conditions in the emergence of the modern world.

Historical Legacies and Landscapes

For most of the world's history, movement, not settlement, has dominated the lived experiences of human societies on Earth.

Long-term movements are an essential part of the human story. As people have migrated from place to place, they have developed and spread new knowledge systems—hunting and gathering, language, religion, and animal and plant domestication. The interactions that have occurred among people from different world regions have had a lasting impact on the landscapes and places we study today.

European Colonialism, Capitalism, and the Industrial Revolution In about 1500 C.E., a rapid reconfiguration of world regions occurred, with implications for the present day. As Europeans circumnavigated the globe, they expanded into the Americas, Africa, and Asia, becoming a driver of new world regional integrations. Europeans dominated how places were connected for several centuries thereafter. **Colonialism** is a political and economic system in which regions and societies are legally, economically, and politically dominated by an external society. In this uneven connection, the colonizer works to reshape the regional economy and politics of the colonized region to suit its own needs.

European colonialism took place in two waves. Each wave had different regional and global implications.

- **The First Wave** Extending from the late 1400s through the early 1800s, the first wave of colonialism was marked by European investment in port cities and trading networks (**FIGURE 1.19**). Led primarily by the Spanish and Portuguese, with most of the extensive activity in South and Central America, this period

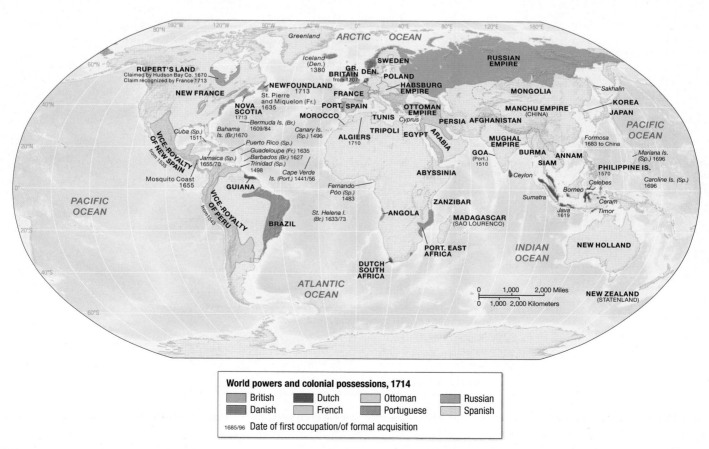

▲ **FIGURE 1.19 European Colonialism in 1714 C.E.** By the middle of the first wave of colonialism, European possessions amounted to less than 10% of the world's land area and only 2% of the world's population.

was characterized by the development of **plantation economies**, the classic institution of colonial agriculture. These economies were extensive, European-owned, operated, and financed enterprises in which crops were produced by local or imported labor for a world market. The massive sugar plantations in the Caribbean were signature examples of colonialism. The enormous demand for labor to work these plantations led to the forced movement of millions of Africans into slavery, creating a connection between West Africa and the Americas that had dramatic implications for the contemporary world.

- **The Second Wave** Beginning in the early 1800s and lasting through the mid-20th century, the second wave of colonialism incorporated the majority of the African continent as well as large parts of Asia, Australasia, and the South Pacific into Europe's direct sphere of influence (**FIGURE 1.20**). The growth of **imperialism**, which is the extension of the power of a state through control of the economic life of other territories, reorganized

the world around a number of colonial powers and their colonies. This period was characterized by heavy exploitation of agricultural and mineral resources and direct and total control over local affairs in colonized regions (**FIGURE 1.21**). European colonial expansion helped fuel the **Industrial Revolution**, the rapid development of mechanized manufacturing that gathered momentum in the early 19th century as well as the evolution of **capitalism**, a system of social and economic organization characterized by the profit motive and individual and corporate ownership of productive goods and resources. Europeans often perceived their colonies as underdeveloped. This perception meant that they saw places in Asia and Africa as ripe for exploitation based on the notion that European culture and economy were more sophisticated and therefore better. The exploitation of resources in Asia and Africa under colonialism led to economic stagnation in many colonies.

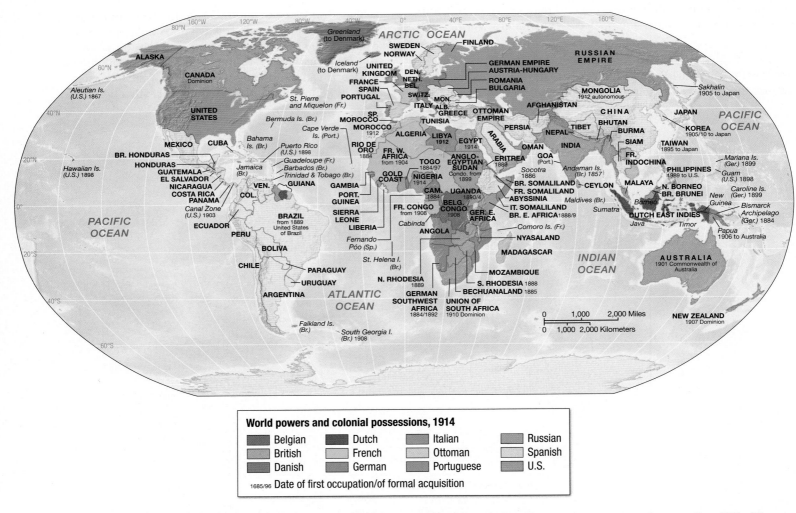

▲ **FIGURE 1.20 European Colonialism in 1914 C.E.** By the middle of the second wave of colonialism, European colonies amounted to more than 55% of the world's land area and 34% of the world's population.

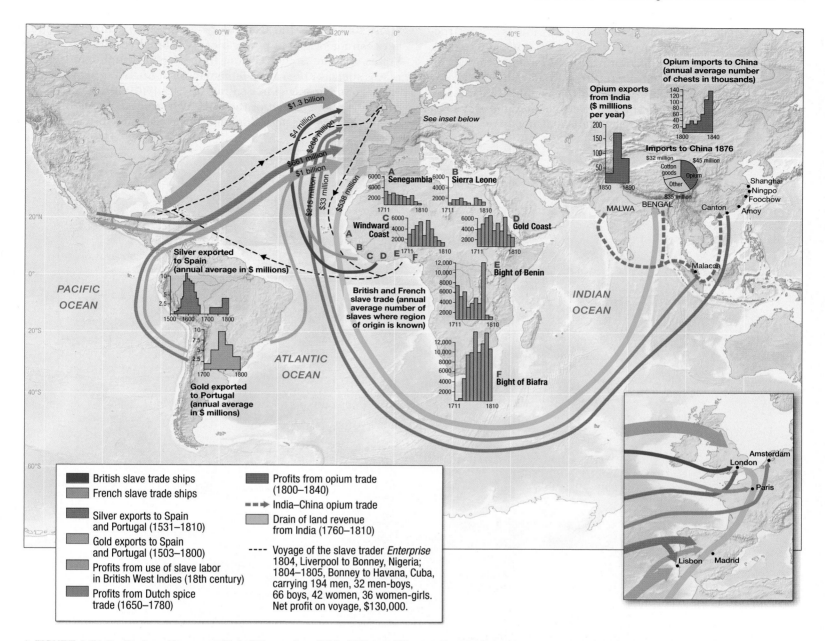

▲ **FIGURE 1.21 Profits from European Global Expansion, 1500–1876 c.e.** This map illustrates the profits generated through European plunder of global minerals, spices, and human beings over more than three centuries. **Describe the pattern of profit flows present in the world from 1500–1873 C.E. What were some of the most profitable commodities during this period?**

The Process of Decolonization In the second half of the 19th century, there was a vast increase in the number of colonies and the number of people under colonial rule. Revolution throughout the Americas led to the emergence of newly formed independent countries in those world regions outside of direct European rule. During the second wave of colonialism, the costs of maintaining this kind of power and influence weakened dominant colonial powers, leading to a fairly rapid period of decolonization following World War II. The legacy of colonialism has lasted into the postcolonial period, as a growing European presence throughout the world restructured the political geographies of many regions through the demarcation of political boundaries around their colonial holdings. Many of the world's political

boundaries and regional formations today have their seeds in the period of European colonial expansion.

Communism and the Cold War New political and economic philosophies emerged in the 19th and 20th centuries, establishing an alternative to the imperial expansion of Europe's capitalist states, leading to the creation of new world regions. Indeed, **communism**, a form of economic and social organization characterized by the common ownership of industry, transportation, agricultural land, and other key economic and social resources, spread in different forms to many parts of the globe. Communism influenced the establishment of a variety of state socialist systems in countries as diverse as China, Czechoslovakia, and Cuba.

The emergence of communism, along with the massive destruction of much of Europe during World War II (1939–1945 C.E.), also facilitated the rapid decolonization of much of the world. This rapid period of decolonization helped usher in a new world regional reorganization, as countries realigned themselves through their connections. The emergence of communism as an alternative economic system to capitalism also ushered in what

many defined as the **Cold War**—a period in which struggles between major global power blocs, centered around the United States and the Soviet Union, took place through conflicts in places such as Korea and Vietnam in Asia, Angola in sub-Saharan Africa, and Nicaragua in Central America (**FIGURE 1.22**).

The global conflict of the Cold War produced new regional configurations. Western Europe became economically integrated

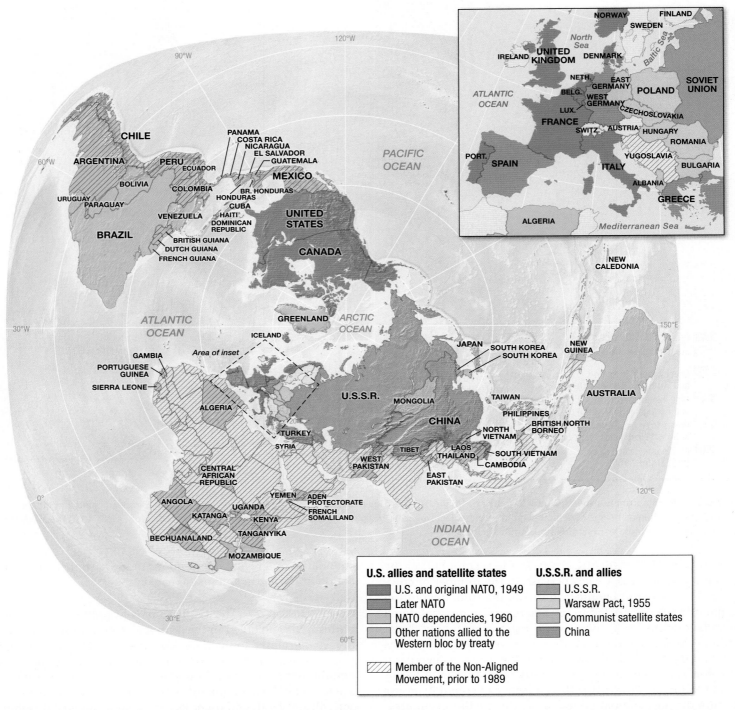

U.S. allies and satellite states
- U.S. and original NATO, 1949
- Later NATO
- NATO dependencies, 1960
- Other nations allied to the Western bloc by treaty
- Member of the Non-Aligned Movement, prior to 1989

U.S.S.R. and allies
- U.S.S.R.
- Warsaw Pact, 1955
- Communist satellite states
- China

▲ **FIGURE 1.22 The Alliances of the Cold War** This map depicts the complex global geography of the Cold War, illustrating how various blocs emerged in this global system of conflict and tension. Illustrated here is also a group of states that declared themselves "not aligned" with either bloc during the Cold War. **Briefly describe the geography of the Cold War. Where were the majority of "non-aligned" countries?**

with the United States and Canada, whereas Eastern Europe was more tightly linked to Russia and parts of Asia. The end of the Cold War, in 1991, resulted in yet another radical reconfiguration of political and economic connections.

Future World Regional Systems The world regional system we know today is not the first such system to have existed. At other moments in the past, different regions held the center of the world stage. Indeed, today's centers of economic power were marginal areas in other periods. Changes in the relative economic and political power of world regions have resulted from changes in their *connections* and *relations.* The rise of Europe over the last 500 years or the United States over the last 75 years was not a product of something unique to European or U.S. culture, resources, or politics. The future regional centers of economic and political power may come to reside in entirely new regions, as is clear from the rise of the **emerging region** known as BRICS, which includes Brazil, India, China, Russia, and South Africa (**FIGURE 1.23**).

Apply Your Knowledge

1.11 Describe the geography of the Cold War. Who were the major players, and how were they aligned in relation to each other?

1.12 How have regional relationships changed in the wake of the Cold War? Do you observe any political legacies that remain from the Cold War era?

Economy, Accumulation, and the Production of Inequality

In spite of the enormous growth of the global economy in the last several decades, there remains a huge gap between the wealthy and poor in the world. These gaps in wealth, particularly at the scale of countries and world regions, are partially a result of the types of economies that exist in different places. Some regions are dominated by extractive industries or agriculture, specializing in farming, fishing, or mining raw materials. Others specialize in manufacturing. Some have little of either of these activities and provide services, like banking, while others specialize in computer software or telecommunication.

Economic Sectors and Regional Economies Economic geographers categorize economic activities into four types: **primary** (extractive), **secondary** (manufacturing), **tertiary** (services), and **quaternary** (information) sectors (**FIGURE 1.24**). When we look at these four sectors, we see regional shifts over time. The United States and Canada were almost exclusively a site of primary sector activities in 1865. By 1960, this world region had become a largely secondary industrial economy, only to become largely a tertiary economy, with few overall jobs in manufacturing by the year 2000.

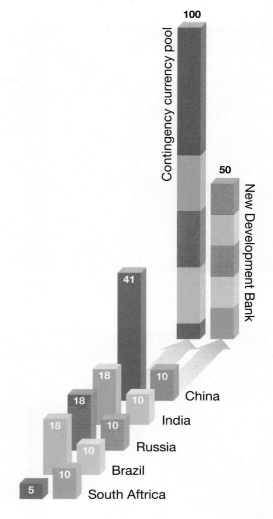

*All contributions
in billion dollars*

▲ **FIGURE 1.23 BRICS Development Bank** The BRICS countries established a development bank in 2014 to counter the global domination of supranational economic organizations such as the International Monetary Fund (see pp. 29–30 for discussion of the International Monetary Fund). The $100 billion dollar fund will provide development monies within and beyond the BRICS member states, creating increased connections between these noncontiguous but increasingly connected places.

This change in the international division of labor is often attributed to the shift of some regions—the United States and Canada, Europe and parts of East Asia, South Asia, Southeast Asia, and Australia, New Zealand, and the Pacific—from manufacturing to service industries. As a result, manufacturing has moved into several new world regions, such as Latin America and the Caribbean, sub-Saharan Africa, Southeast Asia, and parts of South Asia and East Asia. These changes are partially a result of the ability of **multinational corporations (MNCs)**—companies with locations throughout the world—to **outsource** unskilled jobs from places with expensive and highly regulated labor laws to places with cheaper labor costs

Extractive

Primary activities are any form of natural resource extraction. These include agriculture, mining, fishing, and forestry.

Manufacturing

Secondary activities are concerned with manufacturing or processing. These include steelmaking, food processing, furniture making, textile manufacturing, and garment manufacturing.

Services

Tertiary activities are those that involve the sale and exchange of goods and services. These include warehousing, retail stores, personal services such as hairdressing, commercial services such as accounting and advertising, and entertainment.

Information

Quaternary activities include handling and processing knowledge and information. Examples include data processing, information retrieval, education, and research and development (R&D).

▲ **FIGURE 1.24 Four Sectors of the Global Economy** Geographically, the division of labor between people working in these sectors varies, as people in some parts of the world live in areas dominated by primary sector economies, while others live in places that have mostly service sector and information sector jobs. **Can you think of other examples of primary, secondary, tertiary, and quaternary activities? Where do you think you might find the majority of primary activities in the world today?**

and less regulation of working conditions. MNC's have also taken advantage of new service economies in countries with little financial regulations. The result is the **offshoring** of key assets, allowing MNC's to take advantage of tax benefits in host banking countries. These offshoring practices in **global financial centers** often have very little impact on the broader economy of the host country, instead serving the interests of small groups of elites.

In recent decades, other world regions have diversified their economic activities and taken advantage of the global interconnections of production and consumption to drive economic growth. China has taken on a great deal of manufacturing for the global market and is now a significant economic player in other areas of the global economy, including banking. These shifts have altered the playing field of economic power relations both within East Asia and in the world more generally (**FIGURE 1.25**). Despite this change, many parts of the world continue to remain underdeveloped, largely a result of the historical inequalities built into the global economic system. In short, despite the rise of new economies, such as those in Brazil, India, China, South Africa and Russia, the world remains sharply divided between "haves" and "have nots." The key explanation

Gross domestic product, 2014
(% of world total)

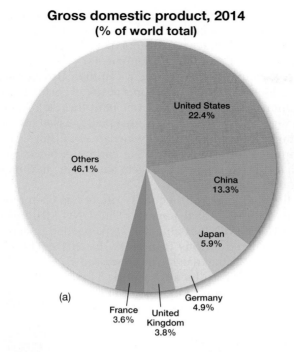

(a)

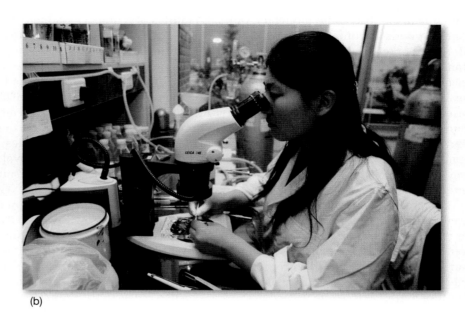

(b)

▲ **FIGURE 1.25 China's Changing Role in the World Economy** (a) China's share of the world's Gross Domestic Product (GDP); (b) A worker in a Chinese biotech firm engages in research.

for this economic difference is that world regions are *connected* through **commodity chains**: a network of labor and production processes that link world regions together in ways that may reinforce economic differences (**FIGURE 1.26**).

The problems of economic disparities and development underline the way regional differences can have dramatic and disturbing implications. People living in grinding poverty and those accumulating wealth are often spatially separated across the globe. But it is often the *connection* of these people that is the driving force of that inequality. Configurations inherited from the Cold War and colonialism often stubbornly persist into the 21st century, with implications for development and the quality of people's lives. The challenge of global development is one of changing the configurations between regions and the people who live in them.

Measuring Economic Development Levels of economic development are often measured using national economic indicators that provide a snapshot of the overall economy. **Gross Domestic Product (GDP)** is one such number. It is an estimate of the total value of all materials, foodstuffs, goods, and services produced by a country in a particular year. To standardize for

Equal Exchange Coffee Chain

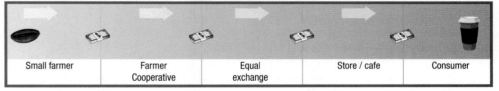

| Small farmer | Farmer Cooperative | Equal exchange | Store / cafe | Consumer |

Conventional Coffee Chain

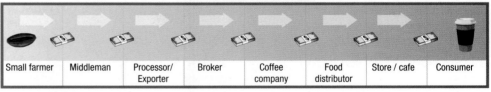

| Small farmer | Middleman | Processor/ Exporter | Broker | Coffee company | Food distributor | Store / cafe | Consumer |

▲ **FIGURE 1.26 Coffee Commodity Chain** A number of intermediaries, buyers, and sellers may exist between coffee producers and consumers. In the conventional coffee commodity chain, brokers and distributors take significant shares of the profits, whereas in equal exchange systems, a shorter and more cooperative chain assures greater returns for small farmers.

countries' varying sizes, the statistic is normally divided by total population, which gives an indicator, *per capita* GDP, which provides a measure of relative levels of overall economic development. **Gross National Income (GNI)** also includes the net value of income from abroad—flows of profits or losses from overseas investments, for example.

In making international comparisons, GDP and GNI can be problematic. They do not include nonmarket goods and services, which lack an observable monetary value. Examples of a non-market goods include wildlife or a coral reef. An example of a nonmarket service is the preparation of food for a family done by a parent. Indicators such as GDP and GNI also obscure inequality within a society. They do not measure differences in health care or differences between women's and men's access to resources or participation in the economy. An alternative indicator compares national currencies based on **Purchasing Power Parity (PPP) per capita**. In effect, PPP measures how much of a common "market basket" of goods and services each currency can purchase locally, including goods and services not traded internationally.

As shown in **FIGURE 1.27**, most of the highest levels of economic development are found in the **global north** as opposed to the **global south**. Some use the population-based term **minority world** to represent the economically wealthier global north and **majority world** to represent the global south, further illustrating the inequality of global economic well-being (**FIGURE 1.28**).

Patterns of Social Well-Being Global inequality is reflected in measures other than income. For example, patterns of life expectancy, a reliable indicator of social well-being, show a dramatic difference between the global south and global north. For adults in industrial countries in the north, life expectancy

is high and continues to increase. In contrast, life expectancy in the poorest countries is dramatically shorter. As explained previously, important development indicators like GDP are not useful for explaining these discrepancies. One useful alternative metric is the UN Development Programme's **Human Development Index (HDI)**, which is based on measures of life expectancy, educational attainment, and personal income (**FIGURE 1.29**). Another is to look more deeply at the drivers of inequality based on social categories, such as gender, whereby we can see differences in how development indicators related to certain experiences of gender, such as life expectancy, access to education, and income equality (**FIGURE 1.30**).

Despite the challenges of inequality, there has been a push by the UN to address global poverty. The UN's Millennium Development Goals (MDG) established steps needed to achieve significant global reductions in poverty by 2015: The eight measure-specific areas that need to be addressed for poverty reduction to be successful. The goals include eradication of extreme poverty and hunger; achievement of universal primary education; promotion of gender equality and empowering women; reduction of child mortality; improvement of maternal health' and combating HIV/AIDS, malaria, and other diseases; ensuring environmental sustainability;

Millennium Development Goals

https://goo.gl/r7S1Gf

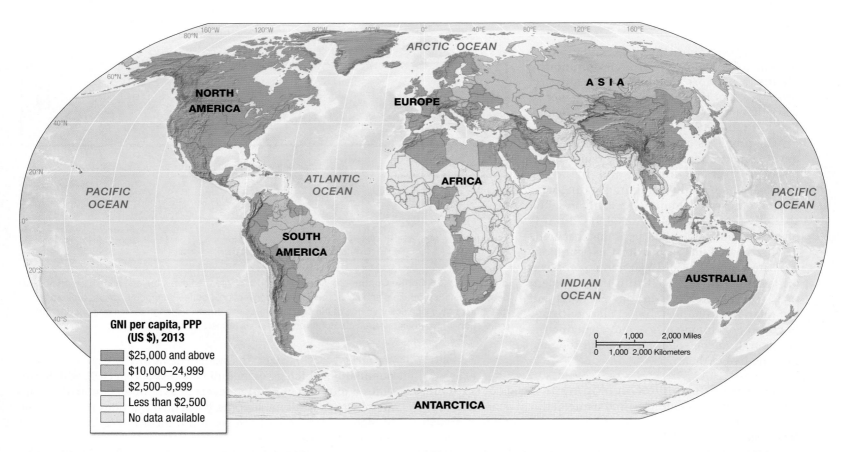

▲ **FIGURE 1.27 GNI per Capita, Purchasing Power Parity (PPP)** PPP per capita is one of the best single measures of economic development. This map, based on 2013 data, shows the tremendous gap in affluence among the countries of the world. The world's average annual GNI per capita PPP is $10,700. The gap between the higher per capita PPPs ($102,610 in Norway, $86,790 in Qatar, and $54,040 in Singapore) and the lower ones ($260 in Burundi, $690 in Afghanistan, and $950 in Cambodia) is huge.

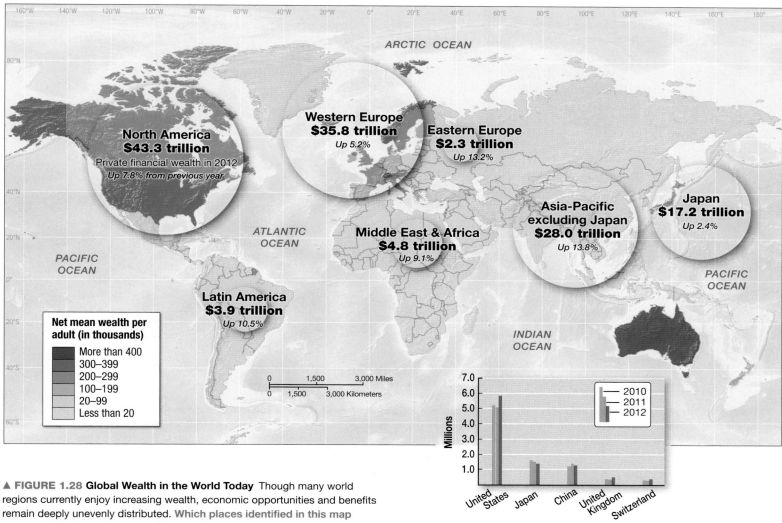

▲ **FIGURE 1.28 Global Wealth in the World Today** Though many world regions currently enjoy increasing wealth, economic opportunities and benefits remain deeply unevenly distributed. **Which places identified in this map witnessed the highest percent increase in net personal wealth in 2012? Which had the least? What do you think explains these differences?**

partnership for development. The 2015 assessment of these goals shows that there has been a lot of positive change in the areas of poverty alleviation, food security, ending debt, maternal mortality, and providing safe water and sanitation. Indeed, all the development goals identified have shown positive progress with the exception of greenhouse gas emissions (see "Welcome to the Anthropocene," Figure 1.2.1). Based on the success of the MDG 2015, the UN Development Program has laid out a plan for achieving continuing development beyond 2015. More generally, the MDG 2015 progress suggests that if we continue to tackle the causes and consequences of poverty and couple those with decreases in fertility, our goals for the year 2050 for a more sustainable planet are relatively feasible (**FIGURE 1.31**).

Explaining and Practicing "Development" There is a general worldwide interest in explaining the abovementioned areas of inequality, so they might be alleviated. Regrettably, there is no universal agreement over the causes of inequality. As a result, there is no consensus over the strategies that should be used to achieve **development**, which is an improvement in people's economic and social well-being as well as their standard of living.

Despite a lack of consensus, many different development experiments have been attempted.

- **Neoliberal Development Model** On one side of this discussions are **neoliberal development** economists and politicians who hold a strong belief that the best development policy is one in which state budgets are reduced, public ownership of industries or utilities is turned over to private parties, and laws regulating working conditions and the environment are minimized. For these observers, it is the *presence* of government subsidies, rules, and interventions that cause poverty. The **International Monetary Fund (IMF)** has championed this approach. The IMF is a powerful international financial institution that provides emergency loans to countries in financial trouble. The IMF sets conditions for its loans that emphasize privatization, limited restrictions on imports, smaller state budgets, and liberalized trade. Many countries around the world have attempted these policies, often with poor results.

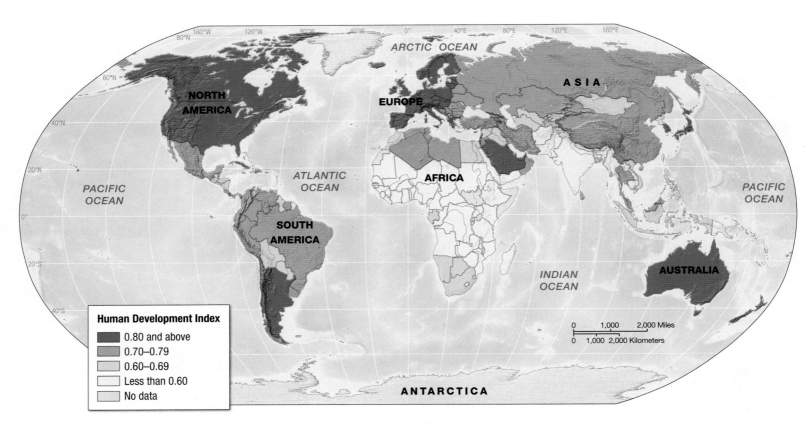

▲ **FIGURE 1.29 The Human Development Index** This index is based on measures of life expectancy, educational attainment, and personal income. A country that had the best scores among all of the countries in the world on all three measures would have a perfect index score of 1.0, whereas a country that ranked worst in the world on all three indicators would have an index score of 0. Most of the affluent countries have index scores of 0.8 or more. The worst scores—those less than 0.6—are concentrated in Africa.

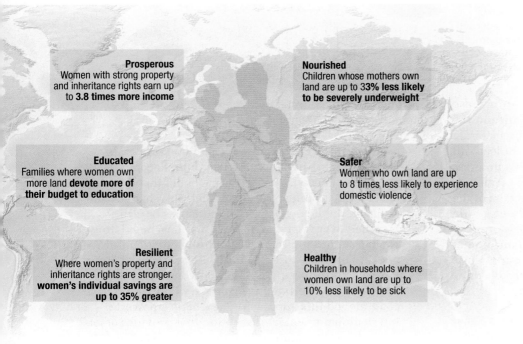

Prosperous
Women with strong property and inheritance rights earn up to **3.8 times more income**

Nourished
Children whose mothers own land are up to **33% less likely to be severely underweight**

Educated
Families where women own more land **devote more of their budget to education**

Safer
Women who own land are up to 8 times less likely to experience domestic violence

Resilient
Where women's property and inheritance rights are stronger. **women's individual savings are up to 35% greater**

Healthy
Children in households where women own land are up to 10% less likely to be sick

▲ **FIGURE 1.30 Women's Land Rights and Development** There is a direct relationship between women's land ownership and human development. When women have access to key resources, rates of educational attainment, health, and fiscal savings are higher and violence against women and the malnourishment of children is lower. **Why do you think there is such a strong relationship between educational attainment and women's and children's health and well-being?**

- **Alternative Development Model** On the other side of the discussions are **alternative development** practitioners, who call into question the basic tenets of **neoliberalism** such as a universal faith in markets, a focus on reducing government programs and roles, and the assumption that development is a purely economic process. Some alternative development organizations empower grassroots movements and promote local knowledge, which they believe can repair the damage done by neoliberal development projects that have focused on the production of wealth (i.e., money) and not on quality of life (i.e., a social good). Alternative development stresses the autonomy of local communities, innovation of technologies rooted in local knowledge, and connections between people and groups around the world.

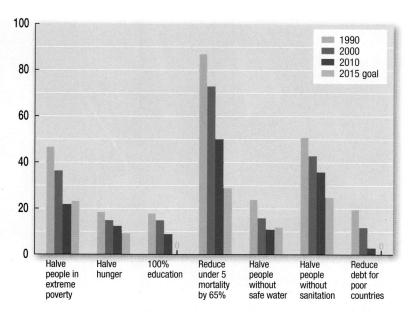

▲ **FIGURE 1.31 Millennium Development Goals** The MDGs are goals that were set out by the UN to end poverty and hunger globally.

Apply Your Knowledge

1.13 Define the term GDP and explain why it is not always the best indicator of economic development.

1.14 What are some other ways to measure and assess development?

Territory and Politics

Like the world economy, the political geography of world regions has its roots in collective global history. Indeed, territories reflect past global configurations and relations. New regional configurations over the last 400 years, for example, are tied to the rise of modern states and nations.

States and Nations A **sovereign state** exercises power over a territory and people and is recognized by other states. The independent power of a sovereign state is codified in international law. These laws give states the right to develop political and economic systems, regulate markets, and invest in the social and economic enterprises of its people. A **nation** is a group of people sharing common elements of culture, such as language and religion. Ultimately, a national identity is built on a common sense of history, geography, and purpose. National identity sometimes compels people to defend the nation and to further the objectives of the state. In rare instances, a state boundary aligns with one national (or ethnic) group. The term **nation–state** is used to describe this ideal state, one consisting of a somewhat homogeneous group of people living in the same territory. Today, there are very few true nation–states. Historical patterns of mobility have resulted in states that govern highly multicultural populations. Even so, **nationalism**, the feeling of belonging to a nation

as well as the belief that a nation has a natural right to determine its own affairs, remains a powerful organizing force of the current state system (**FIGURE 1.32**). Many observers argue that one response to globalization has been a rise in nationalism, which has brought both peace and conflict.

Today, there are over 200 sovereign states that regulate the flow of people, goods, and ideas across world regions as well as across particular national boundaries. States also facilitate greater or lesser regional cohesion, as state members struggle to maintain their position within an ever-globalizing world. Today's political and economic connections are negotiated through international laws and a state system that controls all kinds of movements. Border crossings, citizenship, trade laws, and responsibilities for the environment and resources are all managed through rules and laws within and between states (**FIGURE 1.33**).

▲ **FIGURE 1.32 An Election in India** India is the world's most populous democracy, and people turn out to vote in local, regional, and national elections.

▲ **FIGURE 1.33 International Border in Mae Sai, Thailand and Talicheck, Burma** International borders are often lively sites of everyday exchange. People move back and forth exchanging goods and services. As concerns over violence increase in certain regions, border checkpoints, such as this one, are more highly regulated.

States, like world regions, are not static. They are subject to change and even failure. In 2014, there were 16 states around the world, most of them in Africa, the Middle East, or South Asia, where the central government was so weak that it had effectively no control over its territory. When a government loses control, it is considered to be a **failed state**. At the same time, new states emerge, as was the case in 2011, with the creation of South Sudan, a country that remains tension filled to this day.

Political Globalization There is currently no coordinated system of global governance to accompany economic globalization. That said, **supranational organizations** have emerged to address some of the global issues we face as a world today. These organizations are collections of individual states with a common goal that may be economic or political and that diminish individual state sovereignty in favor of member states' interests.

- **The United Nations** The largest supranational organization is the **United Nations (UN)**, which was founded in 1945, to facilitate cooperation in international law, security, economic development, human rights, and world peace. In the absence of a deliberately organized global governance system, the UN has become something of a *de facto* replacement. With headquarters in New York City, the UN is composed of five administrative organs: the Security Council (**FIGURE 1.34**), which includes as permanent members the United States, Great Britain, China, France, and Russia; the General Assembly, which includes nearly all of the world's internationally recognized sovereign states (Vatican City, Kosovo, and Taiwan are a few exceptions); the Economic and Social Council; the Secretariat; and the International Court of Justice, operating since 1946 in The Hague in the Netherlands.

- **Social Movements** At the same time that formal political organizations like the UN have been expanding their role, informal political organizations and movements have also increased in overall numbers and membership and become more globally connected in the last 50 years. Known as **social movements**—large, informal groups of individuals and organizations that focus on social or political issues—these groups are part of a worldwide array of voluntary civic and social organizations and institutions representing the interests of citizens against the power of formal states and markets. Social movements embrace a spectrum of issues, including women's rights, the environment, human rights, climate change, and others. The issues they choose are ones that affect people very directly and around which people organize to obtain formal political responses.

- **Nongovernmental Organizations** Social movements are sometimes represented by more formally constituted organizations that are managed through some kind of coordinated administration. These **nongovernmental organizations (NGOs)** are among the most visible actors

▲ **FIGURE 1.34 President Barack Obama Chairs United Nations Security Council** The United States is one of the five countries that holds a permanent seat on the UN Security Council. There are 12 members of the Security Council and the Presidency of the Security Council rotates to each member nation for one month a year. In 2015, President Barack Obama held this position in December.

in global politics, their reach is international, and they work across and between regions. Notable examples include Greenpeace (headquartered in Great Britain), Genocide Watch (headquartered in the United States), and the Green Belt Movement (headquartered in Kenya).

The political institutions that make up the system of global politics—including nations, supranational organizations, social movements, and NGOs—are both challenged and strengthened by globalization. As objects move more swiftly and freely through free trade and Internet communication, they cross national boundaries and elude the control of state authorities. On the other hand, the globalization of information has reinforced the power of NGOs and elements of civil society by building new connections between states and regions. New political regions and relationships can potentially emerge out of these kinds of changes.

Apply Your Knowledge

1.15 Compare and contrast the terms sovereign state, nation, nation–state, and nationalism.

1.16 Turn to the Fragile States Index for 2014 (http://ffp.statesindex.org/rankings-2014) and list out the factors that identify states that are highly unstable politically today. Which countries are identified as "very high alert" or "high alert"?

Fragile States Index

https://goo.gl/dJbgWR

Culture and Populations

Economics and politics are not the only aspects of social relations that make up the character of regions. The various cultures of the broader population also define regional dynamics. Broadly speaking, **culture** is a shared set of meanings and practices. The material aspects of culture include such things as dress and house styles, whereas cultural symbols are objects that represent something else, such as a cross as a symbol of Christianity. The shared set of meanings that constitutes culture can include values, beliefs, practices, and ideas about religion, language, family, food, gender, sexuality, ethnicity, and nationality as well as other sets of identities (**FIGURE 1.35**). Cultural values, beliefs, ideas, and practices are—because of globalization and regionalization—subject to reevaluation and redefinition. It is often argued that instantaneous communication is making the world a uniform place, but such arguments don't take into account how new cultures are formed through the regional processes that connect the world's people.

This section traces both cultural and social patterns of people around the globe, as well as population densities, movements, and changes over time, or what geographers call **demographics**.

Culture, Religion, and Language

There are many aspects to culture that are of interest to geographers. Two of the most important are language and religion. The practice of language and religion ties strongly to a sense of place as well as the concept of *nationalism*. Language and religion, therefore, both create connections between people but can also result in divisions.

Geographies of Religion The term **religion** includes belief systems and practices that recognize the existence of powers higher than humankind. Although religious affiliation is on the decline in some parts of the world, it still acts as a powerful influence on daily life, from eating habits and dress codes to coming-of-age rituals and death ceremonies, holiday celebrations, and family practices. Religious beliefs and practices change as people develop or adopt new spiritual influences.

Religious missionizing—propagandizing and persuasion—as well as forceful and sometimes violent conversion have been key elements in changing geographies of religion historically. The Arab invasions following Muhammad's death in 632 C.E. spread Islam. The Christian Crusades of the Middle Ages sought to establish Christian control of Jerusalem. And, from the onset of the modern period in the 15th century C.E., European colonizers brought Christianity to the Americas and elsewhere.

Over time, some religions have become dislocated from their sites of origin, not only through missionizing and conversion but also by way of **diaspora**—the spatial dispersion of a previously homogeneous group—and emigration. Both diaspora and emigration involve the involuntary and voluntary movement of people who bring their beliefs and practices to new locales (**FIGURE 1.36**).

As a result, many religions have spread to so many different places that a global scale map cannot show the complexity of religious geographies. Any global map of religion is inadequate to represent the growing global population of agnostics, who remain unsure of the reality of divine beings and do not adhere to any one faith, as well as atheists, who have no belief in divine beings. Agnostics and atheists—who may make up more than 15% of the global population—share a strong belief in the maintenance of a strict division between church and state. So, **FIGURE 1.37** identifies the contemporary distribution and concentration of the world's largest religions, while largely ignoring the complexities introduced by local religious systems and nonreligious beliefs.

- **Religious Fundamentalism and Identity** The spread and intensification of religious **fundamentalism** is a widespread phenomenon not limited to any particular religion. In general, fundamentalism is a belief in or strict adherence to a

▼ **FIGURE 1.35 Indigenous People Protest Government in Ecuador** Cultures often inform how people imagine their place in the world. In places where a certain culture has been marginalized, people might come together to resist dominate cultural, economic, and political practices. In this image, an indigenous member of the Confederation of Indigenous Nationalities of Ecuador (CONAIE) participates in a demonstration near the Government Palace in Quito, Ecuador, January 2015. CONAIE staged a march demanding that the Ecuadorian Government transfer the ownership of their headquarters building to them. The government only agreed to loan them the building.

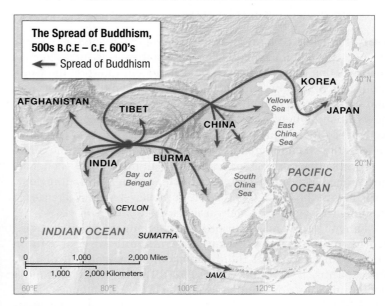

▲ **FIGURE 1.36 Spread of Buddhism, 500 B.C.E to 700 C.E.** Carried by monks, traders, and rulers, as well as the spread of sacred texts, Buddhism diffused out of India more than 2,500 years ago and is now distributed throughout East Asia.

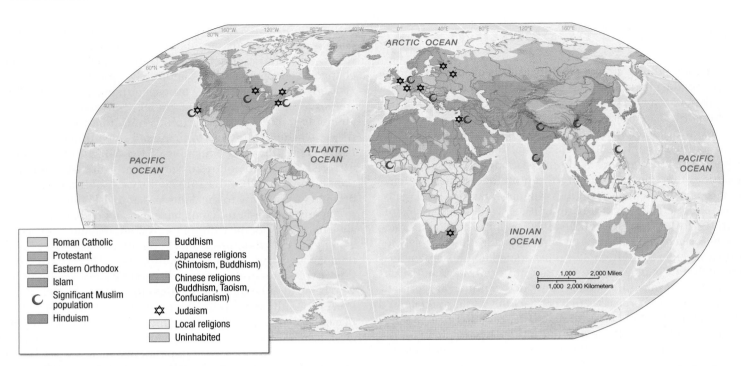

▲ **FIGURE 1.37 World Distribution of Major Religions** Most of the world's people are members of these religions. Not evident on this map are the local variations in practices, as well as the many local religions that are not easily shown on a world map. There is a growing number of people in the world who identify with no religion or as atheists, including large parts of China and Vietnam. **Identify some regions where one major religious abuts another? What kinds of creative interactions or conflicts might this geography produce?**

set of basic principles. Fundamentalism often arises as a reaction to perceived compromises to core values that are in conflict with recent changes to social and political life. An example could be the devaluing of certain religious practices, such as daily prayer, in favor of the responsibilities of work. Religious fundamentalism is an important force in global geographies as it is taken up as a political cause. Though most religious movements are thoroughly peaceful, fundamentalist-inspired terrorism also occurs.

Geographies of Language
Like religion, **language** is a central aspect of cultural identity. Language both reflects and influences the ways that different groups understand and interpret the world around them by providing each group with often-distinct concepts and vocabulary. Through institutions such as schools and courts, governments often promote the use of standard languages, also known as official languages. The global distribution of languages reflects the changing history of human geography and the forces of globalization and regionalization on culture (**FIGURE 1.38**). Languages can also be broken into **dialects**, which represent regional variations of those languages and feature place-based differences in pronunciation, grammar, and vocabulary (**FIGURE 1.39**).

Language diversity can present challenges in a globalizing world that depends on linguistic communication and interconnections. Consider the diversity of languages and dialects in a country like India, where 850 different languages make communication and commerce among different language speakers difficult. In an effort to prevent this, states often designate one national language to facilitate communication and the conduct of state business. Unfortunately, where official languages are put into place, indigenous languages may be lost. Linguists estimate that a language disappears as often as every two weeks.

Although the overall trend appears to be toward the loss of indigenous languages (and other forms of culture), there are movements to revive dying languages by introducing them into schools, offering courses for adults, or establishing native language radio stations (**FIGURE 1.40**). In this way, language has become an important vehicle for new cultural regions.

In other cases, regional formations come to rely on a common language, or **lingua franca**, to communicate across linguistic boundaries. Swahili operates as the *lingua franca* of East Africa, allowing people from different backgrounds to engage each other in various forms of economic and sociocultural exchange. It is quite common to find someone in East Africa who is bilingual, speaking their local language and the regional *lingua franca*.

Apply Your Knowledge

1.17 Look closer at diverse religious traditions in the United States by using the Pew Research Religious Landscape Survey: http://religions.pewforum.org. Select two traditions in the United States and trace their geographic distribution.

U.S. Religious Diversity

https://goo.gl/cvA62w

1.18 What are the basic beliefs, practices, and social and political views of the two religions you selected? Do the religions you selected have certain place-based rules and/or practices (e.g., wearing certain sorts of clothing in public, rules about mobility based on gender, or who has authority to lead and/or speak in religious places such as churches)? Explain.

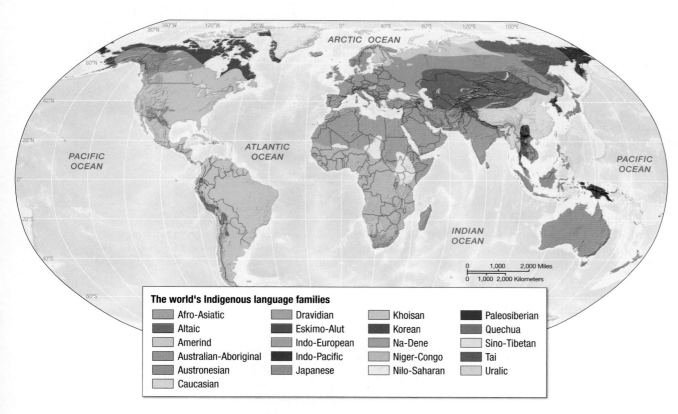

◄ **FIGURE 1.38 World Distribution of Major Language Families** Classifying languages by family and mapping their occurrence across the globe provide insights about important connections across space and time. It is possible through mapping to discover linkages between seemingly unrelated cultures and understand the movement of populations across the globe.

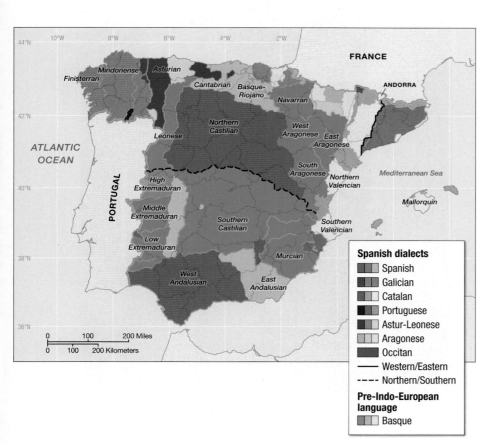

▲ **FIGURE 1.39 Dialects and Languages of Spain** Though Spanish is the common language of most of the Iberian peninsula in Europe, it has many dialects. Basque, which is an Iberian language, it is wholly unrelated to Spanish. **Where do you find Spanish dialects in Spain and where do you find languages other than Spanish? What do you think are the challenges produced by the linguistic diversity of Spain?**

www.kuyi.net

▲ **FIGURE 1.40 The Recovery of Languages Through Technology** KUYI is a Hopi language radio station founded by the Hopi nation, Native Americans living in the United States southwest. "Kuyi" (pronounced KUU-yi) is Hopi for "water," and the explicit goal of the station is to "reaffirm our respect for tradition by preserving our language and culture in a contemporary context."

Cultural Practices, Social Differences, and Identity

Globalization can produce a homogenization of culture through the proliferation of consumer goods. This material culture is predicated on Airbus jets, CNN, music video channels, cell phones, the Internet, Gap clothing, Nike shoes, iPods, Toyotas, Disney, and formula-driven Hollywood movies and fueled by Coca-Cola, Budweiser, and McDonald's. Yet neither the widespread consumption of U.S. and U.S.-style products, nor the increasing familiarity of people around the world with global media and international brand names, adds up to the emergence of a single global culture. The processes of globalization *are* exposing the globe's inhabitants to a common set of products, symbols, myths, memories, events, cult figures, landscapes, and traditions (**FIGURE 1.41**), though consumers of global products often adapt them to their regional cultures. At the same time deep cultural differences are opening up. The more people's lives are homogenized through their jobs and their material culture, the more many of them want to establish a distinctive cultural identity.

Culture and Identity Social categories such as ethnicity, race, sexuality, or class change over time as groups of people interact with each other. Generally, dominant forms of social organization persist for hundreds of years, if not longer. But both subtle and dramatic changes within these forms occur. Global media technologies, such as satellite television and the Internet, increasingly shape new potential social forms and reconfigure old ones. And, some groups use their identities to assert political, economic, social, and cultural claims.

- **Ethnicity** A socially created community identity, **ethnicity** is sometimes based on actual commonalities, such as language or religion, but also on commonalties that are perceived. Ethnicity can be used as a powerful geographic marker of national territory or more informally in ethnic neighborhoods (**FIGURE 1.42**). In some cases, ethnic identities become segregated from the wider society in ghettos or ethnic enclaves.

- **Race** Like ethnicity, **race** is a problematic and illusory classification of human beings based on skin color and other physical traits. Present-day notions of race began to emerge in the 19th century during the era of colonialism. Europeans put forward "scientific" research supposedly demonstrating how people could be classified racially based on physical characteristics. Even though race is a completely social category, with no consistent basis in biological reality, people often insist on the reality of race and racial differences. Social scientists use the concept of **racialization** to describe the practice of creating unequal races where white skin color or some other dominant identity is considered the norm. Cultural perceptions and practices like this can create problematic and discriminatory racialized places—such as in the homelands of South Africa or in the ghettos of Europe and the Russian Federation.

▲ **FIGURE 1.41 Bollywood** Bollywood is the name given to the regional site of film production in Mumbai, India. The Bollywood dance form has been picked up in popular culture across the globe, impacting not just dance but fashion as well.

▲ **FIGURE 1.42 A North African Neighborhood in Paris.** The strong connection between France and its former colonies is represented in the modern-day migration of North and West African peoples to France. These migrations have helped to create new neighborhoods and services like this food shop, catering to new populations.

- **Gender** Like ethnicity and race, groups form around the identity of **gender**. Gender should not be confused with sex, a biological category; gender is a category applied to people and linked to expectations about what different kinds of people can and ought to do. Gender varies across places and times. *Gender* is a category reflecting the socially learned differences between men and women. People may be born as biological males (a physically given condition), but they learn how to act as men (a culturally imposed condition). Certain places are gendered through the inclusion and exclusion of certain people based on their performance of their gender. Consider, for example, the bars or barbershops in many regions frequented almost exclusively by men.

  Gender interacts with other forms of identity and can intensify power differences among and between groups. For instance, women's subservience to men is deeply ingrained within South Asian cultures, and it is manifest most clearly in the cultural practices attached to family life, such as the custom of providing a dowry. The preference for male children in parts of East Asia is reflected in the widespread (but illegal) practice of selective abortion and female infanticide. Women in rich countries, on average, earn less money than men do in the same jobs.

- **Class** Distinctions of **class** are often assigned socially and internalized in the cultures and norms of groups. These distinctions, typically assigned based on wealth and access to resources, are often internalized culturally by groups and communities, creating an often self-conscious sense of class identity. These identities are further reinforced by stratified economic geographies, including neighborhoods and towns occupied by people of specific economic levels, further reinforcing a sense of unified identity (**FIGURE 1.43**).

  The contrast of poverty, environmental stress, and crowded living conditions with the luxurious lifestyles of wealthy elites creates a fertile climate for tension between rich and poor people, countries, and regions. As a result, some metropolitan areas face a prospect of communities polarized between wealthier and poorer people, with outright cultural conflict suppressed through means such as electronic surveillance (**FIGURE 1.44**), security posts, and urban design featuring security fences and gated streets.

- **Sexuality** The increasing attention being paid to human rights issues around the globe includes sexual rights as a political cause. **Sexuality** is a set of practices and identities related to sexual acts and desires. Sexuality is understood to be learned and expressed, or "performed," through practices that most people tend to view as "normal" or "natural." Where gay or straight identities can be performed varies from place to place. In some countries or neighborhoods, the sight of two men holding hands would be unremarkable, whereas in others it might be startling.

▲ FIGURE 1.43 **Working Class Identities and Geographies** Fast-food workers rallying in New York City for higher pay use the April 15, 2015 tax deadline to publicize their claims that they cannot survive on the hourly wages paid by many U.S. corporations.

▲ FIGURE 1.44 **Surveillance Cameras in New York City** In the post-9/11 era, there has been an increased investment in security measures in cities such as New York. These new surveillance systems are designed not only to capture particular illegal practices but to also deter such behaviors. **Where is this security camera possibly pointed? What impact does the presence of a security camera have on the people living and working there?**

Apply Your Knowledge

1.19 Which social categories—ethnicity, race, gender, sexuality, age, class, and religion—are most influential in human activity? Do you think some are more important than others in terms of employment, place of residence, and social life?

1.20 Keep a log of your activities for one full day, including where you go and what you do. Review the log and examine how your experience of particular places intersects with your own social identities (e.g., your gender, race, sexuality, ethnicity, religion, age, or class).

Demography and Urbanization

As the world population density map demonstrates, some areas of the world are very heavily inhabited, whereas others are only sparsely populated (**FIGURE 1.45**). Most of the world's inhabitants live on just 10% of the world's land near oceans or seas, along rivers with easy access to a navigable waterway, and in temperate, low-lying areas with fertile soils. Climate and Earth's topography explain a great deal of population distribution, as harsh, hot or cold places with a dearth of life and resources tend to inhibit large populations.

Within this limited area, human population continues to grow. In 2011 the world's population passed the 7 billion mark. With an ongoing increase of 1.2% annually, the world is projected to contain nearly 9.3 billion people by the year 2050. The distribution of this projected population growth is noteworthy. Population growth is predicted to occur overwhelmingly in the regions least able to support it. Just six countries will account for half the increase in the world's population: Bangladesh, China, India, Indonesia, Nigeria, and Pakistan.

Most of this future population will live in urban areas, since rural to urban migration is ongoing around the world. The creation of **megacities**, huge urban centers, to accommodate this growth may challenge a region's resource base and its ability to provide services. But, urban areas can be among the most efficient and sustainable places for people to live, owing to the efficiencies gained by mass transit, clustered living, and energy transmission (**FIGURE 1.46**).

It is increasingly clear, however, that human population will not continue to grow indefinitely. Analysts suggest that many of the processes associated with industrialization and urbanization will lead to a **demographic transition**, a model of population change that predicts high birth- and death rates will be replaced by low birth- and death rates over time.

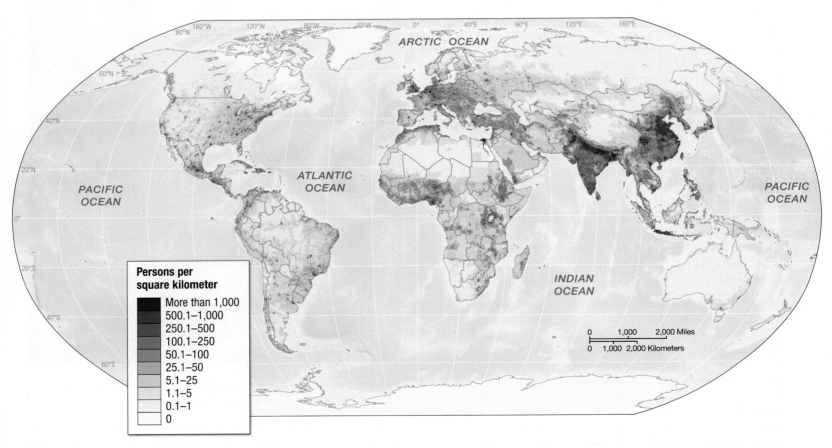

▲ **FIGURE 1.45 Population Distribution** The world's population is unevenly distributed across the globe. **What are the physical characteristics of regions with the lowest density of human population? The highest density?**

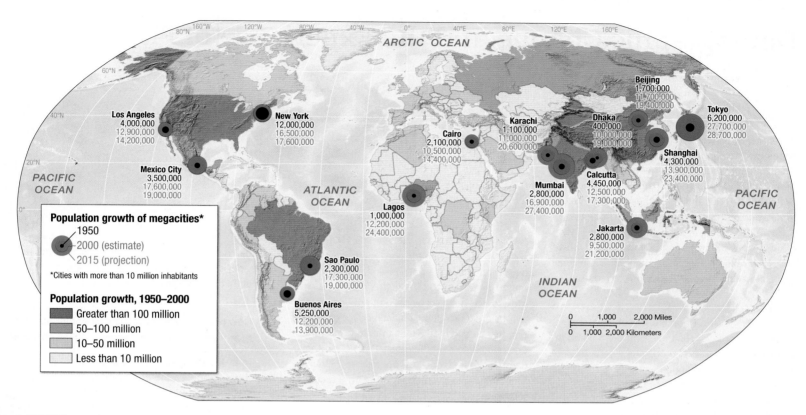

▲ **FIGURE 1.46 Megacities** A megacity is an urban area with more than 10 million people. Approximately 35 such cities exist today and many more are likely to form in the next decade. **Where might we find the world's largest cities in 2050? Can you think of any other cities that are not on this map that might also exceed 10 million people by 2050?**

Birthrates are a measure of the number of births in a population, usually expressed as births per 1,000 people per year or as a percentage figure. High birthrates (as high as 40 births per thousand people or more) are associated with agrarian (farming) societies, where large families are critical for labor, gathering, or subsistence and where birth control options are rudimentary. **Death rates** are measured with a similar statistic. High death rates (as high as 40 deaths per thousand people or more) are also associated with agrarian societies, where medical care may be limited and where work is physically demanding and involves continued exposure to the elements. Conversely, low birthrates and death rates (as low as 10 births per thousand people or fewer) are associated with urban societies, where children are less crucial for labor, birth control is widely available, there is high levels of access to health care, and women are significant participants in the paid workforce.

Modeling Demographic Change

Where birthrates and death rates are roughly the same, little or no growth in a population occurs. Most societies throughout history have not actually experienced high and sustained rates of population growth because their high birthrates were typically countered by comparable death rates. Only when death rates become lower than birthrates does growth occur.

The Demographic Transition The demographic transition model predicts patterns of change over time: from a period of no growth to a period of high growth and back again. **FIGURE 1.47**

shows a hypothetical demographic transition and illustrates how the high birth- and death rates of the preindustrial phase (Phase 1) are replaced by the low birth- and death rates of an industrial and postindustrial period (Phases 4 and 5) and a postindustrial phase of possible demographic decline, after passing through the critical, high-growth transitional phase (Phase 2) and more moderate rates (Phase 3) of natural increase and growth. The transitional phase of rapid growth is the direct result of early and steep declines in death rates while birthrates remain at high, preindustrial levels.

The period of explosive population growth around the world, which is now being followed in some regions by population stability or decline, reflects a very recent change. While the resource demands of growing populations are daunting, the demographic transition model suggests worldwide growth, of an increasingly urbanized population, will end altogether as soon as 2050 or 2060.

Most important, both population geographers and policymakers now recognize that a close relationship exists between women's status and fertility. Women who have access to education, health care, contraception, and employment have fewer children. Success at *damping* population growth in less-developed countries appears to be tied to enhancing the possibility for a good quality of life and to empowering people, especially women. For example, when increased international demand for Mexican products led to a large increase in female factory workers, the result was later marriages and separation between spouses, which unquestionably slowed birthrates in some parts of the Mexican population (See "Faces of the Region" on page 40).

Demographic Change and the Experience of Place

Demographics and demographic change is more than simply statistics. There are real lives that are tied to our changing world. That is why geographers are very interested in the everyday experiences of place and aging. How is growing old in China different today from a generation ago? How have youth in Mexico become politically engaged or disengaged depending on their access to electronic media and spaces for public protests? As you read this book, in each chapter you will meet other "Faces of the Region"—individuals whose daily lives reflect the profound economic, political, and cultural effects of demographic change.

China's Elders: A Change in Social Position
In the People's Republic of China, where older populations have long been cared for in the homes of their children, a new generation of ageing parents and grandparents are finding themselves in nursing homes. These homes didn't exist on a large scale 20 or 30 years ago, but because of China's one child policy and the changing nature of the economy, many aging people are now finding themselves having to live in nursing facilities in their older age (**FIGURE 1.3.1**).

> "Everyone gets older, but how they get older and what it means for them is really different depending on where they are."

Mexico's Youth: Growing Political Engagement
In Mexico, youth are playing a growing role in the political life of many countries. The Yo Soy 132 (I Am 132) campaign in Mexico was initiated by college-age students so disillusioned with the political race for president and the alleged media bias that they supported one candidate over the other. The 132 protest became an internationally supported social movement that brought youth together from around the world (**FIGURE 1.3.2**).

The Changing Life Course
It is important that we understand the different ways that people experience the twin processes of globalization and regionalization across different points in their life course, from their youngest years to their most advanced. Everyone gets older but how they get older and what it means for them is really different depending on where they are.

1. **How have the lives of China's older citizens recently changed? What explains these changes?**

2. **In 2014, Mexican youth protested the disappearance and murder of 43 high school students, a crime that involved corrupt public officials. Research the 2014 protests on the Internet. What were the larger political issues motivating the 2014 protests?**

Mexican Youth Protests

https://goo.gl/QFWyhU

▼ **FIGURE 1.3.1 Nursing Home, Beijing** With more than 160 million people over the age of 60 and the ageing rate gaining pace, China is the world's only developing country facing a rather curious problem: it is greying before it gets rich. The speed at which the number of elderly in China is increasing has alarmed both the government and demographers about the future, with a health care system already straining and many older people without pensions. Nursing homes have emerged to manage China's ageing population, whose one-child policy means that there are far fewer children to care for the elderly.

▼ **FIGURE 1.3.2 Youth Demonstrate as Part of the Yo Soy 132 March in Mexico City, Mexico** Popular protests by 131 Mexican youth was augmented by a larger social movement, Yo So 132 "I am 132," which developed as a protest movement against presidential candidate, Enrique Peña Nieto, and the press, which the protesters suggested were not covering the election fairly.

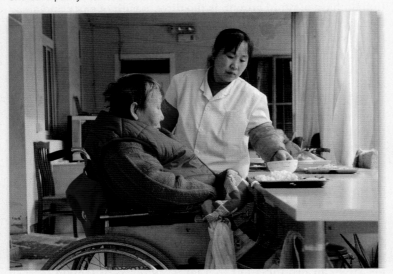

Many factors, such as fertility rates, death rates, and life expectancy, affect the structure of a region's population. In fast-growing regions (like parts of sub-Saharan Africa), a larger proportion of the population is young, whereas in regions where population is level or shrinking (as in parts of East Asia), the proportion of the elderly is higher. **Population pyramids** graph the numbers of people at varying ages in a population. Some countries or regions are dominated by younger populations, reflected by a pyramid with a wide base. These places can be predicted to grow quickly over time. Other regions have a "top-heavy" pyramid, which suggests an aging population, whose care in the future may test the resources of the relatively smaller population of younger people (**FIGURE 1.48**).

Apply Your Knowledge

Global Fertility Rates

https://goo.gl/7woYkj

1.21 The fertility rate for a country is a statistic reflecting the average number of children that women in the country bear during their child-bearing years. Look closer at fertility rates for different countries, by examining total fertility rate data from the World Bank: http://data.worldbank.org/. Using this site, list three countries with relatively high fertility rates (>4) and three with lower rates (<2).

1.22 What might you predict about countries at each end of the scale, in terms of economic development and status of women?

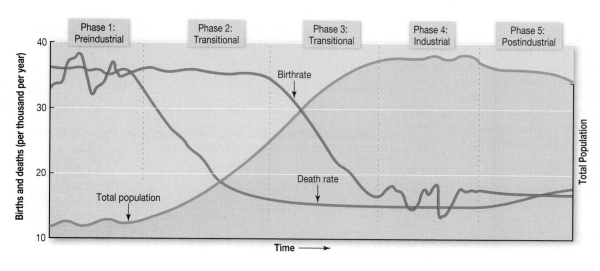

◀ **FIGURE 1.47 Demographic Transition** This model of population change holds that the characteristics of a population (such as rate of growth or decline, changes in birthrate, death rate, and causes of death, etc.) occur in successive stages, in response to changing social and economic conditions. Each stage is seen as leading to the next, though different regions or countries may require different periods of time to make the transition. **At what point does population begin to rise or level off? What causes this inflection point in the curve?**

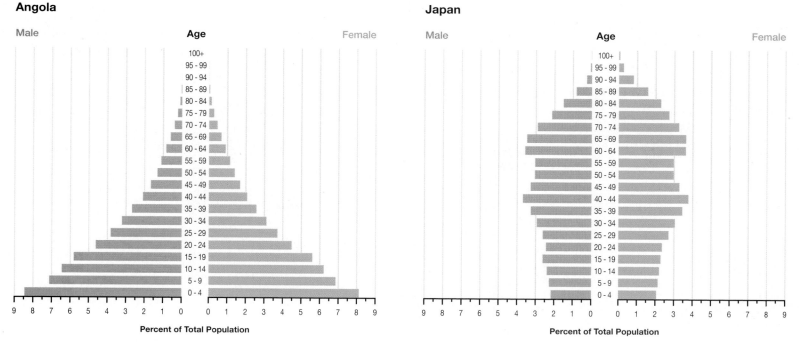

▲ **FIGURE 1.48 Population pyramids** Population pyramids describe the age structure of a population, with age groups stacked from oldest to youngest vertically descending on the chart and the quantity or proportion of people in each group measured horizontally. Angola has a predominantly young population, with many people below the age of 18 years, whereas Japan has an aging population with a large and growing segment of the country over the age of 40 years. **What are the differing prospects of future growth in populations like Angola relative to those like Japan? How might current age structure influence future growth and why?**

Future Geographies

The world is in transition, and the distribution of people, money, resources, opportunities, and crises is necessarily changing as a result. Each of the chapters in this book contains a "Future Geographies" section that speculates about the future of different regions, while "Emerging Regions" sections introduce new regions that are forming now.

Population Boom or Bust?

Many forces are lowering birthrates around the world, leading to leveling off or declining regional populations. In some places, populations may continue to rise, however. **FIGURE 1.49** shows projected population changes according to the UN Estimates up to and 2050.

Emerging Resource Regions

The expansion of the global economy and the globalization of industry will boost the overall demand for raw materials of every kind, and this will spur the development of some previously less exploited but resource-rich regions in Africa and Asia (**FIGURE 1.50**). The emergence of these new resource regions has enormous implications, as in Africa, where new geopolitical, cultural, and economic relationships are forming to rising powers like China and India.

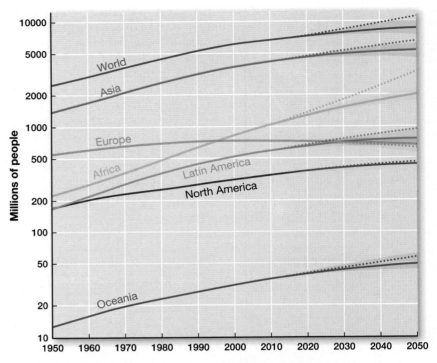

▲ **FIGURE 1.49 Population Geography of the Future** The range of projections is reflected in the widening bars further out in time.

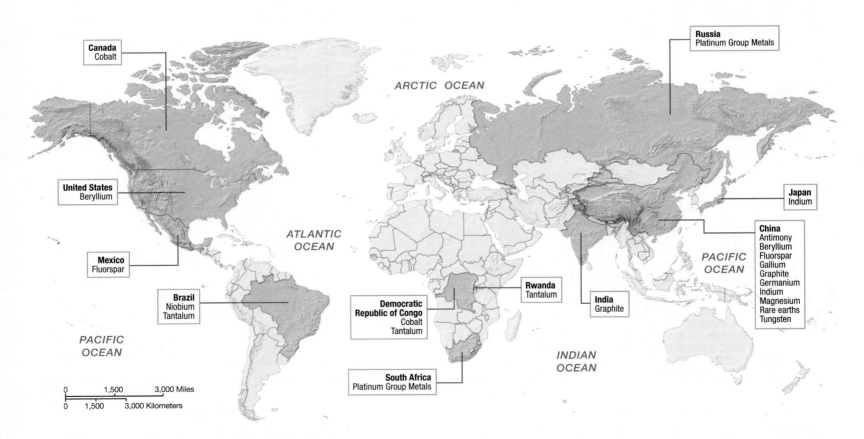

▲ **FIGURE 1.50 Global Resource Production of Important and Rare Elements** As historically peripheral countries develop and use more of the world's resources, demands for scarce materials will increase. **What are some of the applications of rare materials shown on the map (like Cobalt and Platinum)? What makes them important?**

◀ **FIGURE 1.51 The Global Parliament of Mayors** The Global Parliament of Mayors hosts meetings on emerging and future issues and trends, ranging from cultural change, climate change, and migrations, to digital technology and security.

Economic Globalization and Challenges to Regional Governance

Regional and supranational organizations such as the EU, the North American Free Trade Agreement (NAFTA), and the World Trade Organization (WTO) are increasingly attempting to mold the world into seamless trading areas, unhindered by the rules that regulate national economies. The powers and roles of modern states are changing as they are forced to interact with these organizations. Even city governments are in the process of making international connections (**FIGURE 1.51**).

New Regions of Insecurity and Crime

We are arguably living in the most peaceful period in human history. Nevertheless, in some regions, weak governments, lagging economies, religious extremism, and increasing numbers of young people will align to create a "perfect storm" for insecurity and crime. The ongoing integration of the global economy also means the acceleration and interlinking of criminal activities, often associated with violence and human and environmental exploitation (**FIGURE 1.52**).

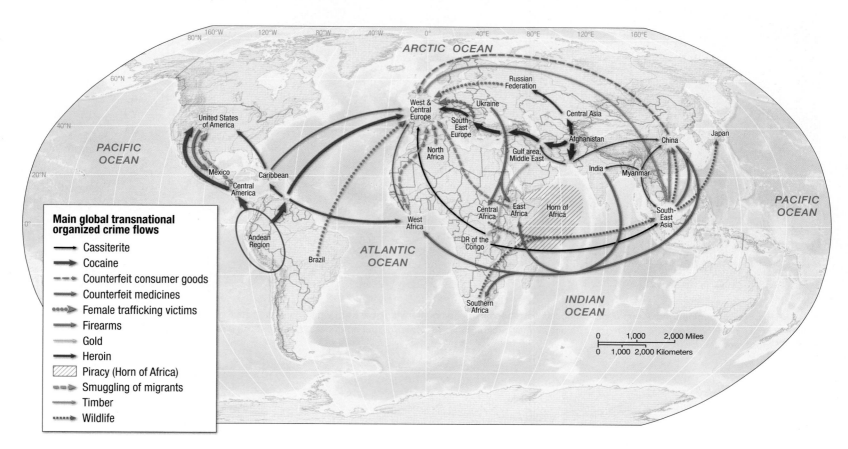

▲ **FIGURE 1.52 The Internationalization of Crime of Important and Rare Elements** Illicit sales and trafficking in a wide range of products, from drugs to endangered species, fuels global crime.

Future Environmental Threats and Global Sustainability

The world currently faces a daunting list of environmental threats: the destruction of tropical rainforests and the consequent loss of biodiversity; widespread, health-threatening pollution; the degradation of soil, water, and marine resources essential to food production; and the effects of greenhouse gas emissions on global climate. During the 21st century, some countries may try large-scale **geoengineering** options to combat climate change, such as blocking incoming sunlight with chemicals or water vapor or trying to store carbon on a massive scale (**FIGURE 1.53**).

Chemicals to save ozone

Aerosols in stratosphere

Giant reflectors in orbit

Cloud seeding

Grow trees

Iron fertilization of sea

Greening deserts

Genetically engineered crops

Pump liquid CO_2 to deep sea

Pump liquid CO_2 into rocks

▲ **FIGURE 1.53 Geoengineering** A number of proposals have been made to counteract global warming through projects that would block or reflect sunlight or even the chemistry of the oceans. **What unintended consequences might large-scaler global engineering of the Earth system have? Which of these efforts might be most promising or most problematic?**

Learning Outcomes Revisited

▶ **Compare** and contrast the concepts of globalization and regionalization.

While globalization suggests that the world is becoming more interconnected, regionalization demonstrates that global relations are organized differently in different regions. Places are different because they are connected.

1. **Describe** the relationship between globalization and regionalization and why world regions remain different despite greater global connections over time.

▶ **Describe** the Anthropocene's global impacts on earth systems and analyze related environmental issues and sustainability choices.

Human impacts in the last two centuries have included the large-scale clearance of forests, the deposition of pollutants in rivers and oceans, and the increase in atmospheric greenhouse gases, resulting in climate change. Sustainable activities reduce the human impact on Earth through greater efficiency and more careful use of resources but also include innovations (like sustainable urban design) that improve economy, human welfare, and the environment.

2. **Briefly** outline ways in which humans can mitigate or adapt to global climate change in the Anthropocene.

▶ **Differentiate** between forms of economic activity and explain why these forms vary around the globe.

Over the last several hundred years, economic activities have diversified away from primary (extractive) and secondary (manufacturing) to include large-scale tertiary (service) and quaternary (information) economies. The distribution of these four sectors is tied to the global international division of labor, which is driven by the processes of capitalist development.

3. **What** are some examples of the different economic activities one would find in the primary, secondary, tertiary, and quaternary sectors of an economy?

▶ **Explain** contemporary economic development trends and describe the main indicators of social and economic advancement.

Contemporary economic development trends indicate that the inequalities within and across world regions have become greater over the last several hundred years. Measures of human development demonstrate that these inequalities are driven by access to health care, education, and income, described by such indicators as GDP and HDI. These indicators can simplify how different communities engage development efforts, and alternative measures provide a complex assessment of global human development.

4. **Compare** and contrast the different measures of economic and social well-being used today to measure development.

▶ **Identify** the global, regional, and national actors that play a vital role in the world today.

Globalization has created new issues for modern states as they have had to manage the increasing flow of not only people but also products, ideas, plants, animals, and diseases. Non-state actors have emerged to grapple with the increasing political and economic interconnectivities of the world and include supranational organizations, such as the UN, social movements, and NGOs.

5. **List** some examples of non-state actors that operate at the global scale today.

▶ **Explain** the implications of globalization and regionalization for world regions and cultures.

Globalization has caused many cultural traditions to interact; access to ubiquitous commodities has created globally familiar brands, symbols, and practices. At the same time, global connectivity has caused the proliferation of new cultures and identities.

6. **Describe** how globalization has produced not only similarity but has perpetuated geographic differences in the world today.

▶ **Provide** examples of how the global distribution of languages and religions is changing.

In some contexts, local and indigenous languages are nearing extinction, whereas other languages—ones that are tied to the global economy, such as English or Chinese—are spreading. Religious geographies continue to change as well, as religions such as Christianity and Islam have used trade, empire, and missions to spread their religious beliefs, and areas that were once homes to certain religions have seen their decline.

7. **Given** your understanding of global religious and linguistic diversity, what are some of the challenges associated with trying to describe religion and language at a global scale?

▶ **Apply** the demographic transition model and use population pyramids to explain how and why regional population growth rates rise and fall.

Population rates rise and fall as economic, political, social, and cultural conditions change in particular places. The rising social and economic role of women in society is closely related to falling birthrates.

8. **Compare** and contrast the factors that lead to both high and low population growth rates in the world today.

Key Terms

Anthropocene (the "human era") (p. *8*)

alternative development (p. *30*)

atmospheric circulation (p. *8*)

biodiversity (p. *16*)

biomes (p. *15*)

birthrates (p. *39*)

capitalism (p. *22*)

class (p. *37*)

climate (p. *8*)

climate change (p. *8*)

climate system (p. *8*)

climate vulnerability (p. *11*)

Cold War (p. *24*)

colonialism (p. *21*)

commodity chain (p. *27*)

communism (p. *23*)

culture (p. *33*)

death rates (p. *39*)

deforestation (p. *18*)

demographics (p. *33*)

demographic transition (p. *38*)

development (p. *29*)

dialects (p. *34*)

diaspora (p. *33*)

ecosystem (p. *16*)

ecosystem services (p. *16*)

emerging region (p. *25*)

emerging world region (p. *7*)

Environmental geography (p. *4*)

ethnicity (p. *36*)

failed state (p. *32*)

fundamentalism (p. *33*)

gender (p. *37*)

geoengineering (p. *44*)

geographical imagination (p. *5*)

geography (p. *4*)

geomorphology (p. *14*)

globalization (p. *6*)

global financial centers (p. *26*)

global north (p. *28*)

global south (p. *28*)

global warming (p. *11*)

greenhouse gases (p. *10*)

gross domestic product (GDP) (p. *27*)

gross national income (GNI) (p. *27*)

Human Development Index (HDI) (p. *28*)

human geography (p. *4*)

imperialism (p. *22*)

Industrial Revolution (p. *22*)

International Monetary Fund (IMF) (p. *29*)

invasive species (p. *17*)

landscape (p. *4*)

language (p. *34*)

lingua franca (p. *34*)

map (p. *5*)

majority world (p. *28*)

megacities (p. *38*)

minority world (p. *28*)

multinational corporations (MNCs) (p. *25*)

nation (p. *31*)

nationalism (p. *31*)

nation–state (p. *31*)

neoliberal development (p. *29*)

neoliberalism (p. *30*)

nongovernmental organizations (NGOs) (p. *32*)

offshoring (p. *26*)

outsource (p. *25*)

physical geography (p. *4*)

places (p. *4*)

plantation economies (p. *22*)

plate tectonics (p. *14*)

Population pyramids (p. *41*)

primary sector (p. *25*)

purchasing power parity (PPP) per capita (p. *28*)

quaternary sector (p. *25*)

race (p. *36*)

racialization (p. *36*)

region (p. *5*)

regionalism (p. *5*)

regionalization (p. *6*)

religion (p. *33*)

rewilding (p. *16*)

Ring of Fire (p. *15*)

scale (p. *5*)

secondary sector (p. *25*)

sense of place (p. *4*)

sexuality (p. *37*)

social movements (p. *32*)

sovereign state (p. *31*)

supranational organizations (p. *32*)

sustainability (p. *20*)

system (p. *5*)

tertiary sector (p. *25*)

United Nations (UN) (p. *32*)

weather (p. *8*)

world regional geography (p. *4*)

world regions (p. *4*)

Thinking Geographically

1. What is geography, and what can studying it provide beyond a description of the world?

2. What is global climate change, and what is the evidence that it may be happening? What might its impacts include?

3. How are the world regions we recognize today different from those prior to 1500 C.E.?

4. How do economic activities differ from region to region? What factors account for these differences?

5. What worldwide demographic trends may emerge over the next 25 years?

6. What impacts might globalization have on the cultural characteristics of a world region or locality?

7. What are supranational organizations and how do they function with respect to formal states?

DATA Analysis

As discussed in this chapter, human activities in the environment directly influence climate change, which affects each region in different ways. Rising sea levels due to climate change directly impact Small Island Developing States (SIDS). Yet SIDS have contributed to less than 1% of the global emissions that cause temperatures and sea levels to rise. Tulun Atoll is one such Pacific island where the population is currently facing forced relocation. Take a deeper look at climate change and SIDS by first reading the 2014 *Guardian* article, "Island nations shouldn't be left to drown from climate change" at http://www.theguardian.com/commentisfree and respond to these following questions:

1. What are the factors causing sea levels to rise? In addition to sea-level rise, how will climate change affect Pacific islands and the surrounding ocean?

2. What is the "call to action" expressed by Tuilaepa Aiono Sailele Malielegaoi, the prime minister of Samoa to other world leaders?

3. How have New Zealand and Australia responded to climate migration of Pacific Islanders?

Going deeper, search Vimeo.com for "Chief Bernard Tunim in Copenhagen / COP15 (2009)" to see this address delivered by Chief Bernard Tunim, a leader and fisherman from Tulun Atoll at the UN Climate Change Conference in Copenhagen, 2009.

4. What does Chief Tunim say about what his people and island are experiencing with regard to food security, changing water levels, and the island's shoreline?

5. In Chief Tunim's view, what and who is destroying Tulun Atoll?

6. How does Chief Tunim rate the relative usefulness of the policies of international organizations in comparison with the everyday experience of his own people? Explain your answer.

7. What does Chief Tunim say about the government's relocation program for his people? Are his criticisms justified? Explain why or why not.

8. Do you agree with Chief Tunim that Europeans and North Americans should curb their consumption of fossil fuels to combat climate change? Explain your answer.

9. Reflecting on both the *Guardian* article and Chief Tunim's address, how much responsibility do you think the high-polluting nations have to the smaller nations who bear the greatest impacts? What percentage of this responsibility should be response, and what part should constitute preventative measures?

Sea-Level Rise

https://goo.gl/yADxc4

MasteringGeography™

Looking for additional review and test prep materials? Visit the Study Area in MasteringGeography™ to enhance your geographic literacy, spatial reasoning skills, and understanding of this chapter's content by accessing a variety of resources, including MapMaster interactive maps, videos, *In the News* RSS feeds, flashcards, web links, self-study quizzes, and an eText version of *World Regions in Global Context*.

Migrants and refugees, en route from Syria wait on the Greek island of Lesbos for a ferryboat to Athens on September 19, 2015.

Europe

The summer of 2015 provided a dramatic example of the interdependence of world regions. Refugees from conflicts in the Middle East and North Africa—mostly from Eritrea, Libya, and Syria—surged into Europe seeking asylum and the opportunity to rebuild their lives. Many of them arrived by taking dangerous land or sea routes after paying their life savings to smugglers. More than 300,000 refugees and migrants sailed across the Mediterranean Sea: nearly 200,000 landing in Greece and 110,000 in Italy. Authorities in Austria found an abandoned truck with the decaying bodies of 71 migrants who had been trapped inside. A few days later, news media carried images of the body of a young boy washed up on a Mediterranean beach. The European public was generally very sympathetic. Soccer fans in Germany and England paraded banners saying 'Welcome Refugees,' and Spain designated 'safe cities' to welcome and accommodate migrants. Germany resettled more than 100,000 refugees from Syria alone, with Sweden taking 75,000 more.

But the scale of the humanitarian crisis soon began to overwhelm the capacity of European countries to deal with the thousands of desperate people arriving daily along their borders. Although international law requires that refugees should stop in the first safe country they reach, the majority insisted on continuing on to Germany or other western European nations with the most prosperous economies and the most progressive social welfare systems. Amid chaotic scenes, Germany suspended its commitment to the open borders shared by members of the European Union, while Greece, Bulgaria, Macedonia, and Hungary put up razor-wire fences along parts of their borders. The inability of European countries to agree on an overall strategy for dealing with the crisis became a serious challenge to the political framework of the European Union.

Learning Outcomes

▶ **Analyze** how Europe's landscapes developed around regions of distinctive geology, relief, landforms, soils, and vegetation.

▶ **Explain** the rise of Europe as a major world region and understand how Europeans established the basis of a worldwide economy.

▶ **Describe** the reintegration of Eastern Europe.

▶ **Summarize** the importance of the European Union and its policies.

▶ **Identify** the principal core regions within Europe.

▶ **Assess** the importance of language, religion, and ethnicity in shaping ongoing change within Europe.

▶ **Describe** Europe's patterns of demographic change and urban development.

Environment, Society, and Sustainability

Two aspects of Europe's physical geography have been fundamental to its evolution as a world region and have influenced the evolution of regional geographies within Europe itself. First, as a world region (**FIGURE 2.1**), Europe is situated between the Americas, Africa, and the Middle East.

Second, the region consists mainly of a collection of peninsulas and islands at the western extremity of the great Eurasian landmass.

The largest of the European peninsulas is the Scandinavian Peninsula, the prominent western mountains of which separate Atlantic-oriented Norway from continental-oriented Sweden. Equally striking are the Iberian Peninsula, a square mass that projects into the Atlantic, and the boot-shaped Italian Peninsula. In the southeast is the broad triangle of the Balkan Peninsula, which

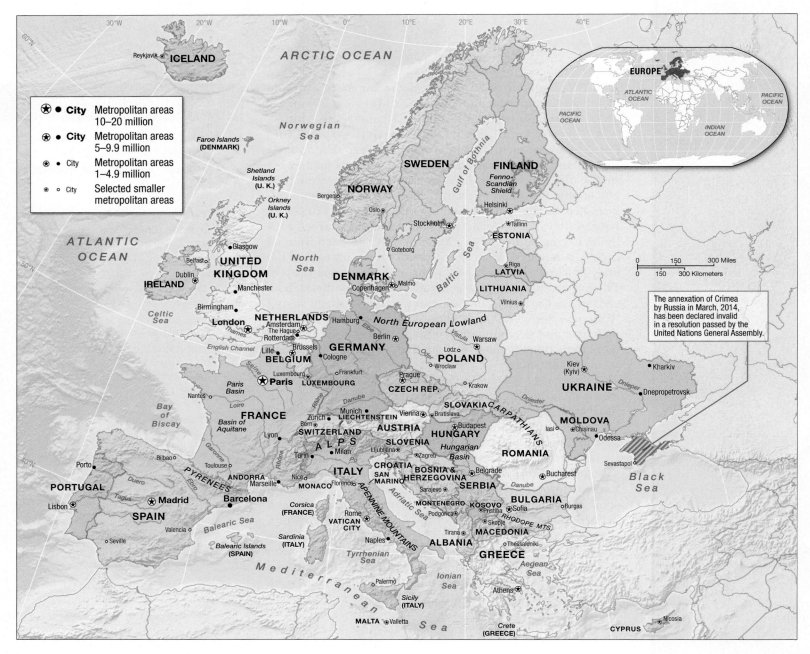

▲ **FIGURE 2.1 Europe: Countries, major cities, and physical features** List all of the diverse bodies of water (seas, rivers, lakes) in Europe and explain how these waterways connected people over time through trade, political contests, and immigration.

"Europeans' relationship to the surrounding seas has been a crucial factor in the evolution of European and world geography."

....

projects into the Mediterranean, terminating in the intricate coastlines of the Greek peninsulas and islands. In the northwest are two of Europe's largest islands, Britain and Ireland.

The overall effect is that tongues of shallow seas penetrate deep into the European landmass. This characteristic was especially important in the premodern period, when the only means of transporting goods were sailing vessels and wagons. The Mediterranean and North seas, in particular, provided relatively sheltered sea lanes, fostering seafaring traditions in the people all around their coasts. The penetration of the seas deep into the European landmass provided numerous short land routes across the major peninsulas, making it easier for trade and communications to take place in the days of sail and wagon. As we shall see, Europeans' relationship to the surrounding seas has been a crucial factor in the evolution of European—and, indeed, world—geography.

Europe's navigable rivers also shaped the human geography of the region. Although small by comparison with major rivers in other regions of the world, some of the principal rivers of Europe—the Danube, the Dneiper, the Elbe, the Rhine, the Seine, and the Thames—played key roles. In addition, the low-lying **watersheds** (drainage areas) between the major rivers of Europe's plains allowed canal building to take place relatively easily, thereby increasing the mobility of river traffic.

Whereas the western, northern, and southern limits of Europe as a whole are clearly defined by surrounding seas—the Atlantic Ocean to the west, the Arctic Ocean to the north, and the Mediterranean Sea to the south—the eastern edge of Europe merges into the vastness of Asia and is less easily defined. Geographers sometimes use the mountain ranges of the Urals to mark the boundary between Europe and Asia, but the most significant factors separating Europe from Asia are human and relate to ethnicity, language, and a common set of ethical values that stem from Roman Catholic, Protestant, and Orthodox forms of Christianity. As a result, the eastern boundary of Europe is often demarcated through political and administrative boundaries, rather than physical features.

Climate and Climate Change

The seas that surround Europe strongly influence the region's climate (**FIGURE 2.2**). In winter, seas cool more slowly than the land and, in summer, they warm up more slowly than the land. As a result, the seas provide a warming effect in winter and a cooling effect in summer. Europe's arrangement of islands and peninsulas contributes to an overall climate that does not have great seasonal extremes of heat and cold. The moderating effect is intensified by the North Atlantic Drift, which carries great quantities of warm water from the tropical Gulf Stream as far as the United Kingdom.

Given its latitude (Paris, at almost 49°N, is the same latitude as Winnipeg and Newfoundland in Canada), most of Europe is remarkably warm. It is continually crossed by moist, warm air masses that drift in from the Atlantic. The effects of these warm, wet, westerly winds are most pronounced in northwestern Europe, where squalls and showers accompany the passage of successive eastward-moving weather systems. Weather in northwestern Europe tends to be unpredictable, partly because of the swirling movement of air masses as they pass over the Atlantic and partly because of the complex effects of the widely varying temperatures of interpenetrating bodies of land and water. Farther east, in continental Europe, seasonal weather tends to be more settled, with more pronounced extremes of summer heat and winter cold. In these interior regions, local variations in weather are influenced a great deal by the direction in which a particular slope or land surface faces and its elevation above sea level.

The Mediterranean Basin has a different and quite distinctive climate. Winters are cool, with an Atlantic airstream that brings overcast skies and intermittent rain—though snow is unusual. Low-pressure systems along the northern Mediterranean draw in rain-bearing weather fronts from the Atlantic. When low pressure over the northern Mediterranean coincides with high pressure over continental Europe, southerly airflows spill over mountain ranges and down valleys, bringing cold blasts of air. These events have local names: the *mistral*, for example, which blows down the Rhône valley in southern France, and the *bora*, which blows over the eastern Alps toward the Adriatic region of Italy. In spring, the temperature rises rapidly, and rainfall is more abundant. Then summer bursts forth suddenly as dry, hot, Saharan air brings three months of hot, sunny weather. There is no rain save an occasional storm; the soil cracks and splits and is easily washed away in the occasional downpours. In October, the temperature drops, and deluges of rain show that Atlantic air prevails once more.

In such conditions, delicate plants cannot survive. The Mediterranean climate precludes all plant species that cannot tolerate the range of conditions—cold as well as heat and drought as well as wet. The result is a distinctive landscape of dry terrain dotted with cypress trees, holm oaks, cork oaks, parasol pines, and eucalyptus trees. These same conditions make agriculture a challenge. The crops that prosper best include olives, figs, almonds, vines, oranges, lemons, wheat, and barley. Sheep and goats graze on dry pastureland and stubble fields. Irrigation is often necessary, and in some localities it sustains high yields of fruit, vegetables, and rice.

Europe and Climate Change Europe has warmed significantly more than the global average: The average temperature for the European land area between 2004 and 2013 was 0.75 °C to 0.81 °C warmer than the preindustrial average; most of those years were among the warmest since 1850. High-temperature

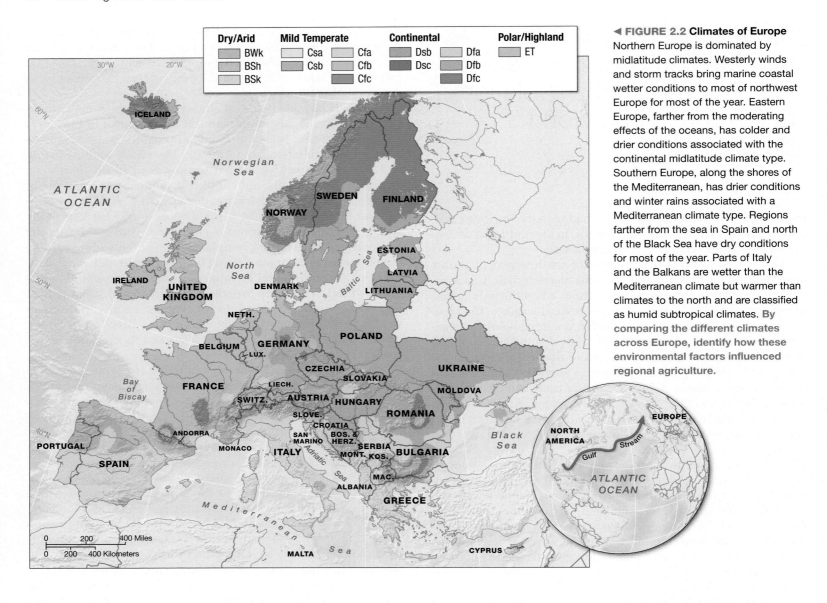

◀ **FIGURE 2.2 Climates of Europe**
Northern Europe is dominated by midlatitude climates. Westerly winds and storm tracks bring marine coastal wetter conditions to most of northwest Europe for most of the year. Eastern Europe, farther from the moderating effects of the oceans, has colder and drier conditions associated with the continental midlatitude climate type. Southern Europe, along the shores of the Mediterranean, has drier conditions and winter rains associated with a Mediterranean climate type. Regions farther from the sea in Spain and north of the Black Sea have dry conditions for most of the year. Parts of Italy and the Balkans are wetter than the Mediterranean climate but warmer than climates to the north and are classified as humid subtropical climates. **By comparing the different climates across Europe, identify how these environmental factors influenced regional agriculture.**

extremes like hot days, tropical nights, and heat waves have become more frequent, whereas cold spells and frost days have become less frequent.

Highly urbanized and industrialized, Europe is a major contributor to the carbon dioxide (CO_2) emissions that drive global climate change. Under the **United Nations Kyoto Protocol**, a legally binding global agreement to reduce greenhouse gas emissions, industrialized nations are obliged to reduce the amount of greenhouse gases being released into the atmosphere. The European Union (EU; 28 countries that account for most of the region; see Figure 2.15, p. 69) has attempted to reduce emissions of greenhouse gases by establishing an **Emissions Trading Scheme (ETS)** that allows energy-intensive facilities (e.g., power generation plants and iron and steel, glass, and cement factories) to buy and sell permits that allow them to emit CO_2 into the atmosphere. Companies that exceed their individual limit are able to buy unused permits from firms that have successfully reduced their emissions. Those that exceed their limit and are

unable to buy spare permits are fined. The system has had mixed success in reducing the total amount of CO_2 released into the atmosphere but can probably be counted as successful in limiting the rate of growth in CO_2 emissions.

Apply Your Knowledge

2.1 Find a European city at the same latitude as a city that is familiar to you. Use the Internet to find climate data (monthly temperatures and rainfall statistics, for example). (Hint: Good sources are http://www.worldclimate.com/ and http://www.bestplaces.net/climate/)

2.2 Describe and provide a reason for the differences between the two sets of data.

Comparing Climates

https://goo.gl/ObzaVP

Geological Resources, Risk, and Water

The physical environments of Europe are complex and varied. It is impossible to travel far without encountering significant changes in physical landscapes. There is, however, a broad pattern to this variability, and it is based on four principal **physiographic regions** that are characterized by broad coherence of geology, relief, landforms, soils, and vegetation. These regions are the Northwestern Uplands, the Alpine System, the Central Plateaus, and the North European Lowlands (**FIGURE 2.3**). Within these broad physiographic divisions, physical differences are often encountered over quite short distances, and numerous specialized farming regions have developed where agricultural

conditions have influenced local ways of life to produce distinctive landscapes. In detail, today's landscapes are a product of centuries of human adaptation to climate, soils, altitude, and **aspect** (exposure) and to changing economic and political circumstances.

Northwestern Uplands The Northwestern Uplands are composed of the most ancient rocks in Europe, the product of the Caledonian mountain-building episode about 400 million years ago. Included in this region are the mountains of Norway and Scotland and the **uplands** (high, hilly land) of Iceland, Ireland, and Wales. The original Caledonian mountain system

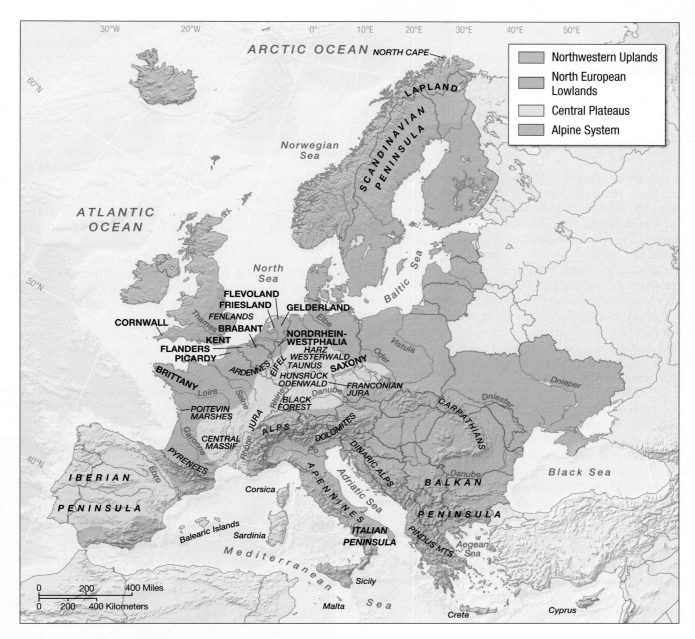

▲ **FIGURE 2.3 Physiographic regions of Europe** Each of the four principal physiographic regions of Europe has a broad coherence in terms of geology, relief, landforms, soils, and vegetation. **Describe three features from each of the four physiographic regions: Northwestern Uplands, North European Lowlands, Central Plateaus, and the Alpine System.**

▲ **FIGURE 2.4 The Northwestern Uplands** The northern parts of the European region are characterized by high mountains and deeply eroded glacial valleys that have been drowned by the sea, creating distinctive fjord landscapes such as this one: Naeroyforden Fjord in Norway.

was eroded and uplifted several times, and following the most recent uplift, the Northwestern Uplands have been worn down again, molded by ice sheets and glaciers. Many valleys were deepened and straightened by ice, leaving spectacular glaciated landscapes. There are **cirques** (deep, bowl-shaped basins on mountainsides, shaped by ice action), glaciated valleys, and **fjords** (some as deep as 1,200 meters—about 3,900 feet) in Norway; countless lakes; lines of **moraines** that mark the ice sheet's final recession; extensive deposits of sand and gravel from ancient glacial deltas; and peat bogs that lie on the granite shield that forms the physiographic foundation of the region. Since the last glaciation, sea levels have risen, forming fjords and chains of offshore islands wherever valleys have been flooded by the rising sea (**FIGURE 2.4**).

This formidable environment is rendered even more forbidding in much of the region as a result of climatic conditions. In the far north, the summer sun shines for 57 days without setting, but the winter nights are interminable.

Much of the Northwestern Uplands is covered by forests. In the southern parts of the region, conifers (mostly evergreen trees such as pine, spruce, and fir) are mixed with birches and other deciduous trees; further north, conifers become entirely dominant, and in the far north, toward the North Cape in Lapland, the forest gives way to desolate treeless stretches of the **tundra** with its characteristic gray lichens and dwarf willows and birches.

The farmers in these mountain subregions depend on **pastoralism**, a system of farming and way of life based on keeping herds of grazing animals, supplementing with a little produce grown on the valley floors. In the less mountainous parts of Scandinavia, as in much of Baltic Europe, with their short growing season and cold, acid soils, agriculture supports only a low density of settlement. Landscapes reflect a mixed farming system of oats, rye, potatoes, and flax, with hay for cattle. Oats, the

largest single crop, often has to be harvested while it is still green. More than half of the milk from dairy cattle is used to produce butter and cheese.

In these upland landscapes, farms and hamlets are casually situated on any habitable site, their buildings often widely scattered. The more temperate subregions of the Northwestern Uplands (in Ireland and the United Kingdom) are dominated by dairy farming on meadowland, sheep farming on exposed uplands, and arable farming (mainly wheat, oats, potatoes, and barley) on drier lowland areas. In these areas, dispersed settlement, in the form of hamlets and scattered farms, is characteristic, and stone is more often the traditional building material.

Alpine Europe The Alpine System occupies a vast area of Europe, stretching eastward for nearly 1,290 kilometers (about 800 miles) across the southern part of Europe from the Pyrenees, which mark the border between Spain and France, through the Alps and the Dolomites and on to the Carpathians, the Dinaric Alps, and some ranges in the Balkan Peninsula. The Apennines of Italy and the Pindus Mountains of Greece are also part of the Alpine System. The Alpine System is the product of the most recent of Europe's mountain-building episodes, which occurred about 50 million years ago. Its relative youth explains the sharpness of the mountains and the boldness of their peaks. The Alpine landscape is characterized by jagged mountains with high, pyramidal peaks and deeply glaciated valleys (**FIGURE 2.5**). The highest peak, Mont Blanc, reaches 4,810 meters (15,781 feet). Most of the rest of the mountains are between 2,500 and 3,600 meters (about 8,200 and 11,800 feet) in height. The Alps pose a formidable barrier between northwestern Europe and Italy and the Adriatic, with only a few passes—including the Brenner Pass, the Simplon Pass, the Saint Gotthard Pass, and the Great Saint Bernard Pass—allowing transalpine routes.

The dominant direction of the Alps and their parallel valleys is roughly southwest to northeast. The major Alpine valleys

▼ **FIGURE 2.5 Alpine Europe** Shown here is the pre-alpine landscape of the Simmental valley in Switzerland, with a sharp peak of the Alpine System in the distance. **What sort of agriculture is best suited to this kind of environment?**

thus have one sunny, fully exposed slope that is suitable for vine growing and a shaded side rich with orchards, woods, and meadows. The mountains and valleys of the Alps proper are surrounded by glacial outwash deposits that provide rich farmland. The limestone of the Alpine region is widely quarried for cement, whereas mineral deposits—lead, copper, and iron—and small deposits of coal and salt have long been locally important throughout the region. In addition, the Alps are a valuable source of hydroelectric power, providing about 36 billion watt-hours in Switzerland (56% of the country's electricity consumption), about 50 billion watt-hours in France (15% of the country's electricity consumption), and about 38 billion watt-hours in Austria (69% of the country's electricity consumption).

The traditional staple of the economy, however, has been agriculture, and farming has given the Alpine region its distinctive human landscape: a patchwork of fields, orchards, vineyards, deciduous woodlands, pine groves, and meadows on the lower slopes of the valleys, with broad Alpine pastures above. In these pastures, which are dotted with wooden haylofts and summer chalets, dairy cattle wander far and wide. Farmers attach bells around the necks of their animals to be able to locate them, and the consequent effect is a resonant pastoral "soundscape" of clanking cowbells. The higher slopes, which receive more rainfall, provide excellent pastures that have made the region famous for its rich cheeses, such as Gruyère.

Central Plateaus Between the Alpine System and the Northwestern Uplands are the landscapes of the Central Plateaus and the North European Lowlands. The Central Plateaus are formed from 250- to 300-million-year-old rocks that have been eroded down to broad tracts of uplands. Beneath the forest-clad slopes and fertile valleys of these plateaus lie many of Europe's major coalfields. For the most part, the plateaus reach between 500 and 800 meters (1,640 and 2,625 feet) in height, though they rise to more than 1,800 meters (5,905 feet) in the Central Massif of France (**FIGURE 2.6**). *Massif* refers to a compact chain of mountains that is separate from other ranges. The Central Plateau landscape is characterized by rolling hills, steep slopes and dipping vales, and deeply carved river valleys.

In central Spain, the landscape is dominated by plateaus and high plains that go on for hundreds of miles, with long narrow mountain ranges—the *cordillera*—stretched out like long cords along the edges. The flat landscape of the region is a result of immense horizontal sheets of sedimentary rock that cover an extensive block, or massif, of ancient rock, with a general appearance of tables ending in ledges—hence the term *meseta* (mesa means table in Spanish), which is generally used to designate the center of Spain. These dry, open lands are covered in grain crops and flocks and herds of sheep and cattle that in summer are driven up to the cooler mountains. Olive trees dominate the shallow valley slopes, and in irrigated valley bottoms vines and market gardens flourish.

Farther east—in the Massif Central in France and the Eifel, Westerwald, Taunus, Hunsruck, Odenwald, and Franconian Jura in Germany—where the climate is wetter, the landscape is dominated by gently rounded, well-wooded hills, with villages surrounded by neat fields and orchards in the vales. These hills rarely rise above 700 meters (2,300 feet), and the landscape includes many attractive and fertile subregions of low hills planted with vines and shallow valleys with orchards and meticulously maintained farms. The hills are covered with beech and oak forests, interspersed with growths of fir. The Black Forest in southwestern Germany is much higher, its bare granite summits reaching 1,493 meters (about 4,900 feet) in the south. The northeastern reaches of the Black Forest form an immense, solid mass of fir trees.

▼ FIGURE 2.6 **The Central Plateaus** Shown here is part of the Odenwald region in Germany, where the River Neckar has cut into the plateau surface.

North European Lowlands The North European Lowlands sweep in a broad crescent from southern France, through Belgium, the Netherlands, and southeastern England into northern Germany, Denmark, and the southern tip of Sweden. Continuing eastward, they broaden into the immense European plain that extends through Poland and Czechia, all the way into Russia. Coal is found in quantity under the lowlands of England, France, Germany, and Poland and in smaller deposits in Belgium and the Netherlands. Oil and natural gas deposits are found beneath the North Sea and under the lowlands of southern England, the Netherlands, and northern Germany. Nearly all of this area lies below 200 meters (656 feet) in elevation, and the topography everywhere is flat or gently undulating. As a result, the region has been particularly attractive to farming and settlement. The fertility of the soil varies, however, so that settlement patterns are uneven, and agriculture is finely tuned to the limits and opportunities of local soils, landscape, and climate.

The western parts of the North European Lowlands are densely settled and intensively farmed, with the moist Atlantic climate supporting lush agricultural landscapes (**FIGURE 2.7**). Surrounding the highly urbanized regions of the North European Lowlands are the fruit orchards and hop-growing fields of Kent, in southeastern England; the bulb fields of the Netherlands; the dikes and rectangular fields of the reclaimed marshland of North Holland, Flevoland, and Friesland along the Dutch coastal plain; the pastures, woodlands, and forests of the upland plateaus of the Ardennes, and the Harz; and the meadows and cultivated fields separated by hedgerows and patches of woodland that characterize most of the remaining countryside.

Farther east, the Lowlands are characterized by a drier continental climate and lowland river basins with a rolling cover of sandy river deposits and **loess** (a fine-grained, extremely fertile soil formed as a result of the deposition of wind-borne silts). The hills are covered with oak and beech forests, but much of the region consists of broad loess plateaus where very irregular rainfall averages about 40 centimeters—approximately 16 inches—a year. There are no woods, and irrigation is often necessary to sustain the typical two-year rotation of corn and wheat. Stone and trees are so scarce that houses are built of *pisé,* a kind of rammed-earth brick. The scarcity of trees forces storks to build nests atop chimneys and telegraph poles. The rich soils produce high yields of wheat and corn, together with hops, sugar beets, and forage crops for livestock.

The area also features residual regions of **steppe**, semiarid, treeless, grassland plains with landscapes that are infinitely monotonous. This land, which is too dry or too marshy to have invited cultivation, was once the domain of wild horses, cattle, and pigs, but today huge flocks of sheep find pasture there.

Apply Your Knowledge

2.3 Name the four physiographic regions in Europe and describe the environment of each region.

2.4 Do an Internet search for landscape images of one European country. Based on the images, describe the physiography and natural features of the landscape and list evidence of how humans use or have modified that landscape.

Ecology, Land, and Environmental Management

Temperate forests originally covered about 95% of Europe, with a natural ecosystem dominated by oak, together with elm, beech, and linden (lime). By the end of the medieval period, Europe's forest cover had been reduced to about 20% of what it was at its peak, and today it is around 5%. Between 1000 C.E. and 1300 C.E., a period of warmer climate, together with advances in agricultural knowledge and practices, led to a significant transformation of the European landscape. The population more than doubled, from around 36 million to more than 80 million, and a vast amount of land was cultivated for the first time. By about 1200 C.E., most of the best soils of Western Europe had been cleared of forest, and new settlements were increasingly forced into the more marginal areas of heavy clays or thin sandy soils

▼ **FIGURE 2.7 The North European Lowlands** Shown here is the flat landscape of East Anglia, in England, with the village of Burnham and its windmill in the distance. **Suggest why the windmill in the photograph might have been built.**

on higher ground and **heathlands**. (Heath is open land with coarse soil and poor drainage.) Many parts of Europe undertook large-scale drainage projects to reclaim marshlands.

Roman Land Improvement The Romans were among the first to demonstrate the effectiveness of drainage schemes by reclaiming parts of Italy and northwestern Europe. Under Roman colonization, land was often subdivided into a checkerboard pattern of rectilinear fields. This highly ordered system was known as *centuriation*, and the pattern can still be seen in some districts today—in parts of the Po Valley, for example. Another important legacy of the Romans is the doctrine of public trust, which asserts public rights in navigable waters, fisheries, and tidelands. It is reflected today in the EU's approach to water management, which features the most progressive overall system of water management, organized by river basin instead of according to administrative or political boundaries.

Medieval Settlement In the 12th and 13th centuries, extensive drainage and resettlement schemes were developed in Italy's Po Valley, in the Poitevin marshes of France, and in the Fenlands of eastern England. In Eastern Europe, forest clearances were organized by agents acting for various princes and bishops, who controlled extensive tracts of land. The agents would also arrange financing for settlers and develop villages and towns, often to standardized designs.

This great medieval colonization came to a halt nearly everywhere in about 1300 C.E. One factor was the so-called **Little Ice Age**, a period of cooler climate that began around 1300 that significantly reduced the growing season—perhaps by as much as five weeks. Another factor was the catastrophic loss of population during the period of the Black Death (1347–1351) and the periodic recurrences of the plague that continued for the rest of the 14th century. The Little Ice Age lasted until the mid-19th century, by which time many villages, and much of the more marginal land, had been abandoned. The Little Ice Age was also responsible for one of the most striking cultural contrasts within Europe: the dominance of wine consumption in southern Europe and beer and liquor consumption in northern Europe (see Geographies of Indulgence, Desire, and Addiction: Beer and Wine on p. 58).

Emergence of the Modern Landscape The resurgence of European economies from the 16th century onward coincided with overseas exploration and trade, while at the same time, domestic landscapes were significantly affected by repopulation, by reforms to land tenure systems, and by advances in science and technology that changed agricultural practices, allowing for a more intensive use of the land. In the Netherlands, a steadily growing population and the consequent requirement for more agricultural land led to increased efforts to reclaim land from the sea and to drain coastal marshlands. Hundreds of small coastal barrier islands were slowly joined into larger units, and sea defense walls were constructed to protect low-lying land. The land was drained by windmill-powered water pumps, and the excess water was carried off into a network of drainage ditches and canals.

The resulting landscape of low-lying areas known as **polders**, reclaimed from the sea and protected by dikes, provided excellent, flat, fertile, and stone-free soils. Between 1550 and 1650 C.E., 165,000 hectares (407,715 acres) of polderland were established in the Netherlands, and the sophisticated techniques developed by the Dutch began to be applied elsewhere in Europe—including eastern England and the Rhône estuary in southern France. Although most of these schemes resulted in improved farmland, the environmental consequences were often serious. In addition to the vulnerability of the polderlands to inundation by the sea, large-scale drainage schemes devastated the wetland habitat of many species, and some ill-conceived schemes simply ended in widespread flooding.

These environmental problems were but a prelude to the environmental changes and ecological disasters that accompanied the industrialization of Europe, beginning in the 18th century. Mining—especially coal mining—created derelict landscapes of spoil heaps; urbanization encroached on rural landscapes and generated unprecedented amounts and concentrations of human, domestic, and industrial waste; and manufacturing, unregulated at first, resulted in extremely unhealthy levels of air pollution. Much of Europe's forest cover was cleared, and remaining forests and woodlands suffered from **acid rain** (rain that contains dilute sulfuric and nitric acids derived from burning fossil fuels). Many streams and rivers also became polluted, and the landscape everywhere was scarred with quarries, pits, cuttings, dumps, and waste heaps. In recent years, however, European cities have been in the forefront of progressive environmental policies (see Sustainability in the Anthropocene: Planning for Green Cities on pp. 60–61).

Apply Your Knowledge

2.5 Describe how polders changed both the economy and environment in the Netherlands, England, and southern France.

2.6 Compare and contrast how Europe has created many environmental problems in the landscape but also sought to be an innovator.

History, Economy, and Territory

Perhaps more than any other world region, Europe has a geography that is the product of its world-spanning economic and demographic systems. Beginning in the 16th century, Europeans became, as University of Wisconsin economic historian Robert Reynolds put it, the "leaders, drivers, persuaders, shapers, crushers and builders"[1] of the world's economies and societies. It was as a result of these processes that the core areas of Europe

---

[1] R. Reynolds, *Europe Emerges: Transition Toward an Industrial World-Wide Society* (Madison: University of Wisconsin Press, 1961), p. vii.

Beer and Wine

Today, Britain is famous for many varieties of ale and beer. But before the cold snap of the Little Ice Age, Britons also produced and enjoyed their own wines. Until the Late Medieval period, viticulture—the cultivation of grape vines for winemaking—was common throughout Europe, extending as far north as Britain and Scandinavia during the Medieval Warm Period that lasted from about 950 to 1250 C.E. Greek civilization had established viticulture by the 6th century B.C.E. Under the Roman Empire, viticulture spread west along the north shores of the Mediterranean and along the valleys of rivers in France and Spain and north to the North Sea and the Baltic. By the 1st century C.E., wine had become a commodity of indulgence, desire, and—for some—addiction throughout Europe. But sharply reduced average tempera-

TABLE 2.1.1 Alcohol Consumption in Europe

Wine Consumption, 2013		Beer Consumption, 2013	
Country	**Amount in 1,000 hL***	**Country**	**Amount in 1,000 hL***
France	28, 81	Germany	85,588
Italy	21,795	United Kingdom	42,422
Germany	20,300	Poland	37,388
United Kingdom	12,738	Spain	35,169
Spain	9,100	France	19,421
Portugal	4,551	Italy	17,504
Netherlands	3,585	Romania	16,300
Greece	3,300	Czech Republic	15,278
Belgium and Luxembourg	3,054	Netherlands	11,690
Switzerland	2,650	Turkey	9,047

SOURCE: *Wine Annual Report* and *Statistics Wine Annual Organisation Internationale de la Vigne du Vin*, May 2014.
*hL = hectoliters; 1 hectoliter = 26.4 gallons

SOURCE: Brewers of Europe, http://www.brewersofeurope.org/uploads/mycms-files/documents/publications/2014/statistics_2014_web_2.pdf
*hL = hectoliters; 1 hectoliter = 26.4 gallons

> "The beer–wine division between northern and southern Europe has persisted since the Little Ice Age."

▼ FIGURE 2.1.1 **The grape harvest in Burgundy, France.**

tures during the Little Ice Age meant that viticulture retreated to Mediterranean Europe, leaving northerners to satisfy their need for alcohol with grain-based beverages, namely beer and spirits. It was a division that has characterized patterns of alcohol consumption ever since (**TABLE 2.1.1**). Incidentally, since North America was populated mostly by northern Europeans, beer and spirits, rather than wine, became characteristic of alcohol consumption in the United States and Canada.

Nevertheless, these divisions are by no means absolute. In the 16th century, Spanish and Portuguese overseas expansion saw the introduction of viticulture to the New World—to Mexico in the 1520s, Peru in the 1530s, Chile in the 1550s, and Florida in the 1560s. The British introduced viticulture to Virginia in the 1600s, and the first vineyards in California were established by Franciscan missions in the 1770s. Meanwhile, in Europe, demographic growth and increasing prosperity rapidly expanded the market for wine.

Today, the exclusivity of Europe's best wines is protected by strict systems of regulation. In France, for example, the Appellation Contrôlée system was introduced to guarantee the authenticity of wines, district by district. Such regulations have been important in reinforcing the appeal of wine as a commodity of indulgence and desire. Fine wines such as those from Burgundy, France (**FIGURE 2.1.1**) denote affluence and distinction. Meanwhile, recent climate change has meant that the frontier of viticulture in Europe has begun to shift northward. At the same time, the consumption of wine in northern Europe has increased significantly, especially among young, middle-class households. Brewers and distillers have responded to the competition by producing specialized cask ales and flavored spirits.

1. **List the environmental factors that created the geographic division between beer and wine in Europe.**

2. **Compare and contrast other examples of indulgent commodities. In your analysis, do you find them to be sourced locally or imported from other places?**

emerged with a strong comparative advantage in the modern world system. For several centuries, other world regions came to play a subordinate economic role to Europe. Nevertheless, the preindustrial trajectories of other parts of the world had often eclipsed that of Europe and sometimes influenced events in Europe itself. Today, Europe is totally integrated into the world economy, with ethnically diverse populations that reflect the region's past economic and political history.

The foundations of Europe's human geography were laid by the Greek and Roman empires. Beginning in about 750 B.C.E., the ancient Greeks developed a series of fortified city–states (politically independent cities, called *poleis*) along the Mediterranean coast, and by 550 B.C.E. there were about 250 of these trading colonies, some of which subsequently grew into thriving cities (e.g., Athens and Corinth), whereas others remain as isolated ruins or archaeological sites (such as Delphi and Olympia). The Roman Republic was established in 509 B.C.E. and took almost 300 years to establish control over the Italian Peninsula. By 14 C.E., however, the Romans had conquered much of Europe, together with parts of North Africa and Asia Minor (**FIGURE 2.8**). Most of today's major European cities had their origin as Roman settlements. In quite a few of these cities, it is possible to find traces of the original Roman street layouts. In some, it is possible to glimpse remnants of defensive city walls, paved streets, aqueducts, viaducts, arenas, sewage systems, baths, and public buildings. In the modern countryside, the legacy of the Roman Empire is represented by arrow-straight roads, built by Roman engineers and maintained and improved by successive generations.

The decline of the Roman Empire, beginning in the 4th century C.E., was accompanied by a long period of rural reorganization and consolidation under feudal systems, a period often characterized as uneventful and stagnant. **Feudal systems** were forms of economic organization where wealth was appropriated though a hierarchy of social ranks by means of institutionalized political or religious coercion. In fact, the roots of European regional differentiation can be traced to this long feudal era of slow change. Feudal systems were almost wholly agricultural, with 80% or 90% of the workforce engaged in farming and most of the rest occupied in basic craft work. Most production was for people's immediate needs; very little of a community's output ever found its way to wider markets.

By 1000 C.E., the countryside of most of Europe had been consolidated into a regional patchwork of feudal agricultural subsystems, each of which was more or less self-sufficient. For a long time, European towns were small, their existence tied mainly to the castles, palaces, churches, and cathedrals of the upper ranks of the feudal hierarchy. These economic landscapes—inflexible,

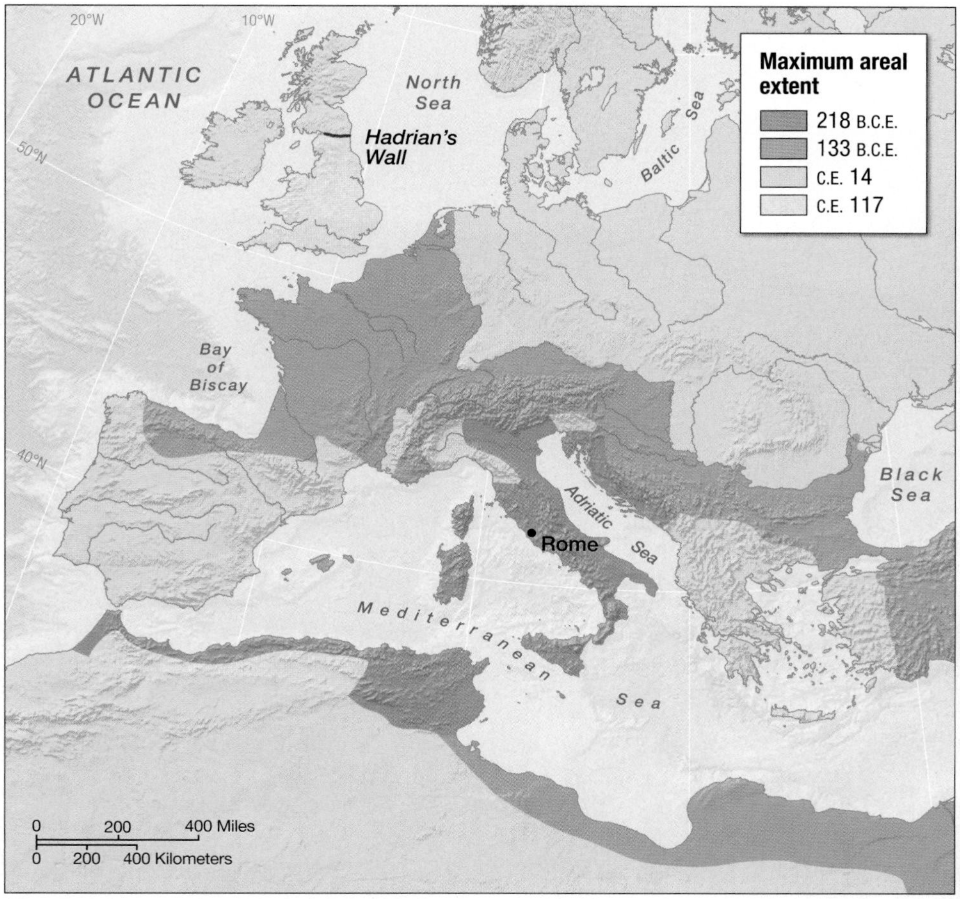
20°W
10°W
ATLANTIC OCEAN
North Sea
Hadrian's Wall
Baltic Sea
Maximum areal extent
218 B.C.E.
133 B.C.E.
C.E. 14
C.E. 117
50°N
40°N
Bay of Biscay
Black Sea
Adriatic Sea
Rome
Mediterranean Sea
0 200 400 Miles
0 200 400 Kilometers
(a)

◄ **FIGURE 2.8 The Extent of the Roman Empire** (a) This map shows the spread of the Roman Empire from 218 B.C.E. to C.E. 117. (b) Hadrian's Wall; (c) The Forum, Rome. **Summarize the geographic conquest of Europe through the movement of the Roman Empire over time. In your summary, describe three ways the Roman Empire shaped the Europe of today.**

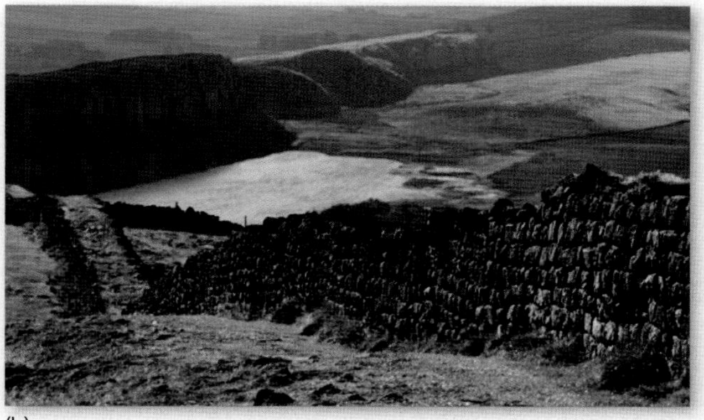
(b)

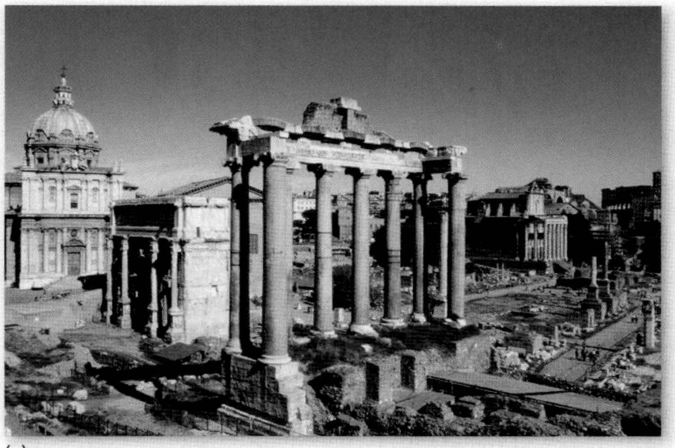
(c)

Planning for Green Cities

What does it take to become a green city? Planning and policies in support of sustainable ways of life are critical. In Europe, the Aalborg Charter (The Charter of European Sustainable Cities and Towns Towards Sustainability) was developed in 1994 to contribute to the EU's Environmental Action Programme. By 2014, more than 2,500 local and regional governments from 39 countries had committed to the Charter. Among these, the German city of Freiburg (**FIGURE 2.2.1**) is widely regarded as one of the most successful in developing green policies and encouraging progressive changes in lifestyles and behavior.

To reach its goal of reducing carbon emissions by 40% by the year 2030, Freiburg employs

"Freiburg's traffic and transportation policies encourage walking, cycling, and public transport."

a unique mixture of environmental, economic, and social policies. The city was one of the first in Europe to establish its own environmental protection agency and pledged to support solar energy after the Chernobyl nuclear power plant disaster in Russia in 1986. Freiburg's traffic and transportation policies encourage walking, cycling, and public

transport. The city's energy policy encourages the use of renewable resources such as solar, wind, and biomass and sets strict standards for the use of energy in new housing developments. As a result of its proactive support of solar energy, Freiburg has become home to many businesses operating in "green" industries such as photovoltaics.

▼ FIGURE 2.2.1 **Freiburg, Germany**

Most notably, the city has developed two new neighborhoods that follow strict ecological standards. Rieselfeld, a 70-hectare (170-acre) district for up to 12,000 residents, is easily accessible by public transport and the houses are built to low-energy standards using photovoltaic and solar thermal technologies that result in 20% less CO_2 emissions than conventional housing developments in Germany. The nearby district of Vauban (**FIGURE 2.2.2**) is home to 5,000 residents, many of them are voluntarily car-free.

Based on H. Mayer, "The Green City," in P. L. Knox, *Atlas of Cities*, New York: Princeton University Press, 2014, pp. 210–225.

1. **Identify the major ways Freiburg has worked to reduce carbon emissions.**

2. **Compare and contrast Freiburg to your city. What are some green policies that your city has employed?**

▼ **FIGURE 2.2.2** The Vauban district of Freiburg boasts 58 "energy plus" homes whose solar energy is fed back into the public grid and generates a profit for homeowners. In accordance with classic solar building principles, the living/dining rooms and bedrooms are to the south, access is in the center, and the service zones on the northern side include kitchens, bathrooms, and building services. Using the principles of Freiburg's energy plus homes, evaluate your own home in terms of energy efficiency.

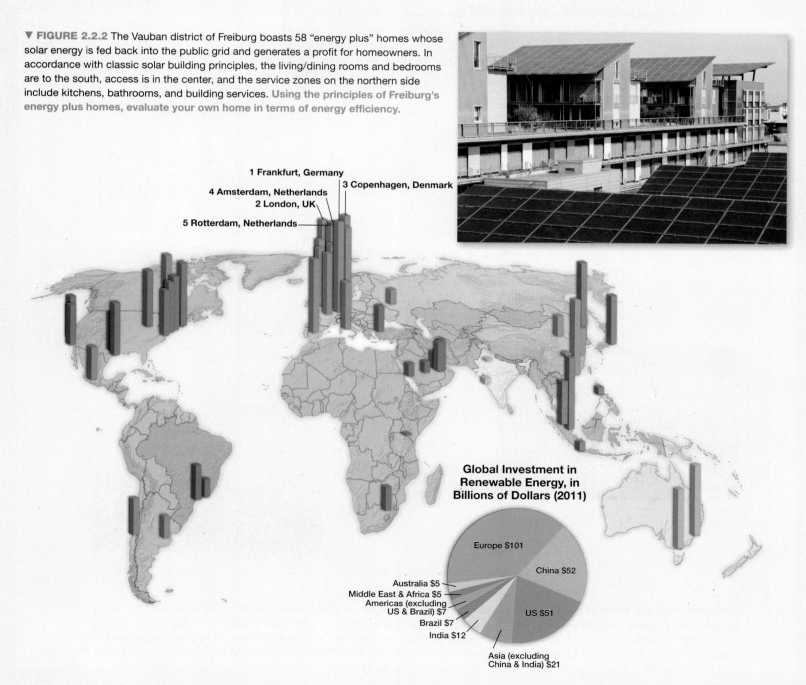

1 Frankfurt, Germany
3 Copenhagen, Denmark
4 Amsterdam, Netherlands
2 London, UK
5 Rotterdam, Netherlands

Global Investment in Renewable Energy, in Billions of Dollars (2011)

Europe $101
China $52
Australia $5
Middle East & Africa $5
Americas (excluding US & Brazil) $7
Brazil $7
India $12
US $51
Asia (excluding China & India) $21

slow motion, and introverted—nevertheless contained the essential preconditions for the rise of Europe as the dynamic hub of the world economy.

Historical Legacies and Landscapes

A key factor in the rise of Europe as a major world region was the emergence of merchant capitalism in the 15th century. **Merchant capitalism** refers to the earliest phase in the development of

capitalism as an economic and social system, when the key entrepreneurs were merchant traders and merchant bankers. The immensely complex trading system that soon came to span Europe was based on long-standing trading patterns that had been developed from the 12th century by the merchants of Venice, Pisa, Genoa, Florence, Bruges, Antwerp, and the Hanseatic League (a federation of politically independent city–states around the North Sea and Baltic coasts that included Bremen, Hamburg, Lübeck, Rostock, and Danzig (**FIGURE 2.9**)).

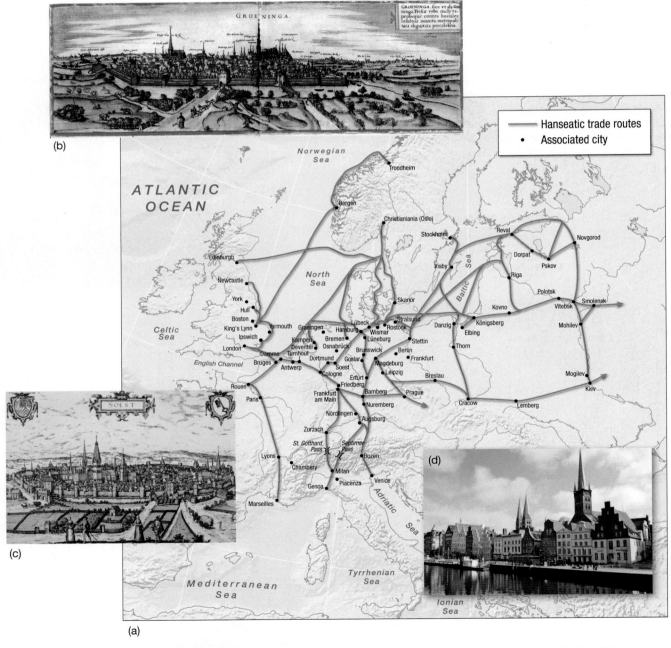

▲ **FIGURE 2.9 Major Towns of the Hanseatic League** The Hanseatic League was a federation of city–states founded in the 13th century by north German towns and affiliated German merchant groups abroad to defend their mutual trading interests. The League, which remained an influential economic and political force until the 15th century, laid the foundations for the subsequent growth of merchant trade throughout Europe. (a) Major trade routes of the Hanseatic League towns, (b) Groningen, (c) Soest. From your analysis of the Hanseatic League, predict how this organization helped lay the foundation for later trade relationships between European powers during the imperial era.

Trade and the Age of Discovery In the 15th and 16th centuries, a series of innovations in business and technology contributed to the consolidation of Europe's new merchant capitalist economy. These included several key innovations in the organization of business and finance: banking, loan systems, credit transfers, company partnerships, shares in stock, speculation in commodity futures, commercial insurance, and courier/news services. Meanwhile, technological innovations began to further strengthen Europe's economic advantages. Some of these were adaptations and improvements of Oriental discoveries—the windmill, spinning wheels, paper manufacture, gunpowder, and the compass, for example. In Europe there was a unique passion for mechanizing the manufacturing process. Key engineering breakthroughs included the more efficient use of energy in water mills and blast furnaces, the design of reliable clocks and firearms, and the introduction of new methods of processing metals and manufacturing glass.

The geographical knowledge acquired during this Age of Discovery was crucial to the expansion of European political and economic power in the 16th century. In societies that were becoming more and more commercially oriented and profit conscious, geographical knowledge became a valuable commodity in itself. Gathering information about overseas regions was a first step to controlling and influencing them; this in turn was a step to wealth and power. At the same time, every region began to open up to the influence of other regions because of the economic and political competition unleashed by geographical discovery.

These changes had a profound effect on the geography of Europe. Before the mid-15th century, Europe was organized around two subregional maritime economies—one based on the Mediterranean and the other on the Baltic. The overseas expansions pioneered first by the Portuguese and then by the Spanish, Dutch, English, and French reoriented Europe's geography toward the Atlantic. The river basins of the Rhine, the Seine, and the Thames rapidly became focused on a thriving network of **entrepôt** seaports (intermediary centers of trade and transshipment) that transformed Europe. These three river basins, backed by the increasingly powerful states in which they were embedded—the Netherlands, France, and Britain, respectively—then became engaged in a struggle for economic and political hegemony.

Colonialism As we noted in Chapter 1, the first wave of European colonialism began in the late 1400s. Europe's growth could only be sustained as long as productivity could be improved, and, after a point, increased productivity required food and energy resources that could only be obtained by the conquest—peaceful or otherwise—of new territories.

This first phase of colonial exploitation could not have been achieved without the development of a remarkable combination of European innovations in shipbuilding, navigation, and naval ordnance. Over the course of the 15th century, full-rigged ship designs were developed, enabling faster voyages in larger and more maneuverable vessels that were less dependent on favorable winds. Meanwhile, the quadrant (1450) and the astrolabe (1480) were invented, and a systematic knowledge of Atlantic winds had been acquired. By the mid-1500s, England, Holland, and Sweden had perfected the technique of casting iron guns, making it possible to replace the bronze cannon with larger numbers of more effective guns at lower expense. Together, these advances made it possible for the merchants of Europe to establish the basis of a worldwide economy in the space of less than 100 years.

Gold and silver plundered from the Americas allowed Europe to live above its means. In effect, the bullion was converted into demand for goods of all kinds—textiles, wine, food, furniture, weapons, and ships—thus stimulating production throughout Europe and creating the basis for a "Golden Age" of prosperity for most of the 16th century. Meanwhile, overseas expansion made available a variety of new and unusual products—cocoa, beans, maize, potatoes, tomatoes, tobacco, and vanilla from the Americas, tea from the Orient—which opened up large new markets to enterprising merchants.

As European traders came to monopolize trade routes, they were able to identify foreign articles and ship them home to Europe, where skilled workmen learned to imitate them. In time, overseas expansion stimulated further improvements in technology and business techniques, thus adding to the self-propelling growth of European capitalism. Ultimately the whole experience of overseas colonization provided a great practical school of entrepreneurship and investment. Most important of all, perhaps, was the way that the profits from overseas colonies and trading posts flowed back into domestic agriculture, mining, and manufacturing. This flow of profits contributed to an accumulation of capital that was undoubtedly one of the main preconditions for the emergence of industrial capitalism in Europe in the 18th century.

Apply Your Knowledge

2.7 Identify and research three cities that have been important entrepôt seaports.

2.8 Use the Internet to create and analyze profiles of the present-day economies of those three cities.

Industrialization In the wake of colonialism, Europe's regional geographies were comprehensively recast once more by the new technology systems that marked the onset of the Industrial Revolution (in the late 1700s). **Technology systems** are clusters of interrelated energy, transportation, and production technologies that dominate economic activity for several decades at a time—until a new cluster of improved technologies evolves. There was, therefore, not a sudden, single Industrial Revolution but three distinctive transitional waves of industrialization, each with a different degree of impact on different regions and countries (**FIGURE 2.10**).

- **The First Wave** The first wave of industrialization, between about 1790 and 1850 C.E., was based on the cluster of early industrial technologies (steam engines,

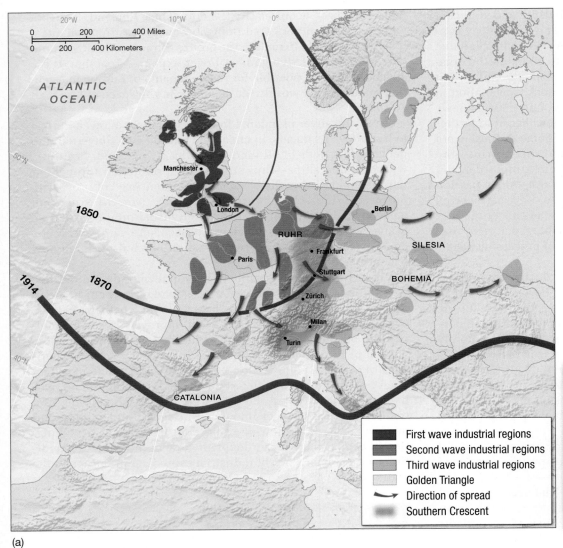

◀ **FIGURE 2.10 The Spread of
Industrialization in Europe** (a) The first
wave of European industrialization began
with the emergence of small industrial regions
in several parts of Britain. As new industrial
and transportation technologies emerged,
industrialization spread to other regions with
the right attributes: access to raw materials and
energy sources, good communications, and large
labor markets. A third wave of industrialization,
after 1870, saw industrial development spread
to southern and Eastern Europe and southern
Scandinavia. (b) Nineteenth-century Manchester.
Note the number of factory smokestacks.
**Explain why the first wave of industrialization
emerged in Britain and how the second two
waves connected the rest of Europe.**

First wave industrial regions
Second wave industrial regions
Third wave industrial regions
Golden Triangle
Direction of spread
Southern Crescent

(a)

(b)

cotton textiles, and ironworking) and was highly localized. It was limited to a few regions in Britain where
industrial entrepreneurs and workforces had first
exploited key innovations and the availability of key
resources (coal, iron ore, and water).

- **The Second Wave** The second wave, between about 1850
and 1870 C.E., involved the exploitation of a technology system based on coal-powered steam engines, steel
products, railroads, world shipping, and machine tools.
This second wave saw the diffusion of industrialization
to most of the rest of Britain and to parts of northwest
Europe, particularly the coalfield areas of northern
France, Belgium, and Germany. New opportunities were
created as railroads and steamships made more places
accessible, bringing their resources and their markets into
the sphere of industrialization. New materials and new
technologies (steel, machine tools) created opportunities
to manufacture and market new products. These new
activities prompted some significant changes in the logic

of industrial location. Railway networks, for example,
attracted industry away from smaller towns on the canal
systems and toward larger towns with good rail connections. Steamships for carrying on coastal and international
trade attracted industry to larger ports. At the same time,
steel produced concentrations of heavy industry in places
with nearby supplies of coal, iron ore, and limestone.

- **The Third Wave** The third wave of industrialization,
between 1870 and 1914 C.E., saw a further reorganization of the geography of Europe as yet another cluster of
technologies (including electricity, electrical engineering, and telecommunications) brought different resource
needs and created additional investment opportunities. During this period, industrialization spread for
the first time to remoter parts of the United Kingdom,
France, and Germany and to most of the Netherlands,
southern Scandinavia, northern Italy, eastern Austria,
Bohemia (in what was then Czechoslovakia), Silesia (in
Poland), Catalonia (in Spain), and the Donbas region

of Ukraine, then into Russia. The overall result was to create the foundations of a core-periphery structure within Europe, with the heart of the core centered on the **Golden Triangle** stretching between London, Paris, and Berlin (Figure 2.10). The peripheral territories of Europe—most of the Iberian peninsula, northern Scandinavia, Ireland, southern Italy, the Balkans, and east-central Europe—were slowly penetrated by industrialization over the next 50 years. More recently, a secondary, emergent, core region—the **Southern Crescent** that straddles the Alps from central Germany to northern Italy—has emerged as a result of the decentralization of industry from northwestern Europe and in part a result of the integrative effects of the EU (see p. 24).

Imperialism and War By the time the Industrial Revolution had gathered momentum in the early 19th century, several of the most powerful and heavily industrialized European countries (notably the United Kingdom, Germany, France, and the Netherlands) were competing for influence on a global scale. Europe's industrialization must be understood in this context: the ascent of Europe's industrial core regions could not have taken place without the foodstuffs, raw materials, and markets provided by the rest of the world. To ensure the availability of the produce, materials, and markets on which they were increasingly dependent, the industrial nations of Europe vigorously pursued a second phase of overseas expansion, creating a series of **trading empires**.

A scramble began for territorial and commercial domination through **imperialism**—the deliberate exercise of military power and economic influence by powerful states to advance and secure their national interests. European countries engaged in expansion to protect their established interests and to limit the opportunities of others. Each power wanted to secure as much of the world as possible—through a combination of military oversight, administrative control, and economic regulations—to ensure stable and profitable environments for their traders and investors. A second phase of colonialism ensued as the major economic and military powers embarked on the inland penetration of mid-continental grassland zones to settle and exploit them for grain or stock production. Britain, with by far the most powerful navy in the world, was able to establish a truly global empire (**FIGURE 2.11**). The detailed pattern and timing of this exploitation was heavily influenced by new transportation technologies—especially railways—and the invention of innovations such as barbed wire and refrigeration.

During the first half of the 20th century, the economic development of the whole of Europe was disrupted twice by major wars. The devastation of World War I was immense. The overall loss of life, including the victims of influenza epidemics and

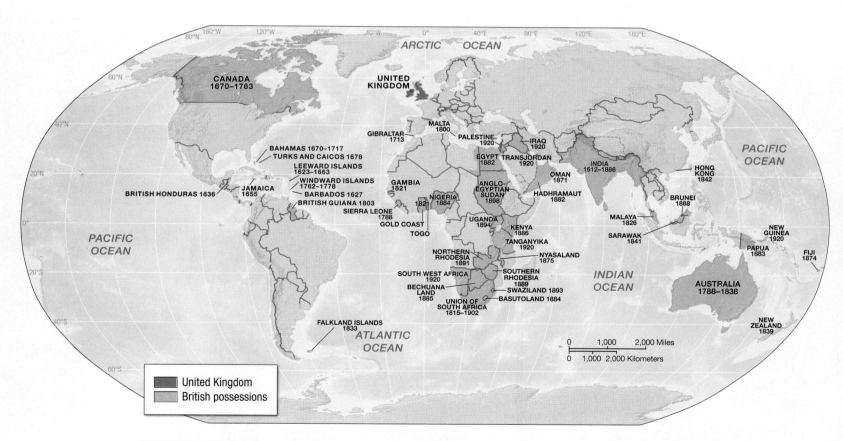

▲ **FIGURE 2.11 The British Empire in 1900** As you consider the geographic expanse of the British Empire, identify one example each of a cultural, economic, political, and demographic result of the imperial project in today's world.

border conflicts that followed the war, amounted to between 50 and 60 million. About half as many again were permanently disabled. For some countries, this meant a loss of between 10% and 15% of the male workforce.

Just as European economies had adjusted to these devastations, the **Great Depression**—a severe decline in the world economy that lasted from 1929 until the mid-1930s—created a further phase of economic damage and reorganization throughout Europe. World War II resulted in yet another round of destruction and dislocation (**FIGURE 2.12**). The total loss of life in Europe this time was 42 million, two-thirds of whom were civilian casualties. The German persecution of Jews and others—the **Holocaust**, Nazi Germany's systematic genocide of various ethnic, religious, and national minority groups—resulted in approximately 6 million Jews being put to death in extermination camps such as Auschwitz and Treblinka, with up to 5 million others, being exterminated elsewhere. The German occupation of continental Europe also involved ruthless economic exploitation. By the end of the war, France was depressed to below 50% of its prewar standard of living and had lost 8% of its industrial assets. The United Kingdom lost 18% of its industrial assets (including overseas holdings), and the Soviet Union lost 25% in the war. Germany lost 13% of its assets and ended the war with a level of income per capita that was less than 25% of the prewar figure. In addition to the millions killed and disabled during World War II, approximately 46 million people were displaced between 1938 and 1948 through flight, evacuation, resettlement, or forced labor. Some of these movements were temporary, but most were not.

After the war, the Cold War rift between Eastern Europe (countries dominated by the Soviet Union) and Western Europe (the rest) resulted in a further handicap to the European economy and, indeed, to its economic geography. Ironically, this rift helped speed economic recovery in Western Europe. The United States, whose leaders believed that poverty and economic chaos in Western Europe would foster communism, embarked on a massive program of economic aid under the **Marshall Plan**, a program of loans and other economic assistance provided by the U.S. government between 1947 and 1952 to bolster European allies whose weakness, it was believed, made them susceptible to communism. This pump-priming action, together with the backlog of demand in almost every sphere of production, provided the basis for a remarkable recovery. Meanwhile, Eastern Europe began an interlude of **state socialism**, a form of economy based on principles of collective ownership and administration of the means of production and distribution of goods dominated and directed by state bureaucracies.

▼ **FIGURE 2.12 London, England, 1940** London sustained heavy bomb damage during World War II. **Compare and contrast the experiences of European cities in WWII to those in the United States and Canada.**

Eastern Europe's Interlude of State Socialism After World War II, the leaders of the Soviet Union felt compelled to establish a **buffer zone** between their homeland and the major Western powers in Europe. The Soviet Union rapidly established its dominance throughout Eastern Europe; Estonia, Moldova, Latvia, and Lithuania were absorbed into the Soviet Union itself, and Soviet-style regimes were installed in Albania, Bulgaria, Czechoslovakia, East Germany, Hungary, Poland, and Romania. The result was what Winston Churchill called an **Iron Curtain** along the western frontier of Soviet-dominated territory, a militarized frontier zone across which Soviet and East-European authorities allowed the absolute minimum movement of people, goods, and information (**FIGURE 2.13**). In addition, Soviet intervention resulted in the complete nationalization of the means of production, the collectivization of agriculture, and the imposition of rigid social and economic controls within the Eastern-European **satellite states** (countries that are under heavy political, economic, and military influence of another state—in this case, the Soviet Union).

The economies of the Soviet Union and its satellites were *not* based on true socialist or communist principles in which the working class had democratic control over the processes of production, distribution, and development. Rather, these economies evolved as something of a hybrid, in which state power was used by a **bureaucratic class** (nonelected government officials) to create **command economies** in the pursuit of modernization and economic development. In a command economy, every aspect of economic production and distribution is controlled centrally by government agencies. The **Communist Council for Mutual Economic Assistance** (CMEA, better known as **COMECON**) was established to reorganize eastern European economies in the Soviet mold and foster mutual trade among the *Soviet bloc*—the Soviet Union plus its eastern European satellite states (**FIGURE 2.14**).

The Reintegration of Eastern Europe Eventually, the economic and social constraints imposed by excessive state control and the dissent that resulted from the lack of democracy under state socialism combined to bring the experiment to a sudden halt. By the time the Soviet bloc collapsed in 1989 (see Chapter 3), Poland and Hungary had already accomplished a modest degree of democratic and economic reform. By 1992, East Germany (the German Democratic Republic) had been reunited with West Germany (the German Federal Republic), and Estonia, Latvia, and Lithuania had become independent states once more.

The reintegration of Eastern Europe added a market of 130 million consumers to the European economy and did not take long for Western-style consumerism to appear on the streets and in many of the stores in larger eastern European cities. On the flip side, it also did not take long for inflation, unemployment, and homelessness to appear. In detail, the pace and degree of reintegration has varied considerably across Eastern Europe. The best-integrated states of Eastern Europe are Czechia, Hungary, Poland, and Slovenia. All have a relatively

▲ **FIGURE 2.13 Berlin Wall** The Berlin Wall was a physical reminder of the separation between eastern and western Europe created by the Iron Curtain after World War II. **What do you think are some of the issues that developed as a result of external powers like the United States and U.S.S.R. dividing Berlin? Draw conclusions on what you think may have happened to people's everyday lives.**

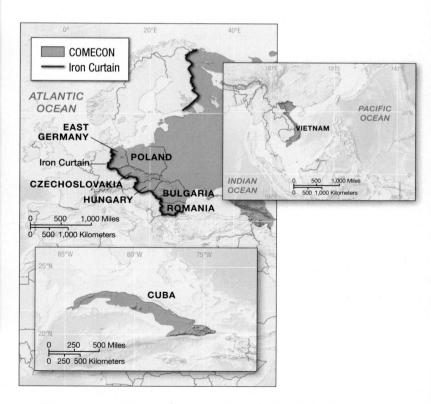

▲ **FIGURE 2.14** Members of the Communist Council for Mutual Economic Assistance (COMECON) trade bloc.

strong industrial base, and Hungary has a productive agricultural sector as well. The Baltic states of Estonia, Latvia, and Lithuania have also been successful in reintegrating with the rest of Europe. Their small size and relatively high levels of education have made them attractive as production subcontracting centers for western European high-technology industries. But ethnic conflict in Bosnia and Herzegovina, Croatia, and the former republic of Yugoslavia has severely retarded reform and reintegration, while Albania, Bulgaria, Macedonia, Moldova, and Romania have suffered from the combined disadvantages of having relatively poor resource bases, weakly developed communications and transportation infrastructures, and political regimes with little ability or inclination to press for economic and social reform. In Ukraine, which has a better infrastructure, a significant industrial base, and the capacity for extensive trade in grain exports and advanced technology, reintegration has been retarded by a combination of geographical isolation from Western Europe and the problem of severing economic and political ties with Russia.

Apply Your Knowledge

2.9 List three ways that Europe has changed since the Cold War.

2.10 Summarize a few of the problems and opportunities that Berlin has faced as a result of reintegration after searching *The Independent* website for articles on Berlin: "The Division and Reintegration of A Nation," at http://www.independent. co.uk.

Berlin Reintegration

https://goo.gl/r1Jias

Economy, Accumulation, and the Production of Inequality

Contemporary Europe is a cornerstone of the world economy with a complex, multilayered, and multifaceted regional geography. In overall terms, Europe, with about 12% of the world's population, accounts for almost 35% of the world's exports, almost 43% of the world's imports, and 33% of the world's aggregate gross national income (GNI). The bulk of Europe's exports are machinery, motor vehicles, aircraft, plastics, pharmaceuticals, iron and steel, nonferrous metals, wood pulp and paper products, textiles/fashion, meat, dairy products, wine, and fish. Levels of material consumption in much of Europe approach those of households in the United States.

Although Europe is a relatively affluent world region, there are, in fact, persistent and significant economic inequalities at every geographic scale. Regional income disparities within many European countries are increasing. In northwestern Europe, this is generally a result of the declining fortunes of "Rust Belt" regions and the relative prosperity of regions with high-tech

industry and advanced business services. In southern and eastern Europe, it is a result of differences between regions dominated by rural economies (generally poorer) and those dominated by metropolitan areas (generally more prosperous). The resulting disparities are significant, with annual per capita gross domestic product (GDP; in purchasing power parity (PPP)) in 2013 ranging from U.S. $4,670 in Moldova and U.S. $8,790 in Ukraine to U.S. $65,460 in Norway and U.S. $90,410 in Luxembourg. Poverty and homelessness exist in every European country, though poverty as measured on a global scale (U.S. $1 or U.S. $2 a day per person) is virtually unknown within Europe.

The development of European **welfare states** (institutions with the aim of distributing income and resources to the poorer members of society) has helped maintain purchasing power during periods of recession and ensured at least a tolerable level of living for most groups at all times. Levels of personal taxation are high, but all citizens receive a wide array of services and benefits in return. The most striking of these services are high-quality medical care, public transport systems, **social housing** (where the dwellings are owned by a government or nonprofit agency), schools, and universities. The most significant benefits are pensions and unemployment benefits.

The European Union: Coping with Uneven Development Contemporary Europe is a dynamic region that embodies a great deal of change. Formerly prosperous industrial regions have suffered economic decline, but some places and regions have reinvented themselves to take advantage of new paths to economic development. Meanwhile, the former Soviet satellite states have been reintegrated into the European world region, and much of Europe has joined in the **European Union (EU)**, a supranational organization founded to recapture prosperity and power through economic and political integration. The EU has its origins in the political and economic climate that followed World War II. The initial idea behind the EU was to ensure European autonomy from the United States and to recapture the prosperity Europe had forfeited as a result of the war. Part of the rationale for its creation was also to bring Germany and France together in a close association, which would prevent any repetition of the geopolitical problems in Western Europe that had led to two world wars.

In the 1950s, several institutions were created to promote economic efficiency through integration. These were subsequently amalgamated to form the European Community (EC), which was in turn expanded in scope to form the EU. EU membership has expanded from the six original members of the EC—Belgium, France, Italy, Luxembourg, the Netherlands, and West Germany—to 28 countries (**FIGURE 2.15**). Member states combined have a population of nearly 500 million, and a GDP (nominal GDP) of U.S. $18.5 trillion in 2014 (compared to the 2013 U.S. GDP of U.S. $16.8 trillion). The EU has developed into a sophisticated and powerful institution with a pervasive influence on patterns of economic and social well-being within its member states. It also has a significant impact on certain aspects of economic development within some nonmember countries.

The origin of the organization was a compromise worked out between the strongest two of the original six members. West Germany wanted a larger but protected market for its industrial goods, whereas France wanted to continue to protect its highly inefficient but large and politically important agricultural sector from overseas competition. The result was the creation of a tariff-free market within the EU, the creation of a unified external tariff, and a **Common Agricultural Policy (CAP)** to bolster the agricultural sector.

The basis of the CAP is a system of EU support of wholesale prices for agricultural produce. This support has the dual effects of stabilizing the price of agricultural products and of subsidizing farmers' incomes. The overall result has been a realignment of agricultural production patterns, with a general withdrawal from mixed farming. Ireland, the United Kingdom, and Denmark, for example, have increased their specialization in the production of wheat, barley, poultry, and milk, whereas France and Germany have increased their specialization in the production of barley, maize, and sugar beets.

The reorganization of Europe's agricultural landscapes under the CAP brought some unwanted side effects, however, including environmental problems that have occurred as a result of the speed and scale of farm modernization, combined with farmers' desire to take advantage of generous levels of guaranteed prices for crops. Moorlands, woodlands, wetlands, and hedgerows have come under threat, and some traditional mixed-farming landscapes have been replaced by the prairie-style settings of specialized agribusiness.

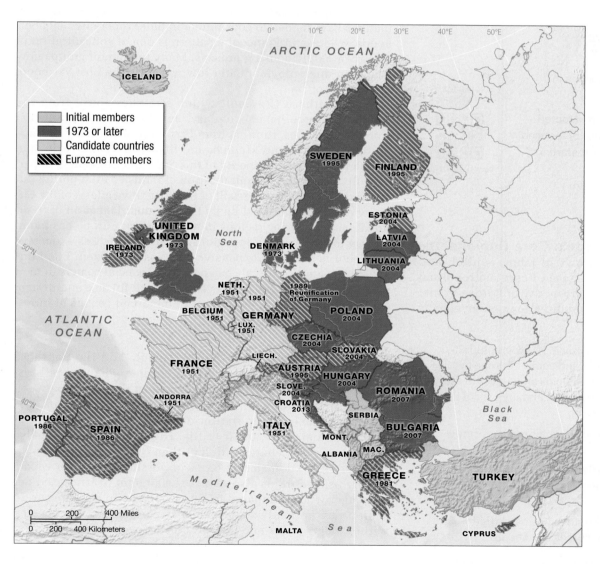

▲ **FIGURE 2.15 The Expansion of the EU** The advantages of membership in the EU led to a dramatic growth in its size, transforming it into a major economic and political force in world affairs. Economic union among very different economies has, however, created significant problems for the member countries of the eurozone, which share the same currency, the euro. Describe three factors that makes membership in the EU attractive. List at least one reason that membership could be a disadvantage.

Meanwhile, the rest of the world economy had changed significantly, intensifying the competitive challenge to Europe's industries. In response, the EC relaunched itself as the EU, affirming the ultimate aim of economic and political harmonization within a single supranational government. The key step was the **Maastricht Treaty** of 1992, which led to the creation of a common European currency, the euro, in 1999. To join the currency, member states had to qualify by meeting strict conditions in terms of national budget deficits, inflation, and interest. Of EU members at the time, the United Kingdom, Sweden, and Denmark declined to join the currency, leaving a "eurozone" that was smaller than the EU as a whole (see Figure 2.15). The 2008–2014 global recession, triggered by an international banking crisis, revealed serious problems in the eurozone, with Greece, Ireland, Portugal, and Spain found to be deeply in debt and having to be given massive loans funded by Germany and other core economies. In addition to the economic and fiscal policies associated with monetary union, the EU has developed a broad range of policies designed to facilitate the integration of member states and improve their competitiveness. One example is the Erasmus program, which provides funding for university students to spend between 3 and 12 months in a university in a different EU country from their own.

Overall, EU membership has brought regional stresses as well as the prospect of overall economic gain. Existing member countries have found that the removal of internal barriers to labor, capital, and trade has worked to the clear disadvantage of peripheral regions and in particular to the disadvantage of those farthest from the Golden Triangle (between London, Paris, and Berlin), which is increasingly the European center of gravity in terms of both production and consumption. As a result, the EU allocates about one-third of its budget—€376 billion (U.S. $487 billion) in its 2014–2020 budget cycle—to projects and policies designed to improve economic and social cohesion within and among its member countries.

Apply Your Knowledge

2.11 Give three examples of inequalities in European economies.

2.12 Summarize an EU policy that is directed toward (a) the environment and (b) focused on regional policy. (Hint: A good starting point is the EU's own website: http://europa.eu/pol/index_en.htm.)

European Union Policies

https://goo.gl/LrTYmG

Regional Development: Europe's Core Regions The removal of internal barriers to flows of labor, capital, and trade has worked to the clear advantage of core metropolitan regions. The character and relative prosperity of Europe's chief core region, the Golden Triangle, stem from four advantages as follows:

- It is well situated for shipping and trade. Its geographic situation provides access to southern and central Europe by way of the Rhine and Rhône river systems and access to the sea lanes of the Baltic Sea, the North Sea, and, by way of the English Channel, the Atlantic Ocean.

- Within it are the capital cities of the major former imperial powers of Europe.

- It includes the industrial heartlands of central England, northeastern France, and the Ruhr district of Germany, and

- Its concentrated population provides both a skilled labor force and an affluent consumer market.

These advantages have been reinforced by the integrative policies of the EU, whose administrative headquarters are situated squarely in the heart of the Golden Triangle, in Brussels, Belgium. They have also been reinforced by the emergence of Berlin, Paris, and, especially, London as world cities, in which a disproportionate share of the world's economic, political, and cultural business is conducted.

The affluence and dynamism of the region have attracted large numbers of immigrants. London, in particular, has become a city with a global mix of populations and subcultures. Almost one-third of London's current residents—2.2 million people—were born outside England, and this total does not take into account the contribution of the city's second- and third-generation immigrants, many of whom have inherited the traditions of their parents and grandparents. Altogether, the people of London speak more than 300 languages, and the city has at least 50 nonindigenous communities with populations of 10,000 or more. Elsewhere in the Golden Triangle, immigrant groups have also added a new ethnic dimension to city populations. Most have settled in distinctive enclaves, where they have retained powerful attachments to their cultural roots. In the United Kingdom, the Netherlands, and Germany, immigrant populations have tended to concentrate in older inner-city neighborhoods, whereas in France the pattern is one of concentration in suburban public housing projects. Many immigrant groups now face high unemployment and low wages as the postwar boom has leveled off.

The Southern Crescent, which stretches south from the Golden Triangle, is a secondary, emergent, core region that straddles the Alps, running from Frankfurt in Germany through Stuttgart, Zürich, and Munich, and finally to Milan, Turin, and northern Italy. The prosperity of this Southern Crescent is in part a result of a general decentralization of industry from northwestern Europe and in part a result of the integrative effects of the EU. Some of the capital freed up by the **deindustrialization** of traditional manufacturing regions in northwestern Europe has found its way to more southerly regions, where land is less expensive and labor is both less expensive and less unionized. Frankfurt and Zürich are global-scale business and financial centers in their own right, whereas Milan is a center of both finance and design, and Munich, Stuttgart, and Turin are important centers of industry and commerce.

High-Speed Rail The Golden Triangle and the Southern Crescent are linked by the trans-European high-speed rail system (**FIGURE 2.16**). With relatively short distances between major cities, Europe is ideally suited for rail travel and less suited, because of population densities and traffic congestion around airports, to air traffic. High-speed trains are also considered more energy efficient than other modes of transit per passenger-kilometer. In terms of possible passenger capacity, high-speed trains also reduce the amount of land used per passenger when compared to cars on roads. In addition, train stations are normally smaller than airports and can be located within major cities and spaced closer together, allowing for more convenient travel. Allowing for check-in times and accessibility to terminals, travel between many major European cities is already quicker by rail than by air. Improved locomotive technologies and specially engineered tracks and rolling stock make it possible to offer passenger rail services on some routes at speeds of 275 to 350 kilometers per hour (180 to 250 miles

▼ **FIGURE 2.16 (a) Europe's High-Speed Rail Network** The expansion of the high-speed rail network marks a major achievement of the EU and is an important factor in the integration of European economies. **(b) High-Speed Train** A third-generation ICE (Inter City Express) train, designed to run at speeds up to 320 km/h (200 mph), stands at the station at Frankfurt International Airport. **(c) Gotthard Base Tunnel** The tunnel, expected to be completed in 2017, is actually a pair of tunnels, each 10 meters (32 feet) in diameter, one for northbound trains and the other for southbound trains. The project will link major economic centers on both sides of the Alps. **Drawing from your study of European geography, why do you think rail is a major form of transportation and connection in Europe?**

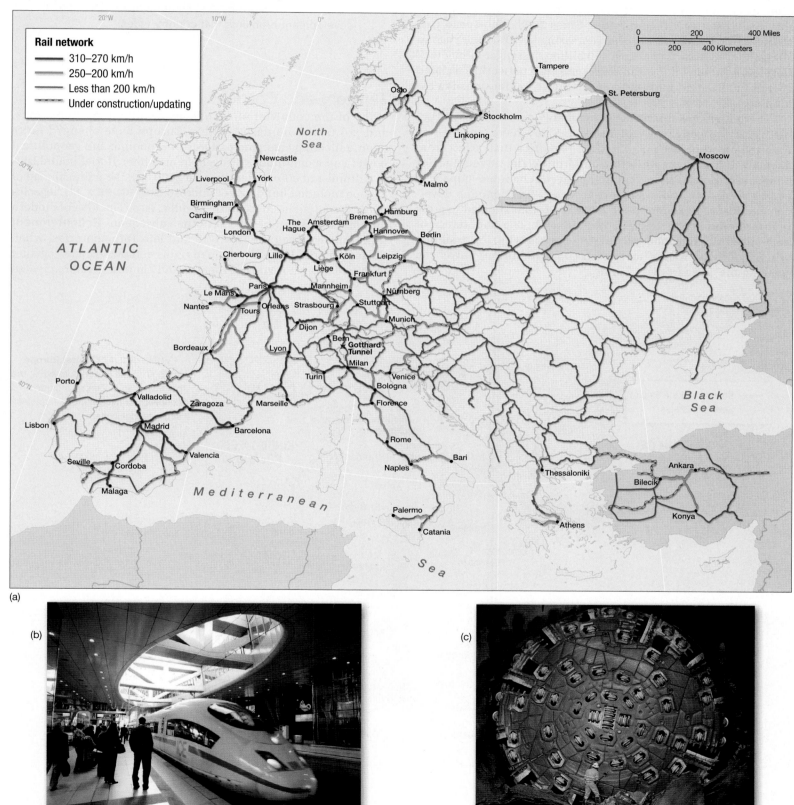

per hour; Figure 2.16b). In addition, new tilt-technology railway cars, which are designed to negotiate tight curves by tilting the train body into turns to counteract the effects of centrifugal force, have been introduced in many parts of Europe to raise maximum speeds on conventional rail tracks.

The Gotthard Base Tunnel, a new 57-kilometer (35-mile) high-speed rail tunnel under the Alps in Switzerland, is part of this overall strategy (Figure 2.16c). Since 2005, high-speed passenger traffic in Europe has been increasing by 10% each year. In 2014, the system carried almost 140 billion passenger-kilometers. Developing a trans-European high-speed rail network is a stated goal of the EU, and most cross-border railway lines receive EU funding.

These high-speed rail routes inevitably cause some restructuring of the geography of Europe. They have only a few time-tabled stops because the time penalties that result from deceleration and acceleration undermine the advantages of high-speed travel. Places that do not have scheduled stops will be less accessible and, then, less attractive for economic development. Places linked to the routes will be well situated to grow in future rounds of economic development. The importance of high-speed rail to economic development is why, for example, billions of Swiss francs (more than U.S. $10 billion) have been spent on the Gotthard Base Tunnel.

Apply Your Knowledge

2.13 Identify the major reasons for the economic prosperity of the Golden Triangle and Southern Crescent regions.

2.14 Analyze one of the geographically peripheral countries of northern or southern Europe and determine the regional development problems that country faces.

Territory and Politics

Many of the countries of Europe are relatively new creations, and political boundaries within Europe have changed quite often, reflecting changing patterns of economic and geopolitical power and a continuous struggle to match territorial boundaries to cultural and ethnic identities. In 1500 C.E., the political map of Europe included hundreds of microstates (**FIGURE 2.17**), legacies of the feudal hierarchies of the Middle Ages. The idea of modern national states can be traced to the subsequent **Enlightenment** in Europe between the late 17th and late 18th centuries, when the ferment of ideas about human rights and democracy, together with widening horizons of literacy and communication, created

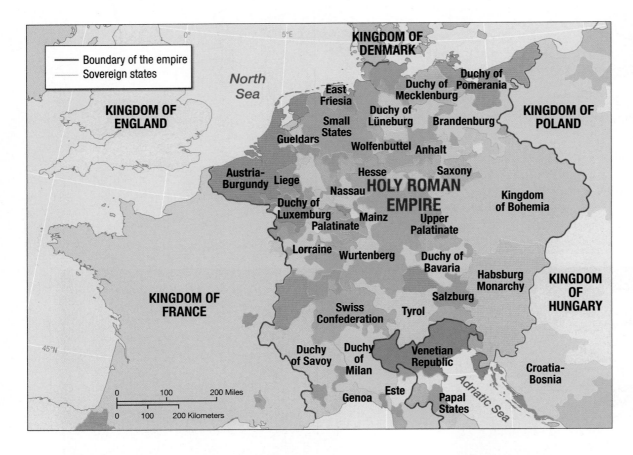

◀ **FIGURE 2.17 Central Europe in 1500** In the late Middle Ages, the political map of Europe included hundreds of microstates—legacies of the feudal system.

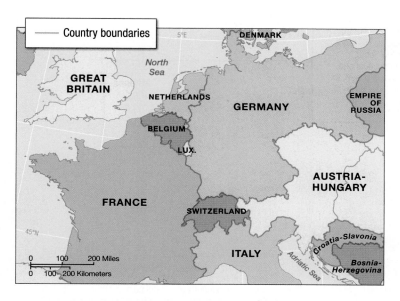

▲ **FIGURE 2.18 Central Europe in 1900** After the French Revolution at the end of the 18th century and the Napoleonic Wars at the beginning of the 19th century, Europe was reordered in a pattern of modern states. **Describe how and why the Napoleonic Wars consolidated European nation–states, impacting the map of Europe today.**

new perspectives on allegiance, communality, and identity. In 1648, the Treaty of Westphalia, signed by most European powers, brought an end to Europe's seemingly interminable religious wars by making national states the principal actors in international politics and establishing the principle that no state has the right to interfere in the internal politics of any other state.

Gradually, these perspectives began to undermine the dominance of the great European continental empires controlled by family dynasties—the Bourbons, the Hapsburgs, the Hohenzollerns, the House of Savoy, and so on. After the French Revolution (1789–1793) and the kaleidoscopic changes of the Napoleonic Wars (1800–1815), Europe was reordered, in the century following 1815, in a pattern of modern states (**FIGURE 2.18**). Denmark, France, Portugal, Spain, and the United Kingdom had long existed as separate, independent states. The 19th century saw the unification of Italy (1861–1870) and of Germany (1871) and the creation of Belgium, Bulgaria, Greece, Luxembourg, the Netherlands, Romania, Serbia, and Switzerland as independent national states. Early in the 20th century they were joined by Czechoslovakia, Estonia, Finland, Latvia, Lithuania, Norway, and Sweden. Austria was created in its present form in the aftermath of World War I, as part of the carving up of the German and Austro-Hungarian empires. In 1921, long-standing religious cleavages in Ireland resulted in the creation of the Irish Free State (now Ireland), with six of the Protestant counties of Ulster remaining in the United Kingdom as Northern Ireland.

Regionalism and Boundary Disputes
The European concept of the nation–state has immensely influenced the modern

world. As we saw in Chapter 1, the idea of a nation–state is based on the concept of homogeneous groups of people governed by their own state. In a true nation–state, no significant group exists that is not part of the nation. In practice, many European states were established around the concept of a nation–state but with territorial boundaries that did, in fact, encompass substantial ethnic minorities (**FIGURE 2.19**). The result has been that the geography of Europe has been characterized by regionalism and boundary disputes throughout the 20th century and into the 21st century.

Examples of regionalism include regional independence movements in the Basque region and in Catalonia (both in Spain), Scotland (in the United Kingdom), and the Turkish Cypriots' determination to secede from Cyprus. Examples of **irredentism**—the assertion by a government that a minority living outside its borders belongs to it culturally and historically—include Ireland's claim on Northern Ireland (renounced in 1999), the claims of Nazi Germany on Austria and the German parts of Czechoslovakia and Poland, the claims of Croatia and Serbia and Montenegro on various parts of Bosnia-Herzegovina and, most recently, Russia's de facto annexation of the Crimean Peninsula region of Ukraine (**FIGURE 2.20**). Some cases of regionalism have led to violence, social disorder, or even civil war, as in Cyprus. For the most part, however, regional ethnic separatism has been pursued within the framework of civil society, and the result has been that several regional minorities have achieved a degree of political autonomy. For example, the United Kingdom created regional parliaments for Scotland and Wales. In response to intensified feelings of nationalism in Scotland, the United Kingdom government provided a referendum in 2014 for Scottish voters to decide whether or not to withdraw from the United Kingdom altogether. The outcome—by a narrow majority—was for continued membership of the United Kingdom.

Ethnic Conflict in the Balkans
In contrast to regionalism, most cases of irredentism have contributed at some point in history to war or conflict. Nowhere has this been more evident than in the troubled region of the Balkans. The repeated fragmentation and reorganization of ethnic groups into separate states within the region has given rise to the term **balkanization**, which refers to any situation in which a larger territory is broken up into smaller units, especially where territorial jealousies give rise to a degree of hostility. In the Balkans themselves, geopolitical reorganizations have left significant **enclaves**, culturally distinct territories that are surrounded by the territory of a different cultural group, and **exclaves**, portions of a country or of a cultural group's territory that lie outside its contiguous land area. These issues came to a head after the breakup of Yugoslavia in 1990 marked the end of attempts to unite Serbs, Croats, and Slovenes within a single territory. Serbian nationalism provided the principal catalyst for violence and conflict in the region, including attempts at **ethnic cleansing**.

The most extreme example of ethnic cleansing was in the Kosovo region. In 1998, Serbia's leader, Slobodan Milošević, initiated a brutal, premeditated, and systematic campaign of murder and violence that was aimed at removing Kosovar

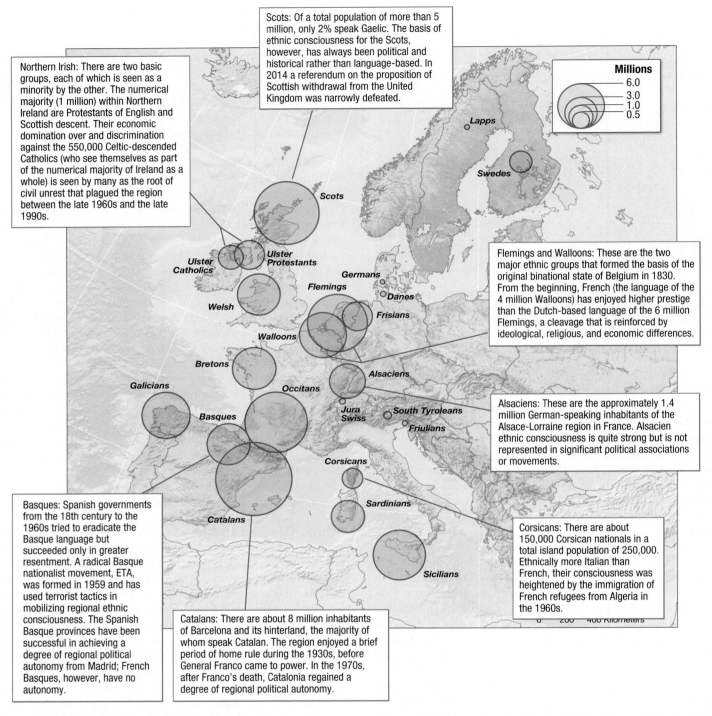

Scots: Of a total population of more than 5 million, only 2% speak Gaelic. The basis of ethnic consciousness for the Scots, however, has always been political and historical rather than language-based. In 2014 a referendum on the proposition of Scottish withdrawal from the United Kingdom was narrowly defeated.

Northern Irish: There are two basic groups, each of which is seen as a minority by the other. The numerical majority (1 million) within Northern Ireland are Protestants of English and Scottish descent. Their economic domination over and discrimination against the 550,000 Celtic-descended Catholics (who see themselves as part of the numerical majority of Ireland as a whole) is seen by many as the root of civil unrest that plagued the region between the late 1960s and the late 1990s.

Flemings and Walloons: These are the two major ethnic groups that formed the basis of the original binational state of Belgium in 1830. From the beginning, French (the language of the 4 million Walloons) has enjoyed higher prestige than the Dutch-based language of the 6 million Flemings, a cleavage that is reinforced by ideological, religious, and economic differences.

Alsaciens: These are the approximately 1.4 million German-speaking inhabitants of the Alsace-Lorraine region in France. Alsacien ethnic consciousness is quite strong but is not represented in significant political associations or movements.

Basques: Spanish governments from the 18th century to the 1960s tried to eradicate the Basque language but succeeded only in greater resentment. A radical Basque nationalist movement, ETA, was formed in 1959 and has used terrorist tactics in mobilizing regional ethnic consciousness. The Spanish Basque provinces have been successful in achieving a degree of regional political autonomy from Madrid; French Basques, however, have no autonomy.

Catalans: There are about 8 million inhabitants of Barcelona and its hinterland, the majority of whom speak Catalan. The region enjoyed a brief period of home rule during the 1930s, before General Franco came to power. In the 1970s, after Franco's death, Catalonia regained a degree of regional political autonomy.

Corsicans: There are about 150,000 Corsican nationals in a total island population of 250,000. Ethnically more Italian than French, their consciousness was heightened by the immigration of French refugees from Algeria in the 1960s.

▲ **FIGURE 2.19 Minority Ethnic Subgroups in Western Europe** Regional and ethnic consciousness now represents a strong political factor in many European countries. Within Spain, for example, the dominant population—some 27 million—is Castilian, but there are between 6 and 8 million Catalans, almost 2 million Basques, and about 3 million Galicians. **Analyze the role of ethnic minority groups in "national identity" throughout Europe. How does ethnic consciousness create a new form of nationalism?**

Albanians from what had become their homeland. International outrage at these human rights violations finally led to the declaration of war against Yugoslavia by the North Atlantic Treaty Organization (NATO) in March 1999. Within a few weeks, Slobodan Milošević and other Serbian leaders were indicted in the International Court of Justice for their roles in human rights violations. Since 2000, many Kosovar Albanian refugees have returned to their homeland in Kosovo to

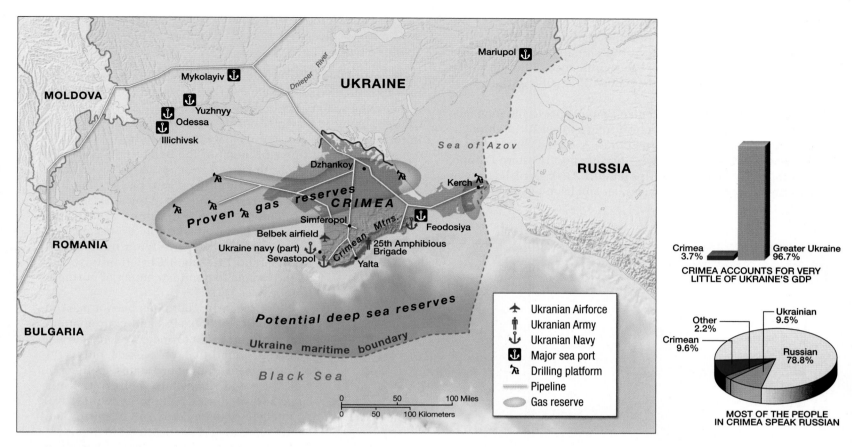

▲ FIGURE 2.20 The strategic importance of the Crimea region.

attempt to rebuild their lives, and in 2008 Kosovo declared its independence—recognized by 22 of the EU's (then) 27 member states, but strongly opposed by Serbia.

Apply Your Knowledge

2.15 Define "irredentism" and "ethnic cleansing" and give another example of each term.

2.16 Summarize the political and cultural conflicts between Russia and the Ukraine by conducting an Internet search.

Culture and Populations

Although its culture is distinctive at the global scale, Europe also bears the legacy of interactions with other world regions through colonialism, emigration, and immigration. Europe's cultures and populations are also characterized by some sharp internal regional variations. In the broadest terms, there is a significant north–south cultural divide, with a lingering legacy of an east–west political and cultural divide. However, contemporary processes of political and cultural change are beginning to modify some of the traditional patterns associated with European geography.

Culture, Religion, and Language

The foundations of contemporary European culture were established by the ancient Greeks, who in turn had been influenced by the ideas and cultural and economic practices of ancient Egypt, Mesopotamia (centered along the Tigris and Euphrates rivers, in modern-day Iraq), and Phoenicia (centered along the coastal regions of modern-day Lebanon, Israel, Syria, and the Palestinian territories). Between 600 B.C.E. and 200 B.C.E., the Greeks built an intellectual tradition of rational inquiry into the causes of everything, along with a belief that individuals are free, self-understanding, and valuable in themselves. The Romans took over this intellectual tradition and added Roman law and a tradition of disciplined participation in the state as a central tenet of citizenship. From the Near East came the Hebrew tradition that, in conjunction with Greek thought, produced Judaism and

Christianity. At the heart of European philosophy, then, are the curiosity, open-mindedness, and rationality of the Greeks; the civic responsibility and political individualism of both Greeks and Romans; and the sense of the significance of the free individual spirit that is found in the Judeo-Christian tradition.

Language Families The Roman Empire spread the Romance family of languages, which evolved from spoken Latin, throughout the western part of southern Europe. A second major group of languages—Germanic languages—occupy northwestern Europe, extending as far south as the Alps (**FIGURE 2.21**). English is one of the Germanic family of languages, an amalgam of Anglo-Saxon and Norman French, with Scandinavian and Celtic traces. A third major language group consists of Slavic languages, which dominate Eastern Europe.

Religious Diversity The Roman Catholic Church emerged in the 4th century under the bishop of Rome and spread quickly through the weakening Roman Empire. The Eastern Orthodox Church, under the auspices of the Byzantine Empire centered in Constantinople (present-day Istanbul), dominated the eastern margins of Europe and much of the Balkans. Meanwhile, Islamic influence began to spread into Southern Spain from North Africa from the 11th century and into parts of the Balkans with the expansion of the Turkish Ottoman Empire in the 14th century. During the 16th and 17th centuries, the Protestant Reformation and wars of religion transformed Europe, and Protestant Christianity subsequently came to dominate much of northern Europe.

More recently, there has been a marked decrease in religiousness and church attendance in much of Europe. A recent survey by the EU found that, on average, only 51% of the citizens of member states "believe in God." An additional 26% believe there is some sort of spirit or life force, while 20% do not believe there is any sort of spirit, God, or life force at all (3% declined to answer).[2] There is, though, a great deal of variation within Europe (**FIGURE 2.22**).

Islam in Europe Meanwhile, immigrants from the Middle East, Africa, and South Asia have reintroduced Islam to Europe, adding an important dimension to contemporary politics as well as culture. Overall, Europe is home to approximately 15–20 million Muslims—between 4% and 5% of the total population of the region. The greatest concentrations of Muslims are in the Balkan countries, where Islam has been important for centuries. In Albania, about 65% of the population is Muslim; in Bosnia-Herzegovina, the figure is 55%; in Macedonia, 29%; and in Montenegro, about 17%. Kosovo's population of 2 million is 93% Muslim. The majority of Europe's Muslims, however, are located in the industrial cities of Western Europe, and they are a relatively recent addition to the population of the region (see Visualizing Geography: Europe's Muslims on pp. 80–81).

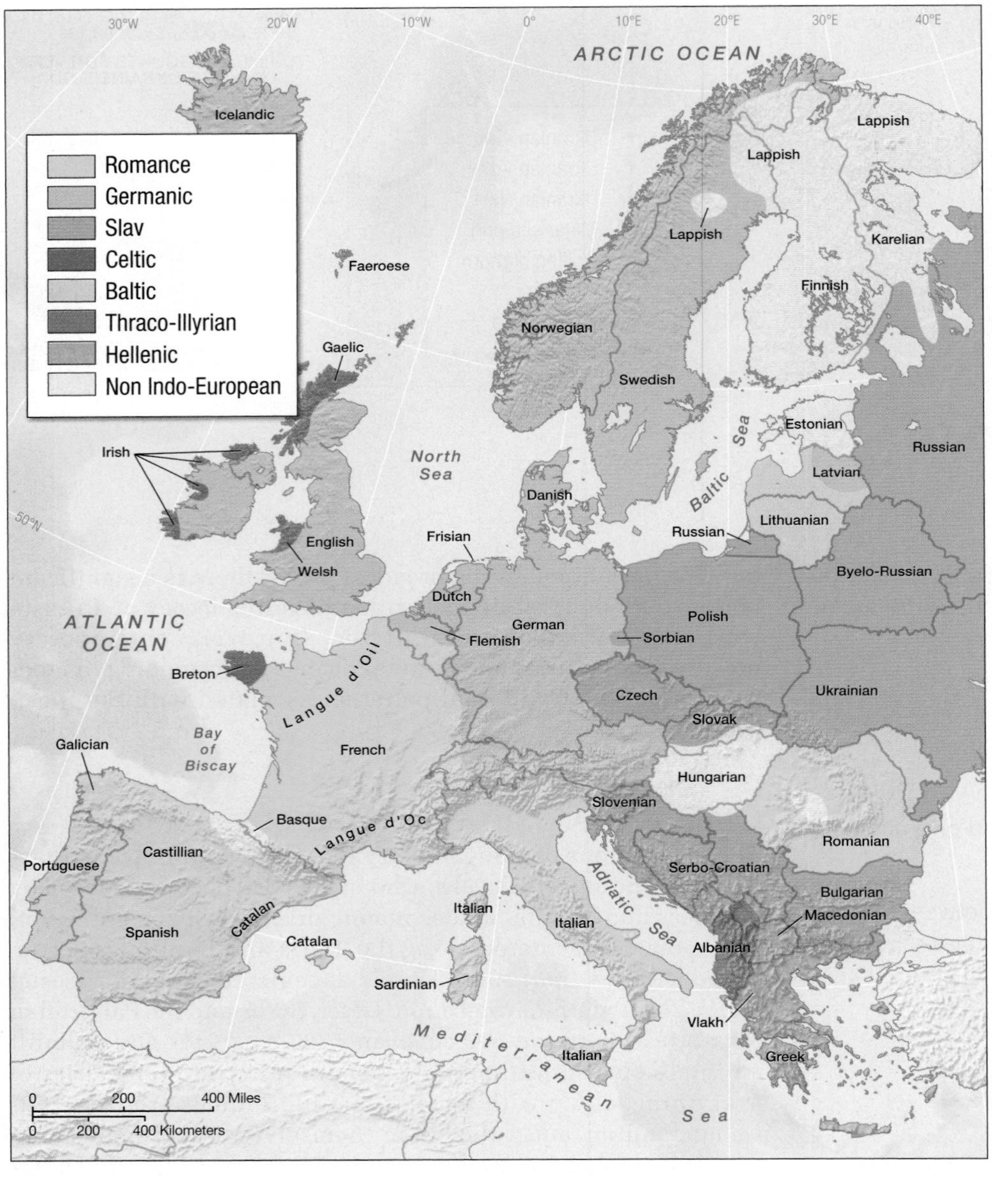

▲ **FIGURE 2.21 Major Languages in Europe** Although three main language groups—Romance, Germanic, and Slav—dominate Europe, differences among specific languages are significant. These differences have contributed to the cultural diversity of Europe but have also contributed to ethnic and geopolitical tensions.

[2]European Commission (2010). "Biotechnology," *Special Eurobarometer 73.1,* http://ec.europa.eu/public_opinion/archives/ebs/ebs_341_en.pdf

Regional Cultures The broad geographic divisions of religion, language, and family life within Europe are reflected in other cultural traits: folk art, traditional costume, music, folklore, and cuisine. For example, there is a Scandinavian cultural subregion with a collection of related languages (except Finnish), a uniformity of Protestant denominations, and a strong cultural affinity in art and music that reaches back to the Viking age and to pre-Christian myths. A second distinctive subregion is the sphere of Romance languages in the south and west. A third is the British Isles, bound by language, history, art forms, and folk music but with a religious divide between the Protestant Anglo-Saxon and Catholic Celtic spheres. A fourth clear cultural subregion is the Germanic sphere of central Europe, again with mixed religious patterns—Lutheran Protestantism in the north, Roman Catholicism in the south—but with a common bond of language, folklore, art, and music. The Slavic subregion of eastern and southeastern Europe forms another broad cultural subregion, though there is considerable diversity beyond the related languages and some common physical traits.

Apply Your Knowledge

2.17 Describe how religion has produced distinct cultural regions in Europe.

2.18 Select three European countries and conduct an Internet search to investigate the rates of religious adherence of their populations. How might you account for the differences in religious beliefs among the three countries? (Hint: A good source of data is the European Commission's survey of social values, see http://ec.europa.eu.)

Religion in Europe

https://goo.gl/4JNUfj

Cultural Practices, Social Differences, and Identity

Although Europe's culture has its roots in the region's ancient civilizations, contemporary Europe has been shaped by ideas that arose over the last few centuries.

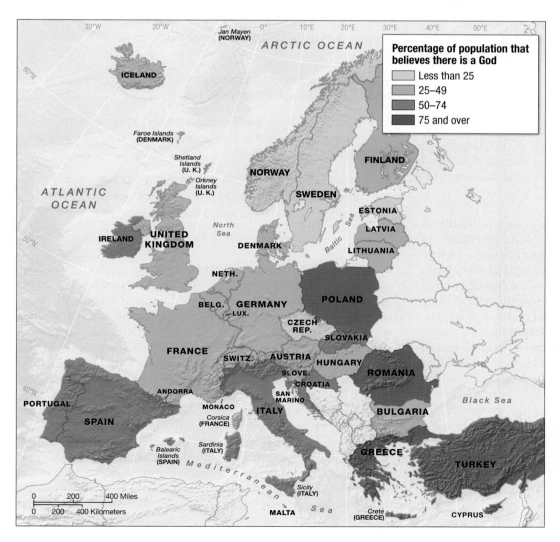

Percentage of population that believes there is a God

Less than 25
25–49
50–74
75 and over

▲ FIGURE 2.22 **Religiosity** What factors might help explain the variations in religiosity within Europe?

The Concept of Modernity One of the single most important developments in European cultural sensibility has been the emergence of the idea of **Modernity**. Modernity is a way of thinking that was triggered by the changing world geography of the Age of Discovery; it emphasizes innovation over tradition, rationality over mysticism, and utopianism over fatalism. As Europeans tried to make sense of their own ideas and values in the context of those they encountered in the East, in Africa, in Islamic regions, and among Native Americans during the 16th and 17th centuries, many certainties of traditional thinking were cracked open. In the 18th century, this ferment of ideas culminated in the Enlightenment movement, which was based on the conviction that all of nature, as well as human beings and their societies, could be understood as a rational system. Politically, the Enlightenment reinforced the idea of human rights and democratic forms of government and society. Expanded into the fields of economics, social philosophy, art, and music, the Enlightenment gave rise to the cultural sensibility of Modernity.

The subsequent dislocations and new experiences introduced by industrialization resulted in still new ways of seeing and new ways of representing things. The places where all this was played out with the greatest intensity were the major cities of Europe. In London, painters, poets, and critics set out to reform art. In Vienna, thinkers met in cafés to discuss new ideas about art, design, psychiatry, and politics. In Milan, Futurists propagated the idea that the past was a corrupting influence on society and celebrating speed, technology, and youth are the keys to the triumph of humanity over nature. Paris became the seat of revolutionary ideas, an unrivaled artistic and cultural scene that included, at various times, the artists Jean-Baptiste Corot, Pierre-Auguste Renoir, Henri de Toulouse Lautrec, Paul Cézanne, Vincent Van Gogh, Georges Braque, Pablo Picasso, and Henri Matisse; and writers Victor Hugo, Honoré de Balzac, and Emile Zola.

Social Critique and the European Dream By the late 20th century, the culture of Europe had become influenced by doubt and criticism. Two terrible world wars, interludes of fascist dictatorships in Germany, Italy, Greece, Portugal, and Spain, repeated episodes of economic recession, and a protracted period of being on the front line of a Cold War that divided European geography in two gave a strong impetus to left-wing critiques that were powerful enough to reshape entire national and regional cultures and, with them, some dimensions of regional geographies.

As a result, contemporary Europe has developed a distinctive set of social values, and the "European Dream" is quite distinctive from the "American Dream."

> The European Dream emphasizes community relationships over individual autonomy, cultural diversity over assimilation, quality of life over the accumulation of wealth, sustainable development over unlimited material growth, deep play over unrelenting toil, universal human rights and the rights of nature over property rights, and global cooperation over the unilateral exercise of power.[3]

A "European" Identity? Globalization has heightened people's awareness of cultural heritage and ethnic identities. As we saw in Chapter 1, the more universal the diffusion of material culture and lifestyles, the more valuable regional and ethnic identities tend to become. Globalization has also brought large numbers of immigrants to some European countries, and their presence has further heightened awareness of cultural identities.

In the more affluent countries of northwestern Europe, immigration has emerged as one of the most controversial issues since the end of the Cold War. Although the economic benefits of immigration far outweigh any additional demands that may be made on a country's health or welfare system, fears that unrestrained immigration might lead to cultural fragmentation and political tension have provoked some governments to propose new legislation to restrict immigration. The same fears have been responsible for a resurgence of popular **xenophobia**—a hate, or fear, of foreigners—in some countries. In Germany, for example, right-wing nationalistic groups have attacked hostels

housing immigrant families while for a long time citizenship laws prevented second-generation *Gastarbeiter* (guest worker) families from obtaining German citizenship. In France, claims that immigrants from North Africa are a threat to the traditional French way of life have led to some success for the *National Front Party*. Asylum seekers, drawn to northwestern Europe from all parts of the globe in such large numbers that they have had to be accommodated in processing centers, have also provoked negative reactions.

All this raises the question, How "European" *are* the populations of Europe? European history and ethnicity have resulted in a collection of national prides, prejudices, and stereotypes that are strongly resistant to the forces of cultural globalization. Opinion surveys show that these stereotypes and prejudices have steadily been countered by a growing sense of European identity, especially among younger and better-educated persons. Much of this can be attributed to the growing influence of the EU (see Faces of the Region on p. 82). On the other hand, the bureaucratization of the EU and its insistence on European-wide standards and policies has provoked a renewed sense of national identity among some populations. In the United Kingdom, for example, a right-wing political party, UKIP (United Kingdom Independence Party) emerged to exploit the single issue of withdrawal from the EU.

Apply Your Knowledge

2.19 How do you think the "European Dream" differs from the "American Dream"? Is it a viable vision for the future—why or why not?

2.20 What factors have led to the emergence of xenophobia in Europe? Will the move to build a "European identity" be enough to counteract xenophobia? Explain your answer.

Women in European Society Another powerful postwar movement deeply critical of the dominant structures of capitalist society was feminism, built on the ideas of Simone de Beauvoir expressed in her 1949 book, *The Second Sex*. Compared with peoples in most other world regions, Europeans have been more willing to address the deep inequalities between men and women that are rooted in both traditional societies and industrial capitalism. Still, patriarchal society and the culture of machismo remain strong in Mediterranean Europe—especially in rural areas—and working-class communities throughout Europe are still characterized by significant gender inequalities.

In northwestern Europe—and especially in Scandinavia—gender equality has improved most, as a result of both the progressive social values of the "baby boom" generation (people born in the years following World War II) and legislation that has translated these values into law. By the mid-1980s, younger men in much of northwestern Europe had acquired a new, progressive collective identity associated with ideals of gender equality—especially as they relate to men's domestic roles.

---

[3]J. Rifkin, *The European Dream: How Europe's Vision of the Future Is Quietly Eclipsing the American Dream* (New York: Tarcher, 2004), p. 3.

In global context, Europe stands out as a region where women's representation in senior positions in industry and government is relatively high. Women in Europe generally have a significantly longer life expectancy than men and have comparable levels of adult literacy. Nevertheless, women in the European labor force tend to earn, on average, 16% less per hour than men. These statistics demonstrate a distinctly regional pattern. Broadly speaking, the gender gaps in education, employment, health, and legal standing are wider in southern and Eastern Europe and narrower in Scandinavia and northwestern Europe. In part, this reflects regional differences in social customs and ways of life; in part, it reflects regional differences in overall levels of affluence.

Apply Your Knowledge

2.21 Summarize the status of women in Europe.

2.22 Compare the gender gaps in southern and Eastern Europe with those in northwestern Europe. What are some of the social customs and economic challenges that perpetuate these gaps?
(Hint: Visit *Women in Europe for a Common Future* [http://www.wecf.eu/] or *United Nations Women* [http://www.unwomen.org/en].)

Europe's Gender Gap
https://goo.gl/xwnG6S

Demography and Urbanization

A distinctive characteristic of Europe as a whole is the size and relative density of its population. With less than 7% of Earth's land surface, Europe contains about 13% of its population at an overall density of nearly 100 persons per square kilometer (260 per square mile). Within Europe, the highest national densities match those of Asian countries such as Japan, the Republic of Korea, and Sri Lanka. On the other hand, population density in Finland, Norway, and Sweden stands at only about 15 persons per square kilometer (39 persons per square mile), the same as in Kansas and Oklahoma (**FIGURE 2.23**). This variation in density reflects a fundamental feature of the human geography of Europe: the existence of a densely populated core and a sparsely populated periphery. Recall that earlier in the chapter we noted the natural and social roots of this core–periphery contrast.

Demographic Change Whereas the population of the world as a whole is increasing

fast, the population of Europe is stable. Europe's population boom coincided roughly with the Industrial Revolution of the late 18th to late 19th centuries. By the early 20th century, Europe began to go through the demographic transition from high birth and death rates to low birth and death rates.

Today, Europe's population is growing slowly in some regions, while declining slightly in others. It seems that conditions of family life in Europe, including readily available contraception, have led to a widespread fall in birthrates beyond what might be expected as a result of the demographic transition. The average size of families has dropped well below the rate needed for replacement of the population (about 2.1 children per family) to about 1.6 per family in 2012. The "baby boom" after World War II was followed by a "baby bust." Meanwhile, life expectancy has increased because of improved health care, medical knowledge, and healthier lifestyles. The effect is not sufficient to outweigh falling birthrates, but it has meant a dramatic increase in the proportion of people over the age of 65 years, from 9% in 1950 to nearly 18% in 2013. The ratio of retirees to workers in Europe is expected to double by 2050, from four workers per retiree to two workers per retiree. By then, one-third of Europe's population will be over 60, compared to 13% who

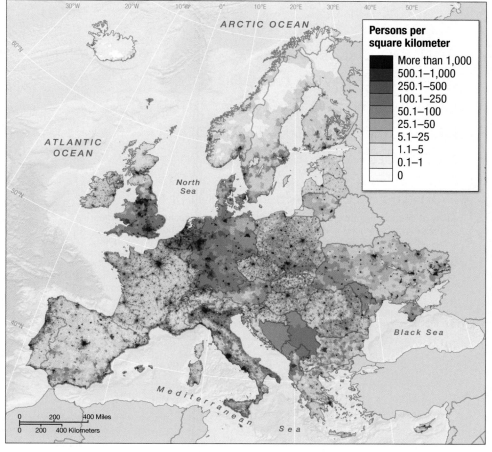

▲ **FIGURE 2.23 Population Density in Europe** The distribution of population in Europe reflects the region's economic history. The highest densities are in the Golden Triangle, the newer industrial regions of northern Italy, and the richer agricultural regions of the North European Lowlands. **List three reasons why Europe's population is dense in these three areas, and more sparse in other areas.**

Europe's Muslims

Government policies in most European countries favor multiculturalism, an idea that, in general terms, accepts all cultures as having equal value. The growth of Muslim communities and their resistance to cultural assimilation, however, has challenged the European ideal of strict separation of religion and public life.

1 Recent Muslim Migration

In the north of Europe, Muslim migration has been dominated by largely legal entry through refugee/asylum applications and employment opportunities motivated by war and civil unrest at Europe's borders, and by associated economic push and pull factors. Muslim immigration has been mostly from Iraq, Somalia, Eritrea, and Afghanistan.

In the south of Europe Muslim migration has been dominated by largely illegal entry (including trafficking in human beings) as a reflection of the geographical proximity of countries with Muslim populations to southern Europe, and motivated by the same factors as migration to the north of Europe.

> "Tensions remain high, and the cultural issues associated with Europe's Muslim population are likely to continue to be an important dimension of European politics."

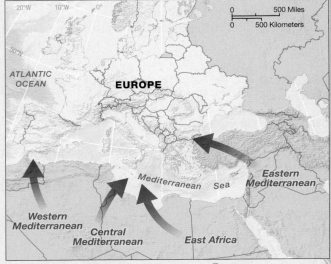

▲ FIGURE 2.3.1 Major migration routes to Europe.

2 Problems Faced by Muslim Immigrants

Many of these immigrant groups, now facing high unemployment and low wages as the postwar boom has leveled off, have clustered together in distinctive urban neighborhoods. In Muslim neighborhoods with high concentrations of unemployment, heavy-handed policing and racial discrimination can easily trigger civil disorder. Islamist terrorist attacks against west European targets have meanwhile heightened fears and tensions between Muslim communities and host populations. Muslim communities across Europe have faced increased resentment and hostility from host populations. Tensions remain high, and the cultural issues associated with Europe's Muslim population are likely to continue to be an important dimension of European politics.

2011 UK Census map

https://goo.gl/px73j0

Estimated net inflow of Muslim migrants to Europe, 2010

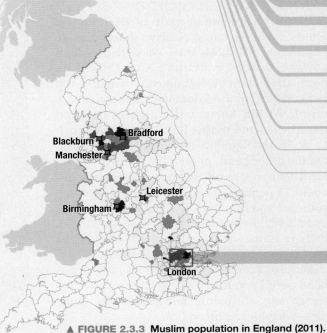

Percent of population that is Muslim

- 20 and more
- 10–19.9
- 5–9.9
- 1–4.9
- 0–0.9

▲ FIGURE 2.3.3 Muslim population in England (2011).

Spain 70,000
In Spain, the Muslim migrants have mostly been from Morocco and sub-Saharan Africa.

France 66,000

United Kingdom 64,000

Italy 60,000
In Italy, large numbers of Muslim migrants have arrived illegally from North Africa and Albania.

Germany 22,000

Sweden 19,000

Belgium 14,000

Greece 12,000 •······ *Greece has experienced migration of mainly Albanian Muslims, but also Muslims from Pakistan, Bangladesh, and Iraq.*

Austria 8,000

Norway 7,000

Ireland 5,000

Netherlands 3,000

Switzerland 2,000

Finland 2,000

Denmark 1,000

Bulgaria 1,000

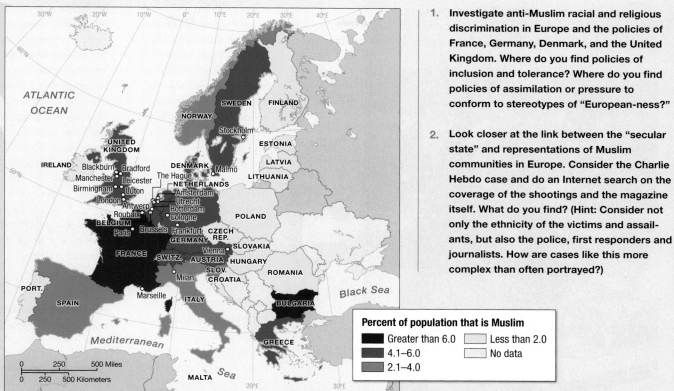

Percent of population that is Muslim

■ Greater than 6.0	▦ Less than 2.0
▦ 4.1–6.0	☐ No data
▦ 2.1–4.0	

▲ FIGURE 2.3.2 **EU cities that are 10% Muslim or more.**

1. Investigate anti-Muslim racial and religious discrimination in Europe and the policies of France, Germany, Denmark, and the United Kingdom. Where do you find policies of inclusion and tolerance? Where do you find policies of assimilation or pressure to conform to stereotypes of "European-ness?"

2. Look closer at the link between the "secular state" and representations of Muslim communities in Europe. Consider the Charlie Hebdo case and do an Internet search on the coverage of the shootings and the magazine itself. What do you find? (Hint: Consider not only the ethnicity of the victims and assailants, but also the police, first responders and journalists. How are cases like this more complex than often portrayed?)

Muslim (Islam) percent by ward, 2011

0 LQ*	0%
Average	2.6%
3.0 LQ	7.8%
5.0 LQ	13.0%
7.0 LQ	18.2%
	50.43%

*LQ is the Location Quotient and describes how far from the national average (LQ=1) the measure is.

▲ FIGURE 2.3.4 **Concentration of Muslims in London and surrounding neighborhoods.**

For the EU's vast economy to thrive, people must feel comfortable working alongside others who differ in language, religion, and culture. The European Commission initiated the Erasmus Programme more than 25 years ago to promote intercultural understanding among the region's young people. In the past 25 years, the program has provided educational exchanges—typically lasting 4–6 months—for over 3 million university students. The program also offers language study and traineeships in companies. For many students an Erasmus Programme grant is a first chance to travel to another country. Top destinations are Spain, France, Germany, the United Kingdom, and Italy (**FIGURE 2.4.1**). But some venture to non-EU countries such as Turkey. This is what one EU student, Emiel, said about her experience in Turkey:

▲ **FIGURE 2.4.1** Maguelone Bastide, an archeology student at the University of Nanterre, France, spent a year at Aristotle University of Thessaoloniki, Greece, through the Erasmus program.

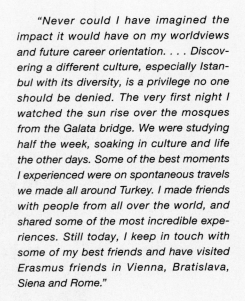

"Never could I have imagined the impact it would have on my worldviews and future career orientation. . . . Discovering a different culture, especially Istanbul with its diversity, is a privilege no one should be denied. The very first night I watched the sun rise over the mosques from the Galata bridge. We were studying half the week, soaking in culture and life the other days. Some of the best moments I experienced were on spontaneous travels we made all around Turkey. I made friends with people from all over the world, and shared some of the most incredible experiences. Still today, I keep in touch with some of my best friends and have visited Erasmus friends in Vienna, Bratislava, Siena and Rome."

1. **Research Erasmus of Rotterdam. Why is it appropriate for the program to be named after him?**

2. **Summarize what Emiel learned from her participation in the program. What would you want to take away from such an experience?**

---

will be under 16. Germany's population (**FIGURE 2.24**) reflects these trends and shows the impact of two world wars.

The European Diaspora The upheavals associated with the transition to industrial societies, together with the opportunities presented by colonialism and imperialism and the dislocations of two world wars, have dispersed Europe's population around the globe. Beginning with the colonization of the Americas, vast numbers of people have left Europe to settle overseas. The full flood of emigration began in the early 19th century, with people from northwestern and central Europe heading for North America and southern Europeans heading for destinations throughout the Americas. In addition, large numbers of British left for Australia and New Zealand and eastern and southern Africa. French and Italian emigrants traveled to North Africa, Ethiopia, and Eritrea, and the Dutch went to southern Africa and Indonesia.

The final surge of emigration occurred just after World War II, when various relief agencies helped homeless and displaced persons move to Australia and New Zealand, North America, and South Africa, and large numbers of Jews settled in Israel.

Migration within Europe Industrialization and geopolitical conflict have also resulted in a great deal of population movement within Europe. Industrial development in the 19th century drew migrants from less-prosperous rural areas to a succession of industrial growth areas around coalfields. As industrial capitalism evolved, the diversified economies of cities offered the most opportunities and the highest wages, thus prompting a further redistribution of population. In Britain, this involved a drift of population from manufacturing towns southward to London and the southeast. In France, migration to Paris from towns all around France resulted in a polarization

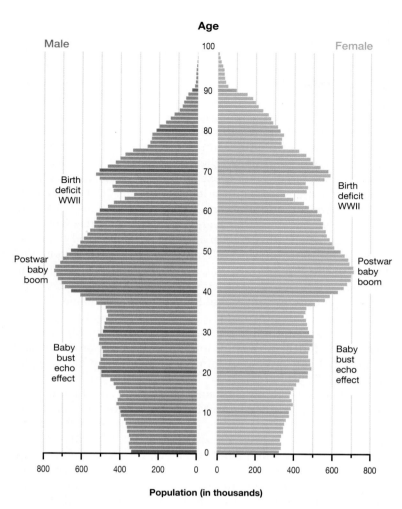

Age

Male · Female

Population (in thousands)

▲ **FIGURE 2.24 Population of Germany, by Age and Sex, 2010** Germany's population profile is that of a wealthy country that has passed through the postwar baby boom and currently possesses a low birthrate. It is also the profile of a country whose population has experienced the ravages of two world wars. **If Germany's economy continues to rise while its birthrate continues to decline, predict what the demographic picture of Germany will look like 20 years from today.**

between Paris and the rest of the country. Some countries, developing an industrial base only after the "coalfield" stage, experienced a more straightforward shift of population, directly from peripheral rural areas to prosperous metropolitan regions. In this way, Barcelona, Copenhagen, Madrid, Milan, Oslo, Stockholm, and Turin all emerged as regionally dominant metropolitan areas.

Wars and political crises have also led to significant redistributions of population within Europe. World War I forced about 7.7 million people to move. Another major transfer of population took place in the early 1920s, when more than 1 million Greeks were transferred from Turkey and a half million Turks were transferred from Greece in the aftermath of an unsuccessful Greek attempt to gain control over the eastern coast of the Aegean Sea. Soon afterward, more people were on the move, this time in the cause of ethnic and ideological purity, as the policies of Nazi Germany and fascist Italy began to emerge. Jews,

in particular, were forced out of Germany. With World War II, further forced migrations occurred that involved approximately 46 million people.

These migrations, together with mass exterminations undertaken by Nazi Germany, left large parts of west and central Europe with significantly fewer ethnic minorities than before the war. In Poland, for example, minorities constituted 32% of the population before the war but only 3% after the war. Similar changes occurred in Czechoslovakia—from 33% to 15%—and in Romania—from 28% to 12%. Southeast Europe did not experience such large-scale transfers, and, as a result, many ethnic minorities remained intermixed or surrounded and isolated, as in the former republic of Yugoslavia. The geopolitical division of Europe after the war also resulted in significant transfers of population; West Germany, for example, had absorbed nearly 11 million German refugees from Eastern Europe by 1961.

Recent Migration Streams More recently, the main currents of migration within Europe have also been a consequence of patterns of economic development. Rural–urban migration continues to empty the countryside of Mediterranean Europe as metropolitan regions become increasingly prosperous. Meanwhile, cities have experienced a decentralization of population as factories, offices, and housing developments have moved out of congested central areas. Another stream of migration has involved better-off retired persons, who have congregated in spas, coastal resorts, and picturesque rural regions.

The most striking of all recent streams of migration within Europe, however, have been those of migrant workers. These population movements were initially the result of Western Europe's postwar economic boom in the 1960s and early 1970s, which created labor shortages in Western Europe's industrial centers. The demand for labor represented welcome opportunities to many of the unemployed and poorly paid workers of Mediterranean Europe and of former European colonies. By the mid-1970s these migration streams had become an early component of the globalization of the world economy. By 1975, between 12 and 14 million immigrants had arrived in northwestern Europe. Most came from Mediterranean countries—Spain, Portugal, southern Italy, Greece, Yugoslavia, Turkey, Morocco, Algeria, and Tunisia. In Britain and France, the majority of immigrants came from former colonies in Africa, the Caribbean, and Asia. In the Netherlands, most came from the former colony of Indonesia. Most of these immigrants have stayed on, adding a striking new ethnic dimension to Europe's cities and regions (**FIGURE 2.25**).

Meanwhile, it is estimated that more than 18 million people moved within Europe during the 1980s and 1990s as refugees from war and persecution or in flight from economic collapse in Russia and Eastern Europe. Civil war and dislocation in the Balkans displaced more than 4 million people in the early 1990s. Wars in Iraq and Afghanistan, dislocations following the expansion of the Islamic State of Iraq and the Levant (ISIL; see Chapter 4), and conflict in North Africa—mainly Libya—have also resulted in significant numbers of refugees entering Europe (**FIGURE 2.26**).

▲ FIGURE 2.25 The Barbès-Rochechouart African Arab quarter of Paris.

Today, most European states are now net immigration countries (the exceptions are Lithuania and Bulgaria), and within Europe as a whole there are more than 56 million immigrants (compared to about 42 million in North America). In the past decade, immigration has become a major electoral issue in Austria, Belgium, Denmark, France, the Netherlands, Switzerland (FIGURE 2.27), and the United Kingdom. The EU has already stepped up the control of illegal immigration across its borders, and some countries have taken steps to impose strict limits on immigration.

Europe's Towns and Cities Europe as a whole is highly urbanized, with a dense network of hundreds of cities and

▼ FIGURE 2.26 Migrants from Sub-Saharan Africa being rescued by the coastguard after their inflatable boat started to sink off the Libyan coast.

thousands of small towns (FIGURE 2.28). There are more than 450 European cities and towns with populations over 100,000. Among the largest are London with more than 8.6 million inhabitants, Berlin with 3.5 million, and Madrid with more than 3.2 million. Because of the pivotal role of Europe in shaping world trade and politics since the 1500s, a number of its cities have been of special significance in organizing and influencing spatial organization well beyond their regional or national boundaries. These are known as **world cities**. In the first stages of the growth of the modern world economy, these cities played key roles in the organization of trade and the execution of colonial, imperial, and geopolitical strategies. The world cities of the 17th century were London, Amsterdam, Antwerp, Genoa, Lisbon, and Venice. In the 18th century, Paris, Rome, and Vienna also became world cities. In the 19th century, Berlin and Manchester joined the ranks of world cities while Amsterdam, Antwerp, Genoa, Lisbon, and Venice became less influential. Today, London, with its financial markets, specialized office space, specialized business services, expert professionals, and high-order cultural amenities, is Europe's preeminent world city. Other contemporary European world cities include Berlin, Paris, Frankfurt, and Milan, control centers for the flows of information, cultural products, and finance that collectively sustain the economic and cultural globalization of the world.

At the other end of the urban spectrum are small towns with fewer than 50,000 inhabitants. Many of these are traditional market towns whose origins were in the trading networks of pre-industrial Europe but which were bypassed—literally—by the canals and railways of the industrial era. Others, in contrast, are specialized mining and manufacturing towns that only emerged during the industrial era but have never grown. Overall, more than one-fifth of Europe's population lives in small towns. Beyond the Golden Triangle, the figure is often closer to one-third and in some regions, such as Scandinavia and southern Italy, at least half of the population live in small towns. Small towns play an important role in the European urban system. They do not have the economic advantages of creativity and productivity associated with metropolitan settings but serve important functions as local and regional service centers, as places that can absorb metropolitan overspill; as specialized fishing, mining, and agricultural processing centers; as centers for tourism, and—in traditional market towns—as centers and symbols of regional culture and identity.

Apply Your Knowledge

2.23 Identify the major waves of immigration in European history and summarize how those waves changed the physical landscape.

2.24 Predict the continued importance of small towns for cultural identity and tourism in Europe, based on demographic changes and economic priorities.

◀ **FIGURE 2.27 The Politics of Immigration** This is a campaign poster (translated as "Stop mass immigration") from the right-wing Swiss People's Party (SVP), the top vote-getting party in the 2011 parliamentary elections in Switzerland. The poster promotes the SVP's anti-immigration policies. **Analyze the increasing support for right-wing parties throughout Europe. What are the factors contributing to this conservative backlash?**

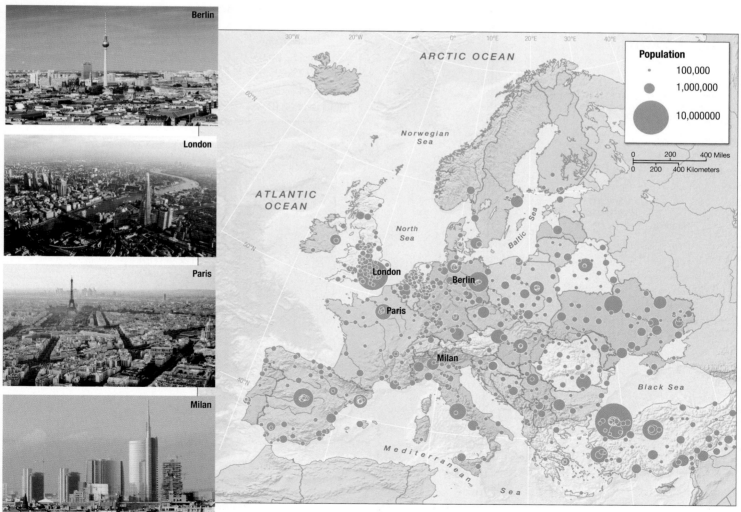

▲ **FIGURE 2.28** Urban Europe. a) Metropolitan areas and large and medium-sized cities, b) Berlin, c) London, d) Paris, e) Milan.

Future Geographies

One of the most predictable aspects of Europe's future geographies concerns its aging populations; in contrast, one of the most speculative is the politics and economics of European integration.

Coping with Aging Populations

Europe already has 19 of the world's 20 oldest countries in terms of average population age. Given current trends, about a quarter of Europe's population will be aged 65 years or above by 2025, and by 2050, about 20% of the population will be aged 80 years or above. This demographic trend will place an immense burden on Europe's working-age populations and their capacity to fund health-care and pension systems at a time when economic growth rates are likely to be modest at best. The **old-age dependency ratio**—the number of people aged 65 years and older compared with the number of working-age people (ages 15–64 years)—is already 28% (**FIGURE 2.29**) and is expected to reach 50% by 2050.

Coping with Immigration

The aging of the European workforce will require more immigration and much better integration of immigrant workers, most of whom are likely to be coming from North Africa and the Middle East. Even if more migrant workers are not allowed in, Western Europe will have to integrate its growing Muslim population, a potential flashpoint for social and cultural conflict that could undermine the European project.

The EU: Costs of Expansion

The EU has put its economic and political stability at risk as a result of its expansion. New EU member countries find

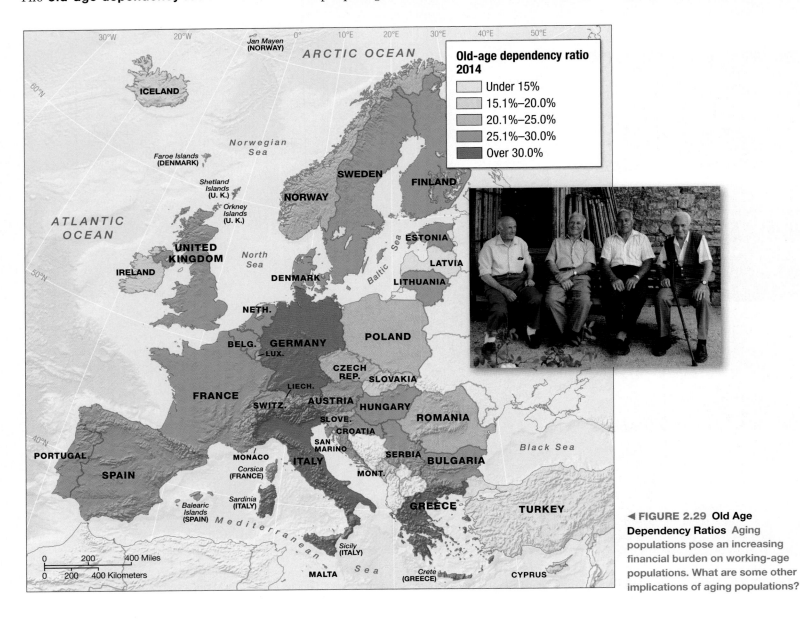

◀ FIGURE 2.29 Old Age Dependency Ratios Aging populations pose an increasing financial burden on working-age populations. What are some other implications of aging populations?

(a)

themselves bound by thousands of "harmonizing" policy directives, such as rules on food preparation and hygiene in restaurants, which are impossibly expensive to implement without help. Meanwhile, the expansion of the EU will remain a costly business; EU subsidies to new member countries currently account for more than 40% of the total budget.

The EU: Problems of Indebtedness

On top of this, the global recession of 2008–2014 highlighted an even greater danger to the European project: the indebtedness of countries like Greece, Portugal, Italy, and Spain, whose uncompetitive and vastly different economies have been tied to the same currency as stronger economies in the "eurozone." (**FIGURE 2.30**) It is possible that one or more of these member states may be forced to leave the Eurozone. Meanwhile, in several of the more affluent countries, increasing numbers of voters support the idea of withdrawing from the EU.

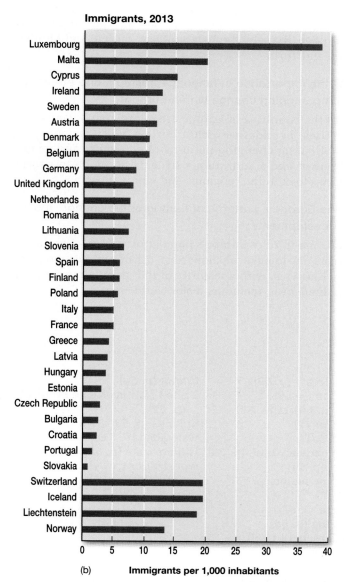
(b) **Immigrants per 1,000 inhabitants**

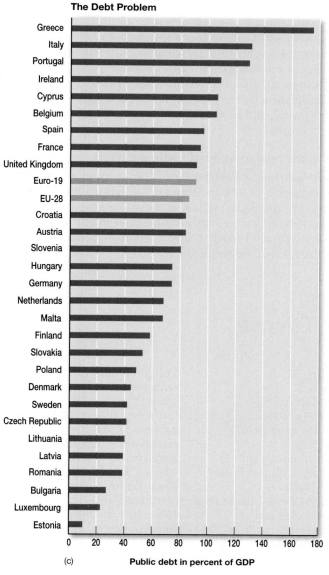
(c) **Public debt in percent of GDP**

▲ **FIGURE 2.30 The challenges of European integration** (a) School students in Petras, Greece, shout slogans against the government's new austerity measures in November 2015; (b) Immigrant populations within European countries in 2013; (c) Indebtedness of European countries, 2014.

Learning Outcomes Revisited

▶ *Analyze* how Europe's landscapes developed around regions of distinctive geology, relief, landforms, soils, and vegetation.

Europe's four key regions are the Northwestern Uplands, the Alpine System, the Central Plateaus, and the North European Lowlands, and the characteristics of these regions are a product of centuries of human adaptation to climate, soils, altitude, and aspect and to changing economic and political circumstances.

▶ *Explain* the rise of Europe as a major world region and understand how Europeans established the basis of a worldwide economy.

The rise of Europe as a major world region had its origins in the emergence of a system of merchant capitalism in the 15th century, when advances in business practices, technology, and navigation made it possible for merchants to establish a worldwide economy in the space of less than 100 years. Since then, Europe's regional geographies have been recast several times: most significantly by the onset of the Industrial Revolution, by two world wars, and by the Cold War rift between Eastern and Western Europe.

▶ *Describe* the reintegration of Eastern Europe.

The reintegration of Eastern Europe has added a potentially dynamic market of 150 million consumers to the European economy, but the region still functions as a group of economically peripheral states, with agriculture geared locally and industry still geared more to heavy industry products than to competitive consumer products.

▶ *Summarize* the importance of the EU and its policies.

The EU emerged after World War II as a major factor in reestablishing Europe's role in the world. Today, the EU is a sophisticated and powerful institution of more than 500 million and combined GDP slightly larger than that of the United States and has pervasive influence on the economic and social well-being of its member states.

▶ *Identify* the principal core regions within Europe.

The principal core region within Europe is the Golden Triangle, which stretches between London, Paris, and Berlin and includes the industrial heartlands of central England, northeastern France, and the Ruhr district of Germany. A secondary, emergent core is developing along a north–south crescent that straddles the Alps, stretching from Frankfurt, just to the south of the Golden Triangle, through Stuttgart, Zürich, and Munich, to Milan and Turin.

▶ *Assess* the importance of language, religion, and ethnicity in shaping ongoing change within Europe.

Many of the countries of Europe are relatively new creations and political boundaries within Europe have changed quite often, reflecting changing patterns of economic and geopolitical power and a continuous struggle to match territorial boundaries to cultural, religious, and ethnic identities.

▶ *Describe* Europe's patterns of demographic change and urban development.

By 2050, one-third of Europe's population will be over the age of 60 years due to medical knowledge, use of contraception, and healthy lifestyles, with a majority of that population living in metropolitan areas, including major world cities.

Key Terms

acid rain (p. *57*)
aspect (p. *53*)
balkanization (p. *73*)
buffer zone (p. *67*)
bureaucratic class (p. *67*)
cirque (p. *54*)
command economy (p. *67*)
Common Agricultural Policy (CAP) (p. *69*)
Communist Council for Mutual Economic Assistance (COMECON) (p. *67*)
deindustrialization (p. *70*)
Emissions Trading Scheme (ETS) (p. *52*)
enclave (p. *73*)

Enlightenment (p. *72*)
entrepôt (p. *63*)
ethnic cleansing (p. *73*)
European Union (EU) (p. *68*)
exclave (p. *73*)
feudal system (p. *59*)
fjord (p. *54*)
Golden Triangle (p. *65*)
Great Depression (p. *66*)
heathland (p. *57*)
Holocaust (p. *66*)
imperialism (p. *65*)
Iron Curtain (p. *67*)
irredentism (p. *73*)
Little Ice Age (p. *57*)
loess (p. *56*)

Maastricht Treaty (p. *69*)
Marshall Plan (p. *66*)
merchant capitalism (p. *62*)
Modernity (p. *77*)
moraine (p. *54*)
old-age dependency ratio (p. *86*)
pastoralism (p. *54*)
physiographic region (p. *53*)
polder (p. *57*)
satellite state (p. *67*)
social housing (p. *68*)
Southern Crescent (p. *65*)
state socialism (p. *66*)
steppe (p. *56*)
technology systems (p. *63*)
trading empire (p. *65*)

tundra (p. *54*)
United Nations Kyoto Protocol (p. *52*)
uplands (p. *53*)
watershed (p. *51*)
welfare state (p. *68*)
world cities (p. *84*)
xenophobia (p. *78*)

Thinking Geographically

1. How has Europe benefited from its location and its major physical features?

2. What key inventions during the period from 1400 to 1600 helped European merchants establish the basis of today's global economy? Why?

3. Which imports from the colonies helped transform Europe? Focus on natural resources and new crops.

4. How did the EU develop? Why is the EU's CAP so important?

5. What migration patterns characterized Europe during the 19th and 20th centuries? Consider movement within Europe as well as movement to and from Europe.

DATA Analysis

Throughout this chapter, we have seen how the geography of Europe changes through flows of people, technology and innovation, political and social movements, and the natural environment and climate. All of these factors impact the "map" of Europe and participation in the EU.

Investigate the economy of Europe by searching for the European Economy Guide and the interactive map of Europe at the *Economist* at http://www.economist.com/ and compare what you find to this chapter and current events.

1. What nations have the highest GDP per person?

2. Where is unemployment the highest? Where is youth unemployment the highest?

3. What correlations between current political events and economic health do you notice?

4. What nations are experiencing the greatest rates of growth? Do any of those countries surprise you? Why or why not?

5. Comparing this map to the text from this chapter, how do you think the population of Europe affects this data? Think about the age of the population where you see the most growth. What are the "young" and the "oldest" countries in Europe based on demographic data?

6. Similarly, which countries experiencing growth are also attracting the most recent waves of immigration? Do countries with large immigration numbers contribute to economic growth? You may need to do additional research on immigration data to compare with this map, use a site like the European Commission's *Eurostat*.

European Economy

https://goo.gl/f2h4P7

MasteringGeography™

Looking for additional review and test prep materials? Visit the Study Area in MasteringGeography™ to enhance your geographic literacy, spatial reasoning skills, and understanding of this chapter's content by accessing a variety of resources, including MapMaster interactive maps, videos, *In the News* RSS feeds, flashcards, web links, self-study quizzes, and an eText version of *World Regions in Global Context*.

Women protest censorship prior to the 2014 referendum in Crimea. Other supporters stand behind the women. Crimea's inclusion in Russia is not recognized by most of the international community.

The Russian Federation, Central Asia, and the Transcaucasus

3

In March 2014, the Russian Federation annexed Crimea, a portion of the sovereign state of Ukraine bordering the Black Sea. While the international community has not recognized the annexation, Russia has formally asserted that Crimea is now part of their federation. Ukraine, meanwhile, more closely aligned with Europe than Russia since the end of the Cold War, struggles to maintain its political integrity as Russian troops support rebels in eastern Ukraine who want a repeat of Crimea. The conflict in Crimea demonstrates how challenging it can be to define world regions and states, whose boundaries are often drawn with little regard for ethnicity.

The Russian government's geopolitical decisions have not come without cost to the federation more generally, which has begun to feel the pressure of international economic sanctions. Though Russia continues to sell its major commodity—oil—on the international market, these sanctions have had an economic impact; the International Monetary Fund suggests they could shrink the Russian economy by as much as 9%. Falling international energy prices have reinforced this effect, and these together reveal the weakness of the Russian economy, which continues to depend on primary commodities for its economic growth. Further compounding Russia's economic problems is the fact that international financiers have become less interested in investing in the Russian economy, further slowing its diversification. Russia has countered some of these trends by increasing interest rates to lure foreign monies to its banks. President Vladimir Putin continues to seek popularity at home by demonstrating strong international presence in Ukraine and the ongoing conflict in Syria. It is clear, though, that Russia wants to continue asserting its regional and global dominance even as the tensions remain along its borders.

Learning Outcomes

▶ **Describe** the physical geography of the Russian Federation, Central Asia, and Transcaucasus region.

▶ **Predict** how climate change and environmental impacts of human activities may influence the region's future sustainability.

▶ **Explain** how people have adapted to and modified the physical landscapes in this world region.

▶ **Describe** how economic transitions in the region are playing out in the post-Socialist era.

▶ **Analyze** the changing political trends in the region in terms of fragmentation and realignment.

▶ **Summarize** the cultural legacies and new developments in the region in terms of high, revolutionary, and popular forms of culture.

▶ **Discuss** and analyze the implications of demographic change in the region.

Environment, Society, and Sustainability

There is only one way to really describe the Russian Federation, Central Asia, and Transcaucasus region in terms of its physical geography. It is *vast* (**FIGURE 3.1**). It takes a full week to cross by train from Vladivostok in the east to St. Petersburg in the west. Despite its daunting size, the region's location has made it a historic crossroads and the site of interaction between Europe and Asia. Although it benefits from vast land area, it also has restricted access to the world's seas and is bounded on the north by the icy seas of the Arctic. Many different people have adapted to and modified this vast landscape, as they have expanded their communities across the region's expanse, creating new environments, economies, and societies.

> "Despite its daunting size, the region's location has made it a historic crossroads and the site of interaction between Europe and Asia."

Climate and Climate Change

Four major climate zones dominate the landscape of this region from the arid and semiarid regions in the southern portion of the region to the continental/midlatitude and polar regions of the

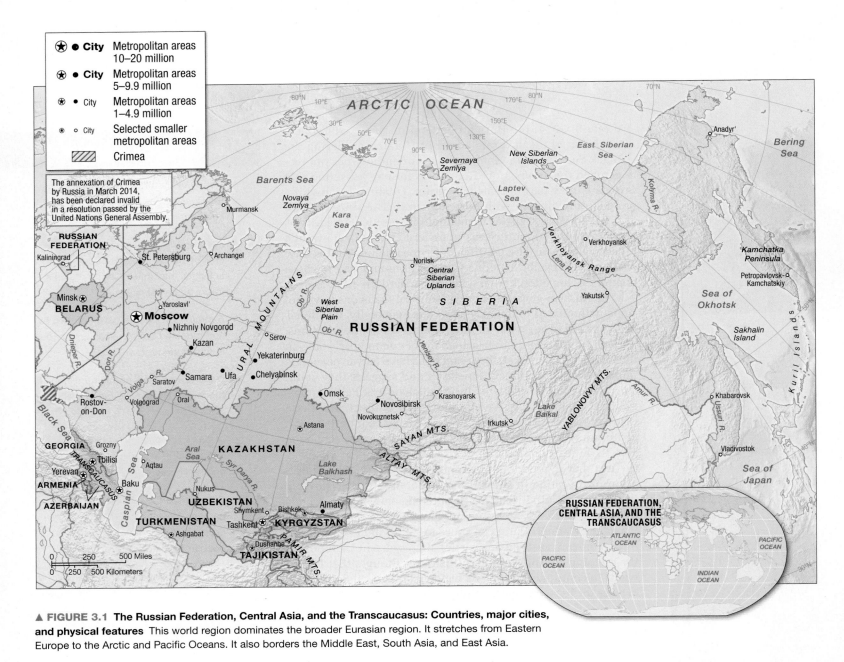

▲ **FIGURE 3.1 The Russian Federation, Central Asia, and the Transcaucasus: Countries, major cities, and physical features** This world region dominates the broader Eurasian region. It stretches from Eastern Europe to the Arctic and Pacific Oceans. It also borders the Middle East, South Asia, and East Asia.

north (**FIGURE 3.2**). What makes this world region unique, however, is its northern position, which results in a severe climate for many of this region's communities. Moscow is approximately the same latitude as Juneau, Alaska, and Tbilisi, Georgia—one of the southernmost cities of the region—is approximately the same latitude as Chicago (42°N). This results in long, cold winters and relatively short, warm summers. Extreme winter cold means that the subsoil is permanently frozen—a condition known as **permafrost**—in more than two-thirds of the Russian Federation. In the extreme northeast, winter conditions can last for 10 months of the year. Consequently, most ports are icebound during the long winter. Murmansk, in the far north, is an important exception. Murmansk benefits from its location near the tail end of the warm North Atlantic current and is open year-round. Some ports, such as Vladivostok, on the Sea of Japan, are kept open by icebreakers.

Summer comes quickly over most of Belarus and the Russian Federation, as spring is typically a brief interlude of dirty snow and abundant mud. Because many rural roads remain unpaved, they are typically impassable until the summer heat bakes the mud. As the landmass warms, low-pressure systems develop, drawing moist air from Atlantic Europe that brings moderate summer rains to the western portions of the region. But, drought is also a frequent problem. In Central Asia, **aridity** (a dry condition where an absence of moisture and rainfall negatively affects plant growth, especially trees) is severe, with desert and semidesert covering much of Kazakhstan, Uzbekistan, and Turkmenistan. The scorching heat in summer is aggravated by strong drying winds that blow on more than half the days of summer, often causing dust storms. In late summer and fall, the increasing temperature range between the hot days and the longer, cooler nights becomes so extreme that rocks exfoliate, or "peel." The intensity of Central Asian climate is offset by the subtropical niche of western Georgia, on the shores of the Black Sea—a unique and striking feature in a region of otherwise severe climatic regimes.

Adapting to a Warming Continent and Melting Sea Ice

The Russian Federation, Central Asia, and Transcaucasus region is already experiencing the effects of global climate change. The

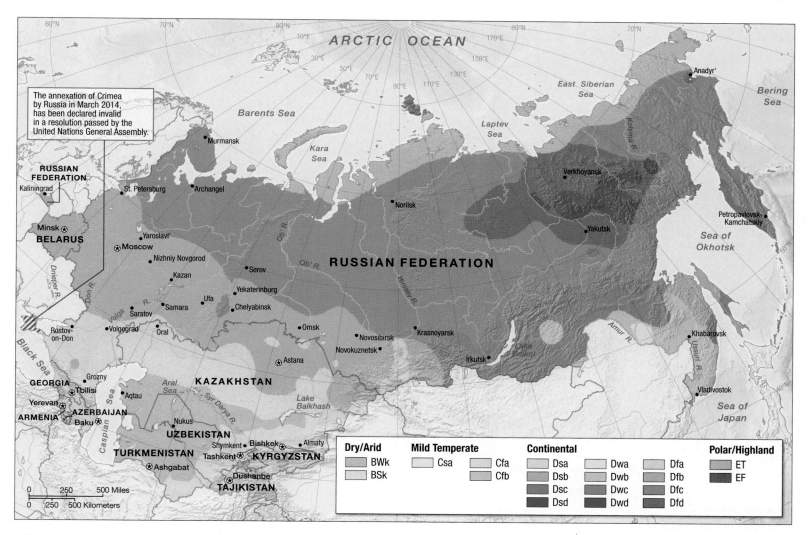

▲ **FIGURE 3.2 Climates of the Russian Federation, Central Asia and the Transcaucasus** The climate of this region features very cold winters and summer droughts are common. Around the Arctic Circle, a very cold and dry polar climate exists. Further south, climates are dry but warmer, typical of semiarid climate types. **In general, how do the climates of the Russian Federation differ from those of Central Asia and the Transcaucasus?**

implications for northern latitudes are mixed. Milder winters mean that growing seasons are becoming longer. It is now possible to use lands for agricultural purposes that previously have been inhospitable. Farmers are also able to raise new crops in some regions. Increased water availability—particularly along the Siberian rivers that can be used for hydroelectric power—should result in increased domestic power production, which may also reduce the demand for oil and gas.

Across vast areas of Siberia, the frequency and intensity of forest fires is increasing as the region warms. In 2010, a Russian heat wave led to numerous explosive forest fires and as many as 14,000 deaths from heat, half of these in and around Moscow. Finally, warming trends across the northernmost parts of the region have begun to result in sea ice melt in the Arctic Ocean. This dramatic transformation of an area previously known only to the hardiest explorers and indigenous Arctic peoples has turned this part of the world into an important player in global commerce and energy production (see "Emerging Regions: The Arctic" on pp. 96–97).

Some of the areas most affected by climate change also have unsettled socioeconomic and sociopolitical relations. For example, the politically turbulent North Caucasus region is becoming drier and hotter and is likely to become less prosperous as an agricultural region. And farther east in areas adjacent to Central Asia, Mongolia, and northeastern China, long-standing cross-border tensions will probably intensify as water availability becomes an increasingly serious challenge, prompting large-scale in-migration to the Russian Federation.

Apply Your Knowledge

3.1 List two serious effects of global climate change in the Russian Federation, Central Asia, and Transcaucasus region.

3.2 Analyze how those changes impact not only the environment but also the wider economy of the region.

Climate Change Knowledge Portal, Russia

https://goo.gl/rAf5J2

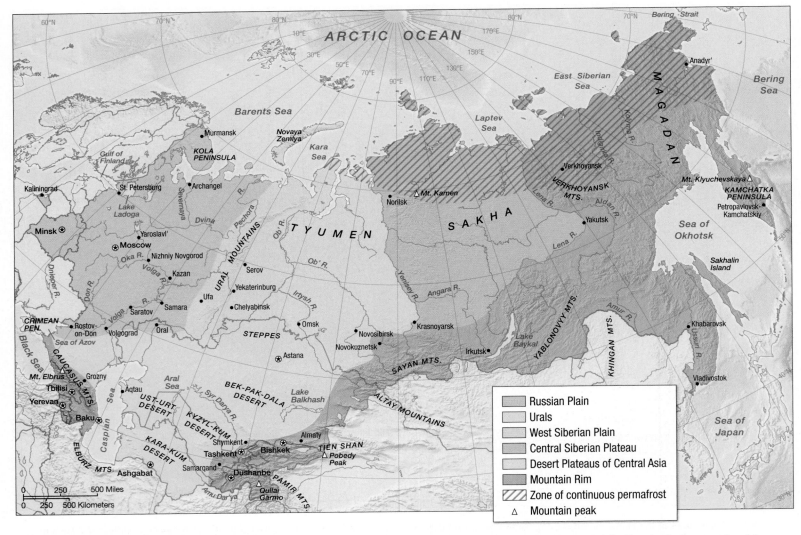

▲ **FIGURE 3.3 Physiographic Regions of the Russian Federation, Central Asia, and the Transcaucasus** The framework for the physical geography of the region consists of two stable shields of the ancient Ural Mountains, with a wall of young mountains that runs along the southern and eastern margins of the shields. **Identify the topographic features that separate the region from the rest of Asia. How were these features formed?**

Geological Resources, Risk, and Water

Remarkably, the entire Russian Federation, Central Asia, and Transcaucasus region sit atop a single tectonic plate. It occupies the eastern portion of the Eurasian Plate (see Chapter 1). The main physical geographic feature that results from this location is the geologically stable monotony of the region's vast plains (**FIGURE 3.3**). Along the southern edges of this expanse, the geological upheaval of colliding plates has pushed up mountains that form a physiographic barrier to the region.

Mountains and Plains This large contiguous region is broken by Earth forces into several subregions. In the west, the Russian Plain, an extension of the North European Lowland, runs from Belarus in the west to the Ural Mountains in the east and from the Kola Peninsula in the north to the Black Sea in the south. East of the Urals lies the West Siberian Plain, the Central Siberian Plateau, and the desert plateaus of Central Asia. The Urals themselves represent their own distinctive region, as does the mountain rim that runs along the southern and eastern margins of the region.

The Russian Plain generally has a gently rolling topography, although the major rivers that drain across these plains—the Dnieper, the Don, and the Volga—have carved a more varied topography along their length. The West Siberian Plain is even flatter and contains extensive wetlands and tens of thousands of small lakes. Poorly drained by the slow-moving Ob' and Irtysh Rivers, the West Siberian Plain (**FIGURE 3.4**) is mostly inhospitable for settlement and agriculture, though it contains significant oil and natural gas reserves.

The Yenisey River marks the eastern boundary of this flat expanse and the beginning of the Central Siberian Plateau (**FIGURE 3.5**), an uplifted region dissected by rivers into a hilly upland topography with occasional deep river gorges. The Urals are a once-great mountain range of ancient rocks that has been worn down over the ages. They stretch for more than 3,000 kilometers (1,864 miles) from the northern frontier of Kazakhstan to the Arctic coast of the Russian Federation. The rocks of the Urals are heavily mineralized and contain significant quantities of chromite, copper, gold, graphite, iron ore, nickel, titanium, tungsten, and vanadium.

The mountain wall that runs along the southern and eastern margins of the region is the product of geological instability. Younger, sedimentary rocks have been pushed up in successive episodes of mountain-building along the edges of the Eurasian Plate, forming a series of mountain ranges of varying height, composition, and complexity. The highest ranges are those of

▲ **FIGURE 3.4 The West Siberian Plain** This photograph of marshland near Primorye, western Siberia, shows the difficult, boggy conditions that prevail in much of western Siberia in summer.

▲ **FIGURE 3.5 The Central Siberian Plateau** The rock shield of the Central Siberian Plateau has been uplifted by geological movements. As a result, the land has been dissected by rivers into a hilly upland with occasional deep river gorges.

the Caucasus (where Mt. Elbrus reaches 5,642 meters or 18,510 feet) and the Pamirs and Tien Shan ranges along the borders with Iran, Afghanistan, and China.

Central Asia connects to the rest of the region to the north across a huge **geosyncline**, a geological depression of sedimentary rocks. The Central Asian geosyncline is of special importance as a source of energy resources. It is home to oil reserves equivalent to between 17 and 49 billion barrels—about 2.7% of the world's proven reserves—plus about 7% of the world's proven reserves of natural gas. Some estimates of the potential oil reserves run even higher—between 60 and 250 billion barrels.

The Arctic

For hundreds of years, explorers and sailors have sought a route through the **Northwest Passage**, an ice-choked waterway spanning the Arctic Sea between the Atlantic and Pacific Oceans, north of the Canadian and Russian mainlands. Recently, global warming has accomplished what generations of explorers and investors have failed to do—it has opened up the Northwest Passage to easier shipping. This opening significantly shortens the shipping distance between Shanghai and New York or Tokyo and London, making a globalizing world all the more tightly connected. In 2014, the Chinese-owned cargo ship "Nunavik" became the first-ever to make the journey without an icebreaker escort, carrying 23,000 tons of cargo to the port of Bayuquan, China. This journey is 40% shorter than going by way of the Panama Canal.

Ice-Free Arctic Summers? As temperatures rise, the extent of ice in the Arctic is decreasing rapidly, especially in the summer (**FIGURE 3.1.1**). Arctic sea ice levels are at their lowest since records have been kept. Projections based on current trends suggest that the Arctic Ocean will be free of summer ice sometime between 2030 and 2080. Far sooner, the route across the Arctic will be reliably open to global commerce, and for the first time, the seafloor will be accessible to extensive resource development involving drilling for oil and natural gas.

Far-Reaching Effects These historic changes will have devastating impacts on the wildlife of the region. Polar bears will effectively be deprived of their natural habitat and ultimately be found only in zoos. And the opening of the ice means the creation of an entirely new world region—a contested prize for key world powers, a novel area for tourism, a critical source of resources, and a connected path between the worlds of the Atlantic and Pacific. The geopolitical contest for the control of this area has already begun, with Russia, Norway, Denmark, Canada, and the United States marking territory and making legal claims on the region (**FIGURE 3.1.2**). In August 2007, the Russian government sent two tiny submarines to plant the Russian flag on the Arctic seafloor (**FIGURE 3.1.3**). In 2015, Denmark, which controls Greenland, hopes to follow suit by making a formal claim to the United

(a)

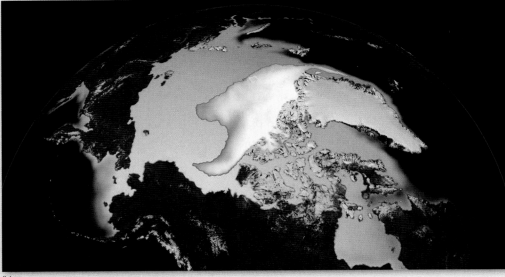

(b)

▲ **FIGURE 3.1.1 The Melting Polar Ice Cap** Shifting patterns of summer ice cover in the Arctic region between (a) 1979 and (b) 2014.

Nations for control over a large portion of the Arctic seabed.

Greenland's Future Although physiographically considered part of North America, Greenland has its own indigenous populations (Kalaallisut-speaking Inuits) and wildlife (polar bear, musk ox, narwhal, and walrus) and is politically an overseas territory of Denmark. The emergence of a new geostrategic region around the North Pole, and the recently established

semi-independent status of Greenland, reinforces this ambiguity. While Greenland is clearly a "victim" of global climate change, through the loss of its ice sheets, wildlife habitat, and indigenous human livelihoods, its position also allows it to assert claims on minerals and oil and gas reserves. This gives Greenland considerable influence and economic opportunity, alongside Canada, Russia, and the United States, though they remain under the nominal control of a small European power, Denmark.

"The geopolitical contest for the control of this area has already begun."

1. For whom does the transformation of the Arctic pose problems and for whom does it present opportunities?

2. What new connections does the opening of the Arctic create between other existing world regions?

Arctic Sea Ice News and Analysis

https://goo.gl/LKQFrR

(a)

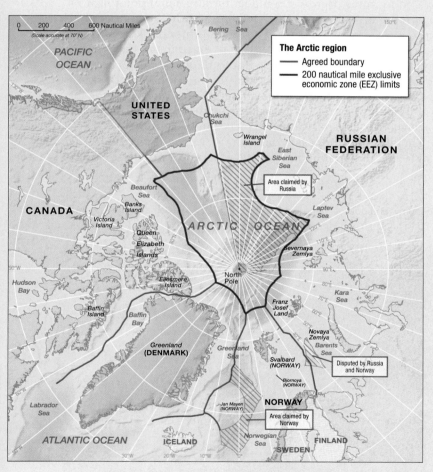

▲ FIGURE 3.1.2 **Claims on the New Arctic Frontier** The national boundaries and the competing claims on this emerging region make this a dynamic place. Each of the five countries bordering the Arctic Ocean has claimed an Exclusive Economic Zone (EEZ), an area where they hold exclusive rights to drill, fish, or mine. Several claims, shown with hash marks, are claimed by nations but not recognized by the international community and depend on contested information about the shape and extension of the continental shelf.

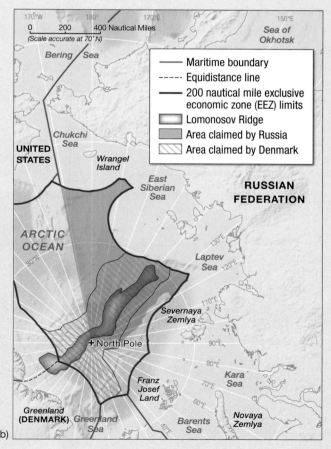

(b)

▲ FIGURE 3.1.3 **Conflicting Claims to the Arctic Seafloor** (a) Russia planted its flag in 2007 on an undersea formation, called the Lomonosov Ridge, a gesture seen by other Arctic powers as a land grab. (b) In 2014, Denmark also claimed the ridge, arguing it is an extension of Greenland's continental shelf.

Rivers and Seas The Russian Federation, Central Asia, and Transcaucasus region sustains several historically important water transport routes, which have allowed for conquest, colonization, and trade. Some of the longest rivers on Earth drain the huge Siberian landmass; the Lena and Ob' Rivers flow north to the Arctic Ocean, while the Amur flows north to the Pacific.

The Black Sea is an inland sea (**FIGURE 3.6**), connected to the Aegean and Mediterranean seas by way of the Bosporus, a narrow strait (see Chapter 4). The many rivers that empty into the Black Sea give its surface waters a low salinity. It is almost tideless, and below about 80 fathoms (478.3 feet; 145.8 meters), it is stagnant and lifeless. The region's other inland sea, the Caspian Sea, is the largest inland sea in the world, at 371,000 square kilometers (143,205 square miles—roughly the size of Germany). There are several lakes of significant size in the region as well, including Lake Baykal (30,500 square kilometers; 11,775 square miles), which, with a depth of 1,615 meters (5,300 feet), is the deepest lake in the world.

Apply Your Knowledge

3.3 Describe the key defining physiographic features of the Russian Federation, Central Asia, and Transcaucasus region.

3.4 Summarize each region's characteristics, and predict how those physical features influence economy and political unity.

Ecology, Land, and Environmental Management

The environments of the Russian Federation, Central Asia, and the Transcaucasus consist of five long, latitudinal zones that run roughly from west to east (**FIGURE 3.7**). These zones are very closely related to climatic patterns and are still easily recognizable to the modern traveler, despite centuries (or, in places, millennia) of human interference and modification. The region's environmental zones are controlled by temperatures that trend upward to the south and precipitation, which tends to trend downward. If one were to walk in a straight line, moving from north to south through the center of the region, dominant patterns would be evident.

The Tundra Fringing the entire Arctic Ocean coastline and part of the Pacific, the **tundra** is an arctic wilderness where the climate precludes any agriculture or forestry. The tundra zone represents almost 13% of the Russian Federation. For nine months or more, the landscape is locked up by ice and covered by snow. Herds of reindeer or occasional polar bear or fox are the only signs of life. During the brief summer, much of the melted snow and ice is trapped in ponds, lakes, boggy depressions, and swamps (**FIGURE 3.8**). In summer, mosses and tiny flowering plants provide color to the landscape, contrasting with the black, peaty soils and the luminous bright skies. Swans, geese, ducks, and snipe arrive on lakes and wetlands for their breeding season, as do seabirds and seals along the coast. Everywhere there are swarms of gnats and mosquitoes.

Though the tundra is a daunting environment, human life is evident. Indigenous people have herded reindeer here for

▼ **FIGURE 3.6 The Black Sea** The Black Sea is a critical component of the ecology, economy, and political geography of the region. It serves as a central passageway for trade and transit between this region and the rest of the world.

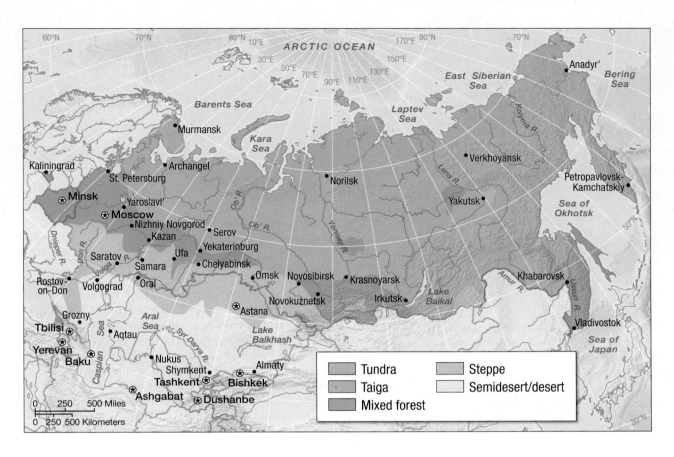

◀ **FIGURE 3.7**
Environmental Zones of Russia, Central Asia and the Transcaucasus The major environments of the region correspond to latitude, and span the region, from west to east, in long bands. **Traveling due north from Ashgabat, what changes would you observe in the environments around you?**

Legend:
- Tundra
- Taiga
- Mixed forest
- Steppe
- Semidesert/desert

centuries, managing their herds' migrations to coincide with changing seasons. The steady warming of the poles in recent years has been devastating for these cultures, however, since these herds must cross areas of seasonal ice to access forage. With delays in the freezing of the sea ice, people and animals remain stranded in overgrazed pastures, subject to starvation (**FIGURE 3.9**).

The Taiga To the south of the tundra, annual temperatures rise and rainfall increases, allowing the growth of larger and sturdier plant life. These conditions give rise to the most extensive zone of all—a belt of coniferous forest known as the **taiga**. The term *taiga* describes the zone of coniferous forest that stretches from the Gulf of Finland to the Kamchatka Peninsula. This makes up more than

25% of the total land mass of this world region (**FIGURE 3.10**). This topography, though broken by some deep river gorges and hills, is uniformly covered in characteristic forests made up of larches: hardy, deciduous, flat-rooted, coniferous trees that grow to 18 meters (59 feet) or more and can establish themselves above the permafrost.

The indigenous inhabitants of the taiga were hunters and gatherers, not farmers. Where the forest is cleared, some cultivation of

▲ **FIGURE 3.8 Tundra Landscape** The tundra landscape is bleak, with sparse vegetation; the terms derives from the Finnish word tunturia: a treeless plain.

▲ **FIGURE 3.9 Indigenous Herders of the Russian Arctic** An ethnic Dolgan herder lassos a reindeer during a roundup in Taymar in Northern Siberia, Russia. Herding animals is an integral part of the life and economy of indigenous people in the Russian Arctic. **What impact might global climate change have on the livelihoods of these herders?**

▲ FIGURE 3.10 **Taiga Landscape** The taiga is a zone of coniferous forest that stretches from the Gulf of Finland to the Kamchatka Peninsula.

hardy crops, such as potatoes, beets, and cabbage, is possible, but the poor, swampy soils and short growing season make agriculture challenging. The central Siberian taiga is one of the richest timber regions in the world. Overall, close to 90% of the territory is covered with forest, and more than a quarter of the Russian Federation's lumber production comes from the region. The taiga is also commercially important for its fur-bearing animals, whose luxuriant pelts are well adapted to the bitter cold.

Mixed Forest Continuing south, annual precipitation begins to decline, leading to a more open region of mixed forest and plains. In this zone, stands of mixed woodland arise, mostly in valley bottoms. This mixed forest was cleared and cultivated early in Russian history, providing an agricultural heartland for the emerging Russian empire.

After decades of relentless exploitation, large swathes of Siberia, once dense and practically impassable, have been cleared. Recently, the forestry sector has become more fully globalized, which has intensified concern over forest resource management. **Privatization** of the timber industry, taking the form of selling government extraction rights to private individuals and firms, has attracted U.S., Korean, and Japanese transnational corporations. The corporations invest in "slash-for-cash" logging operations (**FIGURE 3.11**) in a loosely regulated and increasingly corrupt business environment. The future of central Siberia has become an issue of international concern because the region's vast forests absorb huge amounts of carbon dioxide gas in the process of photosynthesis. This Siberian deforestation significantly contributes to global warming.

The Steppe To the south, the mixed woodlands give way to the **steppe**—the large area of flat grassland or prairie—as rainfall decreases, falling below 400 millimeters (15 inches) annually. The steppe belt stretches from the Don River to the Altay Mountains (see Figure 3.12), covering another 25% of the region. The topography of the region is strikingly flat, punctuated here and there by streams and river valleys. The natural vegetation of tall and luxuriant grasses has matted roots that trap the limited moisture available in this arid region.

▲ FIGURE 3.11 **Logging in Siberia** More than 1.5 million cubic meters of oak, cedar, and ash are illegally logged in the far eastern Primorye region each year. Chinese traders pay U.S. $100 per cubic meter and resell the wood at prices of U.S. $400 to U.S. $500 per cubic meter.

For centuries, the steppe was the realm of nomadic people. In the late 1700s, when the Turkish Empire's hold on the steppes was broken, large numbers of colonists from the wider region began to arrive. Wheat growing rapidly expanded wherever transportation was good enough to get the grain to the expanding world market, but large flocks of sheep dominated most of the colonized steppe until the railway arrived. During the 19th century, Russian settlers began to enter the eastern flat steppe of Kazakhstan in increasing numbers—a million or more by 1900—displacing the Kazakh nomads, thousands of whom died in famines or in unsuccessful uprisings against the Russians.

With the arrival of modern transportation and harvesting technologies in the late 19th century, the steppe region became an enormously productive wheat belt for Russia, as it remains to this day. Since the middle of the 20th century, small farms have been consolidated into large holdings and rivers have been dammed to provide hydroelectric power and irrigation for extensive farming of wheat, corn, and cotton (**FIGURE 3.12**).

▲ FIGURE 3.12 **Intensive Cultivation of the Russian Steppe** Beginning in the 1950s, the eastern steppes were transformed into an extensive wheat farming region through massive investments of modern machinery, fertilizers, and pesticides. **What topographic characteristic of the steppe makes it especially well suited to extensive farming?**

Semidesert and Desert Further south, where the rainfall declines even more and summer temperatures soar to extremes, are zones of semidesert and desert. Making up most of Kazakhstan, Turkmenistan, and Uzbekistan—they continue south of Siberia into Chinese and Mongolian territory. The semidesert is charac-terized by boulder-strewn wastes and *salt pans* (areas where salt has been deposited as water evaporates from short-lived lakes and ponds created by runoff from surrounding hills) and patches of rough vegetation used by **nomadic pastoralists**. These groups herd animals by moving from place to place, carefully and delib-erately following rainfall and plant growth to maintain their flocks. The desert is char-acterized by bare rock and extensive sand dunes, though there are occasional oases and fertile river valleys. Several rivers—including the Amu Dar'ya, the Syr Dar'ya, and the Zeravshan—cross these des-erts toward the Aral Sea.

Challenges to Sustainability in the Region

The environmental challenges in the region are numerous. These include the spectacular over-exploitation of the Aral Sea, where irrigation develop-ment for cotton production led to massive water diversions from sea (**FIGURE 3.13**). Pollu-tion of lakes and rivers remains a constant problem as well. Pollution in Lake Baikal—the world's oldest and deepest lake—has included chlorinated organics that take centuries to biodegrade. Drying and warm-ing may also increase the haz-ard of summer fires in the area (**FIGURE 3.14**). Oil and gas drilling off Sakhalin pollutes the island's waterways, which are crucial habitats for salmon. Offshore oil drilling represents a threat for marine ecosystems.

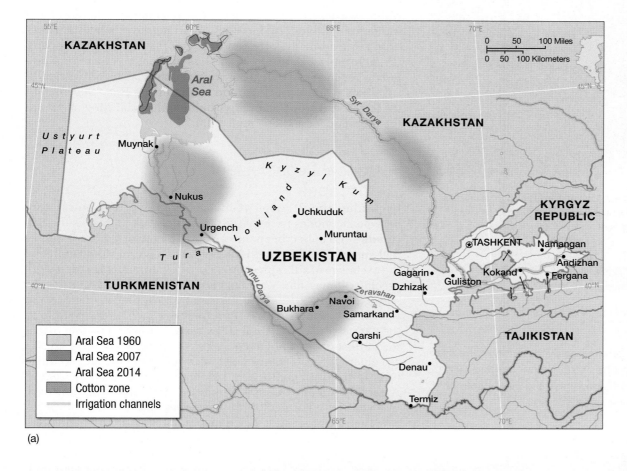

(a)

Aral Sea 1960
Aral Sea 2007
Aral Sea 2014
Cotton zone
Irrigation channels

(b)

◀ **FIGURE 3.13 Destruction of the Aral Sea.** (a) Starting in the 20th century, tributary rivers were diverted to support massive, industrialized, export-oriented cotton development, resulting in its decline. (b) Once the fourth-largest lake on the planet in 1960, by 2008, it had shrunk to 10% of its original size. Legacies of the Lake's reduction can be seen in the abandoned boats that dot the lake's former shores. **Identify a change in land use that might prevent the Aral Sea from further shrinkage. Explain your answer.**

▲ **FIGURE 3.14 Fires burning near Lake Baikal in southern Russia** Lake Baikal—the largest freshwater lake by volume in the world—has seen water levels drop recently. As a result, drying peat reserves around the lake may create more summertime wildfires.

Apply Your Knowledge

3.5 What have the environmental costs of human activity been across the region?

3.6 Compare and contrast the climates and vegetation of the five zones of the physical geography of the region, starting in the north and moving to the south. How have the inhabitants of each zone adjusted to their climates?

History, Economy, and Territory

Although relatively closed off from the rest of the world during the 20th century as a result of Soviet policies, the character of the region was forged by its deep historical connection to Illegal and indiscriminate logging has resulted in massive habitat destruction, threatening the Siberian tiger and the Far Eastern leopard. Illegal logging also results in the loss of as much as a billion dollars in revenue annually, that could otherwise be reinvested in sustainable forest practices and communities (see "Sustainability in the Anthropocene: Environmental Social Movements in the Russian Federation" on p. 103).

Historical Legacies and Landscapes

Beginning almost 2,500 years ago, the towns of Central Asia became key nodes in the vast trading network known as the **Silk Roads**—the collective name given to a network of trade routes that connected China with Mediterranean Europe and facilitated the exchange of silk, spices, and porcelain from the East and gold, precious stones, and Venetian glass from the West (**FIGURE 3.15**). Along the Silk Roads stood ancient cities, places of glory and wealth that astonished Western travelers such as Marco Polo in the 13th century C.E. These cities were east–west meeting places for philosophies, knowledge, and religions. They were known as centers for mathematics, music, art, astronomy, and architecture. In the 14th century, the Mongol empire ruled by Central Asian people built up a vast empire stretching from northern India to Syria. The decline of this empire in the 16th century C.E. saw the rise of smaller kingdoms established by nomadic people and ruled by **khans** (rulers or leaders) in the region. They prospered as independent traders on the caravan routes until the late 19th century.

◀ **FIGURE 3.15 Landscapes of the Silk Road** The Shah-i Zinda complex sits at the center of the city of Samarkand, the third largest city in Uzbekistan today. As a key point along the Silk Road, this city served as an important site for the spread of goods between East Asia and Europe for well over a thousand years. It also became a vital center for scholarly study and Islamic practice in Central Asia. **What is the relationship amongst trade, migration, and urban development in Central Asia?**

Environmental Social Movements in the Russian Federation

In 2013, the Russian government declared a "year of environmental protection," a bold statement in the face of widespread pollution, deforestation, industrial waste, and rising greenhouse gas emissions. In 2011, the World Wildlife Fund (WWF) estimated that only 10% of the government's environmental rules had been implemented, signaling the region has a weak environmental record.

Nevertheless, signs of civic action are unmistakable. A recent Russia-wide campaign emerged, including thousands of protesters in the streets of Moscow, to protect the Khimki Forest from destruction during the construction of a new Moscow–St. Petersburg highway (**FIGURE 3.2.1**). Do it yourself movements have arisen as well, like "Musora.bolshe.net," which translates to "no more rubbish." This initiative has mobilized volunteers to gather and recycle waste from areas around lakes, parks, and forests in the Russian Federation and five former Soviet countries.

These initiatives are also increasingly evident in Russian businesses. The Russian Green Building Council seeks economic benefits that emerge from increasing efficiencies through the use of fewer materials in construction and energy savings from insulation and green roofs. The organization has laid down a "Russia 2030" challenge to make many more buildings throughout the country

> "The Russian Green Building Council .... has laid down a 'Russia 2030' challenge to make many more buildings throughout the country carbon neutral in the next 15 years."

carbon neutral in the next 15 years. The obstacles to success for this effort are daunting, however. Subsidies for cheap energy are a hallmark of Russian domestic construction policy, and codes and requirements in this sector are easily overlooked through the payment of carefully placed bribes.

The overall potential for sustainability in Russia is viewed as enormous, however. The International Finance Group of the World Bank, notably, has forecast highly profitable development of solar and hydroelectric as well as wind power (**FIGURE 3.2.2**). In the northern and western parts of the country, the flat open expanses under high westerly winds suggest a wind-power potential as high as 120,000 terawatt hours (TWh) annually (Russia consumes only 780 TWh per year). Unhappily, the effort was

stalled in 2014 as a result of complex purchasing requirements imposed by the Russian government. The autocratic government of Russia also puts proponents of such efforts at serious risk of government sanction. As in the case of nongovernmental organization (NGO) movements and the Green Building initiative, the challenges to sustainability in Russia are clearly formidable.

---

1. **Describe how citizens are leading the sustainability movement in the Russian Federation.**

2. **Conduct an Internet search to find another example of the environmental movement in Russia. Predict how people's actions may impact government policy.**

▲ FIGURE 3.2.1 **Protesting the Destruction of Khimki Forest, Russia**

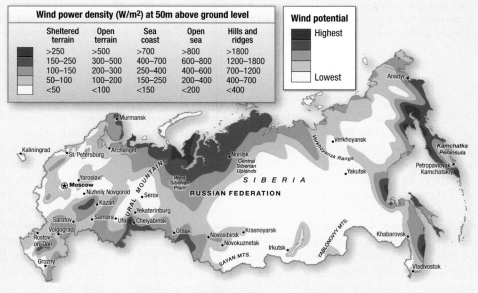

▲ FIGURE 3.2.2 **Highest to Lowest Wind Potential** Wind resources vary across the region but some areas show dramatically high potential.

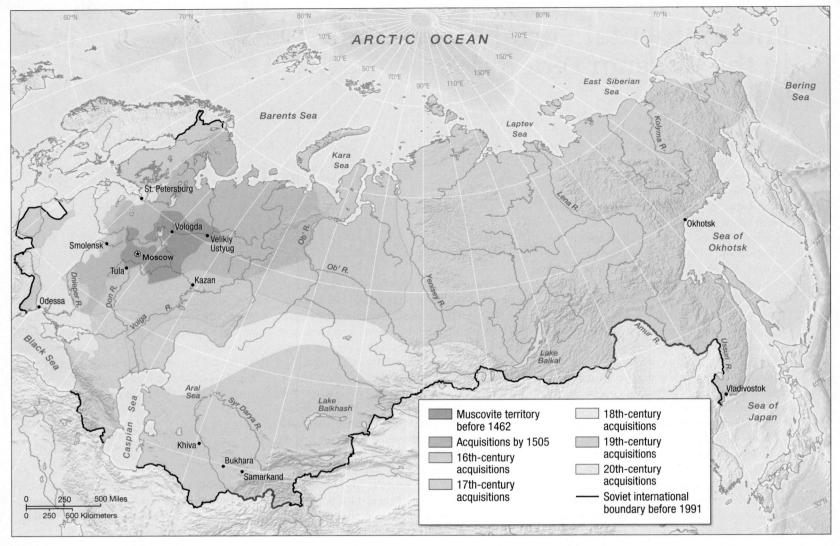

▲ **FIGURE 3.16 Territorial Growth of the Muscovite/Russian State** The Russian Empire was vast and was acquired over the same period (15th century to the late 20th century) that corresponds to the globalization of the world economy. When the Bolsheviks came to power at the beginning of the 20th century, some of the territory was lost. Eventually, however, the Bolsheviks were able to control most of the territories formerly held by the tsars, and it was on this that they built the Soviet state. **When did the Russian Empire gain control of the region that includes Odessa? Why was this acquisition important to the empire?**

The Rise of the Russian Empire

In the mid-15th century, Muscovy was just one principality centered on the city of Moscow. Muscovy was a tributary state to the Mongol–Tatar Empire, whose armies were known as the Golden Horde. In 1552, under Ivan the Terrible, the Muscovites defeated the Tatars at the battle of Kazan—a victory that prompted the commissioning of St. Basil's Cathedral in Moscow. Over a 400-year period, Ivan's successors and the Muscovite state expanded at a rate of about 135 square kilometers (52 square miles) per day so that by 1914, on the eve of the Russian Revolution, the empire occupied more than 22 million square kilometers (roughly 8.5 million square miles), or one-seventh of the land surface of Earth. The expansion of the empire was based on the economic system of **serfdom**—the

practice whereby members of the lowest class were attached to a lord and his land. So, even though early in the 18th century Peter the Great (1682–1725) founded St. Petersburg and developed it as the planned capital of Russia, the country was very much a rural, peasant economy.

The factors behind Russia's imperial expansion were the drive for more territorial resources (especially a warmwater port) and additional subjects. Russian **tsars** (rulers of the Russian Empire) annexed the vast stretches of adjacent land on the Eurasian continent instead of establishing new territories overseas as the European imperial powers had. The Russian empire benefited from its territories' rich resources—ranging from precious metals and timber to furs (see "Geographies of Indulgence, Desire, and Addiction: Furs, Fashion, and Animal

Rights" on p. 106). The extension of political control by the people of the Russian homeland is key to the present-day geography of the entire region. By 1861, when Tsar Alexander II decreed the abolition of serfdom, Russia had built up an internal core with a large bureaucracy, a substantial intellectual class, and a sizable group of skilled workers. The abolition of serfdom was designed to accelerate the industrialization of the economy. In many ways, the strategy seems to have been successful: Between 1860 and 1900 grain exports increased five-fold and collective agricultural production gave way to private land ownership, helping to establish large, consolidated farms in place of some of the many small-scale peasant holdings, as wealthy elites were able to purchase large tracts of agricultural land. Manufacturing activity expanded rapidly. This forced many poorer people to flood to the cities and created acute problems as housing became overcrowded and living conditions deteriorated (**FIGURE 3.16**).

Though the Japanese victory in Manchuria in 1904 brought a halt to Russian territorial expansion, by that time, the Russian Empire contained about 130 million persons, only 56 million of whom were Russian. Russia's strategies to bind together the 100-plus "nationalities" (non-Russian ethnic people) into a unified Russian state were oftentimes punitive and unsuccessful. Non-Russian nations were simply expected to conform to Russian cultural norms. Those who did not were persecuted. The result was opposition and, sometimes, rebellion and stubborn refusal to bow to Russian cultural dominance.

Revolution and the Rise of the Soviet Union
The problems of rural and urban change in the region, to which the Russian tsars remained indifferent, fueled deep discontent among the population. At the turn of the 20th century, Russia was in the grip of a severe economic recession. Inflation, with high prices for food and other basic commodities, led to famine and widespread hardship. Peasants rioted across the countryside and, in 1905, a network of grassroots councils of workers—called **soviets**—emerged to coordinate strikes and help maintain public order. The unrest was eventually subdued by brute force, and the soviets were abolished.

A decade later, World War I intensified the discontent of the population. As casualties mounted, the government's handling of the armed forces and the domestic economy led to the 1917 October or Bolshevik Revolution, named after the **Bolsheviks**, the majority faction of the Russian Social Democratic Party. The Bolsheviks established the Soviet Union and renamed their political party the Communist Party after seizing power. The revolution was seen with great suspicion by its neighbors, who greeted the new government by invading Russia. The conflict ended in a victory for the revolutionary government of the new Soviet Union, but the period set a tone that influenced later global and regional geopolitical relationships.

The Soviet Union's **state socialism**, in which the government (or state) controlled industries and services, was based on a new kind of social contract between the state and the people. In exchange for people's compliance with the system, their housing, education, and health care were provided by the state at little or no cost (**FIGURE 3.17**). By the early 1920s, **Vladimir Ilyich Lenin**, the revolutionary leader and head of state of the new regime, was also able to focus attention on the more idealistic aspects of state socialism. Lenin was an **internationalist**, believing in equal rights for all nations. He wished to break down national barriers and end ethnic rivalries. Lenin's solution was recognition of the many regional nationalities through the establishment of the **Union of Soviet Socialist Republics (U.S.S.R.)** in 1922. Lenin believed that this federal system would provide different nationalities with political independence.

Lenin was optimistic that once inequalities were diminished, the federated state would no longer be needed: Local nationalisms would be replaced by communism. Following Lenin's death in 1924, the federal ideal faded. After eliminating several rivals, **Joseph Stalin** came to power in 1928 and enforced a new

▲ **FIGURE 3.17 The Ideals of the Soviet Union** This painting, titled *The Harvest*, was produced in 1925. It is an example of socialist realist art. This form of art depicted the worker as a hero in the new Soviet society through the literal representation of the everyday labor that supported the people collectively.

Furs, Fashion, and Animal Rights

At the 2015 Australia Fashion Week, animal rights activists interrupted the event to protest the use of animal products in clothing, demonstrating the global tension that surrounds these products (**FIGURE 3.3.1**). While fur, in particular, is still seen by many status-conscious consumers as a fashionable luxury good, a clear marker of material wealth, it is also one of the most controversial luxury products in the market today.

Historical Demands for Fur From earliest times, fur has been a prized commodity. In cold regions, fur coats, hats, and boots are valued for their warmth and as a portable form of wealth. In Russia and Europe, fur has long had royal and aristocratic connotations and, as a result, became a status symbol for all who could afford it. European merchants began trading fur in the Middle Ages, and fur was the commodity that drew Russian trappers and traders to Siberia in the 16th century. Tax revenues from the fur trade were the mainstay of the Russian imperial treasury for the next 300 years. Trade in fur pelts (beavers, muskrats, minks, and martens) fueled Europeans' initial interest in North America.

Fur Trade Today Fur production has spread across the globe. Total mink production worldwide in 2011 was 53 million skins, with the largest share of production occurring in Scandinavia

> "Although Russian farming has declined, global demand for fur is rising: Global retail sales in 2012 totaled more than $15 billion."

(35%), China, (25%), Western Europe (25%), North America (10%), and Russia (5%). Fur farming (raising animals in captivity under controlled conditions) is the principal source of furs for the world market (**FIGURE 3.3.2**). Through controlled breeding, animals with unique characteristics of size, color, or texture can pass on those characteristics to their offspring.

A Slowing Fur Trade? The industry's growth has recently been uneven with a leveling off of sales in the last decade. The reason for this is the result of recent economic downtowns and a shift in attitudes toward furs, led by animal rights activists. Antifur demonstrations in Russia in particular have captured widespread attention. Although Russian farming has declined, global demand for fur is rising: Global retail sales in 2012 totaled more than $15 billion. This shift underlines the way a global cultural phenomenon (the antifur movement) can have a sharp regional impact (on

Russian fur farmers) but also may result in other unanticipated changes in the globalized economy (such as the meteoric rise in Chinese fur farming).

1. **Identify how fur consumption has changed over time due to public taste and awareness.**

2. **Analyze how the global fashion industry influences regional economies by researching the current sales data, the sources of fur, and the major markets on the International Fur Federation website, http://www.wearefur.com.**

Today's Fur Trade

https://goo.gl/DBSmwW

▲ **FIGURE 3.3.1 Protests at Australian Fashion Week** Supporters protesting against the use of animal skin staged protests during Australian Fashion Week in Sydney, Australia, in 2015.

▲ **FIGURE 3.3.2 Russian Fur Farm** Most of Russia's fur farms are located in remote settings, where there are few regulatory controls or inspections.

nationality policy. Although the federal administrative framework remained in place, nations within the U.S.S.R. increasingly lost their independence and by the 1930s were punished for displays of nationalism.

Under Stalin's leadership, a **command economy**, an economy in which all goods, services, prices, and supplies are regulated by the government, was established. Engineers, managers, and *apparatchiks* (state bureaucrats) ran this new economy on a model of five-year plans where clear goals for economic development were set on a rotating basis. The first five-year plan was established in 1928. At this juncture, the Soviet Union chose to withdraw from the capitalist world economy as far as possible, relying on its vast territories to produce the raw materials needed for rapid industrialization. Indeed, Stalin's industrialization drive was founded on severe exploitation of this world region's rural periphery in an attempt to bolster its historic core around Moscow and the central Russian region. Through a process of **collectivization**, peasants were relocated onto collective farms (kohlkoz), where their labor was expected to produce bigger yields. The state purchased harvests at relatively low prices so that the collectivized peasant paid for industrialization by "gifts" of labor.

Severe exploitation required severe repression. Stalin employed police terror to compel the peasantry to comply. Dissidents, along with "enemies of the state" criminalized by purges of the army, the bureaucracy, and the Communist Party, provided convict labor for infrastructure projects to support the industrialization drive. Altogether, some 10 million people were sentenced to serve in the workforce, imprisoned, or shot.

The Soviet economy *did* modernize, however. Between 1928 and 1940 the rate of industrial growth increased steadily. When the Germans attacked the Soviet Union in 1941, they took on an economy that in absolute terms (though not per capita) had industrial output figures comparable with their own.

Soviet Industrialization and Its Environmental Legacy

The entire Soviet Union and the countries that became known as the U.S.S.R.'s "bloc" of satellite states in Eastern Europe continued to give high priority to industrialization after World War II (**FIGURE 3.18**). Soviet economic planners sought to shape

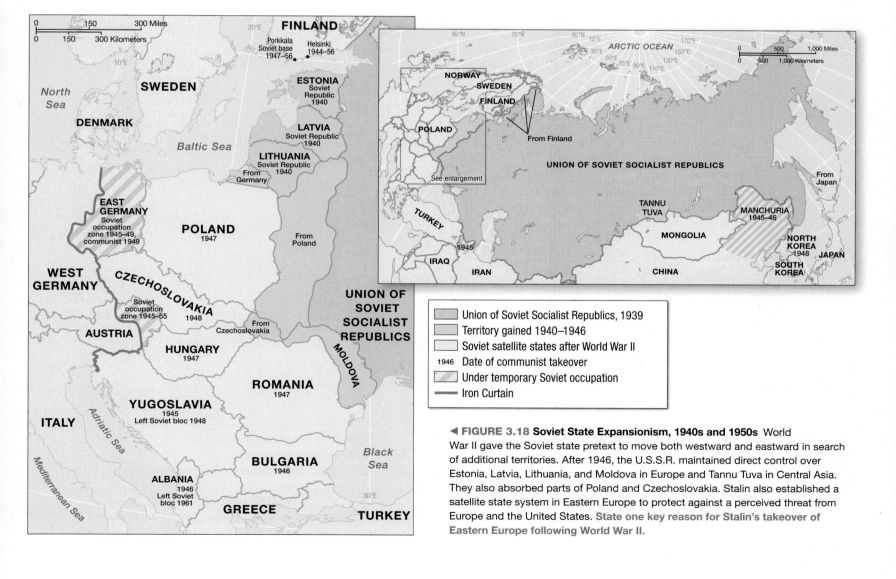

Union of Soviet Socialist Republics, 1939
Territory gained 1940–1946
Soviet satellite states after World War II
1946 Date of communist takeover
Under temporary Soviet occupation
— Iron Curtain

◄ **FIGURE 3.18 Soviet State Expansionism, 1940s and 1950s** World War II gave the Soviet state pretext to move both westward and eastward in search of additional territories. After 1946, the U.S.S.R. maintained direct control over Estonia, Latvia, Lithuania, and Moldova in Europe and Tannu Tuva in Central Asia. They also absorbed parts of Poland and Czechoslovakia. Stalin also established a satellite state system in Eastern Europe to protect against a perceived threat from Europe and the United States. **State one key reason for Stalin's takeover of Eastern Europe following World War II.**

the economic geography of state socialism through strict planning measures. The first was the development of **territorial production complexes**, regional groupings of production facilities based on local resources that were suited to clusters of interdependent industries. Key territorial production complexes were petrochemical complexes, for example, or iron-and-steel complexes (**FIGURE 3.19**). The second was the expansion of industrialization in economically less-developed subregions, such as Central Asia and the Transcaucasus. The third was secrecy and security from external military attack. This criterion led to military–industrial development in Siberia and the creation of so-called secret cities housing military research and production facilities.

The environmental legacies of this form of planning and growth are notable. Every country in the former Soviet region has a legacy of serious environmental problems that stem from mismanagement of natural resources and failure to control pollution during the Soviet era. Soviet central planning placed strong emphasis on industrial output, with very little regard for environmental protection. Today, serious environmental degradation, which is a legacy of Soviet planning, affects all parts of the region (**FIGURE 3.20**).

The 1986 disaster at Chernobyl, in which a nuclear reactor exploded in a power plant in the former Soviet republic of Ukraine, became emblematic of the Soviet nuclear legacy. Stories of radiation sickness filled international newspapers for weeks following the meltdown. Radiation levels at Chernobyl have remained high, and an area surrounding the failed reactor, 31 kilometers (19 miles) in every direction, has been established as a **zone of alienation,** where only a handful of residents and scientific teams continue to reside.

The Chernobyl region has also become a test of how the environment recovers in the almost-total absence of human beings (**FIGURE 3.21**). In the years since the accident, forests have regrown rapidly and many wild animals, including wild horses and wolves, have recolonized the area. Though heavily irradiated, the Chernobyl area has become a unique and emblematic region of the Anthropocene, a place transformed by people and yet still changing and evolving in unexpected and unmanaged ways.

Economic Growth and the Cold War By the 1960s, the Soviet Union had clearly demonstrated its technological capabilities with its first satellite launch, manned space program, and some of the world's most sophisticated military hardware (**FIGURE 3.22**). These successes were paralleled by the Soviet Union's broader geopolitical influence. Armed with nuclear weapons and an economic model that challenged the dominance of capitalism, the Soviet Union posed a very real threat to U.S. power. The resulting tensions between the global power of the Soviet Union and its socialist economic model and the United States with its capitalist economic model resulted in the Cold War.

A **cold war** is not a direct conflict of arms between two states but a war of ideology and proxy conflicts in which other countries are used to fight. The Cold War between the Soviet Union

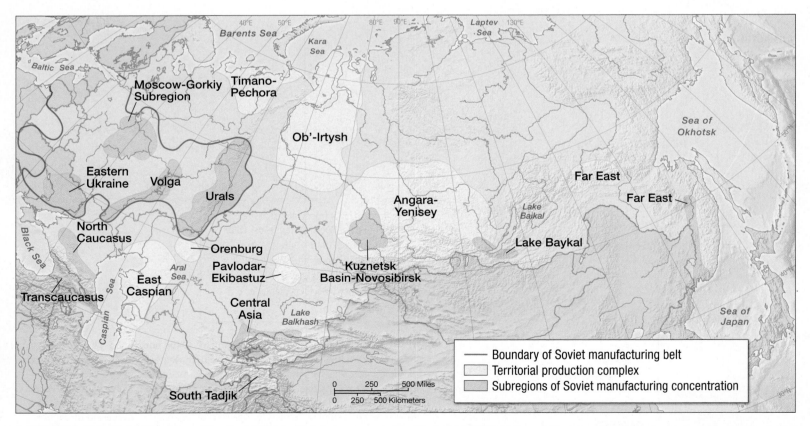

▲ **FIGURE 3.19 Industrial Regions of the Soviet Union** Soviet planners prioritized industrialization and sought to take advantage of agglomeration economies by establishing huge regional concentrations of heavy industry. They also expanded the transportation infrastructure of the Soviet state to further integrate the region's economy.

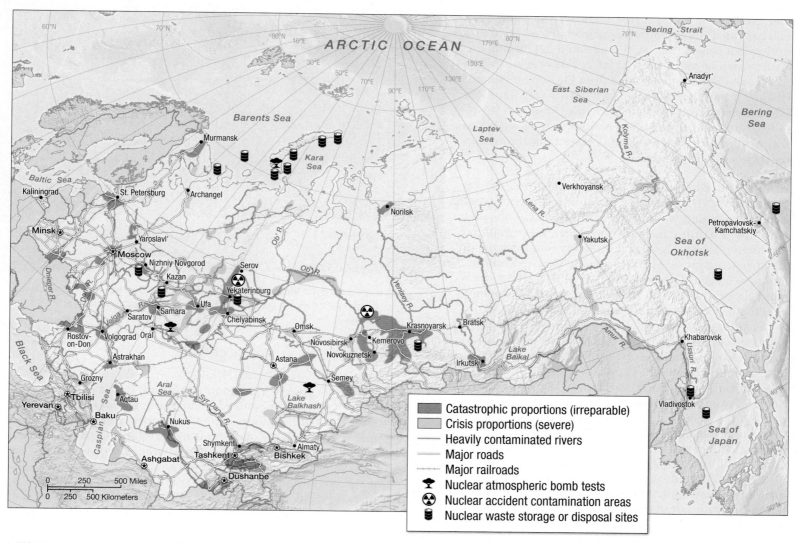

Legend:
- Catastrophic proportions (irreparable)
- Crisis proportions (severe)
- Heavily contaminated rivers
- Major roads
- Major railroads
- Nuclear atmospheric bomb tests
- Nuclear accident contamination areas
- Nuclear waste storage or disposal sites

▲ **FIGURE 3.20 Environmental Degradation** Serious environmental degradation afflicts many parts of the Russian Federation, Central Asia, and the Transcaucasus. **Explain the distribution of the three types of nuclear sites based on a comparison of their distribution with the distribution of the region's population in Figure 3.37.**

◀ **FIGURE 3.21 Environmental Recovery After Chernobyl** These wild buffalo were rare in the region but are now reclaiming the native territory since they were reintroduced to the area following the accident. Though still radioactive in areas, the region surrounding Chernobyl has experienced ecological recovery, largely owing to the absence of human activity.

and the United States lasted between 1950 and 1989 and provided the principal framework for world affairs during this period. The conflict resulted in tensions and a succession of geopolitical crises in many regions of the world (see discussion in Chapter 8 on the Korean War and Chapter 10 on the Vietnam War). In this way, the Cold War simultaneously isolated the Soviet Union and its satellite countries from Western nations as a region while increasing their global reach and connections throughout the less-developed world.

Economic Stagnation and the Decline of the Soviet Union Despite industrial development, peasants in the region worked with primitive and obsolete equipment as they toiled to meet centrally planned production targets. A second economy—an informal or shadow economy—of private production, distribution, and sales emerged. It was largely tolerated by the government because without it the formal **command economy** would not have been able to function as well as it did.

By the 1980s, the Soviet system was in crisis. The crisis resulted from a failure to deliver consumer goods to a population that had become well-informed about the consumer societies of their enemies. Persistent regional inequalities further produced a loss of confidence in the Soviet system. The cynical manipulation of power for personal gain by ruling elites and the drain on national resources from the arms race with the United States also played a role in undermining the Soviet model.

The critical economic failure was state socialism's inherent inflexibility and its consequent inability to take advantage of the new computerized information technologies that were emerging elsewhere. The Soviet authorities had deliberately suppressed investment in the development of computers and new information networks because computers, like photocopiers and fax machines, were viewed as a threat to central control. Surprising even the most astute observers, the Soviet system unraveled rapidly between 1989 and 1991, leaving 15 independent countries—including the Russian Federation as the largest and most powerful of the post-Soviet states—as successors to the former U.S.S.R. (**FIGURE 3.23**). In the process, all local and regional economies were disrupted, leaving many people to survive by supplementing their income with informal activities, such as street trading and domestic service.

▲ **FIGURE 3.22 Victory Day Parade in the Red Square** This Victory Day parade in 2015 marks the 70th anniversary since the capitulation of Nazi Germany.

Economy, Accumulation, and the Production of Inequality

Restoring capitalism in countries where it was suppressed for more than 70 years and reestablishing global connections to other parts of the global capitalist economy have proven to be problematic. The economy of the Russian Federation shrank by 62% between 1991 and 1999. By the end of the 1990s, the Russian Federation's economy was in crisis. About one-half of the government's budget revenue was being absorbed by repaying debts to creditor nations, the rate of inflation reached 100%, and economic output plunged to about half of what it was in 1989.

Building a Market Economy Following the breakup of the command economy, there were no accepted codes of business behavior, no civil code, no effective banking system, no effective accounting system, and no procedures for declaring bankruptcy. Security agencies were disorganized, bureaucratic lines of command were blurred, and border controls between the new post-Soviet states were nonexistent. The absence of these key economic elements provided enormous scope for crime and corruption and fostered regional ethnic **mafiyas** (organized crime groups). State assets were often sold off quickly and at an extremely undervalued price, so that some well-connected entrepreneurs managed to amass huge amounts wealth within the newly created market environment. These new **oligarchs**—business leaders who wielded significant political and economic power—quickly became extremely unpopular among the Russian public because of their undue political influence, extreme wealth, and control over media outlets (**FIGURE 3.24**).

Despite the dominance of these elites, the Russian economy grew dramatically between 2000 and 2014, largely on the strength of its commodities exports and with the active support of the President, Vladimir Putin, who has held top posts—as

Apply Your Knowledge

3.7 Give examples to define these key concepts of the Stalin-era economy: command economy, collectivization, and territorial production complexes.

3.8 Evaluate the loss of confidence in the Soviet system in the 1980s and explain how that influenced the disintegration of the Soviet Union.

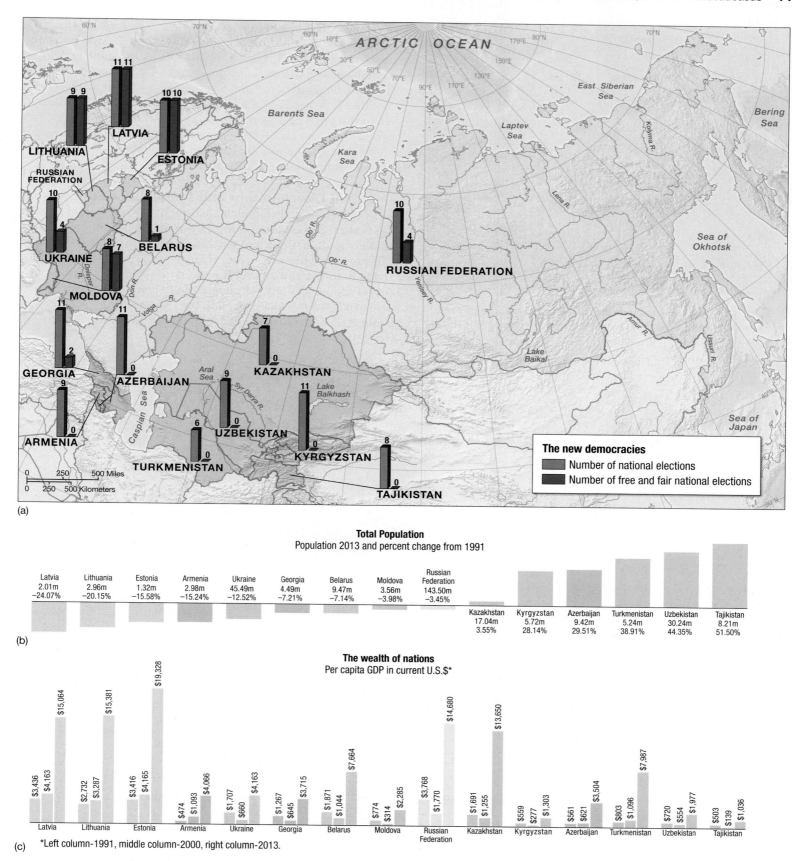

▲ **FIGURE 3.23 The Breakup of the Soviet Union** The breakup of the Soviet Union produced a number of changes in this world region. New countries were formed, which had an impact on the political and economic life of the people in this region. These changes have resulted in (a) a complicated political landscape where some countries have had more free and fair elections than others and (b) dramatic demographic shifts as ethnic Russians and non-Russians moved in and out of former Soviet states and (c) differences in economic development. List the three countries that have fared best in terms of free elections and economic development and the three countries that have fared worst. What might you infer from these lists about how good government and economic growth could be related? Explain.

▲ **FIGURE 3.24 Upscale Housing in the New Russia** A new wealthy elite with the ability to invest in very high end luxury goods has emerged in Russia over the last 20 years.

either President or Prime Minister since 1999. Russia, which was hit hard by the 2008 global financial crisis, appeared to recover relatively quickly. By the end of 2013, Russia's economy was the ninth biggest in the world by gross domestic product (GDP). That said, the dependence of the Russian economy on primary commodities has placed it in a precarious position in 2014 and 2015 when the price of oil and gas dropped quickly. Russia's continued dependence on oil and gas production has made it susceptible to changes in the global supply and demand of these highly valued commodities (**FIGURE 3.25**).

Regional Development and Inequalities Although patterns of regional development are beginning to change, at present, the Central Region around Moscow remains the core of the Russian Federation, Central Asia, and Transcaucasus region. Today, this area is highly urbanized, with about 85% of the population living in towns and cities. Moscow, with roughly 11 million people, is the largest city, although other significant urban centers also exist (**FIGURE 3.26**). Overall, the Central Region accounts for about 20% of the Russian Federation's industrial production.

Russian Exports

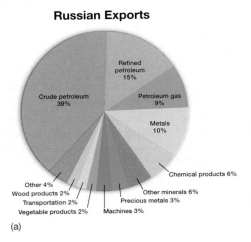

(a)

(b)

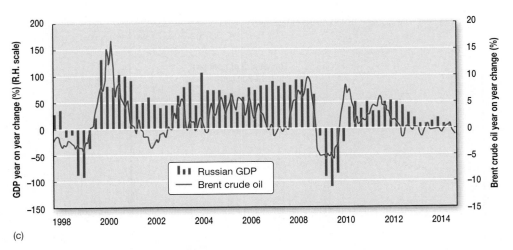

(c)

▲ **FIGURE 3.25 The Changing Economic Fortunes of Russia** (a) Despite attempts to diversify its economy, Russia is still highly dependent on primary commodity production, with over 63% of its exports coming from crude or refined petroleum as was as petroleum gas. (b) The Russian landscape is marked by the needed infrastructure to both extract and move petroleum from the point of production to the sites of refinement and consumption. (c) Russia's GDP is very closely tied to the global price of crude oil. **What are Russia's main exports and how might the country's export profile impact the volatility in its GDP?**

◄ FIGURE 3.26 **Moscow and the New Russian Economy** There has been massive investment in the modern infrastructure of Moscow over the last 20 years, particularly in the development of integrated business, entertainment, and living spaces such as the Moscow International Business Center (also called Moscow City), shown here.

Market forces have introduced a much greater disparity between the economic well-being of regional winners and losers. The unevenness of patterns of regional economic development has been intensified but, after two decades of transition, many of the regional winners are the same as under state socialism. This is partly because of the natural advantages of certain regions and partly because of the initial advantage of economic development inherited from Soviet-era regional planning. However, two different kinds of regions have experienced significantly *decreasing* levels of prosperity. The first consists of regions of armed territorial conflict (such as North Ossetia in Georgia, Ingushetia, and Chechnya in the North Caucasus; Nagorno-Karabakh; and Tajikistan). The second consists of resource-poor regions—mainly in the European north, in Siberia, and in the eastern parts of the region, where conditions are harsh in both rural and urban settings (**FIGURE 3.27**).

Apply Your Knowledge

3.9 List two of the major economic and political changes that have taken place in this region since the 1980s.

3.10 How have these changes provided opportunities or challenges for middle-class individuals or families in this region?

Territory and Politics

Within the Russian Federation, there are approximately 27 million non-Russians across 92 different ethnic groups. Most

▲ FIGURE 3.27 **Marginal Economic Regions and the Periphery** Many of the outlying and rural areas of the region depend heavily on primary resource extraction, especially forestry, drilling, and mining. Working conditions in these industries can be harsh, especially where health and safety regulations are limited. A mining disaster in 2010 killed 90 Russian workers.

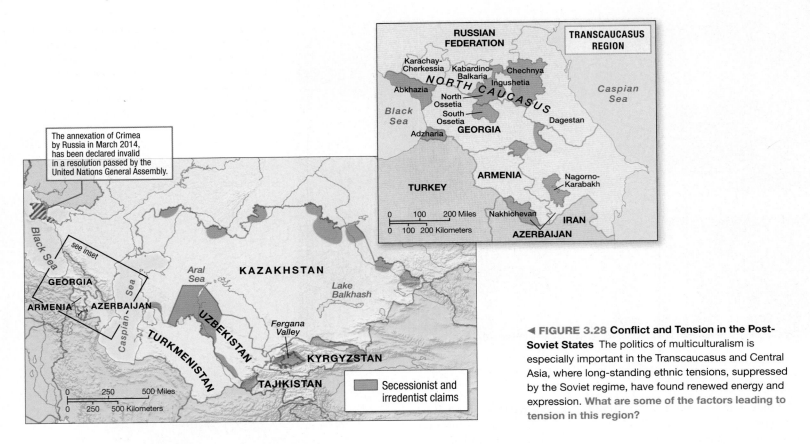

◀ **FIGURE 3.28 Conflict and Tension in the Post-Soviet States** The politics of multiculturalism is especially important in the Transcaucasus and Central Asia, where long-standing ethnic tensions, suppressed by the Soviet regime, have found renewed energy and expression. **What are some of the factors leading to tension in this region?**

of the larger ethnic groups enjoy a fair degree of autonomy within the Russian Federation, although conflicts along border areas throughout the region are numerous (**FIGURE 3.28**). One of the most troubled regions is the North Caucasus, a complex mosaic of mountain people with strong territorial and ethnic identities. Soon after the breakup of the Soviet Union, Ingushetia broke away from the Chechen-Ingush Republic, and Chechnya promptly declared independence from the Russian Federation.

Chechen Independence Of the many movements that surfaced with the breakup of the Soviet Union, the Chechen independence movement has been the most bloody. In the late 1930s, tens of thousands of Chechens were murdered by Stalin as part of his purges against suspected anti-Soviet elements. In 1944, Stalin accused the Chechens of having collaborated with the Nazis and ordered the entire Chechen population—then numbering about 700,000—to be exiled to Kazakhstan and Siberia. Brutal treatment during this mass deportation led to the death of more than 200,000 Chechens.

In 1957, the Soviet Union embarked on a program of de-Stalinization that included the repatriation of Chechens to Chechnya. But when Chechens returned, they found that newcomers had taken over many of their homes and possessions. Over the next 30 years, many of these newcomers withdrew and the Chechen population consolidated and grew to almost 1 million. In 1985, when Mikhail Gorbachev initiated his policy of **glasnost** (an official policy change that stressed open government

and increased access to information as well as more honest discussion about the country's social issues and concerns), Chechens finally saw the possibility for self-determination. With the breakup of the Soviet empire in 1989, Chechens wasted no time in unilaterally declaring their complete independence.

The Russian Federation at first ignored Chechnya's declaration of independence but ultimately could not tolerate the loss of the region because the Russian Federation's major oil-refining centers and significant natural gas reserves reside within Chechnya. In December 1994, Russian troops invaded. Chechen resistance continued, with increased popular support because of the invasion. In 1996 the Russian Federation settled for peace, leaving Chechnya with *de facto* independence. In the summer of 1999, after Chechen rebels had taken the fight to the neighboring republic of Dagestan and to the Russian heartland with a series of terrorist bombings of apartment blocks, the Russians renewed their military effort. Bitter and intense fighting ensued. Hundreds of thousands of Chechens were made homeless and several thousand were killed or went missing. Russian troops took the Chechen capital, Grozny, in February 2000 (**FIGURE 3.29**).

Between early 2000 and April 2009, Russian Federation troops maintained control of Grozny. Human rights organizations documented patterns of abduction, detention, disappearances, collective punishment, extrajudicial executions, and the systematic use of torture by authorities. Chechen rebel guerillas, meanwhile, used terrorist attacks against the regime, including attacks on children. In 2009, Russia formally ended its military campaign

▲ **FIGURE 3.29 Grozny, Chechnya** A Grozny woman sits with her dog in front of her bombed out house in the Staro-Promislovsky district of Grozny, Chechnya. Even though the military campaign has formally ended, the war landscape remains as a reminder of the tension that exists in Chechnya. People remain displaced as a result of the conflict.

▲ **FIGURE 3.30 Resource Development in Azerbaijan** The port city of Baku is the capital of Azerbaijan and its largest city. Located on the shores of the Caspian Sea, the city has long been a cultural crossroads with a rich architectural tradition. It now sits at the center of a network for the production and transport of Caspian oil.

in Chechnya, although terrorists who seek full separation from the Russian Federation have been active in the capital Grozny as late as 2014. In 2014, it was also reported that nearly 40,000 Russian citizens, most of whom are ethnic Chechens, were seeking asylum in Europe as a result of the ongoing tension in the region.

Ethnonationalism in the Transcaucasus and Central Asia

Other ethnicities within the Russian Federation have also pursued nationalist aspirations, including groups in Nagorno-Karabakh, South Ossetia, and Tajikistan.

- **Nagorno-Karabakh** In the Transcaucasus, one trouble spot of **ethnonationalism**—nationalism based on ethnic identity—is the region of Nagorno-Karabakh, in Azerbaijan. The breakup of the Soviet Union brought the opportunity for this region, which was almost 75% Armenian, to petition for secession from Azerbaijan. When the petition was refused, anger was unleashed in Azerbaijan against Armenians and in Armenia against Azerbaijanis. Armenian military forces secured Nagorno-Karabakh and established a militarized corridor as a lifeline from Azerbaijan to Armenia. Russian Federation armed forces were invited to serve a peacekeeping role, and the United States, France, and Russia have been involved in mediating a peaceful settlement. The political situation in Azerbaijan is of broad international interest because of the area's rich natural resources and the 1,760-kilometer (1,094-mile) pipeline from Baku, Azerbaijan, to Ceyhan, Turkey (**FIGURE 3.30**). The pipeline opened in 2005 to carry oil from U.S. and European companies' oil fields in the Caspian Sea to Western markets via the Mediterranean.

- **South Ossetia** Another flashpoint in the Transcaucasus is South Ossetia, which declared its independence from Georgia in 1991. It did so mainly because of strong ethnic ties to North Ossetia-Alania, a republic of the Russian Federation. In 2008, Georgia launched a massive artillery attack on the separatists of South Ossetia, prompting a brief but full-scale war that drew Russian troops into South Ossetia, where they remained at observation and security posts for several months. Since then, South Ossetia has remained independent from Georgia, but its politics remain unstable. The recent presidential election of April 2012 helped Leonid Tibilov rise to power. But he has an uphill battle; the country is marred by a legacy of mistrust and alleged corruption, as the previous government was suspected of misusing Russian aid to the country. The region is also mired in an economic crisis, and as of 2013 the unemployment figures for the region stood at 15%. The region also suffers from an outmigration of its younger populations.

- **Tajikistan** The breakup of the Soviet Union also led to other ethnonational movements in Central Asia. Tajikistan has been beset by conflict between ethnic Tajik and Uzbek insurgents and activists, who define themselves as patriots but assert the primacy of their own individual ethnic groups. A brief civil war ended in 1993, when the ruling government accepted intervention by the Russian Federation. Between 1993 and 2005, Russian troops, in cooperation with the Tajik government, defended Tajikistan's border to counter frequent military incursions by rebels based in Afghanistan. Today, the Russian government maintains a military presence in the country.

Ethnonational movements illustrate the regional dimensions of the problems involved in the new states' ability to guarantee territorial integrity, physical security, and effective governance. In some regions, secessionist and irredentist tensions are clearly undermining the vitality of civil society. **Civil society** involves the presence of a network of voluntary organizations, business organizations, pressure groups, and cultural traditions that operate independent of the state and its political institutions. A vibrant civil society is an essential precondition for **pluralist democracy**—a society in which members of diverse groups continue to participate in their traditional cultures and special interests. Since the breakup of the Soviet Union, civil society is emerging only slowly in this world region, and in some parts—especially in Central Asia—democratic reform has been so limited that the emergence of civil society has been hard to detect.

Geopolitical Shifts and Ongoing Regional Tensions

In an attempt to counter some of the economic disruption caused by the political disintegration of the Soviet Union, several successor states agreed to form a loose association, known as the **Commonwealth of Independent States (CIS)**. The reorientation of the Baltic and eastern European states toward Europe and the expansion of the European Union and North Atlantic Treaty Organization (NATO) membership for some of these states has not only undercut the economic prospects of the CIS (which never really blossomed) but also weakened the geopolitical security of the Russian Federation. Perhaps the final blow to the CIS was Ukraine's decision to leave the body in 2014 and the subsequent annexation of Crimea by the Russian Federation (see Opening Story on p. 91).

The annexation of Crimea has its seeds partially in response to the changes brought about by the breakup of the Soviet Union; the Russian Federation asserted its claims in the 1990s and early 2000s in a special sphere of influence it has called the **Near Abroad**. The Near Abroad consists of the former components of the Soviet Union, particularly those countries that contain a large number of ethnic Russians. Under the leadership of President Vladimir Putin in the early part of the 21st century, in particular, the Russian Federation pressed its authority in this subregion. Russia's interest in maintaining its power in the region is further represented by its role in the Ukranian political crisis, a continuing conflict that most recently re-emerged in 2013 and continues to the present (see Chapter 2, p. 73). Indeed, Russia's role in an uprising of ethnic Russian nationalists in Ukraine, who are

▲ **FIGURE 3.31 Conflict in Ukraine** Pro-Russian gunmen pose at their house position in the North West Oktyaber neighborhood of Donetsk in 2014, following heavy shelling. Russian-backed secessionists forces battled Ukrianian troops to a standstill throughout eastern Ukraine.

fighting the central Ukranian government for independence, has led to harsh sanctions from Europe as well as the United States (**FIGURE 3.31**). As a consequence of the tensions between Russia and Ukraine, the Russian government has tried to intensify its relations with Belarus, a long time ally and strong military partner in the region. Belarus also signed on to the Russian-led Eurasian Economic Union (EuEU) treaty, which takes effect in 2015, drawing Belarus closer to Russia's sphere of influence.

Russia on the World Stage As Russia has focused on its own world region, it has also continued to assert itself on a global scale as well. This is perhaps best seen in the increased attention that Russia has given its relationships with China, Brazil, and India in recent years (**FIGURE 3.32**). Russia's dominant role in the

▲ **FIGURE 3.32 Leaders of five emerging "BRICS" economies** Russian President Vladimir Putin, Indian Prime Minister Narendra Modi, Brazilian President Dilma Rousseff, Chinese President Xi Jinping and South African President Jacob Zuma—pose for ceremonial photos in 2014. The five countries, known as BRICS, agreed on the establishment of a bank to support infrastructure development in developing countries.

BRICS (see Emerging Region, Chapter 7) has been hampered by international sanctions as well as the economic consequences of international sanctions and the rapid decline in global oil prices. That has not stopped Russia and the BRICS from announcing their own development bank initiative in 2014.

Apply Your Knowledge

3.11 Identify three key political conflicts in the region and two factors contributing to each of these conflicts.

3.12 Analyze the role that the Russian Federation played in the regional conflict as well as different ethnonationalist movements.

Russia's Role in the World

https://goo.gl/E8QRez

Culture and Populations

Within the compass of this vast world region, there exists a great deal of cultural and political diversity. The religious and linguistic geographies of the Russian Federation, Central Asia, and the Transcaucasus have been shaped by the influence of past movements of people into the region from Europe, the Middle East, South Asia, and East Asia. Since the region's exposure to global economic, political, and cultural systems in the post-Soviet period, traditional patterns of ethnicity and culture have become increasingly important as the basis for identity.

Religion and Language

Religion and language together form key aspects of regional culture for the people of Russia, Central Asia, and the Transcaucasus.

Religion Russians of all religions have enjoyed freedom of worship since the collapse of the Soviet Union; large numbers of religious buildings that were abandoned or converted during the rule of the officially atheist regime have been returned to active religious use. About 20% of the people in the Russian Federation still profess no religious affiliation, however—a similar figure to Germany, Italy, Spain, and the United Kingdom.

The dominant religion in Russia, professed by about 75% of citizens who describe themselves as religious believers, is Eastern Orthodox Christianity. The emergence of Eastern Orthodoxy dates from the 11th century and the "Great Schism" between Rome and Constantinople, which led to the separation of the Roman Catholic Church and the Eastern Byzantine Churches. Eastern Orthodoxy adheres to ancient traditions and practices rooted in Greek, Slavic, and Middle Eastern traditions.

In Russia today, another 10% of the population self-identify as Muslim. They are concentrated among the ethnic minority nationalities located in the north Caucasus between the Black Sea and the Caspian Sea and in the larger cities of European Russia. Because of higher birthrates, Muslims account for the fastest-growing religious grouping in Russia.

In Central Asia and Transcaucasus, the Sunni branch of Islam is the dominant religious group, dating from the 8th century C.E., when the Arab Caliphate defeated and gained control of the region. In Azerbaijan, about 95% of the population is nominally Islamic. In Georgia, on the other hand, Eastern Orthodoxy is the dominant religion, claimed by about 80% of the population.

Language Dominant today throughout Belarus and the Russian Federation are Slavic people, among whom Russians represent one particular ethnic group. The Slavs are fundamentally defined by linguistic commonalities in the Slavonic group, including Russian and Belarusian, languages written using the Cyrillic alphabet. Slavonic-speaking people correspond to the most densely settled parts of the region, extending along the steppe to the east (**FIGURE 3.33**).

A second major language group is that of Turkic languages, including Azeri and Bashkir, which belong to the Altaic family of languages. The Turkic languages are spoken by the people of Central Asia and parts of the Transcaucasus. Much of northern and eastern Siberia is occupied by people who speak other branches of the Altaic language group.

At the eastern edge of the region, Paleo-Siberian languages are spoken, an inheritance of the indigenous communities of the region, whose presence predates the settlement of Russian people. Finally, there are several smaller areas of Caucasian languages: Abkhaz and Chechen on the northern slopes of the Caucasus and Georgian and Dagestani in the Transcaucasus.

Apply Your Knowledge

3.13 List some of the significant religious, linguistic, and ethnic minority communities of this region.

3.14 What challenges may these communities face in the recently formed nations of the region.

Cultural Practices, Social Differences, and Identity

In the wake of the dramatic transition from state socialism to capitalism, cultural practices in the region have been adapting and contributing to globalization. In some ways, this is simply a resumption of the emerging cultural practices of the pre-Soviet era, when Western, Byzantine, and Eastern influences all contributed to the cultural makeup of the region, and the folkways of Russia and the Transcaucasus found their way into the West through art, literature, and symphonic music.

High Culture The influence of European culture in Russia during the 17th and 18th centuries brought Russian high

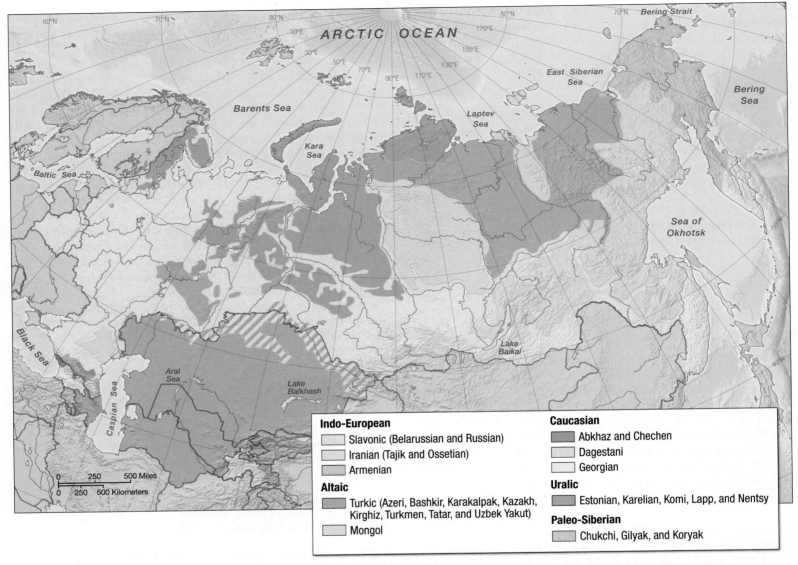

Indo-European
- ▢ Slavonic (Belarussian and Russian)
- ▢ Iranian (Tajik and Ossetian)
- ▢ Armenian

Altaic
- ▢ Turkic (Azeri, Bashkir, Karakalpak, Kazakh, Kirghiz, Turkmen, Tatar, and Uzbek Yakut)
- ▢ Mongol

Caucasian
- ▢ Abkhaz and Chechen
- ▢ Dagestani
- ▢ Georgian

Uralic
- ▢ Estonian, Karelian, Komi, Lapp, and Nentsy

Paleo-Siberian
- ▢ Chukchi, Gilyak, and Koryak

▲ **FIGURE 3.33 Languages of the Russian Federation, Central Asia, and the Transcaucasus** More than 100 languages are spoken in the region, the majority by very small ethnic groups and hence unrecordable on any but the most detailed maps. **To what degree do national borders match or not match the boundaries between major language groups?**

culture closer to the traditions of Western Europe. But uniquely Russian artistic styles also developed, influenced both by liberal social reforms and inspired by Russian rural folk culture. In the performing arts there was the work of composers such as Tchaikovsky, Rachmaninoff, Prokofiev, Stravinsky, and Shostakovich, together with the ballet impresario Sergei Diaghilev and the dancers Vaslav Nijinsky and Anna Pavlova (**FIGURE 3.34**). Equally influential in both Russia and the West were the novels of Dostoevsky and Tolstoy, the stories and plays of Chekhov, and the poetry of Pushkin.

Popular Culture The 1917 Revolution transformed these cultural dynamics. In Russia itself, the trajectories of both high culture and mass culture were incorporated into revolutionary ideology and gave birth to new cultural forms. They produced political posters with a distinctive genre of graphic design and developed a unique form of modernist architecture—Constructivism—that emphasized the importance of industrial power, rationality, and technology.

Popular culture in the post-Soviet era includes a range of practices that together represent the hybrid history of the region. Notably, the food traditions of Russia are a product of its intricate connections both to more western parts of Europe as well as Central and East Asia.

Consider Russian *pirozhki*. Pirozhki are small pies, stuffed with savory meats that come in diverse forms. For example, they are either baked, in the Slavic method, or fried, in a way introduced by the Tatars in the 1500s. Examples such as this one illustrate the ways in which traditional popular culture, though distinctive of the region, also reflects a complex mix

▲ **FIGURE 3.34 Russian Modern Art** Visitors attend exhibits at the Cosmoscow International Contemporary Art Fair in Moscow. Long a source and destination for artistic innovation, Russia has seen a resurgence of art markets in the last three decades.

▲ **FIGURE 3.35 *Pirozhki* Vendor in Moscow** Vendors selling *pirozkhi* in Moscow mix food traditions adapted from the wider region into the city's modern-day street food culture.

of cultures that have been connected through political history (**FIGURE 3.35**).

Since the breakup of the Soviet Union, popular culture has gone through an enormous upheaval, thanks to the pressures of market forces and a flood of cultural imports from Europe, the United States, and elsewhere. The advent of new television programming and different ways of thinking about issues like class, identity, and sexuality opened up post-Soviet society, especially in western Russia and the larger cities. While Russians at first imported a great deal of popular culture from the West, it did not take long before a hybridized, home-grown popular culture began to emerge. No longer do American television shows and Brazilian and Mexican soap operas dominate. People now are more inclined to watch domestic programming, where their own lives are portrayed. Similarly, thousands of rock groups of all kinds—hard, soft, punk, art, folk, fusion, retro, and heavy metal—now flourish in Russia and the Transcaucasus.

The freedom of these media are under constant threat from state control, however. In 2014, Russian President Vladimir Putin signed a law that curtails foreign ownership of Russian media and extends Kremlin control over television and newspapers.

Apply Your Knowledge

3.15 Define high and popular culture in Russia by giving examples of both. How do you see the changing political structure as influencing cultural directions?

3.16 Analyze how Russian arts are 1) an expression of a complicated national identity and 2) a criticism of Americanized global culture by searching and reading "Cultural struggle to define Russian identity," at http://www.bbc.com.

Russian Identity

https://goo.gl/6jVPMq

Social Stratification and Conspicuous Consumption

Clearly, many aspects of life have changed significantly since the breakup of the Soviet Union. Restaurants, for example, were not highly developed under socialism, but the post-Soviet period has seen an explosion of restaurants, cafés, and fast-food places in cities. The majority of people do not frequent restaurants often, mainly for economic reasons, but for the new business classes, dining out is part of a new pattern of conspicuous and competitive consumption. This, in turn, reflects the emergence of new class factions with distinctive identities. Although they had special privileges, most officials in the Soviet system did not accrue wealth. As discussed previously in the chapter, privatization has allowed some of them to become oligarchs by building large fortunes and taking advantage of insider status to acquire a share of direct ownership of state resources and industries. A new

Changing Fortunes for Central Asian Migrants to Russia

On any given day since the breakup of the Soviet Union, people from countries such as Uzbekistan have migrated to Russia seeking work. In the Soviet era these people would have been traveling under one passport and in one country, but in the post-Soviet era the migrants of Uzbekistan are international ones. For many people there are both push factors out of Uzbekistan and pull factors into Russia (**FIGURE 3.4.1**). The poor economy of Uzbekistan forces people to seek work elsewhere, while the growth of the extractive economy of Russia has pulled people there for work. It is estimated that almost 50% of the entire wealth of Uzbekistan results from labor migrants, who remit (or send home) millions of Russian rubles a year to their families (**FIGURE 3.4.2**).

The rapid decline of the Russian ruble, a result of international sanctions and collapsing oil prices in 2015, is having dramatic effects on people like Sabor, who now think of staying home and finding work instead of migrating to Russia. Other Uzbek migrants are leaving because salaries have dropped precipitously over the last several years. Russian policy is another factor impacting the decisions of people to migrate to Russia as guest workers. New regulations are tightening restrictions on international labor migration. Before someone can work in the country, they must obtain permits whose cost has doubled and take exams that demonstrate Russian language fluency and knowledge of Russian history. The international flows of people in this world region are likely to continue to change dramatically, as Russia's economy changes and its policies relative to international migration also change.

1. Create a list of total remittance value from Russia per country from least to greatest. Which countries have the highest remittances from Russia? Which have the lowest?

2. Briefly explain why remittances are so different across the region? For further information go to www.theguardian.com and search for articles on Russia and remittances.

Migrant remittance inflows from Russia

https://goo.gl/mukpUL

"The poor economy of Uzbekistan forces people to seek work elsewhere, while the growth of the extractive economy of Russia has pulled people there for work."

▲ FIGURE 3.4.1 **Russia Migration Policies** Migrant worker from Uzbekistan displays a labour licence issued for him at the Moscow Region's United Migration Centre.

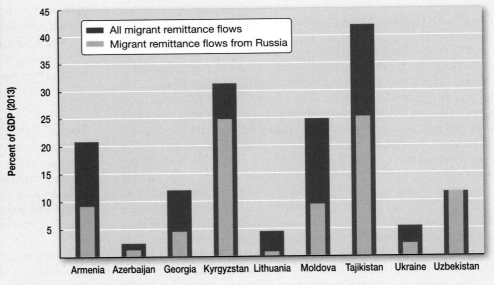

▲ FIGURE 3.4.2 **Measuring Remittances** Remittances, particularly from Russia, remain an important component of the Central Asian economy. Some countries remain heavily dependent on these monies as a percentage of their overall GDP.

entrepreneurial class has also emerged, some of whose members have become significantly wealthy. More slowly, an affluent middle class is emerging in the cities, formed of an educated elite newly employed in business ventures and midlevel management. Most of the rest of the population, meanwhile, remains relatively impoverished.

Gender and Inequality

Amid this upheaval, the role of women has changed significantly. Women, together with all other groups discriminated against under the tsars, were "freed" by the 1917 Bolshevik Revolution, which declared them equal and granted them social and political rights. In theory, women were to be trained for and encouraged to take up what was previously male-only labor, such as operating agricultural machinery, working in construction, and laying and maintaining roads and rail beds. Nurseries and day-care centers were established to liberate women from child rearing. Women's increased participation in medicine, engineering, the sciences, and other fields was supposedly encouraged. In practice, however, the Soviet political system remained male-dominated.

The post-Soviet transition has opened up some opportunities for women but closed down many others (**FIGURE 3.36**). In recent decades, many women have taken up new opportunities presented by the transition to a market economy, venturing into small trade and opening their own businesses. Some women have also taken advantage of opportunities in newly established firms, quickly climbing through the ranks to become managers.

The transition to new market economies has also cost women many of the benefits they enjoyed under state socialism—such as child care, health care, equal pay, and political representation. Legal restrictions have been placed on the kinds of jobs women are allowed to do in the new Russia, for example. As of 2012, there were 460 jobs that were legally off-limits to women, including firefighting and driving metro subway trains. Women's wages are, on average, only 64% of those of men, as of 2010. Between 1985 and 2005, the number of working women in the Russian Federation fell by 24%. Many women have found themselves forced into unskilled work to make ends meet while caring for children and keeping their family together. Often, this means working in the unprotected (and sometimes illegal) realm of the informal economy. A great deal of media attention has been given to the fact that tens of thousands of women have been forced into prostitution, often after being trafficked abroad on the pretense that they would work as maids or waitresses.

Finally, women of the region have clearly regressed in political representation. Under state socialism, quotas ensured that one-third of the seats in parliament went to women. In 2014, only 14% of the seats in the Russian Parliament were held by women (consider, however, that in the United States that figure is only 18%).

Demography and Urbanization

Owing to the enormity of the region, population density is relatively low (**FIGURE 3.37a**). With a total population in 2011 of some 232 million and almost 14% of Earth's land surface, the Russian Federation, Central Asia, and the Transcaucasus contains about 4% of Earth's population at an overall density of only 48 persons per square kilometer (124 people per square mile). Within the Russian Federation, population density is relatively higher in the heavily settled Central Region, falling off to less than one person per square kilometer in the far north, Siberia, and the far eastern parts of the region.

Levels of urbanization reflect this same broad pattern. Most large cities are in the European part of the Russian Federation and in the Urals. These include Moscow, Nizhniy Novgorod, St. Petersburg, Volgograd, and Yekaterinburg. Most of the other cities of any significant size are found in southern Siberia, on or near the Trans-Siberian Railway, a network of railways connecting Moscow with the Russian Far East and the Sea of Japan. Overall, both Belarus and the Russian Federation are quite highly urbanized, with 72% and 76% of their total populations living in cities, according to their respective census counts in the mid-1990s. The populations of the Transcaucasus are moderately urbanized (56% to 69% living in cities), whereas those of Central Asia are more rural (only 30% to 50% living in cities).

Demographic Change This world region has a relatively slow-growing population and even areas of population decline. Throughout the 20th century, there was a general decline in both birth- and death rates (**FIGURE 3.37b**). Since the 1990s, the population began to register a decline as a result of more deaths than births.

▲ **FIGURE 3.36 Women in Today's Russia** Members of Pussy Riot, a Russian radical feminist punk rock group, protest the policies of Vladimir Putin in Moscow in January 2012. In March 2012, three members of the band were arrested for their participation in a protest against the government held in a Russian Orthodox Church. In August of that year they were sentenced to two years in prison. Band members wear masks to draw attention to women's roles in Russia, which they view as more passive and anonymous than those of men.

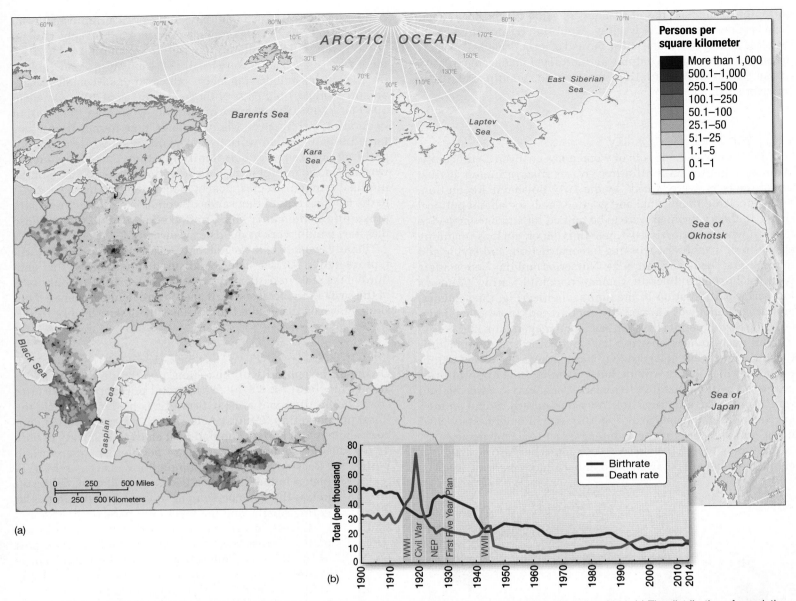

(a)

(b)

▲ **FIGURE 3.37 Population Density and Vital Rates in the Russian Federation, Central Asia, and the Transcaucasus, 2011** (a) The distribution of population in the Russian Federation, Central Asia, and the Transcaucasus reflects the region's economic history, with the highest densities in the industrial regions of the western parts of the Russian Federation and the richer agricultural regions of the Transcaucasus. (b) This graph shows the dramatic drop in the birthrate in Russia that characterized the 1960s and 1990s. It also shows the slight rise in birthrates and rise in death rates after 1990. **How might demographic decline impact the region's economy? What areas might be hardest hit by such change?**

The structure of populations in Kyrgyzstan and the Russian Federation (shown in **FIGURE 3.38**), however, reveals that demographic change varies across the region. While the Russian population pyramid is marked by the massive and distinctive population losses associated with World War I and World War II that of Kyrgyzstan shows an overall smooth widening base more characteristic of a developing country. The Russian pyramid also reflects a declining growth rate in recent years. In Russia, this is a result of a combination of things, including the legalization of abortion, a greater propensity to divorce, deferral of marriage among the rapidly expanding urban population, and a growing preference to trade parenthood for higher levels of material consumption. Kyrgyzstan's socioeconomic conditions are different,

however. Here, almost 40% of the population is under the age of 14 years, owing to more traditional early marriage age and a more rural population.

With the collapse of the Soviet system in 1989 and the subsequent social and economic upheaval that followed, death rates in Russia began to rise. Industrial production in Russia has fallen in the last few decades, hyperinflation has devalued people's savings, and a deep crisis of national finance has followed. The consequences for many parts of the Russian population included rising mortality rates, especially among older men. The higher mortality rates are attributed in part to the stresses surrounding economic dislocation, in part to increasing poverty, and in part to the decline in the provision of health care in post-Soviet

Kyrgyzstan

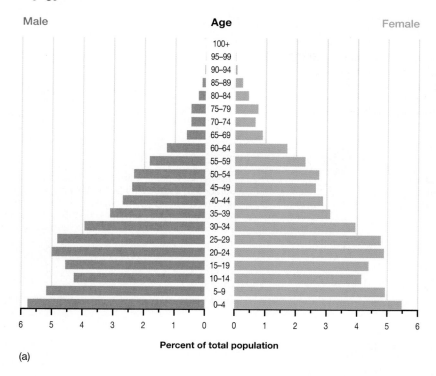

(a)

Russian Federation

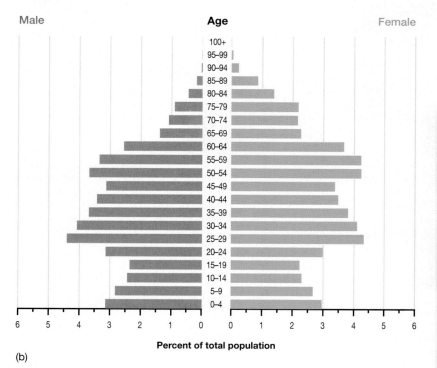

(b)

▲ **FIGURE 3.38 Age–Sex Pyramids (a) Kyrgyzstan and (b) the Russian Federation, 2014** The profile of two countries' populations show many differences, including the effects of World War II (the relative lack of men and women in their early 60s and the reduced number of men aged 70 years and older) and (a) recent reduced birthrates in Russia versus (b) the overall growth trends in Kyrgyzstan. **What other differences stand out and what might account for them?**

Russia. There was also an increase in the rate of industrial accidents and alcohol-related illnesses and a rapid escalation of infant mortality rates.

This has brought the life expectancy of Russians to levels nearly identical to those of Kyrgyzstan, which has historically had much lower rates. According to the World Health Organization in 2012, life expectancy for women in Russia was 75 and for men only 63. This compares to a female life expectancy of 73 for Kyrgyz women and 63 years for Kyrgyz men. In this sense, the two populations are becoming more similar in the post-Soviet era.

Global and Regional Migration Streams

In the late 19th and early 20th centuries, many Russians joined the stream of emigrants headed toward North America. Concentrations of Russian immigrants developed in Chicago, New York, and San Francisco. Over a quarter of these immigrants settled in New York City, the majority in Brooklyn, where distinctive Russian communities, such as the Brighton Beach neighborhood, are vital nodes in the Russian global diaspora (**FIGURE 3.39**).

In the last decade, more people have migrated into Russia than out (see Visualizing Geography: "The Russian Diaspora and Post-Soviet Migration" on pp. 124–125). Migration has come to include a far wider range of nationalities and ethnicities, including not only Slavic speakers, but also members of Central and East Asian ethnicities (**FIGURE 3.40**). This trend is driven by labor demands and opportunities from economic growth in the Russian Federation. The economic changes impacting the Russian Federation may be reversing these trends. For example, between January 2014 and January 2015 the number of migrants to Russia dropped

▲ **FIGURE 3.39 Russian Communities in Brighton Beach, Brooklyn, New York** Brighton Beach is called "Little Odessa" by locals. Several generations of Ukrainians from the city of Odessa have settled there, and the area has numerous Russian markets and restaurants.

The Russian Diaspora and Post-Soviet Migration

I n 2006, President Putin enacted a formal policy encouraging Russians to repatriate from the former republics. This led to a dramatic increase of ethnic Russians migrating back into the Federation. The reason for this repatriation is tied to the history of Russian migration during the periods of both the Russian Empire and the Soviet Union when ethnic Russians had migrated out of their historical core area in and around Moscow to the far reaches of the Russian Empire and the Soviet State.

1 Russian Empire Migration

The spread of the Russian Empire from its hearth in Muscovy took Russian colonists and traders to the Baltic, Finland, Ukraine, most of Siberia, the eastern parts of the region, and parts of Central Asia and the Transcaucasus.

2 Soviet Empire Migration

During the rise of the Soviet empire, many ethnic Russians were directed and encouraged to settle in the Baltic, Ukraine, Siberia, the east of Russia, Central Asia, and the Transcaucasus. This policy was partly to further the Stalinist ideal of a transcendent Soviet people and partly to provide workers needed to run the mines, farms, and factories.

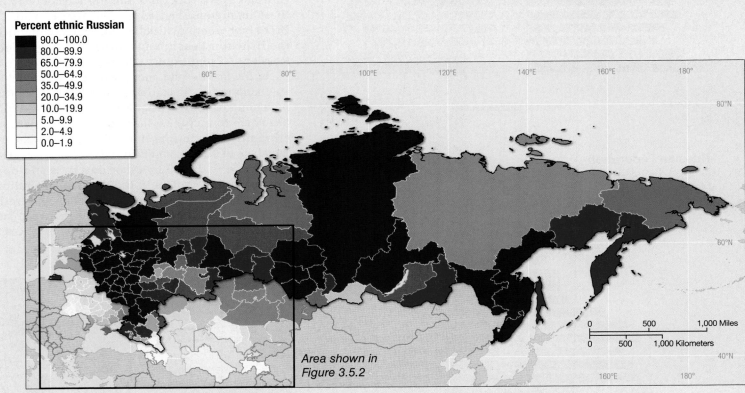

▲ **FIGURE 3.5.1 Russia Diaspora. The percentage of ethnic Russians throughout the former Soviet Union at the time of the last census in each of these countries.**

3 The Post-Soviet Russian Diaspora

By the time of the breakup of the U.S.S.R., 80% or more of the population of Siberia and the eastern parts of the region were Russian, and the Russian diaspora had become very pronounced in most of the Soviet Union's successor states beyond the borders of the Russian Federation (Figure 3.5.1).

"With the dissolution of the Soviet Union, some 25 million Russians found themselves to be ethnic minorities in newly independent countries."

4 Return of the Russian Diaspora

With the dissolution of the Soviet Union, some 25 million Russians found themselves to be ethnic minorities in newly independent countries. The largest group was in Ukraine, where more than 11.3 million Russians made up 22% of the population, with clear political implications for regions like Crimea. In the Transcaucasus, where the proportion of Russians was generally lower than elsewhere, strongly nationalistic governments of the successor states quickly enacted policies that encouraged Russians to leave (Figure 3.5.2). Russians have also left former Soviet states for economic reasons. For example, as the economies of the former

Soviet republics were cut off from the larger Russian state, employment opportunities for ethnic Russians decreased. Moreover, the Russian government has encouraged Russian repatriation for economic reasons – to provide jobs in Russia to Russians first. This has been enforced through new labor legislation, which requires that foreign workers in Russia – except for those in highly valued "expert classes" –speak Russian and can pass a Russian history exam before entering the country to work.

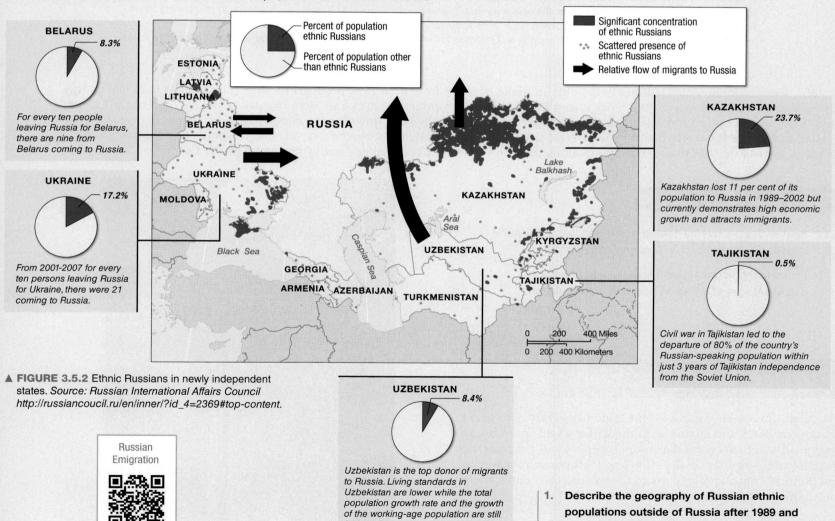

BELARUS *8.3%*

For every ten people leaving Russia for Belarus, there are nine from Belarus coming to Russia.

UKRAINE *17.2%*

From 2001-2007 for every ten persons leaving Russia for Ukraine, there were 21 coming to Russia.

Percent of population ethnic Russians

Percent of population other than ethnic Russians

Significant concentration of ethnic Russians

Scattered presence of ethnic Russians

Relative flow of migrants to Russia

KAZAKHSTAN *23.7%*

Kazakhstan lost 11 per cent of its population to Russia in 1989–2002 but currently demonstrates high economic growth and attracts immigrants.

TAJIKISTAN *0.5%*

Civil war in Tajikistan led to the departure of 80% of the country's Russian-speaking population within just 3 years of Tajikistan independence from the Soviet Union.

UZBEKISTAN *8.4%*

Uzbekistan is the top donor of migrants to Russia. Living standards in Uzbekistan are lower while the total population growth rate and the growth of the working-age population are still fairly high.

▲ **FIGURE 3.5.2** Ethnic Russians in newly independent states. *Source: Russian International Affairs Council http://russiancoucil.ru/en/inner/?id_4=2369#top-content.*

Russian Emigration

https://goo.gl/eUiWE5

1. **Describe the geography of Russian ethnic populations outside of Russia after 1989 and compare that geography to what we see today.**

2. **What are the factors that have motivated Russians to migrate back to Russia in the post-Soviet era?**

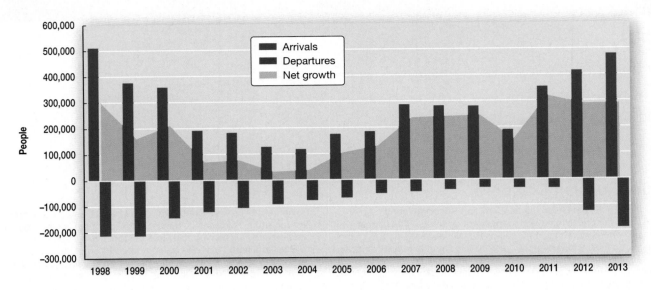

▲ **FIGURE 3.40 Trends in Russian Migration** During the post-Soviet transition of the 1990s, emigration from Russia was enabled by increasing freedom of movement under the new government, whereas immigration to Russia consisted heavily of returning ethnic Russians. After a lull in migration activity during the economic downturn of the early 2000s, Russia has become a destination for labor migrants from across the former Soviet states. **Predict how the fall in oil prices that began in 2013 and continued into 2015 could affect migration trends.**

by 70%, even as President Vladimir Putin puts in place new approaches for attracting migrants to move to and live in Russia. This includes the option for foreign nationals to serve in the Russian military, a relatively lucrative livelihood that is not as severe as some of the jobs in the country's labor sector.

Apply Your Knowledge

3.17 Summarize the overall demographic trends for this world region, especially for the Russian Federation.

3.18 Evaluate the economic and social advantages and disadvantages of an aging or shrinking population.

Future Geographies

The range of possible futures for the region is starkly divergent. On the one hand, interaction with the wider world could bring very positive changes, such as more complex and progressive societies in the region. On the other hand, economic integration and political consolidation portend increasing social and spatial inequality.

Can Russia Conquer Its Energy Dependency?

Russia's recent economic growth has largely been the result of windfall profits from sustained increases in oil and commodity prices. Sixty-eight percent of Russia's total export revenues came

from oil and natural gas in 2013. The freefall in global energy prices in 2013 and 2014 caused Russia's exports to plummet, however, throwing the country into economic distress. Because of Russia's dependence on energy exports, its economic future is linked to global energy demand (**FIGURE 3.41**).

▲ **FIGURE 3.41 Russian Natural Gas Development** Pictured here is a natural gas extraction pipeline located in Tyumen, Russia, which is about 2,200 kilometers (1,367 miles) from Moscow. This pipeline, which is owned by the gas monopoly, Gazprom, is tied into a sophisticated network of lines that bring natural gas to Russia as well as Kazakhstan and other parts of Central Asia and the Middle East.

Democracy or Plutocracy?

As of 2014, the number of Russian billionaires is 111, while Russia's level of income inequality puts it on par with Nigeria. The growth of economic elites presents challenges for a growing democracy. Though many of these oligarchs act to counterbalance traditional political blocks, they also maintain their fortunes through collusion and corruption. The role of this new class of plutocrats in political change is unquestionably critical for deciding the region's future but equally unpredictable (**FIGURE 3.42**).

Will Russia's Pacific Rim Be an Economic Boom?

Russia's leaders and planners ignored the sleepy ports cities of Russia's Far East, between eight and ten time zones east of Moscow, following the breakup of the Soviet Union. With the growth of the Pacific economy (see Chapter 8), however, new investments have pushed this region forward, including a $2 billion casino, $200 million airport, and $30 billion satellite city for auto construction in Vladivostok alone. Whether these will kick-start the economy of Russia's Far East remains unclear (**FIGURE 3.43**).

▲ **FIGURE 3.42   Russian Oligarchs Enjoy Strong Government Connections**   Russian billionaires, Arkady Rotenberg, Boris Zingarevich, Mikhail Zingarevich, play in an ice hockey match with President Vladimir Putin, who was celebrating his 63rd birthday.

Demographic Challenge—Implosion and Decline?

The populations of Russia and Belarus are aging dramatically. Between now and 2025, their populations are expected to decline by as much as 12%. By 2025, between one-fifth and one-quarter of the populations in nine eastern European and former Soviet Union countries will be aged 65 years and older. Ultimately, the labor force may be insufficient for the size of the economy unless immigration is significantly increased (**FIGURE 3.44**).

▲ **FIGURE 3.44   Elderly Russian Women**   Older residents of Moscow discuss the punishing winter weather in a downtown department store. As Russia's population ages, new concerns arise about the size of the workforce, the economic implications of a dependent population, and issues of health and mortality.

◀ **FIGURE 3.43   Vladivostok** Originally founded by Czar Alexander II in 1860 as Russia's Pacific port, massive reinvestments in infrastructure have breathed new life into historically a sleepy industrial city.

Learning Outcomes Revisited

▶ *Describe* the physical geography of the Russian Federation, Central Asia, and Transcaucasus region.

This region's climate is influenced by its mostly northern location, its position at the eastern edge of Europe and inland Asia, its paucity of mountainous terrain (except in the far south and east), and the lack of any significant moderating influence of oceans and seas. Three climate regions dominate the landscape from the arid and semiarid regions in the south to the continental/midlatitude and polar regions to the north. As climate change opens the northern areas of the region, previously unusable resources and transportation routes in the Arctic will become increasingly critical to regional growth and change.

1. *List* the geographic factors that explain the typical climates of Central Asia and Russia.

▶ *Predict* how climate change and environmental impacts of human activities may influence the region's future sustainability.

Climate change is having a dramatic impact on this region's permafrost as well as impacts on water availability in parts of Central Asia. The challenges for the future of the region will be how to manage the decrease in freshwater in Central Asia and the changing landscape of the northern regions, where permafrost decline could lead to real problems for animal life, transportation systems, and economic activity in the region.

2. *Describe* two main problems that climate change may cause in the region.

▶ *Understand* how people have adapted to and modified the physical landscapes of this world region.

People have adapted to the varied environments of this world region by creating economic and social systems that allow them to take advantage of local resources. This includes the people of Siberia, who have adapted to the harsh environments of the tundra by herding animals, and the people of Central Asia, who have learned to manage the large expanse of plains through the domestication of key crops and animals. Recent changes in global climate are having an impact on local practices, however, as the northern reaches of this world region warm and permafrost melts. Moreover, the people of Central Asia must manage the overuse of precious water resources in the region, which include the draining of the Aral Sea. At the same time, changes in climate are creating new opportunities for Russia, in particular, to exploit the resources of the Arctic Ocean.

3. *Explain* the main physical challenges the region's landscapes posed to human settlement.

▶ *Describe* how economic transitions in the region are playing out in the post-Socialist era.

After more than seven decades under the Soviet system, the Russian Federation, Belarus, and the Soviet Union's other successor states are now experiencing transitions to new forms of economic organization and new ways of life. An economic boom has ensued throughout the region propelled by a deep and extensive natural resource base. Growing economic inequality and the rise of economic elites pose challenges to democracy, however.

4. *Compare* and contrast the economies of the former Soviet Union with those of its successor states.

▶ *Analyze* the changing political trends in the region in terms of fragmentation and realignment.

Some of the old ties among the countries of the region have been weakened or reorganized, and patterns of regional interdependence have been disrupted and destabilized. All of the new, post-Soviet states have joined the capitalist world system, and all of them have to find markets for uncompetitive products while at the same time engaging in domestic economic reform. The Russian Federation, as the principal successor state to the Soviet Union, remains a world power with a large standing army and a formidable sophisticated arsenal, including nuclear weapons; a large, talented, and discontented population; a huge wealth of natural resources; and a pivotal strategic location in the center of the Eurasian landmass.

5. *Discuss* Russia's annexation of Crimea in terms of geopolitical "fragmentation and realignment" in countries of the former Soviet Union.

▶ *Summarize* the cultural legacies and new developments in the region in terms of high, revolutionary, and popular forms of culture.

Since the breakup of the Soviet Union, popular culture has gone through an enormous upheaval, thanks to the collapse of censorship, the pressures of market forces, and a flood of cultural imports. New television programming and different ways of thinking about issues like class, identity, and sexuality opened up post-Soviet society, especially in western Russia and the larger cities. But it did not take long before a hybridized, homegrown popular culture began to emerge.

6. *Describe* how official policies and governance under President Vladimir Putin have affected media and popular culture in Russia.

▶ *Discuss* and analyze the implications of demographic change in the region.

Although many parts of Central Asia have growing populations, in the core of the region, which includes the Russian Federation, death rates have risen higher than birth rates. This presents problems for maintaining the labor force, which may depend much more heavily on in-migration into the region in the future. This trend toward increasing immigration into the region reverses the historical trajectory of population mobility, which has historically been dominated by out-migration.

7. *Identify* the factors that have affected birthrates and death rates in the Russian federation since the collapse of the Soviet Union.

Key Terms

apparatchik (p. *107*)
aridity (p. *93*)
Bolshevik (p. *105*)
civil society (p. *116*)
cold war (p. *108*)
collectivization (*Kolkhoz*) (p. *107*)
command economy (p. *107*)
Commonwealth of Independent States (CIS) (p. *116*)
ethnonationalism (p. *115*)

geosyncline (p. *95*)
glasnost (p. *114*)
internationalist (p. *105*)
Joseph Stalin (p. *105*)
khan (p. *102*)
mafiya (p. *110*)
Near Abroad (p. *116*)
nomadic pastoralist (p. *101*)
Northwest Passage (p. *96*)

oligarch (p. *110*)
permafrost (p. *93*)
pluralist democracy (p. *116*)
privatization (p. *100*)
serfdom (p. *104*)
Silk Road (p. *102*)
soviets (p. *105*)
state socialism (p. *105*)
steppe (p. *100*)

taiga (p. *99*)
territorial production complexes (p. *108*)
tsar (p. *104*)
tundra (p. *98*)
Union of Soviet Socialist Republics (U.S.S.R.) (p. *105*)
Vladimir Ilyich Lenin (p. *105*)
zone of alienation (p. *108*)

Thinking Geographically

1. Describe two examples of how recent trade linkages with global commodity markets have created or exacerbated local and regional environmental problems in Russia and Central Asia.

2. What role did the fur trade play in the expansion of Russia?

3. How did the establishment of the Soviet bloc aid development of the Soviet Union following World War II? Discuss technical optimization, industrialization, and military security.

4. What factors led to the breakup of the Soviet empire?

5. Discuss the environmental degradation of the Aral Sea and compare and contrast it to the problems associated with the region's nuclear landscapes.

6. How have national identities been asserted in the decade since the Central Asian republics became independent countries? What cultural factors serve to unify or separate the states in this region?

DATA Analysis

In this chapter, we have examined the political, social, environmental, and economic factors that led to the creation and later dissolution of the U.S.S.R. We have also analyzed the effects of the U.S.S.R.'s legacy on all the countries in this region today. Among these issues, a stable economy has been a consistent challenge for the Russian Federation and many of the nation-states in Central Asia and the Transcaucasus. Compare the current economic prospects for the Russian Federation and two former members of the U.S.S.R., Belarus and Turkmenistan, with data from the World Bank.

1. Begin by accessing the databank for each country: http://data.worldbank.org/country

2. What is the annual "GDP growth," the "poverty head count," the "life expectancy," and the "school enrollment" rate for each country: Belarus, Russian Federation, and Turkmenistan?

Russia's Economy

https://goo.gl/sMjUzL

3. Comparing these numbers, which nation-state is showing the healthiest economy?

4. Using what you have studied from this chapter, what environmental and social factors might explain this data?

5. Scroll to the bottom of each country's data page to the "Projects and Operations" maps. What kinds of projects have been developed in each of the three countries?

6. Finally, examine the infographic from, "Russian Economic Report 33: The Dawn of a New Economic Era?" from http://www.worldbank.org Predict how each of these nation-states is going to deal with projected negative economic growth in Russia.

MasteringGeography™

Looking for additional review and test prep materials? Visit the Study Area in MasteringGeography™ to enhance your geographic literacy, spatial reasoning skills, and understanding of this chapter's content by accessing a variety of resources, including MapMaster interactive maps, videos, *In the News* RSS feeds, flashcards, Web links, self-study quizzes, and an eText version of *World Regions in Global Context*.

The City of Dubai is one of the world's largest ports and a global hub of tourism as well as economic development. Dubai Marina is one of the city's centers of activity even into the early evening.

4

Middle East and North Africa

On the coast of Dubai rests an enclave called Saadiyat Island. The island is a symbol of Dubai's economic growth and diversity. Once completed, it will be home to the first ever Louvre museum outside of France and a branch of the Guggenheim museum. New York University, a leading private university in the United States, has also invested in this new island enclave, building a branch campus there as well. Not unlike the rest of Dubai, this dynamic urban space filled with luxury hotels and high-class amenities symbolizes the transition of both urbanism and economic development in the Middle East. With massive investments in the culture and knowledge economies, cities such as Dubai are becoming global hubs for tourism as well as investment in 21st-century technological change. These emerging cities are also being built with sustainable growth in mind, taking advantage of the latest technologies to increase the efficient use of resources.

The Dubai Plan 2021 for urban growth and development is a utopian vision of a new urban society that is aligned with a global "Smart City" movement focused on inclusiveness, sustainability, and good governance. These visions belie the realities of the global labor workforce that has been hired to build this utopia. While some workers find themselves in state of the art labor housing dwellings, many more workers find themselves living on the outskirts of Dubai's massive urban landscape of wealth and consumption in makeshift housing at the margins of construction zones. As Dubai continues to strive to be a leader in the new urban movement of sustainable and inclusive cities, therefore, it is going to have to come to grips with its exploitative labor practices, which keep international workers from countries such as Bangladesh and Pakistan in substandard living and working conditions.

Learning Outcomes

▶ **Describe** the diversity of physical geographic environments in the Middle East and North Africa.

▶ **Evaluate** the importance of key resources, such as oil and water.

▶ **Summarize** the diffusion of cultural and environmental practices, such as religion and agriculture, throughout the region.

▶ **Explain** the natural and historical factors that contribute to the diverse economies of the Middle East and North Africa, how they vary from country to country, and how they are tied to global and regional processes.

▶ **Analyze** the geopolitical conflicts and tensions that bring instability to the Middle East and North Africa and the relationship between these conflicts and the political history of the region.

▶ **Identify** the differences in practices of kinship, gender, and religion in the region and how these factors affect the lives of women.

▶ **Describe** the patterns of migration in the Middle East and North Africa, the push and pull factors that are involved in migration patterns, and the effects these patterns have on the region's demography.

Environment, Society, and Sustainability

The Middle East and North Africa is an environmentally diverse world region. Pine forests, deserts, and grass plains all can be found in relatively close proximity. The region's plate tectonics has created highland environments and sources of water for some of the world's most historically significant river systems (**FIGURE 4.1**). Human adaptation to the environmental conditions of the region led to some of the earliest agricultural societies in the world, and more recent management of these environments has changed the region's physical landscape in striking ways.

Climate and Climate Change

Although there is variability in the environment of the Middle East and North Africa, the dominant climatic characteristic is **aridity** (**FIGURE 4.2**). Summers in the lowland areas of the region are extremely hot and dry, with daily high temperatures often climbing to 38°C (100°F). That said, highland and coastal areas experience more moderate daily summer temperatures and a predictable influx of visitors escaping the heat elsewhere (**FIGURE 4.3**).

Arid Lands and Mountain Landscapes Except in the coastal mountain areas, precipitation in this world region is generally low and highly variable. Nearly three-quarters of the region experiences average annual rainfall of less than 250 millimeters (10 inches). Scarce rainfall means the soils in the region tend to be thin and deficient in nutrients and most agricultural land must be irrigated. There are exceptions. Agriculturalists along the coastal plains and lowlands of Turkey and the floodplains of the Nile are able to take advantage of fertile soils along rivers. The Central Highlands of Yemen experience abundant rainfall in the summer thanks to the Indian Ocean monsoon system, which brings seasonal rains from the Indian Ocean to the southern part of the Arabian Peninsula.

> "Although there is variability in the environment of the Middle East and North Africa, the dominant climatic characteristic is aridity."

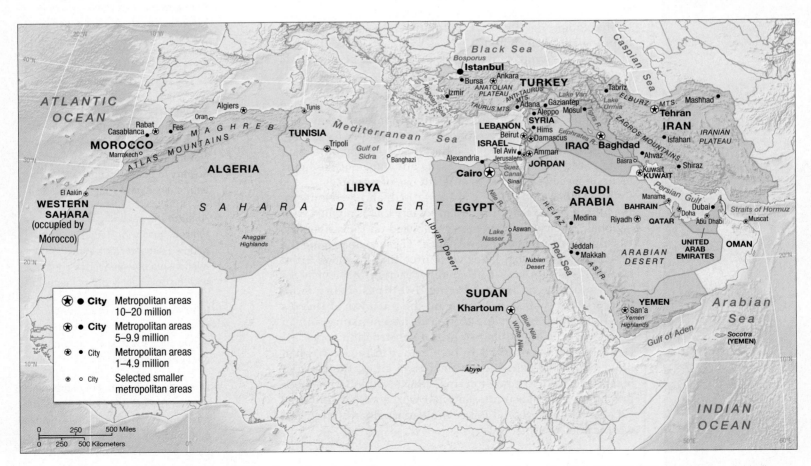

▲ **FIGURE 4.1 The Middle East and North Africa: Countries, major cities, and physical features** This world region stretches from the Atlantic to the Indian Oceans and borders some of the most important waterways in the world, including the Mediterranean, Red, and Arabian seas as well as the Persian Gulf. While many of these countries are quite large in terms of land area, populations tend to cluster near key river valleys, along coasts, and in the highlands of the region.

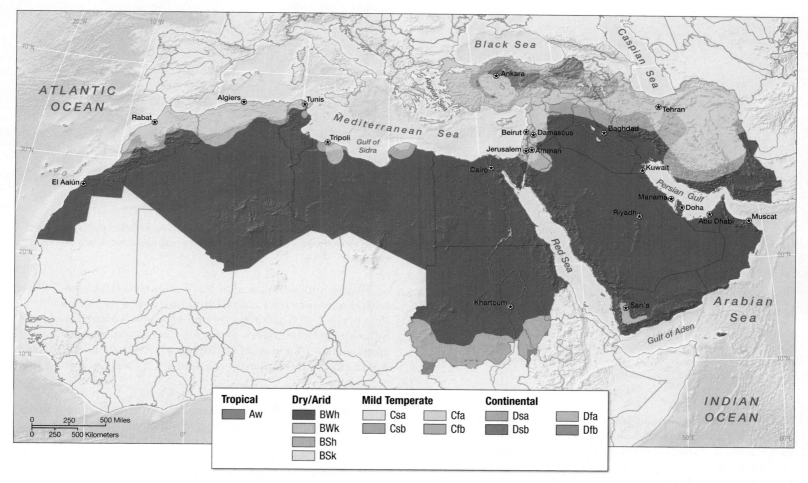

▲ **FIGURE 4.2 Climates of the Middle East and North Africa** The Middle East and North Africa is broad, stretching from the Mediterranean climates of northern Turkey to the wet, tropical climates of southern Sudan, but it is clearly dominated at midsection by a continuous swath of dry lands. **Briefly describe the geography of climate in this world region. What is the dominant climate region? Where do we see greater climatic diversity?**

▲ **FIGURE 4.3 Beach Resort in Israel** There are many coastal cities, such as Netanya, Israel, that cater to tourists from and beyond Israel.

Most of the precipitation in the region falls on the region's mountain ranges. Snowcapped peaks in Turkey, Iran, and Lebanon are the source of snowmelt that provides water for lowland populations (**FIGURE 4.4**). In some mountainous areas where the peaks are especially high, such as the central Anatolian Plateau in Turkey or the Syrian Plateau, a **rain shadow** effect occurs. The mountains cause moisture passing over them to condense and fall as rain before reaching the region's interior deserts. In contrast, coastal areas of the region can experience up to 1,000 millimeters (40 inches) of rain a year. Although most rain falls in the winter and early spring, some areas, such as the Black Sea slope of the Pontic Mountains in Turkey, experience summer rains adequate for **dry farming** techniques, which allow the cultivation of crops without irrigation.

Adaptation to Aridity People have adapted to the aridity and high temperatures characteristic of the region through architecture, dress, and patterns of daily and seasonal activity. Regional architecture features high ceilings, thick walls, deep-set windows,

▲ **FIGURE 4.4 Snowfall in the Atlas Mountains** The Atlas Mountains, with some peaks above 3,000 meters (9840 feet) provide much needed water in the region. These mountains are also home to pastoralists, who benefit from the cooler temperatures and precipitation to raise animals.

and arched roofs that enable warm air to rise. The practice of locating living quarters around a shady courtyard enables residents to move many activities to cooler outdoor spaces that are also very private (**FIGURE 4.5**). Clothing, such as head coverings and long robes, made from fabrics of light color lower body temperatures by reflecting sunlight and inhibiting perspiration while diminishing moisture loss. Some populations, such as the Berber of North Africa, migrate to mountainous areas in the summer and warmer lowlands in the winter to avoid temperature extremes.

▼ **FIGURE 4.5 Traditional-Style Courtyard in Qatar** Open spaces, such as this courtyard, are common throughout the region. They are important spaces for the family and they provide much needed access to cooler outdoor breezes.

▲ **FIGURE 4.6 Native Date Palm** Dates have been grown throughout the region for thousands of years. They are often planted in areas with easy access to water.

Plants and animals also have adaptive strategies to deal with the intense heat and aridity. Native plants store water for long periods of time or survive on very small amounts of water by keeping their leaves and stems very small or developing an extensive root system (**FIGURE 4.6**). Animals adapt through water, salt, and temperature regulation, as well as having light coloration.

Climate Change Recognizing that the arid climate makes the region vulnerable to the effects of climate change, a number of countries in the region have become participants in the UN Framework Convention on Climate Change (UNFCCC) and the Kyoto Protocol (see Chapter 1). Poor environmental protections have left the state of the environment in the region seriously challenged. Rapid urbanization and high levels of migration to cities make solutions extremely difficult to implement. The effects of global climate change may soon be most severe in coastal communities, which may see saltwater encroaching into the freshwater systems of low-lying areas due to sea level rise in this century.

On the positive side, the growing global **carbon market** is highly attractive to countries in this region. The carbon market allows developing countries that have ratified the Kyoto Protocol agreement to receive payments for investments in climate projects that reduce greenhouse gas emissions. Countries can take action to reduce pollution, increase energy efficiency, and participate in global efforts to halt climate change. Internationally supported projects have been launched in Egypt, Tunisia, Jordan, Algeria, Morocco, Iran, and Saudi Arabia.

Apply Your Knowledge

4.1 Analyze how changes in global climate are impacting this world region.

4.2 What are some of the regional responses to global climate change?

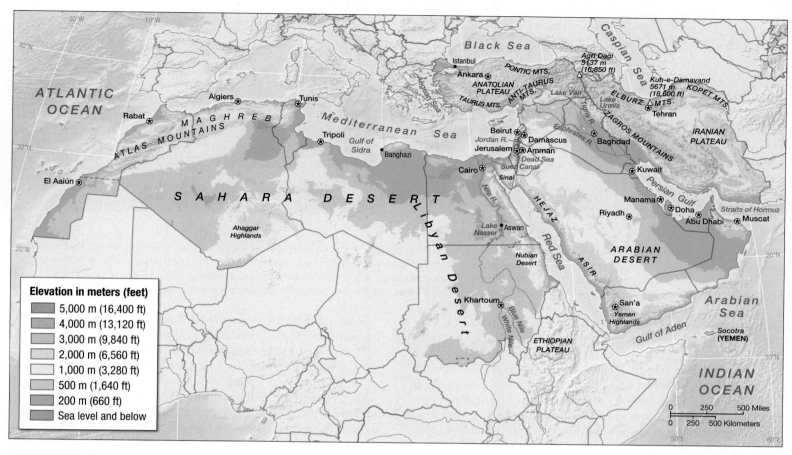

▲ **FIGURE 4.7 Physical geography of the Middle East and North Africa** The physiographic map of the region shows that large portions of this world region are at lower elevations. At the same time, mountains dominate the coast of North Africa, much of Turkey and Iran, play an important role in the coastal areas of Saudi Arabia and Yemen. Rivers and seas also play important, if differing, roles for this world region in terms of agriculture and trade. **What is the relationship between the physical geography of this region and the region's climatic geography?**

Geological Resources, Risk, and Water

The Middle East and North Africa is home to a number of mountain ranges, and two major river systems drain through the region (**FIGURE 4.7**). The region is also interspersed with seas, such as the Mediterranean, and is bordered by the Atlantic Ocean on the Moroccan coast and the Indian Ocean to the south and east.

Mountain Environments and Tectonic Activity
As **FIGURE 4.8** shows, the region is located at the conjunction of several continental landmasses. The separation of the Arabian Plate as it moved eastward millions of years ago developed a rift valley that

▶ **FIGURE 4.8 Generalized Tectonics of the Middle East** The map shows the concentration of tectonic activity in eastern Turkey. The African and Arabian plates are made of ancient rock and are stable; little tectonic activity occurs there. Along the Eurasian Plate, folded and faulted mountains extend from western to eastern Anatolia and then south across Iran and eastward again into the Himalayas. These mountains are the result of active plates colliding.

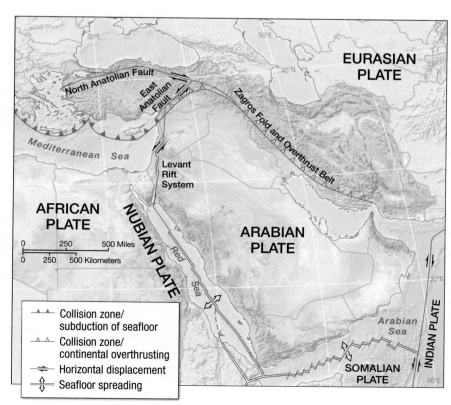

(a)

(b)

▲ **FIGURE 4.9 The Euphrates and Nile River Systems** (a) The Euphrates River is a critical source of freshwater for Turkey, Iraq, and Syria. Shown here is the river as it flows through Dura Europos just north of the border between Iraq and Syria. Ruins in this area date back several thousand years and demonstrate the importance of rivers in the history of this region. (b) The Nile River is essential to human and nonhuman life throughout much of North as well as East Africa. Featured here is a waterwheel, which is used in the Nile Delta region to move water from the river to needed croplands along the river's edge.

filled with water creating the Red Sea. Both the African and the Arabian plates rub against the Eurasian Plate along the Mediterranean Sea and at the mountains that separate the Arabian Peninsula from the Anatolian and Iranian Plateaus. The tectonic activity of this region has also produced three mountain ranges across North Africa, through Turkey and into Iran, and those bordering the Red Sea. Impressive in their beauty and ruggedness, these ranges generate rainfall and are the source of rivers and runoff for this arid region. Some parts of the region is prone to earthquakes, such as the one that shook Turkey in October 2011, killing close to 700 people and injuring thousands more.

Water Resources and River Systems
Water is a precious commodity in the region. This is because there are only two major freshwater river systems—the Nile River and the Tigris and Euphrates Rivers, the latter of which flow together in the lowlands of Iraq.

- **The Nile** With a source in East Africa, the Nile is one of the world's longest river. From the Ethiopian Plateau, the Blue Nile flows northward across the Sahara, where it joins the White Nile at Khartoum, Sudan, to form the Nile River. The river proceeds northward, emptying into the Mediterranean. The Nile flows through some of the driest conditions on the planet, where no additional moisture is added and high evaporation occurs.

- **The Tigris and Euphrates Rivers** These rivers originate in Turkey and flow south through

Iraq and Syria. The Euphrates flows first through Syria while the Tigris flows directly through to Iraq where it forms part of the border between Syria and Turkey. The rivers join together in lowland Iraq and eventually empty into the Persian Gulf. These two river systems have been the lifeblood for the region's inhabitants over thousands of years (**FIGURE 4.9**). It is not surprising that these rivers have also been and remain a source of much conflict in the region—largely having to do with access. The Southeast Anatolia Project in Turkey has produced much regional tension, for example. Turkey is harnessing the agricultural and hydroelectric power of these river systems, but at a cost to downstream regional neighbors, including Syria and Turkey (**FIGURE 4.10**).

▼ **FIGURE 4.10 Map of the Southeast Anatolia Project** This map shows the extent of Turkey's dam system in southeastern Turkey along the headwaters of the Tigris and Euphrates rivers. **What impacts do you think Turkey's dam project has on its neighbors to the south?**

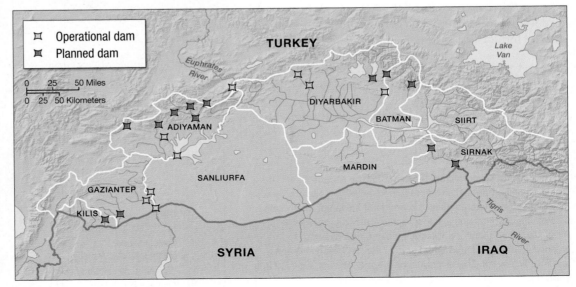

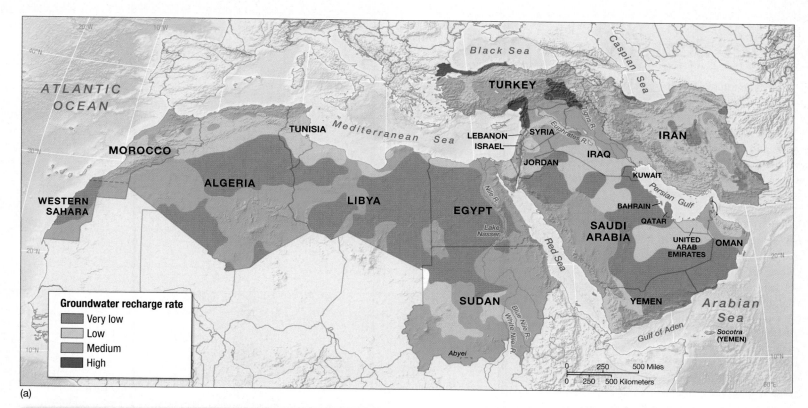

(a)

(b)

▲ **FIGURE 4.11 Water Scarcity in MENA (a) Groundwater Map** This map shows the spatial extent of freshwater aquifers and groundwater in the region. These sources of freshwater are being depleted quickly, a function of drought, declining precipitation, and increased demand. **(b) Sahara Desert Oasis** Geological evidence suggests that between 50,000 to 100,000 years ago, the Sahara possessed a system of shallow lakes and rivers that sustained extensive areas of vegetation. Though most of these lakes had disappeared by the time the Romans arrived in the region, a few survive in the form of oases. Pictured here is the Mandara Oasis and Um El Ma Lake in Libya's portion of the Sahara. **(c) Simplified Diagram of *Qanat* Irrigation** Through a series of low-gradient tunnels, the *qanat* collects groundwater, bringing it to the surface by way of gravity. The diagram shows the shafts drilled into the loose soil and gravel at the foot of the mountain that reach the water table below. The water is drawn up through these shafts and directed out for use through pipes.

Water Resources and Sustainable Environments

Although river systems are critical for providing water to this region, there are other important ways to access freshwater, although many of these are highly undependable (**FIGURE 4.11**).

- **Oases** In the Sahara, runoff from the Atlas Mountains collects underground in porous rock layers deep below the desert surface. In some fertile places in the desert, known as **oases**, underground water percolates to the surface. Because the soils of oases are usually fertile, they support animal and plant life and even agriculture,

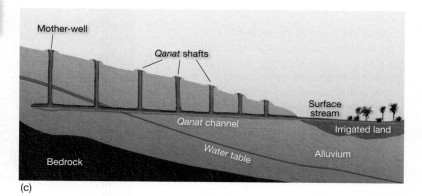

(c)

such as date and other fruit production. Oases play an economic role as stopping points for caravans carrying commercial goods across the vast deserts.

Water Scarcity and Quality

The future of water scarcity and water's availability and quality are serious issues in the Middle East and North Africa. As the region's population grows and the regional economy demands more water for the manufacturing sector as well as for the production of food, the scarcity of water resources will become more acute.

The Causes of Water Scarcity A number of factors contribute to the region's future map of water scarcity (**FIGURE 4.1.1**). In some places, economic conditions make water access difficult, while in other places, changes in global climate and overuse of freshwater resources have pushed many countries in the region to the brink. For example, excessive extraction of water from oasis wells has been occurring for such a long period that oases are dying. The countries of the Persian Gulf region have some of the highest per capita water use rates in the world, nearly double that of Europe. This fact further exacerbates the region's problem of water security.

The Future of Water Quality The expansion of the agricultural economy of the region appears, on the surface, as a positive economic indicator. But, chemical fertilizers for agriculture have a direct impact on the quality of the water available for drinking. Conflict in the region has also had a direct impact on water quality in the region, as fighting directly impacts water quality assurance systems as well as basic access to clean water (**FIGURE 4.1.2**).

> "As the region's population grows and the regional economy demands more water for the manufacturing sector as well as for the production of food, the scarcity of water resources will become more acute."

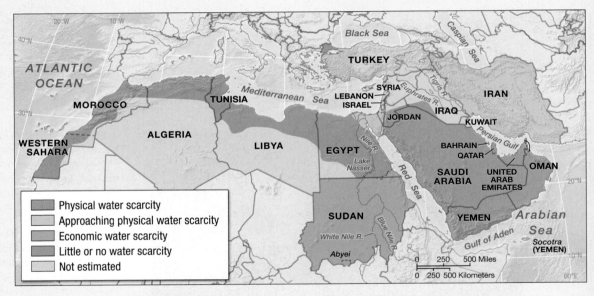

▲ **FIGURE 4.1.1 Map of Water Scarcity** The reasons for water scarcity are tied to the physical use and depletion of water resources, a result of climate change, and economic. What this suggests is that water scarcity is both an ecological and social issue.

Legend:
- Physical water scarcity
- Approaching physical water scarcity
- Economic water scarcity
- Little or no water scarcity
- Not estimated

1. **What are some of the factors that are worsening water scarcity in the region?**

2. **What is degrading water quality in the region? What are the implications for the region's growing population if water quality continues to deteriorate?**

▶ **FIGURE 4.1.2 Child in Syria Seeking Water** A Syrian boy, who fled with his family from the violence in their village in 2012, carries a plastic container as he walks to fill it with water at a refugee camp in the Syrian village of Atma near the Turkish border with Syria.

This process is controversial because it is expensive and has the potential to disturb rich ocean habitats. Despite these concerns, countries in the region have invested in desalinization plants. Qatar gets about 97% of its water from desalinization. Saudi Arabia currently gets about 50% of its water through desalinization. Israel has also produced some of the most efficient desalinization plants in the world (**FIGURE 4.12**).

▲ **FIGURE 4.12 Desalinization Plant, Israel** This plant in Hedera, Israel is one of the world's largest reverse osmosis desalinization plants in the world. Shown here is the return of the brine water, a byproduct of the process, back to the Mediterranean Sea. **What drives the need for desalinisation plants in this world region?**

Apply Your Knowledge

4.3 Identify two examples of how people have applied technological solutions to adapt to the region's water conditions.

4.4 What sorts of limitations and costs do the adaptations involve?

- **Water Mining** One of the most ingenious methods for mining water in the region involves a system of low-gradient tunnels that collects groundwater and brings it to the surface through gravity flow. This **gravity system** of water mining—where holes are drilled into the ground at the foothills of a mountain or hill to tap local groundwater—is known as *qanat* in Iran, *flaj* on the Arabian Peninsula, and *foggara* in North Africa. These systems rely on runoff from rain and snowmelt that percolates below ground level. Under ideal circumstances, the gravity system provides a highly dependable source of water enabling year-round irrigation.

- **Desalinization** The process of converting saltwater to drinking water by removing salt, called **desalinization**, is another way of obtaining freshwater in the region.

Nonrenewable Resources: Petroleum and Natural Gas
Petroleum, more often called oil, and natural gas are **fossil fuels**—deposits of hydrocarbon that have developed over millions of years from the remains of plants and animals and have been converted to potential energy sources under extreme pressure below Earth's surface. **Petroleum** is a liquid compound that can be developed into many unique and useful everyday products. Not only does it burn effectively when converted into fuel, it can be separated into a number of different components—gas and liquid—and converted into lubricants, waxes, and plastics. Petroleum can be used in asphalt and medicine.

The invention of the internal combustion engine and World War I helped establish petroleum as a foundational product of the industrial age. Indeed, in the 20th century the expansion of global military power was fueled by the use of the combustion engine in ships, tanks, and cars. The rise of a petroleum-based economy made the Middle East and North Africa—the region with the highest reserves of oil and natural gas in the world—one of the most economically important and highly scrutinized world regions of the 20th and early 21st centuries (**FIGURE 4.13**).

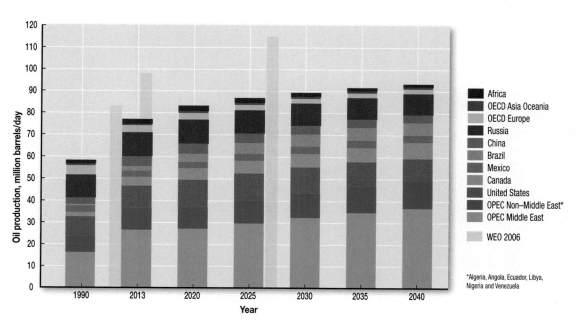

*Algeria, Angola, Ecuador, Libya, Nigeria and Venezuela

◄ **FIGURE 4.13 Predictions of Future Oil Production** This figure illustrates Energy Watch Group's forecast for global oil supply through 2030.

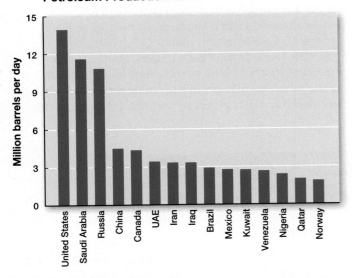

▲ **FIGURE 4.14 Global Oil Production** This graph demonstrates that, while collectively the Middle East and North Africa may be the largest producer overall, countries, such as the US, Russia, China, and Canada have been able to increase production. This has changed the dynamics of petroleum prices worldwide.

- **Early Control of Petroleum and Natural Gas Reserves** In the early 20th century, petroleum deposits in the region were identified as some of the most significant in the world. In the process of consolidating control over global petroleum production, five U.S.-based and two U.K.-based companies began to drill in Iran in early 1900s, Iraq in the 1920s, and Bahrain and Saudi Arabia in the 1930s.

- **The Rise of OPEC** By 1960, enraged by cuts in oil prices made by the seven big oil companies, the major oil-exporting countries in the Middle East and North Africa established the **Organization of Petroleum Exporting Countries (OPEC)**. OPEC's central purpose is to

coordinate the crude-oil policies of its members. When OPEC decided in the 1970s to cut back oil production, raising the price of oil, it produced the world's first "oil crisis". Another oil crisis in 1979 precipitated further concern by oil importing countries about their dependence on OPEC's reserves. Many countries increased fuel efficiency standards for cars and industrial production systems. To cope with their own needs other countries, such as Russia and the United States, increased oil production, driving down prices globally (**FIGURE 4.14**). Regardless of the role that countries such as Russia and the United States now play in the oil market, it is clear that many leaders in the Middle East and North Africa have become very rich from oil production (see "Geographies of Desires, Indulgences, and Addiction: Luxury Cars," p. 141).

- **The Future of Petroleum and Natural Gas** Despite attempts to limit the world's dependence on petroleum products, global consumption continues to rise (**FIGURE 4.15**). As a result, some scientists estimate that we may have already depleted at least half the world's oil reserves, while others have gone even further suggesting that the world's overall oil reserves are exaggerated by as much as one-third. The rapid consumption of natural gas is likely to lead to the eventual depletion of that resource. Some scientists believe we have also hit our "peak natural gas" point with natural gas supply becoming too expensive to produce within 20 to 30 years.

- **Petroleum and Natural Gas: A Blip on the Historical Radar?** It may have taken millions of years for Earth to produce the massive supplies of oil and natural gas we enjoy using today, but peak-oil scientists, as well as many energy-policy experts, believe that the consumption of oil and natural gas as major sources of energy will end up being a very brief affair in global historical terms.

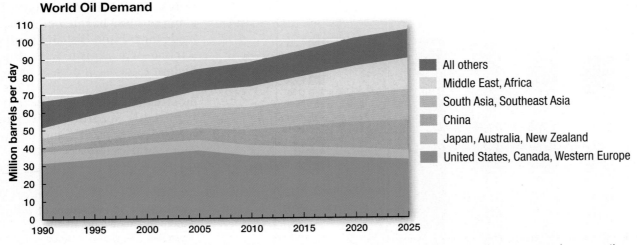

▲ **FIGURE 4.15 Growing Oil Consumption Patterns** The pattern of global oil consumption shows how patterns of consumption will continue to shift from the US, Canada, and Western Europe, to China, South Asia, East Asia, the Middle East and North Africa.

Luxury Cars

Bentley Motors, makers of some of the world's most expensive cars, saw sales rise 23% worldwide in 2014. Those sales were driven by high demand for their product in the Middle East, where luxury car sales are some of the highest in the world.

Elite Status, Elite Cars Luxury cars remain a global status symbol in the age of high consumption of oil and gas. In the Middle East, the makers of Bentley, Jaguar, and Land Rover have found a niche market of incredibly wealthy, high-end consumers ready to drive these new cars off the lot (**FIGURE 4.2.1**). In some countries, such as Saudi Arabia, the privilege of driving is highly gendered as it remains illegal for women to drive.

The Ratio of Cars to People In 2015 the city of Dubai in the UAE surpassed New York City as the city with the world's highest number of cars relative to the total number of people. In 2011, the last year for which data are available, the data show that the Middle East has some of the highest ratios of cars to people in the world (**FIGURE 4.2.2**). As fertility rates decline in this region—as is happening in wealthier countries, such as Qatar, UAE, and Bahrain—the ratio of cars to people will become more skewed toward the car. It is anticipated there will be a car for each person in Dubai, UAE by 2020. In countries where birthrates remain higher, such as Saudi Arabia, SUV demands will likely increase. Such vehicles are not only practical, though. They signal the driver's upward mobility in class and wealth and thus remain central to identities in the region.

The Image of the "Fast Car" The Middle East is now home to many high-end Formula one races as well as company-specific races, such as the Porsche GT3 Cup (**FIGURE 4.2.3**). The popularity of car racing in the region demonstrates just how important luxury cars and the car industry more generally has become in this world region.

---

1. **What are the indicators that luxury cars have become important commodities in the Middle East?**

2. **What role does the car play in the culture of everyday Middle Eastern life? How does the luxury car fit into the image of affluence in the region?**

"In the Middle East, the makers of Bentley, Jaguar, and Land Rover have found a niche market of incredibly wealthy, high-end consumers ready to drive these new cars."

◄ **FIGURE 4.2.1 Car Show in Doha, Qatar** Expos, such as this one in Doha, are becoming more common throughout the region as people invest in high end luxury automobiles.

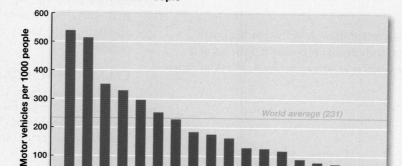

Motor Vehicles to People

Motor vehicles per 1000 people (y-axis: 0 to 600)

Kuwait, Qatar, Bahrain, Israel, Libya, UAE, Oman, Saudi Arabia, Jordan, Turkey, Tunisia, Iran, Algeria, Morocco, Iraq, Syria, Egypt, Yemen

World average (231)

◄ **FIGURE 4.2.2 Number of Motor Vehicles per Person** The distribution of cars to people in this world region is not even and maps onto the overall wealth of this region's member countries.

◄ **FIGURE 4.2.3 The Grand Prix of Bahrain** Bahrain and other countries in the region now host a number of sporting car events.

The rate at which these resources are depleted may increase, as China and India as well as other new global industrial countries increase their consumption of oil and natural gas.

- **Sustainable Futures and Nonrenewable Resources** Climate change related to the increases in carbon dioxide emitted when burning petroleum products and natural gas and the impact of oil and gas pollution on Earth's oceans and rivers is substantial. Increasing amounts of petroleum products are winding up in the ocean, a result of oil spills and petroleum products converted into products such as plastics. As petroleum products break down they are ending up in animal and human food chains; this causes concern among biologists, who worry that the world's fisheries may be irreversibly damaged by petroleum and other industrial wastes in our water.

Apply Your Knowledge

4.5 Using the U.S. Department of Energy website (http://www.eia.gov/beta/international/) briefly describe the geographic pattern of oil reserves, oil production, and oil consumption globally.

4.6 How have production and consumption trends changed over time? Where are future reserves located?

4.7 Go to the following website (http://www.theguardian.com/environment/oil-spills) and identify two recent oil spills and some of the environmental and economic implications of those spills.

Oil Industry Data

https://goo.gl/4Xrmql

Ecology, Land, and Environmental Management

Humans have dramatically altered the landscapes of the Middle East and North Africa, in some cases so much so that entire species of plants and animals have been eliminated. Population increases in recent decades have placed further pressures on existing resources. The region's aridity demands adaptations on the part of all living inhabitants of the region.

Flora and Fauna The current map of plant and animal life belies the impact of human modification to this region's dynamic landscapes. At one time, dense forests existed throughout Turkey, Syria, Lebanon, and Iran. After several thousand years of woodcutting and overgrazing, only a few small areas remain of the forests of Turkey and Syria. Large areas of Lebanon have also been deforested. Sadly, Lebanon's famous cedars continue to grow only in a few high-mountain areas (**FIGURE 4.16**). Iran retains deciduous forests, particularly in the region of the Elburz Mountains. The forests of the Atlas Mountains of Morocco and Algeria are also still being harvested for commercial purposes.

Millennia of human occupation have greatly affected the region's animal life. Several thousand years ago, a wide variety of large mammals inhabited the region's forests, including leopards, cheetahs, oryx, striped hyenas, and caracals; crocodiles also thrived in the Nile; and lions roamed the highlands of present-day Iran. Nearly all these species are now extinct or near extinction. In their place, one can find domesticated camels, donkeys, and buffalo. The highland areas of Turkey and Iran still contain a variety of mammals, including bears, deer, jackals, lynx, wild boars, and wolves. Birds are also plentiful in the region.

Environmental Challenges Because national governments lack the resources or, in some cases, the political will to limit environmental change, there remain intense pressures on the changing environments of this region. Every state in the region except Turkey is experiencing **desertification**—the process by which

◀ **FIGURE 4.16 Cedars of Lebanon** For millennia, this coniferous tree has been significant to the region for trade, medicine, religion, and habitation. The once-abundant cedar forests—the national symbol of Lebanon and shown on its flag—have been almost completely eliminated by overexploitation. Reforestation programs are underway, however, in both Lebanon and Turkey. This photo shows the last remaining cedar forest in Lebanon.

semiarid lands become degraded and less productive, leading to more desertlike conditions. Deforestation has, in some cases, exacerbated the effects of climate changes on the region.

There has been some investment in environmental protection or preservation across the region. Reforestation and **afforestation** programs that convert previously nonforested land to forest by planting seeds or trees are under way. Oman and the United Arab Emirates have begun to take a deliberate stand against desertification through the planting of greenbelts.

History, Economy, and Territory

The first humans in Middle East and North Africa who were able to domesticate plants such as wheat, created semipermanent settlements in the region between 10,000 and 12,000 years ago.

This led to some of the earliest large-scale societies in the world and important inventions, such as writing, mathematics, and cultural-religious systems. These inventions spread throughout the world through global trade, migration, and empire. As the region has influenced the world, the world has also influenced the region. European colonization in the 20th century, global political conflicts, and the thirst for oil have had an enormous impact on the Middle East and North Africa.

Historical Legacies and Landscapes

A remarkable number of sophisticated empires have flourished in the region. The geographical center of these empires was the **Fertile Crescent**, a region arching across the northern part of the Syrian Desert and extending from the Nile Valley to the Mesopotamian Basin in the depression between the Tigris and Euphrates Rivers (**FIGURE 4.17**). Empires in modern-day Iraq, present-day Turkey, the Nile Valley, and the Iranian Plateau as well as Greece

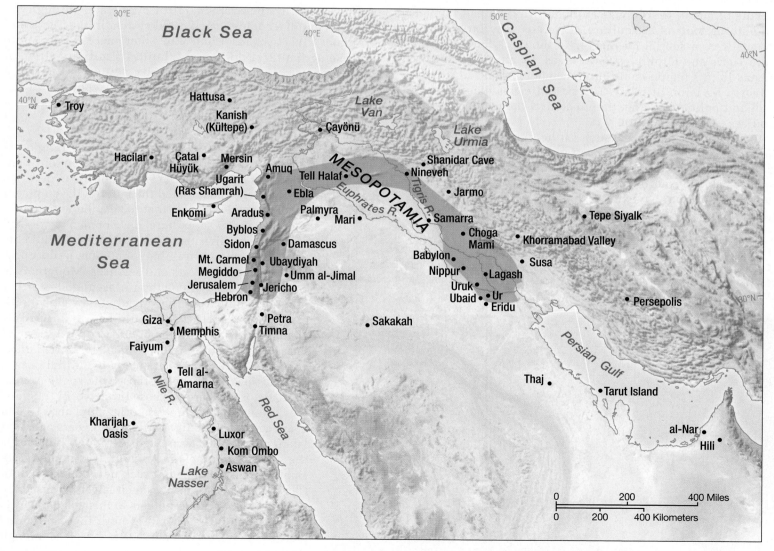

▲ **FIGURE 4.17 The Fertile Crescent** This region is one of the first places in the world where plants and animals were domesticated by humans. Historically, the Fertile Crescent has been an important center for the spread of agricultural technologies as well as writing, mathematics, and religious systems. Many urban-based societies, as highlighted here, developed in this region as a result. **Why does the map of urban development in the Fertile Crescent look the way it does?**

and later Rome all controlled parts of this region over the centuries. The interaction over time and across empires led to an exchange of ideas, goods, people, and belief and value systems that helped tie the region together as well as extend its influence from North Africa to Europe and South Asia.

Plant and Animal Domestication Regional empires developed because people were able to domesticate plants—wheat, lentils, chickpeas—and animals—pigs, sheep, and camels—about 12,000 years ago (**FIGURE 4.18**). The new technologies of domestication signaled the transition from hunting and gathering to agricultural and pastoral (herding-based) societies. The technologies of fire, grindstones, and improved cooking tools and pottery also contributed to these transitions. In about 10,000 to 7,000 B.C.E. Neolithic—New Stone Age—farmers and herders began dry farming, relying exclusively on rainfall. Pastoralists—groups of herders who relied on pack animals, such as camels and later the horse—moved and traded across the region's vast grasslands. Agricultural needs led to the invention of numerous tools, such as the plough, and metallurgy. The working of metal, first bronze and later iron, helped humans build strong tools of agriculture as well as war. Early bronze ploughs were essential for digging in the thick, but rich, soil of the Tigris and Euphrates River areas. By about 4,000 B.C.E. humans in this region could produce enough food to support a diverse division of labor. Towns and cities emerged as a result and political competition to control the region's limited resources increased (**FIGURE 4.19**).

Technological advances, such as the plough, also allowed humans to build systems of irrigation (or **wet farming**). When farmers were able to minimize their dependence on rainfall, they were able to control larger areas and exert dominance over groups who did not have this technology. Archaeologists and other scholars argue that the development of large **hydraulic**

▲ FIGURE 4.19 **Ancient Village of Çatal Hüyük** One of the first sites of sedentary life in the world was in a small village in Turkey.

civilizations emerged as a result of expanded wet farming. These civilizations used the control of water and the food that resulted as weapons of control. Babylon achieved 2,000 years of dominance by increasing control over water through the development of a sophisticated canal system in the Tigris and Euphrates River systems. Control of water enabled Babylon to increase agricultural production, build its military strength (including a walled and fortified city center), establish a long-distance trade network (through extensive port facilities), and organize extensive religious and symbolic political control (**FIGURE 4.20**).

Birthplace of Three World Religions While religion is covered extensively later in the section "Religion and Language," it is important to note that the large-scale urban developments in this world region facilitated the cultural development of **world religions**—belief systems that have adherents worldwide. Extensive agricultural development facilitated an increasing division of labor whereby a small elite had time to pursue training in religion, philosophy, writing, and other arts and sciences. Monotheistic beliefs—the idea of one omnipresent god—further provided context for the spread of new global religions, as a placeless god could be worshipped by anyone anywhere. In this broader context, three of the world's major religions—Judaism, Christianity, and Islam—all developed among the Middle East's **Semitic**-speaking people—those who spoke Arabic, Aramaic, and Hebrew. These three world religions are closely related. Judaism originated about 3,500 years ago, Christianity about 2,000 years ago, and Islam about 1,300 years ago. Although Judaism is the oldest monotheistic religion, it is numerically small because it does not seek new converts. Both Christianity and Islam are much larger because they actively seek to convert others to their religion. This has facilitated the spread of both these religions globally through missionary work as well as through empire and war.

While these three world religions share origins and even core beliefs, there have been tensions

▼ FIGURE 4.18 **Chickpea Processing, Turkey** Chickpeas are processed throughout this world region. Chickpeas are found in many dishes, including in hummus, an important contribution to global food culture.

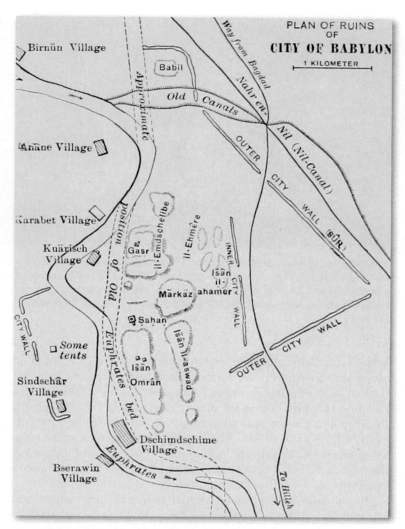

▲ **FIGURE 4.20 Map of Babylon** This map, drawn in 1904, demonstrates the complex urban geography of this city and its relationship to the Euphrates River.

▲ **FIGURE 4.21 Hammurabai's Law Code** As societies grew bigger and became more complex, law codes were created to maintain order and regulate social life. One of the first known written codes was created in the Middle East. **Why did complex societies need law codes? What about the geography of urban life put pressure on leaders to create and enforce such codes?**

between them. This includes a period of intense conflict between Christians and Muslims throughout the Middle Ages (500 C.E. to 1500 C.E.). Much more recently, particularly beginning in the 20th century, there has also been increasing tension between Jews in Israel and some Muslim communities in the region, discussed in the section "Territory and Politics."

Urbanization This world region is home to some of the oldest urban centers, established as humans began to adapt to changing climates by domesticating plants and animals, developing agricultural systems, and building irrigation networks to support larger settlements.

- **Primary Urbanization** In the Middle East and North Africa, the earliest cities developed in the lower valley of the Tigris and Euphrates River areas in the 4th millennium B.C.E. In terms of urban form, early cities, such as Kish, Nippur, and Ur, included three main elements: city walls; a commercial district; and suburbs, including houses, fields, groves, pastures, and cattle folds. These

cities probably contained between 7,000 and 25,000 inhabitants and evidence suggests that they thrived because the major producers in these towns and early cities were fishers and farmers, who supported a non-producing class of priests, administrators, traders, and artisans. These cities were also located at crucial points along natural and well-traveled human routes. It is through these trade networks that the ideas that developed in Mesopotamia—urbanization, writing, mathematics, and law—diffused out from this core region and into other surrounding cultural centers, such as Egypt (**FIGURE 4.21**). Indeed, one of the most important inventions of Mesopotamian culture was the writing system, which was later simplified by Phoenician traders living along the Mediterranean coast. Thousands of years later, this writing system evolved into the alphabet of this textbook.

- **Secondary Urbanization** Urban life spread slowly from Lower Mesopotamia into Upper Mesopotamia—in the areas further north along the Tigris and Euphrates Rivers—by the 3rd millennium B.C.E. and then throughout the entire Fertile Crescent and into areas such as the Nile Valley by 2nd millennium B.C.E. The Egyptians drew on Mesopotamian knowledge—including mathematics, writing, and agricultural technologies—to build large cities centered on monumental tombs, temples, and palaces (**FIGURE 4.22**). Other influences on Egypt included trade with settlements in the northern Red Sea, the upper Nile of sub-Saharan Africa, and the eastern Mediterranean. Egypt imported significant quantities of gold, cedar, ebony, and turquoise for their elaborate rituals and sophisticated body adornments. These trade relations also enabled the transfer of ideas and technologies that enriched all the cultures of the region, including Greece and later Rome. As an example, Greek realist art, which included images of very detailed human bodies, was likely influenced by Egyptian knowledge of anatomy, which resulted from the practice of mummification.

▲ **FIGURE 4.22 Dayr al-Bahri Mortuary Temple, Egypt** In parts of the Middle East, such as Egypt, it was common for rulers to build large and extravagant temples to worship the dead. This unusual temple complex was built in the 15th BCE.

▲ **FIGURE 4.23 Hagia Sophia, Istanbul** This complex has a long history. It was first built as a key seat of the Orthodox Christian Church, was briefly converted to a Roman Catholic Church, and then turned into a mosque in the 15th century CE. It stands today as a museum, although its minarets are also used for the Muslim call to prayer.

- **Greco-Roman Architecture** As the Greek and later Roman empires advanced into the Middle East and North Africa, urban design structures changed, producing new urban landscapes that included great estates, large theaters, and other Greco-Roman architectural forms, including images of Greek and Roman gods. By the 4th century C.E., Christianity, which was supported by a growing Eastern Roman Empire, began to take firmer hold and churches also began to dot the landscape of this world region—forming a critical part of the urban landscape of cities, such as Istanbul, Turkey, and Jerusalem, Israel.

- **Islamic Cities** Successive Islamic empires have had the most visible historical impact on urban patterns in the region, as Islamic architecture and urban form were placed on top of earlier urban forms (**FIGURE 4.23**). At its greatest extent, Islamic rule reached westward as far as Spain; eastward beyond Turkey and the Iranian Plateau into Afghanistan, Pakistan, and India; and southward into North and West Africa. Some of the key features of Islamic-influenced architecture included distinct public and private spaces to parallel Islamic religious practice. Each city also contained a central mosque as well as the citadel or government center (*Casbah*), neighborhoods (*qaraba*) containing clusters of homes of people of the same social status, a market (*suq*), as well as public baths (*hamman*), which were used by men. There is variation in the Islamic city and many have argued that Islamic cities borrowed heavily from their predecessors. So, for example, the Islamic cities of the Magreb in North Africa largely followed Roman form. That said, the influence of Islamic architecture and urban design was widespread, as evinced by examples that can be seen today in places such as Spain.

The Ottoman Empire The Ottomans were Turkish Muslims based on the Anatolian Plateau. They replaced the Christian Greeks as the political power of the region after 1100 C.E. and ruled much of the region for more than 700 years. At its height, the Ottoman Empire extended from the Danube River in southeastern Europe to North Africa and to the eastern end of the Mediterranean. Within the region, only the Persian Empire (on the Iranian Plateau), the central Arabian Peninsula, and Morocco were able to resist direct Ottoman control.

By the mid-19th century, Ottoman rule was under siege from Europe. European occupation of Egypt by Britain and Tunisia by France exposed the region to continental ideas about democracy. **Nationalist movements** emerged based around groups of people who shared common elements of culture—such as language, religion, or history. These movements were problematic for the Ottoman Empire, which had ruled through an imperial legal and administrative structure. By the end of World War I, the Ottoman Empire had been reduced in size, controlling only modern-day Turkey to the north and the coastal regions of the Arabian Peninsula to the south (**FIGURE 4.24**).

The Mandate System After World War I ended, the Ottoman Empire collapsed, leaving modern-day Turkey. The Arab provinces of the Ottoman Empire were divided up and became **mandates**—areas administered by a European power that were to prepare the mandated regions for self-government and future independence (**FIGURE 4.25**). Mandate holders were required to submit to internationally sanctioned guidelines, which required that constitutional governments be established as the first step in preparing the new states for independence. The region's population seriously challenged the political order that was imposed following World War I. People in Egypt, Iraq, Syria, and Palestine all revolted violently against the European presence. By the end of World War II, all the states in the region gained their independence.

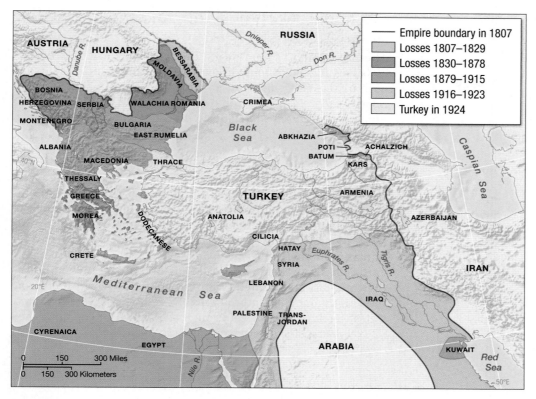

▲ **FIGURE 4.24 Historical Geography of the Ottoman Empire** At its height, the Ottoman Empire controlled a vast range of the Middle East, North Africa, and Europe. Over time, their control of this large empire was eroded.

▼ **FIGURE 4.25 Europe in the Middle East and North Africa in 1920** Relative to the colonial experience in other world regions, the colonial presence of Europe in the Middle East and North Africa region was short-lived—only about 50 years—but very significant. This map shows the European possession of the region in 1920. Theorize the impact that the borders drawn by European powers might have had on the modern-day politics of this world region.

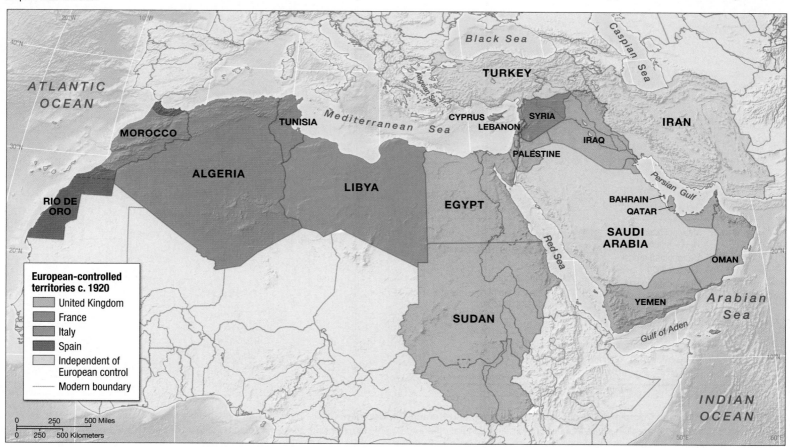

Winning independence from colonizers is not the same as gaining the allegiance of a diverse collection of new citizens. Many challenges emerged for these new states, which had to create national identities from multi-ethnic, religious, and cultural populations. Colonial investments in core urban areas also skewed development toward cities over rural areas throughout the region.

Apply Your Knowledge

4.8 What key technological changes led to social developments in the region?

4.9 How did those technological changes influence urbanization in the region?

4.10 What are the different historical influences on urban form in the Middle East and North Africa?

Economy, Accumulation, and the Production of Inequality

The integration of the Middle East and North Africa into the global capitalist economy in the 19th and 20th centuries brought both increased wealth and poverty to different peoples of this region. This means that today the region can be divided into a number of different subregions with different economic trajectories (**FIGURE 4.26**).

Independence and Economic Challenges After World War II, countries such as Iran, Turkey, Egypt, Syria, Iraq, Tunisia, and Algeria adopted the **nationalization** of economic development, which involved the conversion of key industries from private to governmental operation and control. Despite these attempts, many states began to turn away from nationalization as their economies stagnated, standards of living declined, and debt skyrocketed in the 1970s and 1980s. Pressured by the International Monetary Fund (IMF), the World Bank, and the U.S. Agency for International Development (USAID) many states were forced to privatize their economies and open up their markets to receive external aid dollars.

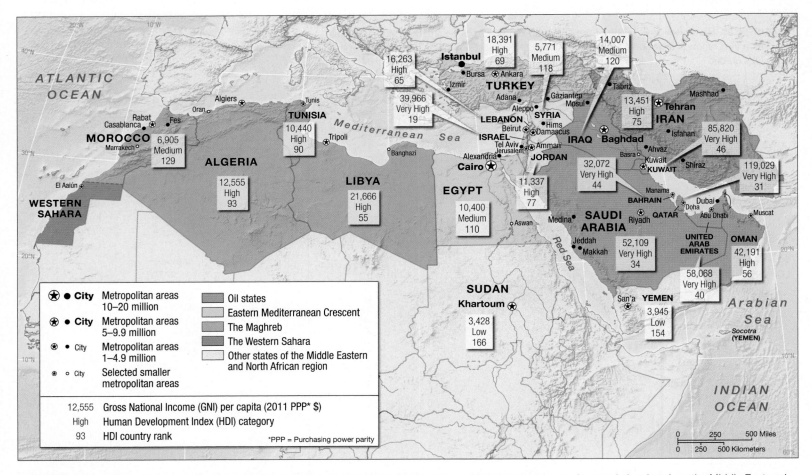

▲ **FIGURE 4.26 Regional Complexity, Gross National Income, and Human Development** While there are many characteristics that draw the Middle East and North Africa together, there are also some stark economic and cultural differences. This map depicts some subregional differences and highlights the broad range of income and human development across the region. **What is the relationship, if any, between HDI and GNI per capita?**

These changes resulted in increasing costs of food and basic necessities, cuts in spending on social programs, and reductions in public sector investment. Urban workers, government bureaucrats, and people on fixed incomes strongly felt the impact of these changes. Corporate farmers were also able to buy up land and use mechanized agricultural practices to reduce labor costs. These changes lowered the living standards of both urban workers and rural peasants. Rural people began to move into the cities to find employment. In cities, these migrants were confronted with decreased public services, such as schools, health care, housing, and clean water. Migrants have found housing in squatter settlements without sanitation. Many now eke out a living in the **informal economy**—economic activities that take place beyond official record and are not subject to formalized systems of regulation or remuneration.

The Oil States

A number of countries in this region derive their wealth from oil. These so-called oil states are pivotal in the global economy. Besides the obvious global dependence on oil, the impact of **petrodollars**—revenues generated by the sale of oil—is significant for the global economy, particularly the economies of Europe, North America, and East Asia, where they are spent, invested, and banked. That said, the high dependence on oil production in the oil states leaves their economies highly vulnerable to fluctuations in the demand for oil.

Recognizing this, some of these states have begun to diversify their economies through tourism, trade, and urban development (see the Chapter Opener on Dubai). Saudi Arabia established a plan to build four new "Economic Cities." One of the planned cities is named "Knowledge Economic City" to attract foreign information industries (**FIGURE 4.27**). These projects are a long way from completion, but they suggest how the region might respond to the changing nature of the energy economy over the next 20 years. Some countries are increasing economic diversification by introducing new industries, such as textile production and food processing plants, while others are developing port facilities to increase their role in global trade. Still others are promoting agriculture, though the scarcity of water makes irrigation an enormous challenge for all of these countries except for Iraq and Iran (**FIGURE 4.28**).

The Eastern Mediterranean Crescent

The Eastern Mediterranean Crescent—home to Egypt, Israel, Lebanon, and Turkey—has the potential to be a different kind of economic success story. Similarities are striking among the four states. All four have sizable middle classes, industrial potential, and rich agricultural traditions and growing agricultural economies (**FIGURE 4.29**). There are also differences among these four countries. Israel's agricultural production is tied to achieving national food security. While Egypt, Turkey, and Lebanon are beginning to encourage more industrial development, Israel possesses a strong industrial base that is fairly diverse.

▲ **FIGURE 4.27 Model of Saadiyat Island, Abu Dhabi, UAE** The new planned region of Abu Dhabi is being planned with a number of international partners. Pictured in the foreground is the model of the new Gugghenheim Museum.

▲ **FIGURE 4.28 Breakdown of GDP by Sector and Composition of the Labor Market in the Gulf Region** (a) the distribution of economic activity by sector varies quite a bit by country in this region, with some countries relying heavily on primary commodity production; (b) the labor market in these countries is divided heavily between people working in both the public and private sectors. In some cases, such as Qatar, the expat labor force (e.g., workers who are not citizens of the coutry) make up the majority of the paid labor force.

▲ **FIGURE 4.29 Hazelnut Farming in Ordu, Turkey** Women play an important part in the agricultural economy of Ordu, an area located on the southern coast of the Black sea. Pictured here are women collecting hazelnuts, a labor intensive job that is not easily mechanized.

▲ **FIGURE 4.30 Beach Resort, Tunisia** Tunisia has long been a tourism destination, particularly for Europeans. Recent bombings in key tourism cities, however, have slowed the flow of tourists. This is having an impact on the overall economy and the livelihoods of thousands of Tunisians.

Turkey also possesses a fairly diversified economy, with a strong agricultural sector and substantial mineral wealth. Egypt also has significant manufacturing capacity across a range of products from food processing to heavy machinery. In contrast, Lebanon possesses a strong agricultural base and historically has been a global banking and financial center. However, civil war in 1975–1990 limited Lebanon's regional and global authority in the area of banking. A 2006 war with Israel further derailed Lebanon's political and economic stability; Lebanon's economy is weak and the country is still deeply fragmented politically.

The Maghreb The Maghreb has a relatively strong economy based on oil and mineral exploitation, agriculture, and tourism. Algeria's oil industry provides nearly 90% of its export revenues. Libya, too, has substantial, high-quality petroleum reserves. Both Tunisia and Morocco have large phosphate industries. All the Maghreb countries are also agricultural producers, though none is self-sufficient.

The Maghreb is also spectacularly beautiful, drawing tourists from Europe and beyond. The region offers a range of tourist experiences, in both luxury and economy style, from beautiful beaches to trekking areas in the Atlas Mountains or the Sahara to ancient archaeological ruins. Algeria and Morocco have the fastest-growing economies because of tourism. Libya and Tunisia were making substantial strides until recent political events curtailed economic development (**FIGURE 4.30**). To promote links across the Mediterranean, the European Union and 15 Mediterranean countries participate in the **Union for the Mediterranean**, which may help to boost all sectors of the Maghreb, from resources to tourism.

Mediterranean Development

https://goo.gl/ePM3zJ

Social Inequalities Across the Region Despite the phenomenal wealth generated from oil production, most of the region remains poor and dependent on an increasingly marginalized agricultural sector. National statistics hide the dramatic variation in wealth between one urban neighborhood and the next as well as between urban and rural spaces. Extreme concentrated wealth in the region comes from oil-based revenues, while states with the largest populations tend to have the lowest per capita wealth. Variation in the Human Development Index (HDI), as measured by life expectancy, educational attainment, and income, varies dramatically (see Figure 4.26 on p. 148). High levels of average wealth do not always translate to a high quality of life. Qatar has the highest GNI per capita PPP in the world, but is only ranked 31st in global HDI.

Apply Your Knowledge

4.11 Compare and contrast the economies of the region's oil states with the states in the region that do not produce large quantities of oil.

4.12 In cases where oil is not available, what sorts of economies have developed over time?

4.13 Which countries have been most successful at diversifying their economies?

Territory and Politics

Although the region has experienced wars and conflict for hundreds, if not thousands, of years, and certainly well before the Europeans arrived, it is generally agreed that most of the present conflicts stem from the colonial period. The colonial imposition of borders has created problems for nation building, while the region's strategic importance and rich natural resources make it a source of global focus. At the same time, the most significant unifying force has been the religion of Islam and, for many of the countries, Arabic language and culture. These unifying forces of religion and language have helped many of the peoples of the region recognize that they have common political and economic goals. Recently, however, economic and class differences and growing interest in democratic reform have shown that these unifying forces may not be enough to overcome some of the region's deep social divisions.

Intra-Regional Conflict and Tension There are various conflicts in the region, each with a unique origin but often stemming from age-old conflicts and the legacies of colonialism.

- **Iran and Iraq** Although there is currently little direct conflict between Iran and Iraq, there has been tension over the last 30 years. One factor is the cultural differences between Persians (Iranians) and Arabs (Iraqis). Iran is a largely Shi'a Muslim country, while Iraq has historically been ruled by a Sunni Muslim minority that persecuted Iraq's Shi'a majority. The British elevated Sunnis in the colonial government in Iraq, giving them power and authority over both the majority Shi'a Muslims as well as the Kurds in the north. This caused tension with Iran's Shi'a population, particularly after 1979, when an Islamist government took over leadership of Iran. Iran's relationship with Arabs in the region has been tested in the last few years, as Iran has been

seen to be supporting conflict in its neighbor Iraq and also in Syria. Some studies show that Iran's favorability among Arabs in the region has been declining, even though they have supported Iraq's government in the conflict with the Islamic State (see Visualizing Geography, "Political Protest, Change, and Repression" on pp. 152–153).

- **Kurdistan** Twenty million **Kurds**, a non-Arab Islamic ethnic group who occupy parts of Turkey, Syria, Iraq, and Iran, have been struggling for either greater regional autonomy within their home countries or an independent Kurdistan state World War I (**FIGURE 4.31**). As a result, the Kurds as an ethnic minority have faced repression and discrimination, particularly in Turkey, where almost 8 million Kurds currently reside. Recent Turkish incursions into northern Iraq have been aimed at stopping Kurdish groups from agitating for regional autonomy within Turkey. A large number of Kurds have left the region and now live in Western Europe, providing an important source of financial support for the resistance movement.

- **Darfur** The brutal violence in Darfur began in 2003. Media reports indicate that the Sudanese government aimed to end rebellion in Darfur by eliminating all tribal Africans in the area. The government supported a militia of African Arabs, who call themselves the **Janjaweed**, to undertake one of the most brutal campaigns of ethnic cleansing that Africa has ever seen.

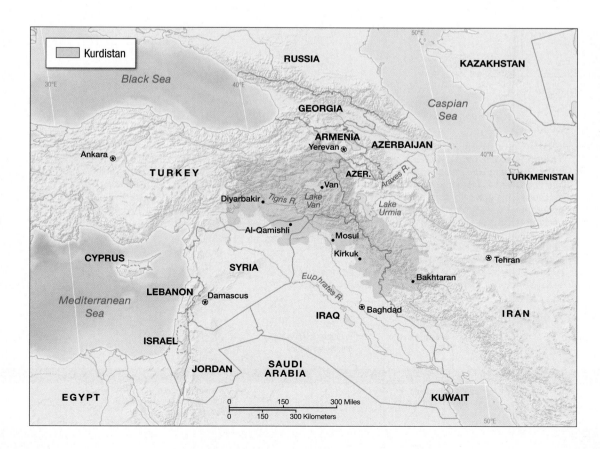

◄ **FIGURE 4.31 Kurdistan, "Land of the Kurds"** Although Kurdistan is not an independent state, the map does provide a sense of the boundaries that (from the Kurdish perspective) it might ideally encompass. **What are the geographic challenges the Kurds face in unifying their diverse geography into one, autonomous country?**

Political Protest, Change, and Repression

In 2009, the Iranian Green Revolution erupted from charges of election fraud in the country. Those protests can be seen as inspiration for what became known as the Arab Spring—protests that occurred across North Africa and parts of the Middle East in Spring 2011 (Figure 4.3.1.).

1 The Arab Spring

▼ **FIGURE 4.3.1** **Arab Spring 2010–2011.**

Tunisia The "Jasmine Revolution" in Tunisia led to end of the 24-year reign of President Zine el Abidine Ben Ali. In the aftermath, the country has held democratic elections. The economic situation in the country remains difficult for many, leading to an increase of Tunisians leaving the country to fight in conflicts in Iraq and Syria.

Egypt In 2012, President Mohammed Morsi of the Muslim Brotherhood won election, replacing Hosni Mubarak, who had been president for 30 years. In 2013, a military coup ousted Morsi from office. Military rule was returned to the country under two caretaker leaders – Chief Justice of the Supreme Court Adi Mansour and then General Abdel-Fattah el-Sisi.

Syria Since a bloody civil war erupted in 2011, the country has seen over 200,000 Syrian deaths and an additional 3.5 million refugees. The country is further destabilized by the rise of ISIS, which controls portions of the country's eastern region.

Rise of ISIS

https://goo.gl/6ukShs

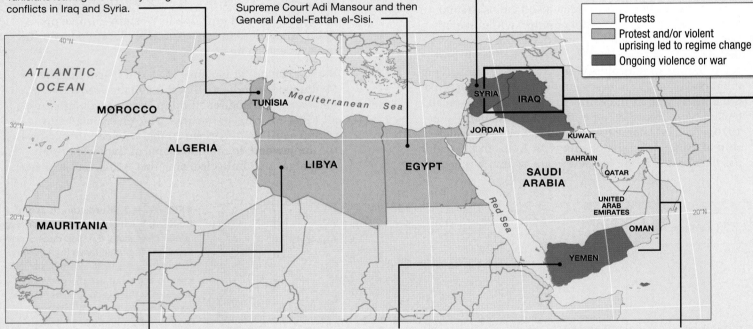

Legend:
- Protests
- Protest and/or violent uprising led to regime change
- Ongoing violence or war

Libya In 2011, Colonel Muammar Gaddafi was killed, ending his 42-year reign over the country. Despite demands for increased political rights and economic opportunities, the country remains divided by conflict and tension. Some cities in the country are said to be in the control of groups loyal to al Qaeda and perhaps ISIS.

Yemen Peace in Yemen was brief after an election in 2012. President Hadi resigned from office in 2015 and eventually fled to his home town of Aden. He has recently reasserted his authority as president, although the government is controlled by another group, the Houthi, seeking to control a unified Yemen. The presence of al-Qaeda in Yemen has also challenged government authority, making the country an increasingly unstable place, while Saudi Arabian backed militias fight to replace Hadi as president of the country. Many observers believe Yemen is in danger of becoming a failed state.

Persian Gulf States While protests broke out throughout the Persian Gulf States in 2011 and 2012, the effects of those protests were limited, as the ruling elite remained in power and little direct political change resulted.

▼ **FIGURE 4.3.3** **Timeline of the Arab Spring and its aftermath.**

Arab Spring begins with Mohdi Bouazizi's self-immolation in Tunisia	Civil war begins in Syria	Gaddafi killed in Libya	President Saleh resigns and leaves Yemen
2011		**2012**	**2013**
Unidentified man self-immolates in Saudi Arabia	Mubarak resigns amid nationwide protests in Egypt	Saudi Arabia grants women the right to vote	Muslim Brotherhood candidate Morsi elected president of Egypt

2 The Rise of the Islamic State of Iraq and Syria (ISIS)

Since 2011, the Islamic State of Iraq and Syria (ISIS) has built support in the region and claimed parts of Iraq and Syria by targeting both foreign forces as well as Muslims who do not follow the "appropriate faith" of Sunni Islam (Figure 4.3.2).

6.4 million
People living in ISIS occupied territory

50,000-100,000
Estimated number of fighters

12,000-20,000 mi²
Territory under ISIS control, including several major Iraqi cities

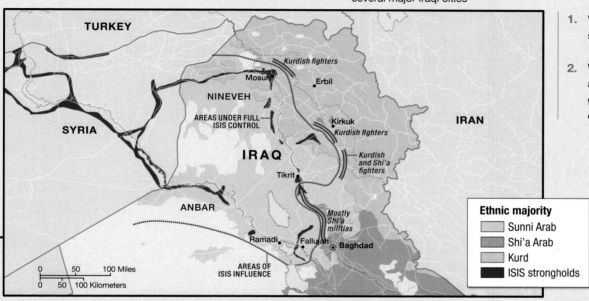

▲ FIGURE 4.3.2 **ISIS Controlled Space in Relation to Ethnic Regions in Iraq.**

1. **What has sparked protest in the region since 2009?**

2. **Why are there different types of protests and different responses by the various governments to them among the many countries in the region?**

"Many observers believe that the region's long-standing political order, consisting largely of military governments or rule by postcolonial dynasties, will remain in control."

3 Economic costs of war in Iraq and Syria

Relative to the levels that could have been achieved if war hadn't broken out:

Syria
over 50%
Unemployment rate

Iraq
−28%
Change in per capita income

$35 billion
Estimated loss of output from the combined economies of Turkey, Syria, Lebanon, Jordan, Iraq, and Egypt, measured by their Gross Domestic Product.

−90%
Change in Imports/exports

−16%
Change in per capita GDP

Economic Impact of the Syrian War

https://goo.gl/lm2nQZ

4 Looking Toward the Future

Many observers believe that the region's long-standing political order, consisting largely of military governments or rule by postcolonial dynasties, will remain in control. The growth of democratic rule—which would include accountability, transparency, and equality—will be very slow. That said, change should continue. Iran appears to have emerged from this period with greater regional and global influence, and minority groups, such as the Kurds in northern Iraq, have gained greater autonomy.

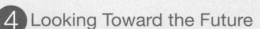

Military coup deposes
Morsi in Egypt

Sisi elected
president of Egypt

Hadi forced to leave
Yemen, civil war begins

2014

2015

Civil war begins
in Libya

ISIS invades
Iraq

Over 650 million displaced Sudanese refugees live outside their country and there are a reported 2.19 million Sudanese **internally displaced persons** (IDPs), individuals who are uprooted but remain in their home country as refugees because of civil conflict or human rights violations (**FIGURE 4.32**).

In May 2006, a peace agreement was brokered but in 2012 violence was still occurring in parts of Darfur. Reports in 2015 indicate that despite years of international calls for peace, violence still reigns in Darfur. Sudanese military personnel reportedly raped hundreds of women in February 2015 as the conflict continued to rage. That most international observers and UN peacekeepers have had only sporadic access to Darfur has only exacerbated such problems.

The Israeli–Palestinian and Israeli–Lebanese Conflicts
The historical geography of the Israeli–Palestinian conflict is

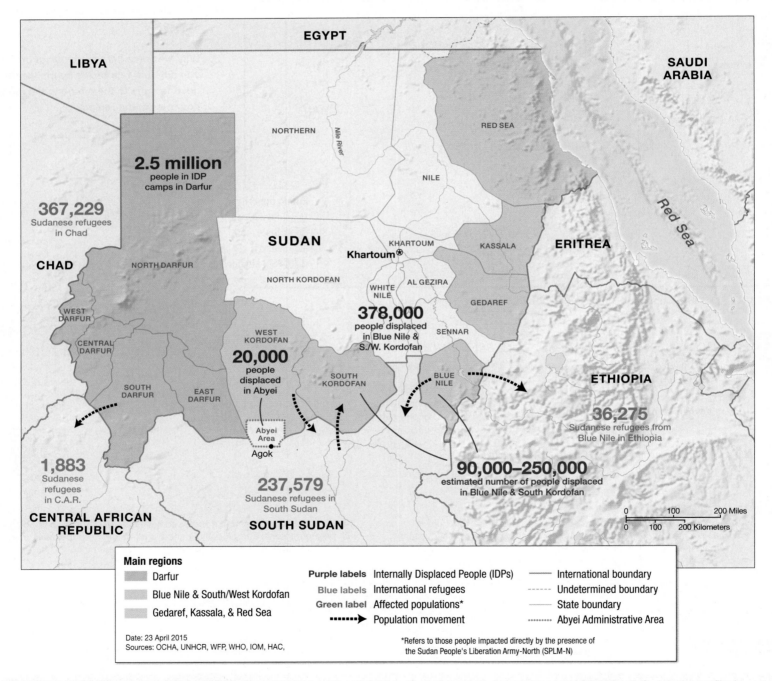

▲ **FIGURE 4.32 Civil Conflict and Human Rights Issues in Sudan** Even though a referendum created a new country in 2011, South Sudan, the conflict in Sudan continues, with many people displaced or without access to international humanitarian support. Some people are fleeing Sudan for the new South Sudan, while others are displaced by conflicts that remain internal to Sudan or South Sudan itself. Complicating matters has been the emergence of the Sudan People's Liberation Movement-North (SPLM-N), which has been fighting the Sudanese government in the Blue Nile and South Kordofan regions of Sudan. **What are some of the challenges related to the movement of so many displaced peoples?**

complex (**FIGURE 4.33**). The Jewish state of Israel is a mid-20th-century construction that has its roots in the philosophy of **Zionism**, a late 19th-century movement that defined the need for an independent homeland for the Jewish people. During World War I, the British issued the **Balfour Declaration**, which supported the legal migration of Jews to Palestine. The Balfour Declaration was highly problematic because Palestinians viewed the arrival of Jews as an incursion into their historical homeland. In 1947, with conflict continuing between Jews and Palestinians,

Britain announced that it would withdraw from Palestine in 1948. The United Nations, under heavy pressure from the United States, responded by partitioning Palestine into Arab and Jewish states. In the plan, Jerusalem was to be an international city administered by the United Nations. The proposed plan was accepted by the Jews and angrily rejected by Arabs, who argued that taking a mandate territory from an indigenous population was illegal under international law. Many Palestinians were passionately against any two-state solution.

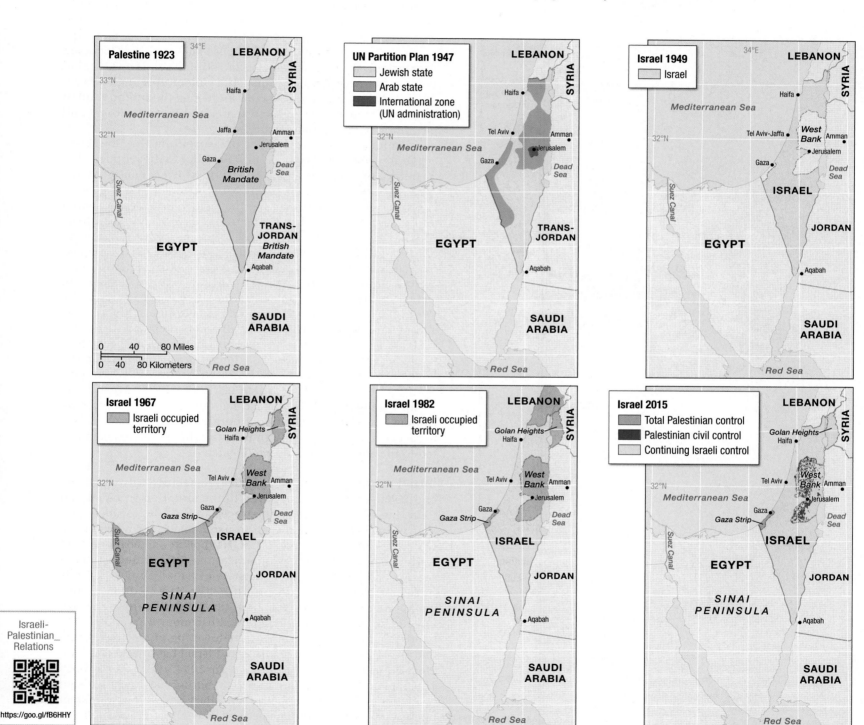

Israeli-Palestinian_Relations

https://goo.gl/fB6HHY

▲ **FIGURE 4.33 Changing Geography of Israel and Palestine, 1923–2015** Since the creation of Israel from part of the former Palestine, the region's political geography has undergone significant modifications. For more on the history of Israeli–Palestinian relations view the timeline at the QR link.

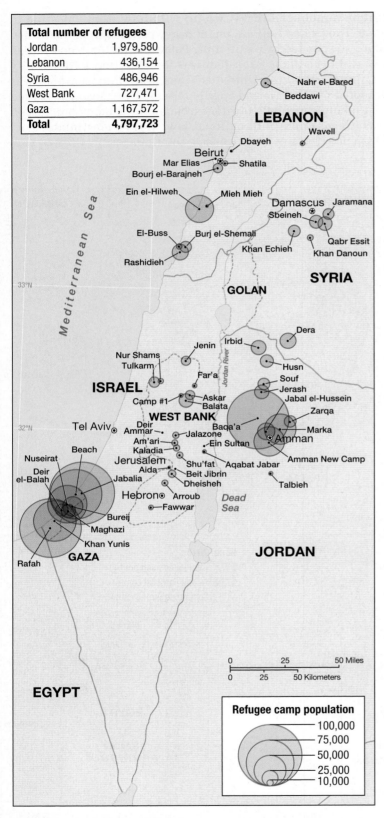

Total number of refugees	
Jordan	1,979,580
Lebanon	436,154
Syria	486,946
West Bank	727,471
Gaza	1,167,572
Total	**4,797,723**

Refugee camp population
- 100,000
- 75,000
- 50,000
- 25,000
- 10,000

▲ **FIGURE 4.34 Palestinian Refugees in the Middle East** This map shows the dispersion of Palestinian refugees, in camps and elsewhere, in the states around Israel and in the West Bank and Gaza. One of the biggest obstacles in the Israeli–Palestinian peace talks has been the question of refugee return and where Palestinians will be allowed to settle.

In 1948, Britain withdrew, Jewish political leaders proclaimed the state of Israel, and war broke out between Israel and the combined forces of Egypt, Jordan, and Lebanon, as well as smaller units from Syria, Iraq, and Saudi Arabia. This war, known as **The First Arab–Israeli War**, ended in the defeat of Arab forces in 1949. Subsequent agreements enabled Israel to expand its territory into Jerusalem. In 1950, Israel declared Jerusalem its national capital, though very few countries recognize it as such.

Israel drove hundreds of thousands of Palestinians from their homeland through their expansion (**FIGURE 4.34**). Many Palestinians were forced to live as refugees in other Arab countries in the region or under Israeli occupation in the West Bank, the Golan Heights, and the **Gaza Strip** (also known as the **Occupied Territories**). By the late 1980s, Palestinians in the Occupied Territories rose up in rebellion. This violent uprising of Palestinians against the rule of Israel, known as the **intifada**, often involved frequent clashes between armed Israeli soldiers and rock-throwing Palestinian young men.

In summer 2005, and under pressure from the UN, Israel began to cede territory back to Palestine. Critics argued that the return of land was a hollow gesture as Israel continued the construction of physical barriers between Israelis and Palestinians (**FIGURE 4.35**). Israel argues that such barriers are essential to protect its citizens from Palestinian terrorism. Palestinians contend that the barrier's purpose is geographical containment of the Palestinians and future expansion of Israeli sovereignty. It is clear, that these barriers uproot Palestinian settlements and separate them from their livelihoods (**FIGURE 4.36**). In October 2003, the UN General Assembly voted 144–4 that the West Bank barrier was "in contradiction to international law" and therefore illegal. The Israeli government called the resolution a "farce." The future of the Palestinian–Israeli peace process remains in doubt today, as tensions remain between the Israel government and the Palestinian Authority, which currently governs some Palestinian territories in the West Bank. Complicating matters is the fact that the Gaza Strip is ruled by **Hamas** (an Islamist Palestinian political party). Israel refuses to recognize or negotiate with Hamas, which remains committed to the abolition of the Israeli state.

Israel and Lebanon The political instability between Israel and Lebanon is very much tied to the situation in Israel and Palestine. The tension can be traced to the 1967 Arab–Israeli War when Palestinians began to use Lebanon as a base from which to launch attacks on Israel. **Hezbollah**—a Lebanon-based political and paramilitary organization—also states as one of their goals the abolition of the state of Israel. Direct conflicts between the two states include the 1982–2000 Israeli occupation of southern Lebanon, the 2006 kidnapping of two Israeli soldiers by Hezbollah and subsequent war, and a 2010 firefight across the border. Despite lowering of tensions, Israel and Hezbollah remain in conflict and Israel has attacked several Hezbollah military supply lines in 2014 and 2015 in both Lebanon and Syria. In January 2015, border tensions escalated again, after Hezbollah attacked a military convoy within Israel and Israel returned fire. The peace between these two neighbors thus remains quite fragile.

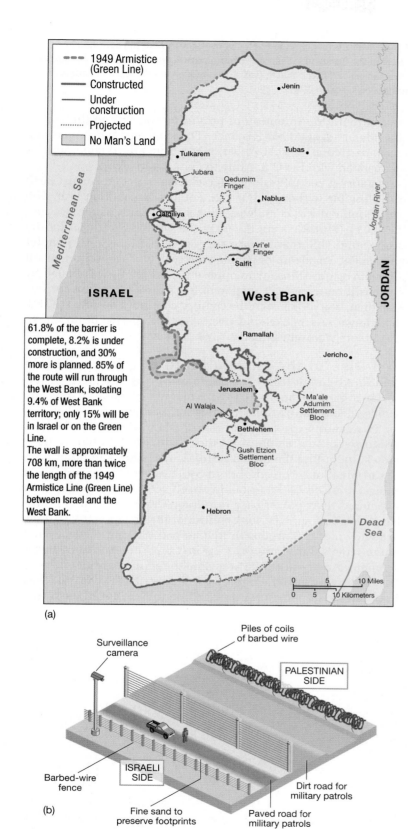

1949 Armistice (Green Line)

Constructed

Under construction

Projected

No Man's Land

61.8% of the barrier is complete, 8.2% is under construction, and 30% more is planned. 85% of the route will run through the West Bank, isolating 9.4% of West Bank territory; only 15% will be in Israel or on the Green Line.

The wall is approximately 708 km, more than twice the length of the 1949 Armistice Line (Green Line) between Israel and the West Bank.

Mediterranean Sea

Jenin

Tulkarem

Jubara

Qedumim Finger

Nablus

Qalqiliya

Ari'el Finger

Salfit

ISRAEL

West Bank

JORDAN

Jordan River

Ramallah

Jericho

Jerusalem

Ma'ale Adumim Settlement Bloc

Al Walaja

Bethlehem

Gush Etzion Settlement Bloc

Hebron

Dead Sea

0 5 10 Miles
0 5 10 Kilometers

(a)

Surveillance camera

Piles of coils of barbed wire

PALESTINIAN SIDE

ISRAELI SIDE

Barbed-wire fence

Fine sand to preserve footprints

Paved road for military patrols

Dirt road for military patrols

(b)

▲ **FIGURE 4.35 Israeli Barriers** (a) Shown are the planned and completed portions of the structures referred to as "security fences" by the Israelis, which are called "the wall" or the "apartheid wall" by Palestinians and other opponents. (b) The physical barriers consist of a network of fences, walls, and trenches equipped with monitoring and surveillance devices.

▲ **FIGURE 4.36 West Bank Fences** The fences built by the Israeli government are designed to isolate and protect Jewish settlements throughout the West Bank. These fences mark more than just a military boundary, they also mark an economic boundary between wealthier Israelis and poorer Palestinians. Palestinians are often cut off from jobs and other key social services by the policed fences.

Apply Your Knowledge

4.14 Compare and contrast the role that ethnic and religious differences have played in various conflicts of the Middle East and North Africa.

4.15 How do ethnic and religious differences complicate the region's image as being unified by one religion (Islam) and one ethnic identity (Arab)?

The U.S. War in Iraq The United States responded to the terrorist attacks of September 11, 2001, by declaring a global war against terrorism. Although the evidence of involvement of Iraq was highly questionable, on March 19, 2003, U.S. President George W. Bush ordered the bombing of the city of Baghdad (**FIGURE 4.37**). The motivation for the war, as expressed by U.S. President George W. Bush, was that Iraq had stockpiled **weapons of mass destruction (WMD)**—chemical and biological weapons capable of massive human destruction. Despite the fact that the UN weapons inspector Hans Blix and his team were unable to locate any weapons, President Bush justified the invasion as part of an escalated "war on terrorism," saying that, "neutralizing" Iraq's leader, Saddam Hussein, was necessary for global security.

On May 1, 2003, President Bush announced the end of major combat operations in Iraq. Over the next 4 years, the United States continued to have a heavy presence in the country. By 2007, the stability of Iraq remained unclear, and President Bush ordered, over Congressional opposition, a *surge* of 140,000 troops

to Iraq with the intention of providing security to the city of
Baghdad and Al-Anbar Province. The surge met with some success, particularly in Al-Anbar, where some local communities
cooperated with the U.S. Army to fight terrorism.

In spring 2009, President Barack Obama set an August 2010
deadline for withdrawal of U.S. combat forces from Iraq. The
actual pullout was completed almost 18 months later, in December 2011 (**FIGURE 4.38**). Even as the United States pulled its
troops out, U.S. intelligence suggested that the global **jihadist**
movement—made up of people who seek to wage war on behalf
of Islam against those who oppose the religion—was fueled by
the coalition forces' occupation of Iraq. It is perhaps not surprising, then, that violence in the country continues despite the U.S.
withdrawal. As an example, over a 5-month period beginning in
2015 there were an estimated 7,000 deaths.

**The Rise of the Islamic State of Iraq and Syria
(ISIS)** Although ISIS became notorious as an emergent terrorist organization in 2014, its origins go back to the late
1990s. The organization's membership adheres to Wahabism,
an extreme interpretation of Sunni Islam, the goals of which
are to purge the region of all those who do not align to their
views, including the region's Shi'a population and other non-
Sunni Muslims. The conflict in Syria, where the group found
refuge, allowed ISIS to establish itself geopolitically. ISIS then
expanded their power base into Iraq. This rise of Shi'a political power in Iraq, who took control of the government, and the
drawdown of U.S. troops in Iraq helped ISIS build resentment
among Iraq's Sunni minority against the Shi'a-controlled government. As a result, ISIS was able to recruit new members in
Iraq relatively easily. As they gained territorial control, they
imposed harsh penalties against those they believed unworthy

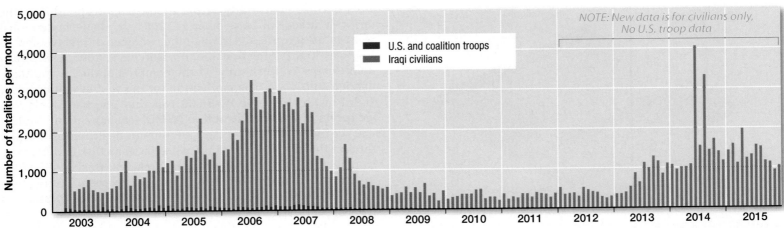

▲ FIGURE 4.38 **Fatalities in Iraq War** Counting the number of people who have died during this war is always a challenge. In this graph, data for civilian fatalities
were compiled by "Iraq Body Count" (www.iraqbodycount.org), which is an independent organization that uses news reports to calculate the total loss of civilian
life since the start of Operation Iraqi Freedom in 2003. U.S. and coalition troop fatalities were gathered from "Iraq Coalition Casualty Count" (icasualties.org), which
bases its numbers on published government reports.

for their state, including beheadings. The fighting also led to the mass displacement of Iraqis from towns that ISIS claimed (**FIGURE 4.39**). ISIS's control of large parts of Syria and Iraq raised questions as to the future viability of a unified Iraqi state.

Nuclear Tensions In 2008, Iran declared, like other countries, its inalienable right to pursue uranium enrichment. Iran's nuclear program precipitated a strong response from many members of the international community, including the UN, which does not support anyone having nuclear weapons. Iran refused to cooperate with the UN, so the organization established trade sanctions to limit Iran's ability to benefit from its oil reserves. Despite UN sanctions, in February 2012, Iran declared that it had developed its own uranium to fuel a nuclear reactor. Iran's first nuclear power plant at Bushehr was reported to be operating at 75% of its full capacity.

Changes in U.S. foreign policy in 2014 and 2015 along with the election of new leadership in Iran in 2013 has provided space for the international community to work more collaboratively with Iran on its nuclear program. This includes the potential for official international oversight of Iran's nuclear program. But, there remains a contentious political landscape in the United States, where some members of the U.S. Congress have pushed against any such deal with Iran. The future of nuclear tensions thus remains unclear.

Regional Alliances Despite regional conflict, many cooperative organizations operate in the region today. Some of these regional alliances have their origins in a political philosophy called **Pan-Arabism**, which was a movement across the region to build cooperation across Arab nations and also ally Arabs against the Ottomans and later Europeans. **The Arab League**, founded in 1945 by Egypt, Iraq, Lebanon, Saudi Arabia, Syria, Transjordan (Jordan, as of 1948), and Yemen, emerged out of this philosophy. Many other countries joined the League after its inception. The stated purpose of the Arab League is to strengthen ties among member states, coordinate policies, and promote common interests. The league is also involved in social outreach, such as literacy campaigns and programs addressing labor issues.

The **Gulf Cooperation Council (GCC)** coordinates political, economic, and cultural issues of concern to its six member states—Saudi Arabia, Kuwait, Bahrain, Qatar, the United Arab Emirates, and Oman. The members of the GCC coordinate the management of their substantial income from oil reserves and address problems of economic development as well as discuss social, trade, and security issues. The GCC has extensively funded Arab countries' economic development as well as military protection during political crises often siding with member governments against protests by their citizens.

▲ **FIGURE 4.39 ISIS Military Fighters** Members of ISIS demonstrate their military power with a march through Raqqa Province of Syria in June 2014.

Apply Your Knowledge

4.16 List three examples of current political and territorial issues in the region.

4.17 What are some of the drivers of these tensions and what are the consequences of the ongoing conflict in the region?

Culture and Populations

The Middle East and North Africa is culturally and demographically diverse. This diversity includes countries where religion and politics are directly aligned as well as countries whose governments are based on **secularism** (nonreligious). Demographically, the region is quite diverse. Some countries produce many migrants, while others are sites of high levels of in-migration. Moreover, the region is crosscut by its ethno-linguistic diversity.

Religion and Language

Unlike in many parts of the globe where societies have become increasingly secular, religion in this complicated region is often a central feature of everyday life for the vast majority of the inhabitants. This is particularly true in countries, such as Saudi Arabia, Oman, Yemen, and Iran, where **Shari'a**, a traditional body of law that comes from Islamic teachings, constitutes both the legal and political system. In countries that apply Shari'a there remain differences—Saudi Arabia is a monarchy whereas Yemen has a two-chamber presidential republic.

Islam In this region, 96% of the population practices some form of **Islam** (FIGURE 4.40). The Arabic term Islam means "submission to God's will." As a religion founded on the revelations of God to humankind, Islam, which is practiced by people called **Muslims**, recognizes the prophets of the Hebrew and Christian Testaments of the Bible, but considers Muhammad to be the last prophet and God's messenger on Earth.

- **The Qur'an and the Sunna** Muslims hold that Muhammed received the word of God from the Angel Gabriel in about 610 C.E. Muhammed's teachings are found in two sources of Islamic doctrine and practice: the **Qur'an** and the **Sunna**. Muslims regard the contents of the Qur'an, the Islamic sacred book, as directly spoken by God to Muhammad. The Sunna is not a written document but a set of practical guidelines for behavior, the body of traditions derived from the words and actions of the prophet Muhammad.

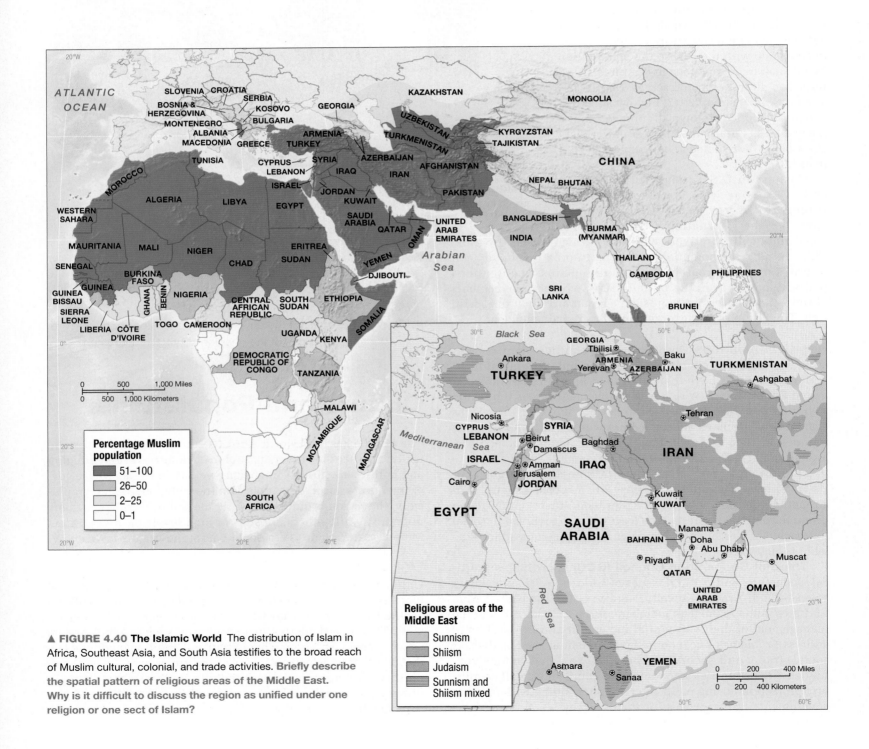

▲ **FIGURE 4.40 The Islamic World** The distribution of Islam in Africa, Southeast Asia, and South Asia testifies to the broad reach of Muslim cultural, colonial, and trade activities. **Briefly describe the spatial pattern of religious areas of the Middle East. Why is it difficult to discuss the region as unified under one religion or one sect of Islam?**

- **The Five Pillars of Islam** In Islam, God has four fundamental functions—creation, sustenance, guidance, and judgment—and the purpose of the people is to worship the one true God. In Islam, it is also held that the actions of the individual should serve humanity, not the pleasures or ambitions of the self. A Muslim must fulfill five primary obligations, known as the five pillars: repeating the profession of the faith ("There is no god but God; Muhammad is the messenger of God"); praying five times a day facing **Mecca** (the city where Muhammed was born in 570 C.E.); giving alms (charitable giving); fasting from sunup until sundown during the holy month of Ramadan; and making at least one pilgrimage, called a **hajj**, to Mecca if financially and physically able (**FIGURE 4.41**). Islam also has strict rules against certain forms of vice, including restrictions on the consumption of alcohol. Islam's moral and ethical code, like Christianity's Ten Commandments, has made it possible for the religion to spread globally and be adopted widely by many different societies.

- **Islam's Sacred Cities** When Islam first emerged, Mecca was a node in the trade routes that connected Yemen and Syria and eventually linked the region to Europe and all of Asia. Because of Mecca's central location, Islam spread through the interconnected trade networks of the region. Today, Mecca is both the most important sacred city in the Islamic world and an important commercial center. A second city, Medina, is also a sacred city; it is the city to which Muhammad fled after he was driven out of Mecca by merchants that felt that he was a threat to the city's commerce. Critical to any Islamic city is the mosque. Mosques serve as the main place of worship for Muslims while also functioning as law court, school, and assembly hall; adjoining chambers often house libraries, hospitals, or treasuries.

▼ **FIGURE 4.41 Mecca, Saudi Arabia** Every year, during the last month of the Islamic calendar, more than 1 million Muslims make a pilgrimage, or hajj, to Mecca. In addition to the required pilgrimage, Islamic traditions require Muslims around the world to face Mecca during their daily prayers.

- **Islamic Sects** Disagreement over the line of succession from the prophet Muhammad occurred shortly after his death in 632 C.E. and resulted in the split of Islam into two main sects, the **Sunni** and the **Shi'a** (sometimes known as Shiite). The central difference revolves around the question of who should hold the *political* leadership of the Islamic community and what the *religious* dimensions of the leadership should be. The Shi'a contend that political leadership must be divine and must derive from descendants of the Prophet. Sunnis argue that the clergy (with no divine power) should succeed Muhammad. Sunni Islam became dominant in the region, although Shi'a Islam as well as other forms of Islam, such as Sufism, a mystical form of Islam, also thrive. Shi'a Islam, for example, is the dominant form of Islam in Iran and also makes up a large portion of Iraq's population (see Figure 4.39). The long-term global networks of Islam have facilitated Muslims' migration out of the region—to Europe and the United States. This movement has resulted in new ideas for Islamic practice that have, in turn, helped shape the practice of Islam globally.

- **The Rise of Islamism** In recent years, the Muslim world has witnessed the rise of **Islamism**, which is more popularly, although incorrectly, referred to as Islamic fundamentalism. Whereas *fundamentalism* represents a strict adherence to the fundamentals of a religious system, Islamism is an anti-Western and anti-imperial political movement. Islamists resist Western forces of globalization—namely modernization and secularization. Very importantly, most Muslims are *not* Islamists. Islamism is a militant movement that wants to create societies that are based in a universal Islamic state—a state that is religiously and politically unified by incorporating principles from the sacred law of Islam into state constitutions. ISIS provides a unique challenge this concept, as they believe in a global post-national society based on Islam.

- **The Meanings of Jihad and Qital** The concept of **jihad** is a complex term derived from the Arabic root meaning "to strive." All Muslims struggle to be better Muslims through their adherence to the Five Pillars of Islam and the broader ethical code of Islam. This means that the current use of the term connotes an inward spiritual struggle to attain perfect faith. For the majority of Muslims, jihad is the peaceful struggle to establish Islam as a universal religion through the conversion of nonbelievers. This can be seen in the struggle of Shi'a Muslims for social, political, and economic rights within Sunni-dominated Islamic states. Jihad has also been used to represent an outward material struggle to promote justice and the Islamic social system. **Qital** (fighting or warfare) is a more focused form of jihad according to the Qur'an. When a war directed against their enemies occurs, it may be interpreted as a holy war. This interpretation has been used by some within Islamist ranks to justify the actions of Islamists, such as ISIS or al-Qaeda. Agressive holy war is, however, not a part of the Islamic tradition or texts.

Christianity, Judaism, and Other Middle Eastern and North African Religions

Although Islam is the most widely practiced religion in the region, it is by no means the only religion of political, cultural, or social significance. There are more than a dozen Christian sects—among them are Coptic Christians in Egypt, Maronites in Lebanon, the Chaldean Catholic Church in Syria, and various other Christian peoples, including Armenian, Greek, Ethiopian, and some Protestant faiths (**FIGURE 4.42**). Generally, Jews in the Middle East and North Africa are secular, observing some Jewish traditions. A small percentage practice **Orthodox Judaism**, which is based on a strict adherence to the religious texts of the Old Testament, or **Hasidism**, which is a mystical offshoot of Judaism.

The three regionally predominant religions—Islam, Judaism, and Christianity—have helped shape the people and the landscape of the region. The most obvious and enduring influences on the landscape have been places of worship and sacred spaces, more generally. Nowhere is the enduring interrelationship of the three religions more apparent than in the ancient city of Jerusalem (**FIGURE 4.43**). The centrality of Jerusalem as an ancient religious space, as well as its contemporary significance as a place of pilgrimage for Jews and Christians, is very much tied up with the Arab–Israeli conflict.

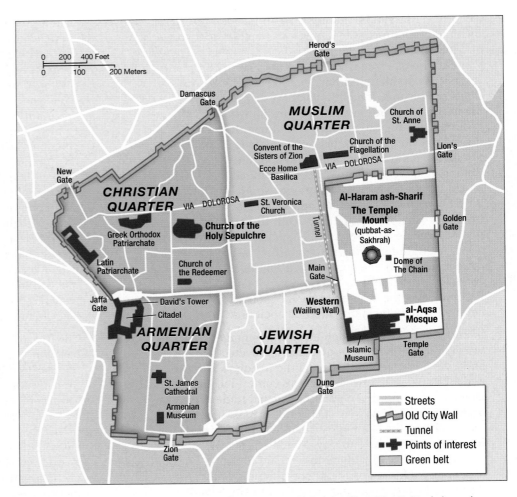

▲ **FIGURE 4.43 Map of the Old City of Jerusalem with Religious Sites Highlighted** Jerusalem has a complex ethnoreligious history because it plays such an important role in all three major world religions. As a long-time international trading city, it has also been an important place for a wide range of ethnic groups, including the Armenians.

Regional Languages

Three major language families dominate in the Middle East: Semitic (including Arabic, Hebrew, and Aramaic); Indo-European (Kurdish, Persian, and Armenian); and Turkic (Turkish and Azeri). These different languages have influenced each other: Persian is written in Arabic script, and Turkish incorporates vocabulary words from Persian and Arabic. All three are spoken in regional dialects that are not always mutually understood and multiple languages coexist within countries. Kurdish, for example, is spoken in five different countries and is a minority language in each of them. In addition to

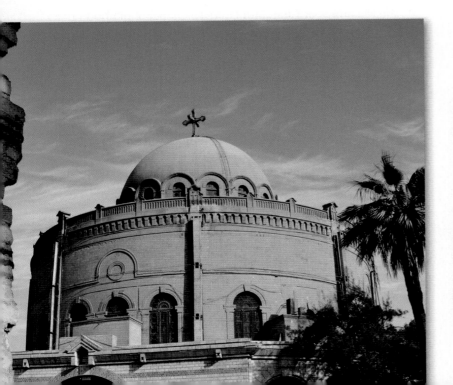

◄ **FIGURE 4.42 Coptic Church of St. George (Mari Girgis) in Cairo, Egypt** The Coptic Church serves between 6 and 11 million Egyptians today. The Church was established around 50 B.C.E. when the apostle Mark came to Egypt. Over its history, the Church has played an important role. It established one of the earliest schools of the Christian religion and was one of the founding members of the World Council on Churches. The Church is led by the Pope of Alexandria. Its membership includes not just Egyptians but about 1 million followers outside of Egypt.

the prominent languages, some ethnic and religious communities have preserved their own languages for religious use, such as Coptic and Greek. Berber, an Afro-Asiatic language, is spoken in the region from Morocco to Egypt. Nubian, a Nilo-Saharan language, is spoken by Egyptians and Sudanese. And there are nearly 1 million Zazaki (an Indo-European language) speakers in southern Turkey. The most common foreign language in the Middle East is English. In North Africa, the colonial languages of French (Morocco, Algeria, and Tunisia), Spanish (Western Sahara), and Italian (Libya) are also spoken, as is English.

Apply Your Knowledge

4.18 Briefly compare and contrast the different religious and linguistic systems found in this region.

4.19 How does this religious and linguistic diversity unify and divide the region socially?

Cultural Practices, Social Differences, and Identity

The culture of the Middle East and North Africa is as diverse as that of any other region of the globe. Global media technologies are facilitating new social and cultural forms and allowing historical ones to be reconfigured. The resulting cultural practices and social issues in the Middle East and North Africa reflect a continually changing mix of both regional and global influences.

Cultural Practices The cultures of the Middle East and North Africa have transformed both places within the region as well as the landscapes of many other parts of the world.

- **Food** The cuisine of the region is available in many large cities in most of the world's regions (**FIGURE 4.44**). The types of food available from this region include *meze* dishes, predominantly subtly spiced appetizers or small plates. Some of the most popular meze dishes include *baba ganoush,* a puree of toasted eggplant, sesame seeds, and garlic; *falafel,* a mixture of spicy chickpeas rolled into balls that are deep fried; *fuul,* brown broad beans seasoned with olive oil, lemon juice, and garlic; and *tabbouleh,* a salad of bulgur wheat, parsley, mint, tomato, and onion. Other regional specialties include grilled meats, especially lamb and chicken, often served with rice.

- **The Arts** The most widespread of the region's dances is the traditional belly dance, a women's solo dance done for entertainment. The dance is characterized by undulating movements of the abdomen and hips and by graceful arm movements. Belly dancing is believed to have originated in medieval Islamic culture, though some theories link it to prehistoric religious fertility rites.

Film industries in the region are also finding their voice, producing some of the most provocative and engaging cinema in the world today. In 2012, the film *A Separation,* directed by Asghar Farhadi, won the academy award for best foreign language film (**FIGURE 4.45**).

▲ **FIGURE 4.44 Foodstuffs from Israel** A spice market in Jerusalem, Israel shows the wide range of ingredients used in Israeli cooking. These same spices, of cousre, are used in cuisine throughout the region.

▲ **FIGURE 4.45 The Oscar Winning Film, "A Separation"** Asghar Farhadi's Oscar Award–winning film traces the conflict that arises when a married woman desires to leave Iran and her husband does not. At the heart of their conflict is the need to care for an aging parent in Iran versus the desire to find better opportunities for their child outside Iran. The film illustrates the ongoing tensions and anxieties found in everyday Iranian society, as people struggle to find their voice in a society with a highly regulated cultural production system. His films are known for their sensitively documenting the human experience in Iran. As a result, while other directors have faced government repression for their work, Asghar Farhadi has thus far avoided government sanction.

▲ FIGURE 4.46 **Berber Shepherd Campsite** The Berbers have lived in North Africa for thousands of years. *Berber* is the name applied to the language and people belonging to many of the tribes who currently inhabit large sections of North Africa. Pictured here is a Berber shepherd campsite in the High Atlas Mountains, where herds of sheep are tended. Increasing numbers of Berbers are raising crops, a practice that signals the erosion of their nomadic practices.

Kinship, Family, and Social Order In this world region, **kinship**, a concept that typically refers to a relationship based on blood, marriage, or adoption, is often expanded to include a shared notion of a relationship among members of a group. This means that neighbors, friends, and even individuals with common economic or political interests can be considered kin. Kinship often determines how people interact with each other.

The concept of **tribe** is also central to the sociopolitical organization of the region. A tribe is a form of social identity created by group members who share a common set of ideas about loyalty, political action, and cultural identity. A shared tribal identity leads to the formation of collective loyalties and a primary allegiance to the tribe over other social groups—nations, for example. Many rural communities, who subsist on the breeding and herding of animals or sedentary agricultural practices, are proudly tribal. This is particularly true for communities that practice **transhumance**, the movement of herds according to seasonal rhythms, as tribe becomes a key defining factor for relations such as marriage (**FIGURE 4.46**).

Where the concept of tribe remains vital, it is a valuable element of local social life. But, tribes are under pressure in this region. Some governments have sought to destabilize tribal leadership to control tribal communities and territories. Tribal communities often exist at border regions and they have strong historical connections with people in another country. Tribal loyalties might destabilize national governments and serve as spaces for transnational connection at the expense of national unity. This may very well be the case in some unstable areas such as the Kurdish region along the Turkey, Iraq, Iran, and Syria borders and along the border of Afghanistan and Pakistan.

Gender Stereotypes of gender relations under Islam in the Middle East and North Africa do not capture the variation of gendered identities across lines of class, religion, generation, level of education, and geography (urban versus rural origins, for instance).

- **Defining Gender Differences** Men and women in this world region, like many others, broadly hold to the ideological assumption that women should be subordinate to men. Many men in the region regard women's subordination as natural. Most women tend to regard their subordination as the product of the society in which they live and therefore as something that can be negotiated and manipulated. What makes the Middle East and North Africa different from many other world regions is how men exercise control over women by restricting female access to public space and secluding women within private spaces (**FIGURE 4.47**).

- **The Diversity of the Veil** If there is one image that tends to stereotypically represent gender politics in the Middle East and North Africa it is the practice of veiling women. But there are differences across the region in both how the veil is worn and what it is called (**FIGURE 4.48**). The veil varies from the all-encompassing full body garment, known as a *burqa* or the *niqab* to a **chador**, which shows the face, to the *al-amira*, which is often worn outside the Middle East

▼ FIGURE 4.47 **Gendered Spaces—Women at the Blue Mosque** A common feature of mosques is distinct spaces of worship for men and women.

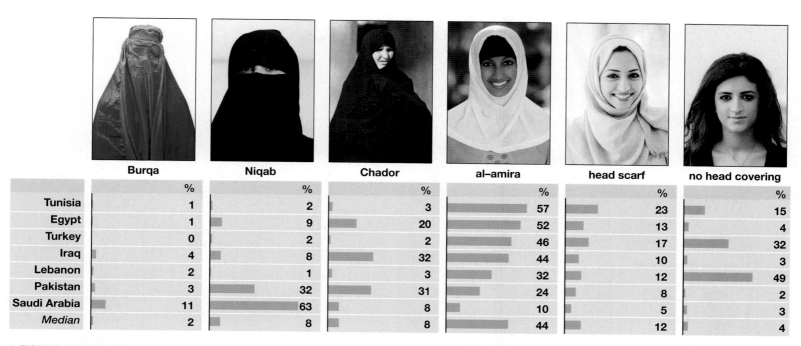

	Burqa		Niqab		Chador		al–amira		head scarf		no head covering	
		%		%		%		%		%		%
Tunisia		1		2		3		57		23		15
Egypt		1		9		20		52		13		4
Turkey		0		2		2		46		17		32
Iraq		4		8		32		44		10		3
Lebanon		2		1		3		32		12		49
Pakistan		3		32		31		24		8		2
Saudi Arabia		11		63		8		10		5		3
Median		2		8		8		44		12		4

▲ **FIGURE 4.48 The Diversity of Women's Coverings** The covering of women's heads and bodies varies dramatically throughout the Islamic world. Compare and contrast at least two countries. What differences in head coverings do you see across those countries?

and North Africa, to a simple headscarf. In some parts of the region, women do not wear any head covering at all. The coverings, which are contingent on national policy and local practice, allow some women to effectively operate in public and yet remain in their personal space. Body coverings allow women physical and social mobility in societies that otherwise restrict women's movements to very private spaces, such as the home.

- **Women's Mobility and Political Participation** National politics related to gender affect how men and women see and represent themselves publicly in the region. Some societies, such as in Yemen and Bahrain, exercise very strict control over women's public movements. In some of the more secular societies of the region, such as Turkey and Egypt, women's public movements are less strictly regulated. Generally, urban women's public movements tend to be more restricted than those of rural women. This is largely because rural village life is typically defined by kinship relations, and women are allowed to operate relatively more freely among kin. In contrast, women in urban areas must move about in a world of both kin and strangers. This means that they must remain covered more often and spend less time in public areas. In some countries, women's rights have recently been expanded, such as their increasing participation in national politics (**FIGURE 4.49**).

- **Women's Economic Realities** Social reality is not always fixed and cultural assumptions and practices around gender are subject to negotiation and change. Although the predominant gender theme in the Middle East and North Africa is that women are subordinate to men,

women can and do exercise significant household as well as political influence and independence across a range of societies in this region. Examples of this include women's control over family resources. That said, women's participation in the workforce is much lower for women 15–24 in this region than for men (**TABLE 4.1**).

- **Women's Rights and Justice** Women's access to political rights and justice is diverse in this world region. In some countries, such as Israel and Turkey, women have full political rights. In countries, such as Saudi Arabia, that have long disallowed women the right to vote, they are now gaining those rights. Women have also been highly active in the political protests that have swept the region, including in Iran and Egypt. Women's rights and access to justice, however, remains precarious in some countries. In Bahrain, marital rape is not a crime when committed against a Shi'a woman, genital mutilation continues to be a serious issue in places such as Egypt, and women's laws regarding who they can marry still limit their flexibility in countries such as Jordan and Kuwait. In some countries, such as Egypt, women must renounce their Muslim faith, a process called **apostasy**, if they are to marry a non-Muslim (**FIGURE 4.50**).

Sexuality The rights of lesbian, gay, bisexual, and transgendered (LGBT) peoples in the region are also quite complex. Homosexuality is illegal (either in civil law or by Shari'a law) in all countries in the region except for Bahrain, Israel, Jordan, Turkey, as well as the West Bank area controlled by the

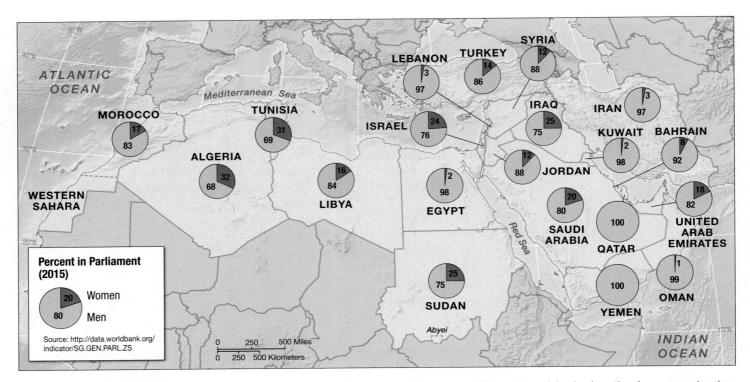

▲ **FIGURE 4.49 The Geography of Women's Political Participation in National Government** Women's participation in national government varies throughout the region. In some cases, parliaments are appointed by the head of state. In Saudi Arabia, for example, in 2013, for the first time, the King appointed women to parliament. In 2015, for the first time, Saudi Arabians can elect women to local offices and some women won. **Given your understanding of this region's complexity, do you see any patterns to the geography of women's political participation?**

TABLE 4.1 Men's and Women's Participation in the Labor Force of the Middle East and North Africa

Country	Labor Force Participation Rate (%), Male 15+	Labor Force Participation Rate (%), Female 15+
Algeria	80	37
Bahrain	85	32
Egypt	75	22
Iran, Islamic Republic of	73	32
Iraq	69	14
Israel	58	49
Jordan	74	23
Kuwait	83	45
Lebanon	72	22
Libya	79	22
Morocco	80	26
Oman	77	25
Palestine, State of	68	17
Qatar	93	50
Saudi Arabia	80	21
Sudan	78	26
Syrian Arab Republic	80	21
Tunisia	71	26
Turkey	75	30
United Arab Emirates	92	42
Yemen, Republic of	74	20

SOURCE: Regional Overview for the Middle East and North Africa: MENA Gender Equality Profile, Status of Girls and Women in the Middle East and North Africa, http://www.unicef.org/gender/files/REGIONAL-Gender-Eqaulity-Profile-2011.pdf

Palestinian Authority (but not the Gaza Strip, controlled by Hamas). Egypt has convicted men for "committing debauchery," while in Morocco one can be jailed for having a sexual relationship with someone of the same sex. In other countries, particularly those that apply Shari'a Law, sodomy is also considered a legal offense. ISIS has publicly executed LGBT people in the areas of the region they control. Israel is sometimes considered more socially open when it comes to the rights of LGBT individuals, but there is currently no state support for such rights like gay marriage in Israel.

Apply Your Knowledge

4.20 Go to http://www.trust.org and search for 'writaw' and create a schematic that highlights some of the major issues for women's rights in this world region.

Arab Women's Rights

https://goo.gl/pniqvE

4.21 Using the same site, what are some of the indicators that women's rights are expanding in parts of this world region?

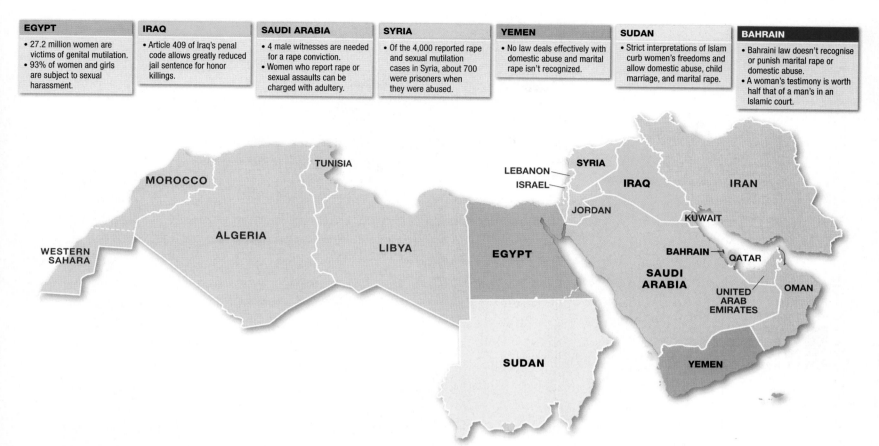

EGYPT
- 27.2 million women are victims of genital mutilation.
- 93% of women and girls are subject to sexual harassment.

IRAQ
- Article 409 of Iraq's penal code allows greatly reduced jail sentence for honor killings.

SAUDI ARABIA
- 4 male witnesses are needed for a rape conviction.
- Women who report rape or sexual assaults can be charged with adultery.

SYRIA
- Of the 4,000 reported rape and sexual mutilation cases in Syria, about 700 were prisoners when they were abused.

YEMEN
- No law deals effectively with domestic abuse and marital rape isn't recognized.

SUDAN
- Strict interpretations of Islam curb women's freedoms and allow domestic abuse, child marriage, and marital rape.

BAHRAIN
- Bahraini law doesn't recognise or punish marital rape or domestic abuse.
- A woman's testimony is worth half that of a man's in an Islamic court.

▲ **FIGURE 4.50 Women's Rights and Justice** Women's rights vary dramatically across the region. But many countries still have strong laws that curtail the rights of women's access to justice in the case of rape or domestic violence and abuse.

Demography and Urbanization

The distinctive pattern of population distribution in the Middle East and North Africa reflects the influences of environment, history, and culture. Though mountain environments can present substantial challenges to human habitation, the availability of moisture means that these environments can support agriculture over a somewhat shortened growing season. The mountains also provided safe havens for minority populations fleeing persecution and discrimination. The Druze in Syria and the Zayidis in Yemen are two groups that sought mountain refuge from their oppressors.

Whereas the highland areas are home to a small portion of the people of the region, the coastal areas, floodplains, and plateaus are the most densely populated (**FIGURE 4.51**). The Iranian and Anatolian Plateaus are also densely populated and the highland plateau of Yemen also contains a sizable population. The coastal areas and floodplains are equally attractive to humans; they constitute some of the most remarkable, highly engineered, and scenic of the region's landscapes. Other population clusters are found in cities, which have grown rapidly since the independence period of the 1950s. However, many of the people of the region still live in rural villages.

Demographic Change The total population of the region is over 500 million, but given the inaccuracy, infrequency, and

inconsistency of national censuses in the region, this number is only an approximation. Recent UN data suggests that the decades-long population boom appears to be coming to an end. Although this means that fertility rates are now stabilizing, there are still challenges in providing for the health, education, and welfare of both children and the elderly as well as providing jobs for those in between. For instance, in several of the most populated countries in the region, including Egypt, Algeria, and Turkey, a large part of the population is younger than 15 years of age. The populations of these countries will continue to grow as these individuals reach reproductive age (**FIGURE 4.52**).

Pull Factors A number of so-called pull factors draw people to this region. The founding of the state of Israel at midcentury drew large numbers of European and Middle Eastern Jews after World War II. Russian Jews arrived in Israel as a result of the end of the Cold War. More recently, Ethiopian Jews have migrated to Israel because of civil war. The oil economy has also drawn people to this region. In the states of the Arabian Peninsula, several factors—small local populations, lack of skill, lack of interest, and cultural resistance to the kinds of jobs made available through the oil economy—have meant a large number imported **guest workers** have been needed.

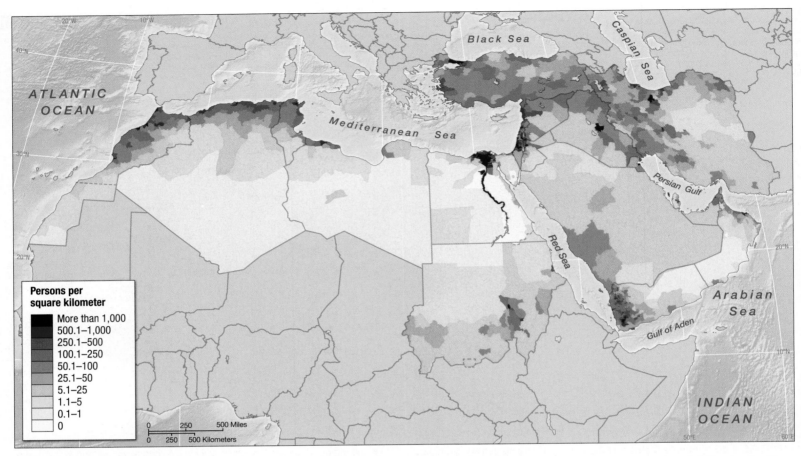

▲ **FIGURE 4.51 Population Distribution in the Middle East and North Africa, 2012** Population distribution is heavily influenced by the availability of water. The intensity and unusual linear pattern of population concentration along the Nile River valley is a perfect illustration of this point. Other concentrations, such as along the eastern end of the Mediterranean Sea, the northern edge of Algeria, and the northern parts of Iran, Iraq, and Turkey, also reflect higher availability of freshwater.

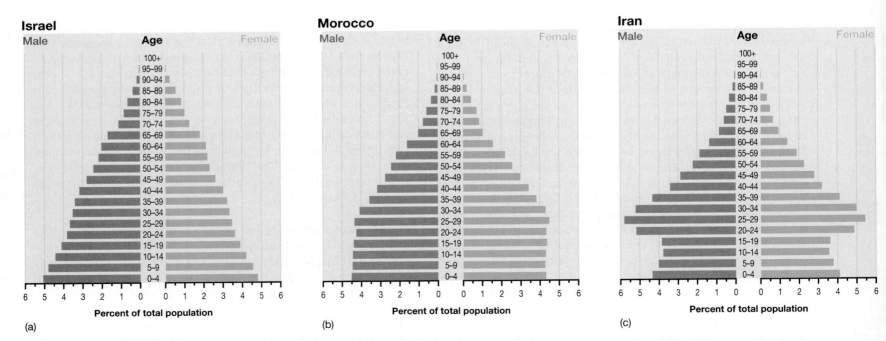

▲ **FIGURE 4.52 Population Pyramids of (a) Israel, (b) Morocco, and (c) Iran** These population pyramids demonstrate the broad differences in the national demographies of the region. Iran, for example, has a shrinking younger population relative to its middle ages population, whereas Israel still has a growing youth population. Morocco appears to have had consistent population growth over the last thirty years.

Refugees Flee the Violence of the Syrian Civil War

Imagine waking up one morning and remembering that you are no longer sleeping in your home country but are now a refugee in a foreign land. For the millions of people displaced by conflict in Syria a few have found their way to countries as distant and different as the United Kingdom. A woman from Syria has managed just this, first by making her way to the Mediterranean Coast and then across the sea to Europe. Another woman, named Nor, was fortunate to find herself in a refugee camp when she was given further refuge in the United Kingdom (**FIGURE 4.4.1**).

For many of those who remain in the refugee camps, however, life is difficult. The camps remain under-resourced, families are crowded together in temporary shelters, and weather conditions make everyday life difficult, particularly in the winter months. For many, particularly the approximately 1.6 million children who have been displaced by the war, the trauma of war can be felt each day in the camp. Many women, children, and men have been victims of torture, rape, or other sorts of violence. Others have seen their family members killed. The camps, while providing some relief from the violence of the war in Syria, do not end the trauma from the war (**FIGURE 4.4.2**). Many want to flee to Europe or other destinations where they hope their lives will be better.

> "Many women, children, and men have been victims of torture, rape, or other sorts of violence. Others have seen their family members killed."

▲ FIGURE 4.4.1 **Syrian Refugess Camp, Qtma, Turkey** Millions of Syrians have been displaced by the conflict in their home country. Many have fled to makeshift camps just outside Syria in countries, such as Turkey.

1. **Where do we find many of Syria's refugees today?**

2. **What are the living conditions of those who remain in the refugee camps?**

Syrian Refugees in U.K.

https://goo.gl/LgygJu

▶ FIGURE 4.4.2 **Syrians Awaiting Asylum** Syrians occupied parts of France's Calais Port in 2013 in an attempt to pressure the French and British government to grant them the right of asylum.

▲ **FIGURE 4.53** **Labor Migrants** Many immigrant workers live in the oil-rich countries of the Middle East and North Africa. In some countries, these workers make up 80% to 90% of the workforce, largely because there are so few local people to fill the jobs. These construction workers from India are at the site of an apartment complex of 40 skyscrapers on the outskirts of Dubai, UAE.

Guest workers have been brought in to work in all aspects of oil production, from exploration and well development to drilling, refining, and shipping (**FIGURE 4.53**).

As oil revenues have been reinvested in economic development projects, more service-sector positions and jobs in the building and construction industry have been created. To lessen the dislocating impact of foreign workers on local social and cultural systems, immigration policy among the oil-producing states of the Arabian Peninsula favors Muslims. Within the region, large numbers of guest workers from Syria and Egypt, as well as Palestinian refugees, have been participants in the Arabian Peninsula oil economy, filling both skilled and unskilled positions. A significant number also come from outside the region, especially from India, Indonesia, the Philippines, Bangladesh, and Pakistan. Most of the workers have been male; female labor migrants participating in the domestic labor industry are also coming from Indonesia and the Philippines (see Chapter 10, p. 382). Some countries, such as Saudi Arabia, have recently tightened their immigration laws to force companies to hire more Saudi Arabian citizens. This has impacted in-migration to the country, and a crackdown by the government on private organizations that employ too many non-Saudi workers.

Many sub-Saharan Africans have migrated from their countries of origin using North Africa as a point of transit into Europe. These migrants are often relatively well educated and from moderate socioeconomic backgrounds. They move because of a lack of opportunity, fear of persecution and violence, or a combination of both. Those who are unable to cross into Europe join growing immigrant communities in North Africa.

Push Factors The most consistent forces pushing migrants out of the region have been war, civil unrest, and the lack of economic opportunity. The most recent and massive out-migration is that of Syrians fleeing conflict and seeking refuge in large numbers in Turkey, Lebanon, and Jordan (**FIGURE 4.54**). Other significant emigration streams are made up of Algerians, Tunisians, and Moroccans, who have migrated mostly to Europe; Turks, who have followed a historical migration route to the Balkans and more recently to Germany; and Egyptians, who have migrated to other Arab countries in the region as well as elsewhere in the world. The migration of better-educated as well as low-skilled Turks and Egyptians are facilitated by European policies that enable temporary workers to take up low-paying, low-skill jobs that are not economically attractive to Europeans.

Cities and Human Settlement Cities have been an important component of this world region for thousands of years, with some of the earliest known cities on Earth emerging here (see "Urbanization", p. 145). Today, the predominant pattern of settlement in the region is a small number of very large cities, a substantial number of medium-sized cities, and a very great number of rural settlements. That said, cities in the region are growing dramatically each year. Only about 50 years ago, most people in the region lived in small rural settlements. Since political independence and the development of the oil economy there has been a dramatic increase in urbanization of the population, although this growth is not even across the region (**FIGURE 4.55**).

What has been most remarkable about contemporary urbanization is that its rapid pace has led to the emergence of one or two very large cities in each country wielding disproportionate

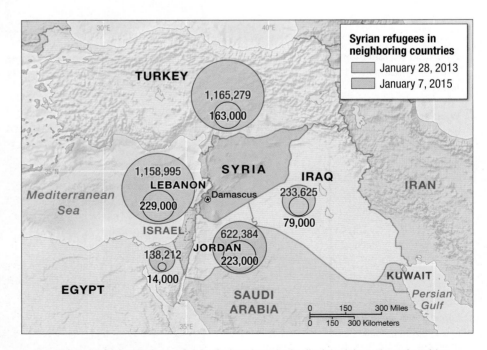

▲ **FIGURE 4.54** **Syrian Refugee Crisis** Syrians have been displaced throughout the wider region as the conflict continues. Which countries are the largest recipients of Syrian refugees? How does this compare to the number of Syrian refugees seeking asylum in Europe?

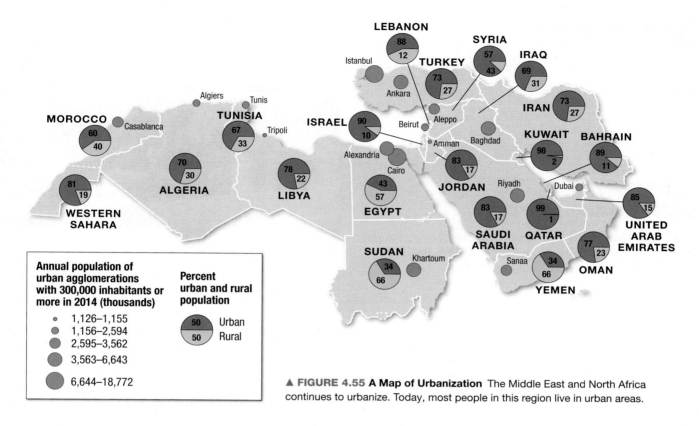

Annual population of urban agglomerations with 300,000 inhabitants or more in 2014 (thousands)

- 1,126–1,155
- 1,156–2,594
- 2,595–3,562
- 3,563–6,643
- 6,644–18,772

Percent urban and rural population

- 50 Urban
- 50 Rural

▲ **FIGURE 4.55 A Map of Urbanization** The Middle East and North Africa continues to urbanize. Today, most people in this region live in urban areas.

political and economic influence in the country. Massive amounts of accumulated wealth means there has also been massive investment in certain urban areas. This was highlighted in the opening to this chapter, but Dubai is just one example. Istanbul, long a hub of connection between Europe and Asia, is also a highly modernized city with a diverse economy. Thus, while Ankara plays a vital political role in the country, Istanbul remains a central cultural and economic hub for the country. Other cities, such as Izmir, along Turkey's western coast, have also grown dramatically. Even as new megacities have emerged and new urban spaces are planned, there remains a massive difference between rich and poor in many countries in this region. This is particularly noticeable in this region's largest cities. One of the most widespread problems is the inability of governments to meet the service and housing needs of growing urban populations. Inadequate and often poor-quality water supplies, electricity, sewer systems, clinics, and schools as well as air pollution and severe traffic congestion plague many cities in the region. Most critically, governments seem unable to provide housing for all who need it, and squatter settlements have sprung up on unclaimed or unoccupied urban land (**FIGURE 4.56**). These problems are compounded as the very largest cities in the region continue to attract migrants, who view the most well-known places as possessing the best opportunities for a better life.

In a few very wealthy oil-producing cities, such as Jubail (Saudi Arabia) and Doha (Qatar), enormous wealth coupled with very low populations in the regions around the cities has made urban growth relatively uncomplicated. Many other cities of the oil-producing region, including Jeddah (Saudi Arabia) and Basra (Iraq), however, have not escaped the erection of shantytowns and the difficult social problems that accompany this type of urban change.

▼ **FIGURE 4.56 Squatter Settlements in Istanbul, Turkey** In Istanbul, squatter settlements are known as *gecekondu,* a Turkish word meaning that the settlements were built after dusk and before dawn.

Apply Your Knowledge

4.22 Describe the various patterns of migration related to this region.

4.23 What produces these patterns?

4.24 How have the urban areas of the region been affected by push and pull factors?

Future Geographies

The future of this region will depend on a number of key issues, including the ability to manage the oil economy and its decline, access to clean potable water, and peace and stability.

Oil

With oil as the foundation of the current global economic system, the region will continue to be important in the global economy for many decades. What is not clear is how new oil production, for example, in the United States and Canada, will impact global prices, which are not expected to reach all-time highs any time soon (**FIGURE 4.57**). Over the long term, climate change and diminishing oil supplies will affect the Middle East and North Africa. Climate scientists have proposed that OPEC establish a production quota system in response to global warming in order to reduce CO_2 emissions. So far, OPEC has resisted this type of external pressure on limiting production, but that does not mean this proposal might not become appealing in the future.

Water

Once the population of the Middle East and North Africa was well-adapted to the extremes of the environment. Today, however, the nations in the region are increasingly home to urban societies that have sought to overcome their environmental limitations through extreme exploitation of resources. This has led to the potential for conflict between countries that share a freshwater resource, such as the Nile River between Sudan and Egypt and the Jordan River between Israel and Jordan. If, instead of consuming water like

▲ **FIGURE 4.58 Wadi Al Mujib Dam, Jordan** Water futures remain a challenge for many in this region. Damming has been somewhat effective in managing water resources, allowing countries to store needed water resources and release as appropriate.

citizens in less arid regions, the people in the region once again adapted to their unique local environment, they could develop sustainability practices that could reshape the region and be applied elsewhere in the world. Experiments like Masdar City in Abu Dhabi indicate that there is a growing understanding of the need for more sustainable approaches to urban development (**FIGURE 4.58**).

Peace and Stability

There are a number of challenges to long-lasting peace and stability in this region. The protests that brought down the leadership of Egypt, Tunisia, and Libya have opened up opportunities for new forms of government that so far remain shaky (Tunisia) or unfulfilled (Egypt and Libya). It is still unclear what will come next. In the context of the ongoing conflict between the Israelis and the Palestinians, there are many challenges to establishing a brokered peace and the Palestinians remain divided between Gaza and the West Bank. The rise of ISIS has also destabilized the region by exploiting the ongoing civil war in Syria and deep political divisions in Iraq (**FIGURE 4.59**). Given all this, it is clear that the Middle East and North Africa will remain an important site for global and regional political negotiations for some time to come.

▼ **FIGURE 4.59 ISIS Destroys Statues, Nineveh Museum, Northern Iraq** ISIS has committed to eliminating artifacts that do not correspond to their vision of a purified Islamic space.

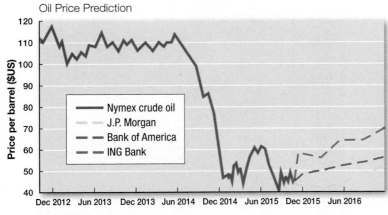

Oil Price Prediction

Legend:
— Nymex crude oil
--- J.P. Morgan
-- Bank of America
-- ING Bank

Y-axis: Price per barrel ($US) — 40 to 120
X-axis: Dec 2012, Jun 2013, Dec 2013, Jun 2014, Dec 2014, Jun 2015, Dec 2015, Jun 2016

▲ **FIGURE 4.57 Price of Oil** Despite predictions, so far the price of oil has stayed fairly low globally.

Learning Outcomes Revisited

▶ *Describe* the diversity of physical geographic environments in the Middle East and North Africa.

While aridity is the defining feature of the region's climate, coastal and mountain environments do exist that bring precipitation to the area. Large river systems, such as the Nile, Tigris, and Euphrates Rivers, remain important sources of freshwater to the region. Mountain ranges, a result of tectonic activity in the region, are important sources of raw materials, such as wood.

1. *What* are the most distinguishing physical geographic features of this world region?

▶ *Evaluate* the importance of key resources, such as oil and water.

The region is resource-rich, particularly as a site of oil production. But, there is debate as to whether or not the world has reached "peak oil" production, suggesting that the development of oil and other petroleum products may diminish in their significance over the next 30 years. Other precious resources, such as water, remain scarce. Innovations are being developed to develop water resources and produce drinkable water from the vast oceans and seas that surround the region.

2. *How* is the region adapting to the global demands for oil and its own internal needs for freshwater?

▶ *Summarize* the diffusion of cultural and environmental practices, such as religion and agriculture, throughout the region.

This region is home to three major world religions: Islam, Judaism, and Christianity. These religions emerged over time in the region, themselves a product of the region's ability to domesticate plants and animals. The Middle East and North Africa has long been a site of innovation, both technological and cultural. From this region, the knowledge of agriculture spread as did other cultural practices, such as writing and mathematics.

3. *Briefly* describe the important contributions this world region has made to global culture.

▶ *Explain* the natural and historical factors that contribute to the diverse economies of the Middle East and North Africa, how they vary from country to country, and how they are tied to global and regional processes.

The Middle East and North Africa is home to some of the world's wealthiest and poorest economies. These stark contrasts have developed over time in relation to the availability of oil. As a rule, the wealthiest states are in the Persian Gulf region; poorer states are outside the belt of oil production. That said, several economies in the region have successfully diversified by exploiting agricultural, manufacturing, and service industries. Most notable of these is Israel, which has strong connections to Europe and a very diverse economy.

4. *Describe* the geography of wealth and poverty in this world region.

▶ *Analyze* the geopolitical conflicts and tensions that bring instability to the Middle East and North Africa and the relationship between these conflicts and the political history of the region.

Conflict and tension have their roots in the geopolitical organization of the region, a product of the colonial experience. In a number of areas, borders imposed under European rule brought different ethnic groups together under larger national banners. The ethnic and religious diversity of the region continues to challenge the stability of the region.

5. *Briefly* explain what geopolitical factors continue to produce conflict in this world region.

▶ *Identify* the differences in practices of kinship, gender, and religion in the region and how these factors affect the lives of women.

Despite images of this region as a uniform space, the practices of kinship, gender relations, and religion vary. For example, women's mobility is quite diverse across different countries in the region. Women's mobility also varies between urban and rural spaces. Kinship—loosely defined by familial and neighborly relations—has a strong effect on the lives of many people in the Middle East and North Africa. Even though the region is predominantly Islamic (almost 96%), other religions, such as Christianity and Judaism, exist throughout the region.

6. *Why* is it inappropriate to say that gender is defined identically in all communities across this world region?

▶ *Describe* and analyze the patterns of migration in the Middle East and North Africa, the push and pull factors that are involved in migration patterns, and the effects these patterns have on the region's demography.

Many regional patterns of migration are tied to economic disparities. People migrate from the poorer countries to the wealthier countries in the region seeking work in the oil industries. Other migrants move to avoid conflict; refugee populations are increasing in the wake of the violence that has plagued the region. Some people with higher levels of education leave poorer countries to seek jobs in places such as Europe. These patterns of migration impact the region's overall demography. Urban areas, in particular, face challenges brought on by rapid and intense population growth.

7. *What* future demographic challenges does this region face?

Key Terms

afforestation (p. 143)
apostasy (p. 165)
The Arab League (p. 159)
Arab Spring (p. 152)
aridity (p. 132)
Balfour Declaration (p. 155)
carbon market (p. 134)
chador (p. 164)
desalinization (p. 139)
desertification (p. 142)
dry farming (p. 133)
Fertile Crescent (p. 143)
The First Arab–Israeli War (p. 156)
fossil fuels (p. 139)
Gaza Strip (p. 156)
gravity system (p. 139)

guest workers (p. 167)
Gulf Cooperation Council (GCC) (p. 159)
hajj (p. 161)
Hamas (p. 156)
Hasidism (p. 162)
Hezbollah (p. 156)
hydraulic civilizations (p. 144)
informal economy (p. 149)
Internally Displaced Person (p. 154)
intifada (p. 156)
Iranian Green Revolution (p. 152)
Islam (p. 160)
Islamism (p. 161)
Janjaweed (p. 151)
Jasmine Revolution (p. 152)

jihad (p. 161)
jihadist (p. 158)
kinship (p. 164)
Kurds (p. 151)
mandates (p. 146)
Mecca (p. 161)
Muslims (p. 160)
nationalist movements (p. 146)
nationalization (p. 148)
oases (p. 137)
Occupied Territories (p. 156)
Organization of Petroleum Exporting Countries (OPEC) (p. 140)
Orthodox Judaism (p. 162)
Pan-Arabism (p. 159)
petrodollars (p. 149)
petroleum (p. 139)

Qital (p. 161)
Qu'ran (p. 160)
rain shadow (p. 133)
secularism (p. 159)
Semitic (p. 144)
Shari'a (p. 159)
Shi'a (p. 161)
Sunna (p. 160)
Sunni (p. 161)
transhumance (p. 164)
tribe (p. 164)
Union for the Mediterranean (p. 150)
wet farming (p. 144)
weapons of mass destruction (WMD) (p. 157)
world religion (p. 144)
Zionism (p. 155)

Thinking Geographically

1. How do countries in the Middle East and North Africa manage their water resources? List an example of an area where water has become a source of conflict in the region.

2. What are some of the factors that limit the possibility of a broader peace in the region? How have recent technological changes, such as the Internet, facilitated dialogue and resistance in the region?

3. How do regional views about gender affect the use of public and private space? Are rules concerning the veiling of women uniform throughout the region? If not, how do they vary geographically?

4. Describe the geographic distribution of refugees and guest workers within the Middle East. What are some current push and pull factors that motivate people to leave one part of the region and enter another?

5. The Middle East and North Africa is becoming highly urbanized. What factors are driving urban growth in this region? How are cities with considerable oil wealth handling their rapid urbanization?

6. Describe the spatial distribution of oil and natural gas production in the region. What are the consequences of the uneven distribution of these resources on this region's politics and economy? What impact has oil and natural gas production had on the local and global environment?

DATA Analysis

Throughout this chapter, we have looked at the political and religious conflicts over the culturally diverse and rich geography of this region. One of the most devastating effects of political and social violence is the creation of large numbers of refugees, people who are forcibly removed from their original geographies and in search of safe spaces. To analyze the current realities for refugees originating in the Middle East and North Africa, use the United Nations High Commissioner for Refugees (UNHCR) and UN Refugee Agency.

1. Begin by reading the United Nations High Commissioner for Refugees (UNHCR) page on the Middle East and North Africa, at http://www.unhcr.org.

2. List three of the main countries of origin for refugees in the region and describe the crises propelling people to leave each state.

3. On the right-hand margin of the web page noted above, download the 2013 UNHCR Global Report on the Middle East and North Africa, from http://www.unhcr.org.

4. How many people have left their home countries in 2013? To what countries did they flee?

5. Using the map on p. 156, compare the amount of refugees to overall population size in the three countries you selected.

6. Skim to the end of the document on the budget.

   a. Where is the budget the largest? Using what you know about this region, explain why that might be.

   b. Who are the donors to the refugee operations in the Middle East and North Africa? List three major donors and one that surprises you.

7. Finally, do a quick Internet news search on refugees who have fled from their country. Describe how the headlines compare to the data you just analyzed. Are there more refugees now than in the past?

8. Where are these refugees going? And can you ascertain their current quality of life?

Middle East Refugees

https://goo.gl/Xglu11

MasteringGeography™

Looking for additional review and test prep materials? Visit the Study Area in MasteringGeography™ to enhance your geographic literacy, spatial reasoning skills, and understanding of this chapter's content by accessing a variety of resources, including MapMaster interactive maps, videos, *In the News* RSS feeds, flashcards, web links, self-study quizzes, and an eText version of *World Regions in Global Context*.

Women play an important role in African economies. These Nigerian women are members of a savings club which offers loans to women at low interest rates and helps them start small businesses.

Sub-Saharan Africa

In the last decade, some of the fastest-growing economies in the world emerged in sub-Saharan Africa. Peace has come to many regions, and health indicators are slowly starting to improve. Individuals have started businesses, campaigned to end civil wars, and labored in remote regions to improve health care and education. Many of these individuals are women.

Born of necessity, opportunity, or both, women's entrepreneurship in Africa has quietly transformed consumer markets, generated employment opportunities, and sparked new philanthropy. Ghana offers an impressive profile of women's business prominence in fields ranging from fair-trade, cooperative chocolate production to high-level public relations. Ghanaian women's businesses have transformed the goods and services produced for African and international markets and fostered schools, programs, and foundations to support the next generation of women entrepreneurs.

For example, Grace Amey Obeng started a skin care company for African skin—Forever Clair—which is now a multibillion dollar enterprise. Her philanthropy includes a training college and a non profit foundation for poor women.

Rural women are also responsible for the growth of sub-Saharan Africa's economies. A Ghanaian farmer's cooperative Kuapa Kokoo led by women includes over 85,999 members who supply raw cocoa for the brand Divine Chocolate. A gender committee works to train women leaders in the cooperative and also conducts a literacy program.

Women's saving clubs and micro loans help women overcome temporary hardship or borrow money to invest in entrepreneurial ventures such as small shops, restaurants or manufacturing companies (Figure 1).

Success stories such as these have led development agencies, charities, and governments to realize that supporting women can be the fastest way to reduce poverty, disease, and conflict and to both enable and inspire future generations to provide their own innovative contributions. As of 2014, women heads of state and government included Catherine Samba-Panza of the Central African Republic, Ellen Johnson Sirleaf of Liberia, Joyce Banda or Malawi, and Saara Kuugongelwa of Namibia.

Learning Outcomes

▶ **Compare** the physical geographies of the diverse regions of sub-Saharan Africa, contrasting the climate and vegetation of the Sahel and the Congo Basin, the East African Rift Valley and Okavango, and the major rivers.

▶ **Evaluate** the role of physical geography, disease, and mineral resources in shaping conflicts, crises, landscapes, and economies.

▶ **Identify** the impacts of human activity on African wildlife and climates and some of the solutions to these threats.

▶ **Summarize** the legacies of European colonialism and the Cold War in the region and how these, together with the diversity of ethnicities, religions, and languages, have created conflict within and between countries.

▶ **Explain** how debt relief, the end of several conflicts, and improving the status of women may increase the chance of positive economic and political futures in sub-Saharan Africa.

▶ **Describe** the main causes of low agricultural production in sub-Saharan Africa, connections of agriculture to desertification and deforestation, and problems that are emerging around land grabs.

▶ **Analyze** factors affecting the characteristics of populations in sub-Saharan Africa, including the demographic transition, HIV/AIDS and other diseases, urbanization, fertility changes, and migration.

▶ **Provide** examples of how African people, goods, and culture have spread around the world as a result of slavery, colonialism, and globalization.

Environment, Society, and Sustainability

Africa is a large, complex, and often misunderstood region. Some people think of Africa as fertile tropical forests and grasslands hosting an idyllic game reserve where rich cultural traditions reach back to the dawn of humanity. Others perceive a harsh desert landscape devastated by war and drought. But Africa is not neatly encompassed by any of these limited descriptions. The continental landmass called Africa straddles the equator, stretching from the Mediterranean Sea to the southern tip of South Africa and from Senegal on the Atlantic to Somalia on the Indian Ocean (**FIGURE 5.1**). The continent's total area is about 30.4 million square kilometers (11.7 million square miles)—three times larger than the United States or Brazil. The physical geography of the continent has shaped ecosystems, the use of land, and the distribution of key mineral and water resources. Human activities within and beyond the region have, in turn, transformed its geography. Perhaps, the most important message of this chapter is that 'Africa' is not a country with a single story but a continent that is home to more than a billion people living in 54 countries that differ from each other and have varied people and landscapes within each of them.

The African continent is often divided into two regions—sub-Saharan Africa and North Africa. In this book, Chapter 4 discusses North Africa as part of the Middle East because of shared Arab ethnicity, and cultural, religious, linguistic, and political links. The current chapter focuses on sub-Saharan Africa (Africa below the Sahara) with an area of 22.64 million square kilometers (8.75 million square miles) and 48 countries, often associated with a predominantly black population. The geographical, racial, ethnic, and cultural basis for dividing Africa between

The danger of a single story

https://goo.gl/epdKSe

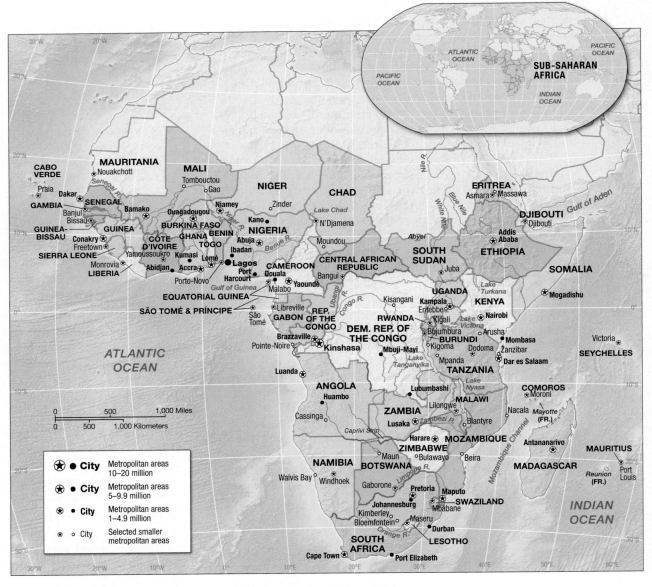

▲ **FIGURE 5.1 Sub-Saharan Africa: Countries, major cities, and physical features**

two world regions is somewhat artificial. The division over-simplifies both the variations within each region and the overlaps between them. For example, the Sahara Desert includes territory from both North and sub-Saharan African countries, and the Nile River links a long fertile corridor that stretches from Egypt in North Africa south into Ethiopia and Uganda. The regional division between Arab and black populations overlooks considerable mixing of race, ethnicity, and religion in many countries; the legacies of colonialism and migration, especially from Europe and Asia; and racial or ethnic constructions of societies that contribute to stereotyping or conflict. Commonly defined regions within sub-Saharan Africa include West Africa, East Africa, southern Africa, central Africa, the Sahel, and the Horn of Africa (**FIGURE 5.2**). Because there are so many physical and human links across the continent of Africa, several sections of this chapter discuss broader patterns across the whole continent.

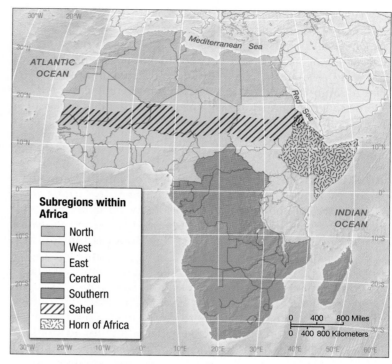

▲ **FIGURE 5.2 Subregions of Africa**

Climate and Climate Change

Most of sub-Saharan Africa has a tropical climate with temperatures higher than 20°C (70°F), except in colder highland climates. Precipitation ranges from heavy rain all year, to a distinct rainy season, or very dry deserts (**FIGURE 5.3**). Two major features of the atmospheric circulation control the region's precipitation—the **intertropical convergence zone (ITCZ)** and the **subtropical high** (see Chapter 1, p. 11). The ITCZ is a low-pressure zone where air rises at the Equator, producing heavy rainfall of more than 1,500 millimeters (60 inches) a year over the Congo Basin. The subtropical high occurs where air sinking over the tropics results in dry conditions over the Sahara and Kalahari deserts. The areas between them, such as the West African coast and East Africa, experience seasonal rainfall as these global circulations shift northward in April and southward in October. These same areas often have unreliable rainfall and frequent droughts that pose challenges to agriculture and water resources management. In July, southwestern trade winds blow onto the coast of West Africa, bringing seasonal monsoon rains to inland countries such as Mali. During December, winds reverse and very dry winds called the **harmattan** blow out of inland Africa carrying large amounts of dust that stress humans and animals.

▶ **FIGURE 5.3 Climates of Africa** Warm tropical conditions dominate the climate of Africa including desert climates (BW/BS), hot climates with rain all year (Af), monsoon seasonal rains (Am) or pronounced dry seasons (Aw). Southern and highland Africa has some milder and cooler climates including Meditteranean with dry summers (CS), temperate with wet summers (Cw) and wetter maritime climates (Cf). See Figure 1.9 for a detailed legend.

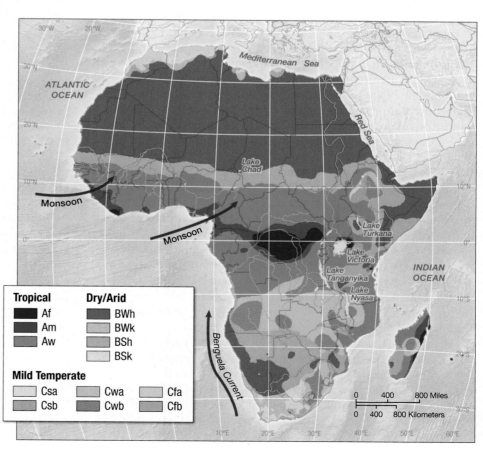

These general climate patterns are modified by the effects of mountains, lakes, and ocean currents. The cold Benguela Current creates cool, dry conditions that suppress rainfall along the coasts of Angola and Namibia and intensifies the desert conditions of the Kalahari. High-altitude areas such as the East African highlands of Kenya and Uganda have higher rainfall and more moderate temperatures, which make them more favorable to agriculture and human settlement. Southern Africa is located at cooler southern latitudes than the rest of the continent with a mild winter and low seasonal rains.

Adaptation to Climate The people of sub-Saharan Africa have adapted to the climate of the continent in various ways. Farmers work floodplains along seasonally variable rivers such as the Niger and implement traditional irrigation and rainwater harvesting schemes in dry areas. The **Sahel** is the southern border of the Sahara Desert (see Figure 5.2) and has highly variable rainfall. The people of the Sahel traditionally depend on **pastoralism**—a way of life that relies on livestock—with some crop production along rivers and at oases (**FIGURE 5.4a**). Herders move their livestock to follow the rains and allow farmers to use the manure from their animals. Despite these adaptations, droughts resulting from variable rainfall—combined with political unrest, the growth of human and animal populations, and changes in land access and land use—threaten food security and can trigger famine in the region.

Rains failed across the Sahel for seven years from 1968 and the drought was blamed for harvest failure, widespread hunger, and the death of 250,000 people and millions of cattle. Images of the drought and starving refugees in Africa began to appear in the international media, resulting in a relief effort and anguished debates among researchers and policymakers about what had gone wrong and what could be done to prevent future tragedy. While some blamed the drought, others pointed out that human factors were also responsible. These factors included land use and political changes stemming from colonialism that prevented migration and promoted crops for export rather than food consumption. Another factor, international assistance for well drilling, concentrated livestock around wells but resulted in overgrazing and starvation.

Land degradation and persistent drought can lead to a process called **desertification** in which arid and semiarid lands become degraded and less productive and desertlike conditions result. The main culprits are seen as climate change, overgrazing, overcultivation, deforestation, and unskilled irrigation. The UN Convention to Combat Desertification seeks to understand and reverse the processes that have led to so much devastation. There are indications that desertification has reversed—a "regreening"—where rains have been adequate, farmers are planting trees and irrigating crops efficiently or where pastoralists have sold their livestock and moved to urban areas (**FIGURE 5.4b**).

Drought still contributes to famine but most recent food crises have been associated with conflict between or within countries. More than half a million people died in each of several conflict-related famines in Biafra (1960s), Ethiopia (1980s), Sudan (1980s), and Somalia (1990s). Drought and insect invasion of crops combined with a global food price increase and conflicts saw widespread hunger returning to Somalia and the Sahel in 2011 and 2012.

(a)

(b)

▲ **FIGURE 5.4 The Sahel** (a) Overstocking of cattle during droughts can result in overgrazing and desertification, especially where cattle gather around water holes like this one in Burkina Faso (b) Some regions of the Sahel, such as along the Niger river, are becoming greener as hand dug wells are used to carefully water vegetables and trees.

Climate Change Africa has been identified as extremely vulnerable to climate change: warmer temperatures, changes in ecosystems, and impacts on the poor have already been observed across the region (**FIGURE 5.5**). In terms of future climate, temperatures are likely to warm by several degrees and although rainfall changes are more uncertain, existing water stresses may increase if drought intensifies. Africa's *vulnerability* (see Chapter 1, p. 13) to climate change is caused as much by poverty as by climate itself. Many Africans are dependent on rain-fed agriculture and livestock, which suffer greatly from the effects of drought and heat waves harming food security and incomes. Many people also lack access to safe water supplies. Scientists are concerned that climate change will expand the range of diseases and insects. Some places where people are better off, resources are shared, and countries are peaceful and well governed are more resilient (able to withstand or recover) than others.

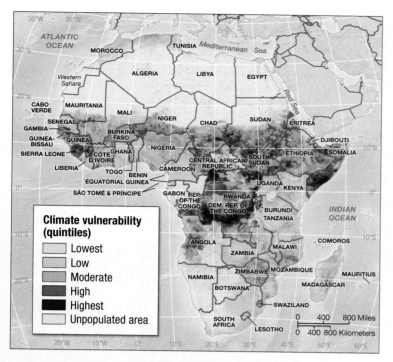

▲ **FIGURE 5.5 Climate Change Vulnerability in Africa** Measures of climate vulnerability in Africa include: exposure to floods, coastal storms, and other hazards; population density; household resilience including literacy, education, health, and safe water; and politics, including conflict and government effectiveness. **Which countries are most and least vulnerable? Any ideas why this might be?**

Although the effects of climate change are predicted to be severe in the region, sub-Saharan countries have some of the lowest per capita greenhouse gas emissions in the world (see Chapter 1 "Visualizing Geography, Causes and Consequences of Climate Change"). This illustrates the inequities of climate change, where those least to blame are often most affected.

Apply Your Knowledge

5.1 What are some of the traditional adaptations to climate extremes in the region and how might these help people adapt to climate change, what would this involve, and what forms could assistance take?

5.2 Many scholars assert that wealthier regions (such as North America and Europe) that have very large per capita greenhouse gas emissions and have caused the climate to change are obligated to help sub-Saharan Africa, which has low emissions, cope with the impacts of climate change. Do you agree or disagree? What, if anything, should wealthier regions do?

Geological Resources, Risk, and Water

The continent of Africa is the heart of the ancient supercontinent called **Pangaea**, the southern part of which broke off to form **Gondwanaland** about 200 million years ago (**FIGURE 5.6**).

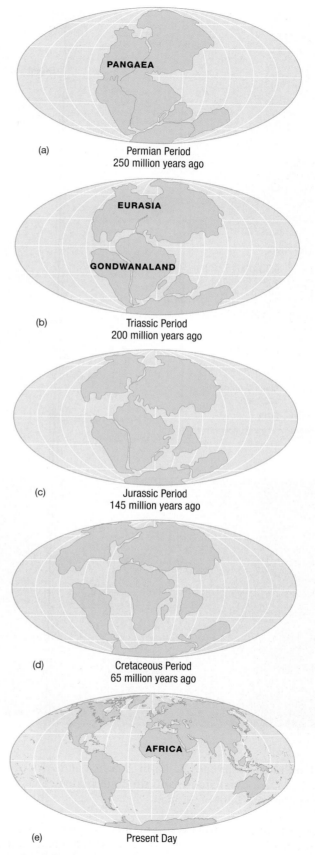

▲ **FIGURE 5.6 Pangaea and the Origins of the African Continent** The continent of Gondwanaland broke off from the ancient supercontinent Pangaea 200 million years ago to become Africa.

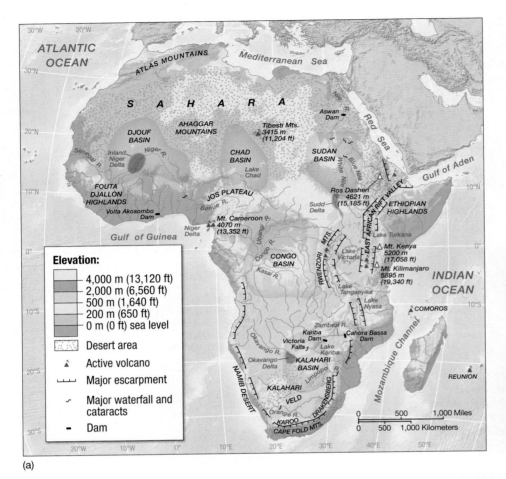

(a)

◀ **FIGURE 5.7 Physical geography of Africa**
(a) Africa is a massive plateau continent with rivers that flow to inland deltas or to the sea over waterfalls at the edge of the plateau. Key physical features include high mountains in eastern Africa and the African Rift Valley with its lakes (b) Lake Bogoria in Kenya is known for its wildlife including pink flamingos and its hot springs. It occupies the Rift Valley (c) As the Zambezi river drops from the African interior to the coast it cascades over the spectacular Victoria Falls—a magnet for tourists to Zimbabwe and Zambia. **What tectonic feature helped form the lakes of East Africa? Explain.**

(c)

(b)

The landmasses that became Latin America and Asia then broke away from Gondwanaland, and the high plateau that remained became the continent of Africa.

Plate Tectonics Africa is still mainly a plateau continent, with elevations ranging from about 300 meters (1,000 feet) in the west, tilting up to more than 1,500 meters (5,000 feet) in the eastern part of the continent (**FIGURE 5.7a**). Steep slopes, especially on the western edge of the plateau, drop to narrow coastal plains.

The higher areas of the plateau, where cooler temperatures and higher rainfall are hospitable to humans, include the Veld of southern Africa, the East African highlands of Kenya and Ethiopia, and the Jos Plateau of West Africa. Volcanic peaks such as Kilimanjaro (5,895 meters or 19,340 feet), Kenya/Kirinyaga

(5,200 meters or 17,058 feet), and the Virungas (4,507 meters or 14,787 feet) rise from the eastern plateau.

A deep trough slices through eastern Africa, where tectonic processes continue to pull the eastern edge of Africa away from the rest of the continent to create a **rift valley**—a large and long depression between steep walls formed by the downward displacement of a block of the earth's surface between tectonic faults. The East African Rift Valley runs more than 9,600 kilometers (6,000 miles) from the Red Sea in the north to Mozambique in the south and is from 50 to 100 kilometers (30–60 miles) wide. It has two major branches and is home to deep lakes, including Lake Tanganyika at 1,473 meters deep (4,832 feet). Lake Victoria, the third-largest lake in the world by area, lies between the two branches. The age, size, and depth of these lakes make them diverse freshwater ecosystems with important fisheries (**FIGURE 5.7b**).

▲ **FIGURE 5.8 Water Access** Women, such as these of the Dogon tribe in Mali, often have to walk miles to wells or rivers to collect water.

Water Resources and Dams

Africa's rivers are extremely important to the people and ecology of the region. The Nile River flows north to Egypt (Chapter 4). It originates in the White Nile that flows from the lakes region of East Africa through Tanzania, Uganda, and South Sudan and the Blue Nile that flows from a source in Ethiopia to meet the White Nile in Sudan. While some rivers flow directly from the high plateaus to the coasts, sometimes dropping over magnificent waterfalls such as Victoria Falls on the Zambezi (**FIGURE 5.7c**), others flow away from the coast and into inland wetlands and deltas before shifting back toward the ocean. Inland wetlands and deltas are important to agriculture and wildlife. Examples include the Inland Niger Delta on the Niger River; the **Sudd**, a vast wetland on the Nile in South Sudan; and the wildlife oasis of the Okavango Delta on the Okavango River in Botswana.

Although rapids and waterfalls pose problems for navigation by boat into the continent, the rivers of Africa also provide considerable potential for hydroelectric development. Several massive dam projects were initiated in the 1950s to harness the energy of the rivers, to provide electricity to industry and cities, and to irrigate agricultural fields. These include the Kariba dam on the Zambezi in southern Africa and the Akosombo dam in Ghana on the Volta River in West Africa. China has financed recent dam projects in Ethiopia, Gabon, and Republic of Congo. While dams foster economic development, they also create problems that include resettlement of displaced people and animals; the loss of sediment, fish, and livelihoods downstream of the dams; and the spread of diseases and insects (such as malarial mosquitoes) in stagnant water.

Water demand is starting to exceed supply in parts of the region, especially where climate is dry and population and irrigation create higher demand. Water quality is bad in many places, such as slum areas of cities, as a result of inadequate sanitation and lack of infrastructure (such as treatment facilities or drainage) contributing to the spread of disease. Women can spend many hours fetching heavy buckets of water for their families (**FIGURE 5.8**).

Soils, Minerals, and Mining African soils tend to not be very fertile because of the great age of the underlying geology and because high levels of rainfall wash out nutrients from exposed soils especially in deforested areas. Some areas have salty, alkaline, and iron- or aluminum-rich soils that are toxic to crops. Soil fertility tends to be higher where volcanic activity has deposited ash, such as the East African highlands, and in wider river valleys, where sediment from floods creates richer soils.

Half the African continent is composed of very old crystalline rocks of volcanic origin that hold the key to Africa's mineral wealth including gold, copper, uranium, cobalt, bauxite, and diamonds. Ancient tropical swamps formed sedimentary rocks containing oil and other fossil fuels, especially in Nigeria and central Africa. Mineral wealth is the mainstay of many African economies, but the distribution of mineral wealth is very uneven between and within countries in the region (**FIGURE 5.9**).

Mineral resources have played an important role in African history. Salt was a key commodity in trans-Saharan trade from the 10th century to the present day. Gold was valued in West Africa from early times; Mansu Musa, the emperor of Mali, carried and traded so much gold on a pilgrimage to Mecca in 1324 that his actions depressed gold prices worldwide. Gold and diamonds spurred European colonial grabs for Africa as well as conflicts between colonial powers. The discovery of diamonds in 1867 at Kimberley and gold in 1886 in South Africa resulted in conflicts between colonizers and indigenous groups. Gold and diamonds, together with oil, continue to incite conflict within Africa and to amplify interest in African economies on the part of other states and multinational corporations (see "Geographies of Indulgence, Desire, and Addiction: Diamonds" on page 188).

183

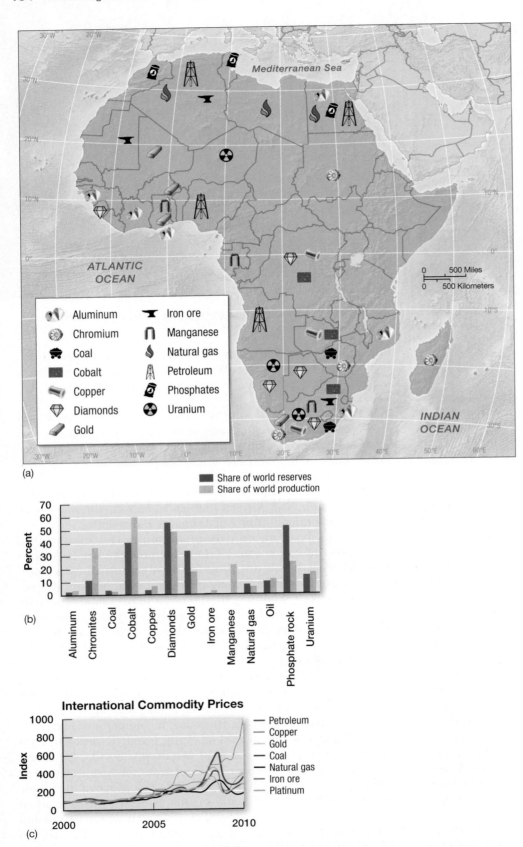

(a)

Share of world reserves
Share of world production

(b)

International Commodity Prices

Petroleum
Copper
Gold
Coal
Natural gas
Iron ore
Platinum

(c)

▲ **FIGURE 5.9 Mineral Resources** Widely distributed across the continent, Africa's mineral resources bring billions of dollars into the region's economy. (a) Map of the distribution of mineral resources. (b) Africa's shares of the reserves and current production for key minerals. (c) The graph shows changes in the global prices of minerals showing how many increased but then dropped creating volatility for those African countries who depended on exporting them. **Where are the critical resources of oil, diamonds, and gold found within the region?**

Rare or high value minerals have fueled conflict in the Democratic Republic of the Congo (DRC). These minerals include cobalt (used for high strength alloys, radioactive imaging, and blue color), tungsten (used for lightbulbs, electrodes, and weapons), and coltan (producing tantalum used for circuits in electronics, especially mobile phones).

Ecology, Land, and Environmental Management

Sub-Saharan Africa hosts some of the world's most fascinating ecosystems, including the habitat of magnificent wildlife such as lions and elephants. But low agricultural yields across the region undermine food security and health and increase pressures for land that result in encroachment on wildlife habitat. Poverty and conflict have increased wildlife poaching. African ecologies also include many pests and diseases that pose serious health threats to humans and agricultural systems. Solutions include efforts to eradicate and cure disease, to increase food production on land that is already being farmed, and to establish parks and conservation programs.

Major Ecosystems African ecosystems are closely tied to climate conditions but also reflect a complex evolutionary history and physical geography that have produced great diversity, unique plants, and the world's most charismatic community of animal species. Large areas are used for grazing and crop production (**FIGURE 5.10**).

The tropical climates of the Congo Basin host Earth's second-largest area of rain forest (after the Amazon), which covers almost 2.2 million square kilometers (780,000 square miles). Other forests are found along the West and East African coasts. Forests make up about 20% of African land area. These forests boast impressive biodiversity. They are home to monkeys and apes, such as chimpanzees and gorillas, as well as tropical hardwoods of significant economic value, such as mahogany. Like many forests around the world, African forests are threatened by demands for timber and firewood, by poaching and conflict, and by conversion to cropland. The highly biodiverse island of Madagascar has lost most of its forests (**FIGURE 5.11**). In addition, recent peace

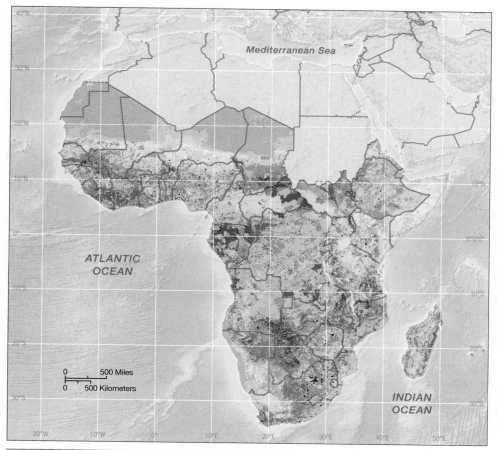

◀ **FIGURE 5.10 Vegetation in Sub-Saharan Africa** This map combines natural vegetation and associated land uses including protected and grazed areas, forests (green), grassland and savanna (orange), shrubs and woodlands (brown), desert (yellow), and cropland (pink). **Where are the main agricultural regions in sub-Saharan Africa (in purple and pink on the map) and where are the protected areas of forest (bright green on map) and grasslands and shrubs (orange and brown)?**

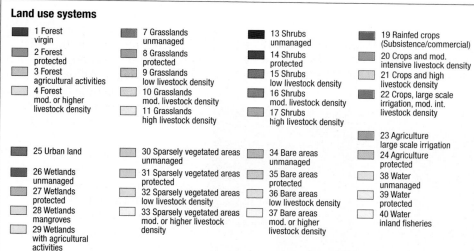

Land use systems

1 Forest virgin	7 Grasslands unmanaged	13 Shrubs unmanaged	19 Rainfed crops (Subsistence/commercial)
2 Forest protected	8 Grasslands protected	14 Shrubs protected	20 Crops and mod. intensive livestock density
3 Forest agricultural activities	9 Grasslands low livestock density	15 Shrubs low livestock density	21 Crops and high livestock density
4 Forest mod. or higher livestock density	10 Grasslands mod. livestock density	16 Shrubs mod. livestock density	22 Crops, large scale irrigation, mod. int. livestock density
	11 Grasslands high livestock density	17 Shrubs high livestock density	23 Agriculture large scale irrigation
25 Urban land	30 Sparsely vegetated areas unmanaged	34 Bare areas unmanaged	24 Agriculture protected
26 Wetlands unmanaged	31 Sparsely vegetated areas protected	35 Bare areas protected	38 Water unmanaged
27 Wetlands protected	32 Sparsely vegetated areas low livestock density	36 Bare areas low livestock density	39 Water protected
28 Wetlands mangroves	33 Sparsely vegetated areas mod. or higher livestock density	37 Bare areas mod. or higher livestock density	40 Water inland fisheries
29 Wetlands with agricultural activities			

drought-resistant plants, such as acacias and woody scrub. The cooler climate of South Africa has produced a unique ecosystem dominated by vegetation (called fynbos) that is characterized by waxy or needlelike leaves and long roots that help plants survive long dry periods. The recent international popularity of the use of protea in flower arrangements and of rooibos tea has increased interest in this ecosystem, which is threatened by climate change.

Diseases and Pests
African pests and diseases can have a devastating impact on human populations and food supplies (**FIGURE 5.13**). The most serious diseases, which kill millions of people, include:

- **Malaria** Transmitted to humans by mosquitoes, **malaria** causes fever, anemia, and often-fatal complications. The disease affected more than 120 million people in the region in 2014, killing more than 500,000 people, many of them children. Although the discovery of treatments such as quinine (in 1820) and chloroquine (1940s) helps fight the disease, several strains of the malaria parasite have developed resistance to these drugs. The use of insecticides, cheap mosquito nets, and control of habitat have helped reduce the risk by almost half since 1990.

- **Schistosomiasis** Also called bilharziasis, schistosomiasis is associated with a parasite that causes gastrointestinal diseases and liver damage. It is passed to humans who are exposed to a snail that is a host for the parasite when people work or bathe in slow-moving water in marshes, reservoirs, or irrigation canals. The disease is not fatal but reduces general health and energy levels. Unfortunately, most cases are not treated and in 2012 more than 200 million people were infected in Africa.

- **River blindness** The bite of a black fly transmits river blindness (onchoceriasis). The bite passes on small

agreements in Central Africa may open up the region to forest exploitation.

The areas of seasonal rainfall in the region have mixed woodlands and grasslands. The **savanna** grassland areas cover about two-fifths of Africa. Savannas have open stands of trees interspersed with shrubs and grasses, vegetation typically found in tropical climates that have a pronounced dry season and experience periodic fires. Savannas provide extended grazing areas for both wildlife and livestock (**FIGURE 5.12**).

Deserts cover another two-fifths of sub-Saharan Africa. Deserts have very sparse and seasonal vegetation and feature

(a)

(b)

▲ **FIGURE 5.11 Deforestation of Madagascar** The forests of the island of Madagascar contain a quarter of all the flowering plants in Africa (including the rosy periwinkle, which is used to treat the disease leukemia) and are also home to unique fauna, including lemurs; 800 species of butterflies; and numerous chameleons, cacti, and corals. Eighty-five percent of the original forest has been cleared for rice production, sugar plantations, cattle ranches, and cutting trees to export tropical hardwoods. (a) A comparison of forest cover in 1950 and 2000 illustrates the loss of many important forest areas. (b) The forests are home to more than 100 species of lemurs.

▼ **FIGURE 5.12 Savanna** Grassland plains, such as the Serengeti of Tanzania and Maasai Mara in Kenya, have some of the densest concentrations of wild, hoofed, grazing mammals in the world. These animals coexist in the region with predators, such as the big cats—lions, leopards, and cheetahs. Larger herbivores in the region include elephants, wildebeest, giraffes, zebras, and rhinoceroses.

worms whose larvae disintegrate in the human eye and cause blindness. At least 35 million people have been infected and more than 700,000 are blind as a result of this disease in sub-Saharan Africa. The eradication of river blindness by controlling fly populations with pesticides and treating victims with drugs has been relatively successful in West Africa.

- **Ebola** Scientists believe bats are host species for the **Ebola** viruses causing illness in primates and other wild animals. Early cases occurred in central Africa, but the most severe Ebola outbreak emerged in

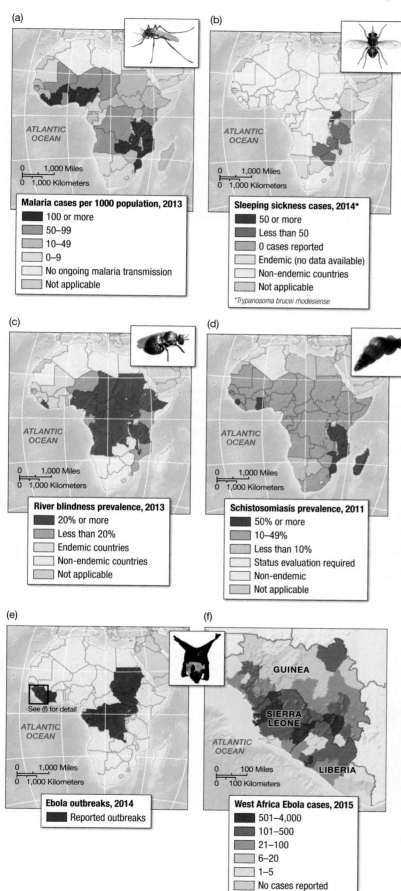

(g)

Diseases in Africa (2011)	Cases/year	% world cases	Deaths/year
Malaria	128 million	65%	525,600
Schistosomiasis	224 million	90%	200,000
River blindness	36.8 million	99%	0 (770,000 blinded)
Sleeping sickness	20,000	100%	No data
Ebola	21,718	100%	8,640

▲ **FIGURE 5.13 Tropical Infectious Diseases and Pests** Africa is home to several serious tropical diseases and their primary nonhuman vectors including (a) malaria across most of the region, (b) sleeping sickness and tsetse in southern and eastern Africa, (c) river blindness in central, west, and east Africa, (d) schistosomiasis throughout the region, and (e & f) Ebola in West and Central Africa. (g) Table of disease impacts. **Pick one of the diseases shown and summarize its distribution, how it is transmitted, and preventive measures to protect people from contracting it.**

West Africa in 2014. The disease is highly infectious, transmitted through bodily fluids including mucus, vomit, and feces. Ebola first manifests as a sudden onset of high fever accompanied with headache, muscle soreness, vomiting, and other symptoms. Infection ends in death in about one-half of reported cases. The 2014 Ebola outbreak was the largest in history and the first to affect West Africa. The 2014 epidemic spread from Guinea to Liberia and Sierra Leone and infected more than 26,000 people and killed more than 10,000 by the end of the year. As of 2015 the number of new cases declined to 100 per week, suggesting efforts to contain the epidemic were taking hold.

- **Tsetse fly** Found in African woodland and scrub regions, the **tsetse fly** is associated with both human and livestock diseases. In humans, the fly's bite causes sleeping sickness (trypanosomiasis). The symptoms of sleeping sickness include fever and infection of the brain that causes extreme lethargy and may end with death of the victim. Sleeping sickness has been reduced to about 20,000 cases a year through drug treatment in early stages and prevented by a variety of pest-control measures. In domestic animals such as cattle and horses, the tsetse fly causes a disease called nagana, which is similar to sleeping sickness and causes fever and paralysis. This disease prevented the introduction of livestock into many low-lying parts of Africa and, as a result, preserved habitats for wild species in an unexpected benefit for conservation.

Many of these debilitating diseases and pests are associated with the tropical climate and diverse ecologies of Africa. Additional pests destroy crops, such as locusts and birds which can swarm across the landscape eating all vegetation in their path. Disease and pest spread may have been facilitated by the expansion of human populations and the transformation of natural environments through deforestation and irrigation. Reducing the human toll from these diseases is a major challenge for scientific research, African governments, charitable organizations, and the UN World Health Organization (WHO), which

Diamonds

Consumers around the world associate diamonds with luxury and love. Larger diamonds are graded, cut, polished, and set in gold or other metals, and then sold for high prices in jewelry stores. About 40% of all polished diamond purchases are made in the United States, but China is the fastest-growing market, now accounting for about 15% of global demand. Diamonds also have industrial value because of their hardness and strong, sharp cutting edge.

More than half of the world's diamonds originate from African countries (see Figure 5.9), with a total annual value of roughly U.S. $8.5 billion. About two-thirds of the global diamond trade is controlled by a South African conglomerate, De Beers, which manages markets to ensure that prices remain high and the supply stable.

Most African diamond production takes place in South Africa, Botswana, and Namibia and contributes significantly to export revenue and local employment. In Botswana, production is guided by traditional leaders, contributing 25% of employment and 33% of the country's Gross

"Diamond profits fueled the purchase of weapons for use in brutal local conflicts and wars."

Domestic Product. But diamonds were also associated with corruption, violence, and warfare in Angola, the DRC, and Sierra Leone (**FIGURE 5.1.1a**). Because of their immense value and small size, diamonds were easily smuggled. Diamond profits fueled the purchase of weapons for use in brutal local conflicts and wars. Some analysts suggest that these so-called **conflict diamonds** (also called blood diamonds) may have accounted for 10–15% of all global trade in diamonds during the 1990s.

The Kimberley Process, involving 80 governments partnered with industry and civil society, was established in 2003 to prevent conflict diamonds from entering the mainstream rough diamond market (**FIGURE 5.1.1b**). Today, the industry reports that less than 1% of diamonds is illegally sourced, and retailers often mention that their diamonds are conflict free in advertisements. In 2013, the World Diamond Council moved for human rights, financial transparency, and economic development to be added to the goal of ensuring rough diamonds were conflict free and legally sourced.

---

1. **Research Internet sites for jewelers and diamond sales and compare how two of them address the issue of conflict diamonds.**

2. **What other minerals found in products that you buy have been associated with conflict in Africa?**

(a)

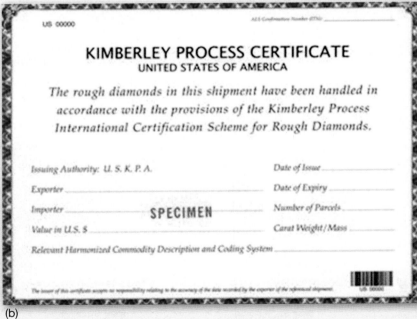

(b)

▲ **FIGURE 5.1.1 Diamonds** (a) Miners dig for diamonds in the Chudja open pit mine in 2009 in north-eastern Congo where proceeds have financed conflict.
(b) An example of a Kimberly process certificate for conflict-free diamonds.

has targeted Africa for extra funds and programs. The Bill and Melinda Gates Foundation has given over U.S. $6 billion to combat disease in developing countries, including more than $1.5 billion for malaria in Africa. The foundation's efforts have included research on a malaria vaccine and the distribution of mosquito nets.

Apply Your Knowledge

5.3 How does Africa's physical environment—including its climate, soils, pests—pose challenges for humans?

5.4 What regional and international steps have been taken to overcome these challenges? How effective have these efforts been?

Land Use and Agriculture

Less than 30% of the soil in sub-Saharan Africa is suitable for agriculture, which is also hindered by harsh climates and pests and diseases. However, Africa, the birthplace of the human species, is also where humans first adapted to the constraints of the physical environment. Early humans survived by hunting wild animals, fishing, gathering plants, and domesticating a number of crop and livestock species including cattle, sorghum, yams, oil palm, and African rice. The highlands of Ethiopia are considered one of the early centers of **domestication**—the adaptation of wild plants and animals into cultivated or tamed forms through selective breeding for preferred characteristics. Ethiopia is the place of origin of coffee, millet, and an important local cereal called *teff* used to make *injera* bread.

Traditional strategies for adapting to low soil fertility include **shifting cultivation**, which involves moving crops from one plot to another. One form of shifting cultivation is **slash-and-burn** agriculture (**FIGURE 5.14a**) where after an area is burned, the ash is used to fertilize crops. After a few years, when the nutrients are exhausted, farmers move on to a new area and leave the previous plot to return to forest or other vegetation. After a long fallow (rest) period, farmers return to clear and burn the land again. An African modification of shifting cultivation is **bush fallow**, in which crops are planted around a village, and plots are left fallow for shorter periods than in the slash-and-burn system. Soil fertility is also maintained by applying household waste to the fields in a practice called "compound farming." **Intercropping**—planting several crops together in a single field—is a common technique for keeping the soil covered to reduce erosion, evaporation, and nutrient loss, and improve soil fertility (**FIGURE 5.14b**). Floodplain farming is used where there are

▼ **FIGURE 5.14 African Agriculture** Adaptations to the poor soils and cost of fertilizer include (a) slash-and-burn agriculture and (b) intercropping shown here in Uganda with maize, beans and pineapple plants. (c) In the savanna and desert regions, pastoralism is the main source of food and income. For example, the Maasai of Kenya and northern Tanzania rely on cattle for meat, milk, and blood. **Which of the three forms of agriculture do you think is most sustainable? Why?**

(a)

(b)

(c)

seasonal floods, such as the inland delta of the Niger River and the Sudd wetlands along the Nile.

Pastoralism—a way of life that relies on livestock—is the agricultural activity best adapted to drier regions of Africa where vegetation flourishes with seasonal rains. Nomads, such as the Bedouin, migrate with their animals in search of pastures in the arid landscapes of the Sahel and North Africa. Other groups, such as the Fulani of West Africa, practice a system of seasonal herd movements called **transhumance**. They move their herds southward to wells and rivers in the dry season and drive them northward in the wet season. Cattle are traded at regional markets and are a family investment for the future in places where there are few secure ways of accumulating capital. East African pastoralists include the Maasai of Kenya and Tanzania (**FIGURE 5.14c**).

Conservation and Africa's Wildlife The rich biodiversity of sub-Saharan Africa is valued by local residents, tourists, and international environmental groups alike. However, differing views about its protection have resulted in many controversies about conservation. Traditional African societies hunted and gathered wild species for food, and while human populations were low, property was unfenced, and hunting technologies were less effective, wildlife populations ranged naturally where climate, vegetation, and terrain were most suitable. As population, technology, and land use changed, especially after colonialism, human activity began modifying habitat. Europeans contributed to the decimation of African wildlife through indiscriminate hunting expeditions, eliminating animals near railroads and farms, and forcing local people off their lands into regions where they came into conflict with wildlife that encroached on their herds and fields.

Currently about 100 million hectares (386,000 square miles) or 5% of sub-Saharan Africa enjoy some sort of protected status, and there are more than 1,000 protected areas—over half in southern Africa. The major parks in East Africa and southern Africa, such as Serengeti in Tanzania and Kruger in South Africa, have become high-profile international tourist destinations, bringing millions of dollars to national economies and employing many local people. But parks have been criticized for providing inadequate benefits to local people who may have been displaced and have lost traditional grazing and hunting rights, or who have had crops destroyed by marauding wildlife. In some parks, too much tourism and high animal densities have destroyed fragile habitats, and poaching has pushed some species close to extinction. Other protected areas have been affected by armed conflicts and by poor people seeking land.

African wildlife is often used by conservationists as a symbol for worldwide campaigns because of threats to key species:

- **Elephants** Appealing because of their size and intelligence, elephants have been hunted for their valuable ivory tusks—known as "white gold"—for centuries. Pressure on herds in East and southern Africa grew with demand from Victorian England and from Asia for decorative ivory. African elephant populations have declined from more than 3 million in the 1930s to fewer than 700,000 in 2014 mostly as a consequence of illegal killing and poaching fueled by the ivory trade. The situation is aggravated by competition between people and elephants for land and by war and civil unrest, especially when firearms are available to unpaid soldiers and desperate refugees in regions where herds have been left unprotected.

  Mounting international pressure resulted in a general ban on international sales of African ivory in 1989. In 1997, elephants were listed under the Convention on International Trade in Endangered Species (CITES) as a species in need of protection. The ban on ivory sales was opposed by countries in southern Africa where elephants were destroying the habitats of other species and where parks and conservation were funded by money earned from legal ivory and hide sales. In 2002, the United Nations granted permission to South Africa, Botswana, and Namibia to sell 60 tons of ivory and in 2008 China was given permission by the United Nations to purchase ivory and is now the global center of trading.

- **Rhinos** Poachers who hope to sell rhino horns have placed rhinos under enormous threats. Rhino horn is sold in Yemen and other countries in the Middle East where it is made into dagger handles or in Asia where it is ground into a highly valued medicinal powder. Protecting the rhino from poachers who can make thousands of U.S. dollars from selling a single horn is a full-time and costly enterprise with individual rhino assigned their own armed guards, often paid for by international conservation groups or tourist revenue (**FIGURE 5.15a**). The newest rhino conservation tool involves the use of *conservation drones*—unmanned aerial machines that can map and observe wildlife and poachers.

- **Gorillas and chimpanzees** These primates are hunted for trophies and **bushmeat** (meat from wild animals hunted for human consumption). Their habitats are being destroyed by deforestation, especially in conflict zones, despite the work of conservationists, such as Dian Fossey and Jane Goodall, who have studied their fascinating social behaviors and campaigned for their protection (**FIGURE 5.15b**).

Sustainability

Sub-Saharan Africa is a focus of concern about sustainable development, including strategies to reduce poverty and improve lives without adding to environmental degradation.

Two-thirds of the region lacks access to electricity, especially in rural areas, and 80% of the population relies on traditional biomass for cooking and heat—wood, charcoal, dung, and crop waste—contributing to deforestation and loss of soil fertility. In East Africa, the loss of forests has disproportionately affected women who are primarily responsible for heating and cooking and have limited access to electricity, gas, or petroleum fuels

(a)

(b)

▲ **FIGURE 5.15 African Wildlife** (a) Guarding endangered black rhino in Zimbabwe.
(b) Tourists watching a gorilla troop in Volcanoes National Park in Rwanda. Foreign tourists
are charged a fee of $750 to see the rare gorillas, funding conservation and benefiting the
Rwandan economy. **Explain the costs and benefits to local people and economies in
protecting African wildlife.**

The best-known social movement of this kind
is the **Green Belt Movement**, which counts 50,000
women as members. Led by Nobel Prize–winning
environmental and political activist Wangari
Maathai, Green Belt planted thousands of trees
around Nairobi and has been the model for similar
groups elsewhere in Africa and around the world.

Hopes for more sustainable energy use rest
on the growth of renewable energy such as wind
and solar (see "Sustainability in the Anthropocene: Renewable Energy in Kenya" on p. 194).
Africa has enormous renewable energy potential
that can reduce already low carbon emissions,
bring electricity to remote areas, and fuel industrial development. Solar panels increasingly
power schools, cell phone and computer charging,
and TVs across rural Africa.

Agriculture in Africa has become more sustainable through the recovery of traditional methods
(such as inter- and mixed cropping, composting
and water harvesting) and also through urban agriculture where city dwellers farm gardens and open
space. There is considerable potential to increase
crop and livestock yields in the region by combining traditional practices with the careful use
of fertilizer, new seed varieties, and fine-tuned
irrigation.

Apply Your Knowledge

5.5 Use the Internet to identify two groups
campaigning on behalf of African wildlife. What
arguments and data do they use in favor of
protection and how do they take the needs of
local people into account?

5.6 What measures do they propose to protect
endangered or threatened species?

History, Economy, and Territory

Sub-Saharan Africa saw the dawn of humanity
on the continent more than 2 million years ago.
Its early global roles continued with the development of major trading societies about 5,000 years ago and the
advent of European colonialism about 500 years ago. Most of
sub-Saharan Africa was under European colonial domination—a
major era of globalization—by 1914. Independence movements, which began around 1950, coincided with Cold War

and therefore rely on wood or charcoal for fuel. Projects to
reduce energy demands by using scrap metal to make more
efficient stoves have complemented the efforts of female-led
nongovernmental organizations to protect trees in and around
Nairobi.

Renewable Energy in Kenya

East Africa is leading the region in embracing renewable energy such as solar, wind, and geothermal energy. In Kenya, low-cost solar panels are providing lights, cell phone charging, water pumps, and TV and radio to both urban and isolated rural communities including schools and hospitals who would not otherwise have electricity (**FIGURE 5.2.1**). Many Kenyans use kerosene lamps that are inefficient and polluting. For a $30 down payment a household can get a 4-watt solar panel, three lights, and a mobile phone charging station that will be paid off after a modest payment of about $15 a month for a year—half the cost of dirty kerosene. The ability to charge mobile phones is increasingly important to sustainability in Africa, allowing people to pay bills and avoid debt, children to learn online, and farmers to get weather and market information. Mobile phones also help to improve response to natural disasters and facilitate democracy through social networks.

Kenya plans to get 50% of its energy from solar in the next few years, but solar is not the country's only source of renewable energy. Steam from tectonic hot spots in the Rift Valley is generating 20% of the country's electricity. A large 300 MW, $800 million wind farm is being built near Lake Turkana.

1. **Research the renewable energy policy of a country in sub-Saharan Africa and evaluate plans for more sustainable energy.**

2. **Use the Internet to find a project where solar energy has improved lives in a community.**

▲ **FIGURE 5.2.1 Sustainability in the Anthropocence: Solar Energy in East Africa** (a) a shopkeeper in Tanzania sells household solar equipment (b) Africa has tremendous potential for generating electricity from the sun. This map shows the intensity of sunlight across Africa.

tensions between the United States and the former Soviet Union, and these superpowers intervened in African political struggles and civil wars as they vied for global domination.

Since the beginning of the 21st century, much of sub-Saharan Africa has continued to struggle with the transition to independent and democratic government. Regional economies still continue to rely on a narrow set of exports to other world regions although some have vibrant and growing economies. Although parts of Africa are becoming highly connected to other world regions through migration, telecommunication, and trade, other places are still isolated and rely on subsistence agriculture.

Historical Legacies and Landscapes

Contemporary African geographies—landscapes, livelihoods, and culture and politics—bear the imprint of prior periods of African history and of Africa's shifting connections to different regions of the world (**FIGURE 5.16**). These connections in turn have shaped other regions of the world. For example, colonialism included worldwide trade in African slaves, resulting in a diaspora (the dispersion of people from their original homeland; see Chapter 1) of African people that continues to influence the culture and societies of other world regions to this day. Colonialism also resulted in regional political boundaries that split ethnic groups across territories or clustered enemies within one territory.

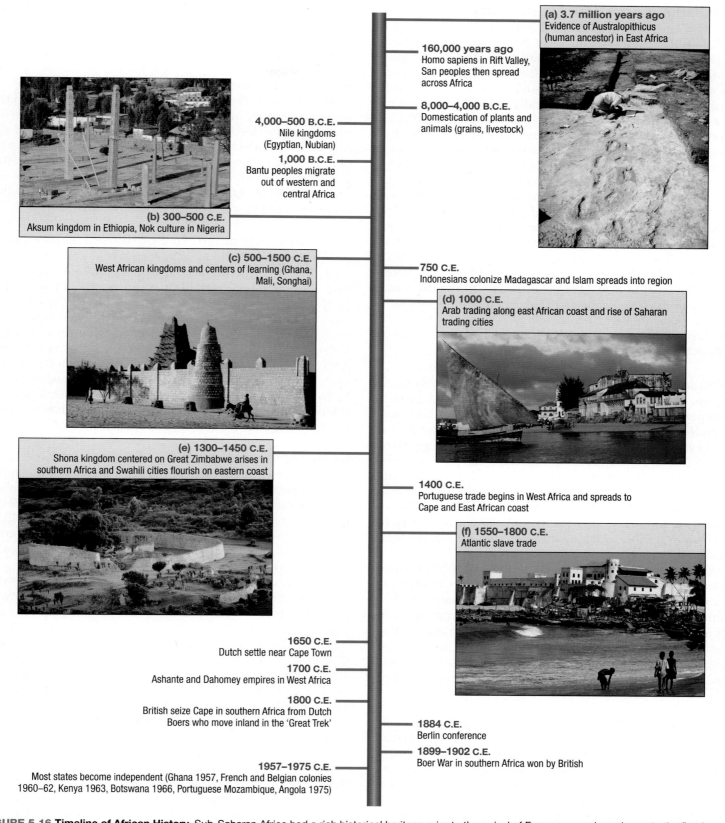

(a) 3.7 million years ago Evidence of Australopithicus (human ancestor) in East Africa

160,000 years ago Homo sapiens in Rift Valley, San peoples then spread across Africa

8,000–4,000 B.C.E. Domestication of plants and animals (grains, livestock)

4,000–500 B.C.E. Nile kingdoms (Egyptian, Nubian)

1,000 B.C.E. Bantu peoples migrate out of western and central Africa

(b) 300–500 C.E. Aksum kingdom in Ethiopia, Nok culture in Nigeria

(c) 500–1500 C.E. West African kingdoms and centers of learning (Ghana, Mali, Songhai)

750 C.E. Indonesians colonize Madagascar and Islam spreads into region

(d) 1000 C.E. Arab trading along east African coast and rise of Saharan trading cities

(e) 1300–1450 C.E. Shona kingdom centered on Great Zimbabwe arises in southern Africa and Swahili cities flourish on eastern coast

1400 C.E. Portuguese trade begins in West Africa and spreads to Cape and East African coast

(f) 1550–1800 C.E. Atlantic slave trade

1650 C.E. Dutch settle near Cape Town

1700 C.E. Ashante and Dahomey empires in West Africa

1800 C.E. British seize Cape in southern Africa from Dutch Boers who move inland in the 'Great Trek'

1884 C.E. Berlin conference

1899–1902 C.E. Boer War in southern Africa won by British

1957–1975 C.E. Most states become independent (Ghana 1957, French and Belgian colonies 1960–62, Kenya 1963, Botswana 1966, Portuguese Mozambique, Angola 1975)

▲ **FIGURE 5.16 Timeline of African History** Sub-Saharan Africa had a rich historical heritage prior to the arrival of Europeans and was home to the first humans. (a) Early human footprints dated to 3.7 million years ago at Laetoli in Tanzania. (b) A 1,700-year-old obelisk in the city of Aksum in Ethiopia. (c) The Sankoré madrassah (center of learning) in the ancient trading city Timbuktu (Mali). (d) The island of Zanzibar, just off the coast of Tanzania, was a key port on the Indian Ocean trading with the Middle East and Africa. The ship shown here is a traditional *dhow* used for trading and transport along the coast. (e) The ruins of Great Zimbabwe. (f) Elmina castle in Ghana was constructed by the Portuguese who processed thousands of Africans for the transatlantic slave trade. **Research one of the early kingdoms or empires mentioned here to learn more about its culture and legacy. How are the places shown in photos above marketed to tourists?**

Human Origins and Early African History

Human Origins and Early African History Africa is often called the "cradle of humankind" because the earliest evidence—tools, footprints, bones—of the human species (*Homo sapiens*) and earlier ancestors such as Australopithecus have been unearthed by archaeologists in East Africa. Remains more than 3 million years old have been found at Laetoli and Olduvai in Tanzania (**FIGURE 5.16a**). Anatomically modern humans, who walked upright and had larger brains, lived at least 100,000 years ago in sites in southern Africa and along the Rift Valley. Genetic evidence shows that these humans are the ancestors of all modern humans—the most basic link between Africa and the world.

Crop and domestic animal production spread from about 8000 B.C.E. with metal smelting dating from 4,000 years ago. Soon after, the societies in the Middle East and North Africa began to influence sub-Saharan Africa with their complex organization, hieroglyphic writing, religions, and hierarchical social organization. Places such as Aksum in Ethiopia (**FIGURE 5.16b**), Djenne and Timbuktu (**FIGURE 5.16c**) in West Africa, and Zanzibar (**FIGURE 5.16d**) in East Africa became familiar to European and Arabic traders and centers of trading and culture. In southern Africa, powerful settlements emerged such as that at Great Zimbabwe (**FIGURE 5.16e**). These sites have become centers of tourism and pride in heritage.

The Colonial Era in Africa

The Colonial Era in Africa With the development of faster and larger ships in the 15th century, contacts with Europe became part of the regular interaction with Africa. The Portuguese traded to get gold from coastal settlements in West Africa, and in 1497, the Portuguese explorer Vasco da Gama rounded the Cape of Good Hope at the southern tip of the African continent, initiating trade with the southeast coasts of Africa en route to India. In return for salt, horses, cloth, and glass, sub-Saharan Africa provided gold, ivory, and slaves to the world via Portuguese and Arab traders. For centuries, African slaves were in demand among Arabs, who used the slaves as servants, soldiers, courtiers, and concubines.

European colonialists took some time to establish control in Africa, and for many years only the coastal ports and trading posts were under European command. One of the main reasons for European reluctance to move inland was the reputation of Africa as the "White Man's Grave." The continent earned this nickname because so many Europeans died from diseases such as malaria, yellow fever, and sleeping sickness, against which they had no natural immunity. In addition, African armies attacked ports and fiercely resisted European attempts to move inland.

Slavery and the Slave Trade

Slavery and the Slave Trade Africa generated enormous profits for European traders through slavery as well as for some coastal African kingdoms who captured and sold their enemies (**FIGURE 5.16f**). The Portuguese took slaves for new sugar cane plantations on the Atlantic islands of Madeira and Cape Verde. In 1530, the first slaves were shipped to the Americas to work on plantations in Brazil. By 1700, 50,000 slaves were shipped each year to the Americas to provide labor on colonized lands and new plantations where local indigenous populations had declined precipitously from European diseases. It is estimated that more than 9 million slaves, who were mostly male, were shipped to the Americas from Africa between 1600 and 1870. At least 1.5 million slaves died during the journey (**FIGURE 5.17**). The conditions of capture and transport were horrific. Hundreds of slaves were packed into the holds of ships with little food and water and were brutally abused by traders. In the Americas, many slaves introduced their own traditions—including ways of cultivating rice, religion, and music—that endure today.

By the end of the 18th century, social movements led to the banning of slavery in Britain in 1772, the end of slave trade in the British colonies in 1807, and the emancipation (freeing) of slaves in the British Caribbean in 1834. Slavery was a divisive issue that played a role in the U.S. Civil War; it was eventually abolished in 1865 by the 13th Amendment to the U.S. Constitution. Some liberated slaves returned to Africa and became elites in the countries now known as Sierra Leone and Liberia.

European Settlement in Southern Africa

European Settlement in Southern Africa Europeans were attracted to settle in southern Africa by the more temperate climate and the strategic significance of the trading routes from Europe to Asia around the southern tip of the continent. In 1652, the Dutch established a community at Cape Town that soon became surrounded with small cattle and wheat farms, which supplied trade ships and local communities. The Dutch settlers spoke a modified form of the Dutch language known as Afrikaans, belonged to the strict puritan Christian Calvinist religion, saw themselves as superior to black Africans, and became known as the Boers (which is the Dutch word for farmer) and, together with French and German settlers, as **Afrikaners**. British immigrants settled around Cape Town and Durban.

European Exploration and the Scramble for Africa

European Exploration and the Scramble for Africa International interest in Africa increased dramatically after the 1850 discovery that quinine could suppress malaria and the discovery of gold and diamonds in southern Africa. Explorers, traders, and missionaries from Britain, France, Belgium, Portugal, and Germany moved to the interior of the continent seeking and competing for new territories, commodities, and souls to convert.

Between 1880 and 1914, European powers aggressively moved to colonize Africa, partly through private trading companies. The British claimed much of coastal West Africa; Kenya, Uganda, and Sudan in East Africa; and southern Africa. The Portuguese claimed Mozambique and Angola and Germany claimed Cameroon and German Southwest (Namibia) and East Africa (Tanzania). France and Belgium split the Congo. Italy took Somalia, Djibouti, and Eritrea and France took most of the remaining territory of West and North Africa.

This hasty scramble for Africa was formalized at the **Berlin Conference** of 1884–1885, a meeting convened by German Chancellor Otto von Bismarck to divide Africa among European colonial powers. Thirteen countries were represented at this conference, but it did not include a single African representative from sub-Saharan Africa. The Berlin Conference allocated African territory among the colonial powers according to prior claims and laid down a set of arbitrary boundaries that paid little respect to existing cultural, ethnic, political, religious, or linguistic regions.

Growing tensions between the British and Afrikaners resulted in the Boer War (1899–1902), which gave control of much of southern Africa to the British. By 1914, almost all of Africa was

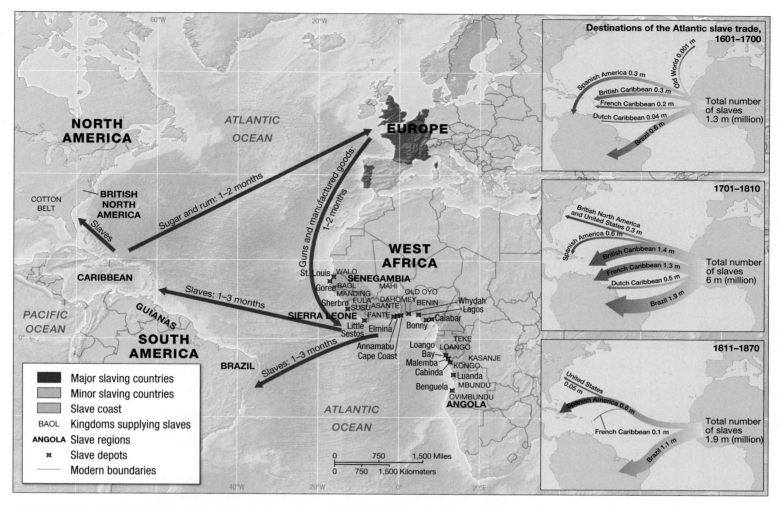

▲ **FIGURE 5.17 The Slave Trade** Millions of slaves were exported from Africa between 1600 and 1870, mainly from the West African coast. Some local leaders acted as suppliers in return for guns and manufactured goods. Slaves were sent to work on plantations in the Americas. The plantations sent sugar, rum, and other products back to Europe in a triangular trade. **Analyze the role of sugar, rum, and guns as drivers of the slave trade. What eventually ended this economic system?**

under European colonial control except for Abyssinia (now Ethiopia), Liberia, and some interior regions of the Sahara Desert (**FIGURE 5.18**). A number of battles were fought in Africa during World War I. Germany's eventual loss in that war redistributed the German colonies to Britain, France, and Belgium.

The Legacy of Colonialism The most general and enduring effects of colonialism—which lasted for 80–100 years from 1880—include the establishment of political boundaries that disregarded traditional territories; a reorientation of economies, transport routes, and land use toward the coast and export of commodities; improved medical care; and the introduction of European languages, land-tenure systems, taxation, education, and governance. Many of the new colonial boundaries divided indigenous cultural groups and in some cases placed traditional enemies within the same country. For example, Nigeria comprised several competing groups, including the Yoruba in the southwest, the Ibo in the southeast, and the Hausa in the north. These groups still struggle over political and cultural differences today.

Colonial mines extracted large amounts of gold, diamonds, and copper for export to Europe. New roads and railways were constructed from inland to the coasts to speed the export of crops and minerals, but few efforts were made to link regions within Africa. The resulting infrastructure still facilitates trade beyond but not within Africa. Colonists also established plantations to produce crops, such as rubber, and used a variety of means, including taxation and violence, to persuade peasant farmers to produce peanuts, coffee, cocoa, or cotton for export. In the temperate climates of the East African highlands and southern Africa—areas that were more attractive to European immigrants—the best land was taken by white settlers for tea and tobacco plantations, livestock ranches, and other farming activities. By 1950, the geography of African agriculture clearly reflected this export orientation. The Firestone Corporation owned vast rubber plantations in Liberia, and the French imposed peanut and cotton cultivation in the Sahel. Cocoa dominated the cropland of Ghana and the Côte d'Ivoire and tea and coffee became important in East Africa (**FIGURE 5.19**). Traditional African land-tenure systems of communal land and flexible boundaries were often subsumed into privately owned and bounded plots, and traditional agriculture, decision-making processes, and legal systems were often replaced with European science, managers, and courts.

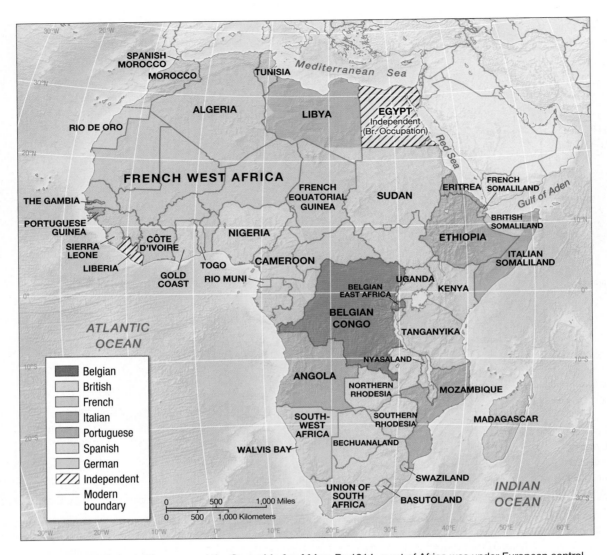

▲ **FIGURE 5.18 Colonial Powers and the Scramble for Africa** By 1914, most of Africa was under European control.

(a)

(b)

▲ **FIGURE 5.19 Export Crops in Africa** (a) A tea plantation thrives below Mount Kenya. Tea was introduced into the Kenya highlands from India in 1903 and grown as a plantation crop on colonial landholdings. There are still 150,000 hectares (580 square miles) under production, often by small scale farmers providing about 10% of the world total. (b) A child carries a peanut plant near Djiffer, Senegal. Peanuts were introduced as a cash crop in French West Africa during the colonial period and are still important in Nigeria and Senegal. **What products made from crops often grown in sub-Saharan Africa do you consume? How do you think increased production of these export crops affects small-scale farmers?**

There were some differences between colonial powers. The British mostly ruled indirectly through preexisting power structures in a decentralized and flexible administrative structure. The British took over land and laws in the areas they settled such as southern and eastern Africa, but maintained some local property rights and laws in other regions such as West Africa. They imposed local taxes that were paid by producing crops for sale to the Europeans. Children learned English in European-style schools. French colonial rule was more direct and militaristic, with top down administration from Paris, especially of agriculture and mining. Some locals became French citizens. The Belgian and Portuguese modes of colonialism in central and southern Africa were much harsher, with direct rule and often ruthless and armed control of land and labor. These authoritarian forms of control provided an unfortunate model for leadership in independent Africa.

Independence

The transition from colonies to independent states in sub-Saharan Africa followed years of struggle but then happened quite rapidly for some countries. There were some relatively peaceful handovers to well-prepared African leadership, but other countries—such as South Africa in 1910—became independent under white rulers. There were some more violent transitions of power to divided or unprepared local elites and militaries. A variety of events inspired independence movements, including Indian and Pakistani independence in 1947; the loyalty of the half-million Africans who fought with the allies in World War II; several foreign-educated activists, such as Kwame Nkrumah of Ghana and Jomo Kenyatta of Kenya; and a Pan-African movement led by black activists in the United States including W. E. B. DuBois and Marcus Garvey.

In British Kenya, for example, white settlers controlled the best land, dominated the government, and set policy in the interests of the white minority, which consisted of only about 60,000 white residents. Many of the white settlers opposed independence, but Kenya eventually gained independence in 1963 in the wake of the violence and repression of the Mau-Mau rebellion. In West and Equatorial Africa, transitions occurred dramatically in 1960 when France suddenly recognized a large group of independent countries. When Belgium abandoned its African colonies in the early 1960s, the countries of Zaire, Rwanda, and Burundi were left with internal ethnic and political conflicts that manifested in later problems. Portugal hung onto its colonies of Angola and Mozambique until 1974, despite independence movements sponsored by the Soviet Union and Cuba, which supported rebel movements as part of the Cold War.

Apply Your Knowledge

5.7 What are the legacies of colonialism for contemporary African landscapes and politics?

5.8 Choose two nations that are former colonies and research evidence of these legacies online and of the process of independence.

Economy, Debt, and the Production of Inequality

Sub-Saharan Africa scores very low on many economic development measures compared with other world regions, but in many countries conditions have improved considerably in the last decade. The region had an average gross national income (GNI) purchasing power parity (PPP) per capita of only U.S. $3,251 in 2013 (but it was double that of 1990) compared to the world average per capita of U.S. $14,338. Many national economies in the region are dependent on just a few low-priced exports, and large numbers of people make a living as subsistence farmers or lack formal employment. Singled out for attention by international agencies, Africa receives the highest amount of development assistance per capita of any world region—almost $56 billion in official development assistance in 2013, $9 billion from the United States. This foreign aid is equivalent to about 28% of the region's total GNI and averages about U.S. $60 per person. The World Bank has identified sub-Saharan Africa as "the most important development challenge of the 21st century."

Dependency, Debt, and African Economies Some blame sub-Saharan Africa's low levels of economic development on the harsh environment or the legacy of colonialism, arguing that colonial powers transformed the political and economic structures of Africa to serve their own interests. The colonial focus on obtaining cheap raw material to export undermined local agriculture and social development. Many African countries emerged from colonialism in the late 20th century with their economies and trade dependent on just a few products (such as minerals or cocoa). The export value of these products declined over time in relation to the price of imports, especially manufactured imports. Because of these deteriorating **terms of trade**, Africans had to sell more and more to purchase the same amount of manufactured goods. Some experts argued that sub-Saharan Africa needed modernization and recommended development programs that included technology transfer, training, and large infrastructure projects, such as hydroelectric dams and roads. Other experts suggested that these actions would reinforce the dependency of the region on high-priced imports and outside expertise and argued that African countries needed instead to substitute local goods for costly imports (import substitution) by subsidizing local industry and setting up trade barriers to foreign imports.

To implement modernization or import substitution, many African countries looked outside the region to borrow money and with poor commercial credit ratings obtained loans from governments such as the United States or through international banks such as the World Bank. As in other regions, increased interest rates in the 1980s led to a financial crisis for severely indebted countries. The International Monetary Fund (IMF) and the World Bank responded with loan adjustments but demanded that African governments cut budgets and remove subsidies and trade barriers in return for debt relief. This required reduced public spending and the privatization of government-held companies sending food prices soaring and increasing unemployment.

By the 1990s some countries were still paying more than 50% of their GNI and most of their export earnings toward external debt (**FIGURE 5.20a**). The extreme poverty of some countries, especially as they emerged from conflict, prompted international agencies and others to try to cushion the impact by providing relief programs through foreign aid. The World Bank and IMF forgave some debt in 33 African countries so long as they showed a willingness to pursue policies of reduced government spending and free trade and to develop poverty reduction strategies. But the **G8** countries controlled much of the debt (the Group of Eight countries that include Canada, France, Germany, Italy, Japan, Russia, the United Kingdom, and the United States) and public outcry (such as the Make Poverty History campaign) lobbied for debt forgiveness. In 2005, a G8 summit agreed to write off the debts of the world's 18 poorest countries, 12 of them in Africa. Overall external debt fell from 66% of GNI in 2002 to less than 25% in 2013 (**FIGURE 5.20b**).

Contemporary African Agriculture
Africa has a larger percentage of its adults working in agriculture than other world regions. For example, the World Bank reports that more than three-fourths of the adult population works in agriculture in Ethiopia, Mozambique, and Tanzania. Agricultural employment is dropping in other, urbanizing countries such as Ghana, Kenya, and Nigeria to much less than half the population. Most farmers in this world region have small plots and grow just enough food to feed their families at a subsistence level. But commercial cash and export crops are also very important where export patterns were established in the colonial period. For example, Ghana and Côte d'Ivoire produce cocoa (for chocolate), Kenya grows tea,

and Ethiopia and Uganda grow coffee for export. Benin, Mali, and Burkina Faso grow cotton; Liberia still produces rubber.

In some countries, there are also new agricultural export sectors. In Kenya, the most rapidly growing export sector is fresh vegetables and cut flowers. Thirty percent of Europe's cut flower imports—mostly roses—come from Kenya in refrigerated planes (**FIGURE 5.21a**). Although these new industries provide employment and higher wages than some other sectors, the strict quality standards, perishable nature of the product, and reliance on air transport make it difficult for many countries and producers to take advantage of export opportunities. There are also concerns about pesticide risks to workers.

Agricultural Challenges and Opportunities

Agriculture in sub-Saharan Africa faces a variety of challenges as well as opportunities. These include increasing crop yields, increasing the food security of urban populations, protecting agricultural land from monopolization by foreign investors, and promoting "fair trade" practices.

- **Yield Gaps** Crop yields are low in many African countries and crop production has not kept up with growing population and urban demands. Agricultural problems in sub-Saharan Africa have been attributed to

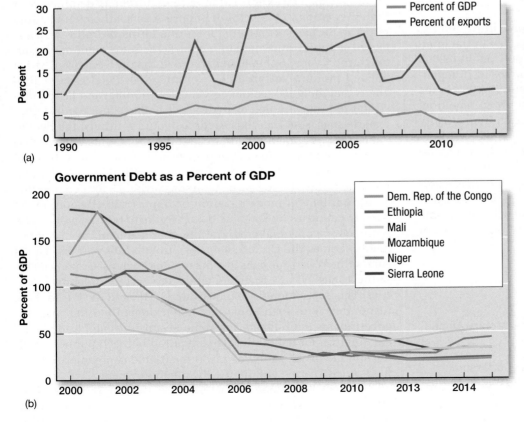

(a)

(b)

◄ **FIGURE 5.20 Declining African Debt** As a result of debt restructuring and forgiveness as well as improved economic conditions, debt across sub-Saharan Africa has declined in the last decade. (a) Debt service payments for the region overall increased as a percent of gross domestic product (GDP) and as a percent of export revenue until 2001 when debt forgiveness started to help economies improve and debt payments decrease. The recession in 2009 caused a short-term reversal. (b) Some of the poorest countries have seen dramatic declines in their overall debt as a percent of GDP. **What else other than debt forgiveness might have contributed to lower debt in these countries?**

(a)

(b)

▲ **FIGURE 5.21 Contemporary Agriculture** (a) Workers cut flowers that will be exported at the Wildfire flower farm near Naivasha, Kenya. (b) Urban market garden in Kigali, Rwanda.

environmental degradation, lack of infrastructure, government policy (especially the control of food prices that can reduce farm income), and unfair international market structures. However, the potential to increase production is considerable and countries like Nigeria are starting to grow wheat, rice, and sugar that they used to import. But increasing production requires access to irrigation, access to cheaper fertilizers, better storage to reduce loss, improved transport to markets, and reduced risks of climate change.

- **Urban Agriculture and Food Security** One often-overlooked African success story is urban agriculture. In cities such as Nairobi (Kenya), Lusaka (Zambia), Kano (Nigeria), and Kinshasa (the DRC), more than half the residents cultivate gardens, either at their homes or on unused

land in the city (**FIGURE 5.21b**). Crops from these urban plots contribute to urban food security and local incomes, as surplus is sold at local markets. More than 200 million people in sub-Saharan Africa are food insecure because they cannot afford to buy food, do not have land to grow food, or their land is used for exports. Some of them are helped with food aid, but many still go hungry.

- **Land Grabs** There is some concern that recent large-scale land purchases, often referred to as **land grabs**, may undermine African food security. This is especially a concern when purchasers plan to use the land for biofuels, mineral or water rights, or private export crop production. The African land observatory reports just over 20 million hectares of land deals to date. Investors and purchasers include China and the Persian Gulf States, major Western banks, and individual U.S. investors. Large tracts of irrigated land that have been important to local food production in Ethiopia and Tanzania have been purchased. Contracts are often for very low cost and tend to generate only a few jobs for local people.

- **Fair Trade** The **fair trade movement** is concerned with ensuring that producers are paid a reasonable wage and that crops are produced sustainably. Products labeled as fair trade from Africa include flowers, vegetables, cocoa, and coffee.

Manufacturing and Services Sub-Saharan Africa is less industrialized than many other world regions, although African manufacturing output has doubled in the last 10 years. South Africa dominates the economy of the region with a substantial manufacturing sector that includes iron and steel, automobiles, chemicals, and food processing. Several major automobile companies produce cars in South Africa, including BMW, Toyota, Volkswagen, Ford, and General Motors. One of the major goals is keeping more of the processing and finishing of goods in Africa rather than exclusively exporting raw materials. A number of countries have important textile and clothing industries often managed by female entrepreneurs.

The telecommunications sector has grown explosively with the rapid adoption of mobile phone technology in sub-Saharan Africa, currently the telecommunications market's fastest growing world region. Mobile phones and the Internet are transforming the economy and culture (see "Sustainability in the Anthropocene: Renewable Energy in Kenya"), even in remote areas, with 608 million internet connections across the region in June 2014 (compared to 4 million in 1998). There are currently 329 million unique subscribers, equivalent to a penetration rate of 38% as compared with the global rate of 92%. Sub-Saharan Africa is forecast to see the highest growth of any region in terms of the number of smartphone connections over the next six years (**FIGURE 5.22**).

The service sector is large in most countries in the region, with many people working as civil servants in government as well as in the tourist sector (especially in southern and eastern Africa). There are also two "hidden" contributions to economic development in most of Africa. The first is the

Mobile Cellular Subscriptions in Sub-Saharan Africa

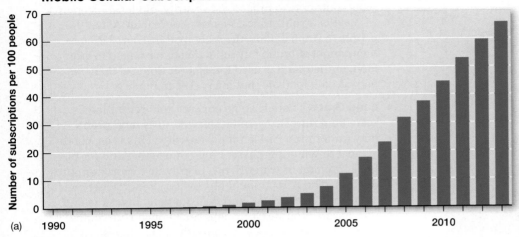

(a)

(b)

▲ **FIGURE 5.22 Cell Phones in Africa** (a) Cell phone use in Africa has grown recently and rapidly. Cell phones are used in many areas where there are no landline phones. (b) Like people in other world regions, Africans like these women picking tea use their phones to obtain information about crop prices, call for medical help, access banking services, and to keep in touch with friends and family.

informal economy; millions of people work as street vendors and maids, for example, and are neither taxed nor monitored by a government. In countries such as Zimbabwe, Tanzania, and Zambia, as much as half the GNI and more than three-fourths of the jobs are associated with the informal economy. The second is from Africans working abroad who sent **remittances** of more than U.S. $32 billion home in 2013.

Since 2000, sub-Saharan Africa has experienced generally strong economic growth, especially from the service,

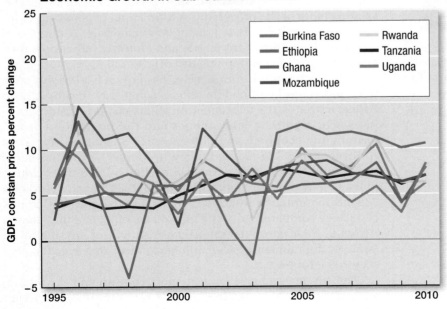

▲ **FIGURE 5.23 Economic Growth in Sub-Saharan Africa** Many African economies have grown rapidly since the mid-1990s—by an average of more than 5% per year. **What sectors have contributed most to the growth in the countries shown?**

agriculture, and infrastructure sectors (**FIGURE 5.23**). In 2014, growth in the region improved for a second consecutive year, rising to 4.5% (and 5.6% excluding South Africa). Strikes in South Africa's mining sector and a decline in oil production in Angola contributed to significantly slowed growth in these countries, along with slowdowns in those countries impacted by the Ebola outbreak. Nigeria, the largest economy in the region, experienced robust growth, as did many of the region's lowest-income countries.

China and the African Economy A deepening trade relationship has developed between Africa and China. Access to raw materials drives Chinese investment and interest in Africa and, as a result, exports of wood, minerals, and foodstuffs from the region to China are growing. China imports more than a quarter of its oil from Equatorial Guinea, Democratic Republic of the Congo, Gabon, Cameroon, and Nigeria as well as timber from the Congo Basin and huge quantities of critical minerals, such as cobalt, manganese, copper, and iron ore.

China also claims a rising share of exports to the region including many low-cost goods. Cheap Chinese textiles and clothes have invaded markets, hurting local producers in areas such as West Africa. China has also made massive investments in infrastructure in Africa (**FIGURE 5.24**).

Social and Economic Inequality Economic and social conditions vary greatly within and among African countries (see p. 204 "Visualizing Geography: The Millennium Development Goals in Sub-Saharan Africa"). People who work in the urban formal sector or are producers of cash crops generally have longer life expectancies, better service access, and higher

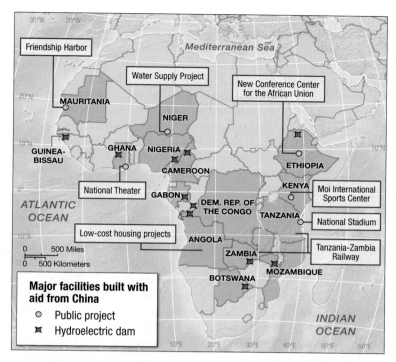

▲ **FIGURE 5.24 China and Africa** China has invested heavily in African countries including in public projects and hydroelectric dams as shown on this map.

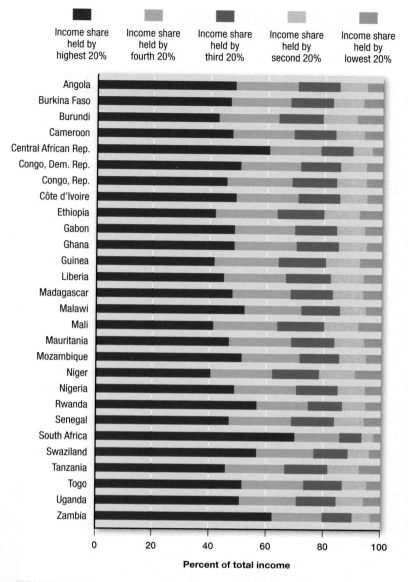

▲ **FIGURE 5.25 Income Inequality** Income is unequally distributed in most countries with the top 20% of earners claiming more than 40% of the income and the poorest 20% receiving less than 5%. **Which countries seem to be most unequal?**

incomes than those who work in the informal and rural subsistence agricultural sectors. Income concentration is high in many parts of sub-Saharan Africa, with the richest 20% of the population receiving more than 50% of overall income in most countries especially in southern Africa (**FIGURE 5.25**). The most recent World Bank data (2011) reports that across sub-Saharan Africa more than 400 million people (46% of the population) lived on less than the equivalent of U.S. $1.25 a day and 70% on less than U.S. $2.

Generally across the region, however, conditions have improved over the last 30 years. From 1970 to 2013, life expectancy has increased from 45 to 56.8 years and infant mortality has dropped from 138 to 64 deaths per 1,000 children born in 2012. Literacy has also shown dramatic changes, increasing from 38% in 1980 to 62% in 2010. Improvements in conditions in specific countries are reflected in life expectancy and infant mortality changes. In Ghana, for example, life expectancy increased from 45 to 64 years since 1970, and infant mortality decreased from 131 to 50 deaths per 1,000 children. But war and acquired immunodeficiency syndrome (AIDS) have also affected Africa. Life expectancy in Botswana and Zambia started to drop in the 1990s as a result of AIDS and in Rwanda as a result of war and genocide. Human immunodeficiency virus (HIV)/AIDS was also responsible for infant mortality increases in Botswana, Kenya, Zambia, and Zimbabwe in the late 1990s.

Poverty rates are reflected in other outcomes. Life expectancy is lower and infant mortality is higher than in other regions. The lowest 18 countries on the Human Development Index are all in Africa. Central African Republic, for example, has a life expectancy of only 41 years for women and 38 years for men and an annual GNI per capita averaging U.S. $588; these figures reflect

the loss of life and economic collapse associated with civil war. Low life expectancies reflect some of the deficiencies in service provision in Africa. Approximately half of the population lacks access to safe drinking water in Ethiopia, Chad, Angola, the DRC, and Sierra Leone.

Regional Organizations
Many regional organizations are working to improve conditions across Africa. **ECOWAS** (Economic Community of West African States), established in 1975 to promote trade and cooperation with West Africa, is an example of a program that promotes economic integration and political cooperation in sub-Saharan Africa. Recent initiatives include a trade and aid agreement with the European Union (the Cotonou agreement), efforts to reduce and recover from conflict, a human rights initiative, and a regional development bank.

Apply Your Knowledge

5.9 Using information from the Internet (e.g., www.worldbank. org or www.undp.org) determine how three major indicators of development and equality have changed in one sub-Saharan country since 2000. Have conditions improved or declined?

5.10 Can you suggest why this may be?

Territory and Politics

Peace and democratic politics in Africa have been hampered by the legacies of colonialism and the Cold War, ethnic rivalries, and the special interests of powerful individuals and sectors. Although some countries in the region were able to create or reestablish a sense of national identity following independence, others have experienced internal struggles or made claims on land beyond their current borders. The economic, human, and ecological cost of wars in sub-Saharan Africa has been a hindrance to investments in development. From 2012–2013 sub-Saharan's military expenditures increased by 7% to $26 billion, and Angola, Botswana, Gabon, and Namibia spent more than $100 per person on their militaries in 2013. In general, however, the region is more peaceful than it has been in recent decades.

Geographers Samuel Aryeetey-Attoh and Ian Yeboah have identified multiple causes for continuing political instability in Africa, including ethnic conflict, poor leadership, outside interference, and the legacies of recent independence struggles and racist governments. For example, Ghana, Nigeria, and Uganda all have experienced at least five coups since independence about 70 years ago. There is also concern about the number of elected leaders who have drifted toward one-party states and dictatorships, imposing accompanying repression and restrictions on freedom of speech. However, there are some signs of optimism in the more recent political geography of Africa as more countries in the region have democratic elections, and political and ethnic tensions have reduced with a decline in the number of authoritarian regimes, coups, and wars in the last 10 years.

South Africa and Apartheid South Africa is often seen as an international economic and political leader in sub-Saharan Africa. But from 1948–1990, it was an international outcast because of **apartheid**, its policy of racial separation. Under apartheid, black, white, and so-called colored (mixed-race) populations were kept apart and the South African government controlled the movement, employment, and residences of blacks. This stemmed from Dutch and British colonization and the settlers who imposed strict racial segregation, native reserves, and required that blacks have permission to enter or live in white areas (the pass laws). After 1931, when the Union of South Africa was given sovereign power, the South African government continued to impose strict racial separation policy, maintained most land in white ownership, and did not allow black people to vote. Regulations prevented social contact and marriage between races, established separate education standards and job categories, and enforced separate entrances to public facilities, such as train stations and hotels. Ten homelands were segregated as tribal territories where black residents were given limited self-government but no vote in national elections for governments that controlled most of their lives.

Protests against apartheid by both white and black South Africans were ruthlessly repressed. Antiapartheid movement leader **Nelson Mandela** was jailed in 1962. International objections to apartheid demanded South Africa's withdrawal from the British Commonwealth. Economic and trade sanctions were imposed and voluntary investment and participation bans were instituted by some major international corporations and sports competitions. Mandela was freed from jail in 1990 (**FIGURE 5.26**) and in 1994, South Africa held the first election in its history in which blacks were allowed to vote. Mandela was elected the first black president of the country as leader of the African National Congress (ANC)—he passed away in 2013. The 1997 postapartheid constitution in South Africa includes one of the world's most comprehensive bills of rights and prohibits discrimination based on race, gender, pregnancy, marital status, ethnic or social origin, color, sexual orientation, age, disability, religion, conscience, belief, culture, language, and birth. The hopes of many South Africans are frustrated by high rates of crime, lack of economic opportunities and access to land, and continued poverty

▲ **FIGURE 5.26 Nelson Mandela** After 40 years of apartheid, Nelson Mandela was freed from jail; President F. W. de Klerk agreed to share power, and Mandela was elected as the first black president of South Africa in 1994. People from all over the world mourned this great African leader when he died in 2013.

with many popular protests against the now black majority government.

Apply Your Knowledge

5.11 Visit the website of the Apartheid Museum in South Africa (http://www.apartheid museum.org/) and find three examples of how apartheid was applied or experienced.

5.12 Research some of the challenges faced by the South African government today in terms of meeting people's expectations for food, water, jobs, housing and health.

Apartheid Museum

https://goo.gl/4okjNn

The Cold War and Africa Independence movements and transitions in Africa following the Second World War coincided with the global tensions associated with the Cold War between the United States and the Soviet Union. Many Africans found communist and socialist ideas of equity and state ownership appealing after the repression, foreign domination, and inequality of colonial rule. For example, in Tanzania in 1967, President Julius Nyerere developed the concept of an African socialism based on the traditional values of communal ownership and kinship ties to extended family expressed as ujamaa (familyhood). Nyerere believed that a socialist system of cooperative production would be more compatible with African traditions than individualistic capitalism.

Africa provided fertile soil for Cold War rivalry as newly independent nations searched for political models, struggled with civil wars and incursions from their neighbors, and sought assistance to develop their economies. In Angola, Mozambique, Ethiopia, and Somalia, revolutionary movements espousing leftist ideals attracted the interest of the then socialist Soviet Union, China, and Cuba, which provided military, economic, and educational assistance. The United States and South Africa supported pro-Western movements, including opposition armies, which fought socialist rule. In many countries, millions of dollars were expended on arms and other military assistance, thousands were killed, and rural areas were abandoned because of land mines.

Civil Wars and Internal Conflicts Recent years have seen some terrible conflicts in sub-Saharan Africa. Some of these conflicts have crossed international borders, but many are associated with internal civil war and unrest.

- **Rwanda and Burundi** Civil wars broke out in 1993–1994 in Rwanda and Burundi when long-standing tensions between the Hutu and Tutsi tribes within each country erupted into violence. A total of 700,000 people were killed and two million refugees ended up in the DRC, Uganda, Kenya, and Tanzania. There are challenges of reconciliation in Rwanda and Burundi where memories

of violence persist, the divisions are deep, and the civil wars were labeled as **genocide**. Genocide is defined as an effort to destroy an ethnic, tribal, racial, religious, or national group. In Rwanda serious efforts have been made to rebuild the economy, tourism, and agriculture. In 2008, Rwanda elected a legislature in which the majority of politicians were women.

- **Liberia and Sierra Leone** In West Africa, the worst civil conflicts were in Liberia and Sierra Leone in the 1990s where simmering resentments between opposition groups were fueled by arms acquired through the sale of diamonds. In Sierra Leone, civil war between rival paramilitaries killed more than 50,000 people and displaced 2 million people. Liberian civil wars killed at least 400,000 people between 1989 and 2003. These wars were notorious for atrocities that included rape and the cutting off of people's limbs and for the forced recruitment of child soldiers. **Child soldiers** are children under 18 years of age who are forced or recruited to join armed groups (**FIGURE 5.27**). Peace was established in Sierra Leone in 2002 and in Liberia in 2003 when UN forces disarmed rebels and militias. Programs of reconciliation were put into place, including efforts to rehabilitate child soldiers and those who had been injured during the conflict. In 2006, U.S.-educated economist Ellen Johnson Sirleaf won the Liberian presidential election and vowed to sustain democracy and rebuild the economy. Sirleaf became Africa's first elected woman head of state. Sierra Leone has also held democratic elections, and conditions were improving until the 2014 Ebola outbreak.

- **Somalia** Some of the most difficult problems in sub-Saharan Africa are in the failed state of Somalia, one of

▲ **FIGURE 5.27 Child Soldiers in Sierra Leone** Thousands of children were forced or recruited to join revolutionary forces during the civil war.

The Millennium Development Goals in Sub-Saharan Africa

African development is increasingly driven by a new set of targets established by the United Nations called the Millennium Development Goals (MDGs); these aim to eradicate extreme poverty and hunger; achieve universal primary education; promote gender equality and empower women; reduce child mortality; improve maternal health; combat HIV/AIDS, malaria, and other diseases; ensure environmental sustainability; and develop a global partnership for development (Figure 5.3.1).

1 Sub-Saharan Africa has traditionally ranked low on most of the MDG indicators. But some of the MDG indicators – such as those for poverty, undernourishment, education, gender equality, child mortality, disease and debt - have shown progress since 1990. But the region is far from meeting most of the goals.

Because of the challenges of data collection in the region, reports on the MDG goals are not produced every year for each country so the dates on some of these figures vary.

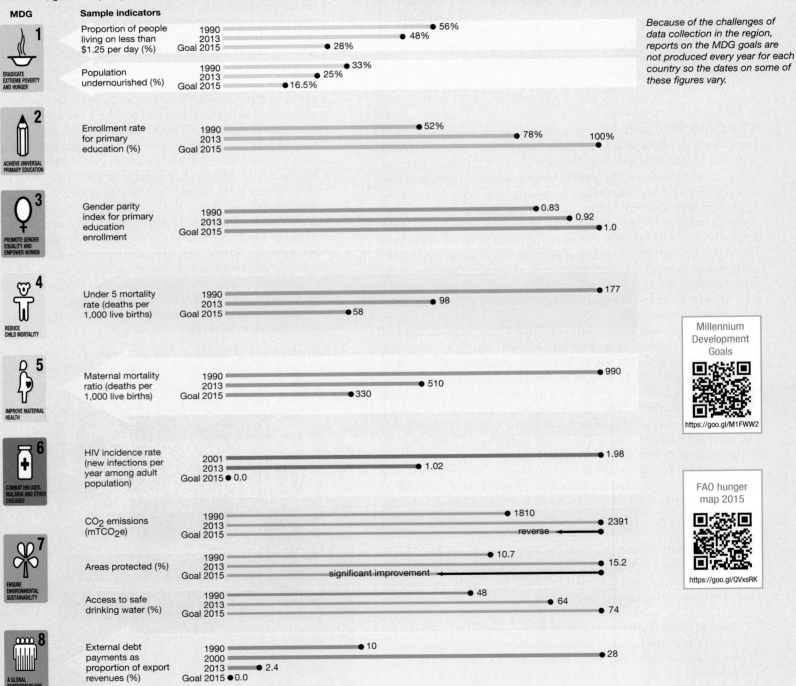

Millennium Development Goals

https://goo.gl/M1FWW2

FAO hunger map 2015

https://goo.gl/QVxsRK

▲ **FIGURE 5.3.1 Millennium Development Goals: Overall Progress in sub-Saharan Africa 1990-2013.**

"Despite some progress toward goals since 1990, there are still large recent differences within and between countries."

2 Maps show the large recent differences within and between countries. In terms of the target of halving hunger some countries are on target to meet the goal and others have made no progress or conditions are deteriorating (Figure 5.3.2a).

South Sudan became the newest country in Sub Saharan Africa in 2011 but also one of the most challenged in meeting the MDGs (Figure 5.3.2b). At the time of independence the country had some of the highest levels of poverty, hunger, maternal and child mortality and lack of access to safe water and sanitation in the world. Civil unrest, a weak economy and underdeveloped infrastructure make it hard for the country to make any progress on the MDGs.

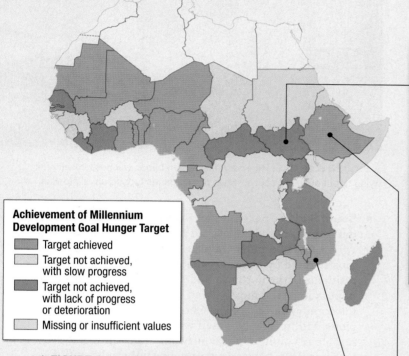

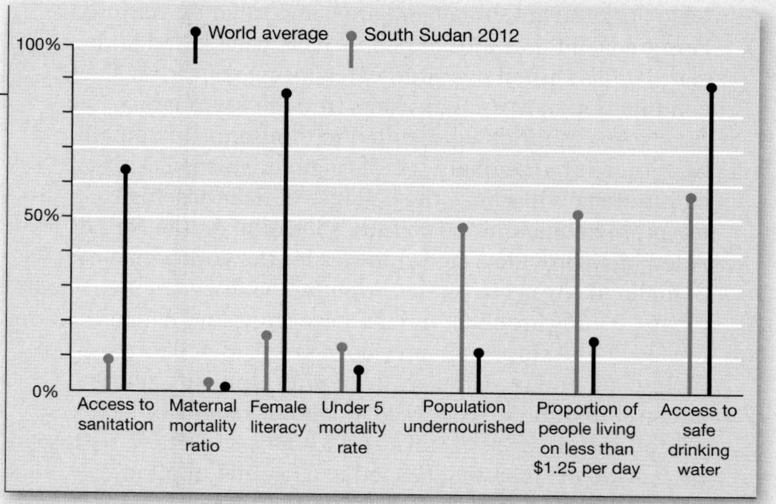

Achievement of Millennium Development Goal Hunger Target

- Target achieved
- Target not achieved, with slow progress
- Target not achieved, with lack of progress or deterioration
- Missing or insufficient values

▲ **FIGURE 5.3.2a** Progress in meeting the Millennium Development target of halving the proportion of under-nourished people between 1990 and 2015. Source: FAO 2015 (cropped from world map).

▲ **FIGURE 5.3.2b** Progress towards selected MDGs in South Sudan.

Some individual countries have made a lot of progress, especially those where conflict has been reduced or there has been a focus on social improvements with the assistance of foreign aid or debt forgiveness such as Mozambique (Figure 5.3.2c) and Ethiopia (Figure 5.3.2d).

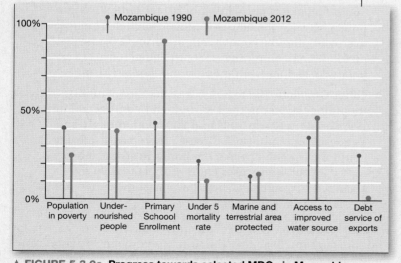

▲ **FIGURE 5.3.2c** Progress towards selected MDGs in Mozambique.

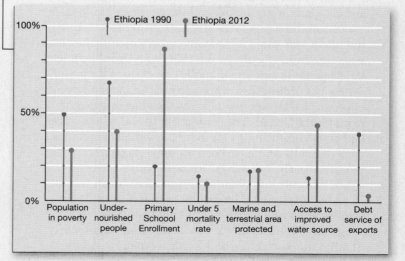

▲ **FIGURE 5.3.2d** Progress towards selected MDGs in Ethiopia.

1. Which country, Ethiopia or Mozambique has made the greatest progress in poverty elimination and other indicators?

2. Why, even though there has been an improvement in the percent of the population living in better conditions might there be just as many people living in poverty, hunger or without water access today?

the poorest countries in the world. A **failed state** is one where government is so weak that it lacks legitimacy and control over its territory, cannot provide public services, and corruption and crime dominate. The boundaries of the state of Somalia, drawn by Italy and Britain in 1960, left out and split some traditional Somali cultures and territories, which were instead granted to Kenya, Djibouti, and Ethiopia. Military governments have failed to regain these areas and lost internal legitimacy as a result. Somalia has been in a state of civil war for much of the time since 1990. Clans in the northeast and southwest have declared autonomy and are being ruled by warlords. Hundreds of thousands of people have been killed and conflict has spilled over into neighboring countries. The combination of natural disasters (droughts, tsunamis, and floods) and conflict has resulted in thousands of Somalis becoming internationally or internally displaced. This continuing civil unrest, the collapse of fisheries, high unemployment, and the country's location on the Gulf of Aden have provided opportunities for the young men of Somalia to engage in piracy. Arguing that they are protecting fishing grounds, pirates seize ships and demand ransoms. The ships that have been attacked or captured include oil and coal tankers, food aid shipments, tourist cruise ships, and private yachts. Various countries have offered antipiracy assistance and formed a maritime patrol in the Gulf of Aden that has led to a decline in piracy.

- **Nigeria** With extensive oil reserves and a population of almost 178 million, Nigeria is one of the most powerful and populous countries in Africa. Nigeria was formed when the British brought together several ethnic groups, including the Yoruba, Igbo, and Fulani, into a large colony, which eventually became independent in 1960 soon after oil had been discovered by Shell-BP. Wealth inequalities have been exacerbated by conflict between ethnic and religious groups and by oil development. In the late 1960s, a million people died from fighting and famine in the Biafra War when Igbo people unsuccessfully declared secession from Nigeria. In the 1990s, the Ogoni people of the Niger Delta suffered oil pollution of their lands and waters but protest was suppressed including the hanging of protest leader Ken Saro-Iwa by the Nigerian government (**FIGURE 5.28**). The conflicts in the oil region have had global implications for international oil companies, such as Shell, who have been accused of complicity with the government and subjected to consumer boycotts.

    Civilian rule and democratic elections, most recently in 2015, have reduced tensions in Nigeria since 1999, and the economy has been growing fast. However, new religious conflicts have arisen based in predominantly Muslim northern Nigeria, where an Islamic extremist movement, Boko Haram, has killed civilians and abducted people, including schoolgirls. A total of 1.5 million people have fled the area. The Nigerian government has hired mercenaries from other countries to fight the terrorism, and a regional West African force has been organized with support from France, the United Kingdom, and the United States.

▲ **FIGURE 5.28 Oil in Nigeria** A resident of Lagos washes himself after trying to put out a fire caused when armed gangs tapped an oil pipeline.

- **South Sudan** Created under British colonial rule in 1947, the country of the Sudan had the largest land area in Africa. Granted independence in 1956, the country was mostly ruled by a military that favored the Arab Islamic residents of the northern Sudan and repressed the black population, who mostly lived in the southern region of the country and practiced Christian or animistic religions. Millions died in civil wars between groups in the north and south of the country and from famine. Peace agreements and the recognition of southern liberation and secession movements resulted in a referendum in which 99% of the 11 million people in the south voted for independence with a new independent country of South Sudan recognized in July 2011. Conflict has continued between the new government and rebel groups over who rules the country, ethnicity, and the unequal distribution of oil revenues. South Sudan remains very poor with serious health, water, and food problems.

- **Other Conflicts** The region has seen other conflicts, including the war in the DRC, which killed several million people between 1998 and 2008 both directly and as a result of famine and disease. There has also been conflict in Uganda, where the Lord's Resistance Army, led by Joseph Kony, uses child soldiers to fight the Ugandan government. There have been recent rebellions in Mali, Niger, and Chad.

Peacekeeping The end of many recent wars in Africa and subsequent peace settlements have been brokered and monitored by the United Nations and regional security forces. The UN Peacekeeping Forces operate under the authority of the UN Security Council and are intended to establish and maintain peace in areas of armed conflict with the permission of disputing parties. In Africa, UN forces, with their distinctive pale blue helmets, have been recently deployed in more than a dozen countries (**FIGURE 5.29**).

▲ **FIGURE 5.29 UN Peacekeeping** The UN peacekeeping forces in sub-Saharan Africa include many soldiers from countries in the region, including South Africa, Nigeria, and Ethiopia. This photo shows a South African brigade in the Congo that includes some female soldiers. **What might be the advantages of having female soldiers as part of a U.N. peacekeeping mission?**

African leadership in promoting peace within the region is growing in importance. These leadership efforts include negotiations led by former South African president Nelson Mandela and a West African peacekeeping force and monitoring group, called Economic Community of West African States Monitoring Group (ECOMOG).

Apply Your Knowledge

5.13 Identify the causes of civil war and conflict within a country, the steps that have been taken to resolve the tensions, and the impacts on residents of the country.

5.14 Consult two or three recent UN or media reports on conflict, peacekeeping, or refugees in sub-Saharan Africa such as http://www.un.org/en/peacekeeping/ or www.unhcr.org. Do these reports seem optimistic or pessimistic in tone? What factors contribute to reasons for hope or greater concern?

> UN Peacekeeping in Africa
>
> https://goo.gl/6ZzjD0

Culture and Populations

Sub-Saharan Africa is a large and extremely diverse region, with hundreds of different ethnic groups who have their own traditional cultures, religions, and languages. In addition to strong cultural legacies, the region has experienced dynamic shifts in culture and attitudes associated with more recent global interconnections. Despite these shifts, the extended family, ties to the land, oral tradition, village life, and music often remain important.

Ethnicity

The multiplicity of religions, languages, and dialects in Africa reflects the large number of distinct cultural or ethnic groups in Africa. Some writers use the term **tribe** to define these groupings and describe Africa as a *tribalist* society. The term tribe describes a form of social identity created by groups who share a collective set of ideas about loyalty and political action. In tribes, group affiliation is often based on shared kinship, language, and territory. Some see the term tribe as negative (related to colonial perceptions of "savagery") and now prefer to use the term **ethnic group**.

The largest ethnic groups in Africa are associated with the dominance of certain languages, such as Hausa, Yoruba, and Zulu, but almost all groups were either split geographically by colonial national boundaries or grouped together with their neighbors, enemies, or others with whom they shared no affinity. This has been a major cause of conflict in contemporary Africa.

Religion

Traditional African religions have been described generally as animist (worship of nature and spirits), but this description simplifies the wide variety of local religious beliefs in Africa. Although natural symbols, sacred groves of trees, and landforms often have significance in local religions, many African religions also include ancestor worship and feature a belief in a supreme being, several secondary gods or guardians, and good and evil spirits. In many of these belief systems, ancestors, priests, or traditional healers mediate and interpret the wishes of the gods and spirits, and rituals ensure the stability of society and relations with the natural world. More than 70 million people (about 10% of the total population) are reported to practice traditional religions on the continent of Africa.

Influenced by migration, colonialism, and politics, Christianity and Islam today dominate in terms of religious practice in sub-Saharan Africa. Of the 360 million estimated Christians in Africa, there are about 125 million Roman Catholics and 114 million Protestants. Evangelical Christianity is growing, attracting more than 10% of the population across Africa. The vibrancy of Christianity in many countries draws more young people to careers in the church than in many other world regions, and Africa-trained pastors are taking leadership positions in European and North American churches.

Islam has more than 300 million adherents in sub-Saharan Africa. Islam is important in the Sahel, North Africa, and in parts of East Africa, such as Somalia and Tanzania; Nigeria and Sierra Leone are now more than 50% Muslim.

Religious differences have fueled numerous political conflicts in some regions of Africa such as those in the Sahel and were one reason for the independence of South Sudan.

Language

The geography of languages in Africa is incredibly complex, with more than 1,000 living languages, 40 of them spoken by more than 1 million people (**FIGURE 5.30**). The dominant indigenous languages, spoken by 10 million or more, include Hausa (the Sahel), Yoruba and Ibo (Nigeria), Swahili (East Africa), and Zulu (southern Africa). Spoken as second languages by many groups to facilitate trade, Hausa and Swahili are examples of a **lingua franca** (see Chapter 9). English, French, Portuguese, and Afrikaans are also spoken in areas that were under colonial control and had European education systems or settlements. Arabic has strongly influenced Swahili along the east coast of Africa. Because most countries have no dominant indigenous African language, many use a European language when they conduct official business and in their school systems. The countries with the most coherent overlap between territory and a dominant African language are Somalia (Somali), Botswana (Tswana), and Ethiopia (Amharic).

Apply Your Knowledge

5.15 How does the geography of language and religion in sub-Saharan Africa reflect the legacies of colonialism?

5.16 Research the linguistic diversity of a sub-Saharan African country (other than Somalia, Ethiopia, or Botswana) and describe how that country's boundaries cut across those of traditional ethnic groups and any conflict this has created.

Cultural Practices, Social Differences, and Identity

Culture in Africa is as varied as religion and language and is changing rapidly as a result of interactions across and beyond the region. Through the slave trade, migration, and global communications, African traditions have influenced the Americas and other regions.

Kinship In many areas, cultural traditions include the importance of the extended family and the kinship ties in social relations and obligations and the widespread respect for elders as sources of wisdom. Kinship ties going back multiple generations define "clan" allegiances in some subregions and sometimes drive primary loyalties. This is the case, for example, in contemporary Somalia, where interclan conflict has dominated recent political events.

Land The tie to the land is connected to traditional forms of land tenure in Africa, where, in many areas, land was viewed as given by the spirits or held in trust for ancestors and future generations. The Elesi of Odogbolu (a traditional leader) in Nigeria has expressed this view in these words: "The land belongs to a vast family of which many are dead, few are living, and countless members are still unborn." This view endows the land with communal nature such that it cannot be bought or sold by individuals. In some cases, land rights are held by extended families or the community, rather than by individuals, and in other societies the chief or king controls land.

◄ FIGURE 5.30 **African Languages** The cultural complexity of Africa is demonstrated by the variety of its indigenous languages. The indigenous languages of Africa can be grouped into larger language families, including the Afro-Asiatic languages of North Africa, such as Somali, Amharic, and Tuareg; the Nilo-Saharan languages, such as Dinka, Turkana, and Nuer in East Africa; and the largest Niger-Congo group, which includes Hausa, Yoruba, Zulu, Swahili, and Kikuyu. Swahili and Hausa are spoken by millions as the trade languages of East and West Africa, respectively. Madagascar is different with Malay-Indonesian languages. **Propose an explanation for the relatively sharp boundary between the Semitic-Hamitic languages across North Africa and the area of much greater linguistic diversity to the south across sub-Saharan Africa.**

Reciprocity Traditions of reciprocity, where a gift is given to obtain a favor, and of helping family members can be a major source of cultural confusion in the region, according to African historian Ali Mazrui. He suggests that these traditions provide an explanation for the way in which some regional leaders have favored family members with jobs in their administrations and for the role bribes play in requests to government officials. Mazrui also notes that under colonialism, local residents viewed stealing from the government as a legitimate form of resistance because they felt that foreigners were robbing Africa of resources and funds through taxation.

Music, Art, and Film Africa has a rich tradition of music and the arts that have had an enormous impact on other regions of the world. For example, traditional music of the West African Sahel influenced the development of American blues, and West African coastal traditions influenced Afro-Caribbean music styles. Slavery was one way in which African musical, artistic, and food customs spread around the world, especially to the Americas. The arts are dynamic in much of Africa, often reflecting international influences and new youth identities.

- **Music** Africa is often associated with music from percussion instruments, especially drums. Other instruments with metal keys that are plucked or tapped, as well as flutes and harp-like stringed instruments, are also commonly used in traditional and popular music. In West Africa, oral traditions are associated with singers and storytellers, some of whom receive the respected name of **griots**. African popular music mixes indigenous influences with those of the West. For example, the highlife music of West Africa is derived from Caribbean calypso and military brass bands; it includes stronger percussion, soul influences, and exchanges between lead and background singers. Juju, or Afrobeat, which is related to highlife music, was made popular by Nigeria's King Sunny Adé and Fela Kuti. One of Africa's most famous vocalists, Youssou N'Dour, has mixed his music career with activism and politics, even making a bid for president of Senegal in the 2012 elections (**FIGURE 5.31a**).

  The diverse audiences for African music mirror its diverse mix of influences. A prime example is the Festival au Désert, held in Mali each year from 2001 to 2012, which brought together musicians from across West Africa with local pastoralists, tourists, and international media (**FIGURE 5.31B**).

- **Art** African art is also richly varied. In traditional African cultures, artists were valued specialists, often working under the patronage of kings and creating works of spiritual value. The masks and wood sculptures of the Dogon and Bambara people of West Africa are collected around the world and still maintain their cultural significance within the region. Kente cloth designs from northern Ghana have become meaningful in African-American identity (**FIGURE 5.31c**). As interest in travel and world culture has grown, local artists have started

(a)

(b)

(c)

▲ **FIGURE 5.31 African Music and Art** (a) Senegalese musician Youssou N'Dour sings with Angelique Kidjo of Benin at a tribute to South African singer Miriam Makeba. (b) The Festival au Désert in Mali features traditional music and popular singers. (c) Kente cloth from Ghana.

to produce items for sale to tourists and international distributors. But of course not all African art is traditional, and galleries in major cities as well as websites promote young modern artists to a global market.

- **Literature and Film** Millions have read the novels or poems of Chinua Achebe, Wole Soyinka, and Chimanda Ngozi Adichie from Nigeria, Doris Lessing of Zimbabwe, and Nadine Gordimer of South Africa. Film is also a growing art form in the region with global influence. Nigeria is home to a booming film industry, known as **Nollywood**, that produces hundreds of recordings for the home video and diaspora market each year and is the third-largest film industry in the world (after the United States and India). Nollywood films are often shot on location for very little money but are incredibly popular, often dealing with moral questions and social issues, such as AIDS, corruption, and witchcraft.

Apply Your Knowledge

5.17 Identify a cultural or ethnic group in sub-Saharan Africa; research traditions of art, music, literature, land ownership, or religion within that group; and describe the traditions' significance for the group.

5.18 Read a novel or view a recent film (or read a review of a book or a film) that is set in sub-Saharan Africa. How does the theme of the work you have chosen reflect the contemporary geography, problems, or culture of the region?

Sport For many young Africans, their heroes are sports stars, especially athletes who make it internationally. Even in remote villages, children play with improvised soccer balls, hoping to be spotted and eventually recruited to play for a European team (**FIGURE 5.32**). Superstars, such as Didier Drogba or Yaya Toure from the Côte d'Ivoire who played for English football clubs Chelsea and Manchester City have become very rich and enormously popular. Another group of African sports stars are the long-distance runners of Ethiopia and Kenya that have produced numerous Olympic gold medalists in both men's and women's events and the cricketers of South Africa.

The Changing Roles of Women Women in sub-Saharan Africa tend to have less education and lower incomes than men, but in most countries, they live slightly longer. Gender differences in Africa reflect the feminization of poverty; more than two-thirds of Africans who join the ranks of the poor are women. The average female annual income (PPP) in sub-Saharan Africa is about U.S. $1,300 less than that of men, and women consistently complete less schooling than men. African women are more likely to be poor, malnourished, and otherwise disadvantaged because of inequalities within the household, the community, and the country. Women are less likely to receive an education, and overall, pay rates are lower in the workplace.

Patriarchy and cultural traditions mean that women may be required to eat less than men and eat only after the men in the family have eaten. Polygamy (men having multiple wives) gave men control over more female labor and is seen as providing both possibilities for women to support each other but also for conflict between them and the spread of disease. Women perform critical functions in rural areas, meeting the basic needs of their families by preparing food and collecting wood and water, generating income with crafts and community work, and comprising at least half the agricultural producers in both subsistence and commercial sectors (**FIGURE 5.33**). The tradition of female circumcision has become a controversial struggle between those who see it as a human rights abuse involving brutal genital mutilation and health risks and others who see it as an important religious and symbolic experience. More than 100 million women in Africa are estimated to have undergone female circumcision.

▲ **FIGURE 5.32 African Sports** African children playing soccer (known as football in Africa) in poor townships aspiring for the international fame and fortune African stars who play for European teams.

▲ **FIGURE 5.33 Women Collecting Fuel** Women carry dung and firewood, which they have collected to use as fuel, in Chandeba, Ethiopia. Women often must walk long distances with heavy loads.

The Ongoing Struggles of South Africa's Youth

Youth organizing played a central role in resistance against apartheid rule in South Africa (see p. 202), from Nelson Mandela's early activism in the African National Congress Youth League to youth activists in townships across the country. During a period of intense upheaval and resistance in the mid-1980s, young activists took up roles on the front lines of struggle. Thousands of children were jailed, and some were killed or wounded in the struggle to overthrow apartheid.

Twenty years after the end of apartheid, substantial inequalities still plague South Africa and with serious gaps between black and white wages and high unemployment for black youth. In a 2014 study entitled "Listening to the 'Born Frees,'" researchers interviewed black youth born after the end of apartheid. The stories of two youth from townships of East London, a city on the country's southeastern coast, showcase the troubling inequalities faced by the postliberation generation struggling to find livelihoods and political voice in postliberation South Africa (**FIGURE 5.4.1**).

Thozama: Twenty-five years old and unemployed, Thozama attended a badly resourced township school and only completed part of her secondary (high school) education. She lives in an

> "Twenty years after the end of apartheid, substantial inequalities still plague South Africa and with serious gaps between black and white wages and high unemployment for black youth."

urban township and her first language is Xhosa. She dreams of going back to school to finish her high school certificate but remains focused on finding work. She feels disillusioned by searching for jobs advertised in the newspapers. She feels that to get work, "you have to have a connection and pay money, and money is something you don't have because you are not employed."

Odidi: Educated at a private school but now working intermittently as a casual laborer, 29-year-old Odidi expresses concern with the lack of jobs available to him. Married with children, his

household monthly income is less than $330 and providing for his family is his primary concern. He expresses a great deal of political disillusionment, stating that although the government makes claims to political representation, from his perspective, "they decide for us."

1. Use the Internet to find stories or blogs about young people in South Africa and summarize their feelings about postapartheid South Africa.

2. Surveys of public attitudes in Africa can be found at http://www.afrobarometer.org/. Review results on views in South Africa, and discuss how they are changing over time and different social and age groups.

Public Opinion in Africa

https://goo.gl/1hm5hJ

▲ **FIGURE 5.4.1 Faces of the Region** Underemployed young men hold placards offering to work in various jobs around Johannesburg, South Africa.

Women are also disadvantaged by many traditional and modern institutions that define property rights. Land in some places may be passed on only to male children, and new land titles are often granted to male heads of household.

Women control distinctly gendered spaces in the African landscape: parts of the home, the market, and the cultivation of certain trees and crops are reserved primarily for women. Women generally have much less access to agricultural inputs such as training, fertilizer, livestock, and land and have less time to farm given their household responsibilities. This limits overall agricultural productivity. Women are often also disproportionately affected by environmental degradation, as deforestation and drought made collecting wood, water, and food more difficult. Although some African governments and development agencies have recognized these disparities and established programs meant to improve conditions for women, poverty reduction among women has been patchy.

Those who understand the role of women's work in African communities and economies have criticized development policies that ignore, undervalue, or displace women. These policies gradually have begun to change to include projects that provide technology, credit, and training to women; that link women's productive and reproductive roles; and that seek to overcome the gender-related differences and barriers to improving the lives of both men and women.

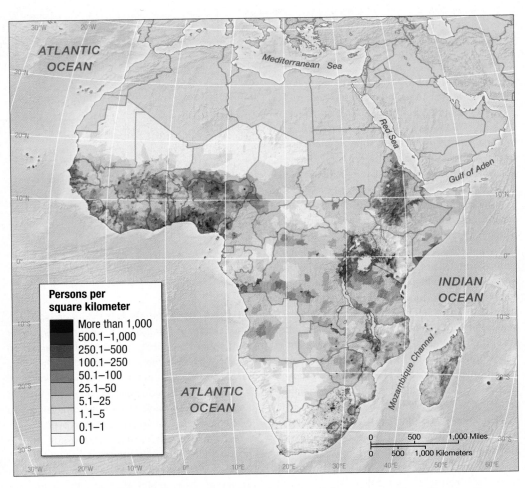

▲ **FIGURE 5.34 Sub-Saharan Africa Population Density** Africa is home to a mostly scattered rural population. Population concentrations tend to be associated with better soils and climate, with colonial centers for mining and export crops, and with coastal ports. **Which regions of Africa have the greatest population densities and can you link any of these to physical geography?**

Demography and Urbanization

The largest numbers of people in the region live in Nigeria, which has a population that is estimated to be over 173 million and in Ethiopia with almost 100 million. Some populations in the region have not been reliably censused for many years, and population estimates are approximate. Population density in the region ranges from very low (e.g., Namibia with just 3 persons per square kilometer) to very high (e.g., Rwanda with 421 persons per square kilometer). The average population density is 36 persons per square kilometer (2014), close to the United States average of 33. However, almost all sub-Saharan African countries have lower population densities than the global average (**FIGURE 5.34**).

Demographic Change

Sub-Saharan Africa is still growing rapidly compared to other world regions. Its population was 936.1 million in 2013 and is projected to reach 1.5–2 billion by 2050. Although fertility has fallen from 5.5 in 2001 to 5.1 in 2013, the annual population growth rate is 2.7%. Fertility rates are much lower in southern Africa with women having three children or less in countries such as South Africa and Botswana. Approximately 43% of the African population is under the age of 15 years although the percent of youth varies by country (**FIGURE 5.35**). This has major implications for future population growth. When this group starts to have children, demands on education systems, health care, jobs, and resources may increase.

Factors Affecting Fertility What are the reasons for high fertility and birthrates in Africa, and what are the prospects for slowing population growth? Population geographers have found that although religious prohibitions of contraception and lack of access to contraceptive devices may play a small role, other factors are much more important. Children are the main source of security for elderly people in countries where there are few pensions or public services for the aged, and it is traditional for younger generations to respect and care for their elders. Children are valued in Africa for many reasons,

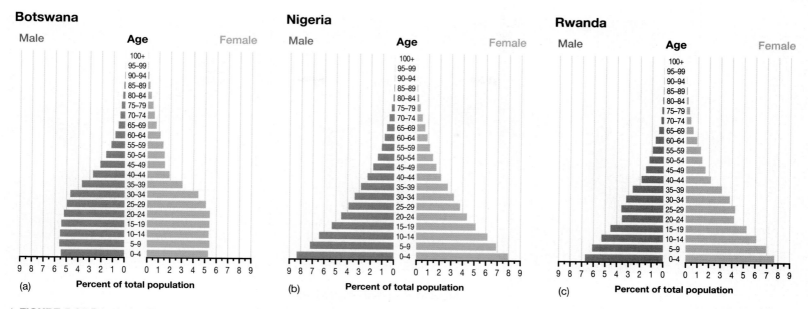

▲ FIGURE 5.35 Population Pyramids for Selected Countries These graphs show the percent of men and women in different age ranges. (a) Botswana has a narrower age pyramid with a smaller percent of younger people (because of lower birth rates and partly to HIV/AIDS) and a population likely to grow slower. (b) Nigeria has a wider base as a result of high fertility rates contributing to population growth in this large country. (c) The Rwandan pyramid reflects the war, genocide, and migration that killed many and reduced births in what is now the 10–30 age range.

including their ability to work in agricultural fields and as herders, to help with household work, and to care for younger siblings. Children are also a possible source of financial or other gain when they marry. Large families are also often perceived as prestigious, a spiritual link between the past and the present, and a way of ensuring family lineage. Even though infant mortality rates have improved with better nutrition and health care in much of Africa, many African families have internalized the need to have a large number of children to ensure that some survive to adulthood.

Many studies have also shown that conditions for women have a strong influence on fertility rates; younger marital ages, minimal female education and literacy rates, and fewer opportunities for female employment all contribute to higher fertility rates. Fertility rates tend to be lower in urban areas with high rates of female education and employment, where women marry when they are older. In Ghana, rural women have higher rates of fertility than urban women, and richer and more educated women have fewer children than poor or less educated women. Statistics such as these indicate that future population growth rates in the sub-Saharan African region and declines in fertility will depend on improved status and conditions for women.

Population Impacts of HIV/AIDS and Other Health-Care Concerns
In sub-Saharan Africa, more people are affected with the **human immunodeficiency virus (HIV)** that causes **acquired immunodeficiency syndrome (AIDS)** than any other part of the world. Three of every four deaths from HIV-related causes take place in sub-Saharan Africa. In 2013, 24.7 million people in Africa were living with HIV; the highest rates were in Botswana, Lesotho, and Swaziland, and the largest absolute numbers in South Africa and Nigeria. More than

15 million Africans have lost their lives to AIDS since the disease was identified in 1981. It has become the leading cause of early death in Africa, killing more people than malaria and warfare combined (**FIGURE 5.36**). Because of the high mortality and infection rates associated with the AIDS epidemic, population projections for several African countries were revised downward in recent years.

The geography of AIDS in Africa varies by country, by regions within countries, and by social groups. In many cases AIDS spread along migration and transport routes. Unlike in other regions, more women than men have AIDS in sub-Saharan Africa, and mothers often transmit the disease to their children. Frequently, married couples are both infected with AIDS and die from it. As a result, there were more than 15.7 million AIDS orphans in sub-Saharan Africa in 2011. Poverty exacerbates the AIDS problem in sub-Saharan Africa because many cannot afford prevention (e.g., through the use of condoms), testing, or antiretroviral medicines that can prolong the lives of people who are infected. According to the World Health Organization, at the end of 2013, 37% of adults in the region affected by HIV were receiving antiretroviral therapy, as compared to just 23% of children. Some governments have low health-care budgets and people have health insurance, diagnosis and treatment are often delayed, cultural stigmas can impede care, and interaction with other diseases, such as tuberculosis, increases mortality rates.

However, there has been considerable success in combating HIV/AIDS in the last 15 years. A combination of domestic policies, foreign assistance, and subsidized treatment helped reduce infection rates by 25% across the region between 2001 and 2009 and deaths by 20% over the same period. Annual deaths from HIV-related causes have further decreased by 22% from 2009 to 2013. Unprecedented international agreements with drug companies in combination with new assistance programs from

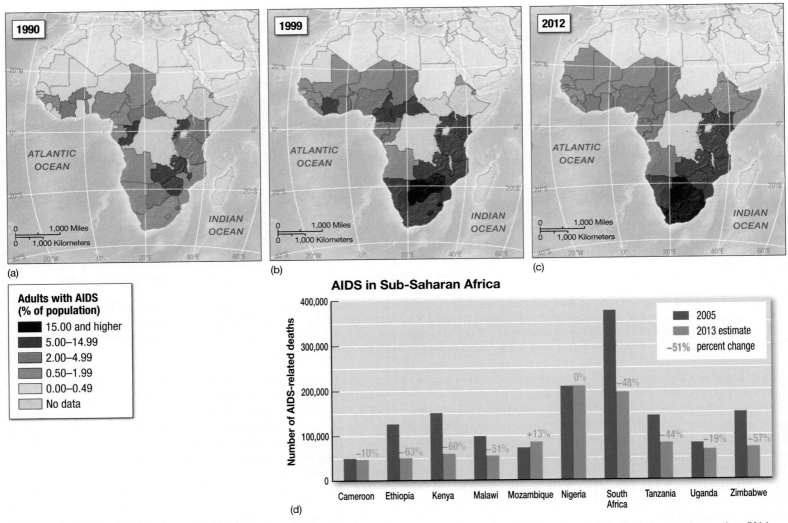

▲ **FIGURE 5.36 AIDS in Africa** (a) In 1990 AID infection rates were highest in Eastern Africa (b) By 1999 rates were over 15% in Zimbabwe and more than 5% in much of east and southern Africa (c) Rates remained high across southern and east Africa in 2012 with at least 1% infection across much of the rest of the region (d) But infection and death rates were starting to fall by 2013. **Which countries had the largest percent drops in death rates from AIDS between 2005 and 2013?**

the World Bank, charities, and donor countries are lowering the cost of AIDS drugs. More than a third of those infected now get antiretroviral medicines compared to only 2% in 2001. But many people with HIV and AIDS are still unable to obtain the drug therapies that will extend their lives.

Apply Your Knowledge

5.19 Research the demography (population, aging, and fertility trends) of a country in the region (e.g., www.un.org/esa/population or www.prb.org) and discuss the implications for environment, economy, and public services.

5.20 List factors that have contributed to the spread of HIV/AIDS in sub-Saharan Africa, describe the impacts of the epidemic, and explain policies that are helping to control it.

African Cities

Although Africa is the most rural of world regions, it has been urbanizing rapidly over the last 40 years. In 1960, the urban population of sub-Saharan Africa was only 17 million people, about 20% of the overall regional population. In 2013, the urban population reached 37.4% of the total population, and the number of people living in larger cities is growing. The level of urbanization varied greatly by country in 2014, Liberia with about 50% of from South Africa, Gabon and Rep. Congo with more than 60% of their people in urban areas to Ethiopia, Niger, Malawi, Uganda and Burundi at less than 20% (**FIGURE 5.37**).

Just like in many other world regions, the major driver of urban growth in sub-Saharan Africa is migration from rural areas to cities. The factors pushing people from rural areas and pulling them to the cities are also somewhat similar to contributing factors in other world regions. People are leaving rural areas because of poverty, lack of services or support for agriculture, scarcity of land, natural disasters, and civil wars. Urban areas are

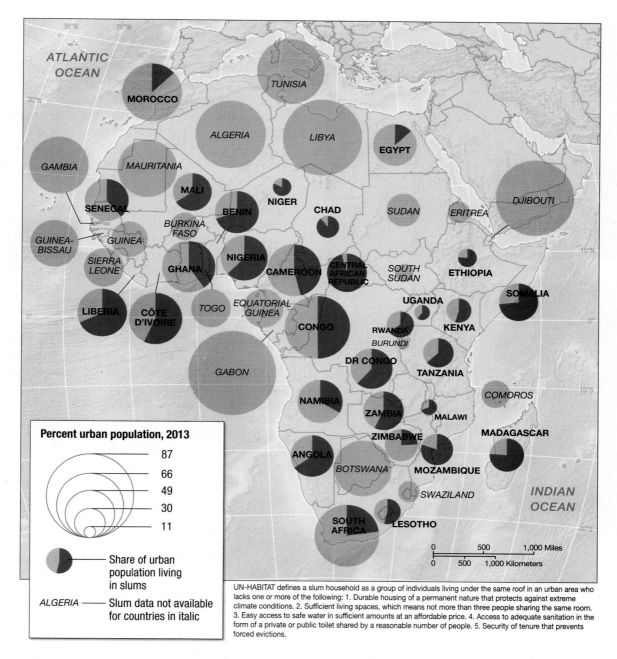

Percent urban population, 2013

87
66
49
30
11

Share of urban population living in slums

ALGERIA — Slum data not available for countries in italic

UN-HABITAT defines a slum household as a group of individuals living under the same roof in an urban area who lacks one or more of the following: 1. Durable housing of a permanent nature that protects against extreme climate conditions. 2. Sufficient living spaces, which means not more than three people sharing the same room. 3. Easy access to safe water in sufficient amounts at an affordable price. 4. Access to adequate sanitation in the form of a private or public toilet shared by a reasonable number of people. 5. Security of tenure that prevents forced evictions.

▲ **FIGURE 5.37 Urbanization and Slums in Africa** The total urban population in countries across Africa has been increasing. In most major cities, however, a majority of the people live in slums. Which three countries have the highest share of the urban population living in slums? **Which have the smallest overall percent of people living in urban areas?**

more attractive because they offer jobs, higher wages, better services (including education and electricity), and entertainment. Urban areas have benefited from the **urban bias** of both colonial and independent governments in Africa, which have tended to invest disproportionately in capital cities that house centralized administrative functions. In addition, food prices have been kept down in the cities to reduce wage demands and to decrease the risk of civil unrest.

Lagos (Nigeria), Kinshasa (the DRC), and Johannesburg (South Africa) are the largest cities in the region, with populations of about 21 million, 8.8 million, and 4.3 million, respectively, and

there are about 40 cities with more than 1 million inhabitants. Life in African cities has great contrasts of wealth and poverty. Many urban inhabitants live in the informal settlements that surround major cities, with inadequate social services, sanitation, water supply, energy, waste disposal, and shelter (**FIGURE 5.38**). Over time, some of the most notorious urban slums, such as Soweto, outside Johannesburg in South Africa, and Kibera in Nairobi, Kenya, have become important to the overall functioning of cities. These slums house many of the lower-paid workers, have received some services, and have become centers of social movements that make political claims for the poor.

(a)

(b)

▲ **FIGURE 5.38 Africa's Major Cities** (a) This view encompasses the poverty of the Alexandra slum township as well as the expensive business and residential area of Sandton in Johannesburg, South Africa. (b) Lagos, Nigeria, street crowded with people, buses, and market.

The most powerful cities in the region in terms of economy and international presence are Johannesburg, Lagos, and Nairobi, which are centers of manufacturing and home to the regional headquarters of major international companies.

- **Johannesburg, South Africa** Sited on the high plateau of South Africa, Johannesburg is surrounded by gold and diamond mines and the townships that housed mineworkers and others who have come to work in the city (**FIGURE 5.38a**). The geography of Johannesburg still reflects the legacy of apartheid. The majority of the white population still lives in wealthy residential areas, such as Hyde Park. In contrast, millions of black workers live in poor conditions in Soweto and Lenasia, where black people and people of Asian descent lived during the apartheid era.

- **Lagos, Nigeria** The colonial capital of Nigeria until the capital was relocated to Abuja in 1992. Lagos has a metropolitan population of more than 21 million people (**FIGURE 5.38b**). Sited on a natural harbor, it was developed by the British as a rail terminus beginning about 1880. It became a leading cargo port, industrial area, and center for production of consumer goods. About 80% of Nigeria's trade goes through Lagos, although some trade is now shifting to the oil regions to the east. The Lagos area includes 53% of Nigeria's manufacturing, 62% of the gross industrial output, and 22 industrial estates. Lagos is also a cultural center, home to many well-known writers and musicians, and Nollywood, the West African film industry (see p. 210). Lagos is infamous for its traffic and crime problems. The average commute to work is more than 90 minutes in polluted air and tangled traffic jams, made worse by inadequate bridges between islands.

- **Nairobi, Kenya** Located in the cooler highlands where English settlers preferred to live, Nairobi still displays its role as a center for globalization in East Africa. It is a major center of commerce and corporate headquarters and hosts several UN agencies, including the headquarters of the UN Environment Program. The Nairobi urban landscape shows some striking contrasts. The central business district is adjacent to Nairobi National Park, which is home to both wildlife and vast slums such as Kibera, where people live in dense, poor-quality housing without adequate access to energy, clean water, or sanitation.

The Sub-Saharan African Diaspora Migration into Africa from other regions is overshadowed by the immense African diaspora and out-migration from Africa to other regions. The descendants of slaves captured and sent away from the region represent a high percentage of the populations in Brazil, the Caribbean, and the United States. A second wave of emigration coincided with the aftermath of colonialism, a time when many Africans who retained British Commonwealth passports or French citizenship moved to Britain and France (or other Commonwealth countries such as Canada) in search of work. Other recent emigrations from Africa include the movements of white populations from South Africa and other countries to Europe, North America, and Australia. There has also been a "brain drain": every year, thousands of African students and professionals, especially doctors, move to study or work in universities or companies in Europe and the Americas. Some are now returning as economic and political conditions improve.

Migration Within Africa Contemporary migration within Africa is mainly associated with the search for employment and with flight from famine, floods, and violent conflict. Traditional migrations included seasonal movements to take advantage of rains, rotate livestock, or for trade. Labor migrations became a factor during the colonial period, when the loss of traditional land to colonists and high taxes forced people to move in search of work, and employment became available in mines and on plantations. For example, residents from the Sahel migrated to work in peanut-, cotton-, and cocoa-producing areas along the West African coast. Labor migration continues from inland West Africa to coastal cities such as Abidjan, Accra, and Lagos. Another migration stream occurs in southern Africa where laborers from Botswana and Zimbabwe moved into South Africa to work in the mining industry. By 1960, more than 350,000 foreign workers were employed in South Africa. These migrations have disrupted family life, but the remittances that are sent back by workers have become an important contribution to local and national economies. Migrants now work in the South African service sector as well.

Refugees and Internally Displaced Peoples (IDPs) The UN High Commission on Refugees (UNHCR) estimated that there were approximately 3 million refugees in sub-Saharan Africa in 2014 (down from 4.5 million in 2002). UNHCR reported that the "total population of concern" in 2014, which takes into account refugees, stateless persons, asylum seekers, and internally displaced persons, was 15.1 million throughout the region. Refugees and IDPs fleeing armed conflict, human rights violations, and drought have originated from the DRC (457,900), Somalia (760,800), Sudan (462,100), Mali (267,000), and Nigeria (650,000).

Refugee populations place serious burdens on neighboring countries that lack the resources to feed and resettle the impoverished starving arrivals. Guinea, for example, absorbed almost a half million refugees from Liberia and Sierra Leone, and Tanzania took in a similar number from Burundi. International organizations and charities support the camps that house most international refugees (**FIGURE 5.39**). Disease spreads rapidly in the crowded conditions of the camps, and food supplies are sometimes interrupted or diverted by military groups or governments. Refugees are often accused of spreading HIV and other diseases but are excluded from HIV/AIDS programs at the same time. Long-standing conflicts and loss of livelihoods in their own countries mean that many refugees spend extended periods in the camps with little hope of returning home. However, more peaceful conditions and carefully monitored repatriation have resulted in the return of refugee populations to countries such as Rwanda and Mozambique.

People forced to move within their own countries are some of the world's most desperate migrants because they may not fall under international definitions of refugees or qualify for assistance. As of 2015 more than a million people have been displaced within Nigeria because of the threat of violence in the north.

Apply Your Knowledge

5.21 How have HIV/AIDS, urbanization, and changes in the status of women altered population projections for countries in sub-Saharan Africa?

5.22 Research a major city in the region and identify the key challenges it is facing and any recent successes in planning.

Future Geographies

In many ways, the prospects for sub-Saharan Africa are improving. Although some Africans maintain rural subsistence lives, disconnected from the world economy, others are working in transnational corporations or producing new exports for the global market. Many countries are experiencing successes as they work to discontinue reliance on foreign assistance, improve living standards, establish democratic governments, and increase food and economic production. What are some of the challenges for the future?

Sustaining Representative and Effective Governments

The transition to peace and democratic governments has occurred in many countries in sub-Saharan Africa, but several governments are still fragile and struggling with internal ethnic and religious tensions, long-term corruption, poverty, and high unemployment. Creating effective governments will require mechanisms for broad-based participation and reconciliation; transparent elections; a professional, reasonably paid, and incorruptible civil service; and reliable public services (**FIGURE 5.40**).

Economic Opportunities

Debt forgiveness, foreign investment, and local entrepreneurship have led to improvements in economic indicators for many African countries in the last decade. This region would benefit from economic diversification—more manufacturing and

▼ **FIGURE 5.39 Refugees** The refugee camps around Dadaab in northern Kenya are linked to the UN High Commission for Refugees and has hosted people fleeing civil war in Somalia and those suffering from the 2011 East African drought. Some of the 400,000 Somalis who came to the camp since the early 1990s are now returning home.

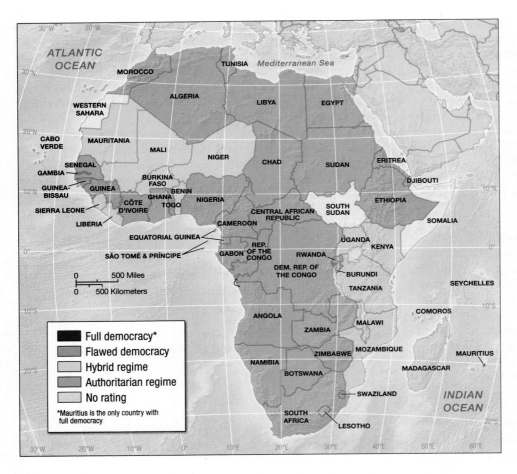

◀ **FIGURE 5.40 Democracy in Africa** The Economist Intelligence Unit has an annual index of democracy based on 60 indicators including electoral freedom, civil liberties, corruption, and political participation. Most of Africa is reported with flawed or hybrid democracies as of 2014 with more than a dozen countries under authoritarian rule. The index is disputed by many countries.

services—and from fairer trading and investment from a broader set of partners. If sub-Saharan Africa can reduce the gap between actual and potential yields, add value to resources through processing or manufacturing sectors, and build an educated workforce that can be employed in national and international services, economic growth can continue creating a middle class that can drive domestic demand (see Figure 5.23).

Food, Water, Health, and Energy for All

As noted earlier in the chapter, many Africans still lack access to safe water, adequate food, clean energy, and basic health care. The new UN Sustainable Development Goals (SDGs) set ambitious goals for development in sub-Saharan Africa (**FIGURE 5.41**).

Sustainable Development Goals
• End poverty in all its forms everywhere.
• End hunger, achieve food security and improved nutrition, and promote sustainable agriculture.
• Ensure healthy lives and promote well-being for all at all ages.
• Ensure inclusive and equitable quality education and promote life long learning opportunities for all.
• Achieve gender equality and empower all women and girls.
• Ensure availability and sustainable management of water and sanitation for all.
• Ensure access to affordable, reliable, sustainable, and modern energy for all.
• Promote sustained, inclusive and sustainable economic growth, full and productive employment, and decent work for all.
• Build resilient infrastructure, promote inclusive and sustainable industrialization, and foster innovation.
• Reduce inequality within and among countries.
• Make cities and human settlements inclusive, safe, resilient and sustainable.
• Ensure sustainable consumption and production patterns.
• Take urgent action to combat climate change and its impacts.
• Conserve and sustainably use the oceans, seas, and marine resources for sustainable development.
• Protect, restore, and promote sustainable use of terrestrial ecosystems, sustainably manage forests, combat desertification, halt and reverse land degradation, and halt biodiversity loss.
• Promote peaceful and inclusive societies for sustainable development, provide access to justice for all, and build effective, accountable and inclusive institutions at all levels.
• Strengthen the means of implementation and revitalize the global partnership for sustainable development.

▲ **FIGURE 5.41 Sustainable Development Goals** In 2015 the United Nations established a new set of goals for poverty alleviation and development around the world. Africa faces many challenges in achieving them despite success with the MDGs (see Visualizing Geography, "The Millennium Development Goals in Sub-Saharan Africa").

Learning Outcomes Revisited

▶ **Compare** the physical geographies of the diverse regions of sub-Saharan Africa, contrasting the climate and vegetation of the Sahel and the Congo Basin, the East African Rift Valley and Okavango, and the major rivers.

The Sahel is the southern boundary of the Sahara with a desert climate with sparse vegetation. The Congo Basin is a vast river basin, where high rainfall creates a tropical wet climate that supports dense tropical forests. The East African Rift Valley is a tectonic feature with steep valley sides containing lakes and the Okavango is an inland delta in Botswana. Major rivers include the Nile, Niger, Congo, and Zambezi.

1. **What** patterns of climate and tectonics created these key physical features?

▶ **Evaluate** the role of physical geography, disease, and mineral resources in shaping contemporary conflicts, crises, landscapes, and economies.

Tropical climates and poor soils limit Africa's agricultural potential as do drought, tropical diseases, and pests. Food crises contribute to unrest and create refugees. Millions of people suffer from tropical diseases, such as malaria, yellow fever, sleeping sickness, and river blindness. Mineral resources such as oil and diamonds are important exports but have fueled serious conflicts across the region.

2. **What** solutions are available to adapt to harsh climates, improve soils and food security, and reduce disease and conflict?

▶ **Identify** the impacts of human activity on African wildlife and climates and some of the solutions to these threats.

Wildlife has been overhunted and poached for ivory, rhino horn, recreation, food, and bushmeat and habitat has been taken for agriculture and cities. Solutions include protected areas and parks, tourism, trade bans, and guarding of key species. Climate change in Africa is mostly caused by other regions and will have serious impacts on people and ecosystems, especially where agriculture is rainfed, water is scarce, and there is competition for resources.

3. **What** are the positive and negative impacts of tourism?

▶ **Summarize** the legacies of European colonialism and the Cold War in the region and how these, together with the diversity of ethnicities, religions, and languages have created conflict within and between countries.

The legacies of European colonialism include trade patterns oriented to the export of primary products, unequal land distribution and apartheid, and political instability associated with contested boundaries and Cold War alignments. Colonialism also resulted in political boundaries that split ethnic groups across territories or clustered enemies within one territory. The region has dozens of different ethnic groups, tribes, and languages; many divided or clustered by colonial boundaries that have split groups or merged traditional enemies. The region includes large populations practicing indigenous beliefs and countries where Christianity and Islam are important and can create political unrest. This diversity poses challenges to governance in independent countries.

4. **What** are some of the contemporary conflicts in sub-Saharan Africa and their origins?

▶ **Explain** how debt relief, the end of several conflicts, and improving status of women may increase the chance of positive economic and political futures in sub-Saharan Africa.

For many years, high rates of foreign debt and poor terms of trade for exports crippled the economies and well-being of residents in many African countries. Campaigns to end poverty contributed to policies that reduced and forgave debt across the poorest countries. This has stimulated higher levels of growth and investment. A number of serious conflicts—mainly over ethnic differences and resources—have ended across the region bringing hopes for peace, poverty alleviation, and better governance.

5. **Which** countries have shown greatest improvement in economic and political futures after debt relief and the end of conflict?

▶ **Describe** the main causes of low agricultural production in sub-Saharan Africa, connections of agriculture to desertification and deforestation, and problems that are emerging around land grabs.

Agricultural production in the region is limited by harsh climates, poor soils, pests, and a lack of fertilizer and other inputs. Higher prices for food and mineral resources are increasing international interest in purchasing large areas of land in sub-Saharan Africa. Chinese investment, in particular, has increased across much of the region, raising local and geopolitical concerns about Africans losing control of their land and economies. Where land is overused by farmers and herders, erosion and vegetation loss may result in desertification or in deforestation if forest land is cleared for agriculture.

6. **What** are some of the solutions to low agricultural production, desertification, and deforestation?

▶*Analyze* factors affecting the characteristics of populations in sub-Saharan Africa, including the demographic transition, HIV/AIDS and other diseases, urbanization, fertility changes, and migration.

Although the region has some of the world's fastest growing populations, fertility rates are falling in most countries as the status of women improves and the region urbanizes. HIV/AIDS spread rapidly through sub-Saharan Africa since it emerged in 1981. The impacts include reduced life expectancy, almost 15 million orphans, vulnerability to other diseases such as tuberculosis, and the loss of a generation of young professionals. Other significant diseases include malaria, river blindness, and Ebola. Millions of people are migrating to major cities such as Lagos and Johannesburg looking for work and many live in slums.

7. *What* changes in the status of women, health care, or economy have led to declines in fertility and diseases?

▶*Provide* examples of how African people, goods, and culture have spread around the world as a result of slavery, colonialism, and globalization.

Slavery forcibly distributed people of African descent across many regions of the world, especially in the Americas. This process shaped the contemporary racial composition and culture of countries such as the United States and Brazil. Migration from the region continues as people flee conflict and poverty. African exports of minerals and agricultural products fuel industrial economies and are important in the production of manufacturing and consumer goods such as cell phones and jewelry. Globalization has brought African music, art, literature, and athletes to the world stage.

8. *Compare* and contrast the impacts of slavery and African culture in the United States and Brazil.

Key Terms

acquired immunodeficiency syndrome (AIDS) (p. *213*)
Afrikaners (p. *194*)
apartheid (p. *202*)
Berlin Conference (p. *194*)
bush fallow (p. *189*)
bushmeat (p. *190*)
child soldier (p. *203*)
conflict diamonds (p. *188*)
desertification (p. *180*)
domestication (p. *189*)
Ebola (p. *186*)
ECOWAS (p. *201*)

ethnic group (p. *207*)
failed state (p. *206*)
fair trade movement (p. *199*)
G8 (p. *198*)
genocide (p. *203*)
Gondwanaland (p. *181*)
Green Belt Movement (p. *191*)
griot (p. *209*)
harmattan (p. *179*)
human immunodeficiency virus (HIV) (p. *213*)
informal economy (p. *200*)

intercropping (p. *189*)
intertropical convergence zone (ITCZ) (p. *179*)
land grabs (p. *199*)
lingua franca (p. *207*)
malaria (p. *185*)
Millennium Development Goals (MDGs) (p. *204*)
Nelson Mandela (p. *202*)
Nollywood (p. *210*)
Pangaea (p. *181*)
pastoralism (p. *180*)
remittances (p. *200*)

rift valley (p. *183*)
Sahel (p. *180*)
savanna (p. *185*)
shifting cultivation (p. *189*)
slash-and-burn (p. *189*)
subtropical high (p. *179*)
Sudd (p. *183*)
terms of trade (p. *197*)
transhumance (p. *190*)
tribe (p. *207*)
tsetse fly (p. *187*)
urban bias (p. *215*)

Thinking Geographically

1. How would you respond to people who seem to see Africa as a single unified region or even a country?

2. How has dependence on mineral resources shaped the economies of countries across Africa and created conflict within and between them?

3. Which subregions and countries of the region have seen the greatest progress in meeting development goals and why?

4. How have colonial and Cold War alignments affected the economic, political, and cultural geography of sub-Saharan Africa?

5. How are women shaping the future of Africa through their fertility choices and economic activity?

6. What are the current major patterns of migration between and within African countries and beyond, and how do they relate to employment opportunities, natural disasters, and conflict?

7. What has sub-Saharan Africa contributed to the demography and culture of other regions?

DATA Analysis

Throughout this chapter we have emphasized the differences between countries and the progress many have made in terms of development, but with more than 50 countries in the region we have not been able to discuss many in any detail. Some of the major challenges for the region include food and water security. Consult the World Bank African Indicators data to answer the following questions about food and water access. Recent data in spreadsheet format can be downloaded from http://data.worldbank.org/.

For each question look for two countries that rank lower, two that rank higher, and two that made the most improvement from 1990 to 2010 or 2000 to 2010 and identify reasons why they may have the ranking or show the improvement (e.g., conflict, economic growth, or debt forgiveness).

1. Which nations in sub-Saharan Africa have low percent of their land in agriculture and which the least in 2010? Why do you think some have less land in agriculture than others?

2. Which nations had agriculture contributing more than 40% to GDP in 2010? Why do you think this might be?

3. Which countries have the highest cereal yields in 2010 (kg per hectare) and which the lowest? Why might this be and what are the implications for food security? Can you identify countries who have doubled yields between 1990 or 2000 and 2010? Any idea about why this might be?

4. Identify countries with less than 20% of the population with access to improved sanitation facilities in 2010. Which countries more than doubled access (100%+ change) between 1990 or 2000 and 2010?

Food and Water Security

https://goo.gl/cQP9r4

5. Which countries had the lowest percent (e.g., 60% or less) of their population with access to improved water sources in 2010? Any ideas why? Which countries showed the greatest improvement in clean water access between 1990 and 2010 (more than 80% improvement)? Which countries show a decline (negative change)?

6. Where are people most undernourished (percent of population) in the region in 2010 (e.g., more than one-third)? Where has undernourishment improved the most (large negative percent change)?

MasteringGeography™

Looking for additional review and test prep materials? Visit the Study Area in MasteringGeography™ to enhance your geographic literacy, spatial reasoning skills, and understanding of this chapter's content by accessing a variety of resources, including MapMaster interactive maps, videos, *In the News* RSS feeds, flashcards, web links, self-study quizzes, and an eText version of *World Regions in Global Context*.

Marchers in 2014 in Washington, DC protest the choking death of Eric Garner by an NYC police officer.

The United States and Canada

I n 2012, 17-year-old African-American Trayvon Martin was fatally shot by neighborhood watch volunteer, George Zimmerman. Martin was walking from a convenience store to the home of his father's fiancé in Sanford, Florida. He was unarmed. His transgression, if it can be called that, was to be "out of place" in the gated community where he was staying with the couple. Following Martin's death, rallies, marches, and protests were held in dozens of cities across the United States and an online petition containing 2.2 million signatures called for a full investigation and the prosecution of Zimmerman. Yet, in 2013, Zimmerman was acquitted of second-degree murder based on a Florida law that allows individuals to use deadly force when they fear death or bodily harm. In response, #BlackLivesMatter (BLM) was founded by three black activists protesting Zimmerman's acquittal.

Black Lives Matter

https://goo.gl/xihGiO

Since 2013, this once online movement has been taken to the streets urging communities across the country to stand up against "state violence against black bodies" especially the high-profile killings of African-American men by police. These deaths included the police shooting of 18-year-old Michael Brown in Ferguson, Missouri, 22-year-old John Crawford, III in Beavercreek, Ohio, 50-year-old Walter Scott in North Charleston, South Carolina, 43-year-old Eric Garner in Staten Island, New York, who died as a result of a police choke hold, and 25-year-old Freddie Gray who died from injuries to his neck and spine suffered while he was in police custody in Baltimore, Maryland. BLM also focuses activism toward police violence against all black bodies by highlighting the wrongful deaths of African-American girls, women, and transgendered and lesbian, gay, and bisexual individuals. The death of 28 year old Sandra Bland, in her jail cell in Waller County, Texas, who was pulled over for failing to signal a lane change, is one such example.

Learning Outcomes

▶ **Describe** the distinctive landscapes of the United States and Canada with respect to landforms and climate.

▶ **Explain** how the United States and Canada are affecting and affected by global climate change.

▶ **Understand** the rise of the United States and Canada as key forces in the global economy.

▶ **Describe** how the two countries, dynamic economies contribute to and shape the global economy.

▶ **Summarize** the economic inequality that exists in both the United States and Canada.

▶ **Compare** and contrast the forms and practice of government of the United States and Canada.

▶ **Summarize** the indigenous and immigrant histories of the United States and Canada.

▶ **Describe** the fundamentals of the geography of religion and language across the United States and Canada.

▶ **Compare** and contrast the urbanization processes that have shaped the United States and Canada.

Environment, Society, and Sustainability

The United States and Canada occupy largest extent of the landmass of North America (**FIGURE 6.1**). This region has played a key role in the unfolding of global environmental processes largely because of its position with respect to climate patterns, plate tectonics, and ocean currents and the adaptations humans have made in response to these factors. People have occupied the region for millennia, though the most profound global environmental impacts and innovations have come about in the last 200 years.

Climate, Adaptation, and Global Change

Complex physical processes characterize the region and shape environment and society relations. Because temperature, precipitation, and terrain patterns combine to influence vegetation, and

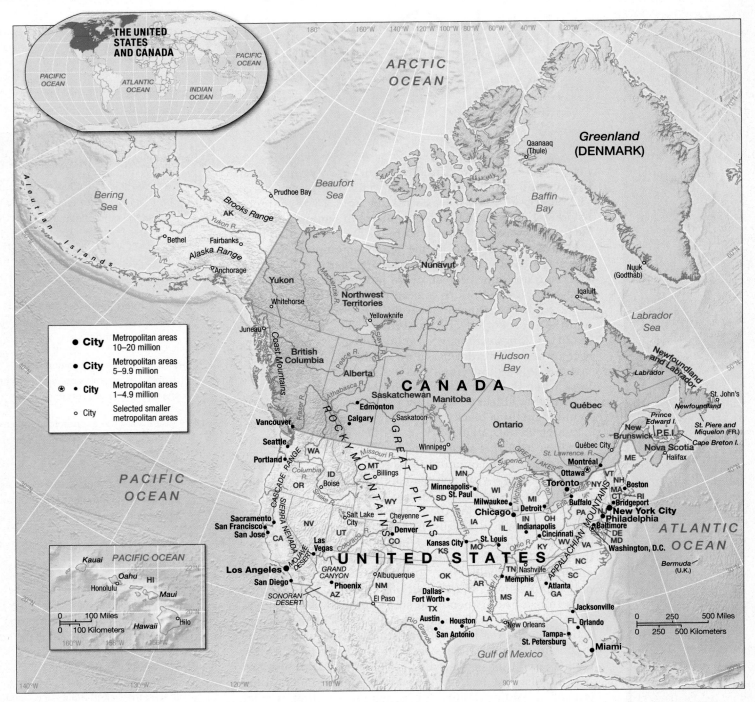

▲ **FIGURE 6.1 United States and Canada: Countries, major cities, and physical features** The landmass these two countries occupy extends from the north polar region to the tropics. While Greenland and Mexico are also shown on the map, this is because geologically, they are part of the North American continent. Greenland, administered by Denmark, is treated in the Europe chapter. Mexico in the Latin America chapter.

to some extent soil, it is important to understand the role that climate plays in this region and the ways people in the region have adapted their practices to climate.

Climate Patterns The United States and Canada contain nearly every type of climate condition possible, with temperatures varying quite dramatically on any one day of the year from north to south. In the Canadian Arctic in the north, it is very cold for most of the year; in the southern United States, it is warm for most of the year. Because of the influence of the oceans on three sides, the Gulf of Mexico, and the interior Great Lakes, as well as the presence of mountains and plateaus, within-region differences also occur.

Variations in precipitation occur throughout the region. (**FIGURE 6.2**). The east and west coastal areas tend to be mild and moderately wet; the interior largely arid as north–south mountain chains prevent moisture-bearing clouds from moving inland to drop their moisture. As a result, the moisture gradient declines slowly but continuously from east to west as far as the three significant mountain ranges—the Rockies, Sierra Nevada,

and Cascades. Once beyond these mountains, the moisture gradient rises dramatically toward the Pacific.

In the southeastern part of the United States, where no significant coastal range exists, moisture-bearing clouds more readily condense into rain. Here, the warm Gulf of Mexico is an important source of moisture and storms, including hurricanes. In the Arctic north, annual precipitation approximates desert conditions because of the dryness of cold air and dominance of very stable air masses with low moisture content. The warmer parts of the region—in the southern United States and Hawaii—experience this precipitation as rain, and the colder, more northerly parts experience snow. In areas around the Great Lakes, the warming effects of these large bodies of water add even more moisture to the mix, bringing especially heavy snowstorms (called "lake-effect" snow) to lake shore places like Buffalo, New York and to Sault St. Marie, Ontario.

Climate Change and Other Environmental Issues The region possesses a bounty of resources: a range of minerals; vast forests; fertile, highly productive land; extensive fisheries; varied and abundant wildlife; and magnificent physical beauty. In addition to this physical wealth, the region has the technological capability and the drive to exploit resources to an extent achieved by few other regions on Earth. Material consumption here results in elevated levels of air and water pollution, soil contamination, solid waste disposal challenges, an ongoing problem of nuclear waste disposal, acid rain, extinct and endangered species.

Certainly the most significant and pressing environmental problem in the United States and Canada today is global climate change (see Chapter 1, p. 12) in which both countries play a major role. Both produce high total and per capita **greenhouse gas** emissions resulting from production and consumption patterns. The **carbon footprint**—the total set of greenhouse gas emissions caused directly and indirectly by an individual, organization, event, product, or place—is a way of understanding how greenhouse gas emissions are contributing to climate change and where their impacts might be reduced. The carbon footprint of both the United States and Canada is much larger than most other nations.

Scientists using complex models that estimate changes in temperature, precipitation, and sea levels at the regional scale have predicted the likely impacts of climate change on the United States and Canada. These include the following:

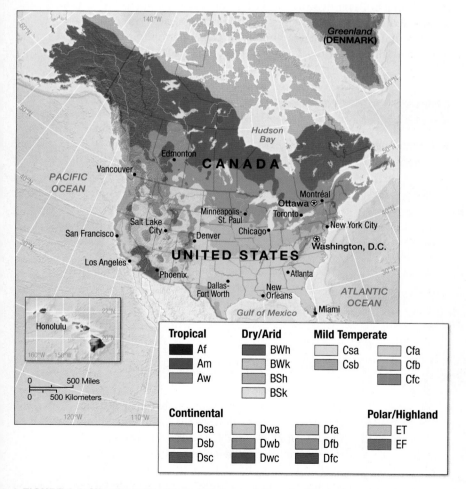

▲ **FIGURE 6.2 Climates of the United States and Canada** This is the only region in the world that contains the full range of climate types that occur globally, stretching from Alaska and the Canadian Arctic to the tropical Hawaiian Islands. **Why is knowing the topography of the region important to understanding its climate?**

"Certainly the most significant and pressing environmental problem in the United States and Canada today is global climate change in which both countries play a major role."

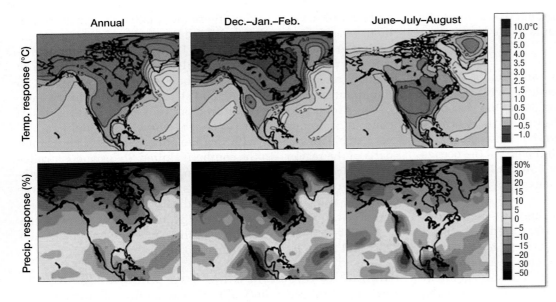

Annual   Dec.–Jan.–Feb.   June–July–August

Temp. response (°C)

Precip. response (%)

10.0°C
7.0
5.0
4.0
3.5
3.0
2.5
1.5
1.0
0.5
0.0
−0.5
−1.0

50%
30
20
15
10
5
0
−5
−10
−15
−20
−30
−50

◄ **FIGURE 6.3 New Temperature and Precipitation Change, 2030** These maps show the anticipated average changes in temperature and precipitation that may occur as a result of climate change in the region. Minimum winter temperatures and maximum summer temperatures are likely to increase more than the global annual mean in the southwestern United States. Projected warming is expected to be accompanied by a general increase in precipitation over most of the United States and Canada except the most southwesterly part.

- An increase in sea levels along much of the coast would exacerbate storm-surge flooding and shoreline erosion, placing at risk those low-lying cities such as that which occurred during Hurricane Sandy (see "Visualizing Geography: Sea-Level Rise in Chapter 1, p. 12).

- Temperature increases could cause adverse health impacts, such as heat-related mortality and mosquito-borne diseases; in more northerly regions, warming might increase agricultural productivity (**FIGURE 6.3**).

- More severe droughts, especially in the western United States, would increase competition among agricultural, municipal, and industrial users and ecosystems around the region's overallocated water resources and could increase risks of wildfire.

Another significant climate change challenge to the region is that posed to Canada's boreal forest (**FIGURE 6.4**). The largest ecosystem in the world, the boreal forest, also known as taiga or "snow forest," is characterized by coniferous stands of pines, spruces, and larches. The Canadian boreal forest is critical not only to Canada and the United States but also to the planet. The forest stores carbon, purifies air and water, regulates the climate, acts as a spring nursery to more than 3 billion migrating birds, and provides habitat to caribou, moose, wolves, and bears. The most all encompassing threat to the forest is rising temperatures, which will cause permafrost degradation, tree-killing droughts, large wildfires, and insect outbreaks.

Apply Your Knowledge

6.1 From the Government of Canada's website on "Impacts of Climate Change" http://www.climatechange.gc.ca/ describe three ways the Canadian natural environment is changing.

6.2 Go to the Nunavut Climate Change Center, http://climatechangenunavut.ca/ and list three ways climate change in the Canadian Arctic region affects human society and economy.

Canada and Climate Change

https://goo.gl/1c50Mw

RUSSIA

Bering Sea

Beaufort Sea

Greenland (DENMARK)

ICELAND

Baffin Bay

Gulf of Alaska

Labrador Sea

Hudson Bay

PACIFIC OCEAN

C A N A D A

Forest–Tundra
Lichen Woodland
Closed Forest

UNITED STATES

0   250   500 Miles
0 250 500 Kilometers

▲ **FIGURE 6.4 North American Boreal Forests** Much of Alaska and Northern Canada is covered in boreal forests, primary among them is the boreal forest. While Alaska is part of the United States, it is included in this map as it is a key element of the boreal forest ecosystem.

Geological Resources, Risk, and Water

The United States and Canada feature a vast central lowland that includes the Canadian Shield, the Interior Lowlands, and the Great Plains. To the east are the Appalachian Mountains, which descend gradually to the Gulf–Atlantic Coastal Plain and become broader as one travels farther south and southwest. To the west are three distinct topographical regions: from west to east are the mountains and valleys of the Pacific

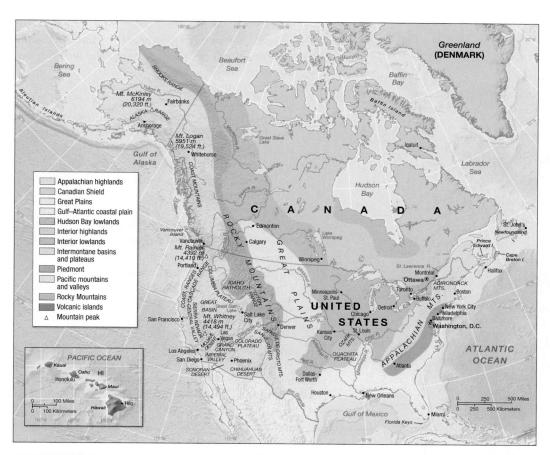

coastal ranges; then an **intermontane** set of basins, plateaus, and smaller ranges that lie between mountains; and finally the great Rocky Mountain range, which rises steeply and imposingly at the western edge of the central lowlands.

Physiographic Regions FIGURE 6.5

provides an overview of the different physiographic regions. (*Physiography* is another term for physical geography, which is the branch of geography dealing with natural features and processes.) The Canadian Shield, shown in light brown, is a geologically very old region and very rich in minerals. The Interior Lowlands, shown in light green, is a glaciated landscape with fertile soils and abundant lakes and rivers. This part of the continental landmass is devoted mostly to agriculture. The third primary physiographic subdivision is the Great Plains, shown in beige, an area that slopes gradually upward as you

▲ **FIGURE 6.5 Physiographic Regions of the United States and Canada**
Mountains, plains, oceans, beaches, fertile soils, and prairies—the region contains the full range of physiographic regions. Images shown include (a) the Idaho Batholith, (b) the Interior Lowlands, and (c) the volcanic beaches of Hawaii.

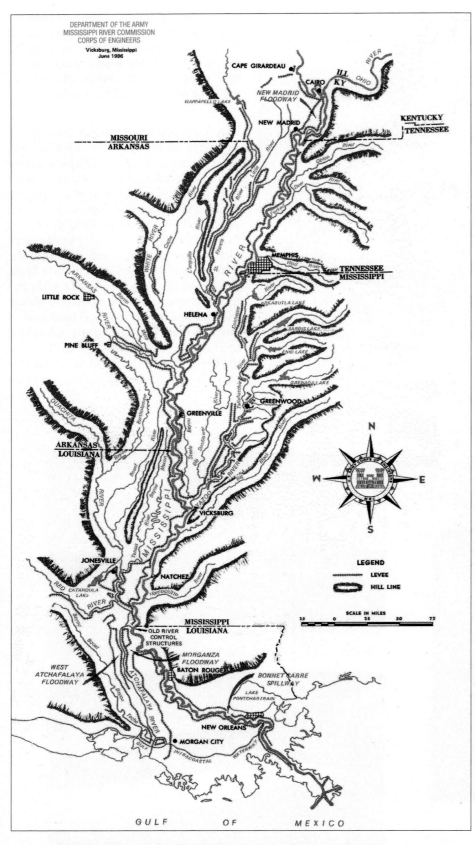

▲ FIGURE 6.6 The Engineered Mississippi River This map from 1986 shows the extent of engineering that was undertaken after the great flood of 1927 along the Mississippi–Missouri river system.

move west toward the Rocky Mountains. This is an area of extensive gently rolling and flat terrain with excellent soils and some of the world's most productive farms.

The major landform feature of the eastern United States and Canada is the Appalachian mountain chain. Surrounding it, on its eastern and southern flanks, is a coastal plain that can be divided into two subregions, the Piedmont area of hills and easterly sloping land and the Gulf–Atlantic Coastal Plain, a lowland area that extends from New York to Texas. The coastal plains are generally level, with soils that are sandy and relatively infertile. Where agriculture does occur, it is intensive and scattered. For instance, the Piedmont is the historic region of plantation agriculture—cotton, tobacco, and corn—but much of the soil in this region has been depleted or eroded from over farming.

The Mississippi–Missouri river system is at the heart of U.S. economic geography. Along with the five Great Lakes (Superior, Michigan, Huron, Erie, and Ontario), it is the most extensive and navigable river system in the world and enabled the early growth of the interior of the United States. However, the engineering that has been applied to the Mississippi–Missouri system (including its major tributaries) over a century or more has altered its ecosystem and contributed to the vulnerability of New Orleans to extreme flooding during Hurricane Katrina. (**FIGURE 6.6**).

Geologic and Other Hazards The west coast of this region sits along the fault line of two active crustal plates, the Pacific Plate and the Juan de Fuca Plate. Both are moving northward, rubbing against the more stationary North American Plate on which the continental landmass sits. As a result of this friction, the coastal area from San Diego through British Columbia to Alaska is subject to frequent tremors. Devastating earthquakes have occurred along this fault line in the past and are likely to occur again in the near future. The extraordinary views of the Pacific Ocean provided by the mountainous topography of the Pacific coastal region have attracted extensive home development and past earthquakes have had destructive effects on many of them. The seasonal rainfall in this area has also wreaked havoc. When the soil becomes saturated with rainwater, liquefaction occurs, and the soil literally moves in one massive slump, carrying very large structures along with it.

In the western United States, the intermontane basin and plateau formation between the Pacific coastal range and the Rockies includes the

Columbia Plateau to the north, which begins at the headlands of the Columbia River, and the Colorado Plateau in the south, which includes the Grand Canyon. In between these two plateaus are the Great Basin in Nevada and Utah, which includes extinct lakes as well as the Great Salt Lake, and the southwestern deserts (Mojave, Sonoran, and Chihuahuan). The region's landscape includes area of dry climates, thin vegetation, and few perennial surface streams. Wildfires, droughts, floods, and landslides are continuing threats in the western United States.

The physical geography of the U.S.-Canada world region is the context for, and an inescapable reminder of, the enormous benefits and sometimes devastating costs of great prosperity. There is no better example of those costs than the accident that occurred when BP's Deepwater Horizon oil rig exploded and caught fire on April 20, 2010, in the Gulf of Mexico, 400 km (250 miles) offshore of Houston. As the world's largest accidental oil spill to date, it will continue to contaminate or damage marine and terrestrial life for decades to come. In 2015, Santa Barbara, California experienced its second significant spill. The first one—the largest in the United States at the time—occurred in 1969.

Mineral, Energy, and Water Resources The United States and Canada possess great energy and mineral resources. The two countries are among the top 20 producers of natural gas in the world. Canada is second only to Saudi Arabia in its proven oil reserves—those where there is a reasonable certainty of oil being recoverable. Canada's oil is mostly located in the province of Alberta. The oil occurs in deposits of sand saturated with bitumen (tar), known as tar sands, whose extraction is toxic. Global Forest Watch, an international consortium of research and environmental organizations, has shown that tar sands development and forest fires in Canada's tar sands have cleared or degraded 775,500 hectares (about 2 million acres) of forest since 2000.

The United States' oil deposits are primarily located in Texas, the Gulf of Mexico, Louisiana, Alaska, and California. In addition to being a leading energy producer, the United States is the second largest consumer, after China, of energy resources. The United States relies mostly on Canadian imports of natural gas, and oil to supplement this thirst for energy.

In addition to these nonrenewable sources of energy, in the western United States and Canada renewable hydropower is available, and solar and wind power are also becoming increasingly prominent on the energy landscape. Both Canada and the United States are also rich in a wide array of mineral and precious metal resources, many of which are integral to new technologies of production from copper to silicon.

Apply Your Knowledge

6.3 Compare and contrast the United States and Canada with two other countries on three global resources: energy, including oil and coal; water; minerals; or foodstuffs.

6.4 Define the relationships between these resources and a country's overall wealth.

Ecology, Land, and Environmental Management

The Europeans who arrived on the Atlantic Coast of the United States and Canada beginning in the late 15th century encountered an environmentally diverse landscape thinly populated by indigenous peoples. It is estimated that around 20,000 years ago, the ancestors of these indigenous peoples began the process of populating the continent and altering (and permanently changing) the environments they encountered (**FIGURE 6.7**). When European explorers arrived, they did not discover a pristine land, but one that had already been transformed by millennia of human settlement.

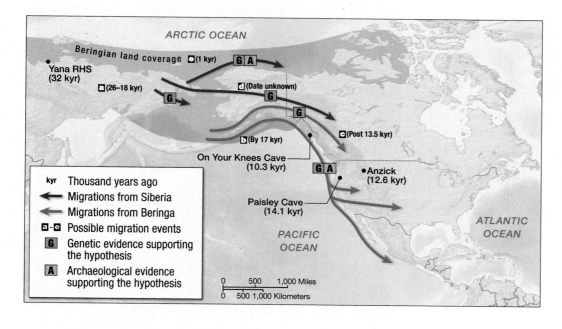

◀ **FIGURE 6.7 Populating the United States and Canada** This map provides hypothetical scenarios for the peopling of the Americas, showing possible migration events, region of origin, some key archaeological sites in North America and Siberia, and the type of evidence—genetic and archaelogical—currently supporting each hypothesized migration. Shading depicts the extent of Beringia—including the Bering land bridge that is thought to have connected Asia and North America during the last glacial period. **How does this map help us better understand the populating of Canada and the United States by early humans?**

Indigenous Land Use These first human inhabitants moved into the United States and Canada, traveling southward along the western edge of the continental ice cap. These Neolithic, or Stone Age, hunters probably originally came from northern China and Siberia. Their descendants gradually moved farther and farther southward into the continental landmass. Abundant evidence suggests they eventually spread throughout North America (as well as Central and South America) and adapted their ways of life to the conditions they encountered. As they began to settle, different groups developed agriculture in different places. With game, fish, and wild and cultivated foodstuffs available, an economic system based on subsistence production and trade emerged. The cultivation of wild foods like maize, tomatoes and potatoes spread northward from Central and South America.

Before European contact, hunting, gathering, and some shifting cultivation existed among indigenous peoples in the region. For example, along the eastern seaboard, hunter–gatherers were mobile, moving with the seasons to obtain fish, migrating birds, deer, and wild berries and plants. Shifting agriculture was organized around planting and harvesting maize, squash, beans, and tobacco. Hunter–gatherers and shifting cultivators all identified and used a wide range of resources. The prevailing practice was to take only what was needed to survive. However, vegetation change did occur as a result of indigenous people's settlement and hunting activities and possibly some species depletion before the arrival of the Europeans.

Colonial Land Use Europeans saw the natural world they encountered in the United States and Canada much differently from how indigenous peoples saw it. Most important, Europeans viewed resources as commodities to be accumulated, not necessarily for personal use but to be sold for profit or export. The arrival of the Europeans in the region meant that pressures on natural resources (especially wood, furs, and minerals) were accelerated. In New England, where European settlement first occurred, there was extreme exploitation of white pine, hemlock, yellow birch, beaver, and whales that led ultimately to deforestation and extinctions.

The arrival of Europeans also meant a dramatic change in prevailing social understandings of land. Native perspectives about the communality and flexibility of land were replaced by European views of land as private and having fixed boundaries. European settlers wanted to own and fence a plot of land, which led eventually to the concentration of land in large private farms, plantations, and estates. Increasingly, the native people of the United States and Canada were forced onto less-productive land or reservations and were prohibited from hunting or gathering on private lands or from moving with the seasons as they had before. A **reservation** (United States) or a **reserve** (Canada) is an area of land managed by an indigenous tribe under the U.S. Department of the Interior's Bureau of Indian Affairs or, in Canada, under the Minister of Indian Affairs.

Contemporary Agriculture and Sustainability Overall in the region, conditions are very good for agriculture as one moves from east to west until the precipitation gradient drops. Most of the agricultural productivity is concentrated in the Interior Lowland and Great Plains. Outside these subregions, soil fertility in western North America is often low, and rainfall is limited and infrequent, although conditions become favorable again in the valleys along the Pacific Coast. Natural conditions favor certain areas for the highest agricultural productivity; other parts of the region are also important agriculturally, largely because farmers have overcome the natural barriers to production through fertilizers, irrigation, pesticides, and other technological applications. This is the case in the Pacific valley areas where irrigation water drawn from the Colorado River enables agriculture to flourish, and in the Southwest where intensive irrigation supports significant cotton and citrus production. In Arizona and New Mexico, northward-flowing air masses bring summer monsoonal precipitation patterns that support unique desert vegetation. The summer and winter rains there allowed the ancestors of contemporary Native Americans to cultivate beans, squash, and corn using sophisticated irrigation systems in a landscape that would otherwise seem inhospitable to subsistence agriculture.

The agriculture of the U.S. Great Plains and Canadian prairie provinces is large-scale and machine-intensive, dominated by a few crops, the most important of which is wheat. Winter wheat, planted in the fall and harvested in late spring, is grown across much of the United States, but predominates in the southern Plains from northern Texas to southern Nebraska. Spring wheat, grown primarily from central South Dakota northward into Canada, is planted in early spring and harvested in late summer or fall. Most of the wheat grown in these former grassland ecologies is not irrigated.

The interior valleys along the western coast produce fruit and vegetables (**FIGURE 6.8**). Rainfall is inadequate to sustain

▼ **FIGURE 6.8 Irrigation in the U.S. West** The agricultural landscapes of the U.S. West include center-pivot irrigation, such as Farmington, New Mexico shown here.

the crops so they are irrigated. More than half of the fruits, vegetables, and nuts consumed in the United States are produced in California. Washington State accounts for 50% of the U.S. apple supply. Fruit and vegetable production in Canada occurs largely in British Columbia, Quebec, and Ontario, with the latter being the leader in vegetable crops and the former dominating the export of cherry plantings around the world. The west coasts of both the United States and Canada are also where marijuana, an increasingly important though still largely illegal cash crop, is produced (see "Geographies of Indulgence, Desire, and Addiction: Marijuana" on p. 246). Many new food movements have begun and disseminated from the west coastal area of these two countries.

Environmental Challenges Decades of federal environmental protection legislation have forced U.S. and Canadian industries to curtail much of their polluting processes. Nonetheless, various subregions of both countries still face serious and persistent environmental challenges. For instance **acid rain**—rainfall made acidic by atmospheric pollution that causes environmental harm, typically to forests and lakes, generated by industrial processes and automobile emissions—continues to pose a challenge in the traditional industrial areas on both sides of the U.S. and Canada border.

The most serious environmental challenge Canadians and Americans face today is their seemingly insatiable appetite for resources, especially energy resources. The impact of the high level of energy consumption not only seriously challenges Earth's supply of renewable energy resources but is also leading to global climate change, as discussed earlier in this chapter.

Popular movements are directing efforts at reducing energy consumption in the region, and a growing consumer interest in sustainability has created a large and growing market for environmentally friendly "green" goods and services in both countries. Leading corporations, such as large automobile manufacturers and energy companies, have made substantial investments in energy efficiency and energy technologies. Individual consumers and smaller businesses are also responding positively. Many greenhouse gas reductions, especially those associated with energy efficiency, from home insulation to fuel-efficient transport, can be made at little or no cost and may even save money. In the absence of serious federal effort to reduce emissions, many cities and businesses across the region have made their own commitments to mitigate climate change.

Several regional alliances have also been created to reduce greenhouse gas emissions. The alliances include the Regional Greenhouse Gas Initiative among the northeastern states in the United States and the Western Climate Initiative between California and British Columbia, Manitoba, Ontario, and Quebec.

Cap-and-trade programs are one way to address emissions reduction. The "cap" is a government-imposed limit on carbon emissions and the "trade" occurs when unused emission permits or quotas from states (or firms) are sold by those who

have been able to reduce their emissions beyond their quota to those who are unable to meet theirs. For example, cities such as Quebec have developed cap-and-trade programs as have industries such as the Cement Association of Canada.

Plans for the mitigation of climate change include coastal protection as well as water management and conservation efforts. There is also a well-established and increasingly municipally organized structure for adaptation to climate change that includes recycling materials and items from aluminum, paper, and glass to appliances, oil, and electronic goods.

History, Economy, and Territory

Today, the United States and Canada are established democracies modeled on European political traditions. Both consolidated their leadership roles in the global economy early in the 20th century through an effective strategy of economic development and global geopolitics. Much of the recent history of these two countries is the result of European colonization (**FIGURE 6.9**). More recent geographies of this region tie into wider movements of global immigration as well as internal migration (discussed in a later section of the chapter). Because of the two countries' comparable economic status, policymakers and government agencies treat them as a world region, although there are certainly differences between them.

▲ **FIGURE 6.9 New France in Quebec** This neighborhood of historic Old Quebec clearly shows the influence of French architecture in these early 18th-century residences with steep gable roofs. French is spoken widely in this neighborhood and in most of Quebec, where English is a second language.

Alternative Food Movements

One form of alternative food is that produced through **organic farming**—any farming or animal husbandry that uses no commercial fertilizers, synthetic pesticides, or growth hormones. Conventional farming, in contrast, uses chemicals such as insecticides, herbicides, and plant fertilizers and relies on hormones for breeding and raising animals. Organic producers reject conventional (also known as industrial) farming practices and the increasing use of **genetically modified organisms (GMOs)** on U.S. and Canadian conventional farms. A GMO is an organism that has had its deoxyribonucleic acid (DNA) modified in a controlled environment rather than through cross-pollination or other forms of evolution (**FIGURE 6.1.1**). Examples of GMOs include a potato that releases its own pesticide or rice that includes more nutrients than non-GMO rice. Although conventional farming practices continue to dominate the region's agricultural sector, organic practices are a growing force not only for small farmers but also among larger corporate entities such as Walmart.

Local food is usually organically grown, but the designation "local" means that it is also produced within a fairly limited distance from where it is consumed. Most interpretations of local food establish a 160 km (100 mile) radius as the limit of what is truly local. Local food movements have resulted in the proliferation of communities who have joined together to support the growth of new farms. **Slow Food** is devoted to a less hurried pace of life and to the true tastes, aromas, and diversity of good food. The alternative food movement also serves as a rallying point against globalization, mass production, and the kind of generic food represented by U.S.-based franchised restaurants.

Slow food, local food, and organic agricultural practices are ones that have been promoted to combat the negative health effects of fast and processed foods, including especially obesity among low-income populations (**FIGURE 6.1.2**).

Fast foods are edibles that can be prepared and served very quickly in packaged form in restaurants or at home.

Processed foods are energy-dense, nutrient-poor foods with high levels of sugar and saturated fats.

Since the 1970s, consumption of fast and processed food has increased dramatically in both the United States and Canada. At the same time, growing concerns about the health effects of these foods are a response to the alarming incidence of obesity (an amount of body fat mass that is a danger to health) and diabetes (a disorder that affects the body's use of food for energy) among North Americans.

A criticism of alternative food practices are largely organized and promoted by white, middle-class people and exclude, simply through cost and associated accessibility, poor people of color. The result is that poor people turn to cheap, easily accessible fast food, more than do members of other economic classes.

▶ **FIGURE 6.1.1 What Is a Genetically Modified Organism?** This infographic provides a snapshot of a range of aspects of GMOs from the science that creates them to popular reactions to them.

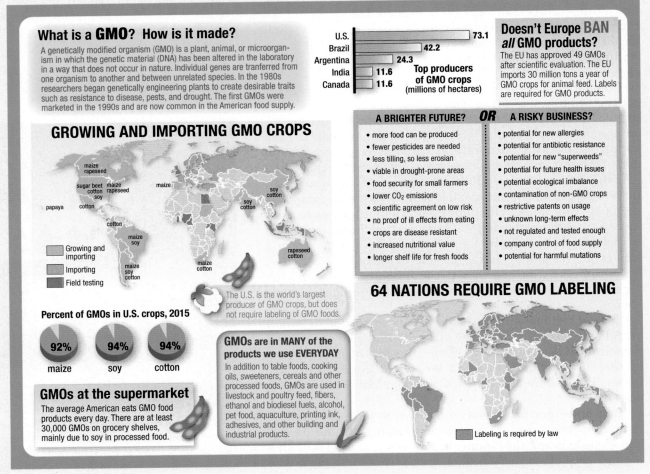

What is a GMO? How is it made?

A genetically modified organism (GMO) is a plant, animal, or microorganism in which the genetic material (DNA) has been altered in the laboratory in a way that does not occur in nature. Individual genes are tranferred from one organism to another and between unrelated species. In the 1980s researchers began genetically engineering plants to create desirable traits such as resistance to disease, pests, and drought. The first GMOs were marketed in the 1990s and are now common in the American food supply.

Top producers of GMO crops (millions of hectares)

- U.S. 73.1
- Brazil 42.2
- Argentina 24.3
- India 11.6
- Canada 11.6

Doesn't Europe BAN all GMO products?

The EU has approved 49 GMOs after scientific evaluation. The EU imports 30 million tons a year of GMO crops for animal feed. Labels are required for GMO products.

GROWING AND IMPORTING GMO CROPS

maize rapeseed
sugar beet cotton soy
maize rapeseed
maize
soy cotton
papaya cotton
soy cotton
cotton
maize soy
maize soy cotton
maize cotton
rapeseed cotton

- Growing and importing
- Importing
- Field testing

The U.S. is the world's largest producer of GMO crops, but does not require labeling of GMO foods.

Percent of GMOs in U.S. crops, 2015

- 92% maize
- 94% soy
- 94% cotton

GMOs at the supermarket

The average American eats GMO food products every day. There are at least 30,000 GMOs on grocery shelves, mainly due to soy in processed food.

GMOs are in MANY of the products we use EVERYDAY

In addition to table foods, cooking oils, sweeteners, cereals and other processed foods, GMOs are used in livestock and poultry feed, fibers, ethanol and biodiesel fuels, alcohol, pet food, aquaculture, printing ink, adhesives, and other building and industrial products.

A BRIGHTER FUTURE? OR A RISKY BUSINESS?

A BRIGHTER FUTURE?	A RISKY BUSINESS?
• more food can be produced	• potential for new allergies
• fewer pesticides are needed	• potential for antibiotic resistance
• less tilling, so less erasion	• potential for new "superweeds"
• viable in drought-prone areas	• potential for future health issues
• food security for small farmers	• potential ecological imbalance
• lower CO_2 emissions	• contamination of non-GMO crops
• scientific agreement on low risk	• restrictive patents on usage
• no proof of ill effects from eating	• unknown long-term effects
• crops are disease resistant	• not regulated and tested enough
• increased nutritional value	• company control of food supply
• longer shelf life for fresh foods	• potential for harmful mutations

64 NATIONS REQUIRE GMO LABELING

Labeling is required by law

"The alternative food movement serves as a rallying point against globalization, mass production, and the kind of generic food represented by U.S.-based franchised restaurants."

1. Provide an example and a description of the local food movement in your city. This might include a product, a market, a restaurant, or a community group.

2. Analyze the impacts of the local food movement by listing three ways that humans and the environment are changed in local food production.

THE AMERICAN DIET: *a recipe for disaster*

INCREASED RATES OF OBESITY

Percent obese, 2010
- More than 29%
- 25% to 29%
- Fewer than 25%

Fewer than 19% were obese in any state in 1996.

COST OF OBESITY

$150 billion *in medical costs*
$73 billion *in lost productivity*

RATES OF OBESITY, 1980 and TODAY

- Children: 7% / 18%
- Teens: 5% / 21%
- Adults: 15% / 35%

Since 1980, the percentage of obese children has tripled, while the percentage of obese adults has doubled.

A NATION OF MEATLOVERS

270.7

Pounds of meat consumed per American per year

Americans can lower their risk of colorectal cancer by limiting their intake of red meat to 18 ounces per week, and avoiding processed meats such as bacon, sausage, hots dogs, and deli meats.

A GRAIN OF SALT

Hypertension (high blood pressure) is partially caused by too much sodium from processed foods, too little potassium from fruits and vegetables, and obesity.

- 60% — adults have hypertension or pre-hypertension
- 30% — will eventually develop hypertension

The average adult consumes twice the amount of sodium recommended for middle-aged or older adults, those with hypertension, or African Americans. Too much sodium also increases risk of stroke, heart failure, osteoporosis, and kidney disease.

Sodium in the American diet
- 12% naturally occurring
- 6% while eating
- 5% home cooking
- 77% processed and restaurant food

OUR DAILY BREAD

Americans spend 17% of their food budget on refined grains and only 1.5% on whole grains. The USDA recommends spending a fourth as much on refined grains and seven times as much on whole grains.

Grains in the American food budget
- refined grains: 17%
- whole grains: 1.5%
- other foods: 81.5%

SKIPPING FRUITS & VEGGIES

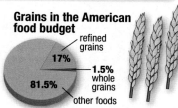

Fewer than 1 in 10 Americans over the age of 12 consume the daily recommended amounts of fruits and vegetables.

HOW SWEET IT IS! *Sugar in the American diet*

Americans consumed 109 pounds of sweeteners per person annually between 1950 and 1959. By 2000, the average American consumed 152 pounds annually as the use of corn sweeteners octupled. Sugar consumption contributes to tooth decay and obesity.

Percent of total calories from added sugars
- 36 million Americans: 25%
- Teens: 21%
- Average American: 14%
- Maximum recommended: 5%

Main sources of sugar are sodas, energy drinks, and sports drinks.

▲ FIGURE 6.1.2 **The U.S. Diet** This infographic provides a summary look at the unhealthy eating habits of many in the United States. It should be pointed out, however, that a survey by the W. K. Kellogg Foundation found that U.S. consumers are beginning to eat more whole grains and fresh fruit and vegetables than they did five years ago.

Marijuana

By Conor Cash

Cannabis is the most commonly consumed illicit substance in the world, with an estimated 125–204 million users in 2015. Developed countries constitute the lion's share of global demand and of those the United States and Canada have the highest per capita use. Perhaps surprisingly, given the current emphasis on global commodity flows, the vast majority of marijuana that is produced worldwide is consumed locally. In the United States and Canada, marijuana produced in both countries travels largely within national boundaries, accruing additional value as it moves across state and provincial lines.

Marijuana production occurs in a wide variety of locales through a variety of vastly different systems of production in the United States and Canada. California produces approximately 75% of the marijuana consumed in the United States. In this state, cannabis is grown on large, illegal plantations on public lands; in medical grow operations, where plants are cultivated specifically for patients in legal medical marijuana buying clubs; and by home cultivators. Production is increasing and changing, however, due to changing marijuana laws across the country.

Marijuana production in Canada is substantial enough to supply the entirety of the country's domestic market (**FIGURE 6.2.1**). While British Columbia remains the largest producer of marijuana in Canada, both Ontario and Quebec have become sites of production in recent years. In Canada, the consumption of medically prescribed marijuana is legal throughout all 10 provinces.

In the United States, the recent spate of state ordinances for medical and recreational marijuana use has changed the national landscape where Alaska, Washington, Oregon, Colorado, and the District of Columbia have legalized it for recreational use.

The legal marijuana market in the United States was worth $2.7 billion in early 2015 and is expected to rise in the next few years to $10.2 billion. Despite its economic impact, the United States and Canada both maintain prohibitionist policies against marijuana at the federal level. Public opinion, however, favors a continued relaxation of prohibition.

---

1. Using the Internet, research the marijuana legalization debate in your state.

2. Compare and contrast your state to the rest of the United States and list the top reasons your state has instituted those policies.

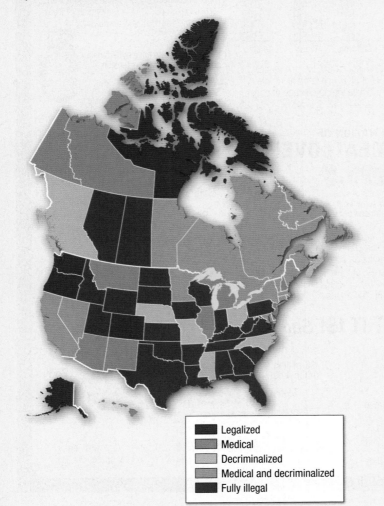

Legalized
Medical
Decriminalized
Medical and decriminalized
Fully illegal

"Despite its economic impact, the United States and Canada both maintain prohibitionist policies against marijuana. Public opinion, however, seems to favor a continued relaxation of prohibition."

▲ **FIGURE 6.2.1 Marijuana Laws in Canada and the United States, 2015** This map shows the variation in the legal status of marijuana use in the two countries where change is occurring rapidly.

Historical Legacies and Landscapes

During the Age of Exploration in the 15th century, European views of the world did not include the existence of the North American landmass. And although 16th-century Spanish missionaries and explorers identified the southernmost section of present-day United States and Mexico to be of interest, most of the rest of North America was considered of little consequence because it presented none of the appearances of grandeur and resource potential like the Aztec, Maya, and Inca empires of Latin America (see Chapter 7, p. 272). This lack of interest in the northern reaches of the continent changed dramatically in the 17th century, with consequences that continue to reverberate today.

Indigenous Histories The distribution and subsistence practices of the indigenous people of the United States and Canada in about 1600 reflected the great diversity of cultural groups that occupied the continent. When Christopher Columbus arrived in the Caribbean, he assumed he had reached the Far East, or the Indies, and he called the indigenous people he encountered *los indios,* or indians. As thousands of native languages existed at the time of European contact, there was no single word of self-description common to the diverse people who occupied the continent. As a result, the word *Indian* has endured as an extremely misleading and sometimes derogatory term for describing a wide range of regional cultural groups. In Canada, the preferred term for describing its indigenous people is "First Nations" and in the United States, it's "Native Americans." The term *indigenous* refers to the status of being the original human occupants of any region around the globe.

Estimates of the size of the indigenous population of the region at the point of European contact range from between 1 million and 18 million. More than anything, these widely differing estimates indicate how little is known about the people who were thriving in this region before the Europeans arrived. What is well known is that there was no common culture, particularly no common language, among the groups who first populated the region. Tribal culture and local environmental conditions were the frameworks within which daily life was lived (**FIGURE 6.10**).

Europeans originally made contact with various individuals and tribes to solicit help with extracting resources, including animal furs; tar and turpentine for shipbuilding; fish; and other primary-sector products exported back to Europe. The experiences of the Dutch, French, and English in North America differed substantially from that of the Spanish in Latin America (see Chapter 7, p. 277). The Spanish conquistadors vanquished sophisticated indigenous civilizations to plunder their gold and silver treasuries and make themselves rich in the process.

The Dutch, French, and English were also interested in improving their financial situations, but they encountered a very different set of cultures with no centralized system of social control that they could exploit as the Spanish did with the Aztecs in Mexico. Instead, the eastern tribes that Europeans first encountered in the United States and Canada were small, autonomous

▲ **FIGURE 6.10 Inuit Share Frozen Meat** The Inuit are known for sharing food caught by one member of the group among the whole group. Shown here are families collecting aged and frozen walrus meat.

groups, possessing an active sense of rivalry and competition with their neighbors.

Colonization and Independence Although the occupation by missionaries and settlers who came to the United States and Canada in the 16th and 17th centuries is widely known as the period of **Europeanization**, the process was actually highly selective and did not involve all of Europe. A few Western European countries—France, Spain, the Netherlands, and Great Britain—dominated colonization in the United States. In Canada, the colonizers were predominantly Britain and France. In both countries, Great Britain was by far the most influential, though others did have substantial impacts, particularly the French in Canada and the Spanish in the United States.

Later, different groups from different European countries settled in different parts of the United States and Canada. As a result, the Europeanization process, as it unfolded along the

▲ **FIGURE 6.11 German Immigrants in the United States** Frankenmuth, Michigan is a town in the east central part of the state with a history of German immigrant occupation. Shown here is the Bavarian Inn of Frankenmuth, also known locally as Little Bavaria. The city's name is a combination of two words. "Franken" represents the Province of Franconia in the Kingdom of Bavaria (in southern Germany), home of the Franks where the original settlers were from. The German word "Mut" means courage; thus, the name Frankenmuth means "courage of the Franconians."

Atlantic seaboard of the United States and Canada through the 19th century, created distinct colonial cultures and societies in different places (**FIGURE 6.11**).

Even though the settlers were not a homogenous group, European settlement routinely resulted in the exploitation and abuse of the indigenous people. The history of colonial settlements, although at first peaceful, over time erupted into disputes over land claims that ended in violence and often outright massacres on both sides. After a time, not only were there conflicts between native people and colonists, but direct conflict also emerged among the various European groups that vied for control over land.

Finally, exposure to the diseases the colonists brought with them had a devastating impact on indigenous populations (see Chapter 7, p. 278). As these groups were decimated, defeated, or pushed onto reservations farther into the interior of the continent, the various European settlers increasingly came into direct conflict, leading eventually to the **French and Indian War** (1754–1763). This war left the British in control of Canada and what is now the United States east of the Mississippi River.

In the two decades following that war, residents of the original 13 U.S. colonies became disillusioned with the taxes they were forced to pay to help Britain recoup the high cost of the war. They then launched their own war, the **American Revolution** (1775–1783), which led to the creation of a new, independent nation in the late 18th century. As a result, an ethos of liberalism, individualism, capitalism, and Protestantism emerged, gained currency, and ultimately came to define a U.S. national character. Canada remained a colony, under British control, composed of both French- and English-speaking settlers. In Canada, a bloodless separation from Great Britain did not occur until well into the 19th century.

The Legacy of Slavery in the United States The impact of European colonization in the 16th and 17th centuries was felt all along the Atlantic seaboard of the United States and Canada. The development of the southern United States following the

end of the revolutionary period, however, differed dramatically from that of its northern neighbors. Before the arrival of the Europeans, the southeastern United States was inhabited by a wide range of indigenous tribes, among them the Cherokee, Choctaw, Chickasaw, Creek, and Seminole people. Early on, the region was occupied by European military personnel living in scattered outposts like Jamestown in Virginia. However, by the mid-17th century, the military outposts had given way to tobacco farms.

At first, **indentured servants**—individuals bound by contract to the service of another for a specific term—from Britain were the primary source of labor on the tobacco and later indigo and cotton plantations. Increasingly, however, servants earned their freedom and were replaced by slaves from Africa. At the same time, disease, armed conflict, and demoralization reduced the indigenous populations that would have been an additional source of laborers.

African slaves had been a well-established commercial staple of the Mediterranean well before the Spanish and Portuguese introduced them to their newly captured territories in Latin America and the Caribbean, thereby establishing the Atlantic slave trade. By the early 18th century, England became the dominant slaving nation in the Atlantic system. As a result, slaves were a part of the social and economic system of the American colonies beginning with their founding. As an institution of formal social and economic organization, slavery endured in the South for more than 250 years, ending officially in 1865, following the end of the **U.S. Civil War** (1861–1865). Notably, Canada never participated to any significant degree in the trade in African slaves.

European Settlement and Industrialization With the creation of new nations and the transformation of colonies into states, the relentless European settlement of the United States and Canada accelerated. In the United States, the movement of the frontier was continuous, at least until settlers reached the Great Plains in the middle of the 19th century and confronted significant mountain ranges at its western edge. In Canada, westward expansion was interrupted early on by the vast, generally infertile, though heavily forested, Canadian Shield, which separated Ontario from the prairies. Many Canadian settlers leapfrogged across the Canadian Shield to the northern midsection of the country to acquire suitable farmland.

The regional economy during this period was oriented to agriculture and trade activities. In short, trading agricultural crops and primary resources, such as fur, fish, timber, and minerals, provided an economic base for the expanding population. Yet by the mid-19th century, a new economy based on manufacturing was rapidly gaining momentum in the United States along the northern Atlantic seaboard, especially in and around southern New England and New York.

As industrialization fueled the economies of the northern United States, the differences between the northern and southern regions of the United States became increasingly pronounced. While the North's economy was more diversified—based on commerce, agriculture, and industry—the South's was more tied to staple crop agriculture but also worked with northern agents—called factors—who found markets for their products. Because southern plantations produced much of their food and clothing needs, and because most of the capital was invested in slave labor (unpaid and thus with no income to stimulate commercial exchange), the economy, of the South was subordinate to the North.

Slavery and the economic issues that surrounded it resulted in a sharp division between the southern and northern states. By 1861, the division was so deep that the South formed a new government, the **Confederate States of America**, and attempted to secede from the United States to protect slavery and the South's agricultural economy (see the previous section). Civil war ensued.

Fearing that the United States, emboldened by its Civil War victory, might launch an invasion of Canada and responding to agitation among the French-speaking minority, the British Parliament passed the **North America Act of 1867**. This act created the Dominion of Canada, dissolving its colonial status, effectively establishing it as an autonomous state with its own constitution and parliament. All the existing Canadian colonies joined the new Canadian confederation except British Columbia, which waited until 1871; Prince Edward Island, which joined the dominion in 1873; and Newfoundland, which remained a colony until 1949.

By the early 20th century, the United States and Canada were occupied from east to west mostly by Europeans and Euro-Americans, and the industrialization of the U.S. economy was well on its way to transforming the country from one of rural agricultural settlement to one of urbanization and industrialization (**FIGURE 6.12**). The 1920 census documented for the first time in U.S. history that there were as many people living in cities as there were in rural areas. From that point onward, the United States and soon after, Canada, became increasingly urbanized. Presently, over 80% of the U.S. and Canadian population live in cities, up from 25% in the mid-19th century.

Apply Your Knowledge

6.5 Summarize the process of Europeanization across both Canada and the United States.

6.6 Using national census data on any three ethnic groups in either the U.S. or Canada (http://www.census.gov or http://www.statcan.gc.ca), construct a map that shows their geographic distribution across the national territory in 1850, 1900, 1950, and 2000.

U.S. & Canada Census

https://goo.gl/X9yJsm

Economy, Accumulation, and the Production of Inequality

By the end of the 19th century, Canada and the United States were fast becoming key players in the global economy. Vast natural resources provided the raw materials for a wide range of industries that could grow and organize without being hemmed in and fragmented by political boundaries. Populations were expanding quickly through immigration, providing an ever-increasing market and a cheap and industrious labor force. Cultural and trading links with Europe provided business contacts, technological know-how, and access to markets and capital for investment in a basic infrastructure of canals, railways, docks, warehouses, and factories. Because of these riches and infrastructural investments, the two national economies in the region took off and expanded.

The Two Economies Canada's path to global economic competitiveness was distinctive. Although most of its population enjoys a high quality of life and high levels of economic productivity, Canada is certainly an atypical economy largely because it was, until very recently, never highly industrialized.

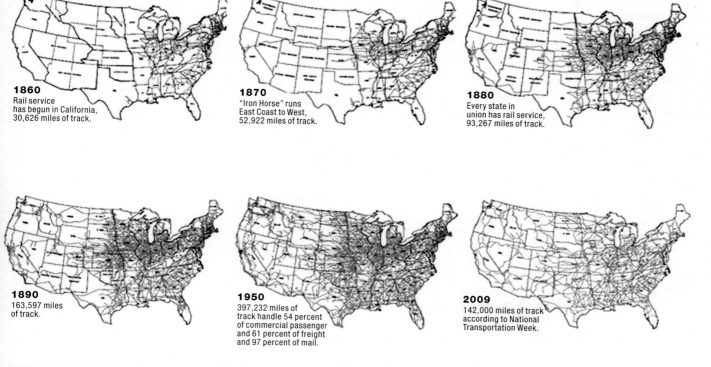

1860
Rail service has begun in California, 30,626 miles of track.

1870
"Iron Horse" runs East Coast to West, 52,922 miles of track.

1880
Every state in union has rail service, 93,267 miles of track.

1890
163,597 miles of track.

1950
397,232 miles of track handle 54 percent of commercial passenger and 61 percent of freight and 97 percent of mail.

2009
142,000 miles of track according to National Transportation Week.

◀ FIGURE 6.12 **History of U.S. Rail Expansion** These six maps provide a sense of how the possibility of goods and people moving from west to east and vice versa was enabled by the fast growing railroad. As the country industrialized, the railroad was key to the spread of both of its processes and the goods it produced. **Look closely at the maps: Has rail coverage simply increased over the years or is there more to the story?**

The primary sector (see Chapter 1, p. 26)—a **staples economy** based on natural resources that are unprocessed or only minimally processed before they are exported to other areas where they are manufactured into end products—was the major pillar of Canada's economic prowess. Although this sector has declined in centrality, it continues to play an important role in Canada's economic structure. When industry did begin to grow and flourish in Canada after World War I, most of it occurred in the midsection of the country along a swath of land at the U.S.–Canada border. A substantial proportion of the industries built there were branch plants of U.S. manufacturers. Canada's major trading partner is the United States, which imports more than half of all Canadian exports.

Over the last decades of the 20th century, Canada's economy shifted. Today, its largest and most dynamic sectors are first, by a large margin, finance, and insurance, real estate (FIRE), and second, manufacturing. **FIGURE 6.13** shows the U.S. economy is also dominated by the quaternary services of FIRE but public administration is its second most important sector. What the graphs do not show is that while the Canadian economy has increased its share in manufacturing, the United States has decreased its share.

Today, the United States and Canada together produce more than one-quarter of the world's gross national income (GNI; see Chapter 1, p. 27). The United States has the world's largest economy, and Canada is the ninth largest. As part of the recent restructuring of the global economy discussed in Chapter 1, pp. 25–30, the various subregions of both the United States and Canada have experienced significant transformations in their economies, societies, political institutions, and even their physical environments.

Transforming Economies In the United States, the most recent wave of internal migration occurred as millions of U.S. residents moved from the **Manufacturing Belt** of North America, the Northeast and the Midwest, to the west and the south. As a result, the original and aged manufacturing area of the region became known as the **Rustbelt**. The area south of the 37th parallel, which experienced massive population growth, became known as the **Sunbelt** because of its inviting climate for people and businesses. Cities like Detroit, Boston, New York, and Buffalo began to experience high labor costs and aging infrastructure, mostly manifested in outdated technology systems. At the same time, once-peripheral areas of the country, the South and West (in cities like Atlanta, Miami, Houston, and Phoenix), began to attract investors. The U.S. military had established bases in the Sunbelt region during World War II. Following the war, the government continued to develop these bases as it built up its military capacity during the Cold War.

As the computer age dawned, the Sunbelt was ripe for civilian investment opportunities. Importantly, its well-educated labor force was not accustomed to unions and high-wage rates and as the profitability of old, established industries in the Manufacturing Belt declined, new industries began to improve their profitability in the Sunbelt.

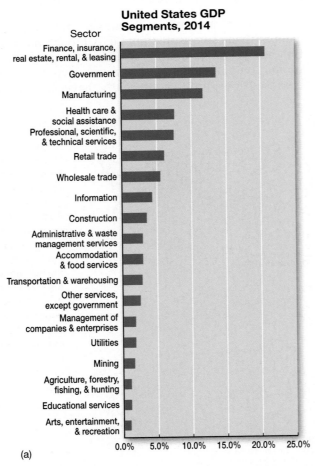

(a)

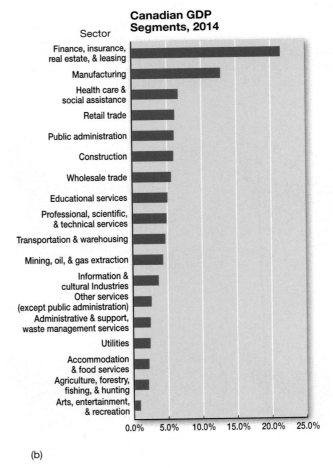

(b)

▲ **FIGURE 6.13 U.S. and Canadian GDP segments, 2014** (a) FIRE, government dominate the U.S. economy. (b) In Canada, manufacturing and services dominate. Despite those differences, the economies of the two countries have much in common.

The process just described to characterize the decline of the Rustbelt and the rise of the Sunbelt occurred through **deindustrialization**, a relative decline (and in extreme cases, an absolute decline) in industrial employment in core economic areas (**FIGURE 6.14**). This process unfolds as firms scale back their activities in response to decreasing profitability. In this case, technological innovations in computerized production systems, transportation, and communication facilitated new industrial applications, and investors and manufacturers began to look around for new places to invest in and build.

In short, old industries and a large proportion of established industrial regions were dismantled to help fund the creation of new centers of profitability and employment. This process is often referred to as **creative destruction**, and it is inherent to the dynamics of capitalism. Creative destruction describes the necessity of withdrawing investments from activities (and regions) that yield low rates of profit to reinvest in new activities (and, often, in new places, domestically or abroad).

The New Economy The new economy that has emerged over the last three decades in this world region has fundamentally transformed industries and jobs through **information technologies (IT)**, the use of computer systems for storing, retrieving, and sending information. This shift to IT has been facilitated by a high degree of entrepreneurialism and competition around the world. But the new economy was born in the United States, sired by the technological changes that emerged from Silicon Valley, California, nearly 50 years ago (**FIGURE 6.15**).

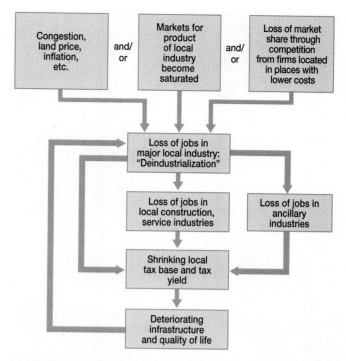

▲ **FIGURE 6.14 Spiral of Deindustrialization** When the advantages of manufacturing regions are undermined for one reason or another, profitability declines, and manufacturing employment falls. This can lead to a downward spiral of economic decline, as experienced by the traditional manufacturing regions of North America during the 1970s and 1980s.

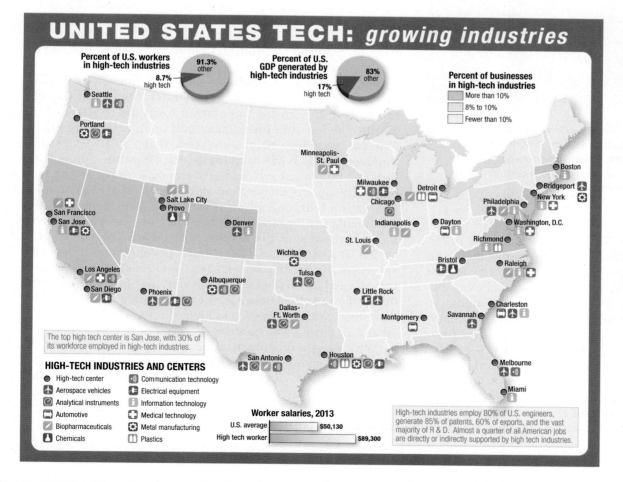

◄ **FIGURE 6.15 High-Tech Industries in the United States** Research, development, and production around high technology are prime movers in the U.S. economy. This map shows those technology hubs that are helping to drive state and regional economies as well as other less technologically dominant cities where related economic activity is occurring. Note that even basic industries such as car manufacturing are considered high tech because they are organized around complex IT systems including robotics, smart materials, and green technologies. **What is it about these cities that make them attractive to high-technology development? Choose a couple of cities and determine why they might be attractive for high-tech investment.**

▲ **FIGURE 6.16 Downtown Toronto** Toronto is Canada's largest employment hub, with one-sixth of the country's jobs. A world banking and finance leader, Toronto also supports a burgeoning variety of high-growth sectors, such as software and hardware design, biotechnology, pharmaceuticals, and telecommunications.

It is generally agreed that the previous economic order, the "old economy," lasted from 1938 (the end of the Great Depression) to about 1974. The year 1974 was a critical year in global economic history: Oil prices were skyrocketing and corporate profits were falling. The old economy's foundation was manufacturing geared toward standardized, mass-market production run by stable, hierarchically organized firms focused on the domestic market. Many regard 1975–1990 as the transitional period from the old economy to the "new economy."

The new economy is about more than just new technology. It is also about the application of new technologies to the organization of work—from the impact of biotechnology on farming to the effect of IT on management hierarchies. Dynamism, innovation, and a high degree of risk are at the center of the new economy.

Canada has been an active participant in the new economy, and part of the shift it has experienced into services and high-tech manufacturing is evidence of this (**FIGURE 6.16**). A particularly dynamic aspect of this change in Canada has been in science-based industries as well as research, development, and manufacturing in communications and transportation technologies.

Wealth and Inequality Although globalization and the new economy helped improve the employment opportunities and level of wealth of many in this region, the Great Recession that began in 2007 seriously set back many people around the globe, including those in the United States and Canada. One result of the great recession was to intensify a trend toward the concentration of wealth in the hands of a small fraction of the population. **Wealth inequality** measures the difference in how much money and other assets combined, individuals have accumulated. **Income inequality** describes the gap in how much individuals earn from the work they do and the investments they make.

In the United States, 75% of all wealth is owned by the richest 10% of people (**FIGURE 6.17**). Among the top 20 developed countries (including among them Australia, Denmark, France, Germany, Singapore, and the United Kingdom), the United States has the most extreme wealth concentration. Another way to look at the statistics on U.S. wealth inequality is that 90% of the population own only about 25% of the private wealth, an imbalance that is more typical of developing countries like Chile, Indonesia, or India. In those countries, 90% of the population own about 40% of the private wealth.

Economists see Canada's wealth distribution as parallel to, though not as extreme as the United States'. In Canada, the richest 20% of the population owns nearly 70% of the wealth; the top 10% own almost half of all private wealth. In Canada, inequality of income means that 1 in 10 people live in poverty, which has been increasing for youth, young families, and immigrant and minority groups. Poverty among First Nations people is higher than among the rest of the Canadian population both on and off reserves. In fact, if the statistics for Canada's indigenous people were viewed separately from those of the rest of the country, they would be ranked 78th on the United Nations (UN) Human Development Index—the slot currently held by Kazakhstan. Over most of the 20th century, Canada provided a strong safety net for poor families, but in the last three decades, the poverty rate among children there has grown to the second highest (after the United States) among all developed countries.

Why does concentration of wealth at the top matter? Because too much inequality reduces opportunity for all and it undermines a vibrant and resilient society, among other reasons. One way to understand wealth inequality's effects on opportunity is to examine the effects of wealth decline on nutritional quality, a basic need for all people, whether rich

MONEY IN THE UNITED STATES: *dollars and sense*

THE WEALTHY FEW

Wealth holdings, by economic class

Bottom 80% (wage and salary workers) **14.9%**
Top 1% **34.6%**
Other 19% (managerial, professional, small business) **50.5%**

34.6% of privately held wealth is held by the top 1% of households, while the bottom 80% holds 14.9%.

HOUSE RICH

9.4%   **61.5%**

House value as a percent of household wealth, for the top 1% and the bottom 90%

TAKING STOCK

Stock ownership, by economic class

Top 1% of households **40%**
Bottom 99% of households **60%**

In 2013, 42% of financial wealth was held by 1% of the population. The bottom 80% held 2% of financial wealth (securities, stocks, trusts, mutual funds, and business equities).

U.S. INHERITANCE

$0 — **91.9%**
$50,000 to $100,000 — **1.1%**
$100,000 or more — **1.6%**

The majority of Americans do not receive an inheritance.

AMERICAN AVERAGE NET WORTH

10.1% in banks
2% vehicles
36.4% investments
51.5% real estate

Household net worth by age

under 35:	$6,682
35 to 44:	$35,000
45 to 54:	$12,175
55 to 64:	$144,200
over 65:	$171,135

GOLDEN STATE

572,000 millionaires in California
(1.5% of the total population)

Connecticut has the highest rate of millionaires, at 2.3%, and Mississippi's rate of 0.3% is the lowest.

INCOME BY ZIP

Highest & lowest average income by zip code

$252,392 — 10022 — New York City
Columbus
$8,965 — 88029

IN THE RED

Consumer debt, by economic class

Top 10% of households **26.6%**
Bottom 90% of households **73.4%**

Average consumer debt profile:
- Mortgage debt: **$156,333**
- Student loan debt: **$32,953**
- Credit card debt: **$15,706**

Student loans and credit card debt are the largest sources of nonmortage debt owed by American families. In 2013, 38.8% of families with a household head under age 40 had student loan debt.

LOCATION, LOCATION, LOCATION

Annual median income by state, 2011–2013

- $60,000 or more
- $50,000 to $59,999
- below $50,000

▲ **FIGURE 6.17 Where's the Wealth in the United States?** This infographic shows the dramatically uneven distribution of wealth according to several variables including debt, inheritance, and geography. **Why is debt so important to understanding wealth?**

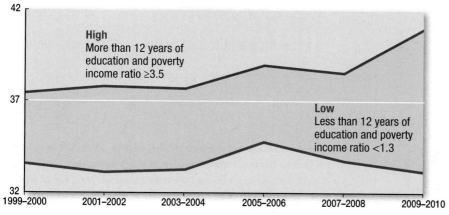

High
More than 12 years of education and poverty income ratio ≥3.5

Low
Less than 12 years of education and poverty income ratio <1.3

1999–2000   2001–2002   2003–2004   2005–2006   2007–2008   2009–2010

▲ **FIGURE 6.18 Diet Quality and Socioeconomic Status** Shown is the Harvard School of Public Health's alternative Healthy Eating Index which measures food and nutrients in relationship to health outcomes. The gap in nutritional quality between rich and poor U.S. diets doubled between 1999 and 2010. A perfect score is 110 points.

or poor. Public health researchers have shown how the diets of low-income people in the United States have gotten worse since 1999, whereas high income people's diets have improved (**FIGURE 6.18**).

Because of the dramatic wealth and income inequality in the United States and Canada, it is not surprising that "Occupy" protestors began to appear in both countries in 2011. **Occupy Wall Street**, the action that appeared first, is part of a popular social movement founded in Canada. With their slogan, "We are the 99%," the Occupy movement signals its dissatisfaction with the growing wealth gap between a small number of superrich individuals and the other 99% of the population. Another aspect of inequality that has been attracting increased attention is the persistent "gender gap" in pay between women and men. See "Visualizing Geography: The Gender Gap in the United States and Canada" on p. 242.

The Gender Pay Gap in the United States and Canada

Income inequality based on gender in both the United States and Canada, as elsewhere in the world, favors men over women. The "pay gap," as it is called, has been an enduring feature of both economies since women joined the work force in large numbers.

"For every one dollar a man earns in the U.S. and Canada, a woman who works the same job full time earns 77 cents."

▼ **FIGURE 6.3.1** The gender pay gap in the United States.

National Women's Law Center

https://goo.gl/R4KUlv

Public and private sector = 77 cents

That 33 cents difference in pay between men and women in the United States adds up. In an average person's life time that's that's enough money to...

Historically and still today, woman tend to be considered secondary earners in a traditional male dominated household. The assumption is that women need less pay as the male breadwinner's paycheck covers family expenses. Unfortunately, then and now, this assumption is only partially true and moreover, reflects acceptance of a belief that an equally qualified female in the same job is somehow worth less than her male counterpart.

Buy a house in the U.S.

+

Put two kids through college at a public school

+

Buy 21,900 gallons of gas

+

Feed a family of four for 6.4 years

▼ **FIGURE 6.3.2** The gender pay gap in Canada.

Conference Board Canada

https://goo.gl/ENBezw

Public sector = 82 cents   **Private sector = 73 cents**

The pay gap is worse for minorities, both men and women, than it is for majority women. In Ontario, minority and Aboriginal women earn much less than white men:

Minority women = 64 cents   **Aboriginal women = 46 cents**

In Canada, gender discrimination in compensation is being addressed both federally and provincially.

Over the course of her career, a typical woman working full time in the U.S. and Canada will lose

$431,000

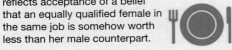

In the United States and Canada over the 20th and now the 21st century, federal legislation has helped to narrow that gap. Solutions offered to close the pay gap include improved policies that support mothers, better pay and benefits, reduce occupational segregation by gender and enforce existing laws.

1. **Compare and contrast the economies of one city in Canada and one in the United States, referring to each city's website.**
2. **Find data on median income levels and cost of living for both cities and draw a correlation between the city and individual economies.**

The United States and Canada and the Global Economic Crisis As discussed previously in this chapter, the global financial crisis that began in 2008 had significant effects on the economies of the United States and Canada. When the failure of several major private U.S. financial institutions in October 2008 prompted panic among international financial markets, millions of households in the region experienced loss: of jobs, wealth due to declines in the value of real estate, homes because of inability to pay mortgages, and value in retirement investments and pension funds. In an attempt to prop up the international financial system and regional, national, and local economies, the governments of the United States and of other leading economies intervened with hundreds of billions of dollars of support for U.S. private financial institutions.

How could the Great Recession have happened? Part of the explanation lies in the steady increase in debt that had been fueling every aspect of the global economy, especially in the U.S., since the late 1970s. Consumer spending had been financed increasingly by credit card debt. The housing boom had been financed by an expanded and aggressive mortgage market. And the wars in Iraq and Afghanistan had been financed by the U.S. government borrowing from overseas. By mid-2007, private debt in the United States had reached U.S. $41 trillion, almost three times the country's annual gross domestic product; the external debts of the United States had meanwhile reached U.S. $13.7 trillion. Everyone, it seemed, was borrowing from everyone else in an international financial system that had become extremely complex and decreasingly regulated.

Because the entire world economy has become so interdependent, the problem in the United States quickly spread. Banks in other countries were caught up in the complex web of loans, and by the fall of 2007, the banking industry was in crisis worldwide. As credit markets seized up, manufacturers and other businesses found it difficult to get the credit they needed to continue operating. Understandably, investors big and small were shaken, stock markets collapsed, and many manufacturers went bankrupt. Consumer confidence also plummeted, prompting retailers to cut back on orders. The sophisticated flexible production system of global commodity chains (see Chapter 7) meant that the impact across the globe was almost instantaneous.

Since the onset of financial calamity in many of the banks that survived have recovered. In spring 2015, U.S. unemployment was 5.5% down from a high of 9.9% in 2009. The economies of both countries in this region continue to grow, as does their manufacturing employment.

The global financial crisis was generally less pronounced in Canada, where the banks are more tightly regulated and more liquid (having more cash on hand to back up the bank's loans). Unlike risk-taking U.S.-based investment banks, Canadian banks tend to operate in a more traditional manner, with large numbers of loyal depositors and a more solid base of capital. But Canada's economy has not been entirely trouble-free. The Toronto Stock Exchange fell significantly in 2008 but recovered in 2010.

Territory and Politics

The United States and Canada are both **democracies**, a form of government in which all citizens possess the capacity to shape the laws and actions of their state. One of the chief requirements in a democracy is that citizens who meet certain qualifications have an equal opportunity to express their opinion through voting. The United States became a federal republic founded on democratic principles in 1789 when it established its constitution. In the United States, the sovereign authority is the people. Canada, also a democracy, is a constitutional monarchy with a parliamentary type of government. The British sovereign is the foundation of Canada's judicial, legislative, and executive branches of government.

States and Government The United States and Canada are both federal states, which means that in each country, political authority is divided between autonomous sets of governments, one national and the others at lower levels, such state/province, county, city, and town. A **state** is an independent political unit with recognized boundaries (though some of those boundaries may be in dispute). **Government**—one element of a state—is an entity that has the power to make and enforce laws. In a **federal system**, many political decisions are made at the local level, but there are differences even among federal states (**TABLE 6.1**).

TABLE 6.1 Differences in U.S. and Canadian Federalism Although both countries are federal states there are important differences.

Canadian Constitution	U.S. Constitution
1) The powers of the provinces are specifically defined and limited.	The powers of the central government are specifically defined and limited.
2) For the most part, the residuary powers were granted to the federal government.	The residuary powers are reserved for states or the people.
3) The Constitution Act, 1867, permits vetoing by the federal government of provincial legislation.	No such clause in the U.S. constitution.
4) All of the upper echelons of the judiciary (federal—such as the Supreme Court—and provincial) are under the control of the federal government.	The federal government controls federal courts; the state governments control the state courts.
5) The chief executive officer of a province (Lieutenant-Governor) is appointed, paid, and removable by the federal government.	The chief executive officer of a state (Governor) is elected by the people of the state.
6) Senators are supposed to protect regional rights and interests and are appointed by the reigning British monarch.	Senators are supposed to protect regional rights and interests and are elected directly by the people of the state.
7) Criminal law is under federal jurisdiction.	Criminal law is under state jurisdiction for state crimes and under federal law for federal crimes.

Adapted from Claude Belanger, 2001, "Comparing Canadian and American Federalism," *Quebec History*, http://faculty.marianopolis.edu/c.belanger/quebechistory/federal/compare.htm)

During the 19th century, the governments of both Canada and the United States made major efforts to develop and regulate commerce. After establishing a strong economic base, both governments began to take an increasingly active and direct role in regulating and supporting all aspects of social and economic life, such as providing for social welfare and developing infrastructure, like dams and highways. In the United States especially, large amounts of tax dollars also began to be transferred to contractors for the buildup of U.S. defense systems, especially during the Cold War.

The view that government's primary responsibility is as a guarantor of social welfare had dominated popular understanding in the United States since the 1930s. In the 1970s and 1980s, as local governments in the Rustbelt were declaring bankruptcy and the federal government was accumulating massive debt, popular opinion changed, and the role of government was reconfigured. Since the late 1980s, it has become routine for local governments in the United States to act more as entrepreneurs than as managers of the social welfare. For example, as deindustrialization accelerated in the Rustbelt, government agencies in the Sunbelt helped lure investment to the region by offering tax breaks, creating needed infrastructure, and providing subsidies for private investment.

Since independence, Canada has fostered a government that has been far more inclined to guarantee social welfare than that of the United States, though Canada also has a tradition of entrepreneurialism (pushing against the status quo to create new opportunities for value creation) in government. In addition to continuing its tradition of providing social welfare, the state in Canada has accelerated its entrepreneurialism by directing support to expanding its tertiary sector (activities involving the sale and exchange of goods and services) and quaternary sector (activities involving the handling and processing of knowledge and information), particularly with respect to high-technology development.

A New Province Perhaps most remarkable with respect to dramatic political geographic change in the region has been the agreement reached in 1999 between Canada and the native Inuit people to cede 787,155 square miles (2,038,722 km²) of land and water combined of the Northwest Territories to the Inuit people. The result was the creation of Nunavut, which means "our land" in Inukitut (the Inuit language). Nunavut has been continuously inhabited by an indigenous population for approximately 4,000 years. The Inuit make up over 80% of the population of Nunavut. Their form of government is consensus based through the Legislative Assembly of Nunavut. Nunavut also possesses representation in the Canadian Parliament of one seat each in both the House and the Senate.

Nunavut Government

https://goo.gl/7Xg4Bu

Apply Your Knowledge

6.7 Analyze how the United States and Canada's electoral systems differ.

6.8 In what ways do these different systems shape elections and policies in both countries.

U.S. Military Influence In combination with its economic power, the United States achieved its place as the world's most powerful country through political and military strength. As the leading global military power, the United States significantly dwarfs Canada. Because of its global military dominance, we focus only on the United States in this section.

As the 19th century came to an end, the United States had established its reluctance to become involved militarily in European affairs. When World War I began in 1914, President Wilson proclaimed a strict policy of neutrality. But when news reached Wilson that Germany refused to suspend unrestricted submarine warfare in the North Atlantic and that it was attempting to entice Mexico into an alliance against the United States, Congress declared war on Germany in 1917.

Stunned by the nearly 5 million war casualties and the horror of the first highly technological war in history and eager to protect its growing economy, the United States had entered a period of relative isolationism after World War I. And although President Franklin D. Roosevelt declared U.S. neutrality with respect to the European war in 1939, in 1941 the country entered World War II. The end of that war in 1945 marked a turning point in U.S. political and economic prowess as U.S. loans helped rebuild war-torn Europe and Japan. U.S. participation in the war effectively solidified the country's status as a world leader. The country's postwar position of leadership and military strength led to its involvement in several subsequent wars, including those fought in Korea and Vietnam.

These latter two wars were fought as part of the Cold War, which pitted the capitalist United States and Western Europe against the socialist Soviet Union for the hearts, minds, and territories of peoples throughout the globe (see Chapters 1 and 4). The Cold War came to an end with the disintegration of the Soviet Union in 1991. A new era of global cooperation was declared, with the U.S. government and U.S. transnational corporations leading the way, though with markedly less militaristic fervor. The recent wars in Afghanistan and Iraq (see Chapters 4 and 9) have halted the brief hiatus in significant U.S. military involvement in global affairs, however (**FIGURE 6.19**).

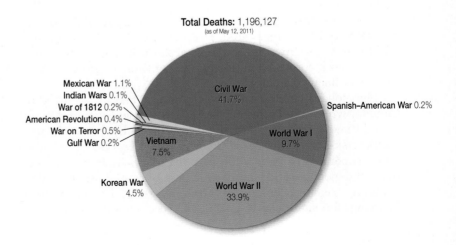

▲ **FIGURE 6.19 U.S. War Dead** This graphic illustrates the dead among military personnel in all the wars in which the United States has been involved over the course of its history.

U.S. War on Terror: Afghanistan, Iraq, and Pakistan

The United States responded to the terrorist attacks of September 11, 2001, by declaring a global war against terrorism and identifying first Afghanistan and then Iraq as the greatest threats to U.S. security. With support from Canada as well as the United Kingdom and Australia, the United States invaded Afghanistan in October 2001 as part of its **War on Terror**. The number of troop fatalities (158) resulting from Canadian military activities in Afghanistan is the largest for any single Canadian military mission since the Korean War. Some 1,500 American lives have been lost, with troops routinely killed by roadside bombs, small-arms fire, and rocket attacks and in unspecified combat operations. The United States officially ended its combat mission in Afghanistan in late 2014.

War in Iraq

Although the evidence of involvement in the September 11, 2001, attacks by Iraq and its leader, Saddam Hussein, was highly questionable, on March 19, 2003, after amassing over 200,000 U.S. troops in the Persian Gulf region, President George W. Bush ordered the invasion and occupation of that country. The Congressional authorization of military action occurred without the explicit authorization of the UN Security Council, and some legal authorities take the view that the action violated the UN Charter. Some of the staunchest U.S. allies (Germany, France, and Canada) as well as Russia opposed the attack, and hundreds of thousands of antiwar protestors repeatedly took to the streets.

On December 15, 2011, the United States formally ended the war in Iraq. During more than eight years of war, nearly 4,500 U.S. service members were killed and 30,000 wounded. Hundreds of thousands of Iraqi insurgents and civilians died, as the United States deposed Saddam's regime and beat down an insurgency and sectarian revenge killings that threatened to destroy the country.

Al-Qaeda and Drone Warfare

A global militant Islamist organization, **al-Qaeda** operates as a network constituted by a multinational, stateless army and a radical Muslim movement. The United States and the United Kingdom, among other countries, as well as the UN Security Council, the European Union, and North Atlantic Treaty Organization (NATO), have identified it as a terrorist organization. Osama bin Laden, who was a member of a wealthy Saudi family, founded al-Qaeda, the organization responsible for the September 11, 2001, attacks on the United States.

- Military activities undertaken in Pakistan by the United States are very much part of the War on Terror. With U.S. troops on the ground in Afghanistan, al-Qaeda moved operations to neighboring Pakistan, where bin Laden was hiding. The U.S. response has been to deploy drones (unmanned aerial vehicles) in Pakistan's tribal areas against suspected al-Qaeda and Taliban targets since 2004 (**FIGURE 6.20**).

- Drones have also been used in the United States and Canada by municipal police departments, federal border protection agencies, and the Federal Bureau of Investigation in the United States for surveillance purposes. As of summer 2015, drones used by U.S. police forces, the Coast Guard, the U.S. Marshals, or any other government entities are prohibited by law from being armed.

▲ **FIGURE 6.20** **U.S. Customs Predator Drone** Unmanned aerial vehicles (drones) have been key weapons in antiterrorist warfare in Pakistan. A drone can be as large as a passenger aircraft, as such as the Customs and Border Protection one shown here, or as small as a hummingbird. Beginning in February 2012, when House Resolution #658 was passed, the U.S. federal government as well as corporations and private individuals were authorized to use drones in U.S. airspace.

Costs of Iraq and Afghanistan Wars

Nobel Prize–winning economist Joseph Stiglitz and coauthor Linda Bilmes argue that the cost of the wars in Iraq and Afghanistan will be between $4 and $6 trillion when the indirect costs are also assessed. These costs include interest on the debt raised to fund the war, the rising cost of oil, health care for returning veterans, and replacing the destroyed military hardware, and degraded operational capacity caused by the war.

War on Terror Continues

The assassination of bin Laden in 2011, in Abbottabad, Pakistan, is seen by many as a symbolic end to the War on Terror. Despite his assassination and the end of major combat operations in Iraq and Afghanistan, the War on Terror continues due to the emergence of the Islamic State of Iraq and Syria or ISIS (also known as ISIL, the Islamic State of Iraq and the Levant). ISIS is a Sunni jihadist militia whose principles include anti-Westernism, strict interpretation of Islam, and promotion of brutal religious violence against Shia Muslims and Christians. The activities of ISIS and other jihadist groups in Nigeria, Somalia, Kenya, and Yemen have led the United States and its allies to extend the War on Terror to new parts of the globe.

Apply Your Knowledge

6.9 Define the "War on Terror."

6.10 Research and summarize the financial and human costs of combating jihadist violence in one of the countries discussed in this section.

▲ **FIGURE 6.21 Protest Against the XL Pipeline** Activists protesting a proposed pipeline to bring **tar sands** oil to the United States from Canada. The Keystone XL pipeline was rejected by President Obama in late 2015 which was seen as a major victory for environmentalists in their fight to address climate change.

Social Movements Two of the most enduring, widespread, and effective social movements in the United States and Canada have been the women's movement and the environmental movement. The women's movement in the United States can be traced to the 1848 women's rights convention held in Seneca Falls, New York. From that point, inspired by women's movements in the United Kingdom, women agitated for the right to vote, to have the same civil rights as men, to have access to birth control, to improve working conditions for themselves, to earn equal pay for equal work, and to end discrimination in the workforce. They also demonstrated in favor of disarmament and peace. Campaigns today for the civil rights of blacks and other minorities, as well as rights for the disabled, gay rights, and children's rights, have derived inspiration as well as tactics and rhetoric from the women's movement. The vice-presidential campaigns of Geraldine Ferraro (1984) and Sarah Palin (2008) and Hillary Clinton's (2015) presidential campaign are testimony to the effectiveness of the women's movement in the United States. Women's rights movements in Canada followed a similar trajectory. Women in Canada received the right to vote in 1918, two years before U.S. women.

The environmental movement in the United States can be traced to the works of men like Henry David Thoreau and George Perkins Marsh, who in the 1850s and 1860s began to lecture and write about the destructive impact of humans on the natural world. The Canadian environmental movement emerged at roughly the same time and has continued to play an important role in Canadian politics since then. Contemporary environmental protection policy and sustainability movements echo the sentiments and commitments of these early activists. Today, citizens in both Canada and the United States are deeply engaged in a wide range of movements that are concerned with environmental issues and the cultural, social, and economic impacts of global warming (**FIGURE 6.21**).

Culture and Populations

The United States and Canada are culturally and politically diverse with influences that extend to remote corners of the globe. From jazz to hip-hop and horror movies to hamburgers, the influence of American culture on the rest of the world has been and continues to be dramatic. The influence of the rest of the world on the region has also been substantial. The region is notably adept at taking what the world has to offer and translating it into something hybrid, popular, and desirable.

Religion and Language

The main language in the United States is English; Spanish is also widely spoken. Although the United States is popularly considered a Protestant country (including large numbers of Baptists, Methodists, Presbyterians, Lutherans, Pentecostals, and Episcopalians), Roman Catholics form the largest single religious group. The largest non-Christian religion is Judaism, and other non-Christian religions, such as Islam, Buddhism, and Hinduism, also have substantial followings.

In Canada, where there are two official languages (English and French), the situation for immigrants has always been somewhat different than in the United States. Instead of **assimilation**, Canadian popular opinion and public policy have advocated **multiculturalism**, the right of all ethnic groups to enjoy and protect their cultural heritage. Multiculturalism in practice includes protection and support of the right to function in one's own language, both in the home as well as in official or public realms.

A significant aspect of the geography of U.S. religion is its regional variation. For example, the **Bible Belt**, which stretches from Texas to Missouri, is dominated by Protestant denominations, many of them fundamentalist and evangelical, as well as Mormons, members of The Church of Jesus Christ of Latter-day Saints, which are concentrated in Utah, where more than 75% of the population are adherents. Large Catholic communities exist throughout the Southwest and in many of the large cities of the Northeast and upstate New York. Sizable concentrations of Jews occur in large U.S. urban centers like New York, Los Angeles, and Miami.

Apply Your Knowledge

6.11 List three cultural contributions of both the United States and Canada to each other and the rest of the world.

6.12 Evaluate your home town's religious identity based on historical and current places of worship.

Cultural Practices, Social Differences, and Identity

Music, art, literature, dance, architecture, film, photography, sports, fashion, journalism, and cuisine, not to mention science, medicine, and technology, have all been shaped by immigrants to the United States and Canada as well as the indigenous populations. The influence of immigrants on music has been

▲ **FIGURE 6.22 DJ Kool Herc** Born Clive Campbell, Herc is widely recognized as the father of hip-hop. This multi-rooted musical form was said to have made its debut at Herc's sister's birthday party in 1973. **What are the musical influences DJ Kool Herc brought to hip-hop?**

particularly impressive. Country, bluegrass, jazz, blues, and rap all originated in the United States but have deep roots in the Old World. From jazz to rap, African Americans have been responsible for musical innovations that have been widely accepted, applauded, and imitated throughout the world.

Arts, Music, and Sports The early 20th-century origins and particularly U.S. expressions of jazz have been influential worldwide. West African folk music, brought by African slaves

to the United States, forms one of the central foundations of jazz, but jazz was also influenced by European popular and light classical music of the 18th and 19th centuries. Rock and roll, country music, hip-hop, bluegrass, folk, gospel, heavy metal, funk, punk, and musicals, all have their source, in part or in full, in the United States (**FIGURE 6.22**).

Native populations in both the United States and Canada have made significant cultural contributions to music, as well as handicrafts (especially basketry, rugs, jewelry, and pottery) and contemporary literature. People from all over the world travel to visit First Nation and Native American sites to view and collect their distinctive handcrafted products such as Navajo rugs from the U.S. Southwest and the wood carvings of the Haida people of Pacific Canada.

The game of baseball is another U.S. innovation, and it, too, has enjoyed widespread popularity beyond the United States especially in Caribbean countries like the Dominican Republic, Venezuela, and Cuba, but also in South Korea and Japan. The diverse nationalities of the players on the major league baseball teams in the United States and the Toronto Blue Jays in Canada demonstrate just how popular this sport has become worldwide. As a high-stakes commercial enterprise, baseball has traveled well. Players on the roster typically hail from Aruba, Australia, Colombia, Cuba, Curaçao, the Dominican Republic, Japan, South Korea, Mexico, Nicaragua, Panama, Puerto Rico, Taiwan, and Venezuela (**FIGURE 6.23**).

Birth of Hip-Hop

https://goo.gl/wquHBZ

U.S. Cultural Imperialism Many scholars argue that "globalization" is really just a euphemism for "**Americanization**," and there is certainly some evidence to support this point of view.

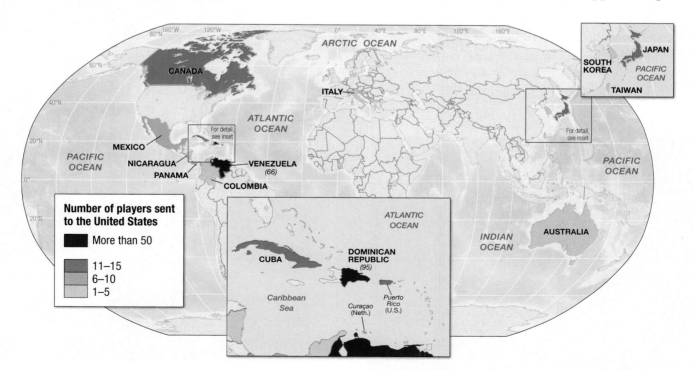

▲ **FIGURE 6.23 The Globalization of U.S./Canadian Major League Baseball, 2015** The old saying "As American as baseball or apple pie" may need revision, given the dramatic transformations that have occurred over the past 10–15 years in the demographics of players on major league baseball teams and the worldwide growth of interest and participation in baseball. **While this is a national map, what states would dominate in a map of the origin of baseball players born in the United States?**

U.S. culture is embraced around the world largely through consumer goods. It seems clear that U.S. products are consumed as much for their symbolism of a particular way of life as for their intrinsic value. Coca-Cola, Hollywood movies, rock and rap music, and National Football League (NFL) and National Basketball Association (NBA) insignia have become associated with a lifestyle package that features luxury, youth, fitness, beauty, and freedom. Neither the widespread consumption of U.S. and U.S.-style products nor the increasing familiarity of people around the world with U.S. media and international brand names, however, necessarily adds up to the emergence of the Americanization of global culture. Instead, the processes of globalization are exposing the world's inhabitants to a common set of products, symbols, myths, memories, events, cult figures, landscapes, and traditions that are variously embraced and employed.

Canadian Cultural Nationalism

Canadian culture tends mostly to represent a mix of immigrant and British and French settler influences. Canadian culture has had to battle the tremendous influence of commercialized U.S. culture, which has been difficult to resist because of its geographical proximity. Canada has developed an extensive and very public policy of cultural protection against the onslaught of American music, television, magazines, films, and other art and media forms.

Government bodies, such as the National Film Board of Canada and the Canadian Radio and Television Commission, actively monitor the media for the incursion of U.S. culture. For example, 30% of the music on Canadian radio must be Canadian. Interestingly, some of the most famous U.S. pop culture icons were either invented by Canadians or are Canadian, such as William Shatner, *Star Trek's* Captain Kirk, Superman (a concept invented by a Canadian and an American), and popular actors, singers and bands, such as Neil Young, Celine Dion, k.d. lang, Pamela Anderson, and Broken Social Scene. Avril Lavigne was identified by researchers at Massachusetts Institute of Technology (M.I.T), through an analysis of Wikipedia, as the most famous Canadian beating out both Jim Carrey and Justin Bieber (**FIGURE 6.24**).

Sex, Gender, and Sexuality

The current distinction between sex and gender has been criticized as misleading by both social and biological scientists. The reason is that it implies that the behavior of an individual can be attributed to either biological or cultural factors. The critics believe that biological manifestations of sex are crosscut with gendered psychosocial and cultural variables—so much so that there is no way to determine that the differences between males and females are due exclusively to biology or to culture.

An example of this widespread understanding of the entanglement of sex and gender is that Facebook, prompted by its users, added more custom categories to their gender option. Recognizing that the simple male/female dichotomy was inadequate in capturing the multiplicity of genders that reside in the space between them, Facebook opened up 58 options. This move by Facebook, seen as excessive by some, is still a fundamental recognition that gender is not something people essentially are (because of a given set of physical characteristics) but something people do (something we enact by the way we present our bodily selves to the world) and something we understand ourselves to be.

With respect to **sexuality**—a person's sexual orientation or preference—certainly the most dramatic transformation in the United States has been the recognition of same-sex marriage first by 38 states, the District of Columbia, 24 Native American tribal jurisdictions, and the federal government. Then, in June 2015, the U.S. Supreme Court decided that same-sex marriage is legal across all states (**FIGURE 6.25**).

▲ **FIGURE 6.25 Freedom to Marry** On June 26, 2015, the U.S. Supreme Court ruled that all states must allow same-sex people to marry. Celebrations of this historic ruling broke out all over the country.

▲ **FIGURE 6.24 The New Pornographers** A globally popular Canadian rock band formed in 1999, the Pornographers, as they are often called, are from Vancouver, British Columbia, Canada.

In Canada, gay and lesbian couples were allowed to marry legally through the passage of the Civil Marriage Act in 2005. These changes reflect the current state of long- and hard-fought battles by gay, lesbian, transgendered, and other queer people to sexual equality.

Freedom to Marry

https://goo.gl/g3oNjG

Apply Your Knowledge

6.13 Identify a cultural phenomenon (music, sports, language, sexual identity, or something else) that is popular in your community and provide a sketch of one or two of its dimensions (e.g., size, duration, or extent).

6.14 How does the local cultural phenomenon connect to its practice or manifestation in communities around the United States and/or Canada who participate in or appreciate it?

Demography and Urbanization

Currently whites (of various European ancestries) constitute nearly three-quarters of the total U.S. population; African Americans, about 13%; Asians and Pacific Islanders, about 4%; and Native Americans, about 1%. Hispanics, who may also be counted among other groups, make up about 14% of the total U.S. population.

Canada consists of primarily two founding ethnic communities, the British and the French. Descendants of these two groups make up the largest piece of the demographic pie at over 50%. Other white European Canadians constitute about 15% of the population with First Nations people contributing 2%; Asians, Africans, and Middle Easterners contributing 6%; and people of mixed background at 26%. Because the United States and Canada ask different questions about ethnicity on their national censuses, it is difficult to compare these numbers. Immigration history provides a more comprehensible picture.

Immigration The United States and Canada have varied and extensive immigration histories. U.S. immigration is frequently discussed in terms of waves, because the numbers and types of immigrants ebbed and flowed over time (**FIGURE 6.26**).

- **First Wave, 1820–1870** At the beginning of the 19th century, the population of the new nation was largely dominated by English colonists and African slaves and small numbers of Irish, Dutch, French, and Germans. The first wave (1820–1870), in which overall immigration rose sharply to 2.8 million individuals, involved large numbers of Irish and German at which time the number of English immigrants declined. The newly arriving Irish were mostly peasants. The Germans who came were mostly skilled craft workers.

- **Second Wave, 1870–1920** In the second wave (1870–1920), in addition to the continuing stream of "old" immigrants from northern and Western Europe, "new" immigrants from southern and Eastern Europe joined the flow. Between the 1870s and the 1880s, the absolute number of immigrants rose dramatically. Widespread economic depression in Europe and North America in the 1890s led to a decline in absolute numbers of immigrants. The numbers rose again in the first decade of the 20th century to an all-time high. This wave included peasants, skilled workers, and successful merchants.

- **Third Wave, 1970–Present** The third immigration wave (1970–present) is substantially different from the other two in that large numbers of recent migrants have been from Asia and Latin America. Although Asians have been part of U.S. immigration history since the mid-19th century, Latin American migration to the United States is a 20th-century phenomenon. By 1990, Mexico had become (and still is) the largest source of immigrants to the United States. Most recent Mexican immigration into the United States has been to California, Texas, and Arizona though southeastern states have also experienced significant Mexican immigration.

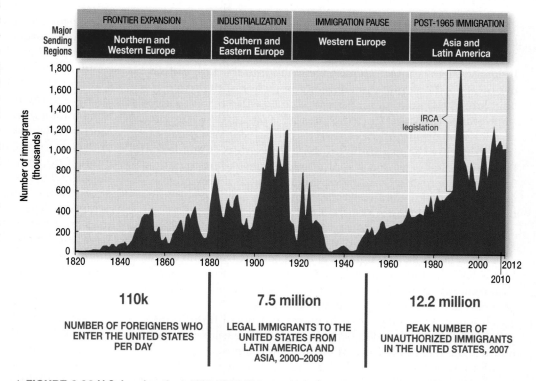

110k

NUMBER OF FOREIGNERS WHO ENTER THE UNITED STATES PER DAY

7.5 million

LEGAL IMMIGRANTS TO THE UNITED STATES FROM LATIN AMERICA AND ASIA, 2000–2009

12.2 million

PEAK NUMBER OF UNAUTHORIZED IMMIGRANTS IN THE UNITED STATES, 2007

▲ **FIGURE 6.26 U.S. Immigration, 1820–2010** This graph shows very dramatic peaks and troughs over the nearly 200 year period. IRCA is the Immigration Reform and Control Act passed in 1986. Among other things, it legalized undocumented aliens who had been in the United States since 1982. **What is an important driver of immigration, whether authorized or unauthorized?**

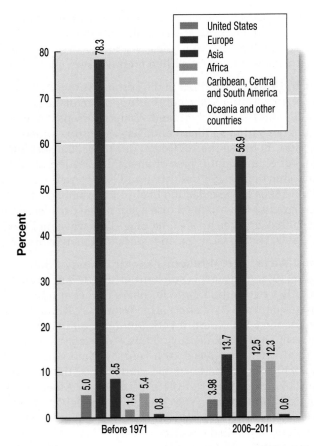

▲ **FIGURE 6.27 Place of Birth of Immigrants to Canada, before 1971 and 2006–2011** While Europeans dominated Canada's immigrants before 1971, Asians are now dominant. In 2013, 20% of Canada's population was foreign born.

Immigration to Canada The immigration history of Canada is very similar to the United States, with one significant difference. French settlers dominated in Canada well into the 18th century, but by about 1750, other immigrant groups from Britain and Ireland joined the immigration stream. Canada also received some immigrants from the United States at least until 1810, when restrictive British policies made it difficult for Americans to immigrate. By the beginning of the 20th century, Canada and the United States had very similar experiences of immigration, including the restrictions that curtailed inflows of new migrants until the late 20th century. Today, Canada is a primary destination for Asian migrants (**FIGURE 6.27**).

Assimilation and Anti-Immigrant Prejudice Although
over time the "new" immigrants of the second wave in the United States have largely been assimilated into mainstream U.S. life, experiencing increasing prosperity and social mobility in later generations, third-wave immigrants continue to confront racism and bigotry. They have been frequent victims of **hate crimes**, acts of violence committed because of prejudice against women, ethnic, racial, and religious minorities, and homosexuals.

Many Central American migrants who come to the United States without visas have been targeted by anti-immigrant legislation such as Senate Bill 1070 in Arizona, which allows for local law enforcement to investigate a person's immigration status without cause. In summer 2014, tens of thousands of young Central Americans, unaccompanied by adults and fleeing economic austerity and political violence in their home countries (Guatemala, El Salvador, and Honduras), arrived along the U.S./Mexico border and were either held in crowded detention centers or allowed to live with relatives already in the United States as lengthy deportation processes took place.

Internal Migration The movement of populations within a national territory—known as **internal migration**—has also played a role in both countries. In the United States, three overlapping waves of internal migration over the past two centuries have altered the population geography of the country. Canada experienced waves of internal migration that differed significantly from the American experience.

- **First Wave: U.S. Westward Expansion** The first wave of internal migration began in the mid-19th century and increased steadily through the 20th century. This wave included both rural-to-urban migration associated with industrialization. Also included was the movement of people from the settled eastern seaboard and Europe through "westward expansion" beginning in the 19th century, when an official settlement policy was created (**FIGURE 6.28**). Despite the emphasis on western expansion and rural settlement, between 1860 and 1920 the United States was also transformed from a rural to an urban society through industrialization drawing agricultural workers and foreign immigrants into the manufacturing cities.

Settlement of Canada The internal migration process was similar for Canada, where urbanization was also the dominant settlement process. Large-scale internal migration westward occurred in Canada, mostly during the early decades of the 20th century. But there were important differences in Canada's immigration history. During the colonial period, most of the immigrants arriving in Canada were British and French, whereas the immigrant stream to the United States was more broadly based. Moreover, whereas immigrants from all parts of Europe migrated to Canada during the late 19th and early 20th centuries, the total number of people immigrating to Canada was smaller than for the United States.

There has long been migration interaction between the two countries, with many Americans immigrating to Canada during the American Revolution. In the early to mid-20th century, Canadian immigration to the United States exceeded American immigration to Canada, as Canadians sought to secure a higher standard of living south of the international border.

- **Second Wave: U.S. Urban Migration of African Americans** The second wave of internal migration, which began early in the 20th century and continued through the 1950s, was the massive movement of African Americans out of the rural South, where they had made livelihoods in agriculture, to cities in the South, North, and West. Although African Americans already

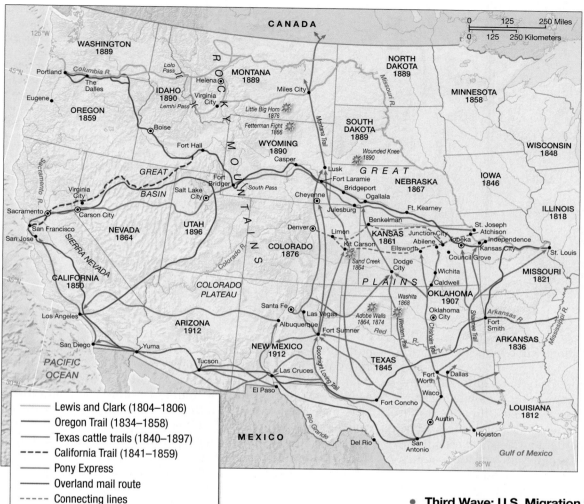

◀ **FIGURE 6.28** **U.S. Frontier Trails, 1804–1897** The western frontier is part of the history and mythology of the United States (and Canada). The various paths individuals traveled suggest the many ways they experienced the passage. The topography of the land west of the Mississippi was intimidating, with vast grasslands and plains, searing deserts, and unimaginable insects and animals.

formed considerable populations in cities such as Chicago and New York, large numbers moved out of the rural areas when agricultural mechanization reduced the number of jobs available. At the same time, pull factors attracted African Americans to the large cities.

In the early 1940s, for example, new jobs in the defense-oriented manufacturing sector became available when many urban white workers left their jobs and entered the military. This second wave of migration can be seen as part of a wider pattern of rural-to-urban migration among agricultural workers as industrialization spread regionally. After the war, a more important catalyst drove this migration: An increasing emphasis on high levels of mass consumption reoriented industry toward production of consumer goods and in turn stimulated large increases in the demand for unskilled and semiskilled labor. The impact on the geography of racial distribution in the United States was dramatic.

- **Third Wave: U.S. Migration to Sunbelt** The third wave of internal migration began shortly after World War II ended in 1945 and continued into the 1990s. Between the end of the war and the early 1980s—and directly related to the impact of governmental defense policies and activities on the country's politics and economy— the Sunbelt, including most of the states of the U.S. South and Southwest, experienced a 97.9% increase in population. During the same period, the Midwest and Northeast, the Rustbelt, together grew by only 33.3%.

Demographic Change By the middle of the 21st century, the population of the United States will increase by about 100 million people for a total of well over 400 million. The population of individuals aged 65 years and older will increase at the same time that the number of working young people will rise and keep doing so growing by 42% by 2050. The large numbers of young immigrants who are part of the United States today have contributed to a fertility rate of 2.1 with a new baby boom expected in the 2020s.

The composition of the U.S. population will also change significantly by the mid-21st century. Today, the United States is 63% white; by mid century it is expected to be 44% white, 29% Hispanic, and 15% Asian. The African-American population is expected to remain at around 12–13%. One implication of these compositional shifts is the emergence of majority–minority

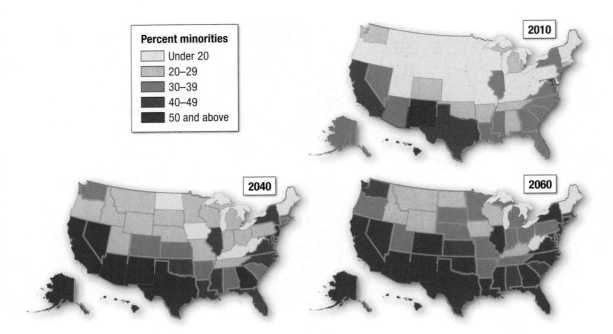

Percent minorities
- Under 20
- 20–29
- 30–39
- 40–49
- 50 and above

◀ **FIGURE 6.29 Minority Eligible Voters Change Political Maps** These maps show the changing demographics of the United States in three snapshots. Majority–minority states will change the composition of the U.S. electorate by mid century and thus the political landscape of the country. **Given the projected age and ethnic composition of future American voters, what kinds of issues might be important to them that are not currently important?**

states—where once minority populations will become the majority (**FIGURE 6.29**).

Canada's demographic future is likely to be quite different from that of the United States. Its population is slowly decreasing. It is expected to continue to experience a low death rate such that the number of births and the number of deaths will be equal even before mid century. In order for Canada to establish a young population able to support its older citizens, the country will need more people and will have to rely on immigration. Within the next 35 years, Canada's immigration rate is expected to increase significantly.

Population experts in Canada believe immigrant populations are likely to settle in the less-populated provinces because cities like Toronto, Montreal, and Vancouver, will be unable to handle increased population pressures. In the less-settled provinces like Saskatchewan, new immigrants are expected to stimulate new businesses, bring about the building of new schools and other public infrastructure, leading to the development of prosperous new cities in these areas.

American Electorate

https://goo.gl/w0xNIS

Apply Your Knowledge

6.15 Identify three negative impacts of demographic growth on the environment and economy.

6.16 Compare those negative impacts to three positive effects of a growing population. Which is a stronger argument for you?

Urbanization, Industrialization, and New Growth Long
before explorers and colonists arrived, indigenous people

built cities in the regions that would become the United States and Canada. The urbanization process began anew during the colonial period as Europeans established settlements as central places to organize commerce, defense, communication, and, later, administration and worship. In Florida and the southwestern United States the Spanish founded cities which became symbols of their political and military authority (**FIGURE 6.30**). French explorers came not to settle but to reap commercial rewards, and they established urban centers to facilitate the exchange of goods. The Dutch also established urban

▲ **FIGURE 6.30 Fort Matanzas in St. Augustine, Florida** Military defense sites were an important part of the early settlement of North America by Europeans, as they fought each other for control of territory. *Matanzas* means "slaughter" in Spanish and refers to the massacre of nearly 250 French Huguenots at the hands of the Spanish at this site, 175 years before the fort was constructed.

The Other Baltimore

Devin Allen is a self-taught photographer whose amateur photograph of the Baltimore community's response to the 2015 homicide of Freddie Grey by city police officers landed on the cover of *Time Magazine*. The image Allen took is of a young African-American man running ahead of a seemingly endless phalanx of police officers in riot gear. The 20-something young man is wearing a baseball cap backward and a bandana covering his nose and mouth, probably against mace and tear gas. The title of the issue reads: "America, 1968 (with the date crossed out and 2015 written over it), What has changed and what hasn't." A statement, not a question, the photograph and its title conjure the continuation of racist social relations where progress since the urban unrest of the 1960s has been slow, especially for young African-American men.

Allen grew up in West Baltimore, Freddie Grey's neighborhood (**FIGURE 6.4.1**). It's a predominantly African-American, low-income neighborhood made famous by the acclaimed HBO television series, *The Wire*. Allen lost two of his childhood friends to gun violence there. He believes he was saved from a similar fate by his art. He has used his camera to tell the story of street life in his beloved city. His goal then and with the protests was to reveal a Baltimore that few people on the outside ever see, a Baltimore where its residents live meaningful lives despite the narratives of the place—where violence is rife and social life is dysfunctional—that circulate through the media.

Devin Allen's photography humanizes the world in which Freddie Grey lived and died. It makes accessible, by extension, the world of so many of the young African-American men who die violently, whether by police brutality or drug deals gone wrong, domestic disagreements, or unfortunate accidents. But his own biography—inescapably intertwined with the larger narrative of Freddie Grey and other young African-American men and women who have died violently—also offers not only a more balanced view of black life in the United States, it also offers inspiration.

1. Go to Devin Allen's Instagram feed and do a survey of the images he has collected there. What is the prominent social message his art conveys to you?

2. Read about the attitudes of adolescents and their perceptions of the relationship between health and physical environment from a 2014 study. How do Baltimore teenagers' attitudes compare to those in Ibadan, Nigeria; Johannesburg, South Africa; New Delhi, India; and Shanghai, China?

"Devin Allen aims to reveal a Baltimore that few people on the outside ever see, a Baltimore where its residents live meaningful lives despite the narratives of the place—where violence is rife and social life is dysfunctional—that circulate through the media."

▶ FIGURE 6.4.1 **Devin Allen** Have a look at Allen's photographs on his instagram page and answer the question: **how does the Baltimore that an insider like Allen pictures compare to the one you've come to learn about, as an outsider, on the news?**

settlements as trading centers for furs and slaves as well as other goods.

With few exceptions, the British played the largest role in shaping U.S. and Canadian urbanization and urban life. Sustained by trade as well as being administrative centers, Atlantic coastal cities such as Saint John, Halifax, Boston, and Charleston were also ports and key nodes in a globally expanding trade system. These urban centers enabled the transfer of resources, goods, and people, not only from the interior hinterland into the cities but also outward to Europe. At the same time, these cities received goods from England and France for consumer markets. Colonists saw their burgeoning cities not only as commercial centers but also as hearths of civilization, where European cultural practices confronted those of the new nation, creating in the process uniquely North American urban places.

Recall from earlier in the chapter, how this dynamic shifted in the 1970s between the North and South in the United States. Cities that had once been sleepy towns began to grow and by the 1980s and 1990s joined the list of some of the largest in the country. In Canada, this period saw the movement of the economy away from overdependence on staples and toward more diversity, including finance, tourism, and high tech. Places that had once been towns in Canada also began to grow and become important cities to the national economy.

▲ **FIGURE 6.31 Okanagan Valley, British Columbia** The Okanagan Valley has become an attractive site for retirement living for Canadians because of its sunny climate and the wide range of available outdoor activities, including skiing, hiking, and fishing. Vineyards and orchards line many of the hills and the area is also a cultural attraction because of its First Nations history.

searching out the good life on the far fringes of the metropolitan core in small towns (**FIGURE 6.31**).

Apply Your Knowledge

6.17 Using the Internet, determine the top 10 most populous Canadian and U.S. cities in each of the following years: 1900, 1930, 1960, and 2000.

6.18 How has the list changed over time? Provide two reasons for any trends you observe.

Urban to Suburban Migration The first evidence of **suburbanization**—the growth of population along the fringes of large metropolitan areas—can be traced back to the late 18th and early 19th centuries. Then, real estate developers looked beyond the city for investment opportunities and wealthy city dwellers began seeking more scenic and less polluted residential locations. Later, residents fled to the suburbs to get away from the new immigrants and their increasing hold over urban machine politics.

Suburbanization was rapidly accelerated in the 1960s as white Americans abandoned racially mixed urban neighborhoods for more racially and class homogenous suburbs. Detroit is an example of white flight and has been much in the news since it filed for bankruptcy in 2013. A largely African-American city, the median household income is around $25,000; median household income in Greater Detroit, which includes the predominantly white suburbs, is between $50,000 and $150,000. As whites left Detroit starting in the 1960s, they took precious tax dollars with them. It is important to appreciate as well that North Americans have most recently chosen to move from cities for many reasons besides race, including retirees who are

Future Geographies

The United States continues to be the most powerful nation on Earth, and Canada is one of its staunchest allies and economic partners. Both countries possess broad resource bases; a large, well-trained, and very sophisticated workforce; and a high level of technological sophistication. The United States has a domestic market that has greater purchasing power than any other single country as well as the most powerful and technologically sophisticated military apparatus.

U.S. Dominance and Its Challenge

The United States' economic dominance is no longer unquestioned. The 2008 collapse of the U.S. financial and housing industry left the country reeling. More important, the globalization of the world economy and years of war have constrained the ability of the United States to translate its economic resources into the firm control of international financial markets that it once enjoyed. Canada, though never a world political power, still maintains substantial economic might. But its past prominence may be overtaken by up-and-coming economic powers like Brazil, Russia, India, and China.

Although the United States still maintains global economic and military dominance, it is currently being challenged by China. The future may witness China surpassing the United States in terms of its economic might. For example, as of 2014, the size of the United States economy (not adjusted for cost of living) is $17.4 trillion; China's, the next largest economy, is $10.4 trillion. But when adjusted for their low cost of living, China's GDP is worth

$17.6 trillion and the United States' $17.4 trillion. It could be argued that the future is already here. It is too soon to tell whether the signs of economic trouble in China beginning in summer of 2015 will have significant effects on the continued robustness of the Chinese economy.

Security, Terrorism, and War

The focus on security by the U.S. federal government is reshaping governing structures as well as economic processes by creating new layers of rules, regulations, and policies. The practices of daily life around increased personal security measures and human rights are also changing as drones become more frequently used domestically. Canada, because of its proximity to the United States as well its dependence on the American economy for its own well-being, is likely to continue to be drawn into U.S. security concerns.

The al-Qaeda attacks of 2001 and the massacre in Paris in 2015 as well as the emergence of ISIS in Syria and Iraq have contributed to a new focus for U.S. geopolitical strategy that is likely to continue into the near future. Also, the invasion of Iraq and Afghanistan, along with U.S. reluctance to lead the way through bold commitments to high-profile international environmental agreements, and its lukewarm cooperation with the UN, has caused concern about the future of the political, cultural, and moral leadership of the United States in global affairs (**FIGURE 6.32**).

U.S. WAR COSTS: *the total is rising*

MILITARY OPERATIONS → **$600 billion**
- Equipment: helicopters, tanks, vehicles
- Combat pay and training costs
- Fuel
- Facilities: bases and embassies

ADDITIONAL MILITARY EXPENDITURES → **$110 billion**
- Costs indirectly attributed to the war

INFLATION
- Adjust for years of inflation

$520 billion ← **OPERATIONAL AND PEACEKEEPING COSTS**
- Costs before and after troop withdrawal

$590 billion ← **VETERAN HEALTH CARE**
- Future medical and disability costs

$280 billion ← **RESTORING THE MILITARY**
- Restore prewar strength, replace worn equipment, matériel, and vehicles

THREE TRILLION DOLLARS ($3,000,000,000,000)

SOCIAL SECURITY COSTS → **$38 billion**
- Payments for veterans who cannot work

INTEREST ON DEBT → **$615 billion**
- The war is funded through borrowed money, increasing the national debt

LOST ECONOMIC OUTPUT → **$370 billion**
- Death and disability of soldiers results in lost potential economic output

$1.9 trillion ← **MACROECONOMIC COSTS**
- Impact of funds diverted from education, infrastructure, scientific research
- High oil prices during the war

In 2003, Secretary of Defense Rumsfeld estimated that the cost of a war with Iraq would be $60 billion. By 2008, the actual cost was more than $600 billion. By the time the United States exits Iraq, it is estimated that the cost of the wars in Iraq and Afghanistan will reach $3 trillion. Some experts believe the total will be even higher, between $4 and $6 trillion.

▲ **FIGURE 6.32 The Cost of the War on Terror** This infographic shows that the cost of war including not only the expense of fighting it but also the expense of recovery from it.

Environmental Change

From observations drawn from the historical record as well as computer simulations, it is clear that the climates of the United States and Canada will continue to change and that this change will have significant effects. Models project warming of 4–9°F for the United States and Canada. While impacts will vary based on place, there is no question that animal and plant life will be affected, precipitation will increase and decrease unevenly, and sea-surface temperatures in the Atlantic and Pacific Oceans will increase (**FIGURE 6.33**). The 2015 Climate Agreement worked out in Paris does provide some hope for a slowing down of global climate change. Among the many things discussed at the talks, both the United States and Canada agreed to a 1.5C max temp long term goal rather than a warmer 2C.

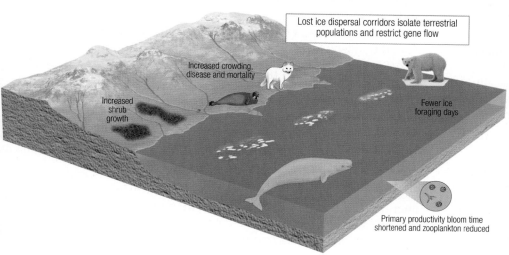

Lost ice dispersal corridors isolate terrestrial populations and restrict gene flow

Increased crowding, disease and mortality

Increased shrub growth

Fewer ice foraging days

Primary productivity bloom time shortened and zooplankton reduced

◀ **FIGURE 6.33 Arctic Ecosystem Change** As sea ice will be melting earlier each year, algae and phytoplankton populations will also peak earlier and decline sooner. The effect is that by the time predators come looking to feed on lower levels of the food chain, much of the food source will be gone. Sea ice loss will also change the physical landscape of the Arctic, limiting the movement of animals who use the ice to migrate for food.

Learning Outcomes Revisited

▶ *Describe* the distinctive landscapes of the United States and Canada with respect to landforms and climate.

The United States and Canada occupy an environmentally rich land mass. Mountain and coastal landscapes are highly scenic and their vast environmental diversity allows for highly productive agriculture. There is also a wealth of natural resources. The climates of the region encompass every possible variation from Arctic to tropical landscapes.

1. *Identify* two different physiographic regions each in the United States and Canada. Describe each of the regions and note the key differences.

▶ *Explain* how the United States and Canada are affecting global climate change.

The United States and Canada play a major role in global climate change. Both countries, but especially the United States, are responsible for producing some of the highest total and per capita greenhouse gas emissions in the world as a result of their production and consumption patterns. These emissions stem from the region's heavy reliance on fossil fuels.

2. *In what* ways are fossil fuels important to the economies of the United States and Canada and how might reliance on them be reduced? Choose one aspect of fossil fuel use in each country to address this question.

▶ *Understand* the rise of the United States and Canada as key forces in the global economy.

The United States' ascension to global dominance has been premised on the richness of its resource base, geopolitical prowess, and the complexity and diligence of its population. Canada has a similar history; however, it is the primary trading partner to the United States and at least some of its success has been due to that relationship.

3. *What* factors contributed to the rise of the United States and Canada becoming global economic forces?

▶ *Describe* how the two countries' dynamic economies contribute to and shape the global economy.

The United States and Canada possess natural resources that are critical to the economic growth of other countries around the globe including highly educated workforces. In addition, their wealth is routinely invested in other parts of the world. The United States' military strength extends to every continent except Antarctica.

4. *On which* measures is the dominance of the United States in the global economy established?

▶ *Summarize* the economic inequality that exists in both the United States and Canada.

In recent years, both superrich and extremely poor populations have become more unequal in the United States and Canada. This inequality is due in part to the recent global financial collapse. The crisis, in a sense, exacerbated a trend that was already in force, however, as the rich have gotten richer and the middle class and the poor have lost wealth.

5. *Why* is inequality of wealth distribution a problem for society?

▶ *Compare* and contrast the forms and practice of government of the United States and Canada.

The United States and Canada are established democracies modeled on European political traditions. Both are federal states in which power is allocated to local government units (provinces in Canada and states, counties, cities, and towns in the United States). The United States is a republic, whereas Canada is a constitutional monarchy.

6. *How* can Canada be both a democracy and a monarchy?

▶ *Summarize* the indigenous and immigrant histories of the United States and of Canada.

The indigenous and immigrant histories of the United States and Canada are strikingly similar since they were inhabited by indigenous groups and settled by the same groups of Europeans at the same time. Settlement of both places largely proceeded from east to west until the 20th century, when large numbers of immigrants arriving from Asia and other parts of the Pacific Rim began settling from west to east.

7. *Why* is the birth of Nunavut as new Canadian province so important for indigenous rights?

▶ *Describe* the fundamentals of the geography of religion and language across the United States and Canada.

There is no official language in the United States though the most commonly used is English; in Canada, it is English and French. And yet, with so many ethnic groups occupying these two countries over their relatively short histories, it is not surprising to find multiple languages spoken and multiple religions practiced on the streets of Toronto and New York and ethnic neighborhoods in large cities as well as in small towns and even in rural areas.

8. *Why* does Canada have two official languages, French and English?

▶ *Compare* and contrast the urbanization processes that have shaped the United States and Canada over the last 100 years.

In the United States and Canada, urbanization first occurred in the Northeast and Midwest of the country, as the processing of agricultural goods and manufacturing drew immigrants in search of jobs. The most recent changes to cities in both countries have come from a period of deindustrialization of old manufacturing areas and urban population growth in the Sunbelt in the United States and in the west of Canada.

9. *Why* are cities important to economic growth?

Key Terms

acid rain (p. *231*)
al-Qaeda (p. *245*)
American Revolution (p. *236*)
Americanization (p. *247*)
assimilation (p. *246*)
Bible Belt (p. *246*)
cap-and-trade programs (p. *231*)
carbon footprint (p. *225*)
Confederate States of America (p. *237*)
creative destruction (p. *239*)
deindustrialization (p. *239*)

democracy (p. *243*)
Europeanization (p. *235*)
federal system (p. *243*)
French and Indian War (p. *236*)
genetically modified organisms (GMOs) (p. *232*)
government (p. *243*)
greenhouse gases (GHGs) (p. *225*)
hate crime (p. *250*)
income inequality (p. *240*)
indentured servant (p. *236*)

information technology (IT) (p. *239*)
intermontane (p. *227*)
internal migration (p. *250*)
local food (p. *232*)
Manufacturing Belt (p. *238*)
multiculturalism (p. *246*)
North America Act of 1867 (p. *237*)
Occupy Wall Street (p. *241*)
organic farming (p. *232*)
reservation/reserve (p. *230*)

Rustbelt (p. *238*)
sexuality (p. *248*)
slow food (p. *232*)
staples economy (p. *238*)
state (p. *243*)
suburbanization (p. *254*)
Sunbelt (p. *238*)
tar sands (p. *246*)
U.S. Civil War (p. *236*)
War on Terror (p. *245*)
wealth inequality (p. *240*)

Thinking Geographically

1. Find an image that shows an aspect of the physical geography of either Canada or the United States (a mountain range, a river system, the prairie, etc.) and discuss the way this particular feature is relevant to agriculture, climate, adaptation, ecology, or resources.

2. The economies of Canada and the United States are critically important to the global economy. Referring to a specific commodity, mineral, or other natural resource, describe how the production of that item in either Canada or the United States shapes another country through its consumption of that particular commodity or resource.

3. Discuss the three main waves of immigration into the United States. When did they occur, and where did the immigrants come from? Discuss internal migration within Canada. How is it the same or different from the United States?

4. Choose a city in both the United States and Canada and describe how and why its fortunes have changed over the course of 110 years—from 1900 to 2010.

5. The future of U.S. global dominance is, of course, unknown. Construct a cogent argument for its continuation and one for its decline and then indicate which future you believe is the most likely to take place and why.

DATA Analysis

Charting voter activity and voting patterns has become a critical component of the electoral process in the United States and other developed countries. For national, state, and local elections, politicians, lobbyists, and pollsters lean on the demographic information to predict voter choices, turnout, and election results. Using the U.S. Census Bureau's website, analyze the following data on "The American Electorate" that provides a snapshot of U.S. voters on the eve of the 2016 presidential election.

U.S. Census Voting Data

https://goo.gl/zwMtC5

1. Go to the U.S. Census Bureau website and download the 2015 report, "Who Votes? Congressional Elections and the American Electorate: 1978–2014."

2. Define the following terms: voting rates, voting population, voting-age citizens, voting-age population, registered population, and nonrespondents. Why are these categories important to elections?

3. Summarize the trends of voter turnout from 1978–2014 on both presidential and congressional elections on the basis of race, ethnicity, and age (Figures 2–4).

4. Using Figures 5 and 6 on page 8, what do you observe about race and age in voter turnout? What do you think influences this data?

5. Reading pages 12–13 on methods of voting, what are the types of alternative voting? What groups use these methods increasingly? Why? How are they shaping elections?

MasteringGeography™

Looking for additional review and test prep materials? Visit the Study Area in MasteringGeography™ to enhance your geographic literacy, spatial reasoning skills, and understanding of this chapter's content by accessing a variety of resources, including MapMaster interactive maps, videos, *In the News* RSS feeds, flashcards, web links, self-study quizzes, and an eText version of *World Regions in Global Context*.

Cubans walk along Callejón de Hamel in Havana with mural and sculpture celebrating African influences on Cuban culture and religion.

Latin America and the Caribbean

Cuba, a large island about 100 miles south of Florida, has been a touchstone for relations between the United States and the Latin American and Caribbean region for more than a century. Claimed by Christopher Columbus for Spain in 1492, the Indigenous Taino were wiped out by European diseases, and the island became a center for sugar plantations worked by West African slaves. Cuba became independent in 1902 following local rebellions and the Spanish–American war, but the United States retained influence over Cuban policy and a U.S. naval base at Guantánamo Bay—now known for the U.S. prison for suspected foreign terrorists. In 1959, discontent with elite and U.S. control of the economy and politics prompted a left-wing revolution led by Fidel Castro. The communist government aligned with the then Soviet Union, nationalized much of the economy, and redistributed land to peasant farmers. Thousands of Cubans fled to the United States, especially to Florida. U.S. reaction included the failed Bay of Pigs invasion in 1961 and an ongoing trade and travel embargo, which made it difficult for Americans to visit or trade with Cuba and made the Cuban economy vulnerable, especially after the collapse of the USSR in 1991. Cuba maintained a good health and education system and embraced sustainable agriculture— Tourists from Canada and Europe drive a vibrant tourist sector. Cuba has a dynamic culture which combines African and Latin traditions into contemporary art, dance, food and a rich musical sector.

U.S.-Cuba relations have improved since 2009 when the United States eased travel restrictions and reduced economic sanctions against Cuba. In December 2014, U.S. President Barack Obama and Cuban President Raul Castro (brother of Fidel Castro) announced a plan to normalize relations between the countries, lifting travel and trade restrictions and reopening embassies. The next decade may bring many challenges and opportunities to Cuba.

Learning Outcomes

- ▶ *Explain* the physical origins and social impacts of natural and human-caused disasters including climate change.

- ▶ *Describe* the ways in which the Maya, Aztecs, and Incas adapted to environmental challenges and the relevance of their experience today.

- ▶ *Identify* sustainability challenges and solutions relating to forests and air in the region.

- ▶ *Summarize* the common impacts and legacies of colonialism and describe the history of U.S. intervention, authoritarianism, revolutions and democratization in the region.

- ▶ *Compare* the economic policies and impacts of import substitution, structural adjustment programs, and free trade and the successes and failures of economic development.

- ▶ *Describe* the history and current distribution of Indigenous peoples, languages, and cultures and their struggles for representation.

- ▶ *Understand* the geography and impacts of the drug trade and international tourism.

- ▶ *Identify* the main migration streams and their causes and list the factors promoting urbanization and associated social and environmental problems.

Latin America is the southern part of the large landmass of the Americas that lies between the Pacific and Atlantic Oceans (**FIGURE 7.1**). The Americas are often divided into the two continents of North America (Canada, the United States, and Mexico) and South America (all the countries south of Mexico). Latin America includes all the countries south of the United States, from Mexico to the southern tip of South America in Chile and Argentina. Latin America is a world region that covers more than 20 million square kilometers (7.7 million square miles). "Latin" America, defined by the shared Latin-based languages of Spanish and Portuguese, may not be the most accurate name for this region, which includes countries where Indigenous languages and English are important. One alternate name for Latin America is Central and South America, where Central America includes Mexico, Belize, Guatemala, Nicaragua, Honduras, Costa Rica, and Panama. This chapter also includes the Caribbean region—the arc of islands that sweeps from the U.S. state of Florida to Colombia across the Caribbean Sea. We include the Caribbean with Latin America because they share histories of colonialism, common cultures, and contemporary problems.

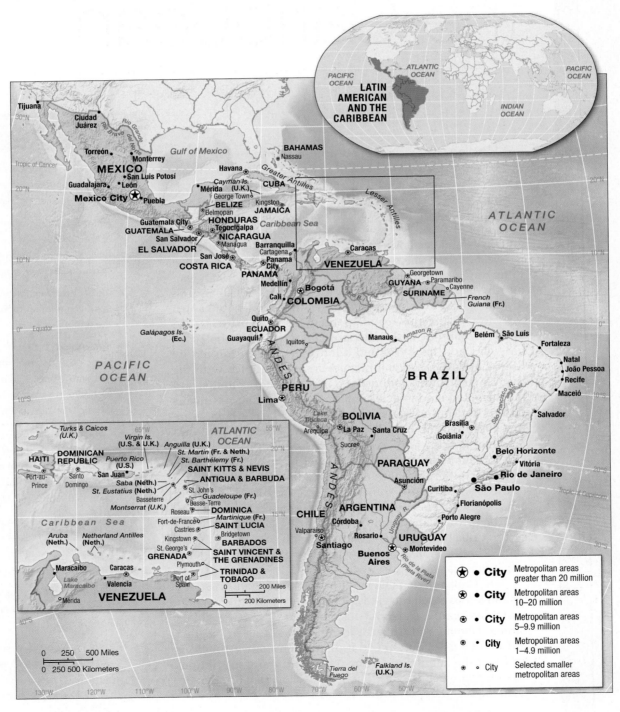

▲ **FIGURE 7.1 Latin America and the Caribbean: Countries, major cities, and physical features**

Environment, Society, and Sustainability

Humans have transformed Latin American and Caribbean landscapes for centuries. Many of the forests were cleared, grasslands burned or grazed, mountains carved into terraces, and the waters of the deserts stored or diverted over the last 1,000 years or more (**FIGURE 7.2**). A major shift occurred, however, when European colonists arrived 500 years ago. Contemporary human settlements and agriculture are vulnerable to natural disasters, and the human geography of Latin America and the Caribbean continues to be influenced by environmental conditions. An understanding of the physical geography, and the ways in which people have adapted to and modified their environment, is integral to appreciating the historical and contemporary human geography of the region and the challenges to its sustainable development.

Climate and Climate Change

The global atmospheric circulation determines the overall climates of Latin America and the Caribbean. The region's climates range from extremely dry deserts such as the Atacama of Chile, to the tropical rainforests of the Amazon, to the cooler steppe climates of the highland Andes, and the midlatitude marine west coast climate of Chile (**FIGURE 7.3**). The Amazon basin is located at the intertropical convergence zone (ITCZ), where moist air heats in intense sun, rises, and cools, forming

▲ **FIGURE 7.2 Andean Terraces** The Incas constructed terraces, not only to reduce soil erosion and provide a flat area for planting, but also to decrease frost risks by breaking up downhill flows of cold air and allow for irrigation canals to flow across the slopes in efficient ways. The deep Colca valley in Peru has thousands of terraces constructed by the Inca on steep hillsides; many are still used by farmers to grow potatoes, quinoa, and barley.

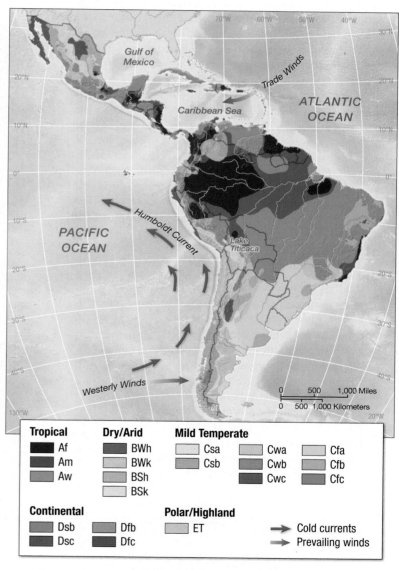

▲ **FIGURE 7.3 Climates of Latin America and the Caribbean** Latin America's climate is influenced by major wind and pressure belts and the configuration of land and oceans. Climates include wet tropical (A), dry deserts and savannas (B), and more temperate highlands and cooler mild zones to the south (C). For more detailed explanation of the legend see Chapter 1 Figure 1.9. **Which countries are found in the wettest and driest climate zones?**

Air Pollution in Mexico City

Air pollution levels have dropped in Mexico City, called the most polluted city in the world in the 1980s. The metropolitan area has a population of more than 20 million people living in a high altitude basin, driving an estimated 4 million cars. Automobiles, trucks, and buses, many with inadequate emission controls, are responsible for about 75% of the air pollution. Dust, fires, industrial plants, and energy production create the remainder. Air pollutants include carbon monoxide, sulfur dioxide, and ozone, that are risky to health, especially to children and people with asthma. Polluted air gets trapped in this basin by the surrounding mountains and by inversions (where warm air traps cold air near the ground). Mexico City is at more than 2,000 meters (6,000 feet), and has a dry climate for much of the year, with rains only washing out the buildup of pollutants in the rainy season. At this altitude, fuel burns less efficiently and humans must breathe more air because of the lower oxygen levels.

Citizen protests and government regulation have improved air quality in Mexico City. Levels of unhealthy carbon monoxide and ozone have fallen considerably since 1986 (**FIGURE 7.1.1a** and **b**). Industry has been closed down or moved out of the basin. Public transport has been expanded and automobiles now have to pass emission inspections. The "Day Without a Car" program requires people not to drive one weekend each month, and one weekday each week, and there are now eco-bikes for rent in many parts of the city. But traffic is still congested, and citizens are calling for even better public transport and low emission vehicles.

1. Use the real time map of air quality in Latin America at http://aqicn.org/map/latinamerica/ to identify cities that have high pollution levels and unhealthy air. Click on a city to see which air pollutants (particulate matter [PM], ozone [O$_3$], nitrous oxide [NO$_2$], sulfur dioxide [SO$_2$]) and carbon monoxide (CO) are in the unhealthy (red) zone. What time of day is the pollution the most serious and when is it the least? How might this relate to people's daily schedules and/or the weather?

Air Pollution in Latin America

https://goo.gl/xEktLS

> "Citizen protests and government regulation have improved air quality in Mexico City."

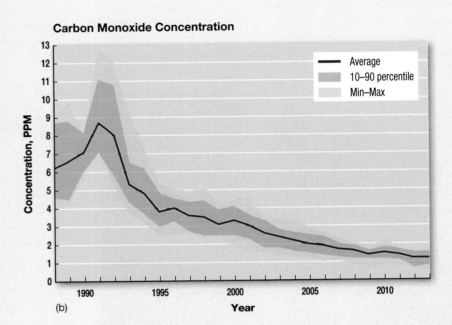

(b)

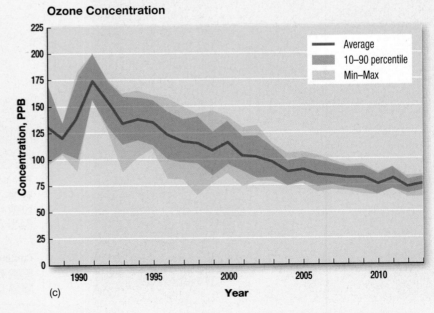

(c)

(a)

▲ **FIGURE 7.1.1** (a) In the mid-1980s Mexico City had some of the worst air pollution in the world with millions of cars and factories emitting pollutants that became trapped by inversions in the high elevation basin of Mexico. Since 1986, air pollutants such as a) carbon monoxide and b) ozone have declined due to aggressive traffic policies and industrial relocation and regulation (c) ozone.

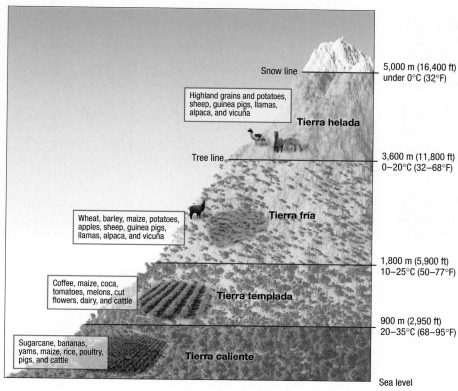

Snow line — 5,000 m (16,400 ft) under 0°C (32°F)

Highland grains and potatoes, sheep, guinea pigs, llamas, alpaca, and vicuña

Tierra helada

Tree line — 3,600 m (11,800 ft) 0–20°C (32–68°F)

Wheat, barley, maize, potatoes, apples, sheep, guinea pigs, llamas, alpaca, and vicuña

Tierra fría

1,800 m (5,900 ft) 10–25°C (50–77°F)

Coffee, maize, coca, tomatoes, melons, cut flowers, dairy, and cattle

Tierra templada

900 m (2,950 ft) 20–35°C (68–95°F)

Sugarcane, bananas, yams, maize, rice, poultry, pigs, and cattle

Tierra caliente

Sea level

▲ **FIGURE 7.4 Altitudinal Zonation** The climate and vegetation in mountainous regions such as the Andes creates vertical bands of ecosystems and provides a range of environments for agricultural production. At higher altitudes, potatoes grow and animals graze; at lower altitudes, grains such as wheat and corn grow; and vegetables and fruit are found at lower levels. The zones shown here are typical of mountain zones near the Equator, but cooler zones start lower down the mountains at latitudes further from the Equator. Global warming is shifting these zones upward in many parts of the tropics.

clouds and high rainfall (up to 2 meters or 80 inches). The dry climates of the region are found where air flowing toward the poles cools, and sinks over the tropics, in northern Mexico and Chile and in the dry **rain shadow** of the Andes, as well as mountains in Central America and the Caribbean. Rain shadows occur where winds blow from oceans over coastal mountains and produce orographic precipitation as the winds cool and release moisture as rain or snow. When the now dry air sinks on the other side of the mountain it becomes even drier, creating arid conditions in the lee of the mountains. In Chile, a cold offshore current exacerbates the extreme dryness of the Atacama Desert. The current cools the air above it, so winds pick up little moisture.

As part of global atmospheric circulation (Chapter 1), the region has zones of prevailing winds that affect climate patterns. Air flows from the tropics to the Equator into an east–west flow called the **trade winds**, and air flows poleward from the tropics into a west–east flow called the **westerlies**. The westerlies bring heavy rains to the west coast of Chile. The Atlantic trade winds bring rain to the Caribbean Islands and east coasts of Central America. As these easterly winds blow across the warm Caribbean they absorb moisture, especially during the fall, when the sea surface is warmest. When storms start to circulate, the warm sea fuels both the moisture and energy of the storms, producing devastating hurricanes.

At the margins of the trade winds and at the edges of the tropical rainfall zone, climates occur with a distinct rainy season. Latin America's extensive grasslands occur where seasonal shifts in wind and pressure belts bring moderate rains, and at the edge of rain shadows or in highland (steppe) climates.

Altitudinal zonation is a classification of environment and land use according to elevation based mainly on changes in climate and vegetation from lower (warmer) to higher (cooler) elevations. In Latin America, these zones are defined as: *tierra caliente* (warm land); *tierra templada* (moderate land); and higher *tierra fría* (cold land). The very high altitudes are called the *tierra helada* (icy land). Each of these zones is associated with characteristic vegetation types and different agricultural activities (**FIGURE 7.4**).

Climatic Hazards Latin America and the Caribbean experience hazardous climate extremes including severe storms, hurricanes, floods, and droughts (**FIGURES 7.5a** and **7.5b**). The last couple of decades have seen several major hurricanes devastate communities in the region. Hurricane Wilma, in 2005, was a category 5 storm (the highest intensity) and caused fatal mudslides in Haiti, flooding in Jamaica and Cuba, and millions of dollars damage to tourist resorts such as Cancun on

> "Climatic variability creates hazards, but it is people who experience *vulnerability* as a result of social inequality, environmental degradation, and unequal access to resources."

Mexico's Yucatán Peninsula. In 1998, Hurricane Mitch left nearly 10,000 people dead in Central America.

Climatic variability creates hazards, but it is people who experience vulnerability as a result of social inequality, environmental degradation, and unequal access to resources. In countries like Honduras and Haiti, which have high levels of poverty and where many people live and farm on steep slopes or in flood-prone areas, it is difficult for the poor to recover from disasters. These vulnerable populations lack the money and resources to rebuild their homes and livelihoods.

(a)   (b)

▲ **FIGURE 7.5 Climate Hazards** (a) Hurricane Sandy, which affected the New York area in 2012, also caused devastation in Jamaica (where it left 70% of the residents without electricity) and Haiti (where it killed 54 people). The Croix de Mission river in Port-au-Prince Haiti is shown flooding in October 2012. (b) The tracks of hurricanes between 1851 and 2010 showing the exposure of the Caribbean and Central America to these damaging storms. The points show the location of storms at 6-hour intervals and the colors show the intensity from Category 5 in red (over 157 mph), to 4 (Orange 130–156 mph), 3 (111–129 mph), 2 (96–110 mph), and 1 (74–95 mph).

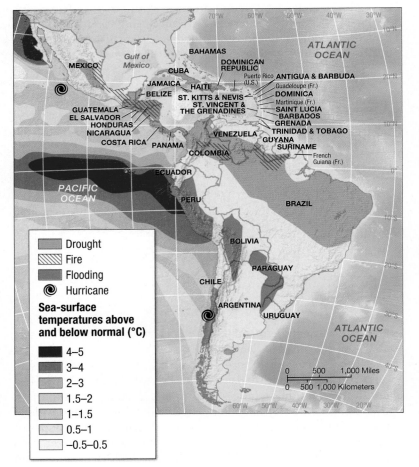

One of the most important causes of climate variability in the region is **El Niño**, the warm phase of the El Niño Southern Oscillation (ENSO). ENSO is a cycle of warm and cold sea surface temperatures that occurs every few years in the Pacific off the coast of South America, that affects climates worldwide, including those in Latin America. El Niño brings rain to the coasts of Peru and Ecuador, which are usually dry because of the cool offshore currents, and is also linked to droughts in northeast Brazil, floods in southern Brazil, and fewer Caribbean hurricanes (**FIGURE 7.6**). The colder phase, called **La Niña**, is linked to floods in northeast Brazil, drought in northern Mexico, and more intense Pacific hurricanes.

Climate Change Impacts Latin America and the Caribbean are experiencing the effects of global warming, especially in the Andes, where glaciers and snowfields are shrinking (**FIGURE 7.7**). Glacial retreat is a problem for communities that rely on ice and snowmelt for irrigation and drinking water, and for hydroelectric power where dry conditions

◄ **FIGURE 7.6 El Niño** When an El Niño occurs, the ocean warms off the coast of Peru and causes heavy rain and flooding along the usually dry Peruvian coast. The Pacific warming is linked to other changes in atmospheric circulation that produce drought in northeast Brazil, the Caribbean, and the Andean Altiplano; floods on the Paraná River; and droughts and fires in Mexico and Central America. Strong El Niños occurred in 1973, 1982, 1997, and 2016.

Average Tropical Andes

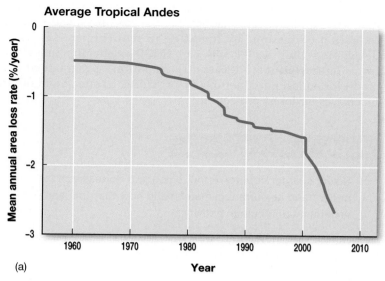

(a)

(b) 1978

(c) 2004

(d) 2011

▲ **FIGURE 7.7 Andean Glaciers at Risk** Climate changes caused by increasing levels of greenhouse gases from fossil fuels and deforestation have already produced higher average temperatures in mountain areas. (a) The graph shows the average annual percentage loss of area for most of the tropical Andean glaciers. (b) The Qori Kalis glacier, which descends from the Quelcaya ice field in Peru, has retreated since 1978, (c) 2004, and (d) 2011. Andean glaciers and snow cover are shrinking, and the rivers that depend on snow and ice melt are drying up, creating problems for irrigated agriculture, urban drinking water, and hydroelectric power generation. **What was the rate of loss of area of glaciers in 1960 compared to that in 2005?**

mean electricity cutoffs in major cities. Although there are many traditional adaptations to climate variability in Latin America—including complex irrigation systems developed by the Aztecs, Maya, and Incas (discussed later in the chapter)—global warming may test the limits of this region if it brings drier and warmer conditions.

The islands of the Caribbean are at risk from climate change and warmer temperatures because of the potential increase in the intensity of hurricanes and the impact of warmer oceans on coral reefs. Sea-level rise threatens coastal ecosystems and settlements in the Caribbean, including economically critical tourist beaches and resorts and water supplies at risk from saltwater intrusion.

Climate Change Causes and Responses The region includes several countries that significantly contribute to greenhouse gas emissions from fossil fuels and deforestation that cause climate change. Brazil and Mexico are among the top 20 emitters worldwide because of their fossil fuel based economies, population size, and land use. Per capita emissions are relatively high in Trinidad and Tobago, Belize, Suriname, Paraguay, and Grenada because of oil development, small populations, and deforestation.

In Latin America, efforts to prevent climate change include reducing greenhouse gas emissions from fossil-fuel burning and deforestation. Mexico and Brazil have promised to cut greenhouse gases up to 40% by switching energy sources to gas, biofuels, and renewables, through energy efficiency, and

by protecting forests (see "Visualizing Geography: Amazonian Deforestation," pp. 268–269). Costa Rica has a goal of becoming carbon neutral (net zero emissions) through reducing use of fossil fuels and planting forests.

Latin American countries also receive financial assistance for reducing emissions through the international **carbon markets**.

(a)

(b)

(c)

▲ **FIGURE 7.8 Carbon Forestry** Programs for Reduced Emissions from Deforestation and Forest Degradation (REDD+) are part of the response to climate change in Latin America and the Caribbean. (a) Programs include protecting existing forests, using others sustainably, reforesting degraded areas, and maintaining forest wetlands and peatlands. (b) The Scolel-te carbon forestry project in Chiapas is used to offset emissions from Formula-1 racing. (c) Some groups are opposed to REDD because they see it as a threat to their traditional rights, taking their resources in a new form of colonialism that turns forests into commodities for the international global carbon market.

In Europe, countries and companies are required to reduce emissions and can earn "carbon credits" toward meeting their reduction obligations by investing in reducing emissions in developing countries (**FIGURE 7.8**). **REDD**—Reducing Emissions from Deforestation and Forest Degradation—programs allow credit for emission reductions by providing financial and other incentives for forest protection.

Apply Your Knowledge

7.1 Select a larger country in the region and do research to understand its current climate(s) and how they influence vegetation and human activity.

7.2 Use the Internet to research and summarize how climate change may impact a country in Central America or the Caribbean.

Geological Resources, Risk, and Water

The varied physical landscape of Latin America includes striking mountain ranges, high plateaus, and enormous river networks (**FIGURE 7.9**). The region's largest-scale physical features are the Andes Mountains and the Amazon Basin and River. The Andes are an 8,000-kilometer-long (5,000-mile-long) chain of high-altitude peaks and valleys that for the most part run parallel to the west coast of South America. Major South American rivers such as the Amazon, the Plata, and the Orinoco provide transport routes, as well as water resources for agriculture and hydroelectricity generation. The two major deserts—the Sonoran and the Atacama—are located along the Pacific coasts of Mexico and Chile.

Other important physical features include the mountainous spines of Mexico and Central America and the high-altitude plateaus between and adjacent to these mountains, including the Andean **altiplano** and the Mexican Mesa Central. These plateaus provide flatter, cooler, and moister environments for agriculture and settlement.

The Caribbean islands and coast of Central America have large areas of limestone geology, where water flowing underground creates large cave and pothole systems, such as the *cenotes* in the Yucatán Peninsula of Mexico and the caves of Puerto Rico. Coral reefs, a beautiful feature of the Caribbean marine landscape, are created when living coral organisms build colonies in warm, shallow oceans. These reefs, hosts to myriad other marine animals, are fragile ecosystems that are easily damaged by boats, divers, pollution, and climate change.

Earthquake and Volcanic Hazards Many of the region's mountain areas and island chains in the region are the result of a long history of tectonic activity (see Figure 1.12). Earthquakes and volcanoes occur near the plate boundaries in parts of the Caribbean, Central America, the Andes, and Mexico. For example, the Caribbean islands of Martinique, Montserrat, and St. Vincent experienced disastrous volcanic eruptions in the 20th century, causing death, destruction, and evacuation: 30,000 people died when Mount Pélee exploded on Martinique in 1902.

▲ **FIGURE 7.9 Physical geography of Latin America and the Caribbean**   The Amazon Basin and the Andes Mountains are the two largest physical features in Latin America, whereas the Caribbean is a region of mostly small islands with the exception of Cuba, Hispaniola, and Jamaica. **Where are oils and minerals found in Latin America and the Caribbean?**

Shifting tectonic plates have produced devastating earthquakes that have ravaged the capital cities of Mexico City, Mexico; Managua, Nicaragua; Guatemala City, Guatemala; and Santiago, Chile. Such natural disasters cannot be blamed solely on geophysical conditions but also on vulnerability. The greatest damages occur when people live in unsafe houses or on unstable slopes because they lack the money or power to live in safer places, cannot afford insurance, or are unable to obtain warnings of impending natural disasters. The Haitian earthquake of 2010 was one of the most devastating in recent history, killing more than 300,000 people and displacing over 1 million residents. A massive worldwide rescue and relief operation followed (**FIGURE 7.10**).

Amazon Deforestation

The fate of the Amazon has attracted global attention, especially from scientists and environmental organizations concerned about the impacts of large-scale forest loss on global biodiversity and climate. The Amazon basin includes some large areas of Brazil, but also parts of French Guiana, Ecuador, Guyana, Suriname, Bolivia, Colombia, Venezuela and Peru. *What are the causes, rates and conse-quences of Amazonia deforestation? What are the solutions?*

1 Deforestation rates in and beyond Brazil

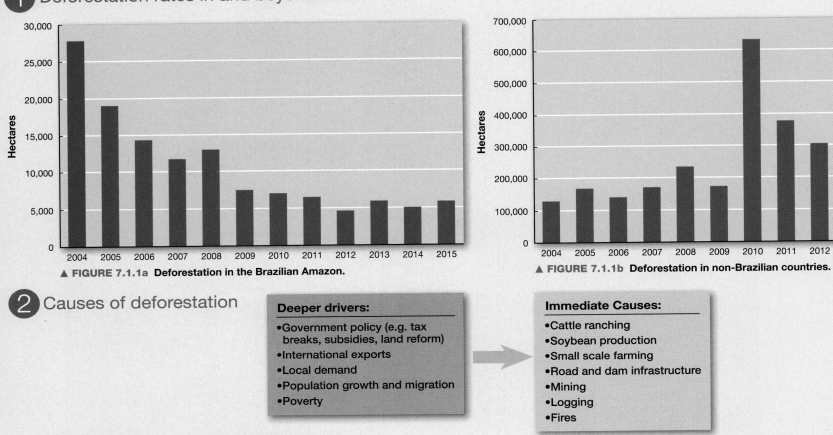

▲ **FIGURE 7.1.1a** Deforestation in the Brazilian Amazon.

▲ **FIGURE 7.1.1b** Deforestation in non-Brazilian countries.

2 Causes of deforestation

Deeper drivers:
- Government policy (e.g. tax breaks, subsidies, land reform)
- International exports
- Local demand
- Population growth and migration
- Poverty

Immediate Causes:
- Cattle ranching
- Soybean production
- Small scale farming
- Road and dam infrastructure
- Mining
- Logging
- Fires

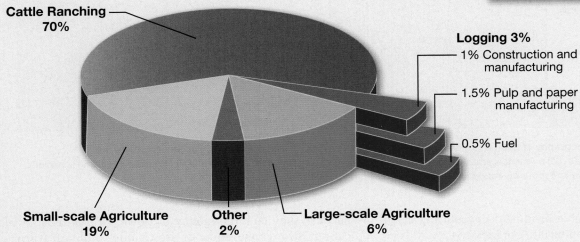

Cattle Ranching 70%

Logging 3%

1% Construction and manufacturing

1.5% Pulp and paper manufacturing

0.5% Fuel

Small-scale Agriculture 19%

Other 2%

Large-scale Agriculture 6%

Amazon deforestation rates

https://goo.gl/wkDrCa

▲ **FIGURE 7.1.2** Causes of Amazon deforestation 2000–2005.

③ Solutions to deforestation

After several decades of high deforestation Brazil was able to reduce forest loss through removing incentives for land conversion, establishing protected areas, and enforcing forest laws. However forest loss remained high in the non-Brazilian Amazon, especially in Bolivia and Peru, and there is a risk that weaker laws and high soybean prices may increase forest loss in Brazil again.

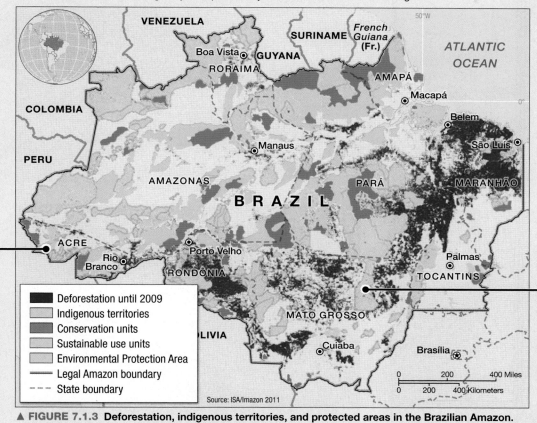

Legend:
- Deforestation until 2009
- Indigenous territories
- Conservation units
- Sustainable use units
- Environmental Protection Area
- Legal Amazon boundary
- State boundary

Source: ISA/Imazon 2011

▲ FIGURE 7.1.3 Deforestation, indigenous territories, and protected areas in the Brazilian Amazon.

1. To what extent do indigenous territories and conservation and protected areas seem to prevent deforestation as shown in Figure 7.1.3 and 7.1.4?

2. Debate whether the US and other countries have the right or obligation to get involved in Amazonian forest issues through pressure on governments or financial incentives.

Indigenous reserves protect the forest

The success of indigenous reserves for forest protection is evident in these satellite photos of the Parakaña indigenous area, an oasis of almost perfectly intact forest surrounded by cleared areas (in yellow and orange).

▼ FIGURE 7.1.4 Deforestation around the Parakaña indigenous area in Brazil.

Protected Areas

Thousands of hectares are now protected, including an areas of the state of Acre named for Chico Mendes, a rubber tapper who organized resistance to deforestation by large ranchers and was murdered in 1988. He pushed for the establishment of areas that were protected for appropriate extractive uses called extractive reserves.

"After several decades of high deforestation Brazil was able to reduce forest loss through removing incentives for land conversion, establishing protected areas, and enforcing forest laws."

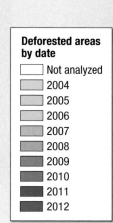

Deforested areas by date
- Not analyzed
- 2004
- 2005
- 2006
- 2007
- 2008
- 2009
- 2010
- 2011
- 2012

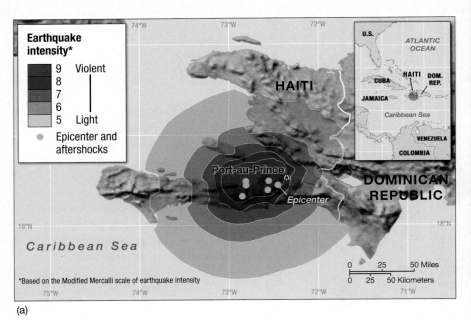

(a)

(a)

(b)

▲ **FIGURE 7.10 Earthquake in Haiti** The 2010 earthquake in Haiti killed 300,000 people and destroyed buildings and infrastructure. (a) The earthquake was centered near Haiti's capital Port-au-Prince. (b) Haiti's population was especially vulnerable to the earthquake because of widespread poverty.

Minerals, Mining, and Oil The mineral wealth of Latin America is typically found where crustal folding brings older rocks near the surface. Major precious-metal mining districts, many developed during the colonial period, include the Peruvian and Bolivian Andes; the silver region of the Mexican highlands; and the gold and iron mines at Carajas on the Brazilian Plateau. Copper is the geological treasure of the southern Andes, especially northern Chile and Peru (**FIGURE 7.11a**), and is also important in northern Mexico.

The other critical resources associated with Latin America's geology are fossil fuels. Coal is found in northern Mexico, Colombia, Brazil, and Venezuela. The earliest oil booms and later gas developments occurred in Venezuela around Lake Maracaibo, and on Mexico's Gulf Coast. Environmental pollution

(b)

▲ **FIGURE 7.11 Minerals and Oils in Latin America** (a) Mining is critical to many national economies such as that of Chile. The largest copper mine is La Escondida. (b) Oil development has been associated with serious pollution in oil regions such as Venezuela, Mexico and Ecuador. The photo shows the aftermath of an oil spill on the Rio Blanco in Ecuador.

has been a serious regional problem, leading to waterways contaminated with waste oil, widespread ecosystem damage, and serious health problems among local residents.

The most recent developments are in the Amazon, where oil and gas were discovered in 1967. In these remote forest areas, however, land rights of Indigenous people are not secure, and as a result, conflicts have erupted between Peru and Ecuador, and among governments, corporations, and Indigenous groups. In Ecuador international oil companies have been accused of causing toxic spills and cancer outbreaks (**FIGURE 7.11b**). In 2010, Ecuador offered not to exploit the oil resources of the Yasuni National Park if the international community will pay to avoid the associated carbon emissions.

Water Resources The three largest river basins in Latin America are the Amazon, the Plata, and the Orinoco, which all flow to the Atlantic Ocean (see Figure 7.1). The rivers in the Plata Basin (including the Paraná, Paraguay, and Uruguay Rivers) originate in the Andes and the Brazilian Highlands. The Orinoco drains the *llanos,* the grasslands of Colombia and Venezuela.

The Amazon tributaries flow downward and eastward from the Andes into an enormous river network that covers a basin of more than 6 million square kilometers (2.3 million square miles) and includes vast rainforests. The Amazon Basin includes the river itself and the surrounding landscape, about two-thirds in Brazil and parts of Peru, Ecuador, Bolivia, Colombia, Venezuela, Guyana, Suriname, and French Guiana. The Amazon River and its tributaries carry 20% of the world's freshwater. They also provide transport, fisheries, and deposit sediment that creates rich agricultural soils.

Water resource development in the region includes complex irrigation systems and major dam projects, which have allowed for the development of intensive agriculture and hydroelectricity. For example, Mexico and Chile have large irrigation districts that grow vegetables for export using water from dams and wells. Brazil's Itaipú dam on the Paraná River is second only to China's Three Gorges dam in terms of installed hydroelectric capacity. Indigenous people who fear the loss of their land often resist

proposals for dams with support from international environmental and human rights groups—such as the enormous dams proposed on the Tapajos and Xingu Rivers in Brazil.

Latin America has several large freshwater lakes including Lake Nicaragua in Nicaragua, and Lake Titicaca at the border of Bolivia and Peru. Major waterfalls such as Iguaçu Falls, (**FIGURE 7.12**) and Angel Falls in Venezuela have become popular tourist destinations.

Apply Your Knowledge

7.3 Do research into a current controversy over mining, oil, or water development in Latin America and identify the different opinions over the value and risks of the project.

7.4 Look up information on a hurricane, flood, earthquake, or volcanic disaster in the Caribbean and explore the role of vulnerability and the response to the hazard.

Ecology, Land, and Environmental Management

The diversity of Latin America's physical environments has produced astounding **biodiversity**—a large number of different species. Latin America's biodiversity is substantial because of the region's size, range in climates from north to south, altitudinal variations within short distances, and long history of fairly stable climate and isolation from other world regions. Many tourists are attracted to the colorful birds, interesting animals, and verdant plants of the tropical environment (**FIGURE 7.13a**).

Desert ecosystems, such as in the Atacama Desert of northern Chile and Peru and the Sonoran Desert in northwest Mexico, are associated with drier climates (**FIGURE 7.13b**). Between the moist forests and dry deserts lie ecosystems where alternating wet and

▼ **FIGURE 7.12 Iguaçu Falls** Iguaçu Falls, on the Paraná River, where Brazil, Argentina, and Paraguay meet, has become a major tourist destination.

(a)

(b)

(c)

dry seasons produce vegetation ranging from scattered woodlands to dry grasslands. Grasslands are also found at higher altitudes, where there is not enough precipitation, or temperatures are too low, to support highland forests. In Argentina, the vast *pampas* grasslands have become important to the cattle economy. Other large grassland ecosystems include the *llanos* of Colombia and Venezuela and the *cerrados* of Brazil. The high grasslands of the Andean Altiplano provide habitat for grazing animals such as the llama, wild guanaco, and vicuña. The long coasts of Latin America and the islands of the Caribbean include about 25% of the world's total area of mangrove ecosystems (**FIGURE 7.13c**).

The wetter climates of Latin America sustain magnificent forest ecosystems, including the tropical rainforests of the Amazon, Central America, and southern Mexico, and the temperate rainforests of southern Chile. The warm and wet climates of much of Latin America, and the large diversity and prevalence of pests and diseases, have limited development in these areas. For example, malaria is endemic in much of the Amazon Basin and lowland Central America.

Plant and Animal Domestication

The most dramatic transformation of nature by early people in Latin America was plant and animal domestication. Starting more than 10,000 years ago, Indigenous people domesticated wild plants and animals into cultivated or tamed forms through selective breeding for preferred characteristics, creating many of the world's major food crops. These include the staples of maize (corn), manioc (cassava), beans, and potatoes as well as vegetables and fruits, such as tomatoes, peppers, squash, avocados, and pineapples (**FIGURE 7.14**). Tobacco, cacao (chocolate), vanilla, peanuts, and coca (cocaine) were also domesticated in Latin America. Latin America has very few indigenous domesticated animals except for the llama and alpaca, tamed and bred for wool, meat, and transport. Dogs (similar to the Chihuahua) and guinea pigs were also bred for pets and meat. The selective breeding of crops and animals for human use has continued, including the **Green Revolution** that increased cereal yields in some regions from the 1950s (see Sustainable Development section).

▲ **FIGURE 7.13 Latin American and Caribbean Ecosystems** Ecosystems in this region range from forests and grasslands to deserts and coastal mangroves. (a) A scarlet macaw flies over a tropical forest in Costa Rica. Costa Rica has more bird species (850+) than are found in the United States and Canada combined. Distinct ecosystems contain more than 6,000 kinds of flowering plants and more than 200 species of mammals, 200 species of reptiles, and 35,000 species of insects (b) The Atacama Desert of Peru and northern Chile is one of the driest locations on Earth. Here the desert meets the sea in the Paracas National Reserve in Peru. (c) Mangrove ecosystems border many coasts in tropical Latin America and the Caribbean, providing protection from storms and supporting important fisheries.

Maya, Incan, and Aztec Adaptations to the Environment

As in other regions of the world, increased yields from domesticated crops in ancient Latin America created a surplus that permitted the specialization of tasks, the growth of settlements, and ultimately the development of complex societies and cultures. Complex societies in this region included the great Maya, Inca, and Aztec Empires centered in Mexico's Yucatán Peninsula and Guatemala, Highland Peru, and Central Mexico. These groups all modified their environments to increase agricultural production and to exploit water, wood, and mineral resources.

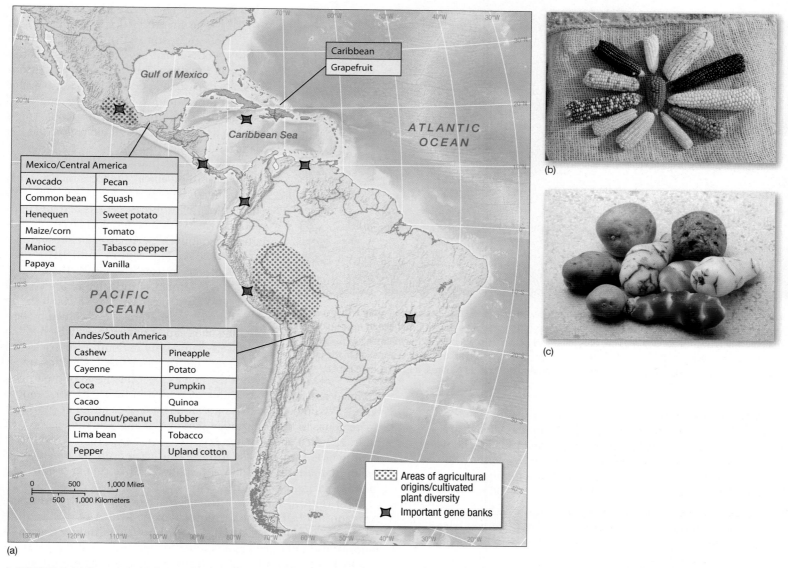

▲ **FIGURE 7.14 Domestication of Food Crops in Latin America** (a) Latin America has two important centers of domestication in Mexico and the Andes, which are indicated on the map by the small red dots. The red bursts on the map denote research centers, where attempts are being made to preserve genetic diversity in crop gene banks. (b) Maize (corn) was domesticated in Mexico from a wild grain called teosinte. Bred for a variety of microenvironments and tastes, traditional maize has many shapes and colors. (c) The various colors and shapes of potatoes grown in the Andes illustrate the diversity of types domesticated in these regions. **Which three food items domesticated in the region do you consider most essential to global diets today?**

- **Maya** In some societies—people placed so much pressure on regional landscapes that environmental degradation may have precipitated social collapse. The Maya occupied the Yucatán Peninsula and parts of Guatemala, Belize, and Honduras from about 1800 B.C.E. and reached a peak of population and political development in about 250 C.E. But in about 800 C.E., the great Maya cities such as Copán, Palenque, Tulum, and Tikal (**FIGURE 7.15**) were abandoned, and overall population declined dramatically.

Many scholars believe that one reason for the Maya collapse was their overuse of the soils. Faced with rapid declines in the fertility of soils after clearing

the rainforest, the Maya adapted by burning forest to capture the nutrients in the ash of burned trees. They moved on to clear another patch, and then another, once the declining fertility of each cleared area resulted in reduced yields. This adaptation to rainforest environments mirrors farming methods in other parts of the world and is called slash-and-burn (see Chapter 5), or swidden, agriculture. The Maya also developed methods for growing crops in wetland areas by building raised fields that lifted plants above flooding, but at the same time took advantage of the rich soils, frost protection, and reliable moisture of wetland environments.

▲ **FIGURE 7.15 Tulum** The city of Tulum in Mexico was one of the last cities of the great Mayan empire and was occupied from about 1200 to 1500. The Maya cleared vast areas of forest inland from Tulum and around other Mayan sites. Soil erosion, droughts, and declining soil fertility in Maya times are believed to have contributed to a decline in the amount of food available to feed the large population in the latter years of the empire, especially when rulers were demanding food as tributes, and thus to collapse.

- **Inca** The Inca Empire, with lands stretching from northern Ecuador to central Chile, dominated the Andean area from about 1400 C.E. The Incas responded to the difficulties of living in a mountain environment in a variety of ways, including constructing agricultural terraces so they could farm the steep slopes of the Andes (see Figure 7.3).

- **Aztec** The Aztecs, who settled in central Mexico in the 1300s, constructed an extensive network of dams, irrigation systems, and drainage canals in the basin of Mexico to cope with the highly seasonal and variable rainfall pattern. They also developed the *chinampa* agricultural system, in which farmers grow crops on islands of soil and vegetation built in lake and wetland environments. Evidence suggests that the Aztecs cleared the forests in the basin of Mexico, which may have contributed to a drop in the water table and to a resulting water crisis that led to the abandonment of some settlements.

Apply Your Knowledge

7.5 How many of the products that you regularly eat are based on crops originally domesticated in Latin America? Did you eat a meal in the course of the past week that *did not* contain foods originally from Latin America?

7.6 List three ways that the Maya, Incas, and Aztecs impacts on and adaptation to their environment might serve as warnings or offer solutions for current environmental issues in the region.

The Fate of the Forests and Biodiversity The forests of Latin America and the Caribbean host a significant share of the world's biodiversity, provide food and shelter for forest dwellers, and provide **ecosystem services** (see Chapter One, p 18), such as water and soil protection, as well as the removal of greenhouse gases from the atmosphere. These services benefit countries in the region and the planet as a whole.

Covering more than a half-billion hectares (about 1.2 billion acres), the forest of the Amazon Basin contains water, forest, mineral, and other valuable resources, yet has had low population density until recent years (see "Visualizing Geography: The Amazon Deforestation," pp. 268–269). The region held little economic attraction until the late 19th century, when development of industry in the United States and Europe caused an explosion in the demand for rubber. Rubber is obtained by tapping the latex sap of rubber trees, which at the time were found only in South America. Local rubber tappers, or *seringueiros,* sold the rubber to middlemen to meet the needs of the industrialized nations. The middlemen traded in turn with the "rubber barons," who constructed enormous mansions and a magnificent opera house in the Amazonian port of Manaus. The rubber boom in Latin America ended when seedlings were exported to Southeast Asia, where they became a successful cash crop. The success of Asian plantations, especially in Malaya, drove rubber production in the Amazon into decline.

The region has many rich forest ecosystems beyond the Amazon, including the rainforests of Central America and the Alerce redwood forests of Chile. There has been extensive deforestation of Caribbean islands for agriculture and fuel. Haiti has lost most of its forest with resulting soil erosion, but other islands such as Anguilla and Dominica are still partly covered by forest, and many other islands have set aside forest reserves.

Sustainable Development

Latin America and the Caribbean are facing many sustainability challenges in the Anthropocene, including those of developing more productive and sustainable agriculture and sustainable cities.

The Green Revolution and Agriculture Increasing yields on existing farmland, rather than clearing new fields, helps prevent deforestation and food insecurity. For the first part of the 20th century, the yields of most agricultural crops in Latin America and the Caribbean were very low. Farmers with small plots of land could not produce enough to feed themselves, let alone to sell in the market. As population and urban-consumption demands increased, countries such as Mexico and Brazil had to import basic food crops, such as wheat and corn.

The **Green Revolution** was an international effort, beginning in the 1940s, to address low productivity and poverty in rural

Agriculture in Mexico

(a)

(b)

(c)

▲ **FIGURE 7.16 Green Revolution** (a) Plant breeding and the use of fertilizer contributed to increases in yields of maize and wheat in Mexico. This graph also shows an increase in production of vegetables in Mexico, mostly for export, and requiring the use of chemicals to control pests and plant diseases. (b) The term the *circle of poison* refers to the chain of events that occurs when imported pesticides are used on crops in developing countries, which then export the contaminated crops back to the regions where the pesticides were manufactured. (c) Some farmers in Latin America and the Caribbean are now protesting the use of agricultural chemicals—such as the Paraguayan farmers shown here—and returning to more sustainable and traditional techniques.

areas through the introduction of new varieties of crops that produced far more food, but also required higher levels of industrial inputs, including fertilizer, water, and pesticides. Countries in the region, such as Mexico, Argentina, and Brazil, also promoted Green Revolution agricultural modernization with higher-yielding seeds of key crops, such as wheat, rice, corn, and soybeans, that were introduced in combination with irrigation, fertilizers, pesticides, and farm machinery.

Although the Green Revolution increased crop production in many parts of Latin America (**FIGURE 7.16a**), it was not an unqualified success. Dependence on imports of chemicals and machines from foreign companies increased, and thus contributed to the debt problem. Benefits accrued to wealthy farmers in irrigated regions who could afford the new inputs, while poorer farmers whose land was watered only by rainfall fell behind or sold their land. In some cases, such as in southeastern Brazil, machines replaced workers, thus leading to unemployment.

Green Revolution technology and training also tended to exclude women, who play important roles in food production. The most serious criticism of the Green Revolution was that it contributed to the worldwide loss of genetic diversity: A wide range of local crops and varieties were replaced with a narrow range of high-yield varieties of just a few crops. Planting single varieties over large areas (monocultures) also made agriculture vulnerable to disease and pests.

Agricultural chemicals, especially pesticides, contributed to ecosystem pollution and worker poisonings, and the more intensive use of irrigation created problems of water scarcity and salinization (the buildup of salt deposits in soil). The increased use of imported pesticides on export crops is associated with damage to ecosystems and workers' health (**FIGURE 7.16b** and **c**). Some farmers are now returning to traditional agricultural techniques that conserve soil, water and biodiversity, and may support a more sustainable food system.

Biofuels Increasingly, both forests and agricultural land in Latin America today are being converted to biofuel production. **Biofuels** are energy sources derived from living matter. Crops, such as sugarcane and corn, can be converted to fuels by various processes, including fermentation. Vegetable oils (biodiesel) or crop and animal wastes can also be used as fuels. Brazil is a major exporter of ethanol, which is produced mostly from sugarcane. Although biofuels can potentially slow climate change by reducing fossil fuel use, there are concerns that biofuel production is driving deforestation, diverting land from food production, and consuming limited water and fertilizer resources. Other sources of renewable energy, such as wind, sun and hydropower may be more sustainable.

Ecotourism and Conservation Latin America and the Caribbean have seen a boom in tourism focused on natural attractions such as beaches and rainforests. Environmentally oriented tourism, or **ecotourism**, aims to protect the environment and provide employment for local people (**FIGURE 7.17**). Costa Rica has won high praise for protecting the rich biodiversity of its ecosystems by setting aside biosphere and wildlife preserves (Figure 7.13a). Ecotourism is the country's main source of foreign exchange, and over 2.4 million tourists visited in 2013. Ecotourism has been a mixed blessing for some in the region. The benefits are not shared equally among residents, and some parks and beaches are becoming so crowded that environmental degradation is occurring.

The ecological diversity of Latin America also encourages biological prospecting—**bioprospecting**—for new medicines and products with commercial uses. Costa Rica has signed agreements with multinational pharmaceutical companies, giving them rights to prospect and develop in return for sharing profits with the government and local people.

Payments for environmental services (PES) are increasingly important across the region where people are paid to protect their local environment because preservation of that environment brings value to others. For example, upstream communities are paid to protect their forests because of the role the forests play in sustaining the flow of rivers for downstream water users, such as plantations and hydroelectric plants.

Apply Your Knowledge

7.7 Consider the ways that the Green Revolution, biofuels, ecotourism, and payments for environmental services have changed environmental and social conditions in Latin America and the Caribbean.

7.8 Keeping in mind the positive and negative impacts, do you think these programs and policies should be expanded across the region and to other world regions?

History, Economy, and Territory

Much of Latin America shared the common experience of Spanish and Portuguese colonialism and their legacies, including the dominance of the Spanish and Portuguese languages, Roman Catholic religion, and associated legal and political institutions. The Caribbean coast and islands had a more diverse colonial experience as a result of British, Dutch, and French colonization. European arrival and settlement also resulted in the collapse of Indigenous populations, and European control of resource extraction, trade links, and other economic activity. Most of mainland Latin America became independent in the 19th century and was drawn into global trade relations, especially with Britain and the United States. In the 20th century, the region experienced rapid integration into global markets and the transition from revolutionary and military governments to democratic ones. Contemporary Latin America has some thriving economies and industrial regions, which serve as international centers of commerce.

Historical Legacies and Landscapes

Archaeological evidence from sites in Chile, Mexico, Peru, and Ecuador suggests a human presence in Latin America from at least 10,000 years ago. Early inhabitants were the descendants of migrants who came from Siberia into North America across the Bering Strait. These Indigenous people of Latin America lived by hunting, fishing, and gathering. As discussed earlier in this chapter, they also eventually domesticated important agricultural crops such as corn and potatoes. Complex societies emerged in the Andes such as the Inca (from about 1200 C.E.), and in Mexico and Central America (such as the Maya, from about 1800 B.C.E.), and the Aztec (from about 1200 C.E.).

▼ **FIGURE 7.17 Ecotourism** The Roatan islands off the coast of Honduras with their coral reefs have become a popular destination for diving.

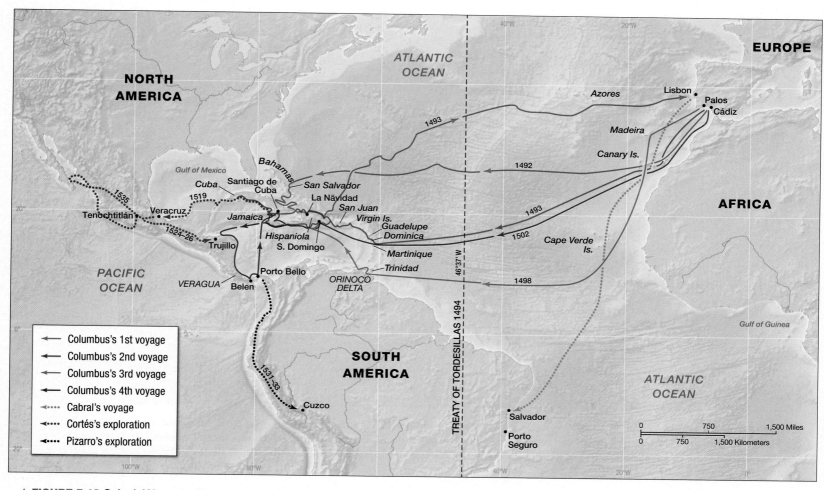

▲ **FIGURE 7.18 Colonial Voyages** The map shows the major voyages and missions of Columbus, Pizarro, Cabral, and Cortés and the division of Latin America between Spain and Portugal under the Treaty of Tordesillas in 1494. The initial line set in 1493 was contested by Portugal and shifted farther west. **How does this map explain why Brazilians speak Portuguese?**

The Colonial Experience in Latin America The integration of Latin America into the global system of political, economic, ecological, and social relationships began more than 500 years ago with the arrival of Spanish and Portuguese explorers at the end of the 15th century.

- **European Exploration** The most famous European explorer was Christopher Columbus, an Italian from the port city of Genoa. Queen Isabella of Spain commissioned Columbus to search for new territory and trading opportunities on a western route to the Indies (as Asia was then called). Although Columbus did not reach Asia, he arrived in the Caribbean in October 1492 and landed on a small island in the Bahamas, to which he gave the name San Salvador (**FIGURE 7.18**). Columbus also visited Cuba and an island that he called Hispaniola (now Haiti and the Dominican Republic). On this and subsequent voyages he left European settlers and expanded his explorations to Trinidad, coastal Venezuela, and Central America.

- **Spanish and Portuguese Claims** The Spanish wanted the new lands to be assigned to Spain rather than to its

rival, Portugal, and negotiated the **Treaty of Tordesillas** (1494) to divide the world between Spain and Portugal along a north–south line about 1,800 kilometers (1,100 miles) west of the Cape Verde Islands. With the approval of Pope Julius II, Portugal received the area east of the line, including much of what is now Brazil and parts of Africa, and Spain received the areas to the west (see Figure 7.18).

- **Conquest and Exploitation** Columbus was followed in subsequent decades by other Europeans seeking gold, fame, territory, and additional resources. His successors included notable explorers (called *conquistadors*) such as Hernán Cortés, who landed in Mexico in 1519 and went on to conquer the Aztec Empire; and Francisco Pizarro, who seized control of the Inca Empire centered in Cuzco, Peru, in 1533. The Portuguese began their colonization with the landing of Pedro Álvares Cabral in 1500 at Porto Seguro in Brazil. Mexico City and Lima became the Spanish administrative centers (called *viceroyalities*) and Salvador da Bahia in Northeast Brazil became the initial Portuguese capital (moving to Rio de Janeiro in 1763).

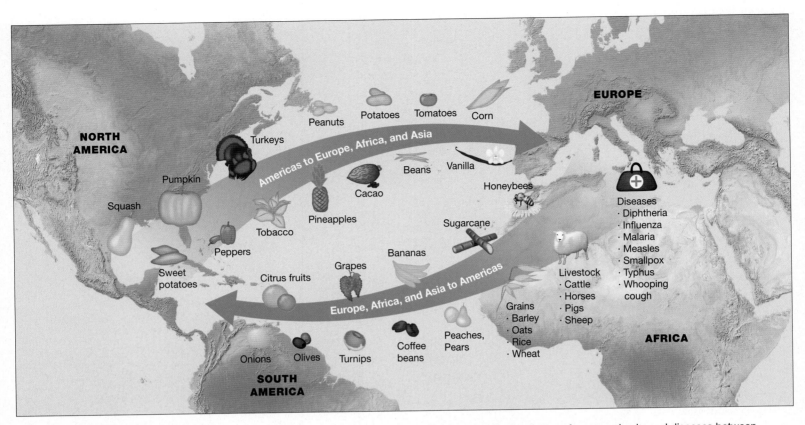

▲ **FIGURE 7.19 Columbian Exchange** The arrival of Europeans in the Americas initiated the extensive exchange of crops, animals, and diseases between continents, including the introduction of European diseases to the Americas, which resulted in the deaths of millions of Indigenous people. The introduction of wheat, sugar, and livestock transformed the landscapes of Latin America, and European and African diets (and health) were altered forever with the arrival of corn, potatoes, and tobacco.

The colonial push into Latin America took place over at least two centuries. Some regions never really came under complete colonial control because of their remoteness (the Amazon) or local resistance (parts of the Andes). Spain demanded that their colonies provide gold and silver for the Spanish crown, convert the Indigenous people to the Catholic religion, and become as self-sufficient as possible through the use of local land and labor. The Spanish crown demanded 20% of all mining profits—the so-called *Quinto Real,* or royal fifth.

Demographic Collapse The search for local labor to work in the mines and fields of the Spanish colonizers was frustrated by one of the most immediate and significant impacts of the European arrival in Latin America—the **demographic collapse**. After about 1500, the Indigenous populations of the Americas began to rapidly die off as a result of diseases introduced by the Europeans to which residents of the Americas had no immunity. Because of the long isolation of the Americas from other continents, Indigenous people lacked resistance and immunity to European diseases, such as smallpox, influenza, and measles. The resulting mortality rates from these diseases were very high: Up to 75% of the population of Latin America died in epidemics in the century following contact.

Columbian Exchange The introduction of European diseases into the Americas is just one example of the interaction between the two continents that historian Alfred Crosby has called the **Columbian Exchange**. The Columbian Exchange refers to the interchange of crops, animals, people, and diseases between the Old World of Europe, Africa and Asia and the New World of the Americas that began with the voyages of Columbus (**FIGURE 7.19**). When the Spanish and other colonial powers arrived in new lands, they brought with them favorite plants and animals to introduce into the new colonies. They also, unintentionally, brought diseases, weeds, and pests (such as rats) that stowed away on their ships. For their return voyages, the explorers and colonists collected species that they hoped could be sold in Europe and elsewhere. The colonists introduced crops and domesticated animals to Latin America and the Caribbean, such as wheat, cattle, fruit and olive trees, horses, sheep, and pigs as well as sugar, rice, citrus, coffee, cotton, and bananas from North Africa and the Middle East (see "Geographies of Indulgence, Desire, and Addiction: Coffee"). The colonizers took corn, potatoes, tomatoes, tobacco, and possibly syphilis back to Europe. For example, potatoes became the foundation of the Irish diet, although the potato blight in the mid-19th century led to famine and Irish migration to the Americas. Corn and manioc (cassava) were introduced into Africa and became new staples, whereas peanuts and cacao became the basis of new African export economies. Tobacco became an addictive and fatal habit across the world.

Coffee

Coffee is one of the world's most popular drinks, stimulants, cultural traditions, and traded commodities. Over 2 billion cups of coffee are consumed daily, and coffee consumption has doubled in the last three decades. Coffee was one of the most important crops introduced into Latin America and the Caribbean by colonial powers. Today the region produces half of the world's coffee, led by Brazil (33% of global production), Colombia, Peru, Honduras, Mexico and Costa Rica (**FIGURE 7.3.1**).

Origin and Diffusion

Coffee comes from the twin seeds, known as beans, inside the fruit of the coffee plant, which are dried and roasted. Roasting and brewing of coffee spread from Ethiopia, where it was first domesticated, to Yemen, and, by the 16th century, across the Middle East. It then spread to Venice, Italy, and the rest of Europe, fostered by the coffee houses where men met to socialize and do business.

The two main varieties of coffee, Arabica and Robusta, contain different amounts of the stimulant caffeine. The stronger (up to 4% caffeine) Robusta coffee is grown at warmer temperatures, and the milder, more popular (1% caffeine) Arabica is grown in cooler, often mountainous areas.

Coffee in Latin America and the Caribbean

Coffee-growing began in the region when the French brought production to what is now Haiti; the Portuguese introduced its cultivation in Brazil; and the British brought coffee to the Blue Mountains of Jamaica.

Coffee is a labor-intensive crop, and slaves produced much of Latin America and the Caribbean's coffee under brutal working conditions. In Guatemala and El Salvador, the growth of coffee production in the 19th century drove Indigenous peoples from their lands and into forced labor on coffee plantations that were controlled by powerful families who eventually supported authoritarian and repressive political systems.

Globalized Coffee Culture

Demand for coffee increased with the spread of specialty coffee stores in the 1970s, including Seattle-based Starbucks, who by 2014 had 21,366 stores in 65 countries. Concerns about justice and environment also began to influence the coffee economy in the 1990s as some consumers preferred coffee from places where people are paid a fair wage, pesticides are banned, and human rights, forests, and biodiversity are protected. "Fair trade coffee" has become a common pronouncement on labels that also boast about qualities such as shade-grown, bird-friendly, or organic coffee. Some countries have developed high value coffee brands such as Blue Mountain coffee from Jamaica.

Commodity Outlook

The coffee industry in Latin America and the Caribbean is at risk from climate change and from the globalization of coffee production. Frosts and hurricanes have the capacity to devastate coffee crops, and climate change threatens production of preferred Arabica coffee grown on cooler, but now warming, mountain slopes (**FIGURE 7.3.2**). The region's coffee production has also been affected by growing competition from coffee production in Southeast Asia, especially Vietnam, which contributed to a drop in prices paid to farmers.

1. **Conduct a survey of 10 friends to assess how much extra they would be willing to pay for assurances that their coffee was fair trade and ecologically produced.**

2. **Consult the website of a major coffee chain or company to note how they present the lives of those who produce the coffee or demonstrate its sustainability.**

"Some consumers preferred coffee from places where people are paid a fair wage, pesticides are banned, and human rights, forests, and biodiversity are protected."

▼ **FIGURE 7.3.1** The coffee produced in Jamaica's Blue Mountains commands a premium price on world markets.

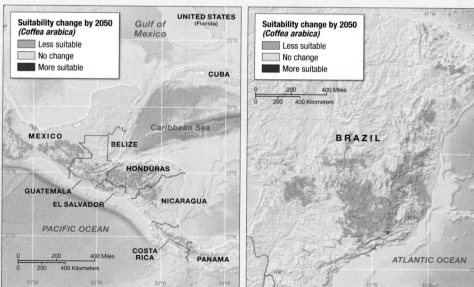

▲ **FIGURE 7.3.2** Climate change will produce warmer temperatures in most of the region by 2050. The area suitable for preferred Arabica coffee will decline in the coffee regions of Brazil and Central America and the Caribbean.

▲ **FIGURE 7.20 Slavery on Plantations** Mexican muralist Diego Rivera portrays the Spanish forcing slaves to do brutal work on sugar plantations in central Mexico. Sugar was introduced by Columbus and other Europeans to both the Caribbean (especially Cuba, Jamaica, and Grenada) and mainland Latin America (especially Brazil and Mexico).

The clearing of land for European crops, such as wheat and sugar, and the overgrazing by cattle and sheep contributed to soil erosion and deforestation in Latin America and the Caribbean. Newly introduced rats, pigs, and cats ate the food that traditionally supported local species, and consumed the local fauna, especially ground-dwelling birds.

Land and Labor The colonial powers introduced several new forms of land tenure and labor relations that still influence contemporary landscapes in Latin America. In areas where Spain wished to directly control the land, they granted land rights over large areas to colonists, often military leaders, and to the Catholic Church, ignoring traditional local uses and establishing fixed property boundaries. Many of these large estates were called **Haciendas** and grew olives, wheat and cattle to supply local mines, missions and cities. Other large estates—variously known as estancias, haciendas, plantations, and fazendas—focused on export crops such as sugar and tobacco. Smaller holdings were termed ranchos, fincas, or quintas.

Plantations are large agricultural estates established in the colonial period that grew crops such as sugar or tobacco for export (**FIGURE 7.20**). The colonists demanded tribute from Indigenous people in labor, crops, or goods; sought to convert communities to the Catholic faith; and taught them Spanish. However, labor remained in short supply, especially in areas where the demographic collapse had devastated local populations. To make up for the labor shortage, colonists imported slaves, mainly from Africa, to the Caribbean, Central America, and Brazilian plantations (see Chapter 5). Slaves worked in the production of sugar and other export crops. In many parts of Latin America, the colonial legacy of land grabs, labor exploitation, and racism still frames contemporary attitudes toward Indigenous people and those of African heritage, who still struggle to regain the land and dignity lost during the colonial period.

Export Commodities The most important export commodities in Spanish colonial America were silver, produced mainly from mines in Mexico and Bolivia; sugar, grown on plantations in Cuba and southern Mexico; tobacco from Cuba; gold from Colombia; cacao (for chocolate) from Venezuela and Guatemala; and indigo, a deep blue dye, from Central America. Spain derived enormous wealth from the bonanza of the silver mines at Potosí (Bolivia), which produced half of the world's silver in the 16th century.

Apply Your Knowledge

7.9 Select one of the crops or animals domesticated in the region (shown on Figure 7.14) and research its role in contemporary food systems including where it is produced and consumed. What role does it play in your own food consumption?

7.10 Visit a website that features tourist or economic information for a Latin American or Caribbean country. In what ways are the legacies of colonialism apparent in the landscapes, cities, or economy of the country as described to tourists?

Independence Over time, merchants and landowners in the Spanish colonies started to resent Spain's strict control of trade and taxation and began scheming to keep the revenue from mines and crops for themselves. Wars in Europe weakened Spain, and revolutionary movements in the United States and France provided inspiration. Pressure built for independence in the colonies, and by the beginning of the 1800s, regional independence movements were calling for liberation from Spanish and Portuguese rule.

Between September 16, 1810, when priest and peasant leader Miguel Hidalgo called for Mexican independence, and 1824, when Simon Bolívar finally led northern South America to independence, regional revolts led to the formation of independent states in Mexico, Argentina, Peru, Colombia, Chile, and Brazil. After 1848, when Mexico lost a war with the United States, large portions of Mexican territory became part of Texas and the future states of Arizona, California, Colorado, Nevada, and New Mexico and Utah.

In the Caribbean, former slaves declared Haitian independence from France in 1804. Cuba remained under Spanish control until after the Spanish–American War in 1898, when limited independence was granted under U.S. influence and frequent intervention (see Chapter opener). Puerto Rico also shifted from Spanish control to a U.S. territory after 1899; it is still currently part of the United States, with some residents desiring independence and others demanding the country be granted full status as a U.S. state.

Independence came late to most of the British Caribbean. Jamaica, Barbados, and Trinidad and Tobago became independent in the 1960s. Other British-controlled islands—and Belize in Central America—did not become fully independent until the 1980s. A number of Caribbean islands remain under colonial control, including Aruba (Dutch protectorate), Martinique and Guadalupe (overseas departments of France), and the British and U.S. Virgin Islands (see Figure 7.1).

▲ **FIGURE 7.21 Mexico's Chemical Industry** Mexico used its oil to develop a domestic refining and chemical industry—shown here in Tampico, Veracruz—including the production of various petrochemicals and fertilizer, and also expanded its automobile and steel industry during the period of import substitution.

Economy, Accumulation, and the Production of Inequality

After independence, the loss of trade with Europe and internal struggles left Latin America economically unstable for the first half of the 19th century. But from 1850, foreign capital and demand helped develop export economies.

Export Dependence A limited number of commodities provided much of the national income for countries in the region. These commodities included nitrate (used to make fertilizer) and copper in Chile; livestock in Argentina; coffee in Brazil, Colombia, and Central America; bananas in Central America and Ecuador; tin in Bolivia; and silver and henequen (a fiber used in making sacks and matting) in Mexico. This export dependence is still evident today (**TABLE 7.1**).

Import Substitution The 1929 stock market crash caused declines in exports, restrictions on investment, and a general economic crisis in the region. In a new economic strategy of **import substitution**, countries such as Mexico, Brazil, and Argentina developed government protected and subsidized domestic industries for goods that used to be imported.

Government nationalization (the process of converting key industries from private to governmental organization and control) and investment in new manufacturing industries fostered production of chemicals, steel, automobiles, and electrical goods in regions such as northeastern Mexico (steel), Mexico's Gulf Coast (petrochemicals—**FIGURE 7.21**), and São Paulo, Brazil (automobiles). But growing criticisms of import substitution emphasized the oversized government bureaucracy and the

high costs of subsidizing industries that were inefficient and produced goods of poor quality.

Debt Crisis After 1960, economic growth in Latin America began attracting international investors and banks. The governments of Mexico, Argentina, Brazil, and Chile were offered the largest loans, but almost all Latin American governments took advantage of the initially low-interest loans to support economic development and other projects. When interest rates rose and debt payments soared in the 1970s, many Latin American governments were unwilling to cut back on popular subsidies and social programs and instead borrowed more money, ran budget deficits, and overvalued their currencies. The resulting runaway inflation and debt reached unprecedented levels. The 1980s have been called Latin America's "lost decade" because of the slowdown in growth and deterioration in living standards that accompanied the financial crisis. Mexico and Brazil owed over U.S. $100 billion and were paying enormous amounts in interest each year. In 1982, Mexico threatened to default on its loans prompting financial institutions such as the International Monetary Fund (IMF) and the U.S. government to find a solution to Latin America's debt crisis. The United States extended the repayment period for debts and lent more money, while the IMF moved to restructure loans on the condition that governments initiate new economic policies.

The new economic policies that were the condition for debt restructuring included requirements that countries curb inflation by cutting public spending on government jobs and services, increase interest rates, control wages, and devalue currencies to increase exports. These **structural adjustment programs (SAPs)** required the removal of subsidies and trade barriers, the privatization of government-owned enterprises such as telephone and oil companies, reductions in the power of unions, and an overall focus on export expansion. These policies, while reducing inflation and debt, had very negative effects on some people and sectors, especially the poor. Food prices increased as subsidies were cut. Health services and education were reduced, and unemployment increased as government jobs were eliminated.

Free Trade and NAFTA Governments in the region were also encouraged to expand free trade through regional agreements

TABLE 7.1 Export Dependence of Selected Latin American Countries		
Country	**Export Commodities**	**Percent of Total Exports**
Chile	Copper	52%
Paraguay	Soybeans	42%
Ecuador	Bananas, crude oil	11%, 50%
Venezuela	Crude oil	96%

to remove barriers among trading partners. The most dramatic step in this direction was taken by Mexico, which in 1994 joined the **North American Free Trade Agreement (NAFTA)** with the United States and Canada.

NAFTA set out to reduce barriers to trade among its three member countries, through reducing customs tariffs and quotas. Advocates of NAFTA argued that free trade would create thousands of jobs in Mexico with higher wages and that these opportunities would reduce migration to the United States. Mexican agriculture would shift to growing high-value fruits and vegetables, where it had a comparative advantage during the winter, and Mexico would be able to reduce food prices by importing low-cost grain from the United States and Canada.

NAFTA has created new jobs in Mexico and increased wages in some industries. *Maquiladoras,* in particular, have generated more than 1.5 million jobs in Mexican cities. *Maquiladoras* or *maquilas* are manufacturing facilities where components can be imported duty-free and assembled for export (**FIGURE 7.22**). NAFTA increased the growth of the *maquiladoras,* including factories producing electronic equipment such as computers, clothing, and appliances.

Unfortunately, the economic crises that followed Mexico's 1994 currency devaluation, and continuing inequality in both urban and rural areas, overwhelmed many of the hoped-for benefits of NAFTA. Imported corn from the US was unpalatable to many Mexicans and caused a drop in prices for Mexican farmers. Prices rose when the US began to promote corn-based ethanol creating a food crisis. NAFTA also created environmental problems as water demand rose for export crops, and export factories and growing border cities caused pollution.

There are several trade and cooperation agreements within the region including MERCOSUR (Chile, Argentina, Brazil, Paraguay, and Uruguay), CAFTA (the Central America Free Trade Agreement between the United States and Central America), and CARICOM (the Caribbean Community). Chile, Peru, and Mexico are expected signatories of the Trans-Pacific Partnership (TPP), a far-reaching trade agreement under negotiation among 12 Pacific Rim countries. The **ALBA** group (*Alianza Bolivariana para los pueblos de nuestra América,* or Bolivarian Alliance for the Peoples of Our America) includes Venezuela, Bolivia, Cuba, Ecuador, and Nicaragua, and opposes neoliberal free trade and U.S. influence in the region.

Apply Your Knowledge

7.11 What were some of the negative consequences of the economic policies associated with debt restructuring in the region? Why did some people in Latin America oppose structural adjustment and free trade policies?

7.12 Search an online news site to find a recent article that evaluates NAFTA and its impacts now, as it is more than two decades since it was passed. What benefits or problems are identified?

▲ **FIGURE 7.22 Maquiladora** The North American Free Trade Agreement increased the number of Mexican export assembly plants, called *maquiladoras,* which employ thousands of workers, especially women, along the border and elsewhere in Mexico. This plant in Tijuana makes implantable electronic medical devices. There have been concerns about pollution from the factories and about poor workplace conditions.

Contemporary Economic Conditions The share of gross national income (GNI) paid as interest on debt has dropped from more than 5% to less than 2% in the last 25 years, and economies in Latin America and the Caribbean have grown and diversified considerably in the last decade (**FIGURE 7.23**). For example, the region's largest economies, Brazil and Mexico, grew rapidly from 2000 to 2013. Brazil is now the world's

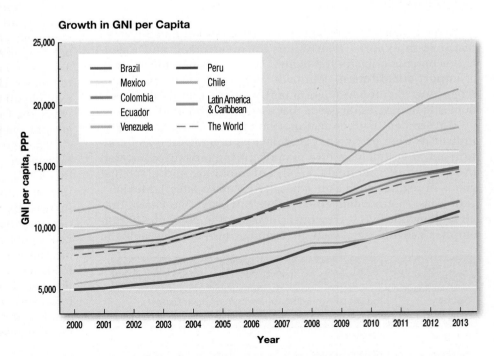

Growth in GNI per Capita

Legend: Brazil, Mexico, Colombia, Ecuador, Venezuela, Peru, Chile, Latin America & Caribbean, The World

▲ **FIGURE 7.23 Economic Growth** This graph shows the growth of major Latin American economies since 2000 in the United States, dollars per capita. **Which countries have the highest GNI per capita now and which have increased the most since 2000?**

eighth-largest economy, and the two countries are home to major manufacturing zones in cities such as São Paulo (Brazil) and Monterrey (Mexico).

- **Income** The World Bank estimated the overall GNI of the Latin America and Caribbean region in 2012 at U.S. $5.1 trillion and per capita GNI at U.S. $9,542. These figures show that this world region is much more prosperous overall than sub-Saharan Africa, South Asia, or Southeast Asia. Total exports in 2012 were valued at about U.S. $1.2 trillion while imports amounted to approximately $1.3 trillion. The Latin American region also received more foreign investment in private capital than any other low- or middle-income region.

- **Current Trends** This all suggests that overall economic conditions in Latin America and the Caribbean are positive and have improved in the last decade, with the overall picture heavily influenced by the success of Brazil. The growth of Brazil's economy merited its inclusion in a new association of emerging economies with Russia, India, China, and South Africa, which are collectively referred to as **BRICS** (see "Emerging Regions: BRICS" on page 285). However, recent trends show a significant slowdown of economic growth in the region, including in Brazil.

Economic Structure

Economies in Latin America and the Caribbean are dominated by the services and manufacturing sector, with a declining share of income and employment in the agricultural sector (**FIGURE 7.24**). While domestic demand supports agriculture, food processing, clothing, and energy development, many economies are still export dependent. Chile, for example, is dependent on copper but also exports fruit and farmed salmon, as well as wine and timber, and has a strong domestic economy based on services, including tourism, retail and informatics.

- **Manufacturing** Domestic demand supports significant manufacturing in countries such as Mexico and Brazil, but smaller countries, such as those in the Caribbean, import most manufactured goods. Mexico's export economy includes oil (13%) and many different manufactured goods such as cars and trucks (17%); machines and electronics (36%) such as computers and telephones; and processed foods. Costa Rica, with a well-educated workforce and stable economy, has lured high-technology companies such as Microsoft, General Electric, and Intel to build factories near San Jose. Textiles and clothing are another very important component of the economy in the region. Countries such as Haiti, El Salvador, and Honduras produce millions of low-cost T-shirts and other garments.

- **Agriculture and Fisheries** The agricultural economy of many parts of Latin America has changed considerably in recent years. Some countries, such as Argentina, have traditionally focused on grains and livestock; others, such as Guatemala, Nicaragua and Honduras, still export colonial crops of bananas, coffee, and

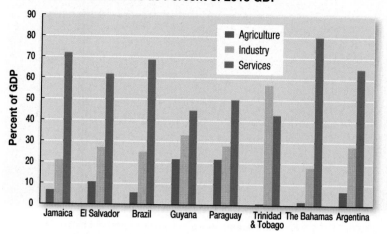

Economic Sectors as Percent of 2013 GDP

▲ **FIGURE 7.24 Economic Structure** Economies in the region are dominated by the service sector and agriculture generates less than 25% of GDP in most countries. **Which countries have a significant manufacturing share of their economy and which have the greatest share in agriculture? Can you do research to explain these differences?**

sugar. Soybeans have become important in southern South America, especially Brazil and Paraguay. Some countries in the region prioritize food self-sufficiency, while others have shifted to export crops such as fruit, vegetables, and flowers. These **nontraditional agricultural exports (NTAEs)** have become increasingly significant in areas of Mexico, Central America, Colombia, and Ecuador, where flowers and vegetables have replaced grain and traditional exports such as coffee and cotton. Rather than grow these and other crops on large company landholdings, the current strategy involves *contract farming* for multinational corporations; farmers sign contracts with companies to produce crops to certain quality standards in return for a promised price.

Fisheries range from subsistence fisheries in small coastal villages to large-scale commercial offshore fisheries and controlled production in fish farming (**FIGURE 7.25**). The UNFAO reports that in 2012, of the total global capture of 91 million metric tons (MMT), the region caught about 1/6th with Peru harvesting almost 5 MMT, Chile 2.5 MMT, and Mexico 1.6 MMT. But offshore catch is down by more than half in Peru and Chile since the early 1990s as a result of overfishing and climate variability.

- **Tourism** In the Caribbean, tourism is the cornerstone of many economies. Almost 24 million international tourists visited in 2011, generating revenues of more than U.S. $150 billion. Mexico's tourist industry, drawing mostly North Americans to resorts such as Cancun, almost matches the Caribbean with more than 20 million visitors. Ecotourism, as discussed earlier in the chapter, is also a growing industry, especially in Costa Rica.

▲ **FIGURE 7.25 Fish and Aquaculture** Aquaculture and mariculture (the cultivation of fish and shellfish under controlled conditions) in coastal lagoons has become an important export sector in countries such as Chile (1 MMT of mostly salmon), Ecuador, and Honduras (shrimp), but has raised concerns about the ecological effects of intensive fish farming. This salmon farm is in Puerto Montt, Chile.

▲ **FIGURE 7.26 Informal Economy** Thousands of people make their living in the informal economy of the region. This street vendor in Mexico City is selling children's toys.

- **Informal Economy** Economic activities that take place beyond the official record and are not subject to formal regulation or remuneration are part of the **informal economy**. The region's informal sector comprises a variety of income-generating activities of the self-employed that do not appear in standard economic accounts. These include street vending, shoe shining, garbage picking, street entertainment, prostitution, crime, begging, and guarding or cleaning cars (**FIGURE 7.26**).

Inequality Despite many economic successes Latin America remains the most unequal region of the world. Wealth and land are often concentrated in the hands of the elite, and many people remain poor and landless. The highest average annual incomes of U.S. $10,000 to U.S. $20,000 per person are reported in Brazil, the Bahamas, Barbados, Antigua and Barbuda, Trinidad and Tobago, Venezuela, Uruguay, Panama, and Chile. In the poorest countries, such as Haiti, Bolivia, Honduras, and Nicaragua, annual incomes average less than U.S. $3,000 per capita. Many of these countries have unequal distribution of incomes within their populations. Across the region, the richest 10% earn 25 times more than the poorest 10%. In Honduras this difference is a factor of 55, while in Uruguay and El Salvador it is a factor of 15.

However, both poverty and inequality have declined across the region as a result of stronger economies, more women in the workforce, and policies that raised minimum wages and provided support for the poor. In Brazil and Mexico, government programs now focus on reducing poverty. Brazil's internationally praised *Bolsa Familia* (family allowance) program gives money to poor families as long as their children attend school and are vaccinated. The funds are usually spent on food, school supplies, and clothing. Mexico has a similar cash transfer program,

called *Oportunidades,* which is conditional on school and clinic attendance and helps the poorest 30% of the population.

- **Social and Health Indicators** Social and health conditions are often considered a better measure of overall inequality within and between countries than economic measures. National improvements in life expectancy, infant mortality, and literacy, for example, reflect improvements at the lower end of the scale rather than for the better-off segments of the population. Latin America and the Caribbean tend to compare more favorably to the rest of the world on social and health indicators than on measures of income and income inequality. For example, life expectancy averages 75 years and adult literacy rates are also relatively high, averaging 92%.

There are wide gaps in social and health conditions within the region. Haiti, Central America, and the Andes tend to have much worse conditions than do southern South America, northern South America, Costa Rica, Mexico, Cuba and the English-speaking Caribbean.

BRICS

The World Bank and other UN agencies have often clustered countries in Latin America and the Caribbean with those in Africa and Asia into the category of less economically well off developing countries. But the utility of this label is breaking down as key countries in Asia, Latin America, and Africa emerge as major economic and political powers. The emerging economies of Brazil, India, and China joined with Russia in 2009 to form an alliance of countries referred to as BRIC. South Africa joined the group in 2010 as a strong representative and economic leader on the African continent. When South Africa joined BRIC, the group adopted a new acronym: BRICS.

The BRICS countries were home to almost 3 billion people in 2013 and include massive economies that collectively rank at the top of the world on several major indicators (FIGURE 7.4.1). When they meet, their agenda includes the global economy, financial reform, and strategies for cooperation in the global arena.

Brazil is the economic leader in Latin America. Brazil benefits from abundant and diverse natural resources, including agricultural land and minerals, and from a manufacturing sector supported by private sector entrepreneurs and the government. A global boom in prices for agricultural commodities produced by Brazil, such as beef and soybeans, has contributed to economic success. Investments in aircraft (such as Embraer commercial planes), automobiles, textiles, and iron and steel have also paid off in terms of exports. Other features of Brazil's economic success include government support for science and technology and social policies that have sustained domestic consumption and a growing middle class. However, there were some signs in 2015 that the Brazilian economy was faltering and some social welfare programs were cut back.

Global Rankings: BRAZIL	
5	Area
6	Population (2014)
7	GDP (2014)
24	Exports (2014)
8	Foreign exchange reserves (2011–2014)
6	Mobile phones (2011–2013)

Global Rankings: SOUTH AFRICA	
25	Area
24	Population (2014)
33	GDP (2014)
37	Exports (2014)
38	Foreign exchange reserves (2011–2014)
28	Mobile phones (2011–2013)

Global Rankings: RUSSIAN FEDERATION	
1	Area
10	Population (2014)
10	GDP (2014)
10	Exports (2014)
7	Foreign exchange reserves (2011–2014)
5	Mobile phones (2011–2013)

Global Rankings: INDIA	
7	Area
2	Population (2014)
9	GDP (2014)
19	Exports (2014)
11	Foreign exchange reserves (2011–2014)
2	Mobile phones (2011–2013)

Global Rankings: CHINA	
4	Area
1	Population (2014)
2	GDP (2014)
1	Exports (2014)
1	Foreign exchange reserves (2011–2014)
1	Mobile phones (2011–2013)

BRAZIL
POP: 190 million
GDP per capita: $10,900
Literacy: 90.3%
Life expectancy: 73
Work force*: 101.7 million

SOUTH AFRICA
POP: 49.99 million
GDP per capita: $10,700
Literacy: 86.4%
Life expectancy: 55.2
Work force: 17.32 million

INDIA
POP: 1.21 billion
GDP per capita: $3,400
Literacy: 61%
Life expectancy: 66.5
Work force: 467 million

RUSSIAN FEDERATION
POP: 139.4 million
GDP per capita: $15,900
Literacy: 99.4%
Life expectancy: 66.2
Work force: 75.55 million

CHINA
POP: 1.33 billion
GDP per capita: $7,400
Literacy: 93%
Life expectancy: 74.5
Work force: 812.7 million

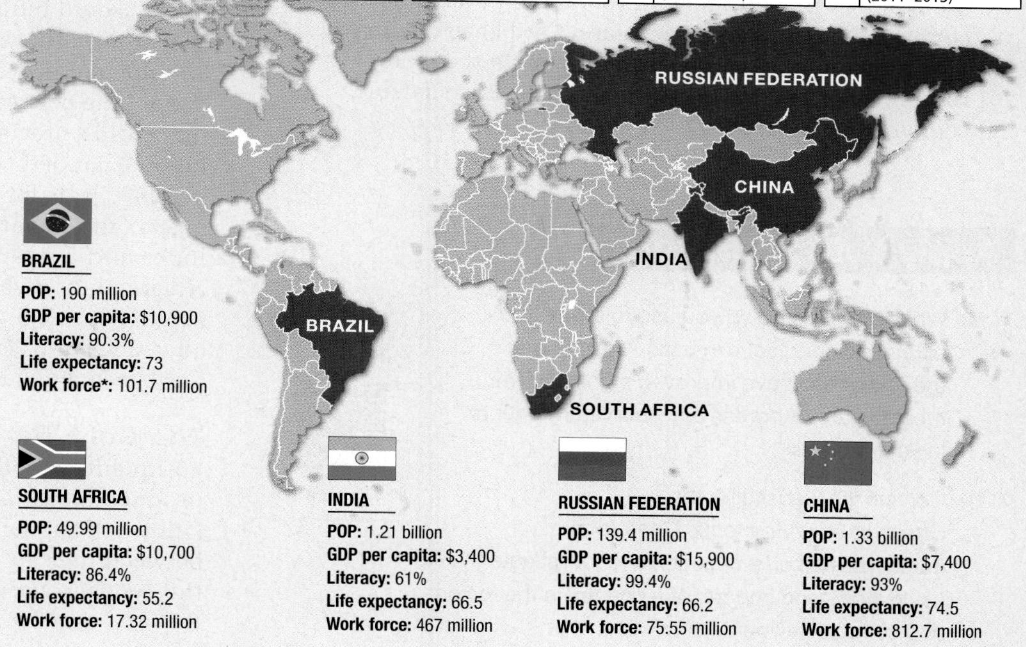

▲ FIGURE 7.4.1 BRICS Characteristics The Brazilian economy is now the 7th largest in the world, along with other emerging economies such as India and China.

Emerging economies such as Brazil and the other BRICS countries represent a shift in global economic power away from North America, Europe, and Japan. Members of the new group have joined to challenge the United States on several issues. The political power of Russia and China is already represented in their permanent membership of the UN Security Council, and Brazil and India are leading possibilities should the Security Council's permanent membership be expanded.

"Emerging economies such as Brazil and the other BRICS countries represent a shift in global economic power away from North America, Europe, and Japan."

1. Consult economic data for Latin America and the Caribbean (e.g., at worldbank.org) and discuss whether any other country in the region can match Brazil for the size and rapid growth of their economy.

2. Consult the MIT observatory of economic complexity (atlas.media.mit.edu) for information on Brazil's imports and exports. Which countries receive most of Brazil's exports and what are they? Where from and what does Brazil import?

For example, in 2013, the average Haitian lived only 63 years and the literacy rate was only 49%, compared to a life expectancy of 80 years in Costa Rica where literacy reaches 97%.

- **Spatial Distribution of Inequality** National indicators also hide large variations in economic and social conditions within the region. In Mexico, the southern regions of the country have lower incomes and life expectancies than do the northern and central areas of the country. In Brazil, the northeastern and Amazon zones have higher infant mortality, lower life expectancy, and lower average monthly incomes than the southern parts of the country. Each Latin American and Caribbean country has its own geography of inequality, with the more rural regions generally experiencing lower social and economic conditions.

Apply Your Knowledge

7.13 What evidence could you use to argue that economic conditions in Latin America and the Caribbean have improved significantly in recent years? Address any counterarguments in your answer.

7.14 Consult the website for the *Observatory of Economic Complexity* (https://atlas.media.mit.edu/) and investigate the key exports and imports of a country in the region and summarize your findings.

International Trade Data

https://goo.gl/jTO36l

Territory and Politics

Latin America and the Caribbean is a dynamic world region. Economic, political, cultural, and social changes have been rapid in the 20th century and have varied in their nature and impact among and within countries, which have taken divergent political paths. The region has included socialist and military governments; authoritarian, single-party, and multiparty systems; and highly centralized and very localized administrations. The challenges of creating functioning national governments and promoting economic growth dominated the postindependence period in the 19th century. The 20th century saw regional factions, the working class and the poor demanding reform through revolution and populist movements, and threats to military and authoritarian rule. One of the most dramatic shifts in the region has been the transition from the military and authoritarian governments that predominated in the 1970s to mostly democratic systems by 2000.

U.S. Influence Latin America in the 20th century became caught up in Cold War politics between the United States and the Soviet Union. The United States has traditionally viewed outside intervention in the region by any country other than itself as threats to its own security. This explains U.S. involvement and intervention in Latin America and the growth of U.S. economic and political dominance in the region (**FIGURE 7.27**). To that end, the United States intervened for their political interests and economic access in Cuba (1896–1922), Haiti (1915–1934), and Nicaragua (1909–1933).

An example of U.S. interests is the Panama Canal, which joins the world's two great oceans—the Atlantic and the Pacific—across the Isthmus of Panama in Central America and shortens the ocean trade route between them by weeks (**FIGURE 7.28**). In November 1903, the Panama Canal treaty specified that in a zone set aside for the purpose, the United States would build a canal, then administer, fortify, and defend it "in perpetuity." In 1914, the United States completed the 83-kilometer (50-mile) lock canal, dramatically cutting shipping travel time between the Pacific and Atlantic. The canal is one of the world's greatest engineering triumphs. A new treaty signed in 1977 handed control of the Canal Zone to Panama at the end of 1999. Tolls for the canal bring in more than U.S. $1.9 billion a year, and Panama is currently expanding the canal with new locks and a deeper channel for larger supersized ships. The Nicaraguan government recently approved Chinese plans to rapidly build a competing 173-mile Nicaragua Canal taking advantage of passage across Lake Nicaragua and connecting the Caribbean and Pacific.

The Cold War and Revolution At the beginning of the 20th century, social and economic inequality in the region produced frustration among the poor, the landless, and opposition or regional factions in several countries. As the Cold War between the capitalist West and the Soviet Union intensified in the 1950s, a series of revolutions in Mexico, Guatemala, Cuba, Bolivia, and Nicaragua reverberated around the hemisphere and the world.

The Cuban Revolution, led by Fidel Castro in 1959, created a socialist state on the largest island in the Caribbean (see chapter opener). After the revolution, the United States took an aggressive stance against the Cuban government. It mounted the Bay of Pigs invasion in 1961, embargoed trade, and closed its embassy. Relations improved in 2014 with the release of political prisoners and changes in U.S. rules for travel and trade.

In Chile and Guatemala, revolutions led to redistribution of land and nationalization of key industries that threatened the local elite and U.S. interests. The United States was implicated in assassinations and military coups that overthrew socialist leaders Jacobo Árbenz Guzmán in Guatemala in 1954 and Salvador Allende in Chile in 1973.

Authoritarianism In the 1960s and 1970s, the threats of economic instability and communist ideas contributed to a rise in nondemocratic authoritarian and military governments in the region. The military took control of governments in Brazil in 1964, Chile and Uruguay in 1973, and Argentina in 1976. Although authoritarian control provided some degree of economic stability and growth, the military governments aggressively kept social order by repressing dissent, especially among students and workers who were branded as having socialist ideals. In Argentina,

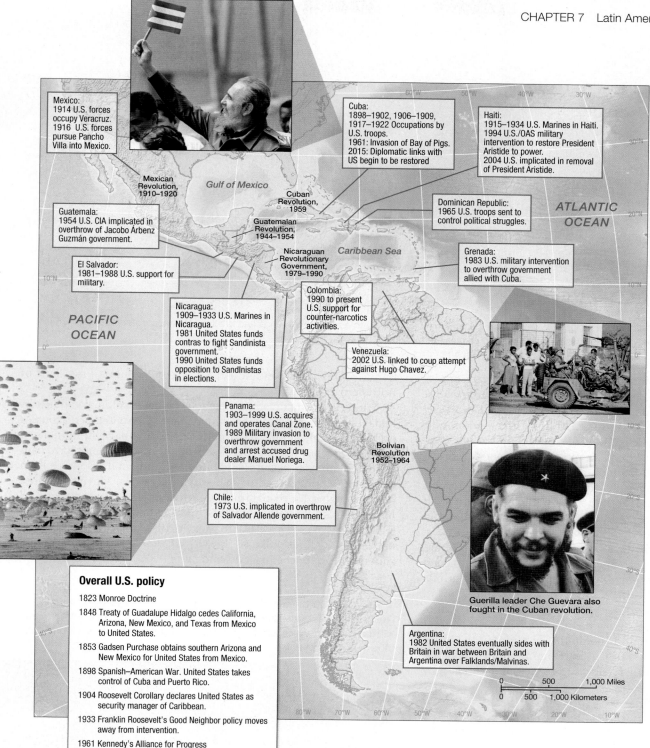

Mexico:
1914 U.S. forces occupy Veracruz.
1916 U.S. forces pursue Pancho Villa into Mexico.

Mexican Revolution, 1910–1920

Gulf of Mexico

Cuba:
1898–1902, 1906–1909, 1917–1922 Occupations by U.S. troops.
1961: Invasion of Bay of Pigs.
2015: Diplomatic links with US begin to be restored

Haiti:
1915–1934 U.S. Marines in Haiti.
1994 U.S./OAS military intervention to restore President Aristide to power.
2004 U.S. implicated in removal of President Aristide.

Guatemala:
1954 U.S. CIA implicated in overthrow of Jacobo Árbenz Guzmán government.

Cuban Revolution, 1959

Guatemalan Revolution, 1944–1954

Dominican Republic:
1965 U.S. troops sent to control political struggles.

ATLANTIC OCEAN

El Salvador:
1981–1988 U.S. support for military.

Caribbean Sea

Nicaraguan Revolutionary Government, 1979–1990

Grenada:
1983 U.S. military intervention to overthrow government allied with Cuba.

Nicaragua:
1909–1933 U.S. Marines in Nicaragua.
1981 United States funds contras to fight Sandinista government.
1990 United States funds opposition to Sandinistas in elections.

Colombia:
1990 to present U.S. support for counter-narcotics activities.

PACIFIC OCEAN

Venezuela:
2002 U.S. linked to coup attempt against Hugo Chavez.

Panama:
1903–1999 U.S. acquires and operates Canal Zone.
1989 Military invasion to overthrow government and arrest accused drug dealer Manuel Noriega.

Bolivian Revolution 1952–1964

Chile:
1973 U.S. implicated in overthrow of Salvador Allende government.

Guerilla leader Che Guevara also fought in the Cuban revolution.

Argentina:
1982 United States eventually sides with Britain in war between Britain and Argentina over Falklands/Malvinas.

Overall U.S. policy

1823 Monroe Doctrine

1848 Treaty of Guadalupe Hidalgo cedes California, Arizona, New Mexico, and Texas from Mexico to United States.

1853 Gadsden Purchase obtains southern Arizona and New Mexico for United States from Mexico.

1898 Spanish–American War. United States takes control of Cuba and Puerto Rico.

1904 Roosevelt Corollary declares United States as security manager of Caribbean.

1933 Franklin Roosevelt's Good Neighbor policy moves away from intervention.

1961 Kennedy's Alliance for Progress

0   500   1,000 Miles
0   500   1,000 Kilometers

◀ **FIGURE 7.27 U.S. Interventions in Latin America and Latin American and Caribbean Revolutions** This map shows U.S. military intervention and those areas where there were major revolutions in the 20th century. In some cases, the United States intervened as a result of Cold War politics and in the interests of U.S. national security to prevent formation of governments allied with the Soviet Union or with socialist orientation. Most revolutionary movements have been inspired by calls for land reform and socialist policies. **Research one of these events and summarize the arguments for and against U.S. intervention.**

the military government's so-called Dirty War against dissenters in the 1970s killed 15,000 people and forced many others to leave the country. In Chile, the military government of General Augusto Pinochet orchestrated similar "disappearances" and human rights abuses from 1973 to 1990. In Guatemala, authoritarian governments suppressed leftist groups and forced Maya populations to flee into neighboring Mexico or farther north, or to retreat into remote mountains, where the military, which did not differentiate between civilians and guerillas, annihilated whole communities.

Apply Your Knowledge

7.15 Use the Internet to research an example of an authoritarian/military government, leftist revolution, or U.S. intervention in Latin America or the Caribbean. (See the examples in Figure 7.23.)

7.16 List three ways the memories and repercussions of the period and events you research are still influencing domestic politics or international relations in the area.

▲ FIGURE 7.28 **Panama Canal** The Panama Canal cuts through the isthmus that joins North and South America and is a critical transport route for both cruise ships and cargo. Large ships must pass through several locks such as these at Miraflores, Panama.

Democracy Public and foreign outrage at authoritarian repression and human rights violations, the inability of military governments to solve economic problems, the end of the Cold War, and international and internal pressures that linked economic globalization to democratic governance, resulted in gradual transitions to democratic governments in most Latin American countries. For example, transitions took place in Argentina in 1983, Brazil in 1985, and Chile in 1989. In Argentina, the departure of the military government was hastened by the loss of a war with Britain when Argentina invaded the Falkland Islands (called the Islas Malvinas in Latin America) in 1982.

Democracy has been fragile and some elections are corrupted or lead to violence. In Argentina and Guatemala trials have been held to prosecute those responsible for human rights violations under authoritarianism (**FIGURE 7.29**). While many countries in the region have elected neoliberal governments who cut public services and opened trade, there has been a backlash against budget cuts, privatization, and inequality in countries such as Brazil, Venezuela, Bolivia, and Chile, bringing more left wing, antiglobalization, and populist leaders to power after 2000.

Social Movements and Indigenous Rights Political opposition and activism in the region often take the form of **social movements** that organize against cuts in government services, for land reform, and for specific resources and issues, such as housing, water, human rights, or environmental protection. New Indigenous social movements have organized to promote language, culture, and land rights.

In Brazil, the Rural Landless Workers Movement (*MST* in Portuguese), supported by an estimated 1.5 million people, organized land occupation, winning titles for 350,000 families, and achieved significant political change when politicians sympathetic to their cause eventually won national elections. In Bolivia, coca farmers organized against the suppression of coca cultivation, and public-sector employees fought against job cuts and the privatization of water. The leader of the coca farmers,

Evo Morales, was elected as the first Indigenous president of Bolivia in 2006.

Latin America's diverse Indigenous cultures, colonial and neocolonial links to Europe and the United States, international discourses on human rights, and Marxist political thought have all influenced these Indigenous emancipatory social movements. Many have consolidated forces to defend land and environmental resources from environmentally-degrading extractive industries. Movements that have struggled for recognition, political power or autonomy, and land rights, include the Ejercito Zapatista de Liberación Nacional (EZLN, known as Zapatistas) in Mexico, the Confederation of Indigenous Nationalities of Ecuador (CONAI), Bolivia's Conamaq organizations, the Mapuche in Chile, and more (**FIGURE 7.30**).

▼ FIGURE 7.29 **Guatemala Human Rights** The 2013 genocide trials in Guatemala included testimony from Indigenous Mayans whose families were killed and communities terrorized during the dictatorships of Rios Montt and other authoritarian rulers.

▲ **FIGURE 7.30 Indigenous Social Movements** Protestors occupy the Confederation of Indigenous Nationalities of Ecuador (CONAIE) headquarters in Quito, Ecuador in January 2015, responding to land evictions and expressing demands for Indigenous autonomy.

Drug Economy Drugs have seriously destabilized politics in some regions of Latin America, especially in Bolivia, Colombia, Mexico, and Peru. Latin America produces drugs that are illegal in many countries, including cocaine (from the coca plant), heroin (from poppies), marijuana, and methamphetamine (meth). Latin American farmers grow drugs because of the high prices they yield in comparison to other agricultural products. In areas where crop yields are low, people have only small plots of land, and market prices for legal agricultural crops do not cover production costs, drug production is an attractive option or necessary evil. Most of the drugs produced in the region are exported to the United States and Europe. The farmers receive only a fraction of the street value of the drugs.

The bulk of drug exports are controlled by powerful families in Colombia and Mexico, who move the drugs from rural Latin America by land, air, and boat into the United States, often through Los Angeles and Miami. The United States has funded and supported efforts to eradicate drug production (e.g., by spraying herbicides) and arrest drug cartel leaders in Latin America. But some analysts argue that the United States should focus on controlling demand within its own borders or legalization of consumption.

In some areas, the drug trade has exacerbated political conflicts. In Colombia, several guerilla movements with links to powerful drug lords controlled large areas and assassinated local officials and police. Working with the United States the Colombian government eventually began to destroy fields and capture the leaders, extradite them to the United States for trial, and provide alternative livelihoods for farmers. By 2010, cocaine production was cut in half, and the area saw a corresponding drop in violence and the political corruption associated with drug production.

As the power of the drug economy declined in Colombia, it rose steeply in Mexico (**FIGURE 7.31**). Mexican **cartels**—organizations and networks that work together to control a product—now control most of the drug flow to the United States, including cocaine from Colombia and meth, heroin, and marijuana from Mexico. Drug traffickers control production areas and distribution networks and also influence the police, the army, judges, and political leaders through intimidation and bribery. The legalization of marijuana in some U.S. states is starting to reduce the power of some cartels.

Apply your knowledge

7.17 Research a social movement in the region and identify what they are asking for and whether they have had any success

7.18 Discuss your views on the solution to the drug crisis in the Americas

Culture and Populations

A mixed racial and ethnic composition of Latin America and the Caribbean influences the cultural heritage and social practices in the region. Indigenous cultures—including traditional dress, crafts, ceremonies, and religious beliefs—persist in regions such as highland Guatemala and the Andes.

Religion

One of the main objectives of Spanish and Portuguese colonialism was the conversion of Indigenous peoples to Catholicism. Although some Indigenous people fiercely resisted missionary efforts, others found ways to blend their own traditions with those of the Catholic Church and create new **syncretic** religions. The term *syncretic* means that practices have coevolved and merged with one another over the centuries into blended religions.

The slave trade brought African religious traditions to Latin America and the Caribbean, and these often merged with Indigenous and Catholic beliefs into syncretic beliefs that are followed by more than 30 million people. These include rituals involving dance, drums and spirit trances, altars with candles and flowers, and traditional herbs, medicine, and chicken sacrifice. Examples include Candomblé and Umbanda in Brazil, Voodoo in Haiti, and Santería in Cuba and other islands. In Jamaica, Rastafari belief has roots in Ethiopia and Christianity, and includes the spiritual use of cannabis, a vegetarian diet, and wearing hair in dreadlocks.

Recent decades have seen the emergence of a new form of Catholic practice, **liberation theology**, which focuses on needs of the poor and disadvantaged and aligned with social movements. But the Catholic Church has experienced a decline in followers, primarily from conversion to Protestant Christian faiths, especially Evangelical groups with messages of literacy, education, sobriety, frugality, and personal salvation. Sixty-nine percent of people in Latin America and the Caribbean are currently followers of Catholicism and 19% are Protestant. Islam, Judaism, and other belief systems account for 2%, while 8% claim no religious affiliation. The election of Argentinian, Pope Francis, in 2013, has revived interest and pride in Catholicism in Latin America.

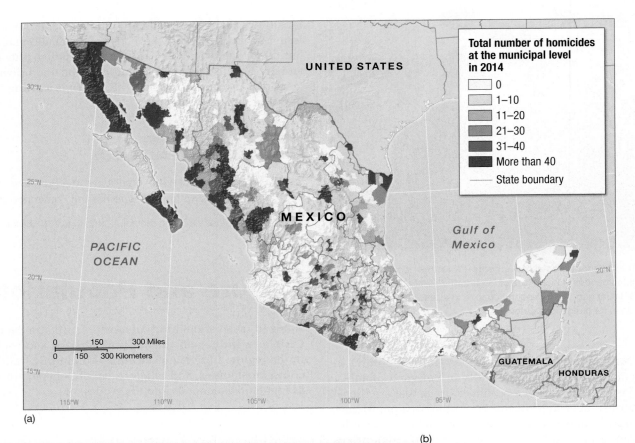

(a)

(b)

▲ **FIGURE 7.31 Drugs in Mexico** (a) Drugs are a major cause of violence in Mexico with thousands of homicides connected to struggles between cartels and including the murders of journalists and local politicians. Human Rights Watch reported deaths of more than 60,000 people in drug-related violence from 2006 to 2012. (b) The Mexican police and army has stepped up attempts to capture leaders of the cartels fueled by hundreds of millions of dollars for military spending from the United States. The Mexican army is struggling to control a deteriorating situation fueled by corrupt police forces, ongoing demand for illicit drugs in the United States, and the flow of U.S. guns into Mexico.

Language Indigenous languages endure in several regions of Latin America (**FIGURE 7.32**). Widely spoken languages include Quechua in the Andean region (more than 10 million people); English Creole mostly in Jamaica, Barbados, and the Bahamas (3.8 million); French Creole in Haiti (10+ million); Martinique (10+ million), Guaraní in Paraguay (4.9 million); Aymara in the Andes (2.8 million); Mayan in Belize, Guatemala, Honduras, and southern Mexico (6 million); and Nahuatl in Mexico (1.5 million).

Spanish is the dominant language across most of mainland Latin America, except for Brazil (Portuguese), Belize and Guyana (English), Suriname (Dutch), and French Guiana (French). In the Caribbean, the dominant languages are English, Spanish, French, and Dutch as well as creoles. (simplification or mixtures of languages)

▲ **FIGURE 7.32 Languages of Latin America** The official language of most of mainland Latin America is Spanish, except for Brazil, whose official language is Portuguese; French Guiana, French; Suriname, Dutch; and Belize and Guyana, English. The dominant European languages of the Caribbean reflect the colonial histories of the islands.

popular media, especially television. The common experience of popular television shows—especially *telenovelas* (soap operas)—and education systems and media that promote modern urban lifestyles, have partly erased differences within the region.

Food The traditional foods of Latin America blend Indigenous crops, such as corn or potatoes, with European influences, especially from Spain (including Arab cuisine). Although Mexico and Jamaica are associated with spicy dishes that include *chile,* the food is quite mild in the rest of Latin America. In livestock-producing areas, such as Argentina, grilled meat is extremely popular, but in much of Latin America, the poor eat simple meals of rice, corn, potatoes, and—for protein—beans. Modified versions of Mexican cuisine have diffused throughout North America and are the basis for many chain restaurants. Foods in the Caribbean reflect the medley of cultures in the region, combining African, Asian, and European influences.

Music, Art, Film, and Sports Latin American and Caribbean art and literature display incredible variety and regional specialization. Traditional textiles, pottery, and folk art are sold to tourists and by import stores in North America and Europe. Literary traditions include the magical realism of Nobel Prize winning authors such as Colombia's Gabriel García Márquez and Peru's Mario Vargas Llosa. Caribbean authors such as Derek Walcott and V.S. Naipaul have won many international prizes for literature. Works of noted Mexican artists such as Frida Kahlo and Diego Rivera are numbered among the masterpieces of global 20th-century art (**FIGURE 7.33**).

Apply Your Knowledge

7.19 What evidence could you marshal to argue against stereotyping Latin America as a region that is Spanish-speaking, Catholic, and comprised of mostly Indigenous populations?

7.20 Use the Internet to research an Indigenous group in Latin America including religion, language, and culture.

Cultural Practices, Social Differences, and Identity

Cultural practices in Latin America have been mostly evolving toward global cultural trends promoted by formal education and

▶ **FIGURE 7.33 Diego Rivera and Frida Kahlo** World renowned Latin American mid-20th century artists Diego Rivera, who focused on murals, and FridaKahlo, who produced iconic, and sometimes disturbing, paintings, are shown in a studio in Mexico City.

Latin American and Caribbean music enjoys worldwide popularity. Caribbean global influences include the reggae of Jamaica and steel drum bands of Trinidad, which resonate with the rhythms of Africa. Latin music includes salsa, samba (Brazil), mambo (Cuba), tango (Argentina), and mariachi and ranchera (Mexico). Also very popular is Latin pop and rock music, often produced in Miami, Latin America's business capital in the United States, and featuring stars such as Jennifer Lopez and Enrique Iglesias. Mexico, Argentina, and Brazil have thriving film and TV industries.

Soccer (fútbol) is the most popular sport in mainland Latin America, but the region's athletes also excel in other sports. The English-speaking Caribbean has produced some of the world's best cricket players such as Antigua's Vivian Richards. Several Spanish-speaking islands, such as the Dominican Republic, are associated with outstanding U.S.–league baseball players, such as Alex Rodriguez and Albert Pujols. Jamaica's Usain Bolt held the world 100-meter sprint record after the 2012 London Olympics and both men and women from the Caribbean often win athletic medals. (**FIGURE 7.34**).

Gender Relations Certain cultural views of the family and gender roles remain powerful in the region. Multiple generations often live and work together, individual interests are subordinated to those of the family. Traditions of **machismo** and **marianismo** define gender roles within the family and the society in parts of Latin America. Machismo constructs the ideal Latin American man as fathering many children, dominant within the family, proud, and fearless. Marianismo constructs the ideal woman in the image of the Virgin Mary; she is chaste, submissive, maternal, dependent on men, and closeted within the family. Latin American and Caribbean society has been generally patriarchal, and institutions have prohibited or limited women's right to own land, vote, get a divorce, and secure a decent education.

These stereotypes are, of course, contradicted by many individual cases and are breaking down in the face of new geographies and global cultures. Family links are weakened through migration and the isolation of many living spaces—from each other and from those of other family members—in urban environments. Men's and women's roles are changing as fertility rates decline, women enter the workforce and politics, and migration and divorce lead to many women-led households. Latin American and Caribbean feminists have organized to obtain the right to vote; to effect changes in divorce, rape, and property laws; to gain access to education and jobs; and to elect women to political office. Women head (or recently led) governments in Argentina (Cristina Fernández de Kirchner), Brazil (Dilma Rousseff), Chile (Michelle Bachelet), and Costa Rica (Laura Chinchilla).

Still, gender inequality is prevalent in Latin America (Figure 7.35). Female literacy, on average, is 2% to 15% lower than that of male populations, although among youth average female literacy now exceeds male literacy. In 2014 women held 27.4% of the seats in Latin America and the Caribbean's national parliaments. The employment ratio for adult women is 50%, while for men this number is 75%. Women tend to earn much less money on average than men and hold traditionally female jobs such as domestic service and food processing. In Latin America and the

(a)

(b)

▲ **FIGURE 7.34 Sports** Jamaican athletes include sprinters (a) Usain Bolt and (b) Shelly-Ann Fraser-Pryce who won three medals each at the London Olympics in 2012.

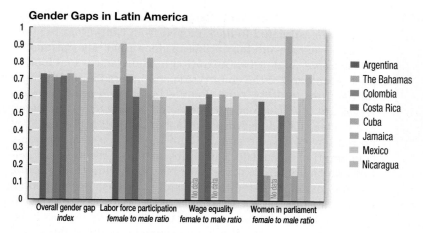

Gender Gaps in Latin America

Legend: Argentina, The Bahamas, Colombia, Costa Rica, Cuba, Jamaica, Mexico, Nicaragua

▲ **FIGURE 7.35 Gender in Latin America and Caribbean** Nicaragua is the most gender-equal country in the LAC region, according to a 2014 report from the World Economic Forum. The report ranks gender gaps in health and survival, education, politics, and economic equality. **Which of the countries shown here has the greatest and least gender gap overall and in terms of employment, wages, and political representation? (the axis shows the ratio of females to males, e.g., women earning less than 0.6 of what men do)**

Caribbean, women earned between 60% and 90% of the male wage in 2013 (**FIGURE 7.35**).

Demography and Urbanization

Prior to the arrival of the Europeans in about 1500, Latin America is estimated to have had a population of approximately 50 million people, including large concentrations within the empires of the Aztecs and Incas and many smaller groups of hunters, gatherers, and agricultural communities. The demographic collapse discussed earlier in this chapter dramatically reduced Indigenous populations, but significant indigenous populations remained in Mexico, northern Central America, and the Andes.

Colonialism also changed the demographic profile of Latin America through the intermixing of European and Indian peoples and the importation of slaves from Africa to the Americas. Racial mixing occurred (by force and by choice) among European, Indian, and African populations, especially in Brazil and Mexico. The resulting mixed-race populations were classified according to their racial mix. The most common category was that of **mestizo** (a mix of European and Indigenous); other categories included *mulatto* (Spanish/African) and *zambo* (African/Indian).

Slave imports to Latin America from Africa totaled more than 5 million people during the colonial period, including 3.5 million to Brazil and 750,000 to Cuba. Many of the Caribbean islands, including Haiti and Trinidad, with very small Indigenous and European populations, had a large number of African slaves working on plantations. African populations also settled along the plantation coasts of Mexico, Central America, northern South America, and Ecuador. Although slavery was not abolished until the mid-1800s (1888 in Brazil), escaped and freed slaves formed communities as early as 1605, most famously the African community of Palmares, which was an autonomous republic from 1630 to 1694 in the Brazilian interior. Escaped and liberated slaves created settlements, also called **maroon communities**, in other regions as well, such as Jamaica.

Recent population censuses that have attempted to record race and ethnicity show some general patterns that correlate with the population history of the region, including where Africans were working as slaves on plantations, where Europeans settled in cooler climates, and where Indigenous groups remained in remote areas. The largest percentages of Indigenous populations are in Guatemala, Bolivia, Peru, and Mexico; the largest black populations are in Brazil, Colombia, Cuba, Venezuela, Panama, and the Dominican Republic; and the largest numbers of self-identified and censused white populations are in Argentina, Chile, Costa Rica, and Uruguay.

Asian immigration to the region began during the colonial period and picked up after the end of slavery when Chinese, Indian, and Japanese workers were brought over to work on plantations and in construction. There are now several million people of Asian descent in Latin America especially people of Chinese and Japanese heritage in Brazil and Peru. Europeans other than the Spanish and Portuguese settled in the more temperate climates in the region, especially in Argentina where many families have Italian, German, Welsh, or British names. Asian Indian migrants have moved to the Caribbean.

Brazil has promoted an image of racial democracy and equality, and musical, religious, and dietary traditions merge into a uniquely Brazilian culture. This myth of racial democracy is contradicted by evidence of continuing racism in Brazil and other Latin American countries. For example, Indigenous people face multiple forms of prejudice across the region. Furthermore, studies show that race and class correlate strongly. Afro-Brazilians are on the whole poorer, less healthy, less educated, and more discriminated against in employment and housing. In Mexico, the media have tended to promote lighter skin as more desirable through the choice of more European-looking actors in commercials and programs.

Demographic Change The overall population of contemporary Latin America totaled 588 million people in 2014, with almost 40 million in the Caribbean. The distribution is clustered around the historical highland settlements of Central America and the Andes and in the coastal colonial ports and cities (**FIGURE 7.36**). The population has grown rapidly since 1900, when the regional total was 100 million, mainly as a result of high birthrates and improvements in health care. Brazil (200 million) and Mexico (122 million) have the largest populations today. Fertility rates have declined throughout much of the region. The 2010–2015 rate approached under two births per woman in Argentina, Brazil, Chile, Colombia, Cuba, Mexico, Uruguay, and many small islands of the Caribbean. Although this rate is slowing and will eventually reverse population growth, a large percentage of the population is under age 15, especially in Central America, and populations are likely to continue to grow as this age group enters its reproductive years (**FIGURE 7.37**).

Higher fertility rates are characteristic of poorer, rural regions, where infant mortality is high, children can contribute labor in the fields, and women do not have access to education, employment, or contraception. Fertility rates in the region have dropped as people move into the cities, health care improves, incomes increase, and more women work and are formally educated.

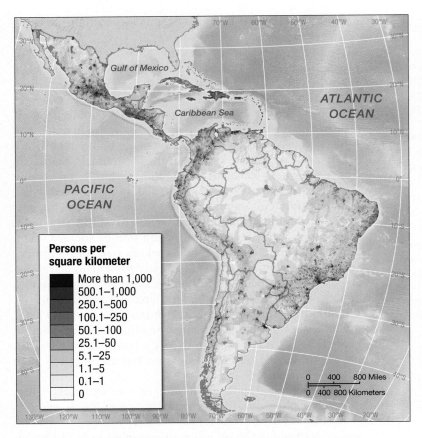

▲ **FIGURE 7.36 Population Distribution of Latin America** The general distribution of population in Latin America includes sparsely settled interiors and high population densities around the historical highland regions of the Andes, Central America, Mexico, and the coastal regions, especially former colonial ports and cities.

Migration Within the Region More than 150 million people are estimated to have moved from rural areas to cities in Latin America and the Caribbean since 1900. The reasons for this massive rural–urban migration include factors that tend to push people out of the countryside and others that pull people to the cities (**FIGURE 7.38**). People leave rural areas because wages are low; because services such as safe drinking water, health care, and education are absent or limited; or because they do not have access to enough land to produce food for home consumption or for sale. Unemployment as a result of agricultural mechanization, price increases for agricultural inputs, and the loss of crop and food subsidies has also driven people from rural areas to the cities. Other push factors have been environmental degradation and natural disasters, as well as long-running civil wars or military repression of rural people.

Cities pull migrants because they offer high wages and more employment opportunities, as well as access to education, health, housing, and a wider range of consumer goods. Governments often have an urban bias and provide services and investment to cities, which are seen as the engines of growth and the locus of social unrest. Social factors that encourage migration to the cities include the promotion of urban lifestyles and consumption habits through television and other media, and long-standing social networks of friends and families that link rural communities with people in cities who can provide housing, contacts, and information to new migrants.

Although most people in the region have migrated to cities within their own countries, there are several other important migration flows within the region. Several countries have encouraged the colonization of remote frontier regions by providing cheap land and other incentives. For example, the building of roads and availability of land in the Amazon

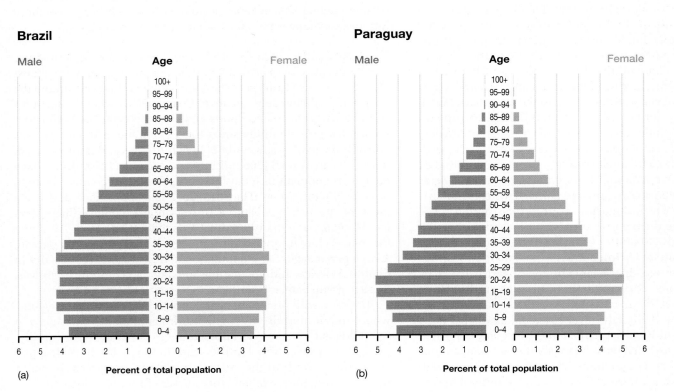

(a)

(b)

Percent of total population

◄ **FIGURE 7.37**
Population Pyramids Fertility rates have dropped considerably in most of Latin America and this is slowly reducing the size of younger age cohorts to create straighter sided population pyramids. Whereas (a) Brazil is approaching replacement fertility, (b) Paraguay still has a large proportion of young people moving into their reproductive years.

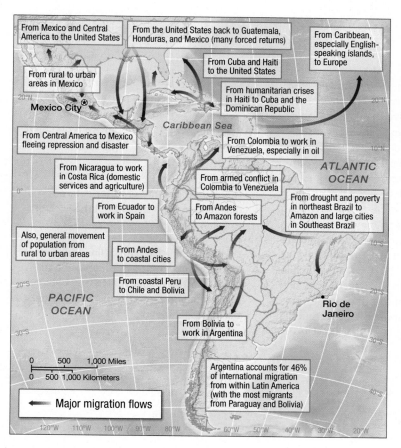

▲ **FIGURE 7.38 Major Migration Streams in Latin America** There are major migration streams within countries and between countries in Latin America and to the rest of the world from Latin America. The most significant overall trend is rural-to-urban migration, but poor people are also moving into frontier regions in the Amazon and southern Mexico; from the Andes to work in Argentina and Venezuela; and away from political unrest and natural disasters. Caribbean people have moved to Europe, especially from the English-speaking islands, and to the United States, especially from Cuba, Haiti, the Dominican Republic, and Puerto Rico.

created a stream of migrants from coastal regions of Brazil to the interior, and the development of irrigation in Mexico and Chile attracted migrants to desert regions. Central Americans have moved to Mexico, sometimes en route to the United States. People have moved out of the Andes to work in mining, agriculture, and oil in Argentina and Venezuela (see "Faces of the Region: Bolivian Youth in Argentina," p. 296), and out of Central America to Mexico either as refugees or workers seeking higher wages. There is a modest flow of migrants seeking work, especially in tourism, between Caribbean countries and refugee flows, such as from Haiti to the neighboring Dominican Republic.

Latin American and Caribbean Migration Beyond the Region
People have also left the region in considerable numbers, creating a global Latin American and Caribbean diaspora. The United States hosts the largest number of people outside the region who define themselves as being of Latin American or Hispanic heritage. Many Mexican families became part of the United States when the land they lived on became a U.S. territory following the U.S.–Mexican War in 1848. They use the phrase "the border crossed us, we didn't cross the border" to

emphasize that they are not migrants but long-standing residents. Between 1900 and 1930, 1.5 million Mexicans (10% of the total population) migrated to the United States to escape the chaos of the Mexican Revolution and to fill labor shortages created by World War I. Although 400,000 Mexicans (some of them U.S. citizens) were deported during the Great Depression in the early 1930s, the growth of the U.S. economy from 1940 and World War II created such a demand for low-cost labor, especially in agriculture, that the U.S. and Mexican governments introduced a formal guest farmworker program. This program distributed 4.6 million temporary permits for Mexicans to work in the United States between 1942 and 1964. Many **braceros** (defined as a guest worker from Mexico given a temporary permit to work as a farm laborer in the United States) never returned to Mexico, and migration continued through social networks after the program ended, even as U.S. immigration restrictions were tightened.

In the past 50 years, Latin American and Caribbean migration to the United States has been dominated by Mexicans (about 40% of the total, and 60% of those defined as illegal or undocumented) but also includes large numbers of people from Cuba (15%) and Central America (10%). Significant Latin American populations can also be found in Canada and Europe (especially in Spain, where large numbers of migrants from Andean countries work in low-paid jobs). The Caribbean diaspora includes migration to the United States. However, because of colonial links to Britain, large numbers of Caribbean people have also migrated there or to Canada, especially from Jamaica and Barbados.

The money that is sent back to Latin America from people working in other countries is called **remittances** and can make a significant contribution to national and local economies. More than U.S. $65 billion was sent to the region from the United States in 2014. Many communities rely on these funds to build houses, purchase agricultural inputs, or educate their children. This is one of the new informal flows of international financial capital in the global economy.

Migration—both with and without documents—from Latin America to the United States has dropped dramatically since 2005 and may have reached a net flow of zero from Mexico (with migration into the United States balanced by those returning home) after 2010. Explanations for the drop-off include recession and unemployment in the United States, improved conditions in Latin America, and stricter border enforcement.

Apply Your Knowledge

7.21 Immigration from Latin America and the Caribbean to the United States is a controversial issue in U.S. politics. Conduct some research (using news websites such as cnn.com or nyt.com) and summarize the factors that promote this migration flow. Then come up with one argument in favor of this migration flow and one against.

7.22 Identify and read a novel or watch a movie about a Latin American or Caribbean migrant from the region to the United States or Europe. What challenges did they face and what traditions did they maintain from their country of origin?

Bolivian Youth in Argentina

In many rural areas of Latin America, labor migration has become a defining aspect of life and social relations. While migration to the United States from Mexico has formed the largest migratory flow in the region in the past century, Argentina is also a destination country for many Latin Americans. The largest groups of migrants to Argentina come from Bolivia and Paraguay.

In the Camacho Valley of Bolivia, located a day's bus ride from the border with Argentina, young men as young as 14 years old leave their communities to work abroad. They migrate, often in pairs or groups, with the goal of earning money, obtaining material goods through their earnings, and seeing the world beyond their towns. Women often migrate once they are in their late teens and early twenties.

After crossing the border, many Bolivians find work in Argentina's commercial agricultural sector. A 21-year-old migrant from Churquiales described working in different areas of Argentina throughout the year: the grape harvest in Mendoza in February, the tomato and pepper harvests in the north from April to September, the tomato and citrus harvests in Corrientes from August onward, and the tomato harvest in Buenos Aires from December to January. Other jobs for Bolivian men include construction and service sector work. Women often work in domestic labor.

Independent agricultural laborers can earn a daily wage in Argentina approximately four times greater than the going wage in the Camacho valley. But they earn less than most Argentinian workers. Migrants often work long hours and live in spartan conditions for 6 to 10 months so as to save money (**FIGURE 7.5.1**), then return for 2 to 6 months dedicated to leisure, socializing, and helping their families before the time comes to migrate again. At home they often struggle to negotiate their migrant identities—marked by different consumption patterns and work patterns—with the experiences and expectations of their parents and of nonmigrant community members.

1. **How might drought or other natural disasters in Argentina affect the lives of Bolivian migrants?**

2. **Consult economic data online to discover the difference in average incomes (and other socioeconomic indicators) between Argentina and Bolivia and consider how this explains the migrant flows.**

"In many rural areas of Latin America, labor migration has become a defining aspect of life and social relations."

◀ **FIGURE 7.5.1 Labor Migrants** Young migrant workers from Bolivia block the street in Argentina with a sign "We are workers not slaves."

Urbanization Eighty percent of people in Latin America and the Caribbean now live in cities (compared to only 10% in 1900), which makes it the most urbanized region in the world. Urbanization levels range from about 71% in most of Central America to more than 90% in Argentina, Uruguay, and Venezuela. The region hosts several of the world's largest metropolitan areas, sometimes called **megacities**, including Mexico City and São Paulo. These two urban areas number among the 10 largest cities in the world, and each has more than 20 million people in its metropolitan areas. Not far behind are Buenos Aires at 13.9 million and Rio de Janeiro at 11.5 million. A major cause of urban growth is migration, although the redefinition of city boundaries (to include metropolitan regions) and internal population growth have also played a role. In many countries, the population of the largest city in an urban system is disproportionately large in relation to the second- and third-largest cities in that system. This so-called **urban primacy** is characteristic of Argentina (Buenos Aires has 33% of the national population), Peru (Lima, 30%), Chile (Santiago, 35%), Mexico (Mexico City, 17%) and many of the larger Caribbean islands such as Cuba and Jamaica. Concentration of population and development in one or two cities within a country can create problems when physical and human resources, political power, and pollution are all focused in one major settlement.

The megacities are global centers of commerce but also have many urban social and environmental problems. The crowding, high land costs, violent crime, poverty, traffic problems, and pollution of the cities are starting to cause people and economic development to shift to smaller neighboring cities across Latin America.

- **Mexico City** The economic, cultural, and political center of Mexico, Mexico City produces 40% of the gross national product and is home to leading Mexican companies and the regional offices of multinational corporations. However, its modern high-rise office buildings and elegant colonial plazas are clouded with air pollution that obscures the snowy peaks of the volcanoes that ring the city (see "Sustainability in the Anthropocene: Air Pollution in Mexico City" on page 262). As in other large Latin American cities, many new migrants to Mexico City cannot afford to rent or purchase homes and settle in irregular settlements, or *barrios,* that surround the city. As much as 50% of Mexico City's housing is defined as "self-help construction" and ranges from cardboard and

▲ **FIGURE 7.40 Rio de Janeiro** Rio de Janeiro has a stunning location with a harbor overlooked by Sugar Loaf Mountain and the beaches of Copacabana and Ipanema. The city hosted the 2012 Rio+20 Earth Summit, the 2014 soccer World Cup, and will host the 2016 Olympics. These global events have caused controversy because of their cost and because facilities have displaced poor people.

plastic shanties to sturdier wood and brick structures with aluminum or tile roofs. Many of these settlements occupy steep hill slopes, valley bottoms, and dry lakebeds that are vulnerable to flooding, landslides, and dust storms. One such example, Neza-Chalco-Itza, houses more than 4 million people on the shores of Lake Texcoco.

- **São Paulo, Brazil** Located on a high plateau about 50 kilometers (30 miles) from the Atlantic coast, São Paulo has wide avenues and many skyscrapers around the central business district. But it is surrounded by neighborhoods of poorer and slum housing (**FIGURE 7.39**). São Paulo is the major financial center for Brazilian and international banks, and has recently developed a large telecommunications and information sector. Brazilian geographer Milton Santos reports that São Paulo, which employs more than 2 million manufacturing workers and produces 30% of Brazil's gross national product, has morphed from a commercial center to a manufacturing hub and then to a service and information core for the global economy. In São Paulo, 28% of residents have no drinking water, and 50% have no sanitation.

- **Rio de Janeiro** The cultural and media center of Brazil (**FIGURE 7.40**), Rio was the capital of Brazil from 1822 to 1960. Its urban structure includes an older city center with a wealthier residential zone and beaches such as Copacabana toward the south, and a poorer, more industrial zone to the north. Rio and São Paulo followed a similar pattern as other Latin American cities, attracting millions of migrants who have settled informally around the urban core. The **favelas**, a Brazilian term for the informal settlements or shanty towns that grow up around the urban core of Rio, lack good housing and services and are home to 11.4 million people.

▼ **FIGURE 7.39 São Paulo** High rise buildings dominate this economic center of Brazil.

Apply your Knowledge

7.23 Research another major city in Latin America or the Caribbean to identify some of its social and environmental problems

7.24 Find a news article about life or urban policy in Mexico City or Rio de Janeiro. What is the focus of the article and how does it reflect your understanding of the geography of the city?

Future Geographies

The region of Latin America and the Caribbean has changed dramatically in recent years as the forces of market liberalization, economic integration, democratization, urbanization, and environmental degradation have transformed the region and changed its relationships with other world regions. Each of these processes has interacted with local conditions to produce a new mosaic of distinct geographies and produced new opportunities and challenges for people and policymakers throughout the region. What are some of the prospects for the future?

Sustainable Development

The future of the region will depend on its ability to manage the resource demands and pollution emissions of new industrial development and farming technologies, and the pressures of new consumption habits. The region faces serious risks; climate change may degrade water supplies and ecosystems, increase disaster losses, and encroach on coastline. Although many countries in Latin America and the Caribbean are moving to new low-carbon energy sources and protecting forests to reduce greenhouse gas emissions, the region is vulnerable to climate change and must adapt to climate changes and sea level rise already underway (**FIGURE 7.41**).

Emerging Economies

While Brazil, Chile, Costa Rica, and Mexico are now seen as emerging rather than developing economies, there are many economic vulnerabilities and problems within the region, especially where growth depends on volatile commodities and demand (such as metals or tourism) or where there is high inflation and unemployment. The region's economic future may ultimately depend as much on markets within the region as beyond.

▲ **FIGURE 7.41 Climate Adaptation in the Caribbean** There are many options for adapting to warmer temperatures and sea level rise in the Caribbean and international assistance is available for poorer countries. But some countries, such as Haiti, are not well adapted even to current climate.

▲ **FIGURE 7.42 Political Representation** Popular protest is increasingly important in Latin American politics. Here people from many different groups march in a people's summit on the occasion of climate negotiations in Peru in 2014.

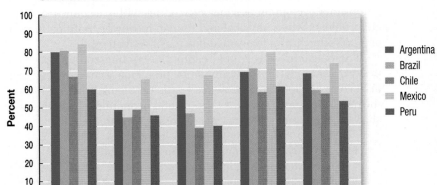

Satisfaction in Selected Latin America Countries

- Argentina
- Brazil
- Chile
- Mexico
- Peru

▲ **FIGURE 7.43 The level of concern about social issues varies in Latin America.** People are generally satisfied with life but for some countries crime, education, and employment are major concerns. **Which of the countries shown here is most concerned about these issues and why?**

Representation and Democracy

Although most of the region has moved toward more democratic governments, given Latin America and the Caribbean's diverse political parties and active social movements, there are many people who feel overlooked and undercurrents—including drugs, corruption, and poverty—that could threaten fragile democracies. Indigenous peoples and youth in particular seek representation and fair treatment (**FIGURE 7.42**).

Public Attitudes

Recent surveys of public attitudes across the region indicate many residents of Latin America and the Caribbean are positive about the future, but their greatest concerns are about economic problems and crime. They generally support democracy and environmental protection. Almost three-quarters of people in the region say they are satisfied with their lives. This region is one of great promise in our global future (**FIGURE 7.43**).

Learning Outcomes Revisited

▶ *Explain* **the physical origins and social impacts of natural and human-caused disasters and climate change.**

The region has mostly tropical climates ranging from rainforests to deserts, with cooler climates in highland regions and in southern South America. The climate and geology of the region produce many extremes, including devastating earthquakes. Changes in ocean currents and temperatures, called El Niño, periodically bring heavy rains and droughts to the Andes and other parts of the region. Poverty and substandard housing increases people's vulnerability to the effects of natural disasters and climate change.

1. *What* **are the main hazards in Latin America and the Caribbean and why are some more vulnerable to them than others?**

▶ *Describe* **the ways in which the Maya, Aztecs, and Incas adapted to environmental challenges and their relevance today.**

The Maya, Aztecs, and Incas adapted to difficult environmental conditions by terracing hillsides to farm steep slopes, developing irrigation techniques for drier areas, and using slash-and-burn farming to clear and fertilize soils in tropical forest environments. They also used raised fields to farm flooded areas and domesticated wild animals. The case of the Maya illustrates how farming that exhausts soils and deforestation that creates local climatic changes can lead to the collapse of a society. Some of their agricultural techniques of terracing and irrigation are still sustainable today.

2. *How* **did the Indigenous people of the region adapt to the constraints of the physical environment?**

▶ *Identify* **sustainability challenges and solutions relating to forests and air in the region.**

The region's tropical forests, most notably in the Amazon, are being cut for timber and agricultural land, causing environmental degradation such as drought, soil erosion, and species loss and disruption of forest communities. Efforts to stem deforestation have been successful in some countries: Brazil set aside reserves for rubber tappers, Indigenous groups, and conservation; and Costa Rica preserved forests to provide tourism and protect biodiversity and watersheds. Air pollution is a serious problem in megacities such as Mexico City, which suffers from too many cars and factories in a high altitude basin that traps pollution. But policies to regulate traffic and relocate industry have reduced pollution.

3. *What* **are some of the policies and responses that seek to reduce deforestation and air pollution in the region?**

▶ *Summarize* **the common impacts and legacies of colonialism and describe the history of U.S. intervention, authoritarianism, revolutions and democratization in the region.**

European colonialism left a legacy of Spanish language and culture from Mexico to Chile and of Portuguese language in Brazil. Colonialism also resulted in widespread practice of the Catholic religion. Other European countries, such as Britain and France, colonized the Caribbean and its coast. Colonial powers oriented regional economies to export minerals and crops and introduced slaves when European diseases decimated Indigenous populations.

U.S. interest in controlling the Americas spurred construction of the Panama Canal. Later, concerns about communist trends led to U.S. interventions in countries such as Cuba, Chile, and Guatemala. Powerful landholding elites and military governments dominated much of the region during the late 19th and 20th century, often violently repressing dissent and Indigenous peoples. Revolutionary movements succeeded in Mexico and Cuba and led to land reform and nationalization, but other such movements were eventually overthrown in Chile, Guatemala, and Nicaragua. In recent years, democratically elected governments have been established across the region, U.S. relations with Cuba have improved, and social movements have achieved some of their goals.

4. *How* **did colonialism shape contemporary Latin America and Caribbean economies and landscapes?**

5. *Why* **did the United States intervene in the region and to what extent is the region now democratically governed?**

▶ *Compare* **the economic policies and impacts of import substitution, structural adjustment programs, and free trade and the successes and failures of economic development.**

Concerns about low export prices and high import prices in the mid-20th century led many countries to prioritize domestic manufacturing and set barriers to imports through import substitution policies. When debt forced many countries to appeal for assistance, structural adjustment programs required countries to sell off publicly owned companies and utilities, cut government spending, and remove trade barriers. Free trade agreements, such as NAFTA, encouraged the opening of Latin American economies and promoted trade.

6. *What* **are the arguments for and against free trade and neoliberalism in the region?**

▶ *Describe* the history and current distribution of Indigenous peoples, languages and cultures and their struggles for representation.

The Indigenous populations of the Americas collapsed rapidly following introduction of European diseases. Before the arrival of Europeans, Indigenous people domesticated plants that included maize and potatoes, which became global food staples, as well as tobacco, which became a worldwide health problem. Today significant Indigenous populations remain in Guatemala, Bolivia, and Mexico. Indigenous social movements have created reserved land, helped elect Indigenous leaders, and gained international support.

7. *Why* did Indigenous populations decline after the arrival of Europeans and which countries now have significant Indigenous populations?

▶ *Understand* the geography and impacts of the drug trade and international tourism.

The region receives millions of international tourists seeking tropical beaches, especially in the Caribbean and Mexico as well as tourists seeking ecotourist experiences in countries such as Costa Rica. Tourism supports economies, but can undermine local culture, overuse resources, and cause pollution. The drug trade, primarily driven by demand in the United States, has promoted the growth of powerful criminal groups known as drug cartels, which have destabilized politics and increased violence in Colombia and Mexico.

8. *Why* do farmers produce drugs in the region and what are the costs and benefits of tourism?

▶ *Identify* the main migration streams and their causes and list the factors promoting urbanization and associated social and environmental problems.

Migration streams within the region include millions moving from rural to urban areas in search of work and education, people from the Andes seeking work in Argentina, from Central America moving to Mexico, and people moving to work in tourism. Large numbers of migrants, especially from Mexico, have moved to the United States seeking employment. Some Andean peoples have migrated to Spain seeking work. Migrants from the Caribbean have settled in Britain, the United States, and Canada. Some migrants flee natural disasters and conflict.

Urban growth is partly driven by economic conditions as people—pushed by unemployment and lack of health, education, energy, and water in rural areas—move to cities in search of work and services. Latin America's largest cities, such as Mexico City, Rio, Santiago, and São Paulo, have serious problems with pollution, congestion, and waste. They are often surrounded by slums that lack services but house millions of people who often work in informal economies.

9. *What* are some of the important migrations within and beyond the region and what is driving people to move?

10. *Why* is the region urbanizing and what are some of the problems and challenges of urbanization?

Key Terms

ALBA (p., *282*)
altiplano (p., *266*)
altitudinal zonation (p., *263*)
biodiversity (p., *271*)
biofuels (p., *276*)
bioprospecting (p., *276*)
braceros (p., *295*)
BRICS (p., *283*)
carbon markets (p., *265*)
cartels (p., *289*)
Columbian Exchange (p., *278*)
demographic collapse (p., *278*)

ecosystem services (p., *274*)
ecotourism (p., *276*)
El Niño (p., *264*)
extractive reserves (p., 269)
favelas (p., *297*)
Green Revolution (p., *272*)
haciendas (p., *280*)
import substitution (p., *281*)
informal economy (p., *284*)
La Niña (p., *264*)
liberation theology (p., *289*)
machismo (p., *292*)

maquiladora (p., *282*)
marianismo (p., *292*)
maroon communities (p., *293*)
megacities (p., *297*)
mestizo (p., *293*)
nontraditional agricultural exports (NTAEs) (p., *283*)
North American Free Trade Agreement (NAFTA) (p., *282*)
Payments for ecosystem services (p., *276*)
plantations (p., *280*)

rain shadow (p., *263*)
REDD (p., *266*)
remittances (p., *295*)
social movements (p., *288*)
structural adjustment programs (SAPs) (p., *281*)
syncretic (p., *289*)
trade winds (p., *263*)
Treaty of Tordesillas (p., *277*)
urban primacy (p., *297*)
westerlies (p., *263*)

Thinking Geographically

1. How did colonization change the ecology and environment in the region through the Columbian Exchange and other processes?

2. Which regions of Latin America are particularly vulnerable to natural hazards, and why and how might climate change and alterations in land use, such as deforestation, increase these risks?

3. How did colonialism restructure the economies of Latin America and what impacts did that restructuring have?

4. Why has the United States played such a prominent role in Latin America and the Caribbean? In addition to the United States, which other world powers have had interests in the region and why?

5. How do trade, migration, and drugs link Mexico and the United States?

6. How have urban–rural contrasts driven the growth of major cities in Latin America, and what are the main environmental and social problems of these urban areas?

7. What has contributed to the recent economic success of Brazil?

DATA Analysis

The Economic Commission for Latin America and the Caribbean is a good source of statistical data on the countries of the region. One useful summary is the Statistical Yearbook: http://interwp.cepal.org/anuario_estadistico/anuario_2014/ that can be downloaded as a pdf file. Select the section of the yearbook on *environmental statistics (section 3 in 2014)*. It is easiest if you rotate the pages you are consulting. Answer the following questions:

1. Which countries have a large proportion of their land in protected area (over 20%)? What benefits do countries gain from protection?

2. Which three countries produce the largest share of their energy from renewables? Use the preceding tables to try and identify which renewable source (e.g., wood, hydro) they rely on the most.

3. Which three countries have the highest levels of overall fish harvest and aquaculture production?

4. Which three countries use the most fertilizer per unit area and which three the least? What might explain the difference and what are the benefits of using it?

Latin America Economic Statistics

https://goo.gl/iOhh7d

MasteringGeography™

Looking for additional review and test prep materials? Visit the Study Area to enhance your geographic literacy, spatial reasoning skills, and understanding of this chapter's content by accessing a variety of resources, including MapMaster interactive maps, videos, *In the News* RSS feeds, flashcards, web links, self-study quizzes, and an eText version of *World Regions in Global Context*.

Fiery Cross Reef, western Spratley Islands, South China Sea, September 2015. This reef has been converted to an island through massive construction by the Chinese government. By building artificial island bases in the region, China seeks to extend its territorial sovereignty.

East Asia

I n October 2015, the United States sent a guided missile destroyer into a section of the South China Sea laced with small islands and atolls, an area China considers its territory, but that the US considers international waters. This follows an event in May 2014 when a Chinese vessel rammed and sank a Vietnamese fishing boat. These incidents, and the accusations and counter-accusations that followed, are only a small part of territorial conflicts in East Asian waters. Countries bordering the South China Sea are drilling the seabed for natural gas and oil; shipping routes for global trade crisscross its length; fishing revenues here are extremely valuable. China is building a series of artificial islands across the sea, to extract resources, secure territory, and support naval operations.

Conflicts over ownership and control of the waters of East Asia are an ancient part of the complex political landscape of the region. These conflicts have become far more acute in recent years as China has expanded as a political and military power and begun to assert its influence into the seas beyond its coastline.

Outright conflict has been limited to a few naval engagements and international legal claims. As the military and naval strengths of East Asian maritime nations have increased, however, the chance of escalation and combat constantly grows. As a result, countries outside the region, including the United States, have made settlement of ocean territorial disputes a high priority. The United Nations Convention on the Law of the Sea (UNCLOS) is the crucial legal framework for sorting territorial rights in the region. UNCLOS came into force in 1994 but has not been ratified by the United States. This failure to ratify the treaty makes peaceful settlement more difficult because the United States lacks credibility as an advocate for standards to which it has itself refused to subscribe.

Learning Outcomes

▶ **Explain** how humans have adapted and modified the landscapes of East Asia over time.

▶ **Describe** how land and water management in East Asia may be impacted by climate change.

▶ **List** the main physical subregions of East Asia and describe their tectonic origins and related hazards.

▶ **Identify** the adverse ecological effects of urbanization, industrialization, and the modernization of agriculture.

▶ **Distinguish** the centers of power in East Asia in the precolonial, colonial, and postcolonial periods.

▶ **Compare** the development of differing economic experiments in the 20th and 21st centuries in China, South and North Korea, and Japan.

▶ **Identify** central geopolitical tensions within East Asia.

▶ **Trace** the linkages of East Asian cultures to other cultures within and beyond the region using specific examples.

▶ **Describe** the forces that have caused significant declines in population growth across East Asia.

▶ **Summarize** and explain the major migration trends in the region such as rapid urbanization.

Environment, Society, and Sustainability

The most striking and geographically significant environmental feature of East Asia (**FIGURE 8.1**) is its position on the east end of the vast Eurasian continent. Its formidably high interior plateaus and towering mountains are the sources of the lengthy river systems. Its peninsular and island coasts create a set of oceanic subregions abutting the continental mass. These characteristics are the product of a long geotectonic history that links the region to other world regions. The entire Tibetan Plateau was uplifted between

65 million and 2 million years ago, and the Himalayans are still rising. The Indian–Australian Plate moved northward and pushed up against the Eurasian Plate, not only uplifting the Tibetan Plateau but also forming the Himalayas. To the east, the movement of the Pacific Plate toward the Eurasian Plate caused the folding and faulting that resulted in the mountains of the peninsulas and islands of the continental margin, including the formation of the Korean peninsula and the Japanese archipelago. The boundary region is therefore part of the tectonically active so-called **Ring of Fire**, a circle of volcanoes and earthquake zones that surrounds the Pacific Plate, stretching from Southeast Asia through the Philippines, north across Japan and Kamchatka and then down the Pacific coast of North America (**FIGURE 8.2**).

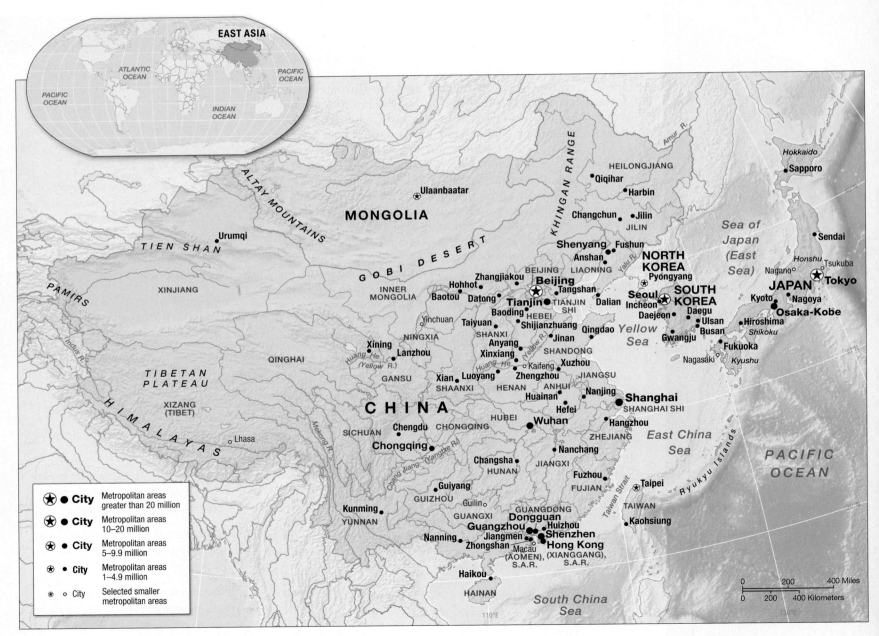

▲ **FIGURE 8.1** **East Asia: Countries, major cities, and physical features** China and Mongolia share the vast continental reach of this region, while the Democratic People's Republic of Korea (North Korea) and the Republic of Korea (South Korea) share the Korean Peninsula jutting towards the island chain of Japan.

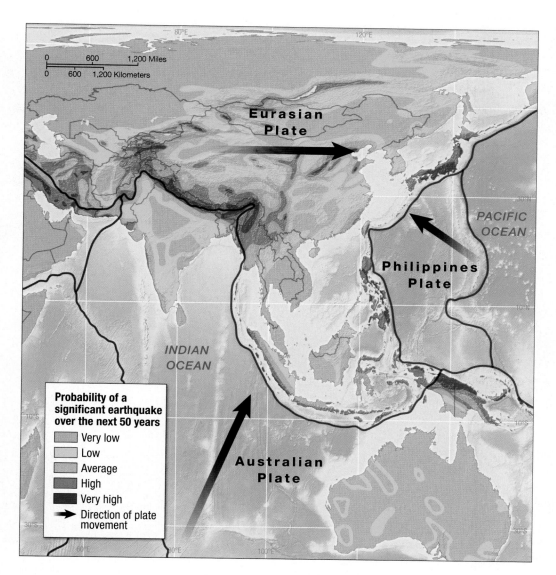

◄ **FIGURE 8.2 The Western Side of the Ring of Fire** East Asia's landscapes are dominated by the impact of the Indo-Australian tectonic plate's movement into the Eurasian plate and the movement of the Eurasian plate into the Philippine and Pacific plates. **Where do these movements create the highest earthquake risks?**

"The boundary region is part of the tectonically active so-called Ring of Fire, a circle of volcanoes and earthquake zones that surrounds the Pacific Plate."

Climate and Climate Change

The climates of East Asia are not wholly unlike those of North America (see Chapter 6), owing to the similarities in latitude, size of the regions, and position relative to oceans in the two regions (**FIGURE 8.3**). As in North America, notably, the western and interior regions of East Asia are largely arid and semiarid. Because of its mountain barriers and sheer distance from the coast, this part of the region averages less than 125 millimeters (5 inches) of rain a year. On the Tibetan Plateau, high elevations make for cool summers and extremely cold winters. Farther north, in Xinjiang, Qinghai, Gansu, and Mongolia, summer temperatures can be extremely hot. In some of these very arid areas, soil erosion and moving sand dunes are a constant hazard. The Chinese government has aggressively sought to control these hazards through large-scale tree planting (**FIGURE 8.4**).

To the east, two distinctive continental climatic regimes prevail. The northern regime is subhumid (slightly or moderately humid, with relatively low rainfall), encompassing the Northeast China Plain, the North China Plain, the northern parts of the Korean peninsula, and the Japanese archipelago. Winters here are cold, while summers are warm, with moderate amounts of rain from the southeasterly monsoon winds. The southern regime (like its counterpart, the southeastern United States) is humid and subtropical. Winters here are mild and rainy, and summers are hot with heavy monsoonal rains. In the arid and subhumid regions of East Asia, drought is a critical natural hazard, causing widespread famine as a result of crop failures in drought years. In addition, the subhumid parts of East Asia tend to be prone to flooding.

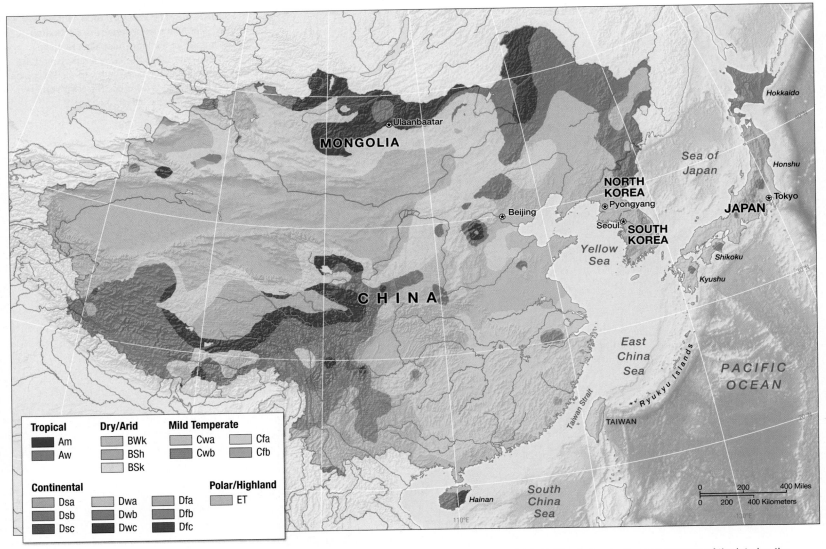

▲ **FIGURE 8.3 Climates of East Asia** East Asia is divided between four main climates, including the variable highland climate, the arid climates of the interior, the continental climates of the north, and the humid subtropical climates of southern China and southern Korea and Japan.

Adapting to Semiarid and Subtropical Climates Each of these distinctive climates influences regional cultures, especially as humans adapt to very different rainfall patterns and temperature ranges. The western deserts and grasslands of Xinjiang and Mongolia, for example, have a long and well-developed herding tradition that mixes settled agriculture with herd mobility to best utilize infrequent and unevenly distributed rainfall. The horse cultures of Mongolia, including the great Mongol Empire of the 13th century but continuing into the present, all evolved within this climate regime (**FIGURE 8.5**).

◄ **FIGURE 8.4 China's "Green Wall"** Large-scale plantation of shrubs and trees in western China represents an effort by the government to control mass erosion and the movement of dunes, made more likely as the climate has become more arid in recent years. **In what parts of the region are Chinese government efforts to control desertification most likely?**

▲ **FIGURE 8.5 The Yurt** The classic dwelling of Mongolian people is still used today. This structure is adapted to life on the arid steppe, where it can be broken down and transported by horseback across the semiarid and grassy plains of interior East Asia.

▲ **FIGURE 8.6 Irrigated Fields and Levees of a Complex Rice System** Controlling the flow of water in paddy agriculture is critically important. Planting and transplanting rice requires careful social coordination for opening and closing gates, managing water levels, and mobilizing labor.

The cultures of southeastern China and Japan, in contrast, have developed systems to capture and control the flow of water to maximize production year-round especially for rice production. The "rice cultures" of Asia have harnessed the waters of major rivers like the Chiang Jiang (Yangtze) into sophisticated systems of irrigated agricultural production (**FIGURE 8.6**).

Climate Change A warming trend over the last 50 years, especially pronounced during the winters, presents serious problems and challenges for this region. The impact of this overall warming is unclear, but it probably will mean decreases in catch in the Pacific fisheries on which most countries in the region heavily depend. It may also do serious damage to the delta ecosystems along the coasts of China and Japan. Rainfall regimes are less easy to predict because

trends over the last century include drier average conditions in some parts of the region (including northeast China) and wetter trends in others (including arid western China). Intense rainfall appears to be on the increase, with more dramatic individual rain events causing increased flooding in western and southern China and Japan.

Equally serious is the threat of melting glaciers in the mountain areas of the interior that feed the headwaters for some of China's major rivers. Glacial melt might ultimately contribute to a lower volume of flow in the Yangtze River, a resource of enormous significance for hydropower generation, farming, and municipal water supplies.

While problematic, these conditions may open onto new regional adaptations. Traditional drought-tolerant crops (including hardy strains of upland rice) and livestock breeds (like Mongolian cattle and Tibetan goats) are already well-adapted for water scarcity and variability and might be further bred or modified to adapt to climate change. Some of the secrets for adapting to climate change in the future likely lie in the adaptive history of East Asian producers of the past.

Given the critical influence of urban activities on the greenhouse gas emissions that drive climate change, urban areas in Japan and China represent an enormous challenge for mitigating climate change. China's 2014 CO_2 emissions (10.5 billion tons) surpass those of the United States (5.3 billion tons). Japan is already a major emitter of greenhouse gases (emitting 1.3 billion tons in 2014). Japan's per capita rate far exceeds that of China, however, due to its smaller population and energy-demanding lifestyle. Where each person in China emits approximately 8 tons per year, the per-person rate in Japan is 10 and in the United States more than 17 tons. Chinese emissions are increasingly problematic and also contribute to urban air quality problems (Figure 8.7). Current negotiations to contain these emissions are promising, however. China and the United States released a joint statement of climate commitments in 2014 and both participated in the 2015 Conference of the Parties in Paris, where nations agreed in the **Paris Agreement** to keep global temperature increases to less than 2°C (3.6°F) by 2100, with an ideal goal below 1.5°C (2.7°F).

Apply Your Knowledge

8.1 How have people in East Asia adapted to weather and climate patterns?

8.2 To what degree might people need to modify this adaptation method in the face of climate change?

Geological Resources, Risk, and Water

Much of East Asia consists of plateaus, basins, and plains separated by narrow, sharply demarcated mountain chains. These broad physical regions contain a great diversity of river systems and elevational gradients, but can be divided into three main subregions: the Tibetan Plateau; the central mountains and plateaus of China and Mongolia; and the continental margin of plains, hills, continental shelves, and islands (**FIGURE 8.8**).

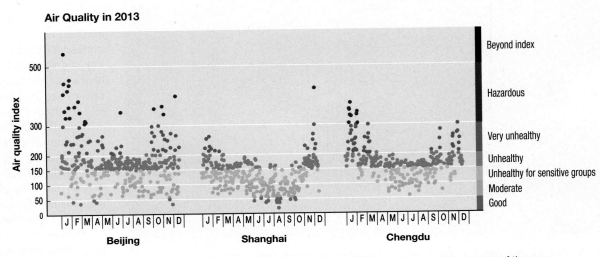

Air Quality in 2013

▲ **FIGURE 8.7 Chinese Urban Air Quality** Daily air quality measures in Chinese cities over the course of the year. Pollution is notoriously bad and getting worse. **During which months/seasons are unhealthy air days especially common? What might explain this pattern?**

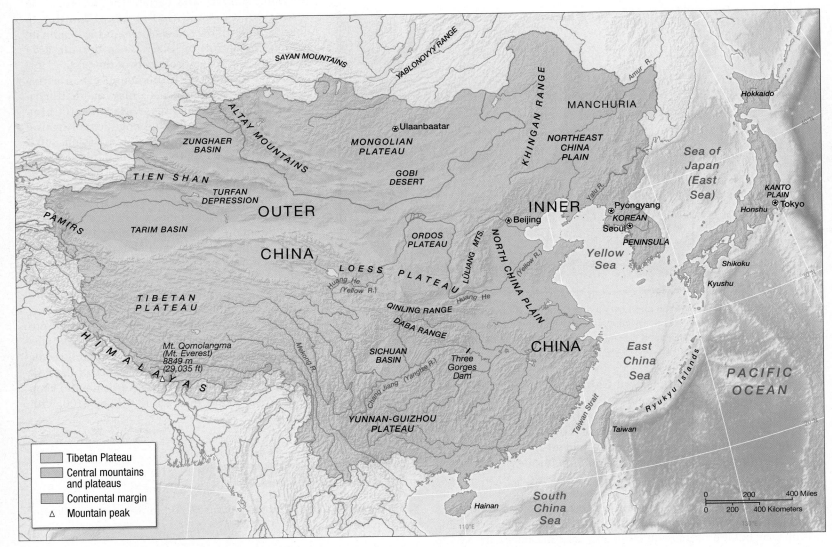

▲ **FIGURE 8.8 Physiographic regions of East Asia** Three broad physiographic divisions characterize East Asia: the Tibetan Plateau; the central mountains and Outer China; and the continental margins in an arc spanning Inner China, the Japanese archipelago, the Korean peninsula, and Taiwan.

The Tibetan Plateau

In the southwest, the Tibetan Plateau is a unique physical environment, a vast area violently uplifted in relatively recent geological time to produce the youngest, highest plateau in the world. The Tibetan Plateau is what geologists call a **massif** (a mountainous block of Earth's crust). Several mountain ranges—including the Himalaya Mountains—with peaks of 7,000 meters (22,964 feet) or more, form the southern rim of the plateau.

Plains, Hills, Shelves, and Islands

The bulk of the population of East Asia lives on the continental margin, which includes the great plains of China: the Northeast China (Manchurian) Plain, the North China Plain, and the plains of the Middle and Lower Chang Jiang (Yangtze River) valley. The Japanese archipelago of Hokkaido, Honshu, Shikoku, and Kyushu forms the outer arc of East Asia's continental margin. A backbone of unstable mountains and volcanic ranges projects from the shallow seafloor and extends to the island of Taiwan.

Earthquakes in a Still-Forming Region

The geological uplift of the Tibetan Plateau described previously, along with the region's position on the western end of the Ring of Fire have made East Asia geologically active. The physical environments of East Asia are, in fact, relatively hazardous. In addition to regular earthquakes along the Himalayan foothills, much of the North China Plain is subject to seismic activity, as is Japan.

Tectonic activity can also create hazards in the form of **tsunamis**. The Pacific Plate, which moves east at a rate of 8 to 9 centimeters (3.1 to 3.5 inches) per year, is subducting under the Eurasian Plate (see Chapter 1, p. 2). This violent process leads to the periodic release of massive amounts of energy. This is what happened in March 2011, when a 9.0-magnitude earthquake occurred just 70 kilometers (43 miles) east of the Peninsula of Tihoku, sending shockwaves through the ocean and creating an enormous tsunami. The multiple resulting waves overwhelmed the heavily populated shore of the country, causing widespread devastation, and contributed to the meltdown at the Fukushima Daiichi Nuclear Power Plant complex (**FIGURE 8.9**).

Flooding, Flood Management, and Hydroengineering

Like earthquakes, inland flooding hazards are an ongoing problem in this region. This is partly because the rivers that flow over long courses across the plains carry a tremendous amount of silt. This silt is deposited in more sluggish lower reaches, building up the height of riverbeds and making river courses unstable. The Huang He (Yellow River) is the largest and most notorious of these rivers, having changed its course several times in the last two centuries. The river's name derives from the color it is given by its heavy silt load: a distinctive brown-gold color.

The people of East Asia have over the millennia developed sophisticated techniques to modify the landscape for water control. Over the centuries, marshes were drained, irrigation systems constructed, lakes converted to reservoirs, and levees raised to guard against river floods. These traditions of water management are

▲ **FIGURE 8.9 The 2011 Tsunami** The devastating waves that hit the Japanese coast in March 2011 were set off by an offshore earthquake.

remarkable examples of coordinated action. This success has caused some researchers to term the regions' historic empires as **hydraulic civilizations**—civilizations that survive, thrive, and expand based on their capacity for controlling water.

The tradition continues into the present. The Three Gorges Dam on the Chang Jiang (**FIGURE 8.10**) is a massive public works

▲ **FIGURE 8.10 Three Gorges Dam** One of the world's largest-ever engineering projects, the dam is located in central China where the Chang Jiang narrows to form the Xiling, Wu, and Qutang Gorges. **What are the possible environmental costs and benefits of dams constructed at this scale?**

The Environmental Price of Industrialization

Industrialization in East Asia has been ongoing since the 20th century, but the recent boom in the Chinese economy has caused spectacular leaps in production, accompanied by unprecedented environmental challenges. Air quality in Beijing is notoriously poor: Residents of this and other major cities take "air vacations" to the countryside to seek relief from smog.

Water pollution is a crisis in many parts of the region. For example, in 2013 a chemical leak of benzene poisoned a tributary of the Huangpu River, a main water supply for many of Shanghai's 23 million residents. After more than 3,000 dead pigs were found floating in the river, local people had to receive emergency water from fire trucks.

These environmental impacts are not localized. Consider the case of the **Asian Brown Cloud**, a blanket of air pollution 3 kilometers (nearly 2 miles) thick. The cloud hovers over a vast area stretching from the Arabian Peninsula across India, Southeast Asia, and China almost to Korea, as well as most of the tropical Indian Ocean. The cloud consists of sulfates, nitrates, organic substances, black carbon, and fly ash. Particulate pollution from Asia can even drift across the Pacific and impact places as far away as North America (**FIGURE 8.1.1**).

A path to sustainability may yet emerge in industrial China, however, as the impacts of all this pollution on citizens and workers are now undeniable. In 2013, the Chinese government itself acknowledged the existence of "cancer villages," settlements with disproportionately high rates of cancer, typically adjacent to sites of industrial production and power generation. At the same time, they set about trying to solve environmental problems. Changes to China's framework Environmental Protection Law (EPL) took effect in January 2015, and include a provision that allows social organizations to bring environmental lawsuits against polluting industries and facilities. In a country famously resistant to public action, the environment is one area where empowerment for sustainability may be slowly emerging.

1. List examples of environmental problems resulting from Chinese industrialization.

2. How has the Chinese government responded to its environmental challenges?

▲ FIGURE 8.1.1 **The Asian Brown Cloud** The Brown Cloud hovers in the left-hand side of this image, just west of the Korean peninsula, partially obscuring the landmass of China beneath. The pollution cloud can be distinguished as a think brown haze spread across the Asian landmass, distinct from the normal clouds, which are smaller and white. In this image, some brown haze can also be seen drifting east over the Pacific.

"A path to sustainability may yet emerge in industrial China, however, as the impacts of pollution are now undeniable."

program aimed at creating hydroelectric facilities, irrigation, and flood control. The dam's 32 massive generators were switched on in 2012; the dam provides almost 10% of China's energy output, bringing electric power to millions of rural households.

Three Gorges is also emblematic of the problematic environmental consequences of large-scale infrastructure development. The river blockage endangers several species indigenous to the region, including the Chinese alligator, the Siberian white crane, and the Chinese sturgeon.

Apply Your Knowledge

8.3 Describe two main physical features of the East Asia region.

8.4 What tectonic forces shaped the landscape into these forms?

8.5 List two hazards associated with these forces.

Ecology, Land, and Environmental Management

The ecologies of East Asia reflect the broad land and climate forms of the region. These natural divisions have been subject to dramatic transformation under successive civilizations that sculpted the physical environment over millennia, creating distinctive regional landscapes.

Tibetan and Himalayan Highlands Nomadic pastoralism remains a chief occupation within the region (**FIGURE 8.11**), though the mixed forests of Gaoshan pine, Manchurian oak, and Himalayan hemlock are increasingly being brought into timber production, accelerating a long history of deforestation in the region.

Steppe and Desert Northwestern China (**FIGURE 8.12**) and much of Mongolia is a territory of temperate desert—grasslands, and scrub-covered hills, and basins—though hundreds of thousands of acres of land have been plowed and irrigated in recent years through agricultural modernization, and Mongolia's extensive mineral resources (including oil, coal, copper, lead, and uranium) have been opened up to exploitation.

Inner China, Korea, and the Japanese Archipelago The remainder of East Asia—"Inner China" plus the Korean peninsula, the Japanese archipelago, and Taiwan—can be divided broadly into the landscapes of the subhumid regions to the north

▲ **FIGURE 8.11 Yak Raising in Tibet** Along with sheep and goats, the yak is a traditional source of livelihood in Tibet, supplying milk and wool as well as meat.

and those of the humid and subtropical regions to the south. Mixed temperate broad-leaf and needle-leaf forests on the hills and mountains dominate the natural landscapes of the northern part of Inner China, along with the northern part of the Korean peninsula and the northern half of the Japanese archipelago, with forested areas accompanied by areas of productive agriculture.

▼ **FIGURE 8.12 Northwestern China** Surrounded by mountain ranges, the vast plains and deserts of northwest China and Mongolia stretch for thousands of kilometers. The Mongolian cultures that emerged in these interior regions are renowned for their skills in livestock management and horse breeding. **How have the landscapes of the highlands, steppe and plains in East Asia influenced human cultures in that past? How have human beings, in turn, altered the landscape?**

▲ FIGURE 8.13 **Japanese Rural Landscape** The historic rural landscapes of Japan, as characterized here by the villages of Gokayama, include elements like traditional, vernacular "gassho" architecture (with steep roofs), neatly maintained rice fields, and carefully stewarded forest plots.

The island of Taiwan, the southern part of the Korean peninsula, and the southern half of the Japanese archipelago are dominated by low hills covered with evergreen forest, though highly intensive cultivation of paddy rice, sugarcane, mulberry trees, and hemp occurs on the remaining 10% of the land (**FIGURE 8.13**).

Around Guilin and in the Zuo River area to the south, near Nanning, erosion has dissolved areas of soft limestone, leaving spectacular pinnacles several hundred feet tall (**FIGURE 8.14**), providing the inspiration for thousands of years of Chinese art.

Revolutions in Agriculture
Several of the world's most important food crops and livestock species were domesticated by the peoples of East Asia, beginning around 6500 B.C.E. Millet, soybeans, peaches, and apricots were domesticated in the northerly subhumid regions. Rice, mandarin oranges, and tea were domesticated in the more humid regions to the south. Chickens and pigs were among the livestock species domesticated and mulberry bushes were domesticated as silkworm fodder. The topography was also altered to suit the needs of growing and highly organized populations, as hills and mountainsides were sculpted into elaborate terraces to provide more cultivable land (**FIGURE 8.15**).

Technological innovations in production have also swept the region since the 1960s, with significant implications for environmental quality and human health. These strategies are a part of the **Green Revolution**, an international effort to introduce and encourage the cultivation of new varieties of crops, which produce far more food but also require higher levels of industrial inputs including fertilizer, water, and pesticides.

The environmental impacts of these changes have been enormous, especially considering the vast areas of East Asia

▲ FIGURE 8.14 **The Guangxi Basin** A unique combination of geomorphology and agricultural development has led to one of the world's most spectacular landscapes. From classic paintings to contemporary tourist posters, this dramatic karst (eroded limestone) scenery serves as an iconic representation of traditional China.

▲ **FIGURE 8.15 Agricultural Terracing, Yuanyang, China** This technique for carving productive farmland out of steep slopes is practiced across the world, but the elegantly worked hillsides of rural China are a testimony to the efficiency and productivity of the region's ancient rice culture. What kinds of labor and human activity might be required to maintain such a system of terraces?

under cultivation. In 2009, farmers in north China used roughly 590 kilograms of nitrogen fertilizer per hectare (526 pounds per acre). This amount is six times more nitrogen fertilizer than farmers use in America on the same amount of land and has resulted in the release of approximately 225 kilograms per hectare (200 pounds per acre) of excess nitrogen into the environment. Released in waterways, fertilizers can radically damage biodiversity and the health of streams and rivers.

More dramatically, the heavy use of herbicides and insecticides, a crucial component of Green Revolution technology, has resulted in widespread illness in rural areas. The recognition of this problem has encouraged the adoption of genetically modified varieties of crops (GMOs) that require fewer pesticide inputs. In a 2011 study, the planting of one type of genetically modified cotton in China was shown to reduce pesticide use by 50% in general and by 70% for the most toxic chemicals (**FIGURE 8.16**).

Apply Your Knowledge

8.6 Describe the characteristics of landscapes and adaptation in East Asia's highlands, the interior steppes, and lowland and coastal regions.

8.7 Compare the advantages and the disadvantages of Green Revolution and GMO technology for people and the environment in East Asia.

Conservation Issues The dense settlement of East Asia holds implications for native fauna. Five important mammal species (including the wolf) have become extinct in Japan. In China, there are more than 385 threatened species, according to the International Union for the Conservation of Nature (IUCN). In China, 90% of the tiger population has vanished in the last century and the iconic Giant Panda has become a threatened species. Pandas have lost half their habitat (upland forest) since the 1970s to logging and farming.

China also hosts a large illegal trade in animal parts, used in medicines, from species like musk deer and bears (see "Geographies of Indulgence, Desire, and Addiction: Animals, Animal Parts, and Exotic Pets"). This trade is international in scope and impacts tiger populations as far away as India (see Chapter 9) as well as elephants in Africa.

East Asia's use of ocean resources is also an area of conservation challenges. For example, despite rules in place since 1986 that ban the hunting of whales around the world, Japanese whalers often violate these restrictions, harvesting minke, fin, and humpback whales in the Southern Pacific Ocean (**FIGURE 8.17**). These harvests are often claimed to be for

▲ **FIGURE 8.16 Pesticide and Fertilizer Application in China** Since the Green Revolution, East Asians have utilized large amounts of insecticides, herbicides, and fertilizers, which endanger farmers as well as environmental quality. This farmer is spraying pesticides on vegetables in Xinlou village, northeast of Guangzhou.

▲ **FIGURE 8.17 Japanese Whalers in the Southern Pacific** Dead minke whales aboard the deck of the Nisshin Maru factory ship. The whaling fleet of Japan, though described as being for "research" purposes only, operates at industrial scale.

▲ **FIGURE 8.18 Wildlife in the DMZ** Inside the heavily fortified demilitarized zone near Chulwon, sixty years of conflict between North and South Korea have left the demilitarized zone less influenced by human activities. **What kinds of species are likely to be returning and thriving in the DMZ?**

"scientific" purposes, though the International Whaling Commission (IWC), who adjudicates such cases, has concluded these whale killings are commercial.

Wildlife in the Demilitarized Zone There may be a location where East Asian biodiversity has thrived over the last few decades, though it remains hard to tell. This area is the so-called **demilitarized zone (DMZ)**, a no-man's-land located between North and South Korea and established at the end of hostilities between the two powers in the 1950s. (There is a detailed discussion of this conflict later in the chapter.) Since that time, the area has been uninhabited by humans and has been heavily strewn with land mines. As a result, wildlife has come to thrive in the long corridor between the two states, including the Amur leopard and the Asiatic black bear.

At the same time, however, the DMZ has become a dumping ground for military-related hazards and waste. The United States Army, who helps patrol that border, has acknowledged extensive use of the herbicide Agent Orange, as well as dumping toxic compounds in and around the DMZ, including carcinogenic perchloroethylene (PCE), pesticides, heavy metals, and dioxin. In this way, the DMZ is a strange accidental wilderness, a product of a half-century of conflict and despoliation, and an example of an **Anthropocene** environment. It is heavily influenced by human activities and also a product of unpredictable natural forces (**FIGURE 8.18**).

Sustainability in East Asia

Though East Asia's recent growth has made it notorious for environmental degradation, innovations for sustainability are also extensive. Taiwan's Declaration on Sustainable Development stresses intergenerational equity and directs the country to track its progress on 24 indicators, including rates of recycling, air pollutant concentrations, and greenhouse gas emissions, as well as quality-of-life indicators like green space and protected areas and monuments. In 2014, Japan held a national summit on "Sustainable Cities," stressing "smart city" solutions to environmental problems like an electric car-sharing system in Yokohama and the development of micro-grids for energy that allow cities like Keihan to maintain power even when the national grid is down. South Korea established a "greenbelt" around the city of Seoul in the 1970s, an undeveloped parkland that has since expanded to more than 1,500 km², or about 13.3% of the total metropolitan area. China, South Korea, and Japan are all members of the Asia-Pacific Partnership on Clean Development and Climate, which fosters cooperation on the development of renewable energy and distributed generation, more efficient power generation, and cleaner production in industries like steel and cement manufacture.

Apply Your Knowledge

8.8 Describe three major conservation or environmental challenges facing East Asia.

8.9 What are some of the main drivers of environmental degradation in the region?

8.10 How are the region's environmental problems being addressed?

Animals, Animal Parts, and Exotic Pets

Trade in animals and animal parts is a worldwide industry, which imperils countless species, harvested for their ivory, organs, and bones. Some investors in China collect and hoard ivory, most notably, investing in the possibility that elephants will go extinct in the wild, making their collection even more valuable. According to the wildlife organization, Save the Elephants, the price of elephant ivory in China in 2014 was $2,100 per kilogram, up from $750 per kilo in 2010. Such collectors are therefore literally banking on extinction.

Roots of Ivory Trafficking
Trade in ivory—a precious commodity hewn from the teeth and tusks of animals—dates back thousands of years and remains a major industry today, with devastating impacts. Where African and Asian elephants numbered in the millions at the turn of the 20th century, current estimates put their populations at 500,000 and 50,000, respectively. Whales and other ivory-producing species have also been ravaged by the trade. In 2011, a major Chinese ivory smuggler was convicted of trafficking 7.7 tons of ivory from Africa to China (**FIGURE 8.2.1**).

Trade in Animal Parts
A thriving global market exists for the parts of other animals as well. The bile of some bears is prized for medicinal purposes, as are parts of tigers, most especially skulls. The incredible demand for tiger parts in Chinese medicine, coupled with the size of the East Asian market and the growing affluence of Taiwanese, Chinese, Japanese, and Korean consumers, has meant an almost unstoppable poaching problem for the dwindling tiger populations in adjacent countries and regions. It is hard to know the size of the global market in these goods, but it is likely at least a U.S. $6 billion-a-year trade. A bowl of tiger penis soup sells for $320 in Taiwan. The humerus bone of a tiger retails for as much as U.S. $3,190 per kilogram in Korea.

Trade in Live Animals
Living wild animals are also valuable and have become part of a global market for exotic pets. The negative consequences of this trade are numerous. Animals themselves suffer from mishandling, abandonment, and abusive conditions. The overharvesting of animals contributes to the destruction of their populations in the wild. Animals like the Slow Loris (**FIGURE 8.2.2**) of Southeast Asia and Yunnan China have become popular pets in Japan. Their value in the pet trade has contributed to making them an endangered species.

International Countermeasures
Dealing with this damaging trade has proven difficult for governments, since these luxury items are easy to smuggle and highly valuable. One effort has been the **Convention on International Trade in Endangered Species (CITES)**, a global treaty first signed in 1990 that restricts trade in rare or endangered animals and their parts between member countries. Many observers credit CITES with reducing the rates of decline for high-profile species, like elephants and tigers. Critics suggest that banning trade in objects like ivory altogether, however, results in artificially high prices, actually serving to increase demand, desirability, and the incentive to poach wild species.

1. **Where are some of the sources and markets for rare animal parts?**

2. **How might a total ban on ivory lead to economic speculation, hoarding, and an increased value for ivory?**

> "Dealing with this damaging animal trade has proven difficult for governments, since these luxury items are easy to smuggle and highly valuable."

▲ **FIGURE 8.2.1 Elephant Tusks** Customs officers in Osaka, Japan, display a record-setting seized cache of ivory, which totaled 2.8 tons. Osaka is a top black market destination for elephant tusks.

▲ **FIGURE 8.2.2 Asian Slow Loris** The Asian Slow Loris is a popular pet in Japan, as well as the United States. This exotic animal is now endangered.

History, Economy, and Territory

East Asian civilizations are indisputably some of the oldest in the world, and the mark of these cultures on the contemporary landscape is indelible. East Asia also played an important later role as the historical eastern anchor for a global trade system established long before the colonial era. During the postcolonial and contemporary eras, the region has been home to highly divergent economic and political experiments.

Historical Legacies and Landscapes

State societies, with organized bureaucracies, consolidated militaries, and unified legal systems, appeared in East Asia millennia ago. The geographic footprints of these societies, in the form of canals, irrigation systems, road networks, and fortresses, remain visible in the landscape to this day.

Empires of China China has had a continuous agricultural civilization for more than 8,000 years. The first organized territorial state was that of the Xia dynasty, a Bronze Age state that occupied the eastern side of the Loess Plateau and the western parts of the North China Plain (northwestern Henan Province) between 2206 and 1766 B.C.E. The succession of early dynasties that followed included the Shang dynasty and Qin dynasty, culminating in the rule of Emperor Shih Huang-ti (221–209 B.C.E.), who established a unified empire, an imperial system, and a centralized bureaucratic administration.

Shih Huang-ti began construction of the Great Wall (**FIGURE 8.19a** and **8.19b**) to protect China from "barbarian" nomads.

(b)

▶ **FIGURE 8.19**
The Great Wall The Qin dynasty began construction of the Great Wall from 316–209 B.C.E. (a) The Wall that visitors see today dates from the Ming period and was built between the late 14th and mid-16th centuries. (b) This is only part of a much more vast series of walls and towers constructed over centuries on China's northern perimeter.

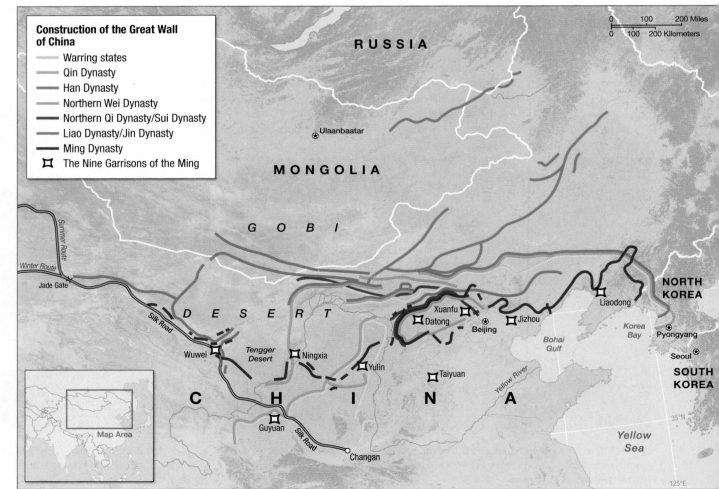
(a)

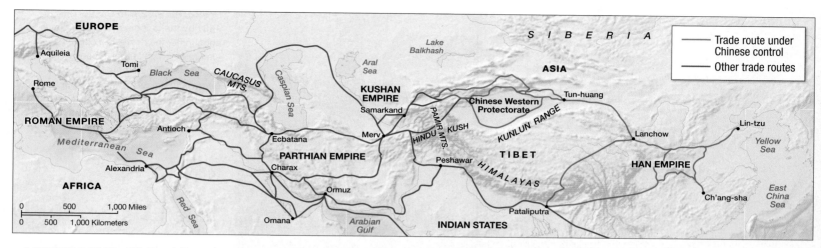

▲ **FIGURE 8.20 The Silk Road** The political and economic stability created by successive Chinese empires fostered a trade corridor connecting the region to North Africa and Europe and the rise of thriving trade cities along its length.

Really a series of walls and outposts constructed between the time of the Qin and the much-later Ming dynasty of the 16th century, the Great Wall is a monumental architectural and logistical achievement, reflecting the genius of imperial administration, the massive labor power it could mobilize, and the scale at which Chinese central authority planned its infrastructure.

The stability of the Qin dynasty was also a precondition for the success of the **Silk Road** as an economic and cultural link among the civilizations of China, Central Asia, India, Rome, and, later, Byzantium (**FIGURE 8.20**). This trade route made East Asia the effective center of gravity for the global economy for centuries, a role the region is reestablishing in the present era.

The Han dynasty extended the empire westward from 206 B.C.E. to 220 C.E., allowing the Chinese to control more trade routes along the Silk Road. In 1279, the Mongols, under Kublai Khan, conquered China, establishing imperial rule over most of both Inner and Outer China. The Mongol dynasty, known as the Yuan dynasty, was expelled after less than 100 years and was succeeded by the Ming (1368–1681) and the Qing (1681–1911) dynasties (**FIGURE 8.21**).

Japanese Feudalism and Empire

Japan's historic civilization formed around the introduction of Buddhism (covered in more detail later in this chapter), which arrived from India via China and Korea, coupled with a distinctive form of feudalism. The centralization of power in an imperial clan in Japan dates to the 6th century C.E. From this period forward, a succession of Japanese rulers maintained a rigid system in which a subjugated peasantry sustained

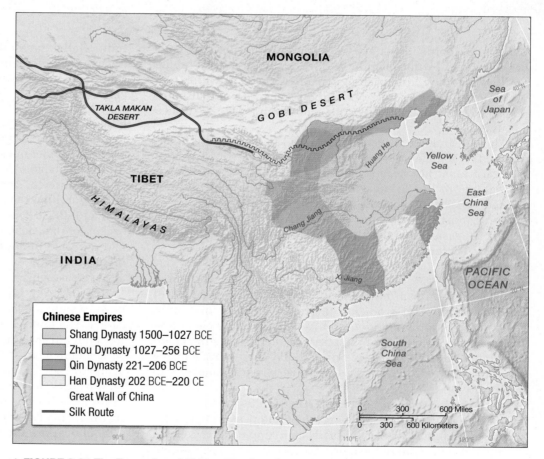

▲ **FIGURE 8.21 The Expansion of Chinese Empires** From the earliest Xia dynasty to the time of colonialism and the last Qing Emperor, the Chinese empire largely expanded through a succession of cultural, economic, and military conquests. **Which geographic features bound the areas of earliest Chinese Imperial power, and what are are the characteristics of the later empire's frontiers and boundaries?**

a sophisticated but inward-looking civilization. In 794 C.E., the ruling elite established a capital in Kyoto, which became the residence of Japan's imperial family for more than 1,000 years and the principal center of Japanese culture (**FIGURE 8.22**). The culture and politics of this feudal system were governed by a dynasty of imperial emperors, on the one hand, and a set of

◄ **FIGURE 8.22 Nijō Castle** Built in 1601 by the founder of the Tokugawa shogunate, the castle is a classic example of high feudal Japanese architecture. The ornate interior gardens and courtyards are organized hierarchically, so visitors would have access only to what their status dictated.

feudal noble lords (or daimyos), one of whom—the **shogun**—was appointed by the emperor to rule the country. These noble families governed from castle towns, the largest of which was Edo (known today as Tokyo), which reached a population of around 1 million by the early 19th century under the Tokugawa shogunate.

While an industrial system developed in the western hemisphere, the Tokugawa shogunate (1603–1868) strove to maintain traditional Japanese society. To this end, the patriarchal government of the Tokugawa family excluded missionaries, banned Christianity, prohibited the construction of ships weighing more than 50 tons, closed Japanese ports to foreign vessels (Nagasaki was the single exception), and deliberately suppressed commercial enterprise.

Imperial Decline The dynasties of Imperial China and Imperial Japan both eventually succumbed to a combination of internal and external problems. Internally, the administration and defense of growing populations began to drain the attention, energy, and wealth of the imperial regimes. As the economy stalled, peasants were required to pay increasingly heavy taxes, driving many into grinding poverty and thousands to banditry.

By the early 19th century, both China and Japan had moved into a phase of successive crises—famines and peasant uprisings. The imperial courts of both China and Japan suppressed the spread of knowledge of modern weapons because they feared internal bandits and domestic uprisings. Though the crossbow, guns, and gunpowder were innovations of East Asia, the 19th century found the Japanese and Chinese empires relying on antiquated military technology compared to the Europeans and Americans.

External problems and the power of western colonial force ultimately undermined both imperial regimes. European and American powers began to impinge on weakened rule in these empires in the 18th and 19th centuries, eventually seizing control or indirect authority over parts of China and catalyzing the

industrialization and militarization of Japan (see "Visualizing Geography: Colonialism in East Asia"). The stage was set for dissolution of imperial power and revolutionary changes during the 20th century.

Revolutions, Wars, and Aftermath Revolution came to China in 1911, when the Qing dynasty was overthrown and replaced by a republic under the leadership of Sun Yat-Sen's Nationalist Party (or Kuomintang). After a lengthy civil war and a struggle against the Japanese in World War II, Communist forces under **Mao Zedong** unified the country in 1949 as the People's Republic of China (or PRC). These forces came to control almost all of China except for the island of Taiwan, where the Nationalist leadership had retreated under U.S. protection. China would emerge from these struggles as an isolated nation, but one with enormous unrealized economic potential.

In 1931, the Japanese army advanced into Manchuria to create a puppet state and, in 1937, began a full-scale war with China. The Japanese military leadership ultimately overplayed its hand, however, by attacking the United States at Pearl Harbor in December 1941. The resulting sustained U.S. offensive against Japan, culminating in the dropping of atomic weapons on the cities of Hiroshima and Nagasaki (resulting in hundreds of thousands of deaths), led to Japan's defeat in World War II in 1945. Japan's industry lay in ruins. In 1946, output was only 30% of the prewar level.

Apply Your Knowledge

8.11 List some of the characteristics of Chinese and Japanese imperial systems that made them successful.

8.12 How did the aftermath of the colonial era differ in China and Japan?

Economy, Accumulation, and the Production of Inequality

Resistance to colonialism and the formation of new nations in East Asia during the postcolonial era brought with it turbulence and violence as well as economic experiments, starting in the years immediately following World War II. There are several contrasting systems of economic management in the states of the region today, each with its own pattern of accumulation and development. Japan, China, and South Korea, most notably, all have large economies, but their road to economic development has led down three very different paths (distinguished below).

These paths to economic development do share some interesting similarities. All the states in the region have geographically uneven patterns of development and accumulation. Agriculture and rural areas, most notably, have suffered in both Korea and China. All of these nations, moreover, are moving toward greater regional integration. Mutual investment and integration of firms between states are hallmarks of East Asian economies in the 21st century. This has led to the emergence of a new kind of economic region in East Asia. The new region is *externally* more competitive internationally and shifting toward increasing accumulation and centrality within the world system, but *internally* is increasingly competitive and fractured, with some subregions expanding while others contract. These three countries also form important hubs in the **Pacific Rim**, a loosely defined region of countries that border the Pacific Ocean (see "Emerging Regions: The Pacific Rim").

Japan's Economy: From Postwar Economic Miracle to Stagnation

Though World War II left Japan in ruins, incredibly, within 5 years, the Japanese economy recovered to its prewar levels of output. By 1980, Japan had outstripped the major industrial core countries in the production of automobiles and television sets, and had become an economic giant.

Several factors account for this period of spectacular growth in Japan. These included exceptionally high levels of personal savings, rapid acquisition of new technology, and extensive levels of government support for industry. The Ministry of International Trade and Industry (MITI), a state bureaucracy, guided and coordinated Japanese corporations organized in business networks known as **keiretsu**, making "Japan, Inc." the envy of global capitalism (**FIGURE 8.23**).

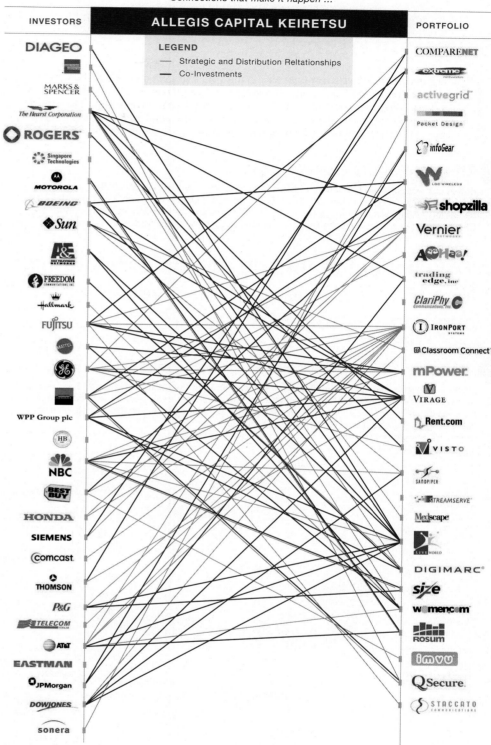

Connections that make it happen ...

▲ **FIGURE 8.23 Keiretsu** Built around strategic partnerships between firms, shared capital, and mutual investment, Keiretsu are an engine of economic organization that long characterized "Japan, Inc."

Since the late 1980s the Japanese economy has suffered significantly, however. Stagnant economic growth and recession began in the 1990s and the government was unable to pull the economy from its nosedive. In 2014, Prime Minister Shinzo Abe set about defeating the long recession through a stimulus campaign, in which the

Colonialism in East Asia

The legacy of Colonialism in East Asia in the 19th and 20th century still has an influence over the region today. The impact of foreign rule on China would lead to underdevelopment, revolution and ongoing mistrust of the West, while in Japan it would lead to the emergence of military empire in the early 20th with enormous implications for geopolitics after its dismantlement after World War II.

1 The Role of the Opium Trade

British colonizers forced the opium trade (based on opium grown in India for export by the East India Company) on China in the 19th century. The so-called Opium Wars (1839–1842) ended with defeat for the Chinese and the signing of the Treaty of Nanjing, after which, the opium trade to China expanded enormously; the British profited from Chinese opium addiction (Figure 8.3.1).

"The impact of foreign rule on China would lead to underdevelopment, revolution and ongoing mistrust of the west."

2 Opening Chinese Ports to Colonial Powers

The Treaty of Nanking also ceded the island of Hong Kong to the British and allowed European and American traders access to Chinese markets through a series of treaty ports. These ports were opened to foreign trade as a result of pressure from the major powers, and allowed spheres of increasing control over mainland China (Figure 8.3.2).

▲ FIGURE 8.3.1a **Artist's depiction of a Chinese opium den.**

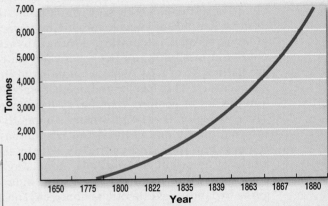

▲ FIGURE 8.3.1b **Opium imports into China 1650–1880.**

Spheres of influence, c. 1900
- British
- French
- German
- Japanese
- Russian

Treaty Ports
- ○ Original port opened by Treaty of Nanjing (1842)
- ⚔ Treaty port opened by 1900
- ● Major city

Was colonialism good for Asia?

https://goo.gl/k7q6Nv

◀ FIGURE 8.3.2 **Colonial possessions, treaty ports and spheres of influence in China 1800–1911.**

In the wake of the Opium Wars, the way was clear for European and American powers to exert control over fragmented China, setting unfair trade relations, and extracting goods and resources.

3 Japan's Reaction to Western Influence

Japanese trade, long-closed by the country's rulers, was forced open by Western foreign powers as well. Reacting to this humiliation in 1868, the Meiji imperial clan and a clique of daimyo set out to industrialize Japan as quickly as possible (Figure 8.3.3). Going from colonized to colonizer, the Japanese government promoted industrial development and military aggression over China (1894–1895), Russia (1904–1905), Taiwan (1895), Korea (1910), and Manchurian China in (1931) (Figure 8.3.4).

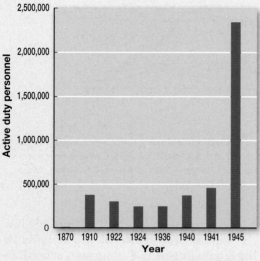

▲ FIGURE 8.3.3a **Growth of the Imperial Japanese Army 1870-1945.**

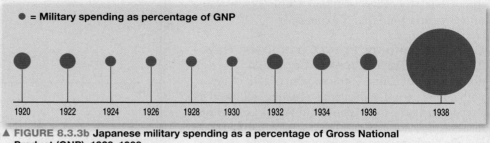

▲ FIGURE 8.3.3b Japanese military spending as a percentage of Gross National Product (GNP), 1920–1938.

From 1870 to 1945, Japan went from an isolated island state to a global military power. At the height of its power in 1942, the Empire of Japan ruled over a land area spanning

7,400,000 km² (2,857,000 mi²)

making it one of the largest maritime empires in history.

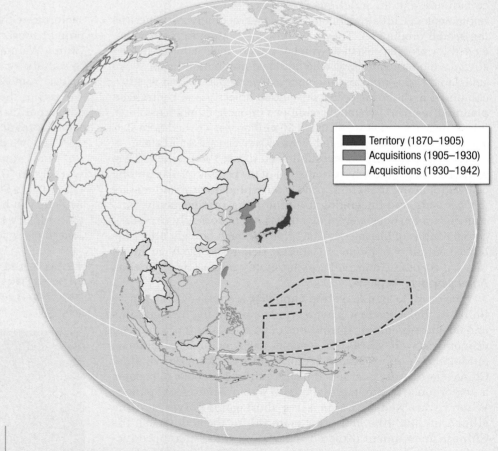

Legend:
- Territory (1870–1905)
- Acquisitions (1905–1930)
- Acquisitions (1930–1942)

▲ FIGURE 8.3.4 **Japanese Expansionism.**
In the late 1920s and early 1930s, Japan pursued a policy of military aggression to secure a larger resource base for its growing military–industrial complex.

1. **How did European colonial powerscome to dominate China in the late 19th century?**

2. **What was the role of opium in British colonial strategy?**

central Bank of Japan created money by buying government bonds. By 2015, the country had just barely emerged from recession, however, with GDP growth of 2.2%, which remains a lackluster performance.

Critically, Japan has been forced to compete for foreign investment with China. Several Japanese electronics giants, including Toshiba Corp., Sony Corp., Matsushita Electric Industrial Co., Olympus, Pioneer, and Canon, Inc., have expanded operations in China even as they have shed tens of thousands of workers at home. China's gains have often occurred at Japan's expense.

China's Economy: From Communist Revolutionary to Revolutionary Capitalist

The economic story of China is the reverse of that of Japan, with a long period of postwar stagnation, followed by a recent meteoric rise. After World War II, Communist leader Mao Zedong faced the task of reshaping society after two millennia of imperial control. In 1958, in what became known as the **Great Leap Forward**, Mao launched a bold scheme to accelerate the pace of economic growth. Land was merged into huge communes, and an ambitious **Five-Year Plan** was implemented. China's planners were concerned only with increasing overall production. They paid little or no attention to whether a need for products existed, whether the products actually helped advance modernization, or whether local production targets were suited to the geography of the country. Several years of bad weather, combined with the rigid and misguided objectives of centralized planning, resulted in famine conditions throughout much of China. It is estimated that between 20 and 30 million people died from starvation and malnutrition-related diseases between 1959 and 1962.

Radical economic changes followed in the 1980s when China embarked on a thorough reorientation of the economy, dismantling central planning in favor of private entrepreneurship and market mechanisms, and integrating itself into the world economy. Special free trade zones were opened and the historically capitalist, colonial cities of Macao and Hong Kong, though brought under the official authority of China in 1999 and 1997 respectively, maintain relative economic and political autonomy under law as **Special Administrative Regions**.

The result is a unique **state capitalist** economy. China is capitalist in the sense that privately owned manufacturing plants and businesses employ its millions of workers, but it is a planned economy as well, since state agencies and functionaries endorse, guide, and support certain sectors and markets. For example, China strategically positioned itself in the early 2000s to become a powerhouse in the manufacture of photovoltaic solar panels. When confronted with a massive shortage of polycrystalline silicon in 2007 (the main raw material for these devices), the Chinese government directed investment into silicon production and procurement from foreign sources. The result has been a boom in Chinese solar panel production and exports, making China the world leader in this market by 2015 (**FIGURE 8.24**).

This model has shown other remarkable results. The Chinese economy has been growing at double-digit rates for most of the past 30 years. In the early 21st century, China's manufacturing sector grew by almost 15% annually. In 2014 the United States bought more than U.S. $466 billion worth of goods made in China. Walmart was the single largest U.S. importer, according to a study by the Citizens Trade Campaign, procuring at bargain prices from China everything from T-shirts to car stereos. The success of the Chinese export boom has created a burgeoning consumer class in China and led to the emergence of big-box retail stores across the country (**FIGURE 8.25**).

The low cost of labor underpins the preponderance of "Made in China" apparel labels in U.S. stores. Whereas skilled clothing workers in Toronto's garment district, for example, are paid U.S. $350 to $400 a week, plus benefits, in China, skilled garment workers can be hired for U.S. $30 for a 60-hour week. Studies of the industry show that rural girls as young as 12 years old are sent to work as sewing machinists in city workshops. They sleep

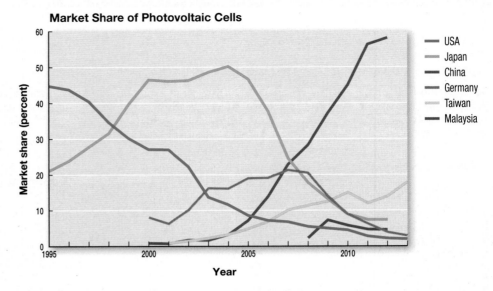

Market Share of Photovoltaic Cells

▲ **FIGURE 8.24 Market Share of Photovoltaic Cells Worldwide** After a deliberate effort to turn their state capitalist system toward the production of photovoltaic solar cells, China is now a world market leader and exporter. As China comes to take a larger share of the solar market, which nations contribute relatively less to photovoltaic production?

▼ **FIGURE 8.25 Walmart China** The sale of inexpensive consumer goods in the United States at stores like Walmart has long fueled the Chinese economic boom. Now, a growing consumer class makes China itself a target for big-box retailers.

The Pacific Rim

Over the last few decades, the direction of flows in international trade reversed. Before the liberalization of the Chinese economy and the rise of the so-called **Asian Tigers**, the exports and imports of the United States typically looked *eastward* toward Europe and *northward* toward Canada. Canada still remains the largest importer of U.S. goods and services, but China is a close second, and imports from China dominate all U.S. markets, as do those from Japan and other thriving Pacific economies. Trade among East Asian countries is also strong, including imports and exports that move between South Korea, Malaysia, China, and Japan. Added to this is the economic power of intermodal freight transport, involving the movement of massive quantities of goods cheaply in large containers (sometimes called "cans"), transported on giant trans-Pacific ships (**FIGURE 8.4.1**).

Ties between smaller economies are also strong. Peru has deep historic ties to Japan. Chinese investment in Nicaragua is extensive, including underwriting the massive "Nicaragua Grand Canal" that will connect the Pacific and Atlantic. South Korea and Australia have close trade ties and 150,000 people of Korean origin live in Sydney and Melbourne.

The resulting network of connections around the Pacific Ocean has drawn together the countries at its edges to form a new region: the *Pacific Rim*. Defined as the 47 nations and territories bordering the Pacific, the emerging Pacific Rim region accounts for 48% of world trade and 58% of world GDP and includes Japan, China, and the United States, the world's three largest economies (**FIGURE 8.4.2**).

This emerging region is marked by its coastal urban economic centers, which are tied far more closely to one another than they are to their respective interior areas. Among these, the trade centers of China are the fastest growing. As of 2015, three coastal Chinese export cities were numbered among the 10 fastest-growing cities in the world: Fuzhou, Xiamen, and Hangzhou. These municipalities are also becoming some of the most cosmopolitan. A visitor to the financial and trade center of Shanghai, for example, hears dozens of languages. Shanghai is as closely tied to Singapore as it is to Beijing.

Much the same can be said of cities on the eastern edge of the Pacific. Vancouver is home to Canada's largest port, ranks first in North America in total foreign exports, and trades more than U.S. $43 billion of goods every year, with most of these moving west to Asia. After English, the most commonly spoken languages in Vancouver are those of the Pacific Rim: Cantonese, Mandarin, Korean, and Tagalog.

1. **Describe the main factors behind the integration of countries around the Pacific Rim.**

2. **What are the social and cultural implications of these economic connections across the Pacific?**

> "The emerging Pacific Rim region accounts for 48% of world trade and 58% of world GDP."

▼ **FIGURE 8.4.1 Intermodal Freight Transport** By packing and transporting freight and goods in containers, which can be transferred from rail to ship to truck, goods can traverse the planet quickly and cheaply without ever being handled by workers.

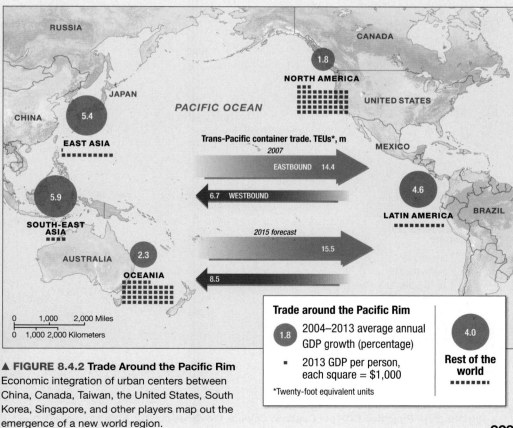

Trans-Pacific container trade. TEUs*, m

2007

EASTBOUND 14.4

6.7 WESTBOUND

2015 forecast

15.5

8.5

0   1,000   2,000 Miles
0   1,000 2,000 Kilometers

Trade around the Pacific Rim

1.8   2004–2013 average annual GDP growth (percentage)

■   2013 GDP per person, each square = $1,000

4.0   **Rest of the world**

*Twenty-foot equivalent units

▲ **FIGURE 8.4.2 Trade Around the Pacific Rim** Economic integration of urban centers between China, Canada, Taiwan, the United States, South Korea, Singapore, and other players map out the emergence of a new world region.

eight to a room and sew seven days a week from 8 A.M. to 11 P.M. As a result of these conditions, the retail margin (gross sales profit) on clothing made in workshops in China and sold in Europe and the United States is 200% to 300%, compared to a margin of only 70% or so on clothing made in the West. The success of these industries has resulted in rising wage demands by Chinese garment workers. Competition from new labor markets—like the Philippines—means that the continued growth of the garment industry in China is questionable; firms may outsource garment work to even poorer places in the future (**FIGURE 8.26**).

This new pattern of accumulation in China occurs at a sacrifice. The success of industrial modernization has intensified the gap among different regions and between urban and rural areas, and this disparity creates the potential for political tensions within China (**FIGURE 8.27**). Low wages and difficult working conditions have also created controversies for manufacturers in China. Finally, GDP growth has begun to decline in China, falling below 10%

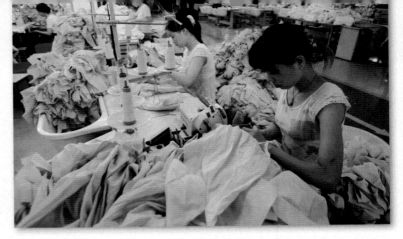

▲ **FIGURE 8.26 Textile and Garment Production in China** The growth of exports from China has depended on low wages for industrial labor, especially women's labor. Pictured here is a garment factory in Shenzhen.

between 2011 and 2015. This last fact portends serious ongoing economic and political difficulties.

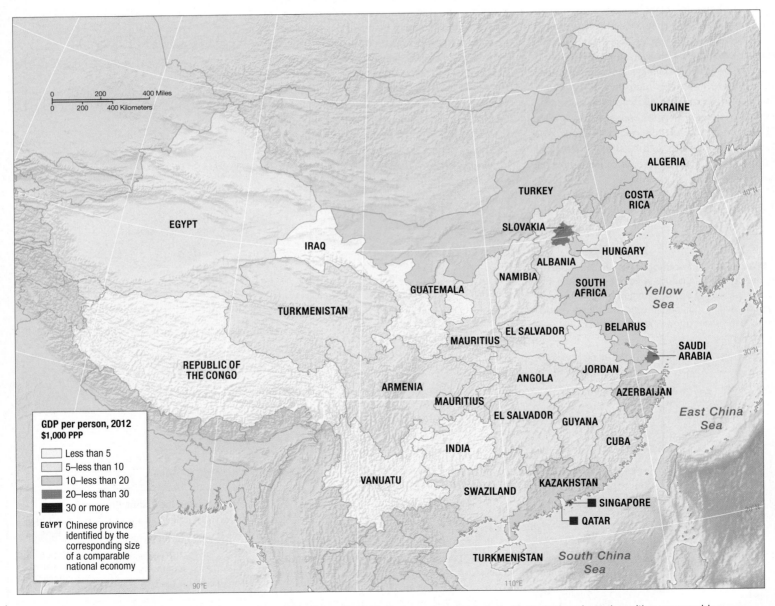

GDP per person, 2012
$1,000 PPP

- Less than 5
- 5–less than 10
- 10–less than 20
- 20–less than 30
- 30 or more

EGYPT Chinese province identified by the corresponding size of a comparable national economy

▲ **FIGURE 8.27 Projected GDP per Capita in Chinese Provinces in 2012** Each province is labeled with the name of a nation with a comparable per capita GDP; the lag in rural and western provinces suggests that China is becoming many different economies. **What accounts for the economic disparities between the coastal and inland regions of China?**

Even so, China's massive international mobilization of capital has included the acquisition of significant debt from other nations. As of 2014, China, which owns more than U.S. $1.3 trillion in United States bills, notes, and bonds, is the largest foreign holder of U.S. debt (followed by Japan with $1.2 trillion). This financial position gives China great political power in negotiations with the United States, but again makes China vulnerable to the changing status of the U.S. economy. As the Chinese economy has become more globalized, it has become more exposed to distant regional economic patterns.

Apply Your Knowledge

Follow this link (which shows the United States' Treasury's list of countries holding US debt in securities) to explore how debt contributes to interdependence in the globalized economy

8.13 Which countries around the world own U.S. debt?

8.14 Do some world regions hold more U.S. debt than others? What are the economic and geopolitical implications of this relationship between debtor (United States) and debt-holder (China)?

Two Koreas, Two Economies The end of World War II brought about an important geopolitical change to Korea as well. In 1945, the Soviet Union occupied the north of the peninsula while the United States occupied the south, their zones of occupation divided at the 38th parallel. A key site in the Cold War (see Chapter 3), Korea suddenly found itself divided into two parts. After a series of revolts and following separate declarations of independent statehood for the south and then the north, North Korean troops carried out a major attack on South Korea in June of 1950. The United Nations intervened, largely led by U.S. forces, and rapidly rolled back North Korean forces all the way to the Chinese border. This prompted China to enter the conflict on behalf of the North Koreans, and there followed a devastating war. Casualties were horrendous, with more than one million killed and much of the country destroyed. In 1953, the Demilitarized Zone (DMZ) was established as the de facto border, but the two Koreas remain bitter rivals and the war has never officially ended.

The division resulted in two very different economic experiments on the Korean peninsula. In South Korea (the Republic of Korea), the government crafted a state-led development effort that carefully controlled imports, managed internal competition between firms, and took out enormous international loans to invest heavily in private industrial development. The initial emphasis was on **import substitution**, in which the South Korean government facilitated the development of export industries by providing incentives, loans, and tax breaks to firms. By the mid-1980s, manufactured goods accounted for 91% of total exports. More than half of these were in the form of ships, steel, and automobiles. South Korea is now one of the world's leading exporters and is the home of major international conglomerates in areas such as electronics (Samsung) and automobiles (Hyundai-Kia).

Like Japan's, South Korea's economy has been adversely affected by the recent growth of Chinese manufacturing. The city of Busan, once the center of the South Korean footwear industry, experienced an enormous decline in sales of manufactured goods and is now full of deserted factories. Most of that export sector lost out to new Chinese factories. On the other hand, this city has developed tourism, shipping, and luxury residences, leveraging growth in the tertiary and quaternary economic sectors (see Chapter 1) to offset declines in secondary production.

North Korea (officially the Democratic People's Republic of Korea) has long followed the path of a closely planned, centrally guided economy, a model pioneered by the Soviet Union and China in the era before economic reform. This form of central planning often retards capitalist investment and adaptation to changing market conditions. North Korea remains highly economically isolated. A major economic barrier for the country is the incredibly disproportionate fraction of the economy dedicated to military spending, perhaps as much as 25% of GDP. This, coupled with sanctions that periodically isolate the country further from global trade, means a North Korean economy that is largely stagnant, small, and marked by shortages and hunger.

Apply Your Knowledge

8.15 Compare and contrast the economic experiments in China, South and North Korea, and Japan.

8.16 What have the costs and benefits of each strategy been for the economic prominence and power of each country? For their people?

Territory and Politics

East Asian political geography is marked by competing historical trends. The war in Korea remains, for all intents and purposes, unresolved. The fate of Taiwan also remains in question more than a half century after it was seized by Chinese Nationalists. On the other hand, the expansion of Chinese geopolitical power has escaped the bounds of the region. Chinese influence is now being felt not only across North America and Europe but also in the resource frontiers of Africa, even as the Chinese government struggles to maintain its own internal control.

An Unfinished War in Korea The partition of North and South Korea in 1953 (described earlier in the chapter) left North Korea behind the world's most heavily militarized border. The first period of leadership in that country set the tone for politics and development for the decades that would follow. The nation's first leader, **Kim Il Sung**, governed the country according to a philosophy called *juche*: a mixture of centrally planned economics, self-reliant nationalism, and the cult of the national leader. From its inception, North Korea's dynastic leadership has retained tight relations with China and a posture of continued hostility toward South Korea.

To carry out his strategy of national development, Kim Il Sung imposed an austere regime directed through central planning, a regimented way of life dominated by a totalitarian government, and the development of military power, including nuclear power capability that could lead to nuclear weaponry. In 1994, Kim Il Sung died and was succeeded by his son **Kim Jong Il**. Estimates of the total number of deaths because of North Korea's food shortages between 1995 and 2005 range between 1 and 3 million. By 2011, after 58 years of sacrifice and austerity, personal incomes in North Korea were only one-tenth of those in South Korea, and infant mortality rates were five times higher. North Korea remains not only one of the world's most impoverished societies but also one of the most closed and rigid. Listening to foreign radio broadcasts is punishable by death, citizens can be detained arbitrarily, and there are an estimated 150,000 political prisoners.

In 2012, Kim Jong Il died and was succeeded by his own son, **Kim Jong-un**, an enigmatic figure and the world's youngest head of state (he was born in 1983). Kim Jong-un governs one of the most highly militarized countries in the world. North Korea has the world's fifth-largest standing army (after China, the United States, Russia, and India), even though it has a relatively small population (an estimated 25 million in 2015; **FIGURE 8.28**). For its part, North Korea occupies an embattled position, with US and South Korean troops on its border and a general state of isolation under crippling economic embargo. North Korea's weapons and missiles are sold indiscriminately in world markets to raise sorely needed foreign exchange. Its own arsenal—including nuclear weapons—is pointed at South Korea and Japan.

Most recently, North Korea has been suspected of mobilizing **cyberwarfare** technology to hack or sabotage websites. The matter became highly publicized in 2014 when North Korea was alleged to have launched a massive hack of the Sony Pictures website in retribution for their planned release of the comedy movie, "The Interview," which portrayed the Democratic People's Republic in a humorous way and depicted the assassination of Kim Jong-un.

Unresolved Geopolitics in Taiwan When the Communist revolution created the People's Republic of China (PRC) in mainland China, the ousted Nationalist government of China established itself in Taiwan as the Republic of China in 1949. Granted diplomatic recognition by Western governments but not by the PRC, Taiwan immediately became a geopolitical flashpoint in the Cold War. Large amounts of economic aid from the United States helped prime Taiwan's economy. By the early 1960s, the nation's political stability and cheap labor force provided a very attractive environment for export processing industries.

Taiwan lost its full international diplomatic status in 1971, when U.S. President Richard Nixon's rapprochement with the PRC led to China's entrance into the United Nations. In 1987, Taiwan began a phase of political liberalization and relaxed its rules about contact with the PRC (also referred to as mainland China). Mainland China still claims Taiwan as a province of the PRC (even as Taiwan has symmetrically declared the People's Republic a "breakaway" from true China) and has offered to set up a Special Administrative Region for Taiwan, with the sort of economic and democratic privileges that the PRC has given Hong Kong. Geopolitical tensions periodically flare up in conflicts over Chinese or U.S. naval exercises and other symbolic acts. So, while it is extremely unlikely that any real conflict will determine the future of Taiwan's status, meaningful diplomatic relations have not emerged despite a long period of effort.

◀ **FIGURE 8.28 North Korean Military** A significant proportion of the North Korean gross domestic product is dedicated to supporting the military, rendering the economy anemic. This photograph shows North Korean female soldiers marching in a military parade in Pyongyang on April 15, 2012, to mark 100 years since the birth of the country's founder, Kim Il-Sung. **What strategies characterize North Korea's development and how do they differ from those of their neighbors?**

Apply Your Knowledge

Go to the home pages of the Democratic People's Republic of Korea (North Korea): http://www.korea-dpr .com/, the Republic of Korea (South Korea): http://www.korea.net/, and Taiwan (Republic of China): http://www.gio.gov.tw/

Politics in Two Koreas and Taiwan

https://goo.gl/mSEODA

8.17 How do the political visions of these states differ?

8.18 As presented in the text on these sites, what distinct geopolitical vision and history have they chosen to communicate?

The Contested Periphery of Tibet As early as the 7th and 10th centuries C.E., Tibet had developed a distinctive subregional culture based on Tibetan Buddhism. Far away from the center of Chinese power, though ostensibly part of a few Chinese empires, Tibet was divided among feudal chieftains ruling repressively and autocratically.

In 1950, China seized on this fact as justification for its invasion of the strategically important region. Between 1950 and 1970, the Chinese destroyed much of Tibetan cultural heritage and caused the death of an estimated 1 million Tibetans. In 1959, the Tibetans rose in an attempted revolt, and their spiritual leader, the **Dalai Lama**, fled into exile. In 1965, Tibet was granted the status of an autonomous region by the PRC, called Xizang, but by that time, large numbers of ethnic Han Chinese had flooded in, often taking positions of authority and leaving Tibetans as disadvantaged, second-class citizens. A significant, though hard to determine, proportion of the area's population is now of Han Chinese descent, which the Tibetans in exile describe as "demographic aggression." In 2015, the Chinese government proposed a policy to evaluate the "patriotism" of

Buddhist monks and nuns and to install national flags, telephone connections, and Chinese newspapers and reading rooms in monasteries. These developments have given rise to significant ethnic tensions within the region (**FIGURE 8.29**).

For their part, the Chinese cannot understand the ingratitude of the Tibetans. As the Chinese see it, this is a process of modernization, which has saved the Tibetans from feudalism and built roads, schools, hospitals, and factories.

China has expressed some interest in naming the Dalai Lama's successor. Since that successor is considered by Tibetan Buddhists to be the reincarnation of the last Dalai Lama, this puts the athiest Chinese Communist Party in the awkward position of interpreting, intervening, and tacitly acknowledging matters of Buddhist faith.

The Geopolitics of Globalized Chinese Investment The success of state capitalism in China has resulted in new global geopolitical problems. These problems have arisen because the enormous production systems of the industrial zones of China have tested the limits of available raw materials and energy resources within the PRC. This lack of resources has led to a search for new sources of inputs and has led China abroad into new resource frontiers, especially into Africa. Chinese investments overseas, especially in underdeveloped areas have grown. As of 2012, China purchases more than one-third of its oil from Africa and is the continent's largest trade partner.

China's appetite for resources makes it a large and quickly growing presence in the developing world, with implications for international conflict. In a notable case, Chinese development money has flowed heavily into many countries in Africa, including Sudan, a country under international scrutiny for high levels of uncontrolled and state-tolerated violence and genocide (see Chapter 5). From the Chinese point of view, its supply of development capital to countries like Sudan provides a net benefit to Sudan's citizens and the mode of governance a nation chooses is not the business of other members of the world community. From the point of view of the United States and the European Union, on the other hand, Chinese support for the Sudanese regime subverts efforts to bring a disastrous situation under control. Whatever the outcome in the case of Sudan, the rise of China as an international development and investment actor on a global scale is a key factor in geopolitics and the formation of new regions.

▼ **FIGURE 8.29 Chinese Development in Tibet** A Chinese tourist takes a photo of a Lama monk outside the Shigatse railway station in Shigatse city, Tibet. The influences of Chinese culture, economy, and political control on the Tibetan landscape are now unmistakable.

Apply Your Knowledge

8.19 Describe the political changes of fortune for the great powers of East Asia (especially Japan and China) before colonialism, under colonialism, and in the last few decades.

8.20 How has the center of power shifted across the region?

Culture and Populations

Traditional societies in East Asia have centered around family, kin networks, clan groups, and language groups. These societies developed in largely rural contexts, with a strong bureaucracy enforcing social order. In China and Japan, society was historically hierarchical, and individuals were subsumed within the family unit, the village, and the domain of the local lord. In these environments, important social values include humility, understatement, and refined obsequiousness, with particular deference being shown to older persons and those of superior social rank.

There are, however, marked regional variations in cultural traditions within East Asia, including key regional differences in religion, language, diet, dress, and ways of life. The changing relations of the region to other global cultures, moreover, have transformed local ways of living.

Religion and Language

The cultural traditions of East Asia, especially its religions and languages, show signs of merging and hybridizing with one another, while they remain diverse and independent. Japanese and Chinese languages share some symbols, for example, and Buddhism is found in many parts of the region. However, the languages are also entirely distinct, and Buddhist practices in Mongolia differ from those of Japan.

Religions Chinese culture found spiritual expression in the philosophy of **Confucianism**. Unlike formal religions, Confucianism has no place for gods or an afterlife. Confucianism's emphasis is on ethics and principles of good governance and on the importance of education as well as family and hard work. Formalized religions in China include Daoism (throughout much of the country), Tibetan Buddhism (in Xizang), and Islam (in large parts of Inner Mongolia and Xinjiang).

For a large number of Chinese, however, folk religions are more important than any organized religion, especially **ancestor worship**. This practice is based on the veneration of the dead and that the dead spirits, to whom offerings are ritually made, have the ability to influence people's lives. Offerings include food or burning paper money, to heal the sick, appease the ancestors, and exorcise ghosts at times of birth, marriage, and death.

Japanese Indigenous culture was expressed in **Shinto**. Shinto is not a distinctive philosophy but, rather, a belief in the nature of sacred powers that can be recognized in every individual existing thing (**FIGURE 8.30**). Shinto traditions can be thought of as the traditions of Japan itself. Seasonal festivals elicit widespread participation in present-day Japan, regardless of people's religious affiliation. These festivals usually entail ritual purification, the offering of food to sacred powers, sacred music and dance, solemn worship, and joyous celebration.

Buddhism is also important in Tibet, Mongolia, China, and Japan. Imported from India in the 1st century C.E. (see Chapter 9), the religion changed drastically from its traditional form and took on a distinctive set of regional forms. Tibetan Buddhism stresses the monastic tradition and the holy leadership of monks, whereas Zen Buddhism in Japan is known for its distinctive material culture, including practices of painting, ceramics, and tea-drinking ceremony.

The official belief system of the Chinese Communist Party is 'Marxism-Leninism with Chinese characteristics', and an avowed **atheism**. Enforcement of this belief was especially strong during the **Cultural Revolution** in China during the 1960s, when traditional beliefs were punished and religious institutions vandalized and destroyed. Millions of people were displaced, tens of thousands lost their lives, and much of urban China was plunged into a terrifying climate of suspicion and recrimination. The party now formally recognizes and tolerates five religions: Buddhism, Taoism, Islam, Protestantism, and Catholicism, though religious practitioners report coercive state controls and limits on their practice.

Feng Shui Throughout East Asia there is widespread adherence to **feng shui** (pronounced "fung shway")—not a religion, but a belief that the physical attributes of places can be analyzed and manipulated to improve the flow of cosmic energy, or qi (pronounced "chi"), that binds all living things. Feng shui today involves strategies of siting, landscaping, architectural design, and furniture placement to direct energy flows.

Languages *Chinese* is a somewhat imprecise linguistic term. In the region of East Asia dominated by Chinese people, residents speak not only many Chinese dialects but also an enormous range of diverse minority languages (like Uyghur in the west, Sibo in the northwest, and Chuang, Yi, Nakhi, and Miao in the south). The dominance of Han people within China is reflected in the geography of language, however. Mandarin, the language of the old imperial bureaucracy, is spoken by almost all Han people as well as Hui and Manchu people (and the people of Taiwan). As many as

▲ **FIGURE 8.30 Shinto Gate** The classic archway of the Shinto shrine (Torii) in Miyajima, Japan, is emblematic of Japan's deepest religious traditions. The gate marks a location of passage between the profane and the sacred. **What characteristics do East Asian spiritual traditions share? Which are distinctive and in what ways?**

53 ethnic minorities in China, however, maintain their own languages.

Chinese writing dates from the time of the Shang dynasty (1766–1126 B.C.E.). It features tens of thousands of characters, or **ideographs**, each representing a picture or an idea. A single dictionary contains 50,000 distinct characters. Ideographs are enormously complex, especially in their studiously artful calligraphic form; a single Chinese character might take as many as 33 brush strokes to complete (**FIGURE 8.31**). More recently, the government of China has facilitated literacy by simplifying characters and reducing the number of characters in use. The PRC has also adopted a new system, **pinyin**, for spelling Chinese words and names using the Latin alphabet of 26 letters.

Though the Japanese adopted some Chinese pictographic characters in the 3rd century C.E., the two main languages are largely unrelated. Japanese ideographs—symbols representing single ideas or objects (*kanji*)—number more than 10,000. These are joined by numerous other symbols (*kana*) that represent phonetic word fragments.

The affiliation of Korean to other languages in the region is uncertain. Although it contains Chinese words, its structure and grammar are far closer to Japanese. Even so, the linkage between Korean and Japanese is unclear. The Korean alphabet, unlike Japanese and Chinese, is entirely phonetic (characters represent independent sounds), though the script—wholly unlike that of other languages in the region—is written in clustered groups. The fundamental differences between the languages of East Asia remind us that the region is a product of interactions, not an inevitable coherent whole.

Cultural Practices, Social Differences, and Identity

In China, since the death of Mao Zedong in 1976 and the end of the Cultural Revolution, traditional arts and practices have reemerged strongly, including new museums, folk art

▼ **FIGURE 8.31 Chinese Calligraphy** A Chinese calligrapher at work writing on a scroll. **How do the languages and scripts of China, Japan and the Koreas differ?**

celebrations, and espousal of commitments to traditional culture (**FIGURE 8.32**). This may in part reflect an interest in selling a managed form of Chinese culture to foreign governments and tourists, or it may be an intentional effort to utilize certain art forms, language, and music in the service of Chinese nationalism. Whatever its motivation, the back-and-forth struggle over traditional practices is a key feature of the East Asian cultural landscape, as can be seen in the changing state of gender roles, conflict over ethnic identity, hybrid food traditions, and new cultural traditions that have come to travel worldwide.

Gender and Inequality Traditional East Asian cultures insist on distinct and hierarchically ordered relationships between men and women. There are numerous dramatic historical examples of patriarchy. The Chinese tradition of foot binding, practiced mainly by elites, involved the often painful disfiguring of women's feet. The Japanese Meiji Civil Code of 1898 denied women many basic legal rights. Neo-Confucian

▼ **FIGURE 8.32 Traditional Opera Culture in China** This performance in the Beijing Opera includes an appearance by Monkey, a favorite character from Chinese classic literature.

Emerging Women's Power Amid Patriarchy and Corruption

More women in China are in the workforce than perhaps ever before and women occupy positions of power in industry and government. Nonetheless, traditional patriarchal norms can mix with the disproportionate power of officials amidst a booming economy. When traditional patriarchal authority attempts to exert control, however, China's new information and knowledge economy can push back.

Deng Yujiao learned this in 2009 when she fought back against assailants at her place of work in Hubei province. The 21-year-old was doing laundry when she was forced to defend herself with a small knife against three male assailants, who had attempted to bribe and threaten her for sex and then assaulted her. The attacker she stabbed, who did not survive his wounds, was a local official. The police arrested Yujiao, restrained her in a mental institution, and began swift proceedings to find her guilty of murder (FIGURE 8.5.1).

It is at this point that Deng Yujiao became a critical figure in China. Shortly after her arrest, people from across the country became familiar with the case in chat rooms, blogs, and reportedly more than four million posts across the Internet. A circulated public letter, signed by hundreds of journalists and academics, circulated widely online. Despite efforts of the Chinese state to censor these sites, Yujiao's case came to take on national significance, representing for many the immunity and corruption of China's emerging elite class of officials and, for others, the persistent lack of power for women, even in the "New China." Before the tide of public opinion pushed back, reporters covering the case suffered beatings and regional officials threatened Yujiao's family. The outcome was a highly public trial where a local lawyer (rather than the one assigned from Beijing) represented Yujiao, whom the court eventually found guilty of a far lesser charge: excessive self-defense. In the wake of Deng Yujiao's experience, columnists and reporters from across China have used the case to criticize both the system of justice that is tilted toward public officials with increasing money and power, as well

> "Yujiao's case came to take on national significance, representing for many the immunity and corruption of China's emerging elite class of officials and, for others, the persistent lack of power for women, even in the "New China.""

as the difficult and vulnerable position of women. Her face began to appear on T-shirts across the country. A new generation of Chinese women are simultaneously facing new opportunities and changing public feelings about the role of women, while at the same time confronting an increasingly powerful group of male elites.

1. How does Yujiao's case reflect the changing nature of China's economy and society?

2. To what degree does information technology, the Internet, and media impact the way China is governed?

▲ FIGURE 8.5.1 Deng Yujiao

culture in Korea, as elsewhere in the region, focused on the woman's role of providing a male heir.

Much of this tradition has been overturned since the end of World War II. Independent, working, professional women have been ubiquitous in Japan since the time of its postwar economic miracle. Even so, patriarchy in Japanese work culture persists and Japan consistently rates near the bottom 1/3 of the world's countries in the gender gap report the World Economic Forum publishes yearly. In China, traditional gender roles were disrupted during the Communist period, when society was officially gender-neutral and images of female beauty and ideas of romance were suppressed. Today, the participation of women in political and economic activities remains widespread. In China, competitive markets in education have increased the number of college-educated women. As of 2014, women made up 48% of university enrollment. These better-educated women have made their mark in China's labor markets, earning high salaries and challenging traditional gender boundaries for top jobs. A startling paper by Forbes reports that in 2013, women held 51% of China's senior management positions, compared to the 24% global average.

Even so, traditional gender stereotypes have resurfaced. This has played out in labor markets, as older and poorer women have borne the brunt of layoffs in the state sector and have been left with lower-paid occupations. As a result, despite successes for women in management, there is a widening income gap between men and women in China. Women in China earned 19.6% less than their male counterparts in 2000 but earned 40.7% less in 2013. Other countries in the region harbor similar disparity. In South Korea in 2007, the employment rate for women with college degrees was 61.2%, at the very bottom among advanced economic countries. Women in South Korea receive 66% of men's wages for the same work, according to the Korean Employment Information Service.

Ethnicity and Ethnic Conflict in China

Paralleling changes in gender roles, many traditional ethnic divisions have been both challenged and reinforced in recent years. Although there are few immigrant populations of any significant size in the region, there are numerous important long-standing minority ethnic communities, especially in the People's Republic of China, which officially recognizes 56 ethnic groups (**FIGURE 8.33**). The dominant group in China is the Han, who make up more than 91% of the population. The Han people originally occupied the lower reaches of the Huang He and the surrounding North China Plain. They spread gradually inland along river valleys, into present-day Korea, Manchuria, and toward the humid and subtropical south. Today there are more than a billion Han in China. The remaining 55 minority groups add up to less than 110 million people, though they are dominant populations in their home regions. They are mostly groups of minority ethnicities—including the Hui, the Manchu, the Uyghurs, the Tibetans, the Zhuangs. Indigenous peoples, such as the Miao, the Dong, Li, Naxi, and Qiang live in the remote border regions, and are relatively economically disadvantaged.

Relationships between these minority communities and the dominant Han people are complex and

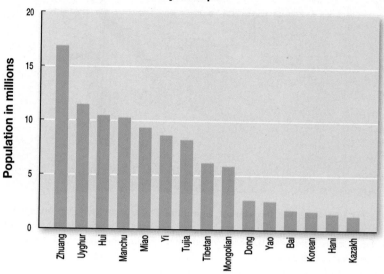

China's Ethnic Minority Groups

▲ **FIGURE 8.33 China's Ethnic Minority Groups** 55 ethnic minority groups are recognized by the People's Republic of China (PRC). Some, like the Uyghur, are numerous and concentrated, while others, like the Bai, are relatively few.

vary from community to community. The Naxi people (pronounced: na-shi), who live in the hill country of the far southwest, have a growing prominence, owing to their distinct folk culture and unique musical traditions (**FIGURE 8.34**). Chinese authorities have recently come to celebrate distinct Naxi traditions, in part to develop the Naxi's Autonomous County in Lijiang, Yunnan Province, for tourism.

Not all relationships are so mutually advantageous; tensions often exist between several of the larger minority groups and the dominant Han. The ethnic Uyghurs, who predominate in westernmost China (and in many post-Soviet central Asian states),

▼ **FIGURE 8.34 Naxi Musical Orchestra** A unique folk tradition of the Naxi people, Naxi music celebrates local tradition but also playfully engages dominant Han traditions. **How have minority communities adapted to, and survived amidst, dominant Han Chinese culture. In what cases have there been conflicts?**

▲ **FIGURE 8.35 Muslim Market, Shanghai** The Muslim Market in Shanghai highlights minority food traditions of China, especially including Uyghur culture and cuisine. Held near the Huxi Mosque every Friday, the stalls sell lamb and rice dishes, dumplings, buns and noodles. **How are Chinese food traditions a product of inter- and intra-regional relationships?**

▲ **FIGURE 8.36 Chinese Street *Bao Zi* (Dumplings) in Shanghai** *Bao zi* are dough dumplings filled with vegetables. This classic Chinese street food is common all over China.

have faced suppression from the PRC. Starting in 2009, violence broke out between Uyghurs, ethnic Han people, and state police in the city of Ürümqi in far western China. The Chinese state looks upon Uyghur activism as hostile to the interests of the PRC and has branded some Uyghurs as terrorists. Uyghurs cite ongoing economic marginalization and restrictions on the practice of Islam, a product of both ethnic marginalization and the history of religious repression leftover from the Cultural Revolution.

Even so, China has also managed to absorb and make space for Muslim minorities in some places. The Friday Muslim market near the Huxi Mosque in Shanghai has been closed and reopened by authorities from time to time, but remains a valued destination by locals and tourists alike (**FIGURE 8.35**).

Food For people outside East Asia, Chinese food or Korean food may conjure up images of an undifferentiated set of globalized generic dishes (from Kung Pao chicken to Mongolian barbecue). The diverse and nuanced food traditions in East Asia, however, emphasize the regionally unique character of food practices and their dependence on traditional food sources, as well as the complex influences from other regions.

In terms of regionally distinct foods, for example, within China there are fundamental differences between the cuisines of Inner and Outer China. In Outer China, milk-based dishes—yogurt, curds, and so on—are common and herds of sheep, goats, cows, horses, camels, and yaks also provide meat dishes. Inner China, by contrast, has little stock-raising and, consequently, few milk-based dishes; its principal meat dishes feature ducks, poultry, and pigs. Within Inner China, the humid and subtropical south has developed a cuisine based on rice, while in the subhumid north, noodles are the staple.

More generally, the distinctive character of East Asian cuisine is heavily influenced by historical linkages with the rest of the globe. Many Chinese dishes, for example, are marked by the creative use of hot chilies. This Central American domestic plant only became part of Asian cuisine after 1492. The meat-filled dumpling, a standard fundamentally "Chinese" staple on Asian menus in Britain, France, and the United States, was likely introduced to

China in the 3rd century C.E. from elsewhere in Asia (**FIGURE 8.36**). Similarly, the vegetarian traditions in East Asian cuisine thrived under the influence of imported Buddhist teaching (originally from South Asia—see Chapter 9), and had to compete against meat-loving habits of people from different regions and faiths.

East Asian Culture and Globalization In addition to food, other East Asian cultural products have become popular beyond the region. East Asian cultures have long been part of the process of globalization. Chinese art and artifacts have been popular in Europe and the Americas since the 18th century. Chinese and Japanese cuisines have been introduced to cities throughout the rest of the world. Japanese architecture and interior design have influenced Modernist design. Religions such as Tibetan Buddhism and Zen Buddhism have a growing following in both Europe and North America, and feng shui has found adherents in many Western countries.

Two stylized artistic Asian entertainment forms that have swept the globe in recent years are manga and anime. **Manga** refers to Japanese print comic books (*komikku*), which date at least to the last century. Contemporary manga novels cover topics from fantasy to history, usually with an easily recognized and common style, typically involving characters with large eyes and childlike features (**FIGURE 8.37**). **Anime**, in a general sense, refers to all Japanese animated cartoons. More specifically, however, the term *anime* is associated with a distinctive style of animated film. Though stylistically anime can vary from highly realistic to more abstract, internationally it is most closely associated with the styles of manga. Chapters, settings, and stories draw heavily from manga print cartoons and feature human forms with large eyes and exaggerated expressions.

Of these newly globalizing East Asian cultures, Chinese culture in particular is entering a period of global ascendance. Chinese music and products are increasingly familiar throughout Asia and internationally. China's global cultural profile is reinforced by its growing status as a major destination for tourism. In 2012, China received 57 million international visitors, behind only France and the United States. In 2011, foreign exchange

▲ FIGURE 8.37 **Manga** These manga books (Japanese comics), which are enjoyed by a global audience, are on sale at the annual Paris book fair.

revenue from tourism to China generated U.S. $35 billion. The hidden costs of modernizing infrastructure to accommodate tourists are notable. Stringent police controls remain common, free expression is quashed, and many small traditional neighborhoods, or **hutongs**, are demolished to make way for construction (and ongoing urban development).

Apply Your Knowledge

8.21 Identify some East Asian cultural traditions that have elements or ingredients introduced from other regions.

8.22 Does describing these cultural phenomena as traditional, local, or regional have economic or political implications?

Demography and Urbanization

The most striking demographic characteristic of East Asia as a whole is the sheer size and density of its population. The region's total population of 1.59 billion in 2015, as **FIGURE 8.38** shows, is distributed along coastal regions and in the more fertile valleys and plains of Inner China. In contrast, population densities throughout Inner China and in Japan, North and South Korea, and Taiwan are very high: between 200 and 500 per square kilometer (518 to 1,295 per square mile) on average. The great majority of Japan's 127.5 million people live in dense conditions in the towns and cities of the Pacific Corridor, where space is so tight and expensive that millions of adults, unable to afford places of their own, live with their parents. The average dwelling in metropolitan Tokyo is about

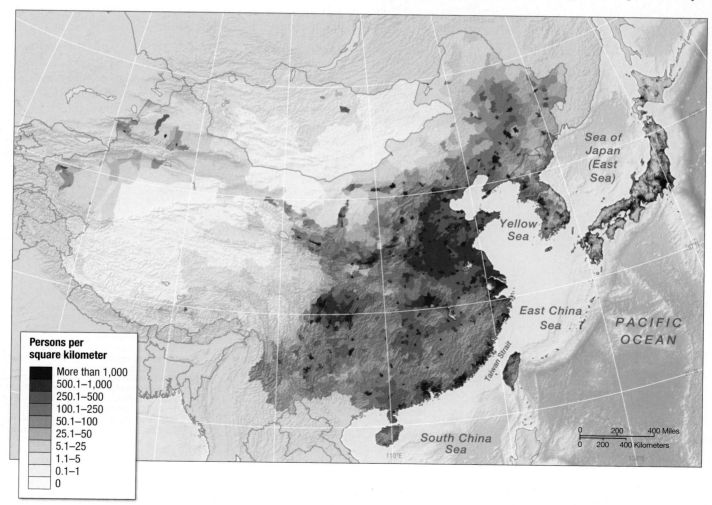

Persons per square kilometer: More than 1,000 / 500.1–1,000 / 250.1–500 / 100.1–250 / 50.1–100 / 25.1–50 / 5.1–25 / 1.1–5 / 0.1–1 / 0

▲ FIGURE 8.38 **Population Density in East Asia** The density of population is very high throughout most of the continental margins, as well as the fertile basin and river valleys. Arid and mountainous interiors are sparsely populated.

▲ FIGURE 8.39 Small Home, Tokyo, Japan Scarcity of space makes elegant but efficient home design an architectural goal in Japan.

60 square meters (646 square feet; **FIGURE 8.39**). (By comparison, the median dwelling size in metropolitan Los Angeles is just over 150 square meters, or 1,615 square feet.)

People's behavior and urban development patterns also follow from a need to use space efficiently. Japanese daily life is filled with rules, and many of them have evolved because of space shortages. Japanese commuters are used to tight conditions in subway trains during rush hour and sharing public spaces such as parks with large numbers. As a result, transportation and daily life in Japan are relatively free of hassles, despite the high overall density of people.

Demographic Change Although the region was long known for high rates of population growth, the current population profile of East Asia is radically changed from its historic trends. Population pyramids for Korea, Mongolia, Japan, and China all show a common trend: decreasing populations of young people, owing to falling birth rates (**FIGURE 8.40**). Japan led this trend by decades, as reflected in the rectangular shape of the pyramid, while Mongolia's trend is only apparent in recent years.

The drivers of these changes in each case are somewhat different. Japan's early economic boom led to smaller families and its lengthy recent recession (described above) reinforced this trend by making child-raising and education expensive. For Mongolia, the last few years have been ones of economic change, where historically agricultural families have become increasingly settled, with a predictable fall in birthrates.

Population Control in China This radical transition is most notable in China, however. There, a fertility rate decline from more than 6 to less than 2 accompanied industrialization and urbanization. At the same time, however, dramatic and invasive policy experiments also contributed to the decline.

During the 1960s, China's population grew at a rate of 2.6% per year, owing to improving medical care and public health, while birthrates remained high, encouraged by the political leadership.

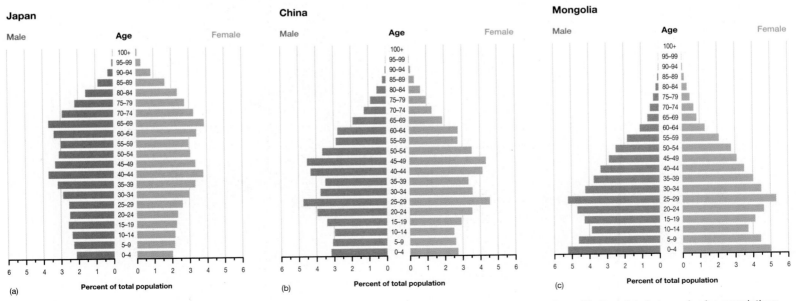

▲ FIGURE 8.40 Population Pyramids: Japan, China, and Mongolia Though there is some variability, the overall trend in East Asia is toward aging populations and demographic stagnation. **How do the youngest cohorts in these populations compare to those of middle-aged people? What does this portend for population growth and change?**

By the early 1970s, the average family size was 5.8 children, and it became clear that China could not sustain such a large population. In response, China's Communist Party instituted an aggressive program of population control. A sustained propaganda campaign was reinforced in 1979 by strict birth quotas: one child for urban families; two for rural families; and more for families that belonged to ethnic minorities. The policy involved rewards for families giving birth to only one child, including work bonuses and priority in housing. In contrast, families with more than one child were fined and penalized in annual wages and made ineligible for education and health-care benefits. This **one-child policy** was rigorously enforced until 2015, when it was discontinued, owing to an increasingly evident dearth of working-age young people.

In terms of reducing population growth, the policy was very successful. Fertility rates have fallen, and zero population growth is on the horizon (**FIGURE 8.41**). Without its aggressive population policy, China's population today would have been more than 300 million larger.

Nevertheless, in addition to the personal and social coercion involved, the one-child policy led to social stresses, specifically

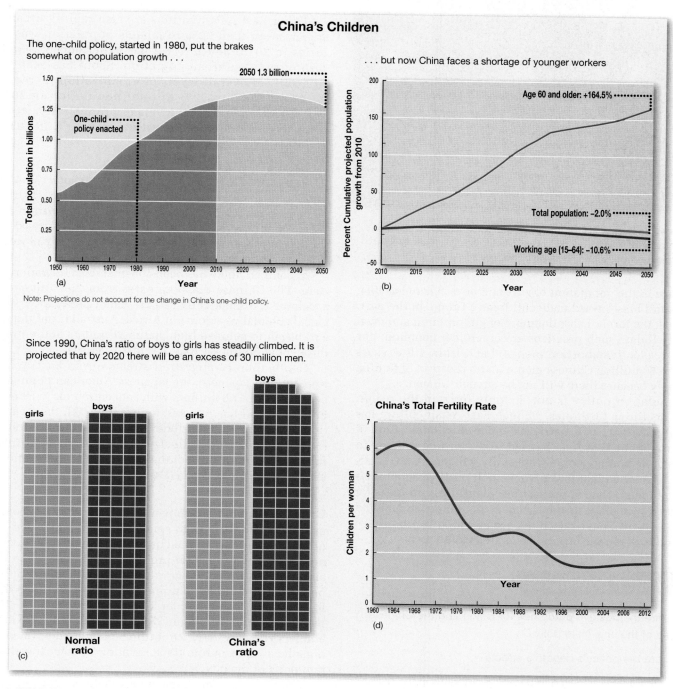

▲ **FIGURE 8.41 China's Demographic Transition:** Changes in Chinese population includes: a) a leveling off of population growth, b) a decline in working age people, c) a growing gender imbalance, and d) a plummeting fertility rate. **What are the potential impacts of China's demographic transition to low and no growth for its economy and society?**

▲ **FIGURE 8.42 Japanese Professional Women** Urban, single women in Japan are marrying later and working longer. **What impact do these social changes suggest for population growth or decline?**

an imbalance in the population of girls and boys. There has long been a cultural bias toward male children in China. In the past, this bias took the form of abandonment of girl children and even infanticide. Today, such practices are much less common, but widespread selective abortion means that within a few years there will be 50 million Chinese men with no prospect of finding a wife, simply because there will not be enough women.

The population policy is not the only force in the region leading to declining birthrates. Though East Asian culture has traditionally favored early marriages, the professionalization of women throughout East Asia has led to decreasing interest in marriage among young people, especially women. A 2011 survey in Japan revealed that fewer women felt positive about their marriage than men. New opportunities, identities, and ideas for young women in society, as they become more independent and publically assertive, have raised the age at which families have children and lowered the birthrate overall (**FIGURE 8.42**).

Apply Your Knowledge

8.23 What are the positive outcomes for China and the Chinese people of the one-child policy?

8.24 What are the policy's negative effects?

8.25 What other forces have led to declining birthrates in the region?

Migrations For decades, internal migration within East Asia has been dominated by the movement of people from peripheral, rural settings to the towns and cities of more prosperous regions. This has been especially dramatic in China where 35% of the population that is officially registered as urban, but at least 20% more of the population now resides in cities. These latter migrants are officially unregistered as urban dwellers under China's **Hukou** system, which attempts to control and rationalize the rate of population migration. In cities like Shanghai, these unregistered "floating people" are key to the economy, but live as second-class citizens (**FIGURE 8.43**).

Other East Asian nations are also rapidly urbanizing. In Japan, more than 92% of Japan's population was living in cities in 2015. South Korea has experienced a similar pattern of rural-to-urban migration. Seoul has been the focus of a shift of overall levels of urbanization increasing from 21.4% in 1950 to 92% of total population in 2013. Even North Korea, with poor levels of productivity, experienced a steady increase in urbanization: from 31% in 1950 to 60% in 2013.

Diasporas East Asia is distinctive insofar as it has few immigrant populations of any significant size. Distrust of foreigners of all kinds has been long-standing within East Asia. The Communist regimes of China and North Korea contributed to this mind-set as well by maintaining tightly closed borders for several decades.

In contrast, there has been considerable emigration from East Asia. The Chinese diaspora dates from the 13th century, but accelerated dramatically as East Asian populations joined the global colonial workforce in Africa, Australia, and the Americas, and began to achieve high status as traders and entrepreneurs throughout Southeast Asia and the Americas.

Contemporary migration streams from East Asia include Koreans and Japanese; the Japanese-American population numbers more than 1.3 million, with concentrations in Honolulu, San Francisco, and Los Angeles, for example. But the greatest migration flow is from China. Today, Chinese living abroad number in the tens of millions, with significant and economically powerful groups throughout Southeast Asia and large numbers in the United States and Canada (**FIGURE 8.44**).

Great Asian Cities Japan and Korea have become highly urbanized in the years since World War II, and China is urbanizing quickly in the current era. The resulting cities have numerous environmental problems and stresses, but they have also radically propelled new forms of architecture, public transportation, and urban planning and design. Through innovative logistics and technology in Tokyo, for example, its 35 million people experience urban design marked by flexibility and efficiency. Every day, 27 million people ride Tokyo's highly organized rapid public transportation system, traveling to distant parts of the city.

The new cities of China are all the more spectacular for the recentness of their growth. The city of Guangzhou, for example,

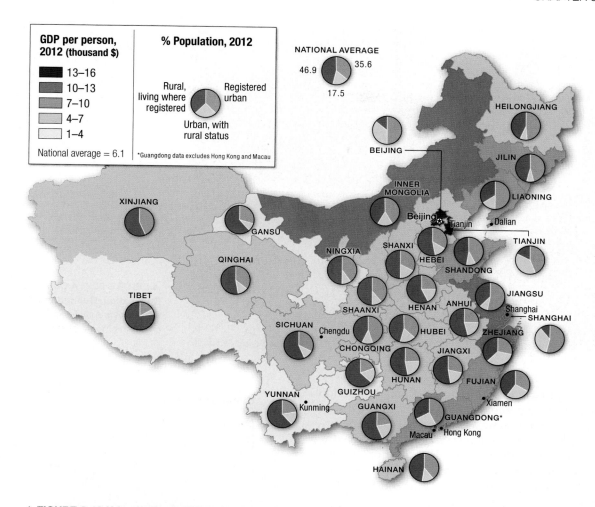

▲ **FIGURE 8.43 Urbanization in China** Chinese cities are booming, with populations moving in large numbers to urban areas, especially in the more affluent East. In the west, where rural populations still prevail, economic growth is far lower.

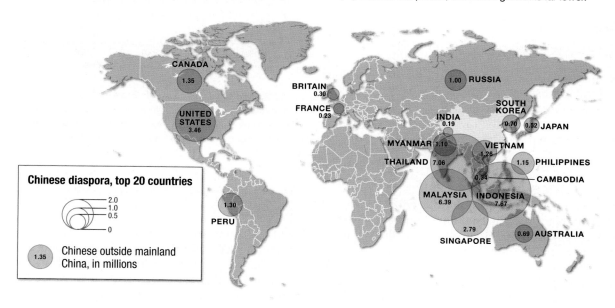

▲ **FIGURE 8.44 Chinese Migrants Abroad** Chinese emigration began on a large scale during the 19th-century Industrial Revolution, when Chinese laborers found opportunities for employment in newly colonized lands. **Which world regions host significant Chinese populations?**

◄ **FIGURE 8.45 A Construction Site in Guangzhou, Guangdong, China** Dozens of new skyscrapers now form the skyline of this bustling city at the mouth of the Pearl River in southern China.

recently grew to a population of more than 12 million people (**FIGURE 8.45**). The bursting economic activity of the city is embodied in a construction boom, where innovative structure, curvilinear design, and expressive experimentation are creating a whole new skyline. It is critical to recall that impoverished migrants from the countryside construct these glistening new cities. More than 150 million laborers live in China's cities as temporary and typically second-class citizens.

Future Geographies

The future geographies of East Asia rest on the trajectory of four issues with uncertain paths: the transition of the Chinese economy, the role of environmental degradation and sustainability in Chinese development, the fate of the Korean peninsula, and the role of China in international geopolitics.

A Nation of Chinese Consumers?

The Chinese leadership has begun an ambitious effort to turn the country's economy into a postindustrial one dominated by consumers. By developing educational institutions, offshoring manufacturing jobs, and investing in infrastructure, the government hopes that its middle class will number in excess of 630 million people (what will be half of the total population) by 2022. This is occurring, unfortunately, precisely as the economy cools off; in 2014, China's economy grew at its slowest pace in 24 years, logging only 7.4% growth (**FIGURE 8.46**). Some observers are skeptical that China can accomplish its middle class revolution.

China GDP Growth

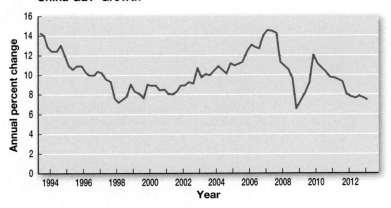

▲ **FIGURE 8.46 China's GDP Growth: 1993–2015** The 1990s were a period of meteoric economic growth in China. The last decade, however, has been marked by decline growth and more erratic returns. **What are the potential social and political implications of this economic slowdown?**

Environmental Rebirth or Collapse in China?

China's environmental degradation, air and water quality, and mounting contributions to climate change are now all matters of official Communist Party concern. China has actually shown a dramatic decrease in greenhouse gas emissions per unit of energy generated, notably, owing to a diversification of energy sources into renewable energy (**FIGURE 8.47**). The power of central planning and state capitalism may yet be brought to bear on environmental issues in China.

CO₂ Emissions in China

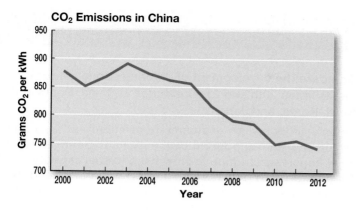

▲ **FIGURE 8.47 Declining Carbon Intensity in Chinese Energy Production** China is a massive greenhouse gas emitter, but the efficiency of energy production continues to rise, as does the use of alternative energy sources. The result is a falling level of emissions per unit of energy produced.

Emerging Korean Conflict or Reconciliation?

The ascendance of Kim Jong-un as supreme leader of North Korea in 2011 ushered in intense uncertainties. A recent failed missile test, coupled with the emergence of North Korea as a potential cyberwar force, suggest that the world can expect ongoing aggressiveness. On the other hand, efforts at diplomacy between the Koreas are ongoing and the Korean people share an interest in unification. Many observers predict a unified Korea will emerge in the 21st century, which would hasten regional disarmament and bring East Asia back from the brink of a conflict that has simmered for 50 years.

Emerging Chinese Hegemony

As we have seen, the growth of the Chinese economy has placed the country in a new political position relative to the rest of the world. China is now the major trade partner with Africa and Chinese development has come to compete with U.S. and European foreign aid in Africa (**FIGURE 8.48**) and it uses its power selectively. For example, in the wake of recent extreme violence on the part of the Syrian state against its citizens (resulting in hundreds of thousands of civilian deaths), China joined Russia in vetoing two UN Security Council resolutions condemning the violence. How, where, and when will China project power?

▼ **FIGURE 8.48 Chinese Development in Africa** Chinese workers from the Zhongyuan Petroleum Exploration Bureau (ZPEB) celebrate the completion of the major part of an oil pipeline with Sudanese workers. China gives more loans to underdeveloped countries than the World Bank. Most of the money ends up in Africa. The influence of China over African politics parallels this rise.

Learning Outcomes Revisited

▶ **Explain** how humans have adapted and modified the landscapes of East Asia over time.

People in East Asia have utilized arid regions through herding and more humid areas through farming systems, including the creation of complex irrigation works and sophisticated "rice cultures."

1. **Which** adaptations and modifications of the landscape have had unintended risks or negative consequences?

▶ **Describe** how land and water management in East Asia may be impacted by climate change.

The management of waterscapes in the region has long involved large and sophisticated irrigation systems, often overseen by expert centralized authority. This continues to be the case, especially in China, where hydroelectric investment and water diversion are key aspects of development.

2. **What** effects are likely to result from climate change that might be expected to test the capacity of the region's water management systems?

▶ **List** the main physical subregions of East Asia and describe their tectonic origins and related hazards.

East Asia is distinguished by the vast continental interior of China and the peninsular and island coasts formed at the edge of the Ring of Fire. Risks associated with the region's physical features include flooding from the enormous drainages across China and earthquakes such as the one that triggered the recent 2011 Japanese disaster.

3. **List** some of the distinctive land features of East Asia.

▶ **Identify** the adverse ecological effects of urbanization, industrialization, and modernization of agriculture.

Widespread urban growth has led to a decline in wildlife and placed stress on open space, leading to "vertical" urban development in the region. Traffic and industry have led to serious air quality problems and a persistent "Asian Brown Cloud." Green Revolution agriculture has resulted in high levels of fertilizer use, but new genetically modified crops may have caused a decrease in pesticide use and poisonings.

4. **What** are some of the major threats to wildlife in East Asia?

▶ **Distinguish** the centers of power in East Asia in the precolonial, colonial, and postcolonial periods.

The historical geography of the region is characterized by the core economic power of China, which was at the center of global trade networks for more than a 1,000 years prior to colonization. The colonial era and the upheavals of World War II drove China from its historically dominant position in the region, allowing for a reconfiguration of the region dominated by trans-Pacific trade between Japan and the United States. China has resumed its central position, however, through globalization in the 21st century.

5. **What** are some current concerns about economic conditions and change in East Asia, especially China?

▶ **Compare** the development of differing economic experiments in the 20th and 21st centuries in China, South and North Korea, and Japan.

Japan's economy is conglomerate-oriented and based on a strong and state-supported banking sector. South Korea utilized import substitution to grow its industrial capacity. North Korea continues to feature an authoritarian and centralized planning approach to economic development. Chinese development, however, dramatically switched from central planning to state-led capitalism. All of these developments have led to spatially uneven patterns of development.

6. **Describe** some of the economic characteristics of the emerging Pacific Rim region.

▶ **Identify** central geopolitical tensions within East Asia.

The geopolitics of East Asia revolve around the shifting balance of economic power between Japan and China and the unresolved problem of the Korean peninsula, where the Korean War has never officially come to an end.

7. **How** are the region's geopolitical tensions reflected in conflicts in the South China Sea?

▶ **Trace** the linkages of East Asian cultures to other cultures within and beyond the region.

In this region, many of the dominant languages (including Japanese and Korean), religious and philosophical traditions (including Confucianism, Buddhism, and Shintoism), and food traditions (including iconic local dishes like dumplings) have distinct regional roots but display shared characteristics, feature some elements from outside the region, and are often the object of political contestation.

8. **How** does the status of ethnic minorities differ between China and other countries in the region?

▶ **Describe** the forces that have caused significant declines in population growth across East Asia.

The one-child policy in China led to a major decline in the fertility rate, although with some social problems, including the imbalance in the populations of girls and boys. Cultural change and professionalization of women, in countries including Japan, Taiwan, and South Korea, have also contributed to later marriage ages and declining fertility rates.

9. **How** do the growth rates in East Asian countries compare to European countries and the United States?

▶ **Summarize** and explain the major immigration trends in the region such as rapid urbanization.

East Asian populations have long migrated within and beyond the region. The region has also seen a rapid pattern of urbanization as a result of industrialization, especially in trade centers along the Pacific Rim. Urbanization has led to a growing disparity of wealth and power between cities and rural areas.

10. **What** are the status and impacts of East Asian emigrants in other world regions?

Key Terms

2015 Conference of the Parties (p. *307*)
ancestor worship (p. *328*)
anime (p. *332*)
Anthropocene (p. *314*)
Asian Brown Cloud (p. *310*)
Asian Tigers (p. *323*)
atheism (p. *328*)
Buddhism (p. *328*)
Confucianism (p. *328*)
Convention on International Trade in Endangered Species (CITES) (p. *315*)

Cultural Revolution (p. *328*)
cyberwarfare (p. *326*)
Dalai Lama (p. *327*)
demilitarized zone (DMZ) (p. *314*)
feng shui (p. *328*)
Five-Year Plan (p. *322*)
Great Leap Forward (p. *322*)
Green Revolution (p. *312*)
Hukou (p. *336*)
hutongs (p. *333*)
hydraulic civilizations (p. *309*)
ideograph (p. *329*)

import substitution (p. *325*)
keiretsu (p. *319*)
Kim Il Sung (p. *325*)
Kim Jong Il (p. *326*)
Kim Jong-un (p. *326*)
manga (p. *332*)
Mao Zedong (p. *318*)
massif (p. *309*)
one-child policy (p. *335*)
Opium Wars (p. *320*)
Pacific Rim (p. *319*)
pinyin (p. *329*)

Ring of Fire (p. *304*)
Shinto (p. *328*)
shogun (p. *318*)
Silk Road (p. *317*)
Special Administrative Regions (p. *322*)
state capitalist (p. *322*)
Treaty of Nanking (p. *320*)
tsunami (p. *309*)

Thinking Geographically

1. How has water management been an important part of the civilizations of East Asia? To what degree has water management continued to be a factor into the present?

2. How and to what degree was East Asia economically linked to the rest of the world in the precolonial era? What is the Silk Road?

3. Explain how the central government involves itself in corporate activities in China, Japan, and South Korea.

4. What are the driving forces creating the emerging Pacific Rim region? Is this region a fully coherent or unified one? Why or why not?

5. How and why is the status of women changing in East Asia?

6. What are some dominant East Asian cultural traditions? How have East Asian beliefs and traditions affected other world regions?

7. How has the state been involved in population control in East Asia? What other forces have influenced the region's birthrates?

DATA Analysis: Millennium Development Goals in East Asia

The Millennium Development Goals (MDGs) are international development goals established by the United Nations in 2000 that include (among others) eradicating extreme poverty and hunger, achieving universal primary education and gender equality, and reducing child mortality, HIV/AIDS, malaria, and other diseases. Go to the Asian Development Bank's 2015 Basic Statistics table (go to http://www.adb.org and search for Key Indicators for Asia and the Pacific in 2015) and compare performance in some of these areas for China, South Korea, Mongolia, and Taiwan (identified as Taipei China here). Japan and North Korea are not represented in these data.

1. Which East Asian countries have the lowest percentage of people living below the poverty level? What factors described in this chapter might account for the relatively higher or lower success in this area?

2. What is the proportion of seats held by women in these countries' national parliaments? What cultural and historical factors might account for these relatively low proportions?

3. Infant mortality rates per 1,000 live births are indicators of the quality of women's health care. Which East Asian counties have the lowest rates, and what are the implications of having a lower rate of infant mortality?

4. Tuberculosis is one of the world's deadliest diseases, and its prevalence rate (per 100,000 people) in a population is one indicator of overall health and quality of health care. Which East Asian countries have higher rates of tuberculosis?

5. Examine the rates of telephone lines, cellular subscriptions, and Internet usage. Which counties are the most tech savvy? What factors, apart from overall economic development, might influence the rates of each of these three measures of "connectivity"?

Asia Development Statistics

https://goo.gl/vk0yLH

MasteringGeography™

Looking for additional review and test prep materials? Visit the Study Area in MasteringGeography™ to enhance your geographic literacy, spatial reasoning skills, and understanding of this chapter's content by accessing a variety of resources, including MapMaster interactive maps, videos, *In the News* RSS feeds, flashcards, web links, self-study quizzes, and an eText version of *World Regions in Global Context*.

Middle class families have emerged as a typical part of the street life, economy, and politics of India.

South Asia

I n a recent poll of 70,000 people in India, a remarkable half of the respondents reported that they were members of the "middle class." In a country most Americans and Europeans associate with poverty, a social and economic revolution is clearly afoot. This is because, though the economy has stalled in the last few years, the area of fastest growth remains in the service sector, including everything from technology companies, to retail chains, to administration. This trend has been accompanied by booming urbanization and a rise in participation in higher education. Many Indians, only one or two generations away from a rural farm, now work in high tech or marketing and often shop in malls, maintain two cars, and seek better schools and tutors for their children.

The opportunity for such lifestyles has even engendered a small but important "reverse brain drain," where western-educated professionals have returned to India. The numbers of these return migrants is still small, but this represents a remarkable shift in the flow of global talent.

This new class of consumers is also having an impact on politics, supporting candidates who promise to improve infrastructure, like roads and public transit, as well as reduce corruption. These priorities have arguably turned attention away from traditional areas of concern, including poverty reduction and environmental justice.

Whether this emerging population of consumers represents a "middle class" is a more complex question. The term "middle class" in a country like India can include people at the margins of poverty as well as those with standards of living as high as anywhere in the world. At the same time, not all self-identified "middle class" Indians share the same political priorities, aspirations, or sense of community. In a country with vast disparities and diverse regional and caste communities, an emerging middle class that transcends these differences still may be elusive. Roughly 180 million people still live below the poverty line (on less than $1.25 per day), moreover, making the rise of the middle class only one part of the ongoing economic drama of the subcontinent.

Learning Outcomes

▶ *Describe* the pattern of the monsoon climate, explain its origins, and identify the key adaptations people in the region use to cope with change and uncertainty.

▶ *Describe* the physiographic regions, environmental hazards, and mineral resources of South Asia and explain how some land uses contribute to environmental destruction, while others promote sustainability.

▶ *Provide* examples of rulers, governments, and economic systems that historically put South Asia at the center of global trade, and explain how this position was later exploited and overridden by colonial governance.

▶ *Identify* the major engines of recent rapid South Asian economic expansion and discuss the uneven benefits of expansion.

▶ *Describe* the key relationships and borders that present the greatest current geopolitical challenges for South Asia.

▶ *Summarize* the characteristics of the region's major religions and languages and explain how global connections have interacted with the region's traditional cultures and emerging consumer cultures.

▶ *Explain* where and under what conditions South Asian populations have begun to stabilize or decline and where they continue to grow relatively rapidly.

Environment, Society, and Sustainability

Two aspects of South Asia's physical geography help define it as a world region (**FIGURE 9.1**). Each of these features also highlights the region's connections to other regions. The first striking feature is the **monsoon**, seasonal torrents of rain on which the livelihoods of South Asia depend. The monsoon system is propelled by the annual heating and cooling of the landmass of Central Asia to the north and the constant feeding of moisture from the Indian Ocean to the south. This signature physical condition is a result of South Asia's position between other world regions. Forbidding mountain ranges also set South Asia apart from the rest of Asia, creating a porous boundary around the people of the region. Like the monsoon, this distinctive feature is born

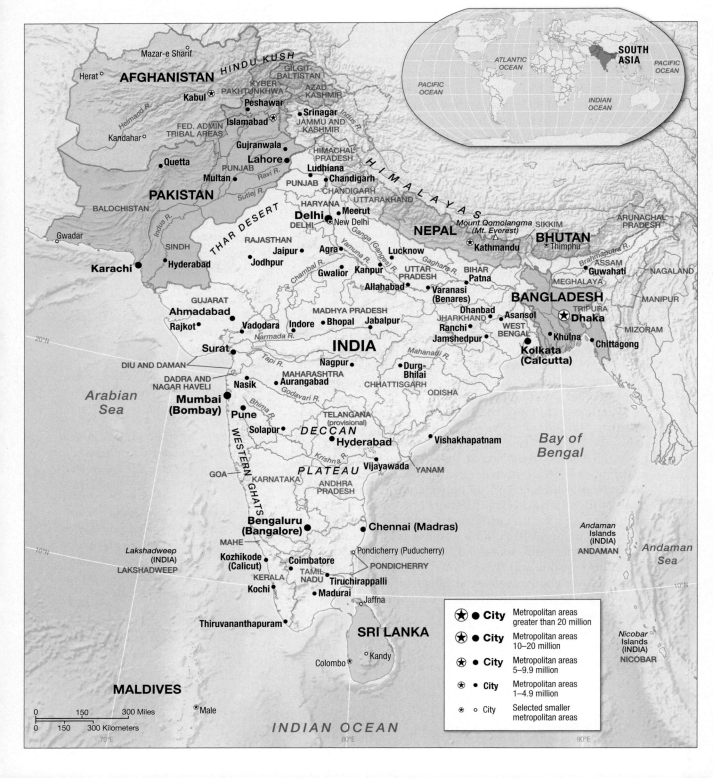

◀ **FIGURE 9.1**

South Asia: Countries, major cities, and physical features India dominates the subcontinent, occupying the central majority of the peninsula, with Pakistan, Afghanistan and Bangladesh in the northwest and northeast. Nepal and Bhutan sit abreast the high Himalayas and the island nations of Sri Lanka and Maldives lay in the Indian Ocean.

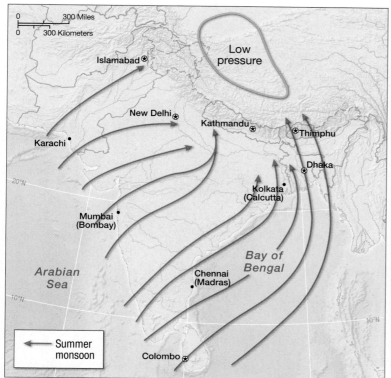

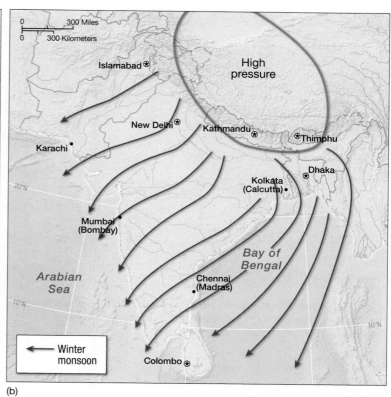

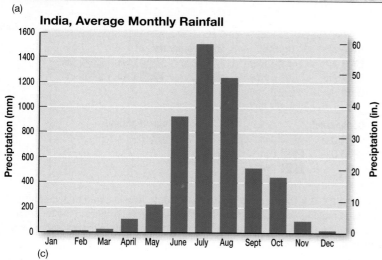

India, Average Monthly Rainfall

(c)

▲ **FIGURE 9.2 Summer and Winter Monsoons** (a) The wet summer monsoon arrives in June and lasts into September or later. (b) In winter, the prevailing dry monsoon winds are northeasterly. (c) This results in dramatic precipitation differences across the course of the year.

from the interaction of regions, as the South Asian Plate collides with the Asian Plate, creating massive uplift and ongoing mountain growth.

Climate and Climate Change

The climate of South Asia is distinguished by the dramatic effect of the southwesterly summer monsoon, a deluge of rain that sweeps from south to north across the region. Although now widely used to describe any windy, rainy season, the term

monsoon was originally applied only to the distinctive seasonal winds in the Indian Ocean.

The Monsoon In most of South Asia, the seasonal pattern of climate consists of a cool and mainly dry winter, a hot and mainly dry season from March into June, and a wet monsoon that bursts in June and lasts into September or later. The engine behind the monsoon is the heating and cooling of the Asian continent to the north of the region. During the early summer, the interior parts of Asia begin to heat more rapidly than the areas to the south, creating low-pressure convection as hot air rises all across Asia. By midsummer, this pressure gradient becomes strong enough to draw moisture-laden air from the Indian Ocean and propel it northward across the subcontinent. This wet monsoon moves inland bringing drenching rains. The rains slowly progress from the island of Sri Lanka, across the southern coastal zones of India, across Bangladesh and the plains of India, and finally into the northwestern parts of India and Pakistan (**FIGURE 9.2**). The northern hills and uplands exert a strong **orographic effect**, causing moist air from the sea to lift, condense, and produce heavy rainfall. The arrival of the wet monsoon season is announced by violent storms and torrential rain (**FIGURE 9.3**). In winter, this system reverses, as relative low pressure over the Indian Ocean draws air outward from the interior of Asia. The result is northeasterly winds, which blow from the interior toward the sea. These are the dry monsoon winds that typically do not bring rain.

The southern part of Sri Lanka, the Maldives, and the Nicobar Islands are far enough south to remain within the Intertropical Convergence Zone year-round, so these locations rarely experience a dry month all year. The result is the overall pattern

▲ **FIGURE 9.3 Monsoon in South Asia** People and cycle rickshaws moving through the flooded streets of the densely populated neighborhood of Paharganj in Delhi after a heavy monsoon rainfall.

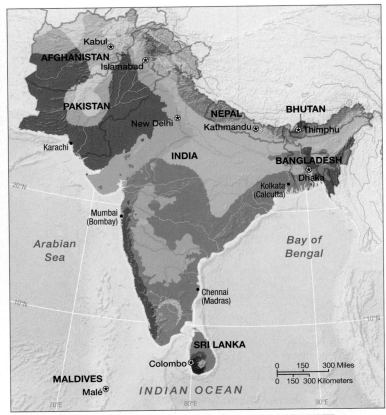

▲ **FIGURE 9.4 Climates of South Asia** South Asia is home to a variety of climate regions, with desert in the northwest and wet tropics in the far south. **Which counties in South Asia are dominated by arid climates and which by tropical ones?**

of climate across the region, with tropical climates hugging the coasts and the islands, a broad belt of humid subtropical climate following the Ganga (Ganges) and abutting the Mountain Rim, and arid climates in the desert northwest of India and Pakistan and in the interior of the Indian Plateau (**FIGURE 9.4**).

Rainfall, Drought, and Agricultural Adaptation For much of South Asia, there is a significant risk of drought if there is a late or unusually dry monsoon season. In early 2015, nearly 9 million farmers in the Indian state of Maharashtra lost their winter crop, exacerbating rural poverty in a state where rural poverty, debt, and farmer suicides are a serious problem. Even where the monsoon does not fail completely, it can be highly uneven, so that villages in one region may receive sufficient rainfall to plant and harvest, while nearby settlements do not.

The fierceness of the monsoon rains, as well as their unpredictability, spatial unevenness, and occasional failure, has led to a range of ingenious adaptations. In the mountain areas of the Himalayas and the Karakoram, **terracing** of slopes over generations has created a distinctive landscape and allowed sustained agricultural yields amid torrential rainfall and steep slopes.

"For much of South Asia, there is a significant risk of drought if there is a late or unusually dry monsoon season."

In more arid parts of South Asia, agricultural systems maximize **intercropping**, the mixing of different crop species that have varying degrees of productivity and drought tolerance. In bad rainfall years, farmers salvage a harvest by relying on the heartiest part of the crop. This high agrodiversity helps absorb climate shocks. Traditional grains like millet can grow to harvest after soaking up only two or three light rainstorms.

Livestock production has also adapted to the region's climates. In Afghanistan, the Himalayan regions of Nepal and India, and the arid desert districts of India and Pakistan, local communities practice **transhumance**, an animal herding migration scheme that allows members of families to be away many months of the year tending animals (**FIGURE 9.5**). It also accommodates modern technology. Herders now transport their animals by truck and use radios and computers to access up-to-date weather information.

▲ **FIGURE 9.5 Itinerant Herding in India** The goatherds shown here are migrating with their flocks in arid Rajasthan, India. Livestock keepers adapt to the vagaries of climate by strategically moving animals to find forage.

Since the 1960s, efforts have been made to boost agricultural production through the use of modern technologies as part of the **Green Revolution**. These innovations include high-yielding varieties of wheat and corn, which can double or triple production per hectare of land. These crops require more inputs than their traditional counterparts, however, including more water, fuel, fertilizer, and pesticides. The result has been an increase in productivity, but also increased dependency on purchased farming products, decreased native seed diversity, and exhausted soils and water supplies.

Implications of Climate Change
It is too soon to tell precisely what global climate change might portend for South Asia. The region is undeniably vulnerable to possible effects of climate change, ranging from more intense storm events to increased incidence of disease.

- **Storms, Floods, and Sea-Level Rise**
  Recent years have actually seen an increase in the intensity of rainfall, which has led to flooding. There have been record-breaking and recurring floods in Bangladesh, Nepal, and northeast India. Typhoons (hurricanes), which are endemic to the Bay of Bengal and a serious environmental hazard in Bangladesh, have grown less frequent in recent history but more intense. These storms are now harder to predict and have increasingly devastating impacts. Coastal areas are vulnerable to the sea-level rise that will accompany melting of polar ice as well. Should sea-level rise by 1.5 meters (5 feet) in the next century, as many as 22,000 square kilometers (8,494 square miles) could be lost to the sea in Bangladesh and seasonal storms in the Bay of Bengal will become all the more serious (**FIGURE 9.6**).

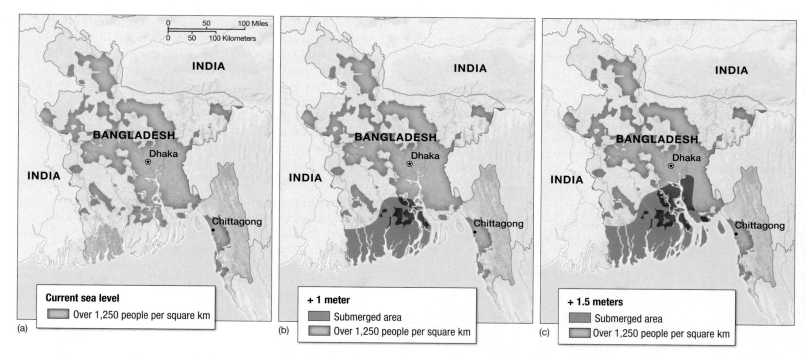

▲ **FIGURE 9.6 Forecasted Hazards of Sea-Level Rise for Bangladesh** With only 1.5 meters (5 feet) of sea-level rise, much of Bangladesh could be impacted. Map (a) shows the present-day sea level. The blue areas on maps (b) and (c) show the areas that will be partially submerged if sea level rises. **What are the possible implications for vulnerable people in the region? Where might coastal migrants go?**

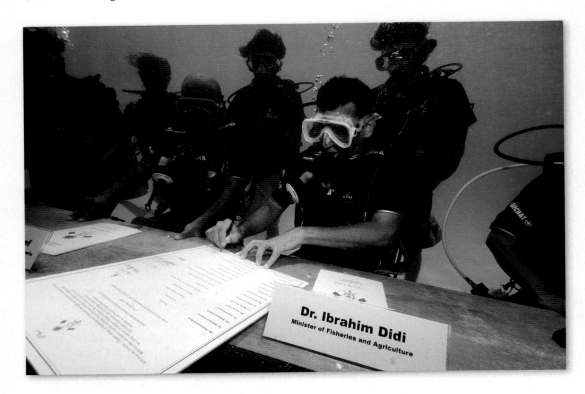

◀ **FIGURE 9.7 Threat of Sea-Level Rise to Low-Lying Islands** The leaders of the Maldives held an underwater cabinet meeting in 2009 to draw attention to the very real threat of sea-level rise for island nations resulting from climate change.

These risks are even more serious for island nations. Located on islands formed from coral reefs, the Republic of Maldives, in the Indian Ocean, is the flattest country on Earth; 80% of its land area lies below 3.3 feet (1 meter). Most sea-level rise projections therefore suggest inundation for this island nation (**FIGURE 9.7**).

- **Heat and Drought** Conversely, climate change may also increase drought events in some areas; crop yields in many regions have already suffered in recent years

▼ **FIGURE 9.8 Drought in South Asia** A village boy leads his goat through a parched pond on the outskirts of the eastern Indian city of Bhubaneswar, in the month of May. Huge swathes of rural farmland turn dry during this period before the annual monsoon rains. If these rains do not arrive, difficult conditions will continue indefinitely.

from unexpected high temperatures and heat waves. More frequent and recurring failures of the monsoon are possible, especially for arid northern India and Pakistan, where the monsoon often fails to reach (**FIGURE 9.8**).

- **Glacial Melting and Water Supplies** One of the most uncertain impacts of global climate change on the region has to do with the status and condition of Himalayan glaciers. As the Intergovernmental Panel on Climate Change (IPCC) reports, these glaciers, which cover 17% of the mountain areas of the Himalayas, are receding faster than in any other area of the planet. Should these glaciers experience serious decline, the major rivers of the region (Ganga, Indus, and Brahmaputra) will lose all but their seasonal (monsoonal) water sources. This threatens agriculture as well as the drinking water supplies for more than a half-billion people, especially during the lengthy dry season.

- **Disease Ecology and Health** Another poorly under-stood but serious problem posed by climate change is disease ecology. Of the many diseases that present a hazard to people in the region, those spread by mos-quitoes are especially problematic. These include malaria, which India's National Institute of Medical Statistics estimates to kill more than 40,000 people per year in India alone. With overall global warming, it is reasonable to anticipate that some cities will experience increasingly lengthy mosquito breeding seasons and a significant, if not exponential, escalation in rates of dis-ease. Devastating viral fevers, such as Chikungunya and dengue, may also increase.

Apply Your Knowledge

9.1 How has adaptation to the monsoon rains prepared the people of South Asia for the vagaries of climate change?

9.2 What new hazards may be too unprecedented to be readily addressed based on past experience and traditions?

Geological Resources, Risk, and Water

In geological terms, South Asia is a recent addition to the continental landmass of Asia. The collision of the South Asian Plate with the southern edge of Asia sets the stage for the current landforms and hazards of the region (**FIGURE 9.9**).

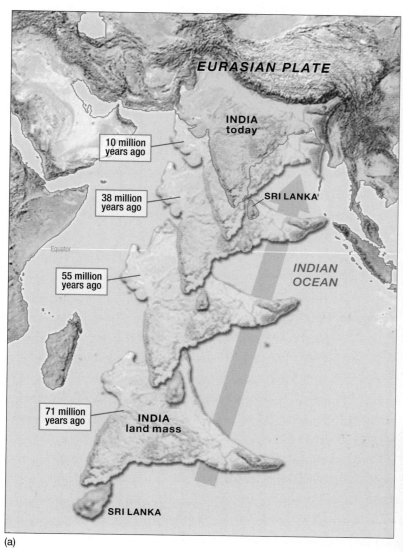

(a)

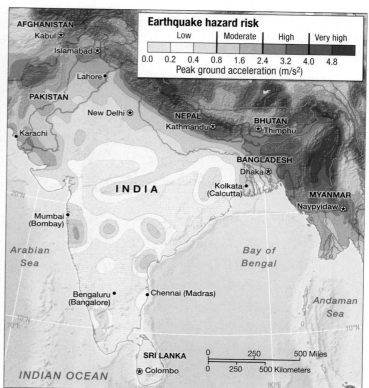

(b)

(c)

▲ **FIGURE 9.9 Plate Tectonics in South Asia** (a) The movement of the Indian Plate into the Eurasian Plate 10 million years ago created a zone of turbulent contact and uplift, forming the Himalayas and setting the stage for critical earthquake hazards (b). The 2015 earthquake in Nepal, which killed more than 8,000 people, also destroyed many ancient buildings in the city of Kathmandu (c) and devastated the regional economy, which depends heavily on tourism.

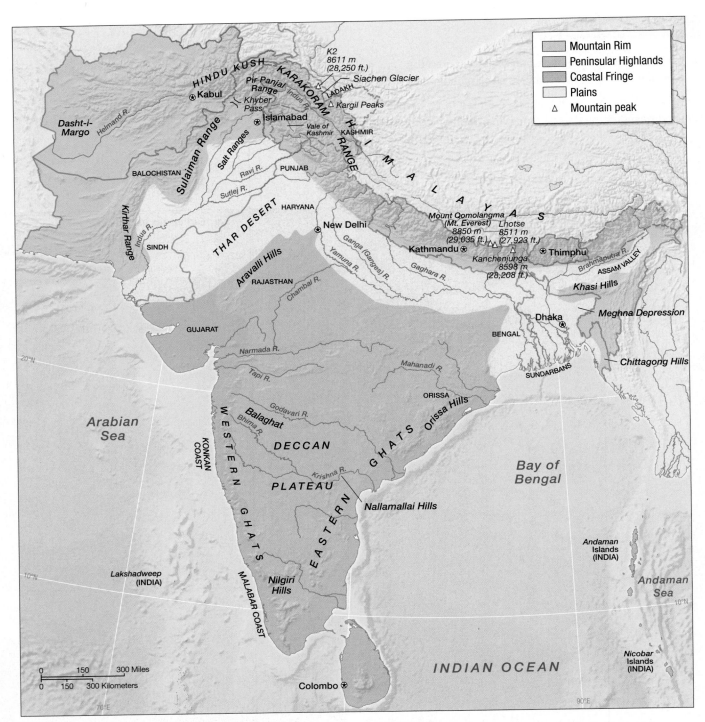

▲ **FIGURE 9.10 Physiographic regions of South Asia** The physical geography of South Asia is framed by the highland plateaus of an ancient continental plate and a young mountain rim; these are separated by broad plains.

The principal physiographic features of South Asia reflect this major geological event. They include the Peninsular Highlands, the Mountain Rim, the Plains, and the Coastal Fringe (**FIGURE 9.10**).

- **Peninsular Highlands** The Peninsular Highlands of India form a broad plateau flanked by two chains of hills. The highlands rest on an ancient layer of volcanic rocks; between 65 and 55 million years ago, immense

eruptions of lava buried parts of the peninsula beneath dense volcanic rock, creating the distinctive Deccan Traps formation (**FIGURE 9.11**).

- **Mountain Rim** A vast region of spectacular mountain terrain, the Mountain Rim is known for its remote valleys, varied flora and fauna, ancient Buddhist monasteries, and fiercely independent tribal societies.

The physical geography consists of several mountain ranges, including the young Himalayan and Karakoram ranges and the world's highest peaks, including K2 and Mount Everest (known as Sagarmāthā in Nepal and Chomolungma in Tibet) (**FIGURE 9.12**).

▼ **FIGURE 9.11 Deccan Traps** The distinctive geology of the Deccan plateau is the product of massive volcanic eruptions that covered much of northwestern India with thick layers of lava.

- **River Systems, Plains, and Bengal Delta** These mountains also serve as the source of the three great river systems of the region—the Indus, the Ganga (Ganges), and the Brahmaputra. The Plains region has long been widely irrigated, has supported a high population density, and has been the site of many of the region's great cities (**FIGURE 9.13**). The Bengal Delta, which covers a large portion of Bangladesh and extends into India, is the product of the Ganga and Brahmaputra Rivers.

- **Coastal Fringe** The narrow Coastal Fringe is the product of marine erosion that has cut into the edge of the mountainous Peninsular Highlands. During the rainy monsoon seasons, the Coastal Fringe is filled with luxuriant vegetation, especially along the southwest Malabar Coast of India and the southwest coast of Sri Lanka (**FIGURE 9.14**).

The Hazard of Flooding Where the summer monsoon hits hills and mountains, the rains are especially heavy. The peninsula typically receives between 2,000 and 4,000 millimeters (79–158 inches) of rainfall where the southwest monsoon winds meet the steep slope at

▼ **FIGURE 9.12 Mt Everest (Sagarmāthā in Nepali and Chomolungma in Tibetan)** The Mountain Rim of South Asia consists of the highest peaks and valleys on Earth. **How do the geologic histories of the Deccan Plateau and the Himalayas differ? How is that evident in their form and shape?**

▼ **FIGURE 9.13 Ganga (Ganges) River Flowing Through Varanasi** A set of steps (or *ghats*) descend into the Ganga (Ganges) in the sacred city of Varanasi. Some of the greatest and most ancient cities of the region were built on the banks of this important river.

▲ **FIGURE 9.14 The Coast of Sri Lanka** Traditional stilt fishermen practice their craft at Kogalla, Sri Lanka. The Coastal Fringe of South Asia is home to distinctive fishing cultures and rich trading civilizations.

▲ **FIGURE 9.15 Stonecutters' Labor** The marble and sandstone quarried in India and Pakistan have been a crucial component in the construction of the great works of classical architecture, including the Taj Mahal, but their mining takes a terrible toll on workers who suffer from silicosis and injury.

the edge of the Peninsular Highlands. Similar levels of annual rainfall fall in the central and eastern parts of the Mountain Rim. The outflow from river flooding from these highlands, coupled with heavy local rainfall, can have a doubly hazardous effect.

In 2011, extremely heavy monsoon precipitation had devastating effects. Dams overflowed in the state of Gujarat, and more than 100 people died or went missing in India and Nepal. Over 5 million people were affected and almost 1 million were left homeless. In 2014, flooding in Pakistani-controlled Kashmir and the downstream state of Punjab forced more than 700,000 villagers to flee their homes.

Earlier, in 2010, floods in Pakistan affected 20 million people and killed at least 2,000. That event was the product of an unusually rainy year. The engineering of the floodplain with extensive levees or embankments, however, made the floods worse. These levees allow intensive irrigation and some protection from typical rainfall but also accelerate and concentrate floodwaters during more extreme events. The geography of the Pakistan floods shows how climate events and human decision making can combine to lead to tragedy on a vast scale.

Energy and Mineral Resources
The geological resources of the South Asia region are modest by global standards, but there are several important deposits. The Peninsular Highlands region contains valuable deposits of iron ore, manganese, gold, copper, asbestos, and mica. Coal and bauxite are found in workable quantities in eastern parts of the parts of the South Asian peninsula. New oil deposits have recently been discovered in the Indian state of Rajasthan. The geology of the region also yields large quantities of sandstone and marble, construction materials that have long been a fabric of the region's architecture, including monumental works like the Taj Mahal in Agra and numerous major structures in Delhi (**FIGURE 9.15**).

Sustainability: Energy Innovation in South Asia
Despite setbacks in environmental quality, efficiency, and clean energy, recent efforts in sustainability show progress throughout South Asia. In 2015, the government of India set in place a plan to develop 60 new "solar cities," in which at least 10% of municipal power would be generated by renewables. The first city slated for retrofitting is Agra, home of the Taj Mahal, whose façade has suffered from the air quality problems associated with coal-fired power plants. Though Bhutan imports some energy from fossil-fuel sources, the country is actually carbon-neutral, owing to its enormous hydropower potential; 95% of commercial activities are powered by the country's rivers. Experiments in Pakistan to implement solar power have focused heavily on remote villages in places like Baluchistan, where many villagers still live entirely off the grid. In Bangladesh, biogas digesters are being brought on line, which consume waste and dung and generate energy. These generated five gigawatts of power in 2013 (**FIGURE 9.16**).

Arsenic Contamination
A dramatic case of geology's impact on humans in the region was the discovery in 2000 that millions of **tube wells** in Bangladesh were drawing arsenic-contaminated water. Tube wells are lined with a durable and stable material, usually cement. This makes it possible to sink them deeper than traditional water wells. These wells were installed in Bangladesh as a result of a campaign in the 1970s by the United Nations Children's Fund (UNICEF) intended to provide drinking water free of the bacterial contamination that was killing more than 250,000 children each year.

Unfortunately, the well water was never tested for arsenic contamination, which occurs naturally in the groundwater. By the 1990s, high rates of certain types of cancer throughout much of Bangladesh led researchers to investigate and identify the cause as arsenic-contaminated water. According to a 2010 study by the British medical journal *Lancet,* up to 77 million people in Bangladesh are exposed to toxic levels of arsenic, taking years

▲ **FIGURE 9.16** An Indian worker checks the flame from a biomass gasifier power plant in Gosaba, a town south of the city of Kolkata. 1,200 families are served with power from this power plant.

off their lives. Tube wells that test positive for unsafe arsenic concentrations are now painted red as a warning to residents (**FIGURE 9.17**).

Apply Your Knowledge

9.3 How did efforts to improve irrigation in Pakistan and provide access to drinking water in Bangladesh have unintended hazardous impacts?

9.4 Describe how these cases, along with others from South Asia, show that hazards are a combination of both natural and human causes.

Ecology, Land, and Environmental Management

The ecologies and landscapes of South Asia are diverse and support a huge range of life. The ecosystems of the region include deep deserts, thick forests, isolated mountains, and a critical coastal zone. These natural ecosystems have undergone heavy

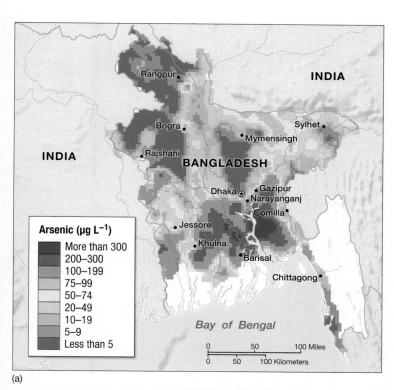

Arsenic (µg L⁻¹)

	More than 300
	200–300
	100–199
	75–99
	50–74
	20–49
	10–19
	5–9
	Less than 5

(a)

(b)

▲ **FIGURE 9.17 Water Not for Drinking** (a) Arsenic contamination impacts much of Bangladesh, though its distribution is uneven; (b) a Bangladeshi boy collects drinking water from a green-painted arsenic-free tube well at a village near. Wells known to pump arsenic-contaminated water in Bangladesh are marked red as "undrinkable," while usable wells are painted green. An estimated 30 million Bangladeshis—out of population of 138 million—are at risk of arsenic related diseases.

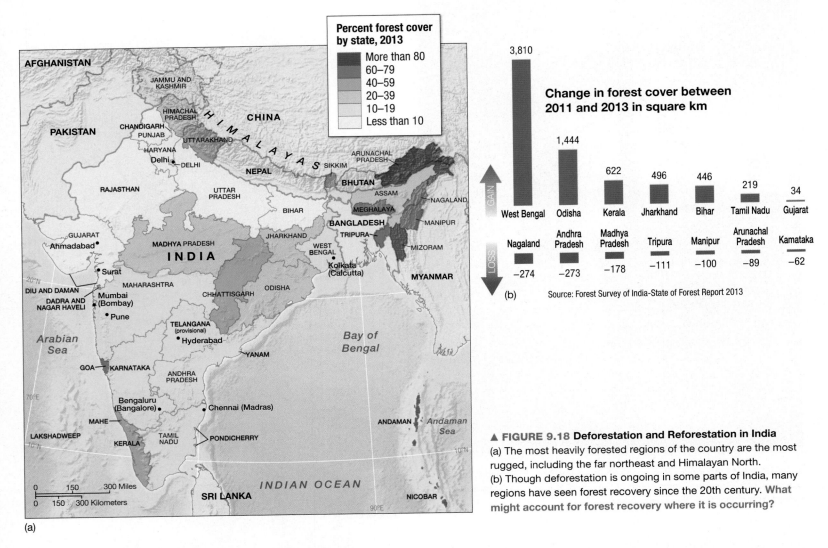

▲ FIGURE 9.18 Deforestation and Reforestation in India
(a) The most heavily forested regions of the country are the most rugged, including the far northeast and Himalayan North.
(b) Though deforestation is ongoing in some parts of India, many regions have seen forest recovery since the 20th century. **What might account for forest recovery where it is occurring?**

human influence, which has both harmed the ecosystem, but also created new habitats and environments for other species.

Land Use Change Humans have inhabited this region for millennia and intensive development of the landscape has left its mark. For example, residents have cleared large areas of natural forest to cultivate land from earliest times. For the most part, however, subsistence farmers, herders, fisherfolk, and artisans have historically drawn on local resources only for their food, traditional medicines, housing materials, and fuel. Their activities, although resulting in some **deforestation**, were managed in such a way that many critical habitats were preserved.

The arrival of European traders, the era of British rule, and the subsequent period of independence (described next) coincided with accelerated and persistent environmental transformation. In 1750, when the British were beginning their imperial conquests, more than 60% of South Asia was still forested. British imperial rule brought systematic clearing of land for plantations and the methodical exploitation of valuable tropical hardwoods. By 1900, only 40% of South Asia remained forested. The most rapid period of change has been the past 50 years, as the independent countries of South Asia have sought to modernize and expand

their domestic economies. Only 21% of India remains forested today, and less than half of that is intact, natural forest—the rest consists of forest plantations.

Though forest plantations often consist of nonindigenous species, these areas, coupled with natural regrowth in others, have led to reforestation in many regions (**FIGURE 9.18**). According to recent satellite surveys of the region, some large areas are still covered in mixed forest, with tree cover thriving in agricultural areas, wastelands, and even in and around cities. The proportion of land area that has at least a density of 10% forest cover on it is 44% in India, 71% in Nepal, 73% in Bhutan, and 91% in Sri Lanka. This compares favorably to the roughly 50% of the area worldwide that has at least some tree cover. In these areas, important native animal and bird species also thrive. South Asia is ultimately a region with high levels of **anthropogenic** forest cover (tree cover created or retained by human beings), which can serve as wildlife habitat.

Agriculture and Resource Stress Environmental resources are increasingly under stress, which sets in motion a destructive cycle. Rural people, even those disconnected from the booming global economy, are forced to use their limited resources in increasingly unsustainable ways, depleting

▲ **FIGURE 9.19 Biomass Fuel** A majority of households in South Asia continue to burn raw biomass—wood and dung—for cooking.

▲ **FIGURE 9.20 Water Tanker Truck in India** Urban dwellers are shown collecting water from a tanker in New Delhi. An unprecedented water crisis looms large over this national capital. Though a municipal tanker is shown here, shortfalls are often met by water salesmen, who draw off the public water supplies to sell to water-starved local neighborhoods.

sources of fuelwood, exhausting soils, and draining water resources. Though development and agricultural modernization are ongoing, notably, in rural areas a large number of people live directly off the land, at subsistence levels. According to a 2014 United Nations study, over two-thirds of India's 1.3 billion people continue to rely on gathered fuelwood and collected dung as a fuel for cooking. Not only does this degrade local forests and contribute to greenhouse gas emissions, the burning of these fuels creates a health hazard for women cooking in enclosed spaces (**FIGURE 9.19**).

Water is also under stress. In parts of Punjab and Haryana—the "breadbasket" of India, where almost a third of the country's wheat is grown—73% of crops are dependent on groundwater. Unfortunately, the water table has fallen by an average of 55 centimeters (more than 2.5 feet) per year in the last decade. Because 86% of water resources are used for agriculture in India, cities must compete with farming for very scarce water resources. Poor water distribution systems contribute to local urban scarcity, even where water is more plentiful. Chennai and New Delhi are just two examples of the many large cities in India that are heavily dependent on supplementary water supplies hauled in by tanker (**FIGURE 9.20**).

Environmental Pollution Water is often polluted in addition to being scarce. In India, about 200 million people do not have access to safe and clean water, about 690 million lack adequate sanitation, and an estimated 80% of the country's water sources are polluted with untreated industrial and domestic wastes. Only 10% of all sewage in South Asia is treated. In India, 1,769 tons of organic water pollutants are discharged *daily* into local waters. As of 2011, Indian cities alone were estimated to generate about 68.8 million tons of solid waste each year, most of which is disposed of in unsafe ways: burned, dumped into lakes or seas, or deposited into leaky landfills.

In 1984, a toxic gas release at the Union Carbide chemical plant in Bhopal, India, killed thousands of local residents and left many thousands more permanently disabled. Workers there had been encouraged to ignore safety regulations and, in the aftermath, foreign investors were never held fully accountable for horrific loss of life. That event reverberates into the present, with ongoing calls throughout India for greater accountability, more stringent safety regulations, and more careful oversight of foreign investment.

Air pollution has also become a serious issue. According to the Tata Research Institute in New Delhi, air pollution in India causes an estimated 2.5 million premature deaths each year. In addition to air pollution from coal-fired power plants across the country, motor vehicle emissions are a major contributor to urban air pollution. As of 2010, 40 million passenger vehicles were driving on Indian roads, an all-time high. The increasing congestion of city streets and highways has resulted in a corresponding rise in respiratory diseases. Studies show these effects are unequally distributed. A study comparing urban men showed that the poorest men, who work disproportionately outdoors, have far worse respiratory problems than wealthier ones, who tend to work in offices.

Apply Your Knowledge

9.5 What are key drivers of environmental change in South Asia? To what degree are environmental stresses and challenges the result of unregulated economic change, concentration of wealth, population growth, or technology? Provide evidence for your answer.

9.6 What signs are there of environmental recovery in South Asia? Name three examples of environmental innovation and sustainability efforts.

Conserving Wildlife

Extinction of species is another ongoing crisis in South Asia. According to the World Conservation Union's "Red List," as of 2012, 132 species of plants and animals in India are listed as critically endangered. These include many magnificent and charismatic large animals, such as elephants and tigers. Cheetahs disappeared from the wild in India more than 50 years ago. The Bengal tiger (FIGURE 9.1.1) and the Asian elephant are emblematic of critically endangered species in South Asia. Pakistan is also home to a number of iconic species, including the snow leopard of the Himalayas, which are declining for similar reasons.

"People are learning to live with wildlife in South Asia; there are signs of endangered species recovery."

In Bhutan, the golden leaf monkey and the red panda are endangered.

Equally dramatic, though regrettably greeted with less global concern, is the decline of a large number of vulture species. These enormous birds of prey are crucial to the functioning of natural ecosystems, as they scavenge and recycle carcasses and waste. Nesting in high rocks and thriving on the enormous garbage dumps of India and Pakistan, vultures were found in huge numbers only a few years ago but are now in fast decline. The Indian vulture (*Gyps indicus*) experienced a cataclysmic 97% population decrease between 2000 and 2007 (FIGURE 9.1.2). There are many reasons for their decline, but the major culprit is the veterinary drug *diclofenac*, widely administered to livestock to reduce pain. When ingested by vultures, which scavenge the carcasses of dead cows and bulls, the drug causes kidney failure and death.

People are learning to live with wildlife in South Asia in the Anthropocene, however; there are signs of endangered species recovery. A 2014 census of India's tigers showed the population was up by 33% since 2011, owing to extensive conservation measures and efforts to decrease conflicts with rural people. For vultures, a recent ban on diclofenac may have contributed to some recovery as well; a 2014 study showed that diclofenac-related vulture deaths had declined by 65% since the ban was put in place. This may be too little, too late, however.

Visit the World Wildlife Fund (WWF) web page, and search for the organization's work in the Himalayas: http://www.worldwildlife.org.

1. **What conservation efforts and activities does the organization emphasize?**

2. **To what degree do their programs address the full range of threats to wildlife in South Asia?**

South Asia Wildlife

https://goo.gl/vOvkQZ

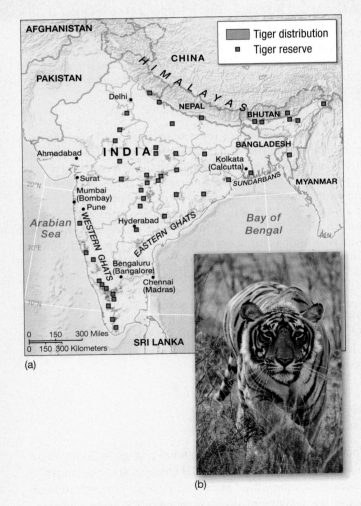

(a)

(b)

▲ FIGURE 9.1.1 **Bengal Tiger** (a) Areas of highest tiger distribution in India occur in the hills of the southwestern part of the country as well as across the central belt up to the coastal mangrove "Sundarbans." Parks have been founded in all these regions to protect these animals. (b) The Bengal tiger is one of the most seriously endangered species.

▲ FIGURE 9.1.2 **Indian Vulture** While less charismatic than other endangered species, the vulture is among the most ecologically important; the decline of numerous vulture species in Pakistan and India in the last decade is viewed as a conservation crisis.

History, Economy, and Territory

South Asia has long been at the center of global cultural exchange and commerce. Sophisticated cultures and powerful political empires have emerged and spread across the region, British colonialism led to the region's profound underdevelopment, and the subcontinent's emerging economics and geopolitics have made it a center of world attention.

Historical Legacies and Landscapes

South Asia appears to sit, in terms of physical geography, in an isolated position, yet nothing could be further from the truth. The seas to the west and east of the jutting peninsula connect the subcontinent to lively cultures and great economies of Africa and Southeast Asia while traders and invaders have constantly traversed the region's rugged mountains.

Harappan and Aryan Legacies   The first extensive imprint of human civilization in the region dates from at least 5,300 years ago, when the people of the **Harappan Civilization** began to irrigate and cultivate large areas of the Indus Valley. Between 3,000 and 2,000 B.C.E. this Indus Valley culture utilized sophisticated agricultural techniques to produce enough surplus, primarily in cotton and grains, to sustain an urban civilization. The Harappans built enormous, organized cities at the sites of Mohenjo Daro and Harappa (located in present-day Pakistan). Distinctively manufactured Harappan trade goods, including jewelry and fine crafts, are found in archaeological sites as far as the Mediterranean (**FIGURE 9.21**).

The disappearance of the Harappans, likely following a shift in both climate and the course of the Indus River, was followed immediately by the rise of the **Aryan** civilization (1500–500 B.C.E.). This group of livestock-keeping peoples from the northwest populated the plains of the Ganga (Ganges) and developed a series of kingdoms. Their advent was accompanied by farming and new trade links, and they slowly transformed the plains of the northern part of the subcontinent from a green wilderness to a heavily agricultural zone.

Early Empires   The **Mauryan Empire** (320–125 B.C.E.) was the first to establish rule across greater South Asia. By 250 B.C.E., the emperor Ashoka had established control over all but present-day Sri Lanka and the southern tip of India. Securing control wreaked such havoc and destruction, that Ashoka renounced armed conquest and adopted Buddhist teachings, which also included Buddhist principles of vegetarianism, kindness to animals, nonacquisitiveness, humility, and nonviolence.

After Ashoka's death in 232 B.C.E., the Mauryan Empire fell into decline, and northern India soon succumbed to invaders from Central Asia. After more than four centuries of division and political confusion, the **Gupta Empire** (320–480 C.E.) united northern India. The Gupta period is generally regarded as a great classical period. It produced the decimal system of notation, the golden age of Sanskrit and Hindu art, and contributions to science, medicine, and trade.

▲ **FIGURE 9.21 The Indus Valley Civilization**   This King Priest figure, carved from steatite (soapstone), demonstrates the enormous artistic talents of the Harappan Civilization, also known as the Indus Valley Civilization. Urban, organized, and with trade linkages across West Asia, the Harappans were the first to put South Asia at the center of global trade, more than 5,000 years ago.

Mughal India   Toward the end of the 15th century, a clan of Turks from Persia (now Iran), known as the **Mughals**, moved east into the region. Led by Babur, they conquered Kabul, in what is now Afghanistan, in 1504. By 1605, Babur's grandson, Akbar, had established control over most of the Plains, and in the next century Mughal rule extended to all but Sri Lanka and the southern tip of India from 1526–1707 (**FIGURE 9.22**).

Akbar synthesized the best of the many traditions that fell within his domain. Traditional kingdoms and princely states were kept intact, but they were integrated in a highly organized administrative structure with an equitable taxation system and a new class of bureaucrats. Over time, Islam proved attractive to many, especially in the northwest (the Punjab) and the northeast (Bengal). By 1700, mosques, daily calls to prayer, Muslim festivals, and Islamic law had become an integral part of the social fabric of South Asia. Spectacular architecture became a signature of Mughal rule. The Mughals built lavish mosques, palaces, forts, citadels, towers, and gardens, including the iconic Taj Mahal. Most important, under Mughal rule, South Asia became an economic powerhouse, and textiles and other goods were traded widely across the Indian Ocean and around Southeast Asia.

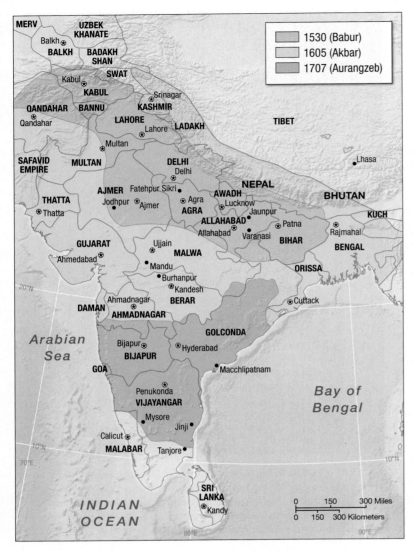

	1530 (Babur)
	1605 (Akbar)
	1707 (Aurangzeb)

▲ **FIGURE 9.22 Mughal India** Two centuries of Mughal rule began in 1504, with the conquest of Kabul, in present-day Afghanistan. By 1707, Mughal rule extended almost to all the subcontinent. **To what degree do the boundaries of the Mughal Empire differ from those of the modern countries of South Asia?**

The British East India Company Internal strife during the rule of the last Mughal, Aurangzeb, coupled with conflict with neighboring Hindu kingdoms, left South Asia open to the increasing interest and influence of European traders and colonists. By the 1690s, Dutch, French, Portuguese, and British trading companies had established a permanent presence in several ports, though they had little initial interest in establishing settler colonies or exerting political authority at first. The British East India Company (established in 1600) was the most successful of these, and in 1773 the British government transformed the company into an administrative agency.

Soon afterward, the British pushed ahead with aggressive imperialist policies in South Asia. The Company began direct acquisition of Indian territory in pieces, using a strategy of annexing small kingdoms, deposing their rulers, and winning pitched battles over other colonial powers (notably including

the French) and their Indian allies. Some princely states were adopted for control through complex treaties. Still others fell under "indirect rule" where local rulers remained autonomous on paper but answered to Company agents. In this way, a single company, the East India Company, managed to control and rule a vast country for a century from 1757 to 1858.

Toward the end of the 1700s and on into the 1800s, the focus of the Company shifted beyond trade to economic imperialism. The central goal of British policy was to harness the agricultural wealth of the region and to export raw materials, like tea, opium, and cotton, while dismantling native industry, creating a condition of **dependency** of the region on the import of finished goods like textiles (made with Indian cotton!) from industrializing England (**FIGURE 9.23**). During this period, the East India Company itself, which kept its own standing army, ruled the region—not the British Crown.

Rebellion and the Raj The inevitable anticolonial reaction came to a head in 1857, when an Indian Army unit rebelled. The incident quickly spread into a yearlong civil uprising, known as the Indian Mutiny (known in India as the First War of Independence). The British Crown put down the rebellion with enormous and brutal force (mutineers were strapped to discharging cannons) and, in 1858, established "the **Raj**," assuming direct control over India.

Under the Raj, western institutions of property and law were extended to secure and maintain rights and control, while modern industrial infrastructure—railroads, roads, bridges, and irrigation systems—led to increased efficiencies in production and transportation. In an effort to create and maintain a trained cadre of local managers and bureaucrats, Western educational curricula flourished and British-style public universities were established.

British imperial rule over South Asia extended to the border of present-day Afghanistan by 1890. Well into the 20th century, the British worked to extend political control further northward, dueling against Russia for political and military control beyond the Khyber Pass. The British and the Russians fought real and proxy wars over this border region, which includes both Afghanistan and parts of Western China. Armies, ambassadors, and spies were engaged in a conflict that became known as the **Great Game**. That conflict was never fully resolved, and the British were never able to maintain control of Afghanistan during their long reign in South Asia (**FIGURE 9.24**).

Apply Your Knowledge

9.7 Which trade connections and governance systems became stronger under British rule? Which became weaker? Explain.

9.8 Do some research on foreign activities in and around Afghanistan in the 20th century, including American and Soviet intelligence efforts and military activity. In what way do these more recent events mirror earlier elements of the "Great Game"?

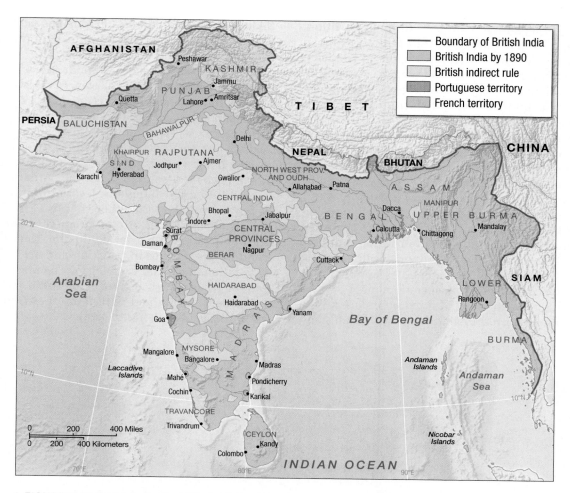

▲ **FIGURE 9.23 British India** By 1890, the British had come to control, directly or indirectly, most of South Asia, stretching north towards the border of Afghanistan, East to Burma, and south to Ceylon (today Sri Lanka).

◀ **FIGURE 9.24 The Great Game in Afghanistan** This 1878 political cartoon shows Afghanistan asking to be saved from the colonial ambitions of its "friends," Russia (Bear) and Britain (Lion). Rivalry between Russia and Britain over political control of Afghanistan in the 1900s resulted in instability and warfare in the region for decades. This conflict has contemporary implications as well.

Independence and Partition Long-standing grassroots resistance to British imperial rule had gained momentum through the **Indian National Congress Party**, formed in 1887 to promote greater democracy and freedom. A leader and the inspirational figure of this movement was **Mohandas Gandhi**, who advanced a vision of social justice and independence. Gandhi and his followers emphasized methods of nonviolent protest, including boycotts and fasting. Under Gandhi's leadership, the case for national independence became irrefutable. Soon after the conclusion of World War II, the British set about withdrawing from South Asia altogether.

In creating new, independent countries through **partition**, some independence leaders sought to follow the European model of building national states on the foundations of ethnicity, with a particular emphasis on language and religion. As a result, a separate Muslim-majority country called Pakistan was formed. Administrative districts under direct British control that had a majority Muslim population were assigned to Pakistan, together with several princely states that were joined to Pakistan rather than India, which was to be home to a predominantly Hindu population. Pakistan was created in two parts, East Pakistan and West Pakistan, one on each shoulder of India, separated by 1,600 kilometers (994 miles) of Indian territory (**FIGURE 9.25**).

The results of partition were geopolitical chaos and humanitarian disaster. At the moment of partition in 1947, when Pakistan and India were officially granted independence, millions of Hindus found themselves as minorities in Pakistan, whereas millions of Muslims felt threatened as minorities in India. Religious violence erupted. In desperation, more than 12 million people fled across the new national boundaries—the largest refugee migration ever recorded in the world. Hindus and Sikhs moved toward India and Muslims moved toward Pakistan. As many as 1 million people were killed in the resulting confusion and violence.

In some ways, the states of South Asia are still adjusting to the 1947 partition of India and Pakistan. In Pakistan, divergent regional interests in East and West Pakistan quickly developed into regionalism, and East Pakistani leaders called for secession. As a result, the country was split into two independent states in 1971: West Pakistan became Pakistan, and East Pakistan became Bangladesh.

The patchwork of borders that emerged amidst these periods of chaos is also notable. In the borderlands between what is now India and Bangladesh, are more than 250 pockets—*enclaves*—ruled by the other country. Though these complex political geographies are the product of both precolonial and pre-independence border confusion, they have been exacerbated by the politics of partition and are only now beginning to be rationalized and exchanged. The most extreme of these is Dahala Khagrabari, a pocket of India, within a pocket of Bangladesh, contained within another Indian enclave, itself within Bangladesh (**FIGURE 9.26**).

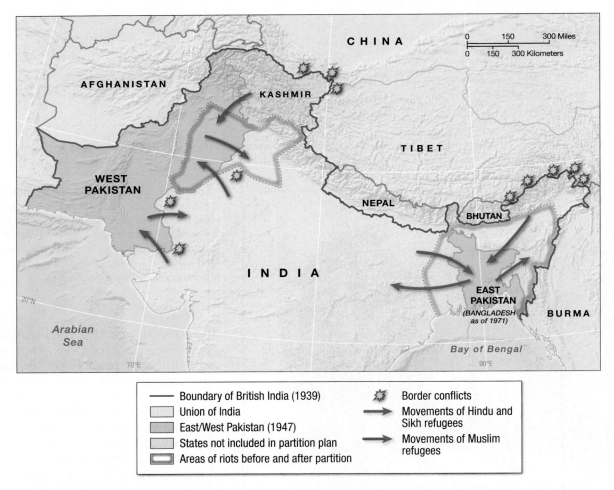

Boundary of British India (1939)
Union of India
East/West Pakistan (1947)
States not included in partition plan
Areas of riots before and after partition
Border conflicts
Movements of Hindu and Sikh refugees
Movements of Muslim refugees

▲ **FIGURE 9.25 Partition of India and Pakistan, 1947** The decision to create two states from British India had both immediate and long-term consequences. In the period just following partition, mass flight of refugees, riots and border conflicts occurred, resulting in the deaths of as many as 1 million people. Eventually, the independence of Bangladesh from Pakistan would set the contemporary boundaries of the region into place.

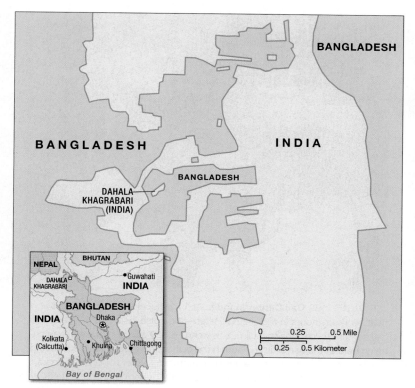

▲ **FIGURE 9.26 The Political Geography of Enclaves in India and Bangladesh** Enclaves are sections of a sovereign nation contained within the territory of another nation. Dahala Khagrabari represents the most complex case, a 3rd order enclave: a state within a state within a state within a state.

A more critical case regards the kingdom of Kashmir, which at the time of partition became a conflict zone. Here, a Hindu maharaja in a Muslim majority region had elected to join India in 1947, but Pakistani forces intervened to protect the majority Muslim population, splitting the territory in two, with

implications that have never been resolved. Neither India nor Pakistan can agree on the status of Kashmir, and the two countries have repeatedly gone to war over the region.

Apply Your Knowledge

9.9 What problems have emerged in South Asia owing to the partition of the region into separate nation states?

9.10 How do contemporary issues of political geography, like enclaves, reflect the decisions and priorities of 1947?

Economy, Accumulation, and the Production of Inequality

In the decades since independence, South Asia—and India, in particular—has come to play a large role within the economic world system. Though between 1950 and the 1990s many key institutions in India—including banks, utilities, airlines, railways, radio, and television—were government owned and operated, most all of these have since been privatized, and foreign investment has flowed into the country.

The New South Asian Economy At points over the past 20 years, India has been a country with the second-fastest rate of economic growth in the world—after China—averaging well above 6% growth per year. This booming economy, moreover, has shown terrific resilience in the face of recent global financial turmoil. By 2012, India had become the 19th largest exporter and 10th largest importer in the world, a fully international economy linked closely to the growth of China as well as the economic power of Gulf States in southwest Asia, especially the United Arab Emirates (**FIGURE 9.27**).

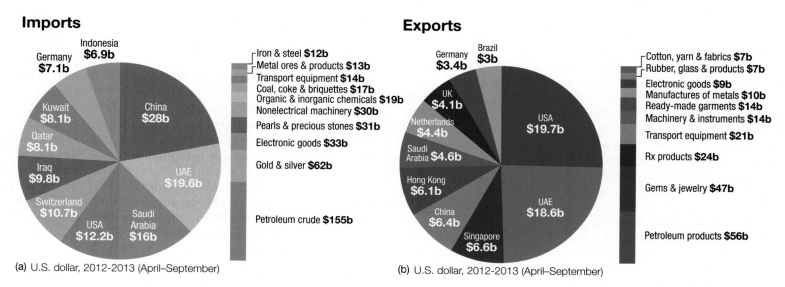

▲ **FIGURE 9.27 India's Imports and Exports** India's economy is rapidly growing, linked globally through (a) imports and (b) exports worldwide. **Which world regions figure especially prominently in trade into and out of India?**

▲ **FIGURE 9.28 Consumer Culture in South Asia** Brigade Road in Bangalore, India, is one of the country's largest commercial and shopping centers. Elite restaurants, stores, and clubs line the road, representing the booming consumer culture and economy of an emerging Indian middle class.

▲ **FIGURE 9.29 Call Center in India** At 1 A.M. local time in Bangalore—peak workday hours in the United States—the noise inside a 24/7 customer call center crescendos to a climax as nearly 1,300 phone conversations are being held at the same time. Though the call-center industry is a successful signature component of the regional economic boom of the past 20 years, young workers in this industry work long hours for poor pay and benefits with little chance of advancement.

Though exports from India are still dominated by petroleum and pharmaceuticals, one of the most dramatic examples of growth is that of the software industry in south India. Bangalore, in particular, has become a thriving industrial and business center as a result of a tech boom. Altogether, Bangalore boasts more than 500 high-tech companies that employ 100,000 people. These relatively affluent employees of the tech industry have contributed to the city's progressive and liberal atmosphere and its lively commercial centers, featuring fast-food restaurants, theme bars, and glitzy shopping malls (**FIGURE 9.28**).

Not all jobs made possible by high technology are those in software design. Another major growth industry is **call centers**, where workers respond to sales and customer service phone calls from throughout the world, keeping long hours for only modest pay. Educated Indian workers are especially attractive to foreign companies for this work, since their typically excellent English skills suit the needs of phone conversation with customers in North America and the United Kingdom, two major markets (**FIGURE 9.29**).

The region's economic growth has not come exclusively through external foreign investment. Some of the most prominent household names in India are the heads of industrial families whose firms emerged before independence and are now global. The Tata Group, for example, was originally a hotel company founded in the 1860s by entrepreneur Jamsetji Tata. It is now a multinational conglomerate company headquartered in Mumbai, whose operations extend to materials (Tata Steel); automobiles and trucks (Tata Motors); information (Tata Consultancy Services and Tata Technologies); as well as chemicals, energy, and communications. The reach of these South Asian conglomerates is international. Tata has acquired, for example, English tea companies, South Korean auto plants, and Anglo–Dutch steel firms.

This growth in the regional economy is not restricted to India, moreover. The South Asian Association for Regional Cooperation (SAARC) was established in 2010 as an economic and political cooperation group that includes India, Pakistan, Nepal, Bangladesh, Sri Lanka, Afghanistan, and the Maldives. Together, the SAARC nations constitute the third largest economy in the world in terms of gross domestic product (GDP). Efforts are ongoing to open a free trade zone between the member states and to ease visa restrictions.

The Tourism Boom Tourism is another growing sector in the South Asian economy. Tourism is India's largest service industry, employing approximately 9% of the total workforce. The central thrust of these industries is a combination of environmental amenities, including mountain trekking and beaches, along with cultural destinations and monuments, like the ancient city of Kandy in Sri Lanka and the "golden triangle" of tourism in India, which includes the Taj Mahal and the "pink" city of Jaipur.

There are downsides to being tourism dependent. When a tourism-dependent country experiences apparent instability or even a minor disease outbreak, tourists are quick to stay away, making the sector vulnerable to rapid downturns. Even so, much of the recent growth in South Asian tourism has been *internal,* with emerging wealthy classes in India and Pakistan beginning to tour their home region in large numbers. Though apt to spend less than foreign tourists, their numbers are far higher than that of foreigners; Indian tourists outnumbered international tourists by more than 41 times in 2010.

The most quickly growing part of the tourism sector is **medical tourism**, where international visitors come to South Asia—especially India—to undergo medical procedures at reduced cost, as much as 85% cheaper than what they might pay at home.

Trekking

In the country of Nepal, tourism rules the economy. Nepal's boundaries enclose 8 of the 10 highest mountains in the world, including Mount Everest, making the country a destination for mountain climbers, trekkers, and nature enthusiasts from around the world. In 2013, Nepal earned $420 million from trekking. The country's travel and tourism sector directly supported 504,000 jobs that year, around 3.2% of the total employment in the country. This does not count secondary receipts and jobs that the sector fuels, and tourism may account for as much as 20% of the economically active population.

The urge to hike the Himalayas for fun or a challenge dates to the colonial era, when British officials would summer in "hill stations" and take advantage of the underpaid armies of local people to serve as porters and guides. During this era in Nepal, the Sherpa community became iconic in this role.

Though the term is sometimes used to describe any local guide, the Sherpa people are a specific, small, Nepali ethnic group (perhaps 150,000 people) from the most rugged part of the Himalayas. Made famous for their support of grueling ascents of Mount Everest in particular, the Sherpa economy depends almost entirely on serving foreign trekkers, all of whom are willing to pay for guiding, and many of whom are dangerously inexperienced. Sherpas earn up to $5,000 a year, far higher than the average annual earnings of $700 in Nepal. They also die in startling numbers, as their community has borne a disproportionate

> "The Sherpa economy depends on serving foreign trekkers, all of whom pay for guiding, and many of whom are dangerously inexperienced."

share of the risk associated with foreign climbers. When George Leigh Mallory climbed the Tibetan side of Everest in 1922, an avalanche killed seven Sherpas. Sir Edmund Hillary and Sherpa Tenzing Norgay would ultimately scale the peak in 1953. In 2014, an avalanche at Everest base camp killed 16 Nepalis, most of whom were Sherpa. The Sherpa community has reacted defiantly, with some guides pulling their teams from the mountain. The 2015 earthquake in Nepal also resulted in 19 deaths at the base camp, 10 of whom were Nepalese Sherpas.

Risk for local guides is only one part of the problem that foreign climbing brings. The growth of visitors has also meant the deforestation of hillsides to collect campfire fuel. The industry is notoriously dirty as well, with rubbish and human waste becoming major problems. Writing for *National Geographic* in 2013, mountaineer Mark Jenkins described "pyramids of human excrement befouling the high camps." For all these reasons, the dependence of the Nepali economy on fulfilling the trekking dreams of foreigners has proven a double-edged sword (**FIGURE 9.2.1**).

1. Navigate to http://data.opennepal.net/ and search for the table (under datasets) on the purpose of tourist visits. What proportion of all tourist visits are based on trekking and mountaineering? What other tourist activities draw people to Nepal? Do you think these activities draw people from the same parts of the world? Why or why not?

   Nepal Tourism

   https://goo.gl/hJlycH

2. How does a heavy dependence on tourism benefit Nepal's economy, relative to other possible economic activities (e.g., industry)? Under what conditions might it make the economy more vulnerable?

◀ **FIGURE 9.2.1 The Trekking Economy** Two western trekkers smiling with a large group of porters and Sherpas on a Himalayan pass—Bodhi Himal, in Nepal.

▲ **FIGURE 9.30 Medical Tourism in South Asia** An Indian doctor in Mumbai consults with a foreign couple, showing a slide with a two day old embryo. Lack of regulation surrounding fertility services, and lucrative returns, have turned India into a popular hub of "In Vitro Fertilization tourism". Childless couples from overseas are attracted by the relatively low-cost treatment. There are some 400 IVF clinics in the country, providing an estimated 30,000 assisted reproductive treatments a year.

Popular procedures for foreigners include eye surgery, in vitro fertilization, hip replacement, and heart surgery (**FIGURE 9.30**). The industry draws on the long-standing and excellent medical training from educational institutions. The city of Chennai is a well-known destination, with modern facilities and more than 12,000 hospital beds, half of which are unused by locals. Though it is hard to track the overall value of medical tourism, in India, it is estimated to currently be a $2 billion annual industry, comprising as many as 3 million patients.

Inequality and Poverty Not all economic change in the region has been positive. The rapid growth of India's affluent middle class serves to highlight the desperate situation of a larger group: the extremely poor. As of 2012, around 60% of India's rural population lives on less than 35 rupees (50 U.S. cents) per day and roughly the same proportion of urban dwellers live on less than Rs 66 (U.S. $1; **FIGURE 9.31**).

Economic Change and Farming The liberalization of India's economy has had other unpredictable consequences. Lifting export controls has enabled farmers with access to large amounts of capital to reorganize their production toward lucrative overseas markets, with the result that domestic consumers have to pay more for traditional staples. For example, many farmers are turning to the cultivation of flowers and strawberries, which are airfreighted abroad. And now that a global market has become aware of high-quality local specialties, such as the fragrant basmati rice of the Himalayan foothills and the short-season Alphonso mangoes of Maharashtra, their prices within India have made them luxury items, out of reach of many consumers who traditionally regarded them as staples.

At the same time, the modernization of agriculture has made it more capital-intensive since farmers require more

▲ **FIGURE 9.31 Poverty in India** A homeless man in Kochi in the state of Kerala seeks goods for resale in the garbage. Poverty is crushing in the region, but drives people to creative and informal economic activities.

cash to pay for expensive inputs and equipment. This has had a steep cost in lives. Over the past decade, roughly 15,000 farmers in India annually committed suicide. Though many observers are quick to blame the adoption of genetically modified crops for this outcome, a 2014 study suggests that suicide risks were highest amongst any small-holding farmers, growing cash crops, with debts in excess of 300 rupees (around U.S. $5).

Opium In war-torn Afghanistan, finding a crop that can sustain its value despite erratic global markets is essential for rural producers. The opium poppy (see "Geographies of Indulgence, Desire, and Addiction: Opium and Methamphetamine" in Chapter 10) fulfills this profile. Harvested in large quantities, the poppy is the source of a resin rich in morphine, which is refined and sold for export. Further refinement into heroin occurs in-country or after export. This means there is a steep markup in price, but farmers still make reliable profits. In 2002, one kilogram (2.2 pounds) of opium was worth U.S. $300 for the producer, a value far exceeding that of cut flowers. Exports from Afghanistan amount to perhaps U.S. $4 billion, with farmers retaining a quarter of that

▲ **FIGURE 9.32 Opium Production in Afghanistan** Two men work in a field of opium poppies, which grow in abundance in Afghanistan. Despite efforts by U.S. and Afghani authorities to stamp out opium production, poppies remain among the most reliable source of revenue for poor farmers in the region.

value before officials, insurgents, and traffickers take their cut (**FIGURE 9.32**).

Living Conditions and Informal Settlements In rural South Asia, illiteracy remains common, and even the most basic services and amenities are lacking. Life expectancy is low; hunger and malnutrition are constant facts of life. In urban areas, poverty is compounded by crowding and unsanitary conditions. Clean drinking water is limited, and most poor households do not have access to a latrine of any kind. Much of this poverty results from the lack of employment opportunities in cities that are inundated with people. To survive, people who cannot find regularly paid work resort to various ways of gleaning a living. Some of these ways are imaginative, some

▼ **FIGURE 9.33 Dharavi in Mumbai** One of the largest slums in Asia, Dharavi teems with economic activity and energy as a population of more than 1 million people live in dense, improvised housing.

desperate. Examples include street vending, shoe shining, craftwork, street-corner repairs, and scavenging on garbage dumps.

Life is especially difficult for those living as squatters, or slum-dwellers, in the growing cities of Pakistan and India. Squatters and slum-dwellers, who have no formal rights to the land they may have occupied for years or decades, face a further challenge of poverty in that they have no legal standing and little political voice.

Consider Dharavi, one of Asia's largest slums, located in Mumbai, a city where slum-dwellers constitute perhaps half of the 12 million citizens (**FIGURE 9.33**). Despite the hardships these residents face in living in this slum, its location between two suburban rail lines is convenient for local workers and rents are low enough (about 185 rupees or U.S. $4 per month) for poor families to afford. Residents have rigged electricity and often have televisions. They have created a range of small, informal industries, such as sewing, leatherwork, and pottery. And yet the land these Dharavi squatters occupy, many for decades, lies in the near center of the city, on prime property for development. As a result, the citizens face constant threats of eviction. While residents have organized to oppose evictions, they possess little in the way of formal rights.

Self-Help and Microfinance The picture is not entirely negative, however, and one of the most significant developments has been the emergence of self-help movements. The best known of these is the **Grameen Bank**, a grassroots organization formed to provide small loans—**microfinance**—to the rural poor in Bangladesh. The Grameen Bank, which claims to be a financially sustainable, profit-making venture, runs completely against the established principles of banking; it lends money to poor borrowers who have no credit. The average size of a Grameen loan is about U.S. $120, typically enough to purchase a cow, a sewing machine, or a silkworm shed. Studies have shown that the bank's operations have lifted more than one-third of all borrowers above the poverty line. The most distinctive feature of the Grameen Bank is that 97% of its borrowers are women (**FIGURE 9.34**).

Child Labor Throughout South Asia, the informal labor force includes children. In environments of extreme poverty, every family member must contribute something, and children are expected to do their share. Industries in the formal sector often take advantage of this situation. Many firms farm out their production under subcontracting schemes that are based not in factories but in home settings that use child workers. In these settings, labor standards are nearly impossible to enforce.

The International Labour Office has documented the extensive use of child labor in South Asia, showing that many children under 10 years of age are involved in manual labor: weaving carpets, stitching soccer balls, making bricks, handling chemical dyes, mixing the chemicals for matches and fireworks, sewing, and sorting refuse. Most of them work at least 6 and as many as 12 hours a day.

▲ **FIGURE 9.34 Rural Enterprise** Microcredit programs such as those pioneered by the Grameen Bank have enabled tens of thousands of rural women to begin small businesses. This photo shows women signing up for loans in Gazipur, Bariali, Bangladesh.

Apply Your Knowledge

Navigate to http://www.worldbank.org and search for the summary report on the prospects for economic growth in South Asia.

South Asia Economic Growth

https://goo.gl/gjEOXF

9.11 How do countries in the region differ in their prospects and strategies for joining the new South Asian economy? What constraints do India, Bangladesh, Pakistan, and Sri Lanka face?

9.12 What implications does rising economic growth have for ameliorating inequality and poverty as described in this chapter?

Territory and Politics

As it does in other world regions, the colonial imposition of arbitrary borders continues to have an enormous impact on governance and conflict in South Asia. Cultural diversity and uneven economic development have also given rise to regionalism, separatism, political tension, social unrest, and, occasionally, outright conflict.

Indo-Pakistani Conflicts and Kashmir

South Asia was of great interest during the Cold War, in part owing to its location south of the Soviet Union. During the late 20th century, both the Soviet Union and the United States vied for favor with South Asian states, leading to off-and-on diplomatic relationships between Pakistan and the United States. South Asia

has become even more of a geopolitical hot spot since the end of the Cold War.

Part of the reason for this is the unresolved question of Kashmir (introduced earlier in the chapter). As a result of the de facto borders resulting from armed conflicts after partition, Pakistan controls the northwestern portion of what India claims as Kashmir, and China controls the northeastern corner (**FIGURE 9.35**). Since 1989, more than 30,000 people—separatist guerillas, policemen, Indian army troops, and civilians—have died in a conflict pitting an ongoing, pro-Pakistan guerrilla campaign against a repressive and violent Indian Army response.

Tensions mounted further after November 2008, when roughly a dozen carefully coordinated attacks were carried out in the dense business and tourist district of Mumbai, over the course of a few hours. More than 160 people were killed in the attacks, and hundreds more were wounded (**FIGURE 9.36**). In the period since, there has been some easing of hostility, with the opening of borders and increased cultural exchange between the two nations.

Geopolitics of Afghanistan

Indo-Pakistani tension is exacerbated by the international troubles surrounding Afghanistan. Afghanistan has long been of particular geopolitical significance because it is situated pivotally between Central and South Asia. Control of its mountainous terrain and mountain passes—such as the Khyber Pass—has frequently been disputed. During the 20th century, Afghanistan had economic and cultural ties to the Soviet Union and began to pursue a Soviet-style program of modernization and industrialization. This provoked resistance from a zealous group of fundamentalist Islamic tribal leaders

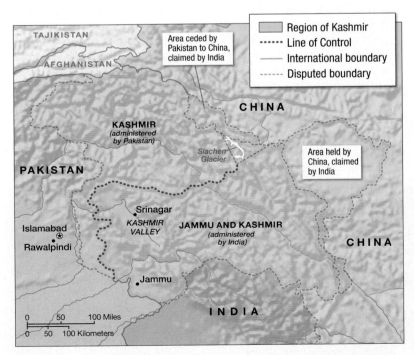

▲ **FIGURE 9.35 Kashmir and the "Line of Control"** Established by a brief war following partition, the unsettled boundary that splits the region of Kashmir has become a problem for both Pakistan and India.

▲ **FIGURE 9.36 Mumbai Terrorist Attacks** Smoke rising from the Taj Hotel building in Mumbai. The 2008 terrorist attack on luxury hotels and properties in Mumbai by armed gunman left 164 people dead and at least 300 wounded. India holds Pakistan responsible for the attacks.

Trade Center in the United States. Consequently, as a core component of the **War on Terror**, Afghanistan became the focus of a U.S. military operation, Enduring Freedom, which resulted in the defeat of the Taliban and installation of a U.S.-backed government in Kabul in December 2001.

Since the declared end of U.S. combat presence in Afghanistan in 2014 (U.S. and NATO military activity continues in the region), some progress has been made toward democracy, including the election of Ashraf Ghani as president in 2014. Even so, the governance of the country remains weak and chaotic. The global organization Transparency International ranks Afghanistan as the third most corrupt country in the world. Moreover, the Taliban has managed to survive in the border regions between Afghanistan and Pakistan.

The Afghan–Pakistan Frontier Between 2008 and 2015, the unresolved conflict in Afghanistan spilled over into Pakistan. Several of that country's frontier districts lie adjacent to the long, remote, mountainous border with Afghanistan. These are borders that tribal communities have freely traversed throughout history and where government control has been difficult to impose. Because Taliban forces operate in these areas, U.S. forces have conducted military operations in the area. These operations in the War on Terror include the use of **drones**, unmanned aerial vehicles flown out of bases in Afghanistan (and Pakistan) into Pakistani territory that are used to spy and strike enemy targets. Drone strikes have resulted in collateral civilian casualties, triggering both popular ire and official Pakistani outrage. Pakistan views these actions as transgressions of its territory.

Tensions increased in May of 2011, when U.S. Navy SEAL commandos made a raid on the hiding place of al-Qaeda leader Osama bin Laden in Abbottabad, Pakistan. Bin Laden had been hiding, essentially in plain sight, in a suburban area occupied mostly by retired Pakistani military officers. The fact that bin Laden was found and killed in Pakistan was an embarrassment and source of frustration for Pakistan's government, and Pakistan–U.S. relations deteriorated further as a result. The United States transports matériel and personnel into Afghanistan through Pakistan, and Pakistan is crucial to the maintenance of any peace in the future. As a result, problems along the border of Afghanistan–Pakistan translate into broader unsettled geopolitical relations.

Apply Your Knowledge

Examine the map of South Asia in Figure 9.1.

9.13 What areas and features on the map do you think are of most strategic concern from India's point of view?

9.14 Now consider the map from Pakistan's point of view. What areas and features do you think Pakistan considers to be the most strategic?

called the **mujahideen**, who were armed and trained by Pakistan. A total of more than 120,000 Soviet troops invaded Afghanistan to confront the resistance; however, like British colonizers before them, they were unable to establish authority in Afghanistan outside the capital city of Kabul.

The Soviets withdrew in 1989, but the conflict and suffering did not end. Differing factions continued to fight violently for control and by 1994 Kabul had been rocketed into rubble by forces supported by both the United States and Pakistan. By 1996, a hardline Islamist faction, the **Taliban**, was able to rise to power amidst the chaos, gaining control of Kabul and most of Afghanistan. The Taliban regime harbored Osama bin Laden and his al-Qaeda terrorist network, who were responsible for the September 11, 2001, attacks on the Pentagon and the World

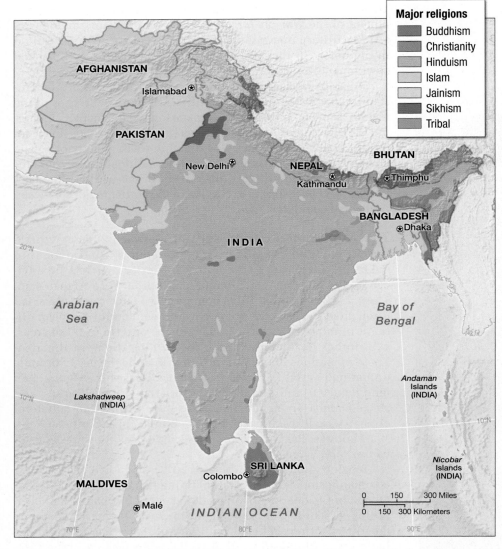

Major religions	
	Buddhism
	Christianity
	Hinduism
	Islam
	Jainism
	Sikhism
	Tribal

▲ **FIGURE 9.37 The Geography of Religion in South Asia** Partition between India and Pakistan in 1947 resulted in mass migrations of Hindus from Pakistan to India and of Muslims from India to Pakistan. However, more than 123 million Muslims still live in India. **What parts of the subcontinent have concentrations of significant minority religious communities?**

Culture and Populations

South Asia is a region with deep cultural roots that are tangled with those of neighbors and the broader world. Even where regional traditions have not been mixed or hybridized, there are differences in the extent traditional cultures have accommodated or resisted globalization.

Religion and Language

Religious and linguistic traditions in South Asia possess elements that have evolved fully within the region, such as Hinduism and the Dravidian languages. At the same time, many elements are wholly imported products of invasions and interactions from across the world, such as Islam and the English language.

Religion The two most important religions in South Asia are Hinduism and Islam. **Hinduism** is the dominant religion in Nepal (where about 90% of the population is Hindu) and India (about 80%). Islam is dominant in Afghanistan (99%), Bangladesh (more than 80%), the Maldives (100%), and Pakistan (about 80%). **FIGURE 9.37** shows the broad regional patterns of religion in South Asia; this geography is much more complex when considered in detail.

- **Hinduism** Roughly a billion people practice the dominant religion of the region, Hinduism. Hinduism is not a single organized religion with one sacred text or doctrine; it has no unifying organizational structure, worship is not congregational, and there is no agreement as to the nature of the divinity. Rather, Hinduism exists in different forms in different communities. It features both formal texts and major common beliefs, as well as diverse regional practices, idiosyncratic deities, and local legends. These traditions originally derive from the Vedas, a collection of oral poems that date from the 10th century B.C.E. There are several other important Hindu texts, including the epic tales of the *Ramayana* and the *Mahabharata*. The key common aspects that emerge from these are an acceptance of polytheism; belief in **karma**, the idea of cosmic responsibilities for actions and deeds visited upon eternal souls throughout an endless cycle of reincarnated lifetimes; faith in **dharma**, the actions aligned with a right way of being and doing in the universe; and the practice of *puja*, worship and prayer, typically at a temple or shrine (**FIGURE 9.38**). Nonviolence is also an important tenet and many, though by no means all, Hindus are vegetarians.

- **Islam** A sizable minority population in the region adheres to religions other than Hinduism. The most significant of these other religions in terms of numbers is **Islam**. Although several million Muslims migrated from India to Pakistan at the time of partition, more than 123 million Muslims still reside in India today. Islam is practiced in South Asia in a manner similar to its practice elsewhere in the world (see Chapter 4). Certain unique regional traditions also exist. South Asian mosque architecture can be distinctive, especially

▲ **FIGURE 9.38 Prayer at a Hindu Holy Site** Hindu worshippers are shown here immersed in the waters of the Ganga (Ganges) River at the sacred city of Varanasi, performing *puja* (prayer). Hindu rituals take place in temples, in homes, and outdoors at sacred sites such as this one.

Mughal architecture, which fuses Hindu and Muslim styles.

- **Buddhism, Jainism, and Sikhism** Numerous other religions were founded in South Asia and persist. **Buddhism** originated in South Asia, although it is currently followed by only about 2% of the region's population (see Chapter 8). Buddhism is the predominant religion in Bhutan and Sri Lanka, and an enclave of Buddhism is found in Ladakh, the section of Kashmir closest to China. Buddhists trace their faith to Prince Siddhartha, a religious leader who lived in northern India prior to the 6th century B.C.E. and who came to be known as Buddha (the Enlightened One). Buddhism stresses nonviolence, moderation, and the cessation of suffering.

  Jains are another distinctive religious group whose origins are in South Asia. Jains, like Buddhists, stress self-control and nonviolence, and their roots also can be traced back to the 6th century B.C.E.

Guru Nanak founded the monotheistic *Sikh* religion in the 16th century C.E. in Punjab. Sikhism remains largely associated with the people of the Punjab, where its sacred Golden Temple sits in the city of Amritsar (**FIGURE 9.39**).

- **Christianity** There are also almost 20 million Christians in India. The Portuguese brought Roman Catholicism to the west coast of India in the late 1400s, and Protestant missionaries, under the protection of the British East India Company, began to work their way through the region in the 1800s. Christianity is most widespread in the state of Kerala, in southwest India, where nearly one-third of the population is Christian.

Religious Identity and Politics Historically, the religions of South Asia have displayed a high level of syncretism, meaning that practices have coevolved and merged with one another over the centuries. Recent changes in identity and geopolitics, however, have exacerbated religious antagonisms and fostered social movements oriented around religion. In Afghanistan and Pakistan, notably, powerful Islamist movements have attempted to promote traditional culture as the basis of contemporary social order. There is strong adherence in both countries, for example, to traditional forms of dress and public comportment.

In India, political movements founded in a conservative interpretation of Hinduism have also been on the increase since the 1950s. **Hindutva**, a social and political movement that calls on India to unite as an explicitly Hindu nation, has given birth to a range of political parties over the years, including the Bharatiya Janata Party (BJP). The BJP attracted popular support starting in the late 1980s. Given the number of non-Hindus in the country and the explicitly secular nature of the national constitution, many people view the rise of Hindu nationalism with suspicion and fear. In 2014, the BJP won the national general elections in a landslide victory, launching Narendra Modi into place as the 15th prime minister of the country, and the first one born after Indian independence. Modi's election was controversial since, during the time that he served as Chief Minister of the State of Gujarat in 2002,

◄ **FIGURE 9.39 The Sikh Golden Temple at Amritsar** The Golden Temple at Amritsar is the holiest site of the Sikh religion, where the sacred text of the faith is housed. The temple is a destination for hundreds of thousands of pilgrims each year.

horrific anti-Muslim riots went unchecked for several days, resulting in the deaths of 1,000 people, mostly Muslims (**FIGURE 9.40**).

Islam and Hinduism both provide cultural symbols and practices that conservative and intolerant groups and parties have rallied around and used to organize. The continued success of secular government in India, however, and, to a lesser degree in Pakistan, shows that these parties and groups are neither universally embraced nor consistently successful.

Language A great diversity of languages is spoken in South Asia. In India alone there are approximately 1,600 different languages, about 400 of which are spoken by 200,000 or more people. There is, however, a broad regional grouping of three major language families:

1. Indo-European languages prevail in the northern plains region, Sri Lanka, and the Maldives. This language family includes Hindi, Bengali, Punjabi, Bihari, and Urdu.

2. Dravidian languages (including Tamil, Telegu, Kanarese, and Malayalam) are spoken in southern India and the northern part of Sri Lanka.

3. Tibeto-Burmese languages are scattered across the Himalayan region.

Written languages also vary owing to the diverse populations that have settled in the region over millennia. Most notably, Hindi and Urdu are in the same language family and are extremely similar in structure and vocabulary, but while the former is written in the Devanagari script derived from ancient Sanskrit writing, the latter is written in Nastaliq, a modified version of the Persian script (**FIGURE 9.41**).

Though Urdu is the national language of Pakistan, most Pakistanis speak another first language, including Punjabi and Pashto. Similarly in India, no single language is spoken or understood by more than 40% of the people. There have been efforts to establish Hindi, the most prevalent language, as the Indian national language, but this has been resisted by many states whose political identities are closely aligned with different languages. English, the first language of fewer than 6% of the people, links India's states and regions; English is the **lingua franca** (a widely recognized common second language) throughout the region for commerce, government, and travel.

Apply Your Knowledge

9.15 What are the major religions of South Asia, and which have origins outside the region?

9.16 How does language diversity present a challenge for nationalism in the region?

Cultural Practices, Social Differences, and Identity

The cultural practices of South Asia are diverse and represent a merging of indigenous traditions with a range of customs from other cultures. Some of the traditions most widely associated with social identity in the region include caste and religious identity, but new cultural phenomena are also widespread.

Caste, Marginality, and Resistance A very important—and often misunderstood—aspect of the region's cultural traditions is **caste**, which is a system of kinship groupings, or **jati**, that are reinforced by language, region, and occupation. Though associated with Hinduism, and therefore most common in India and Nepal, forms of caste distinctions are not unknown in Pakistan. There are several thousand separate jati in India, most of them confined to a single linguistic region. Many jati are historically identified by a traditional occupation, from which each derives

▲ **FIGURE 9.40 Indian Bharatiya Janata Party (BJP)** An election rally for Prime Minister Narendra Modi in Ranchi, India, during 2014. Modi was elected Prime Minister in 2014 and while many are enthusiastic about his modernization of the economy, others are deeply concerned about his partisan record of anti-Muslim chauvinism.

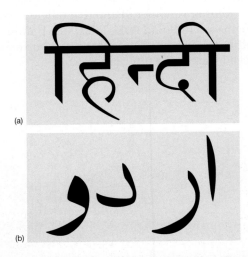

▲ **FIGURE 9.41 Devanagari and Nastaliq Scripts** The word "Hindi" is written (a) above in Devanagari and the word Urdu is written (b) below in Nastilaq script. The two are closely related Indo-European languages and are often orally indistinguishable; however, their respective written scripts derive from entirely different cultural lineages.

Aziz Royesh

Rudimentary content and rote memorization often characterize elementary education in the poorer parts of South Asia. In Afghanistan, the situation is far worse. Decades of war and the years of Taliban rule ripped apart the country's educational infrastructure. Schools are scarce, quality teachers are lacking, and girls typically have no access at all.

The Marefat School is different. Here, in a set of small buildings in the Afghan capital, Kabul, instructors teach through discussion, question and answer, and debates of major social and economic themes, including democracy and society. Girls make up half

▲ **FIGURE 9.3.1 Aziz Royesh** Founder of the revolutionary coeducational Marefat School in Kabul, Royesh was a finalist for the $21 million Global Teacher Prize in 2015.

> ## "Setting up school in a ruined building in Kabul's westernmost neighborhood, Aziz lured families to class by providing heat."

of the school's 3,500 students. The school holds classes on everything from history to music and has its own radio station and arts department. The students come from the poorest families, especially from the Hazara community, a Shia Muslim minority that has suffered especially in recent years.

Marefat's founder was Aziz Royesh, an energetic, determined, and imaginative instructor. The Soviet Union's invasion of Afghanistan in 1979 cut short Aziz's own education in the fifth grade. After that he labored in factories, a bakery, tailoring, and carpentry before he founded Marefat after the fall of the Taliban in 2002. Setting up school in a ruined building in Kabul's westernmost neighborhood, he lured families to class by providing heat (where most households lack even basic heating). Since then, the school has grown dramatically under Aziz's nurturing, gaining international attention and global support.

Aziz's teaching philosophy is especially notable. He stresses the ability to question. As he told an interviewer for National Public Radio (NPR) in 2015: "All these students around me, they can easily come and they can challenge me. They can reject me. They can oppose me. They can laugh with me. Sometimes even they can laugh at me, you know, they can."

Marefat High School

https://goo.gl/TCWINk

1. **What impact might the long-standing conflicts in Afghanistan have for the next generation of Afghan children? What other implications, beyond education, might the war have on the future population?**

2. **Visit the Marefat High School web page: https://marefat.wordpress.com. What are some of the surprising features about the educational program and its social facilities?**

---

its name: *jat* (farmer), for example, or *mali* (gardener), or *kumbhar* (potter). Modern occupations such as assembly-line operators or computer programmers, of course, do not have a traditional jati and people doing these jobs do not cease to be members of the jati into which they were born. Jati are also **endogamous**, which means that families are expected to find marriage partners for their children among other members of their jati.

In each village or region, jati exists within a locally understood social hierarchy—the caste system—that determines the norms of social interaction. Each individual person's jati is fixed by birth and the broad structure of caste systems always places certain groups at the top and others at the bottom. Caste systems tend to be held in high esteem by those who are religious and those who are especially learned. Those who pursue wealth or hold political power are typically less well regarded, and those who perform menial tasks are accorded the lowest status. Priestly jatis—known as Brahmins—are at the very top of the caste hierarchy. At the bottom end of all caste systems are the so-called untouchables—whose members dispose of waste

or dead animals. In an effort to eliminate the demeaning term *untouchable,* today most people in these communities prefer to be referred to as Dalits, meaning "the oppressed."

Traditionally, these lower-caste groups were forced to live outside the main community because they were deemed to be capable of contaminating food and water by their touch. They were denied access to water wells used by other communities, refused education, banned from temples, and subjected to violence and abuse. Although these practices were outlawed by India's constitution in 1950, discrimination and violence against lower-caste communities is still routine in some areas (**FIGURE 9.42**).

In modern, urban areas, and in recent years, practices relative to caste are far less regularly observed. Most urban Indian newspapers, however, still contain a lengthy "matrimonial" section, where advertisements are placed for "suitable" matches for arranged marriages, most often within the same specific caste group. Discrimination is also experienced by groups based on criteria other than caste. **Adivasi**, or tribal people, in India represent a significant part of the manual labor force in the regions

▲ **FIGURE 9.42 Caste and Poverty** A lower caste woman holds her child outside her house in the village of Bhaddi Kheda in the northern Indian state of Uttar Pradesh. This is among the most poverty-plagued states of India (its per-capita income is just above 50 percent of the national average) but marginal caste communities are amongst the poorest populations of the state.

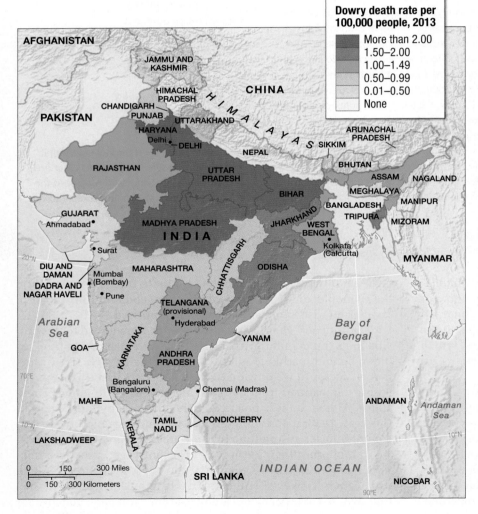

▲ **FIGURE 9.43 Dowry Deaths Across India in 2012** Though dowry murders occur occasionally across South Asia, they are concentrated in some of the most poor and rural parts of the country and far more common in the north than the south of India. **What might explain this uneven distribution?**

where they predominate. Tribals tend to hold the least land, have the poorest job prospects, and live in the most difficult of conditions.

Increasingly, marginal castes and adivasi communities have become more politically active, and some violent resistance movements are associated with these groups. So-called **naxalite** insurgencies, violent armed uprisings, have occurred across India, especially in the poorest states: Chhattisgarh, Jharkand, and Bihar. Because they attack landlords and police stations, the naxalites are considered terrorists by the government of India, but they have a considerable following in rural areas and among students.

Status of Women Across South Asia, the status of women continues to be a problem. Though women participate in business, government, and education, they face several traditional constraints. These include the practice of **purdah**, where women in some traditional households are segregated from men entirely, as well as the tradition of **dowry** (called Jahez in Pakistan), where the marriage of a girl child requires the bride's parents to make significant payments to the groom's family. This latter practice has, on occasion, led to "dowry murder," where the husband's family harasses or assaults the bride to extort further payment. In 2012, there were 8,233 reported dowry death cases in India (**FIGURE 9.43**). Women's subservience to men is manifest most clearly in the cultural practices attached to family life. The widespread (but illegal) practice of selective abortion reflects the preference for male children. Within marriages, husbands routinely neglect and maltreat many (but by no means all) poor women. In Afghanistan, women's education remains a serious challenge; according to government statistics as of 2013, only 26% of the country's population is literate and women's literacy is only 12%, among the lowest in the world.

Popular Culture Traditional cultures are deep and abiding in South Asia, but there are very few places where the impact of contemporary global culture is not heavily felt. The growth of a large and affluent middle class in India has brought Western-style materialism to larger towns and cities. Fast-food outlets, automated teller machines (ATMs), name-brand clothing, consumer appliances, video games, luxury cars, pubs, clubs, and shopping malls abound. Indian consumers have an increasing appetite for jewelry, electronics, and fine dining, wines, and spirits—markets that are entirely new to the economy.

South Asian popular media is also highly distinctive. As of the most recent census, approximately half the homes in India have a television. While foreign programming is popular, the top 10 Indian television shows of 2013 (a mix of reality shows, game shows, and soap operas) did not include a single foreign program. The Hindi-language film industry, known as **Bollywood** since it is centered in the city of Mumbai (formerly Bombay), is huge, with 1,000 films released each year reaching an audience of more than 3 billion people,

▲ **FIGURE 9.44 Krrish 3** A shot from a 2013 Bollywood superhero science fiction film. Made for only 17 million dollars (U.S.), the film grossed this within its first week.

▲ **FIGURE 9.45 Bhutanese Children** Junior high school students smile for the camera in Thimphu, Bhutan. Though one of the least developed and economically active countries in South Asia, tiny Bhutan has led a worldwide effort to revaluate development by stressing gross national happiness rather than material gain.

which includes domestic audiences and Indians abroad. Bollywood movies thrive on local markets and regional traditions, though most modern Bollywood movies address contemporary themes and characters in ways that relate directly to the lives of viewers. Although India also produces avant-garde movies that have been recognized for their artistic and dramatic content, most Bollywood products are exuberant, musical, spectacle-driven entertainment: melodramatic fantasies that mix action, violence, romance, song, dance, and moralizing into a distinctive form (**FIGURE 9.44**).

New forms of popular consumer culture have not gone without contestation in South Asia. Many local groups, nongovernmental organizations, blog authors, and cooperatives criticize consumerism as the predominant culture. Operating on a national scale, as a remarkable example, the government of Bhutan has worked to recognize **gross national happiness**, rather than economic output, as the measure of national success (**FIGURE 9.45**). See "Visualizing Geography: Gross National Happiness."

The Globalization of South Asian Culture Culture also flows outward from South Asia. Mysticism, yoga, and meditation found their way into Western popular culture during the "flower power" era of the 1960s. South Asian methods of nonviolent protest developed in the 20th century by Gandhi, such as boycotts and fasting, have spread around the world. Contemporary South Asian literature from writers such as R. K. Narayan, Aravind Adiga, Satyajit Ray, and Salman Rushdie has found a global readership.

One of the most visible global influences of South Asia is its food. Consider chicken *tikka masala,* a dish possibly conceived in Punjab, which has become global in scope owing solely to its popularity abroad (**FIGURE 9.46**). The dish has many fully regional South Asian components; roasted chicken (tikka) and spicy sauce (masala) are signatures of Pakistani and Punjabi cuisine. But the specific mix of yogurt, cream, and spices into a kind of gravy is novel, and it has become the most popular dish

in British restaurants. Britain's foreign secretary, Robin Cook, declared the meal "a true British national dish" in 2001.

Apply Your Knowledge

9.17 Describe two key cultural traditions from South Asia.

9.18 To what degree have modern media and consumer culture accommodated and absorbed these traditions?

▲ **FIGURE 9.46 Chicken Tikka Masala** This dish is by no means traditional, and may have been invented far from South Asia, but its unmistakable Pakistani and Punjabi elements are a testimony to the linkages and influences South Asia has across the globe.

Demography and Urbanization

South Asia's population is large and rapidly growing. India alone accounts for just over 1.25 billion of the world's 7 billion people; almost one in five people on Earth live in India. The rate of natural growth varies, with Afghanistan growing at a relatively high rate of 2.8%, while India is growing at a far lower 1.5%. Patterns of population density (**FIGURE 9.47**) further reflect patterns of agricultural productivity. Rich soils support densities of more than 500 people per square kilometer (1,300 per square mile) in a belt extending from the upper Indus Plains and the Ganga (Ganges) plains through Bengal and the Assam Valley.

Demographic Change

Despite the high rate of overall growth, population growth is slowing or halting in many parts of the region. Although some rural areas have fertility rates as high as six children per woman, a rate consistent with historic averages, others have declined to far lower figures, some to as few as two births per woman. Though it is an extremely populous country, India's births are declining, as are those in Sri Lanka, even while those in Afghanistan continue at historically high rates (**FIGURE 9.48**).

Factors Affecting Fertility Rates Little of this change resulted from the explicit and harsh population policies favored by regional governments in the past. In the 1960s and 1970s, notably, the government of India opened "camps" for the mass insertion of intrauterine devices (IUDs) and others for sterilization by vasectomy. Not surprisingly, a popular backlash put an end to these programs, which did little to curtail population growth.

Instead, these demographic changes have come about as a result of infrastructural, educational, and economic changes, associated with Demographic Transition (see Chapter 1). The unevenness of this effect can be seen across the region. **FIGURE 9.49** shows the average fertility rates by state in India. Notably, many of the states of southern India have rates of roughly two births per woman or fewer, effectively a condition of **zero population growth**, or ZPG. In the state of Kerala, often taken to be an example of demographic success, this change appears to have been the result of a combination of factors. First, rural health care in the state is well-developed and well-distributed. With access to good maternal and infant care, infant mortality rates are low, which discourages families from having more children. Good health care also means access to birth control, either in the form of prophylactics or sterilization. This is coupled with high levels of female literacy, high levels of participation in the labor force by women, as well as a local matrilineal tradition, which gives women significant control over property. Conversely, states in India where population growth remains high, like Rajasthan and Uttar Pradesh in north India, typically have low women's employment, low women's literacy, and a poorly developed health infrastructure.

Male–Female Sex Ratios An additional population concern in the region centers on male–female sex ratios (the proportion of women to men). From 1951 and 2011, the number of women in the population per 1,000 men declined from 983 (near even) to 918. The specific causes are numerous and include female feticide (observed also in East Asia, see Chapter 8) and higher rates of female infant mortality. The driving factor, however, is a continued preference for boy children and a stubbornly persistent system of gendered discrimination.

Urbanization South Asia is still very much a land of villages. Indeed, India has the largest rural population in the world. In 2011, approximately 65% of Pakistan's population and 71% of India's lived in rural settings; in Afghanistan, Bangladesh, and Sri Lanka, the rural population

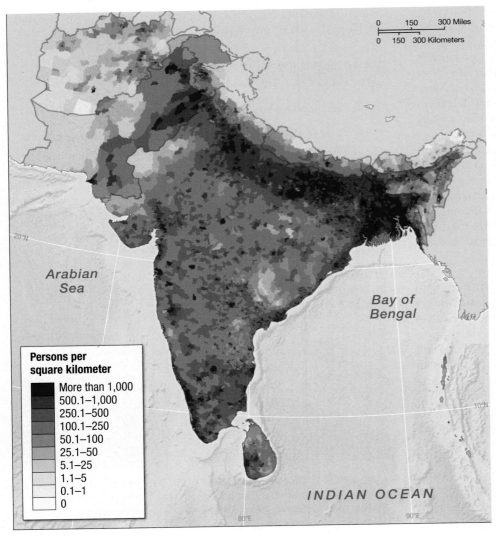

Persons per square kilometer

- More than 1,000
- 500.1–1,000
- 250.1–500
- 100.1–250
- 50.1–100
- 25.1–50
- 5.1–25
- 1.1–5
- 0.1–1
- 0

Arabian Sea

Bay of Bengal

INDIAN OCEAN

◀ **FIGURE 9.47 Population Density in South Asia, 2010** The density of population is very high throughout most of South Asia, but especially so in the plains and in subregions with good soils and humid climates.

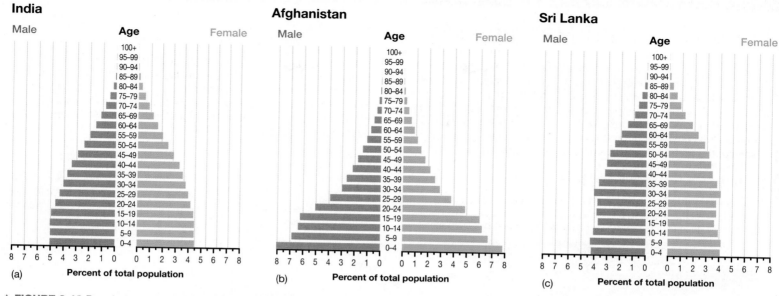

▲ **FIGURE 9.48 Population Pyramids for India, Afghanistan, and Sri Lanka** Like many around the world, (a) India's population pyramid, like that of (c) Sri Lanka, has a base that is leveling off. **What might account for continued growth in (b) Afghanistan?**

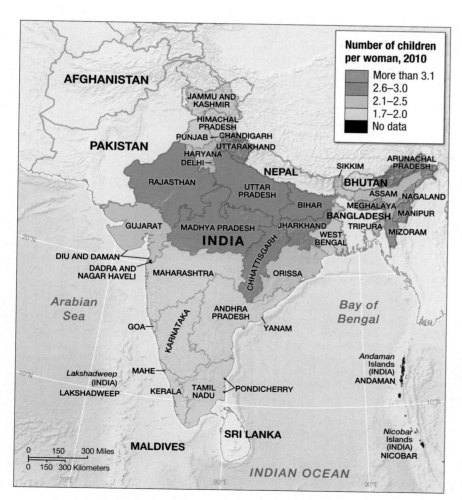

▲ **FIGURE 9.49 Fertility Rates Across the States of India** Although high fertility and large family sizes are the norm in some parts of the country, many states, like Kerala, have achieved nearly zero population growth.

is about 77%; the tiny state of Bhutan is 67% rural, with Nepal at 82%.

Rural-to-urban migration is shifting the balance toward towns and cities, however. Delhi is perhaps the third largest city in the world, with almost 25 million residents in its metropolitan area, while Karachi's 23 million puts it among the ten largest. The explosion of cities, middle-sized urban areas, and large towns has fundamentally changed the character of South Asian politics, landscapes, and culture. Urban areas are characterized by intense mixing of ideas, dress, and traditions, including many religions, classes, and castes. The dynamism of cities extends beyond the signature urban areas of Mumbai and Bangalore to the tens of thousands of middle-sized towns, where housing construction is booming. The competition for land at the edge of cities is intense, as agricultural land prices have risen dramatically through speculation, and formerly quiet rural areas have been quickly incorporated into bustling metropolises.

The South Asian Diaspora and Counter-Diaspora

The population dynamics of South Asia are highly influenced by its relationship to the rest of the world. The South Asian diaspora—the movement of Indians, Pakistanis, and Bangladeshis abroad—has involved between 5 and 6 million people, most of whom relocated to Europe, Africa, North America, and Southeast Asia. The origins of this diaspora can be traced to the British Empire when demand for cheap labor on plantations and railways was filled in part by emigrants from British India. In the mid-19th century, thousands of Indians left for the plantations of Mauritius (in the Indian Ocean), East Africa, the West Indies, and South Africa.

Gross National Happiness

Bhutan is the smallest and most isolated of the countries of South Asia, with a population of less than a million people, a low level of urbanization, and a highly traditional, rural, Buddhist society. The country's king began to open the country to international influence and trade in 1972, but chose to do so on local terms, prioritizing developments and economic change that would bring serenity and satisfaction, rather than mere prosperity. He coined the term "gross national happiness" (playing on the economic indicator of gross national product) and sought to devise a measure of the quality of life of the population as opposed to per capita economic activity. The gross national happiness concept has since expanded to a globally recognized metric: The global mean in a UN ranking system is 5.1 out of 10.

1 Satisfaction with Life Index

A World Happiness Report from the United Nations in 2013 ranked the world's countries. Happiness is hard to map, but it is widely variable from country to country and depends on a number of factors. The index used in the map is calculated from answers people gave to a Gallup poll about law and order, food and shelter, institutions and infrastructure, work conditions, well-being, and education.

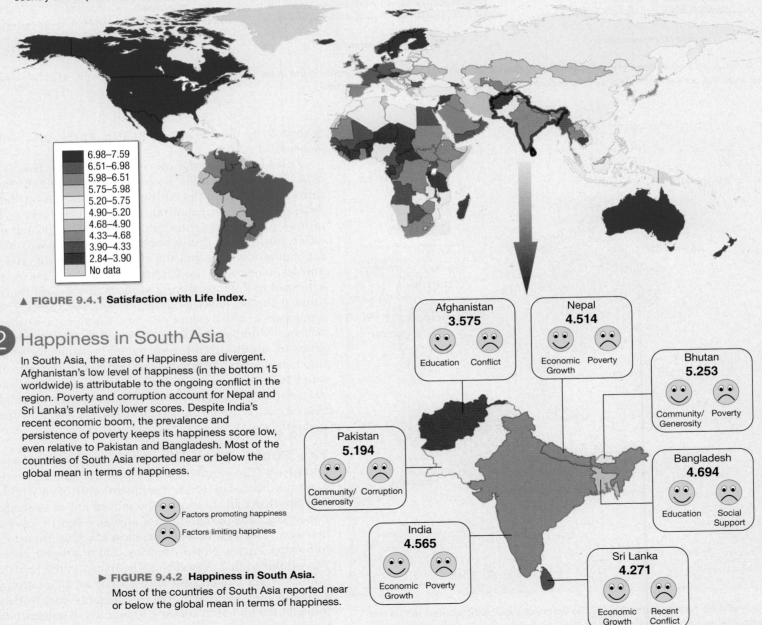

6.98–7.59
6.51–6.98
5.98–6.51
5.75–5.98
5.20–5.75
4.90–5.20
4.68–4.90
4.33–4.68
3.90–4.33
2.84–3.90
No data

▲ FIGURE 9.4.1 **Satisfaction with Life Index.**

2 Happiness in South Asia

In South Asia, the rates of Happiness are divergent. Afghanistan's low level of happiness (in the bottom 15 worldwide) is attributable to the ongoing conflict in the region. Poverty and corruption account for Nepal and Sri Lanka's relatively lower scores. Despite India's recent economic boom, the prevalence and persistence of poverty keeps its happiness score low, even relative to Pakistan and Bangladesh. Most of the countries of South Asia reported near or below the global mean in terms of happiness.

😊 Factors promoting happiness

☹ Factors limiting happiness

▶ FIGURE 9.4.2 **Happiness in South Asia.**
Most of the countries of South Asia reported near or below the global mean in terms of happiness.

Afghanistan
3.575
😊 Education   ☹ Conflict

Nepal
4.514
😊 Economic Growth   ☹ Poverty

Bhutan
5.253
😊 Community/ Generosity   ☹ Poverty

Pakistan
5.194
😊 Community/ Generosity   ☹ Corruption

Bangladesh
4.694
😊 Education   ☹ Social Support

India
4.565
😊 Economic Growth   ☹ Poverty

Sri Lanka
4.271
😊 Economic Growth   ☹ Recent Conflict

③ Happiness Trends

While happiness increased in some world regions, including East Asia, it fell in others, including South Asia. Poll trends suggest that though continuing economic growth and greater generosity were reported by people in South Asia, people also complained of declining social support, and less perceived freedom to make life choices.

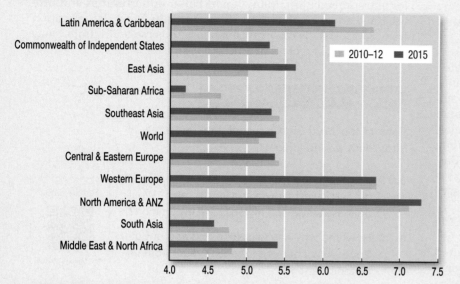

▲ FIGURE 9.4.3 **Happiness Trends worldwide 2010-2012 and 2015.**

1. **Happiness trends for South Asia show downward trajectory even amidst overall economic growth and the rise of a new middle class. What might explain this divergence?**

2. **The Happiness Index is not the only measure of regional wellbeing. Explore the data at http://www.happyplan-etindex.org/data/ and compare the map of "experienced wellbeing" with that of "ecological footprint". Do the happiest countries have the smallest ecological footprint? What does this suggest about the relative environmental costs of happiness and the acceptable environmental price of happiness?**

"Happiness is hard to map, and is widely variable from country to country."

Happy Planet Index

https://goo.gl/tz5rlu

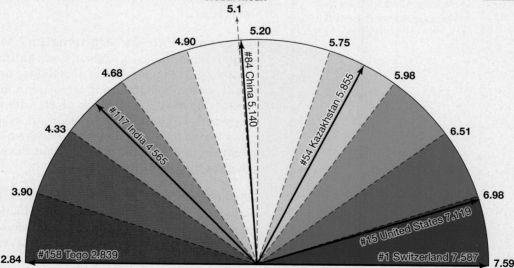

Negative variables:
- Corruption
- Poverty
- Lack of generosity
- Conflict
- Poor social support
- Short life expectancy
- Less freedom to make life choices

Positive variables:
- No corruption
- Prosperity
- Generosity
- Peace
- Good social support
- Healthy life expectancy
- Freedom to make life choices

▲ FIGURE 9.4.4 **2015 World Happiness Report rankings of selected countries.**

After World War II, the pattern changed significantly. Britain received more than 1.5 million South Asian immigrants, whose permanent presence not only filled a gap in the labor force, but has also served to enrich and diversify British urban culture. About 800,000 South Asians moved to North America, mainly to larger metropolitan areas, where most found employment in service jobs. From the 1970s onward, there has been a steady stream of South Asian immigrants to the oil-rich Persian Gulf states, recruited on temporary visas to fill manual and skilled jobs.

South Asia has experienced a **brain drain** over the past several decades, as some of the most talented and well-educated young people have emigrated to Europe and the United States. The brain drain began with the emigration of physicians and scientists to Britain in the 1960s and accelerated as South Asian students, having completed their studies in British and American universities, stayed to take better-paying jobs rather than return to South Asia. The idea of living abroad gained popularity among India's cosmopolitan and materialist middle classes as newspaper and television features publicized the global successes of Indian emigrants.

In the last two decades, the most distinctive aspect of the brain drain from South Asia has been the emigration of computer scientists and software engineers from India to the United States and parts of Europe. For example, 86% of Indian students receiving doctorates in the United States in 2002 did not return to India by 2007, second only to China (**FIGURE 9.50**).

Remarkably, a small tide of return migrants has begun a counter-migration from the United States and Britain. Though difficult to quantify, a large number of foreign-educated South Asians, especially from the technology sector, have chosen to move back to their native countries. This is likely a result of the economic slowdown in the West and the meteoric rise of the Indian economy and its growing middle class. A study from the Harvard Law School reports that half of those nonresident Indians who have returned to India in recent years have done so for entrepreneurship. Many report better opportunities and professional advancement there. So even while most out-migration from South Asia is one way, a modest "reverse brain drain" may be underway for a class of elite professionals.

Apply Your Knowledge

Refer to the Word Bank's recent data on fertility at http://data.worldbank.org/

9.19 Which countries in South Asia have the highest and lowest growth and fertility rates? What do these states and regions have in common?

9.20 How and why has migration into and out of South Asia changed in recent years from its historic pattern?

Fertility Rates

https://goo.gl/J99BBE

Future Geographies

How the shape and character of South Asia will change over the next 50 years depends on the outcome of several tensions and trends. While the end of population growth and an increase in the use of genetically modified organisms (GMOs) seem likely, peace and economic growth are more uncertain.

A Peaceful Fate for Afghanistan? Violence, war and US military operations in Afghanistan continue. The recent elections suggest the possibility of stability, despite ongoing fighting. If orderly governance is achieved in the next decade and a semi-stable regime remains in place, with normalized relationships with Pakistan, it is far easier to imagine a peaceful outcome for the region. In that case, for example, India and Pakistan might resume their discussion about the state of Kashmir and other disputes (**FIGURE 9.51**).

Zero Population Growth—When? Parts of South Asia have actually achieved zero population growth and overall demographic growth in the region may halt before 2050; however, how much sooner growth ends depends on several factors, including women's education, health-care availability, women's participation in the labor force, and the degree of urbanization. If conditions in these areas improve even modestly, as they have in Bangladesh where the fertility rate has fallen from seven births per woman in 1975 to 2.2 births per woman in 2013, the prospects are good (**FIGURE 9.52**).

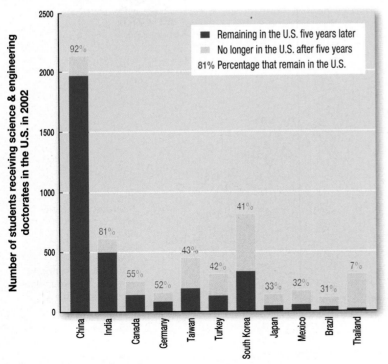

Staying in the U.S.

Legend:
- Remaining in the U.S. five years later
- No longer in the U.S. after five years
- 81% Percentage that remain in the U.S.

Y-axis: Number of students receiving science & engineering doctorates in the U.S. in 2002

China 92%, India 81%, Canada 55%, Germany 52%, Taiwan 43%, Turkey 42%, South Korea 41%, Japan 33%, Mexico 32%, Brazil 31%, Thailand 7%

▲ **FIGURE 9.50 Brain Drain** A vast majority of Indian PhDs from American schools do not return to India.

▲ **FIGURE 9.51 Afghan Elections** Women wait to cast their ballots at a polling station in Mazar-i-sharif in 2014. Voting was largely peaceful during Afghanistan's presidential election that year, the first democratic transfer of power since the fall of a Taliban regime in 2001.

▲ **FIGURE 9.53 Protesting Genetically Modified Crops** Though genetically modified cotton is grown in Pakistan and India and the Green Revolution introduced many crossbred crops into the region, consumers have balked at the production of genetically modified food crops, especially eggplant.

India 2050

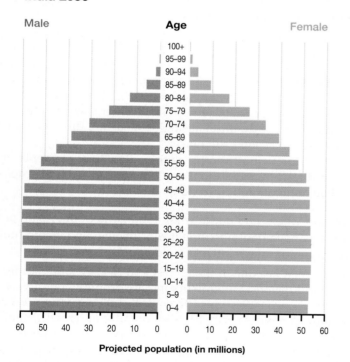

▲ **FIGURE 9.52 Projected Population Pyramid for India 2050** Exactly when global population growth will end depends heavily on the status of women in South Asia and their influence on the population pyramid.

New Food Technologies?

South Asia has experimented with the adoption of **genetically modified organisms (GMOs)** in agriculture. To date, no genetically modified food crops have been introduced in either Pakistan or India, where opposition to genetically modified crops has been vehement. Both countries, however, do grow *Bt cotton*—a variety of the cotton modified to

resist pests without the use of pesticides. In 2012, Bangladesh began movement toward allowing the cultivation of genetically modified eggplant, and Pakistan has begun to consider genetically modified maize. The future of food production in South Asia likely includes GMOs (**FIGURE 9.53**).

Can India's Economy Catch China's?

In 2015, India released a projected expansion of GDP growth of 7.4%. This would make it seem that India is on the heels of China and might overtake that giant sometime soon. The sudden jump in growth, however, was an artifact of a new calculation of GDP by the Indian government. Corrected in current U.S. dollars, India will probably lag behind China for some time to come (**FIGURE 9.54**).

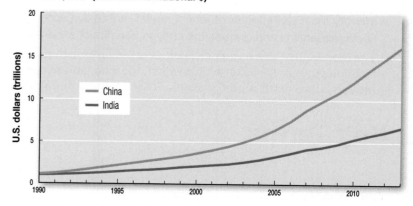

▲ **FIGURE 9.54 GDP in India and China** Though India's growth has been strong and China's economy is cooling off, it is unlikely that India will overtake China in the near future.

Learning Outcomes Revisited

▶ *Describe* the pattern of the monsoon climate, explain its origins, and identify the key adaptations people in the region use to cope with change and uncertainty.

The monsoon climate is caused by the relative pressure between the Asian interior and the Indian Ocean, which results in a band of intense summer rains and winter dry winds. People of the subcontinent have adapted to monsoon conditions with a range of agricultural practices, put at risk by some Green Revolution technologies.

1. *What* are some of the specific adaptations people have used to maintain agricultural production in a monsoon climate?

▶ *Describe* the physiographic regions, environmental hazards, and mineral resources and explain how some land uses contribute to environmental destruction, while others promote sustainability.

The major rivers of the region flow from Himalayan mountains and glaciers. Periodic heavy rainfall and steep drainages can contribute to flooding. When these natural events are exacerbated by human engineering of rivers and deforestation of hill slopes, the landscape is vulnerable to rare but catastrophic flooding.

2. *How* has South Asia's unique tectonic position created hazards, and where are these located?

▶ *Provide* examples of rulers, governments, and economic systems that historically put South Asia at the center of global trade, and explain how this position was later exploited and overridden by colonial governance.

Both the Harappan Civilization and the Mughal Empire demonstrate the region's linkages between South Asia and the world through trade links that reach west to Africa and north toward China. Seeking to exploit the region's location, colonial powers, especially the British, slowly dismantled the productive capacity of South Asia's economy.

3. *What* are some of the major political and economic legacies of colonialism in South Asia?

▶ *Identify* the major engines of recent rapid South Asian economic expansion and discuss the uneven benefits of expansion.

The new South Asian economic boom is based on growth in high technology, service-sector employment, tourism, and exporter industrialists with a global reach. Though an affluent and booming middle class has emerged, a significant majority of the region's population has not been touched by economic growth.

4. *Which* South Asian populations have been left out of the recent economic boom and why?

▶ *Describe* the key relationships and borders that present the greatest current geopolitical challenges for South Asia.

Long-standing antagonism exists between Pakistan and India, who since the time of partition have contested the status of the region of Kashmir. The U.S.-led NATO war in Afghanistan fueled further antagonisms between the two states. The outcome of that conflict will either lead to opportunities for further diplomacy or continued conflict between these two nuclear states.

5. *How* do the independence-era borders of South Asia influence present-day regional political relationships?

▶ *Summarize* the characteristics of the region's major religions and languages and explain how global connections have interacted with the region's traditional cultures and emerging consumer cultures.

The dominant religions of the region, Hinduism and Islam, as well as the popular languages, Hindi, Urdu, and Bengali, exist amidst enormous local diversity. Key elements of traditional South Asian culture include caste, gender-specific social roles, and religious identity. As consumer culture has increased, media and luxury goods have taken center stage. South Asian culture has greatly influenced global culture through cultural aspects such as food and music.

6. *Describe* some specific traditional cultural elements of South Asian culture that continue to exist, or are amplified, amidst modernization.

▶ *Explain* where and under what conditions South Asian populations have begun to stabilize or decline and where they continue to grow relatively rapidly.

South Asia has one of the largest and densest populations of all world regions, but growth rates are dropping rapidly, especially where health care, birth control, and education are available to women. Areas with high population growth contrast sharply with areas with far lower growth in these aspects.

7. *What* subregions of South Asia have the lowest population growth and why?

Key Terms

adivasi (p. *371*)
anthropogenic (p. *354*)
Aryan (p. *357*)
Bollywood (p. *372*)
brain drain (p. *378*)
Buddhism (p. *369*)

call centers (p. *362*)
caste (p. *370*)
deforestation (p. *354*)
dependency (p. *358*)
dharma (p. *368*)
dowry (p. *372*)

drone (p. *367*)
endogamous (p. *371*)
genetically modified organisms (GMOs) (p. *379*)
Grameen Bank (p. *365*)
Great Game (p. *358*)

Green Revolution (p. *347*)
gross national happiness (p. *373*)
Gupta Empire (p. *357*)
Harappan Civilization (p. *357*)
Hinduism (p. *368*)

Hindutva (p. *369*)
Indian National Congress Party
   (p. *359*)
intercropping (p. *346*)
Islam (p. *368*)
jati (p. *370*)
karma (p. *368*)

lingua franca (p. *370*)
Mauryan Empire (p. *357*)
medical tourism (p. *362*)
microfinance (p. *365*)
Mohandas Gandhi (p. *359*)
monsoon (p. *344*)
Mughals (p. *357*)

mujahideen (p. *367*)
naxalite (p. *372*)
orographic effect (p. *345*)
partition (p. *360*)
purdah (p. *372*)
Raj (p. *358*)
Taliban (p. *367*)

terracing (p. *346*)
transhumance (p. *346*)
tube well (p. *352*)
War on Terror (p. *367*)
zero population growth
   (p. *374*)

Thinking Geographically

1. How do the Himalayas and the monsoons help define the character of South Asia?

2. How did Britain transform the physical and human geographies of South Asia in terms of agriculture, infrastructure, and economy?

3. Compare poverty in rural and urban South Asia. With malnutrition and illiteracy rates high, what strategies do the poor use to survive in cities and in the countryside?

4. What areas or sectors of economic growth in South Asia are most vibrant?

5. As South Asians migrated worldwide, which jobs did they take? Which cultural traditions did they bring with them?

6. Why do India and Pakistan both make a claim to the Kashmir region?

7. How are key external political and military linkages between South Asia and the rest of the world affecting internal relationships within the subcontinent?

DATA Analysis: UN World Food Programme

In this chapter we began looking at food security issues, as well as tracing the history of agriculture and the impact of global food economies for South Asia. Take a closer look at food security and analyze these data for today's demographic realities in India. Begin by going to the United Nations World Food Programme website, and search for the Indian page on Food Security at https://www.wfp.org/.

1. Read the "Overview." What is the current headline on food security in India?

2. Navigate to the side bar and click on the "Food Prices." Go to the "Annual Data" and compare the fluctuation of prices on the following commodities: atta (wheat flour), lentils, and rice. Use all the markets and all the years in your data selections.

3. What do you notice in comparing the prices of the three commodities over the time and geographic span of the data?

4. Go to the "Reports and Bulletins" page, and download the "Food Security Atlas of Rural India, 2011" at https://www.wfp.org/. Read the "Executive Summary" and "Introduction."

5. In the Executive Summary (pp. i–iv), the writers rate certain regions in India as "food insecure." Explain why they call this a "geography of hunger."

6. List the three factors at the household level that contribute to food security, discussed in the Introduction (pp. 1–6).

Food Security
in India

https://goo.gl/6iz9b6

MasteringGeography™

Looking for additional review and test prep materials? Visit the Study Area in MasteringGeography™ to enhance your geographic literacy, spatial reasoning skills, and understanding of this chapter's content by accessing a variety of resources, including MapMaster interactive maps, videos, *In the News* RSS feeds, flashcards, web links, self-study quizzes, and an eText version of *World Regions in Global Context*.

A woman retrieves her possessions amongst the debris in the aftermath of Typhoon Haiyan in Tanauan, Leyte in central Philippines November 14, 2013.

Southeast Asia

In November 2013, Typhoon Haiyan made landfall in the Philippines. The storm was one of the most powerful ever recorded, with sustained winds topping 170 miles per hour. Like all typhoons, Haiyan began over the Pacific Ocean, picking up speed and water as it made its way toward the Philippine coast. On November 8th, the typhoon made landfall in the central parts of the Philippine islands, causing mass destruction. More than 6,300 people lost their lives in the storm, with over 1,700 of those deaths occurring in the provincial capital of Tacloban. The flooding was so extensive that regional airports were inoperable, making support flights impossible for several days. Houses were flattened and thousands were reported missing immediately after the storm passed the country.

Because of their dependence on the fishing economy, many people of the region hardest hit by typhoon Haiyan cannot move inland to safer areas. They are dependent on the economy of the seas that surround this island nation. But, global climate change continues to intensify the storms that build each year in the Pacific and storm surges as well as high winds may become a more regular occurrence for the people of this region. There are more immediate concerns as well. Poor housing conditions and a lack of resources for people to invest in storm shelters mean that the world's most vulnerable people economically also remain the most vulnerable to the vagaries of the storms that are driven to greater intensity each year with warming oceans.

Moreover, while the Philippines has intensified its efforts to reforest the country, deforestation allows for floods caused by typhoons, such as Haiyan, to create even greater chaos. Mudslides are commonplace in the parts of the Philippines where forest cover is the least dense. The local changes to the environment thus have both local and global affects. Fewer forests impact changing global climate and put local populations at more immediate risk.

Learning Outcomes

▶ **Describe** the region's climate pattern, geology, and physical landscapes and explain how people have adapted to the region's physical geographies.

▶ **Describe** the relationship among local economic development, Southeast Asian regional economic cooperation and competition, and global economic issues and trends.

▶ **Assess** the history of political tension in the region, noting the role that local understandings of the region's political geography and global conflicts have played.

▶ **Trace** the unique historical development of social life in the region, explaining how and why differences and similarities between places emerge over time.

▶ **Identify** the major social issues in the region today and describe how they are the result of historical, geographical, and cultural patterns.

▶ **Explain** the patterns of demographic change and migration in the region and their effects.

Environment, Society, and Sustainability

The relationship between Southeast Asia's physical geography and human geography is complex. This region's position in relation to global climatic patterns and large rivers and oceans has fostered rice cultivation and fishery development for over 5,000 years (**FIGURE 10.1**). The people of Southeast Asia have adapted by developing large-scale agricultural systems and harnessing biodiversity for medicinal purposes. Tectonic activity also makes this region susceptible to volcanic activity and earthquakes. Major earthquakes can produce **tsunamis**, very large ocean waves that can cause great destruction along coastlines.

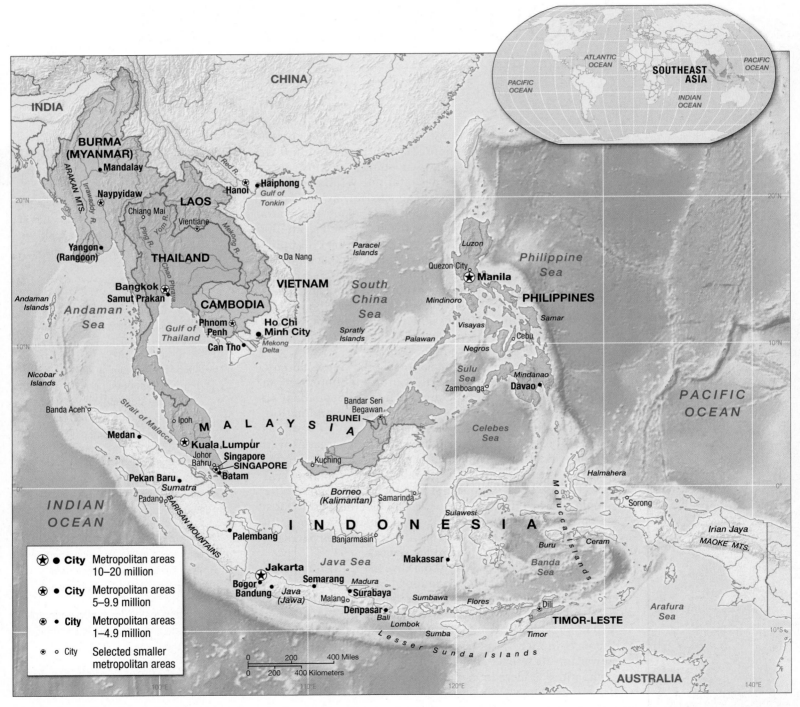

▲ **FIGURE 10.1 Southeast Asia: Countries, major cities, and physical features** Southeast Asia is an expansive region stretching from the Pacific to the Indian oceans. It borders East Asia and South Asia and surrounds a large portion of the South China Sea.

Climate and Climate Change

Although most common images portray Southeast Asia as a uniformly tropical region, it has diverse climatic patterns stretching across mainland Southeast Asia and island (or insular) Southeast Asia (**FIGURE 10.2**). The mainland consists of Burma, Thailand, Laos, Cambodia, Vietnam, and the peninsula of Malaysia, while the island region consists of Singapore, Indonesia, Timor-Leste, and the Philippines. The region's climatic diversity is a function of its relation to the intertropical convergence zone, or ITCZ (see Chapter 1, p. *9*), which produces the **monsoon**—a wind system that reverses directions periodically and produces seasonal rain patterns throughout the region.

Climate Factors of Mainland and Islands On mainland Southeast Asia, the November–March period brings cooler temperatures as air flows south from the large highlands and plateaus

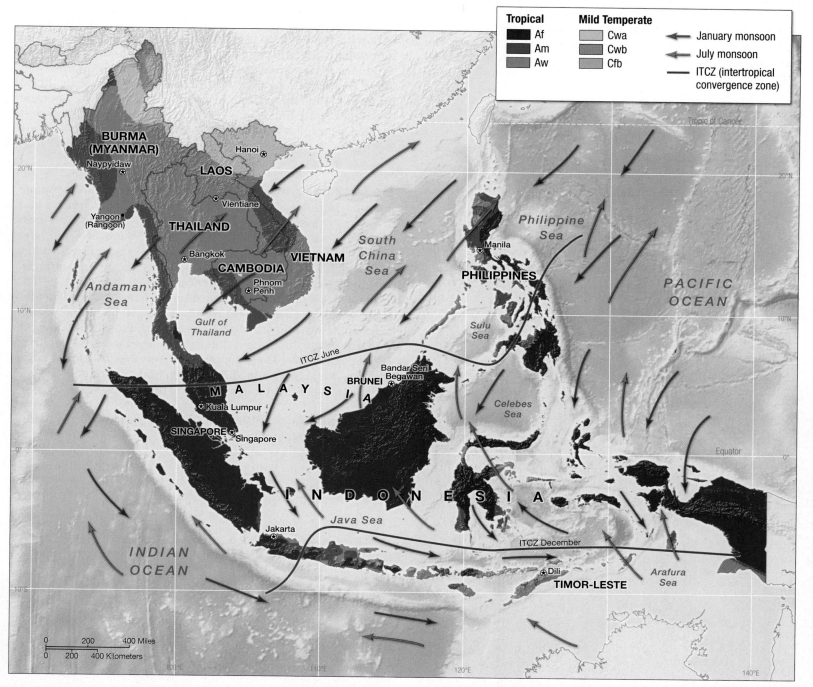

▲ **FIGURE 10.2 Climates of Southeast Asia** Southeast Asia has a hot, wet climate and seasonal monsoon winds that bring heavy precipitation. The mainland is drier than the islands, with more seasonal rainfall. For more detail on the legend see Chapter 1, Figure 1.9. **How does the January monsoon of Southeast Asia's islands differ from the same monsoon on the Asian mainland?** (*Hint:* **Study the wind pattern on the map for the January monsoon.**)

of East Asia. During this period, the monsoon winds blow out of the continental interiors and across the ocean. On the mainland, the November–March period also brings drier conditions, resulting in lower overall annual rainfall totals. During this same season, the islands of Southeast Asia receive monsoon rainfall, this time on the north-facing slopes. The Indonesian island of Sumatra receives the bulk of its rain on north-facing slopes during the December-to-February period as winds blow southward from mainland Southeast Asia across the South China Sea. Between June and August, rain falls on south-facing slopes when monsoon winds sweep northward from the Indian Ocean.

In places where rainfall happens only part of the year, the onset of the monsoon is a momentous event (**FIGURE 10.3**). It brings some alleviation from very high temperatures as

> "The onset of the monsoon is a momentous event. It brings some alleviation from the very high and oppressive temperatures as rains cool the air, and it brings water for crops and drinking."

▲ **FIGURE 10.3 Water Festival, Chiang Mai, Thailand** Thai performers celebrate the Songkran festival, marking the traditional Thai New Year. The throwing of water was historically a sign of respect and well wishing during the festival.

rains cool the air, and it brings water for crops and drinking. The monsoons can also cause severe floods, and the constant heavy rain can promote the growth of molds and fungus, which may have an impact on the health of people living in this region (see Sustainability in the Anthropocene: Regional Effects of Climate Change on p. 388).

Tropical Cyclones and Typhoons The Southeast Asian islands are located in the equatorial region relative to the ITCZ; they experience daily tropical thunderstorms throughout the year. Warm ocean temperatures from August to October combine with trade winds to produce large storms often called as **typhoons** in Pacific Asia and **cyclones** in the Indian Ocean. Typhoons most commonly develop east of the Philippines and move westward into the South China Sea, whereas cyclones begin in the Indian Ocean and sweep into Southeast Asia from the west. The combined effect of monsoons, the ITCZ, and typhoons/cyclones means that the islands of Southeast Asia are among the wettest regions in the world with lush forest vegetation and fairly consistent annual temperature.

Trade Winds and Trade Networks The region's unique position in relation to wind patterns has brought traders and migrants to Southeast Asia for thousands of years, and with them their religious beliefs—first Hinduism and Buddhism and later Islam. The shift of winds in winter carried traders back home, which globalized Southeast Asia's spices and other products. Imagine a person in 16th century Spain buying nutmeg in their local market, a product that came from the famed "**Spice Islands**" of the Molucca Islands (**FIGURE 10.4**). Even though that person in Spain knew little or nothing about the islands and the people who lived there, his or her tastes were changed as a result of these global connections. These connections were expansive. The spice trade encompassed parts of Africa, the Middle East, and Europe and the transoceanic trade networks of the Indian Ocean.

Climate and Agriculture The dynamic climatic forces in this region have provided humans a unique opportunity to adapt to the conditions of this region. About 5,000 years ago, the selective breeding of a grass with edible seeds produced rice. Rice is one of the few major crops that can grow in standing water and is

▼ **FIGURE 10.4 Spice Market, Jakarta, Indonesia** A vendor sells spices traditionally used in Indonesian cooking in one of Jakarta's many markets.

suited to the flooding that accompanies heavy monsoon rains. **Swidden farming**—the practice of clearing small forest areas for agriculture—has historically been the main form of agriculture in the highland areas of the region. The extensive shifting of cultivation involved in swidden farming has been sustainable under two conditions: consistent rainfall and sufficient land area. Where widely scattered villages controlled large areas of land, farmers could use fields for a few years and then leave them for 10 to 30 years to recover. As climate has changed, access to land has decreased, and population and consumption has increased, it has become more difficult for swidden farmers to produce strong agricultural yields.

Irrigated rice requires the control of large river systems and labor throughout the season to prepare and maintain fields, transplant seedlings, weed, and harvest each stalk by hand. Facilitated by the gendered division of labor in some Southeast Asian societies, women have historically been responsible for the planting and care of rice cultivation as well as other domesticated crops, while men have been responsible for clearing fields, felling trees, and burning plots in preparation of planting. This has changed as mechanization of agricultural production and communal systems of farming (where villages participated collectively in rice production) have given way to large-scale industrialized agricultural production systems. Changes in agricultural systems have given rise to a class of landless people—both men and women—who must sell their labor to wealthier landowners during the planting and harvesting seasons.

Humans have dramatically modified the region's physical landscape for rice production to manage the rainfall produced by the monsoons. Agricultural adaptations include the construction of terraces, paddies, and irrigation systems (**FIGURE 10.5**). Terraces cut into hillsides provide level surfaces that control water flow and reduce erosion. The construction of dikes (ridges) allows fields to be flooded, plowed, planted, and drained before harvest in a system called **paddy farming**.

Apply Your Knowledge

10.1 Identify and describe the patterns of rainfall across Southeast Asia.

10.2 What are the global processes that leave parts of Southeast Asia wet all year and other parts of the region seasonally dry?

▼ **FIGURE 10.5 Terrace Agriculture in Luzon, the Philippines** Terracing affords a number of advantages in hilly areas that experience heavy rainfall. The practice allows people to control the flow of water through irrigation systems between different parts of the field; it protects topsoil from eroding during periods of heavy rainfall, sustaining the nutrient levels of the field; and it allows farmers to divide the land into paddies, and grow fruits and vegetables to accompany this staple crop. **Will terraces like these remain competitive as Southeast Asian societies move toward mechanized, large-scale industrial agriculture? Explain.**

Regional Effects of Climate Change

According to the Intergovernmental Panel on Climate Change (IPCC) Assessment Report, global climate change will impact agriculture, coastal systems, broader ecosystems, and water in Southeast Asia. Broader global dynamics, such as deforestation in other world regions, also affect global climate patterns. These kinds of dynamics may have a long-term impact on rainfall and the monsoon patterns in the region.

Agricultural Impacts Increased rainfall is likely to occur in the region as a result of changing climate. Large-scale rice agribusiness, which now keeps rice under cultivation year-round in parts of Southeast Asia, may also be affected by temperature increases related to global warming. This could place increased stress on crops and lower yields. Increased rainfall, also a result of climate change, in some parts of the region may be coupled with extended periods of drought in other parts of the region, such as northeastern Thailand (**FIGURE 10.1.1**).

"According to the IPCC Assessment Report, global climate change will impact agriculture, coastal systems, broader ecosystems, and water in Southeast Asia."

Coastal Systems Increases in global temperature also affect fisheries in the region. The long-term viability of many low-lying island communities are subject to problems associated with increasing sea levels.

Ecosystems Climate change will have an impact on the future of land cover in the region, as forests may become more susceptible to pests that will thrive in the warmer, wetter climates of the region. One of the most fragile and yet most important ecosystems in the region, mangroves, may come under further threat from sea-level rise and human encroachment on their habitats (**FIGURE 10.1.2**).

Water In coastal areas, sea-level rise stresses the delicate balance between fresh- and saltwater by allowing saltwater to seep into groundwater used for drinking. Changes in snowfall and melting in upstream mountain areas throughout East and Southeast Asia will impact broader water flow throughout the region, impacting not only agricultural development but also the production of renewable energy resources, such as hydroelectric power (**FIGURE 10.1.3**).

▲ FIGURE 10.1.2 **Mangrove Replanting, Philippines** Many people in Southeast Asia are trying to mitigate the impacts of human-induced environmental change. Mangroves protect coastal landscapes and provide much needed space for a variety of fish species.

1. **What are some of the major environmental challenges facing Southeast Asia today?**

2. **What opportunities exist for people in Southeast Asia to address the changes to regional climate over the next decade?**

▲ FIGURE 10.1.1 **Drought in Northeastern Thailand** Changing global climate patterns have different impacts on local regions. In the case of Northeastern Thailand, changes in global climate are increasing the duration and intensity of droughts.

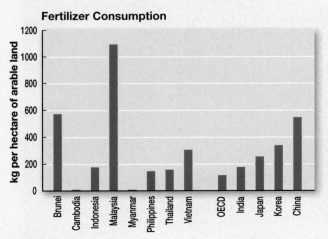

Fertilizer Consumption

(kg per hectare of arable land — bar chart for: Brunei, Cambodia, Indonesia, Malaysia, Myanmar, Philippines, Thailand, Vietnam, OECD, India, Japan, Korea, China)

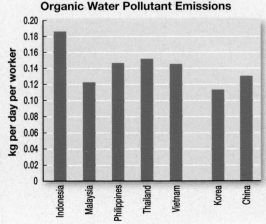

Organic Water Pollutant Emissions

(kg per day per worker — bar chart for: Indonesia, Malaysia, Philippines, Thailand, Vietnam, Korea, China)

◄ FIGURE 10.1.3 **Fertilizer Consumption and Organic Water Pollutant Emissions in Southeast Asia** Fertilizer consumption and organic water pollutant levels are relatively equivalent to other parts of the world. But some countries in Southeast Asia, such as Malaysia, are seeing growing consumption of fertilizer as increasing pressure on farming is produced by larger populations. Other countries in the region may follow, as food demands change both regionally and globally.

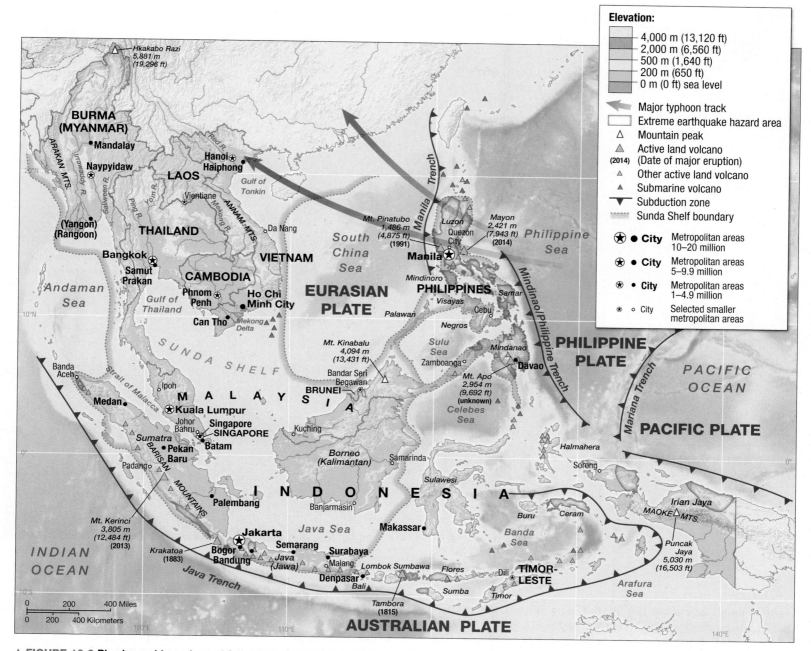

▲ **FIGURE 10.6 Physiographic regions of Southeast Asia** Southeast Asia, one of the most geologically active regions of the world, experiences many earthquakes and volcanoes because it lies at the intersection of active tectonic plates. The region is very mountainous except for the major valleys of the Irrawaddy, Chao Phraya, Red (or Song Hong), and Mekong Rivers and some coastal plains. **Identify and explain two different hazards associated with the region's subduction zones.**

Geological Resources, Risks, and Water Management

Plate tectonics have shaped the physical geography of Southeast Asia and influenced the historical configuration of land and oceans, highlands and lowlands, and geological hazards and resources (**FIGURE 10.6**). The collision zones between tectonic plates are associated with mountain building and volcanic activity and have created thousands of islands that form the Indonesian and Philippine **archipelagoes**—a series of islands that are co-located within a large body of water.

Indonesia has more than 13,600 islands (only half of them inhabited), and the Philippines is made up of more than 7,000 islands.

The economies and lifestyles of insular Southeast Asia have been oriented to the sea through ocean fishing and maritime trade for centuries (**FIGURE 10.7a**). It has often been easier to use ocean or river transport than overland routes. The complexity of the many islands of some nations, together with the economic significance of ocean fishery and oil resources, has created conflicts over maritime jurisdictions. Countries within and outside Southeast Asia, including the People's Republic of China, lay

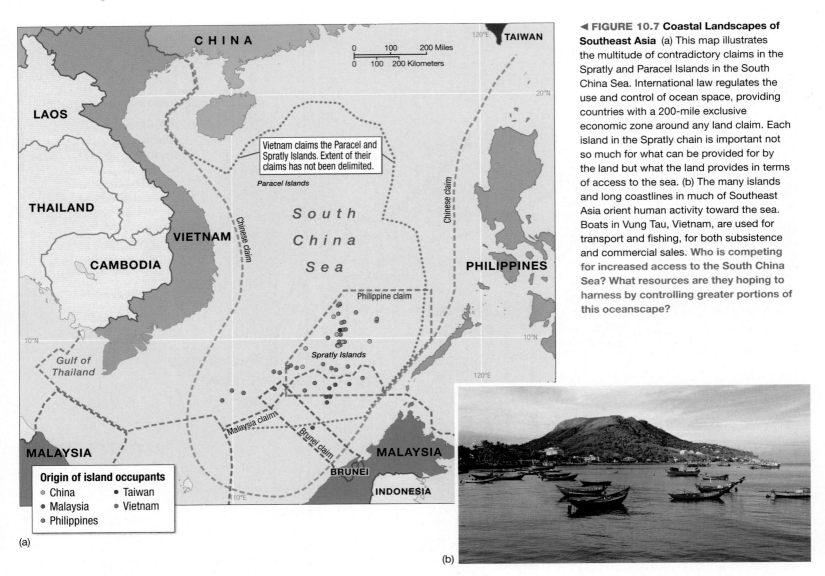

◄ **FIGURE 10.7 Coastal Landscapes of Southeast Asia** (a) This map illustrates the multitude of contradictory claims in the Spratly and Paracel Islands in the South China Sea. International law regulates the use and control of ocean space, providing countries with a 200-mile exclusive economic zone around any land claim. Each island in the Spratly chain is important not so much for what can be provided for by the land but what the land provides in terms of access to the sea. (b) The many islands and long coastlines in much of Southeast Asia orient human activity toward the sea. Boats in Vung Tau, Vietnam, are used for transport and fishing, for both subsistence and commercial sales. **Who is competing for increased access to the South China Sea? What resources are they hoping to harness by controlling greater portions of this oceanscape?**

claim to small islands in the South China Sea and the resources that surround those islands (**FIGURE 10.7b**).

Mainland Southeast Asia is geologically older but still very mountainous. Peaks in the highlands of northern Burma reach beyond 5,000 meters (16,400 feet), and ridges link to the Himalayas. The countries of mainland Southeast Asia feature fragmented and elongated geographies that have created challenges to national integration, transportation, and economic development. Burma, Thailand, and Vietnam all include long narrow segments, although rivers cut through these countries, providing networks for transportation and large-scale agricultural developments.

Energy and Mineral Resources Southeast Asia has mineral and energy resources, such as tin and oil, although oil from Southeast Asia only contributes about 5% of global production. Indonesia, Malaysia, and Brunei are the principal oil producers, exporting mainly to Japan from the north coast of Borneo, Sumatra, and Java.

With foreign assistance, Vietnam is expanding its oil production, and Burma is reinvigorating its energy sector through exploitation of gas reserves for domestic use and sale to Thailand.

From colonial times, tin has been a major export from Malaysia, but exports have dropped as deposits have become exhausted and competition from Thailand and Indonesia as well as outside the region have lowered prices. The Philippines has a wide range of mineral resources, including copper, nickel, silver, and gold, and Indonesia is developing new gold, silver, and nickel mines in Irian Jaya, in the face of opposition from environmentalists and indigenous groups.

The newest energy sources to be developed in Southeast Asia are **biofuels**, especially palm oil, which can be converted into biodiesel. Indonesia and Malaysia accounted for over 80% of world crude palm oil production in 2015, and Thailand is increasing its production as the world's third-largest producer. Southeast Asian countries export a large portion of this production, sending it to India, China, and the European Union. Concern is growing about the conversion of tropical forests to palm oil plantations across the region, particularly when expansion of this crop intrudes on **megafauna**—large-bodied animals—such as orangutans (**FIGURE 10.8**). As global demand for palm oil increases, it will continue to be an important export crop for these countries.

Risks and Benefits of Volcanic Activity The active volcanoes and earthquake zones of Southeast Asia pose great risks to the human populations of the region. The Indonesian and Philippine volcanoes are part of the Ring of Fire (see Chapter 8, p. *304*), which surrounds the Pacific Ocean. The explosion of Mount Tambora in Indonesia in 1815 is one of the largest eruptions ever recorded on Earth. The eruption was so powerful that it shot ash into the stratosphere and noticeably cooled global climate for several years following the eruption. More recently, the tsunami of 2004 that struck off the coast of Indonesia and impacted Thailand and Burma, as well as Sri Lanka and India, represents the modern-day risks of tectonic activity in the region (**FIGURE 10.9**). The risks of volcanoes are balanced by the benefits they provide in terms of soil nutrients. Volcanic eruptions deposit ash that contributes to soil fertility and high crop yields. In Java, the rich volcanic soils have contributed to productive agriculture and dense populations.

River Systems Between the mountain ridges of Southeast Asia lies a series of river valleys and deltas, all of which have been important to the growth of societies in this region (**FIGURE 10.10**). The Mekong River is the longest waterway in the region, flowing from the highlands of Tibet to its delta in Vietnam and Cambodia. The Mekong forms the main transpor-

▲ **FIGURE 10.8 Palm Oil Collateral Damage** Conservationists believe that one of the largest populations of wild orangutans on Borneo will reach extinction unless drastic measures are taken to stop the expansion of palm oil plantations.

▲ **FIGURE 10.9 Tsunami Damage in Banda Aceh, Indonesia** The city of Banda Aceh was the hardest hit by the tsunami that struck Indonesia in 2004. An estimated 387,000 people were left homeless in the city and surrounding areas.

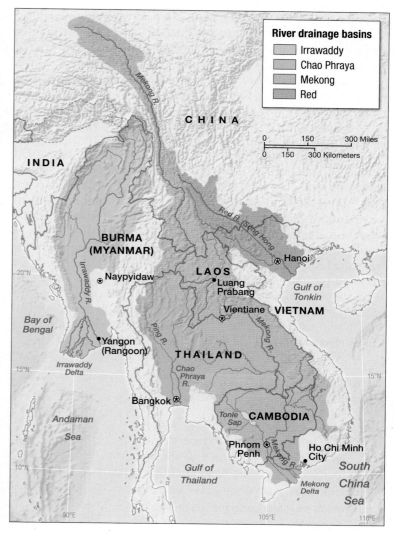

▲ **FIGURE 10.10 River Systems in Mainland Southeast Asia** The large river systems of this world region have played an important role in the historical, political and, economic development of this region. These river systems have supported large-scale urban societies for several thousand years.

◄ **FIGURE 10.11** **Tonle Sap Lake, Cambodia** The Tonle Sap Lake is a very important part of the region's ecosystem and economy. In the dry season, the lake shrinks to under 3,000 square kilometers, but it grows to over 16,000 square kilometers in the rainy season. The lake provides freshwater for farming and an abundance of fish to the people of Cambodia.

tation and settlement corridor for Laos and Cambodia, while also serving as a key trade network between these two countries and China, Burma, and Thailand (see "Emerging Regions: The Greater Mekong" on pp. *396–397*). The river also provides water for irrigation, hydroelectric power, and fishing. One of the most unusual physical features in the Mekong Basin is Tonle Sap in Cambodia. This large lake acts as a safety-valve overflow basin for flooding. During the dry season, water flows out of the lake along the Tonle Sap River into the Mekong; during flooding on the Mekong, water flows back up into the lake, quintupling the lake's area (**FIGURE 10.11**).

Soil fertility is very high in the river valleys and deltas that receive regular replenishment of river sediment and nutrients during annual floods. These complex river systems serve a vital role in the food production and transportation systems of the mainland Southeast Asian countries. River systems in the region are also contested sites of local and regional politics, as people struggle to control precious water-related resources. Dam projects along river systems have met with local resistance for extended periods of time. Regions farther from rivers in Southeast Asia experience soil limitations typical of tropical climates around the world; in these areas, heavy rain and warm temperatures wash nutrients through the soil, and organic material is broken down and recycled rapidly into forest vegetation.

Apply Your Knowledge

10.3 To explore recent volcanic activity in the region, visit http://www.independent.co.uk/ and search for articles on the February 2014 eruption of Mt. Kelud. What are some of the factors that make Southeast Asia susceptible to so many earthquakes and volcanic eruptions?

Southeast
Asia Volcanoes

https://goo.gl/rkCFKc

10.4 What are the risks related to high levels of tectonic activity in the region? Are there any rewards?

Ecology, Land, and Environmental Management

The vegetation and ecosystems of Southeast Asia reflect its wet tropical climate. A natural land cover of dense forests historically dominated the region. Today, the two major types are evergreen tropical forests in the wetter areas and tropical deciduous, or monsoon, forests where rainfall is more seasonal or less intense and trees lose their leaves in the dry season. Much of this historical land cover has been changed by land use patterns in the region today (**FIGURE 10.12**). In drier regions, forests are less diverse and vegetation cover changes to savanna and grasslands. **Mangrove forests**—groups of evergreen trees that form dense, tangled thickets in marshes and along muddy tidal shores—are found along coastlines. Mangrove forests, with their deep, strong root structures, serve a vital function, protecting areas from erosion while providing habitat for myriad animal species.

Southeast Asia's Biodiversity Climate change and plate tectonics together have produced a fascinating division in the ecology of the Southeast Asia region. During the last ice age the Sunda Shelf was exposed between the mainland and Indonesia until about 16,000 years ago. Many megafauna, including tigers, elephants, and orangutans, migrated across this **land bridge** to the islands of Indonesia. To the south, animals such as kangaroos and opossums moved north across the land bridge formed by the Australian Plate to New Guinea and other islands.

Between Bali and Lombok is a deep ocean trench that remained ocean, even during ice ages, and prevented these two very different types of species communities from mixing. This created an ecological division called **Wallace's Line**, named after naturalist Alfred Wallace (**FIGURE 10.13**), who traveled extensively in the region in the 19th century. Wallace's theory about biodiversity, evolution and natural selection, was theorized in tandem with Darwin. The quandary for Wallace and others was to explain why two islands in the same climatic zone had such dramatically different flora and fauna. From there, Wallace argued that the deep ocean separating the islands prevented the

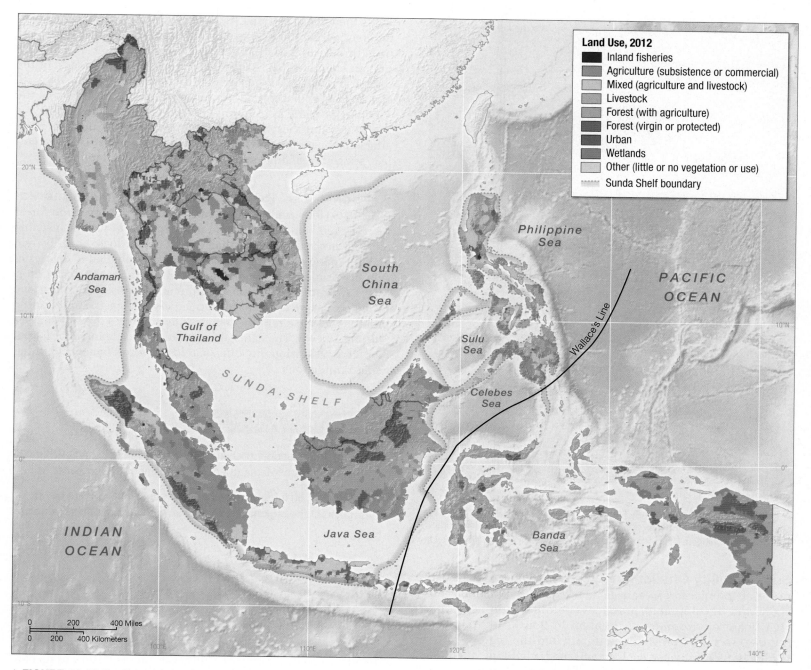

▲ **FIGURE 10.12 Southeast Asian Land Cover Land Use** The land cover of most of Southeast Asia is historically forest—evergreen in the wetter areas and tropical deciduous in drier regions. This map shows agriculture and other land uses in the region. Wallace's Line follows the deep oceans east of the Sunda Shelf. This line represents a distinct break between the islands to the east and west of the line, as each developed its own distinct biodiversity as a result of the separation produced by the deep ocean. **Describe the landscapes of Java and Borneo based on their land use. Which island do you think has the larger population?**

migration of certain species. The result was different evolutionary pathways for the animals that existed on either side of the line.

European Colonial Influences on Regional Ecosystems

In addition to the naturally occurring vegetation of this region, ecological systems have been introduced via European colonialism. Rubber production grew rapidly after 1876, when plants grown in Britain from seeds smuggled out of Brazil were introduced to Malaya. Rubber plantations grew as the explosion of automobiles in North America and Europe increased demand for rubber tires. Even though there was a ban on production by local people to protect the profits and monopoly of European planters, rubber quickly became popular with local Malayan farmers because of the high price it brought and the limited labor its production required.

▲ **FIGURE 10.13 Alfred Wallace** Although less known than Charles Darwin in popular culture, Wallace offered significant contributions to theories of natural selection and evolution. His work in Southeast Asia is enshrined in Wallace's Line.

Sustainability Challenges and Deforestation

In part a result of colonial forest management practices and modern-day economic development practices, deforestation is the most significant environmental problem in Southeast Asia today (**FIGURE 10.14a**). Forests once dominated the region, providing habitat for a diverse ecology and food, medicine, fuel, fiber, and construction materials to local peoples. But, widespread deforestation began in the late 1800s with the expansion of rice production and export of tropical hardwoods under European colonial control. Most of the deforestation during this period was in lowland regions, such as the Irrawaddy Delta. After World War II and decolonization, timber extraction expanded in the highlands, especially teak cutting in Thailand and Burma for export to the furniture industry. More recently, the growth of oil palm plantations, the pulp and paper industry, and cutting trees for plywood and veneers has placed even greater pressure on forests. The timber industry provides thousands of jobs and a significant amount of foreign capital in Southeast Asia, making it difficult for governments to eliminate.

The impact of deforestation includes loss of species habitat, flooding and soil erosion, and smoke and pollution from forest burning. Local communities have also been marginalized from lands they once occupied, as logging concessions are given over to larger corporate or state entities. Deforestation has destroyed or is threatening the habitat of many species and indigenous human communities. At the same time, the deep forests of the region are still relatively unexplored by biologists. New species are being discovered, including large animals such as the Vu Quang ox and giant muntjac deer, identified as recently as the 1990s. But these isolated regions are coming under further attack, as poachers move into once-isolated forest lands to collect prized megafauna, such as tigers (**FIGURE 10.14b**).

Responses to Deforestation and Environmental Change Governments have responded by setting aside forest reserves and banning logging in many regions as well as insisting on local processing rather than export of raw logs. Local people and international environmentalists have responded to deforestation by forming social movements and alliances to protect forests. Throughout the region, environmental conservationists have called on the government to stop large-scale logging and mining projects that have rapidly depleted the country's forest resources (**FIGURE 10.14c**). The damage from forest clearing includes loss of biodiversity, erosion, flooding, and loss of forest benefits to indigenous populations.

Apply Your Knowledge

10.5 List and describe three impacts that deforestation has on local indigenous human and animal populations in Southeast Asia.

10.6 Conduct an Internet search to find out what steps are being taken at the regional and global level to address these impacts.

History, Economy, and Territory

The historical geography of Southeast Asia is a story of its connections to other world regions, from East Asia (Chapter 8) to South Asia (Chapter 9) to Europe (Chapter 2) and the Americas (Chapters 6 and 7). Because of these connections, the politics and economies of this region are related to the trade in and out of India and China and the changing nature of global financial markets, for example. Southeast Asia's diversity emerges from processes of regionalization, as local practices have adapted to the flows of people, ideas, and biological species.

Historical Legacies and Landscapes

Maritime trade brought merchants from the Middle East, China, and India to Southeast Asia about 2,000 years ago, together with new religions, crops, and technologies. Southeast Asian **entrepôt** (commercial trading) towns, such as Oc Eo in modern-day Vietnam, became points of connection between this region and

(a) Forest cover

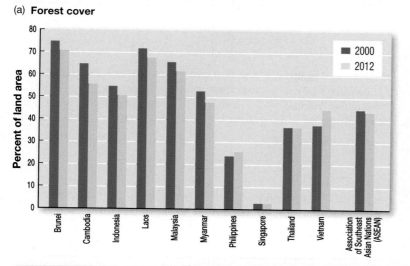

(b) Biodiversity

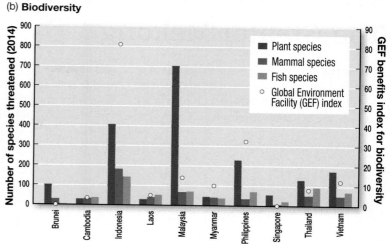

◀ **FIGURE 10.14 Deforestation in Southeast Asia** (a) the member of countries of the Association of Southeast Asian Nations (ASEAN) has collectively witnessed decreasing forest cover throughout the region, with the exceptions of the Philippines and Vietnam; (b) the plant, mammal, and fish biodiversity of this region varies, as does the potential for each country to contribute to the world's global environment facility (GEF) benefits (Indonesia and Philippines representing potentially high global biodiversity benefits if species are preserved); (c) monks in Cambodia seek to protect precious forests in that country's Aoral Wildlife Sanctuary. **If you were a botanist searching for medicinal plants, which countries in the region might you want most to visit? Explain.**

India and China. For some Southeast Asia societies, early interaction with India, the Hindu and Buddhist religions and the related forms of theocratic governments were attractive to local chiefs who saw advantage in the divine privilege they could be granted as god-kings (*deva raja*).

Building on the cultural and political practices of Hinduism and Buddhism, large kingdoms emerged in the region, including the mainland **Khmer** (or Cambodian) Empire and the island **Srivijaya** Empire of Sumatra. The 9th-century Khmer kingdom near Tonle Sap Lake later built the magnificent 12th-century temple compound of **Angkor Wat** (see "Visualizing Geography: Angkor Wat" on pp. *400–401*). Srivijaya ruled Sumatra and the southern Malay Peninsula from the 7th to the 12th centuries by controlling the region's maritime trade. This powerful state built numerous Buddhist monuments, such as Borobudor (**FIGURE 10.15**).

On the mainland, an increasingly powerful Thai kingdom conquered the Khmer Empire in the 14th century and maintained control of parts of the Malaysian Peninsula for four centuries. The Thai competed with different regional empires from the 14th to the 19th centuries for control of the people and

▲ **FIGURE 10.15 Ceremonial and Religious Space in Southeast Asia** Borobudor Temple, Indonesia, is one of the best examples of Buddhist architecture in the world. This site dates from the 8th and 9th centuries and represents the life of the Buddha as well as other important Buddhist ideals.

The Greater Mekong

The Mekong River is a truly international river, as it moves goods and people back and forth between China and all the countries of mainland Southeast Asia (**FIGURE 10.2.1**). In fact, one can view from its shores three countries at once—Burma, Thailand, and Laos. Known as the **Golden Triangle**, this meeting point has been critical to the historical trade of opium, although today it is more important as a tourism site.

On any given morning in and around the Golden Triangle area, a visitor can see Chinese

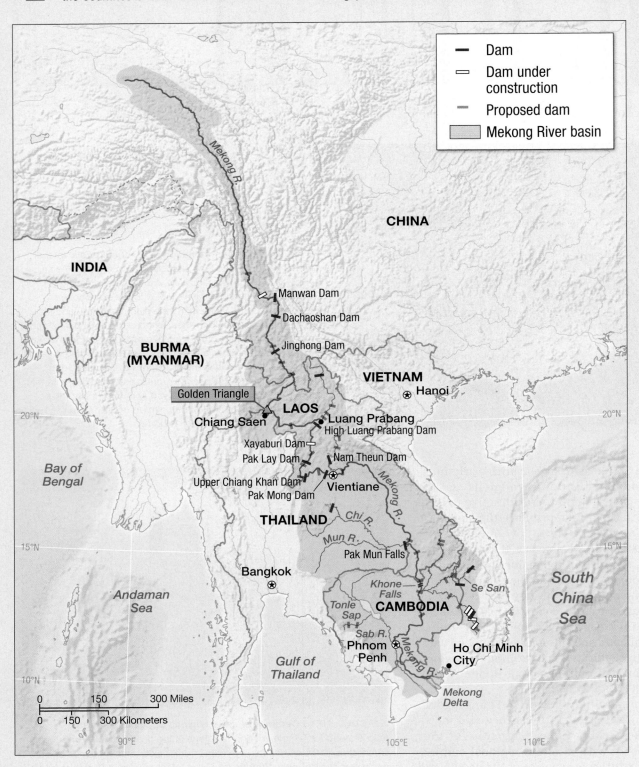

◄ **FIGURE 10.2.1 The Mekong River** The Mekong River links China in East Asia to mainland Southeast Asia, as it flows from a source on the Tibetan Plateau to the delta in Vietnam and Cambodia.

"On any given morning in and around the Golden Triangle area, a visitor can see Chinese boats parked at port towns along the Mekong in Thailand, Burma, or Laos."

▲ FIGURE 10.2.2 **Chinese Trading Boats in Chiang Saen, Thailand** Every day, boats travel up and down the Mekong River moving goods between China and Southeast Asia. The simple boat structure and docking area hides the reality that this place is an important site of global trade and connection.

boats parked at port towns along the Mekong in Thailand, Burma, or Laos. One town in Thailand, Chiang Saen, is a critical port for Chinese goods making their way into mainland Southeast Asia (**FIGURE 10.2.2**). As boats arrive, Thai customs officers greet them, examine the manifest, and sign off on the goods to be brought into the country. The fresh fruits and vegetables that come in from Yunnan Province in China are welcomed commodities in local markets.

Even as boats move goods back and forth across the Mekong and through the two world regions of East Asia and Southeast Asia, tension remains about the river's development. The **Mekong River Commission (MRC)** coordinates flood control, dam projects, and studies of the basin among its four member nations—Cambodia, Laos, Thailand, and Vietnam. The MRC is also a vehicle for resolving disputes between member states and their two "dialogue partners," China and Burma.

While dams can be found throughout the basin, they have not been erected without controversy. Often local interests are in conflict with global organizations that seek to profit from the dam ventures. The most recent tension revolves around the Xayaburi Dam, which is being built in Laos. The large dam is expected to put increasing pressure on downstream communities and threatens species, such as the famous freshwater Mekong dolphins (**FIGURE 10.2.3**). China also controls the headwaters of this river system and has engaged in both dam projects and sand removal projects, the latter of which feeds the growing and quite lucrative sand economy necessary for building materials, such as glass and concrete.

1. What defines the Golden Triangle, historically and today?

2. What factors suggest that the Greater Mekong River is becoming more integrated into an emerging region over time?

◄ FIGURE 10.2.3 **Freshwater Dolphins of the Mekong River** These unique dolphins can be found throughout this world region. Dams and intensification of fishing in rivers have impacted their habitats. The dolphins pictured here are not in their native habitat, but are instead shown in captivity.

resources of the mainland. These empires privileged **Theravada Buddhism**—a form of Buddhism that focuses on the monastic order and adherence to rituals written in Pali, a language that originated in South Asia. As Theravada Buddhism spread, these empires were consolidated around important city-states, such as Bangkok, Thailand, and Pagan, Burma.

By the 11th and 12th centuries, new religious systems, including Islam, entered Southeast Asia through global trade from South Asia. By the 15th century, the political geography of insular Southeast Asia was restructured around Islamic-based **sultanates**—Muslim states ruled by supreme leaders or sultans—such as Malacca and Brunei. Malacca's strategic location gave it commercial power, and after the rulers enthusiastically adopted Islam, Malacca disseminated Islamic beliefs and institutions throughout different parts of Southeast Asia.

European Colonialism in Southeast Asia Southeast Asia has had two overlapping but somewhat distinct colonial periods. The mercantile (or trade) period spanned from about 1500 to 1800 C.E. The industrial (or export-oriented) period lasted from about 1800 to 1945 C.E. (**FIGURE 10.16**). The external powers were from Europe as well as the United States.

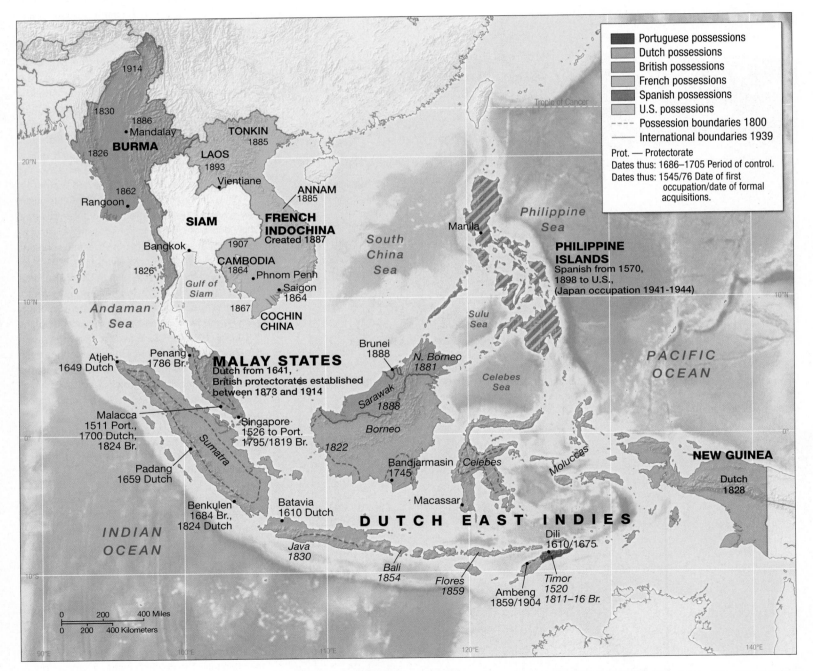

▲ **FIGURE 10.16 European Expansion into Southeast Asia** This map shows the areas controlled by major colonial powers in Southeast Asia. The dates on the map represent the time at which a particular city or region within a country came under direct colonial control. Based on the dates of colonization, which two colonizing nations arrived in Southeast Asia first? By 1900, which colonizers controlled most of the region?

- **Portugal and Spain** Portugal dominated early mercantile trade, establishing headquarters in Goa, India, and then at the strategic ports of Malacca in 1511 and Timor-Leste in 1515. To obtain commodities such as cloves, nutmeg, and pepper from across the Indonesian archipelago, they relied on indigenous producers and merchants. The Spanish sailed across the Pacific, arriving in the Philippines in the 1500s as well. The Spanish spread Catholicism and expanded their global empire by gaining access to trade in spices and other commodities in Southeast and East Asia. The city of Manila, founded in 1571, became a center of trade with Latin America.

- **The Netherlands** The third colonial power to move into Southeast Asia in the mercantile period was the Netherlands. Their initial focus was on the Molucca Islands of Indonesia—the famous Spice Islands. The formation of the Dutch East India Company in 1602, with its headquarters in Java, consolidated commercial interests in Indonesia by restricting the production of valuable spices such as pepper and by destroying communities that ignored restrictions or participated in smuggling.

- **Britain** The British colonial effort in Southeast Asia was an extension of their activities in India. They focused on Burma, Malaya, and Borneo and on the control of strategic ports. Like the Dutch, the British were led by a trading company, the British East India Company, formed in 1600 and committed to British domination of South Asia (see Chapter 9). The British made a move into Southeast Asia by claiming the strategic ports of Penang (1786), Singapore (1819), and Malacca (1824) and fighting wars with the Burmese while also claiming the Malay Peninsula from the Dutch in 1873 (**FIGURE 10.17**). The British reoriented colonial economies to exports of tin, rubber, and tropical hardwoods and controlled trade routes between India and China.

- **France** The French first entered the region in the 18th century but consolidated their holdings in the 19th century in response to rivalry with Britain over commercial links to China. The French governed Cambodia, Laos, and the districts of Tonkin, Annam, and Cochin China in Vietnam as the Union of French Indochina. They created a new writing system for Vietnamese, which is still used today, and instituted French language and cultural practices for local elites.

- **United States of America** The United States was a colonial power in Southeast Asia for less than 50 years, acquiring the Philippines in 1898 after victory in the Spanish–American War. The United States lost control of the islands to the Japanese from 1941 to 1944 and then granted independence to the Philippines in 1946, it continued to treat the islands as a strategic location, maintaining several military installations, such as Clark Air Force Base and Subic Bay Naval Base. Both bases

▲ **FIGURE 10.17 Raffles Hotel in Singapore** The Raffles Hotel is named after Sir Thomas Stamford Raffles, who secured control of Singapore for the British in 1819. This building stands as both a symbol of British colonialism and European architectural influence in the region.

were closed after the eruption of the volcano Mount Pinatubo in 1991.

- **Independent Holdout: Thailand** Unlike other parts of Southeast Asia, Thailand was able to maintain its political independence throughout the colonial period. It provided a buffer between British and French interests even as they lost territory to the British in Malaya and Burma and to the French in Cambodia and Laos. The ability of the Thai to adapt themselves to the changing geopolitical dynamics of the region also benefited their autonomy. Thailand created linkages to colonial trading systems and adopted European policies in the region to accommodate their trade.

Colonial Legacies of the 19th Century During the 19th century, Southeast Asia was fully integrated into a European-centered global trading system and land use practices in the region were dramatically altered. The Dutch introduced the **Culture System** into Java in 1830, which required farmers to devote one-fifth of their land to export-crop production, especially coffee and sugar, with the profits going to the Dutch. The British and French promoted rice production to feed laborers and growing populations. Rice production grew dramatically in the Irrawaddy Delta of Burma and the Mekong Delta of Indochina. Colonial authorities invested heavily in urban areas and, as a result, Singapore, Jakarta, Bangkok, and Manila developed as global and regional gateways. Export agricultural regions were oriented to these cities, while remote rural peripheries remained in subsistence livelihoods with little investment in education or other services. Most of Southeast Asia remained under colonial control during the first part of the 20th century.

Angkor Wat

Throughout the history of Southeast Asia, religious, social, and political life has been intertwined and religious practice has often reinforced social and political authority. These systems of authority were symbolically represented in the complex landscapes of religious architecture throughout the region.

1 The Khmer Empire

The greater Khmer Empire centered in modern-day Siem Reap, Cambodia controlled many parts of present-day Cambodia, Burma, Thailand, Laos, and southern Vietnam (Figure 10.3.1). One can find the influence of Cambodian architecture in many of these places, where Khmer-style stupas—mound-shaped monuments that hold sacred Buddhist relics—can still be found today.

1,200,000 km² (463,323 mi²)

were controlled by the Khmer Empire at its peak in 1150 with an estimated population of

4 million people.

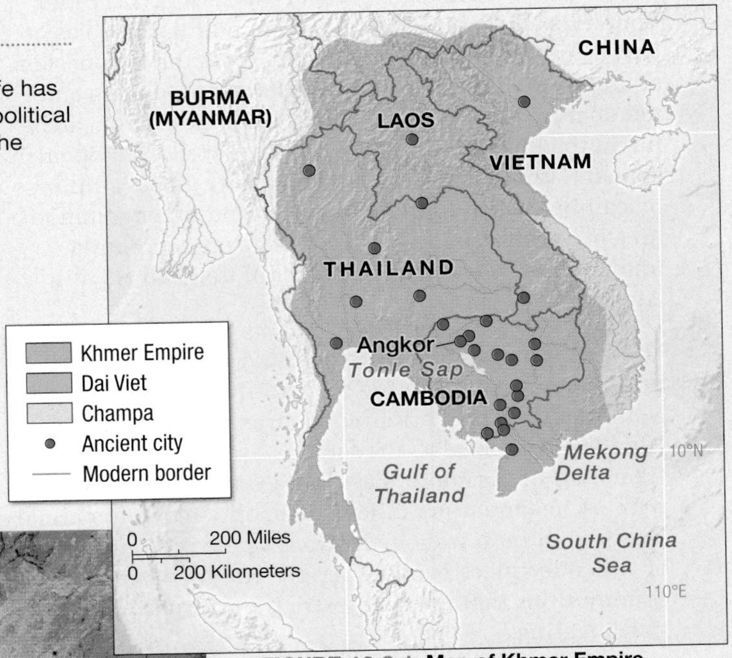

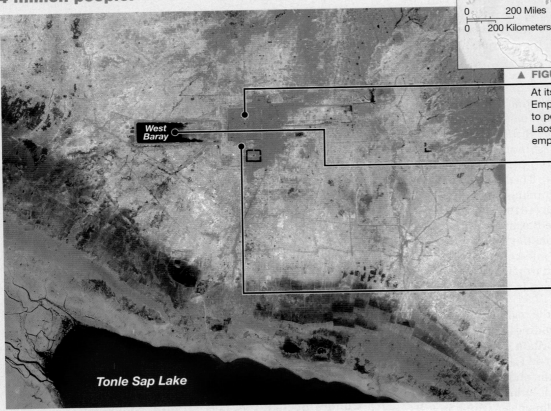

▲ FIGURE 10.3.1 **Map of Khmer Empire.**

At its height in the 12th century C.E., the Khmer Empire had either direct control or strong linkages to people in what is modern-day Burma, Thailand, Laos, and Vietnam, making it one of the largest empires in the region's history.

▲ FIGURE 10.3.2 **Aerial view of the Angkor Wat temple complex.**

At its height, the Angkor complex in modern-day Cambodia was a highly productive urban space. Agricultural production and other economic activity tied to Tonle Sap in the rural regions surrounding the city supported Angkor's growth.

2 Angkor Wat Temple Complex

At the center of the Khmer Empire from the 9th to the 15th centuries C.E. was the larger Angkor Wat complex (Figure 10.3.2), which included hundreds of sites and buildings dedicated to the Hindu religious systems and the notion that Khmer kings were "god-kings" or deva raja, a concept borrowed from South Asian religious practice (Figure 10.3.3). Like many complexes in other parts of Southeast Asia, the Angkor Wat complex was much more than just temples. It was a vibrant and robust urban society that influenced peoples across the region. It may have been one of the largest cities in the world at its zenith, outstripping smaller cities that could be found in Europe or the Americas during the same period.

"It may have been one of the largest cities in the world at its zenith, outstripping smaller cities that could be found in Europe or the Americas during the same period."

▲ FIGURE 10.3.3 **Angkor Thom, Cambodia.**

By the 10th century, there were large kingdoms influenced by Indian culture and religion throughout mainland Southeast Asia. This is exemplified by the 12th-century temple of Angkor Wat and sacred city of Angkor Thom, which was constructed by the Khmer Empire just north of Tonle Sap Lake in Cambodia. The temple design features symbols from the Hindu cosmos, including mountains, artificial lakes, and sculptures.

③ Water Management and the Sustainability of the Angkor-Based Kingdom

A large reservoir and canal system collected water to be used in agriculture during the dry seasons. Khmer leaders were able to extract loyalty and labor from its citizens in exchange for consistent harvests and access to needed resources for the largely agricultural economy. The demise of the kingdom is not linked to one distinct process, but scientists have surmised that an extended period of drought affected the ability of kings to manage the needs of the people that lived in and around the city.

④ Preserving Angkor Wat and the Heritage Tourism Economy

Today, Angkor Wat is a United Nations Educational, Scientific, and Cultural Organization (UNESCO) heritage site. Massive investment has been made in the complex's revitalization. Cambodia, which was subject to a long period of violent civil war in the 1970s and 1980s has slowly built up its tourism economy based, in part, on the heritage tourism of the Angkor complex (Figure 10.3.4).

20%
increase in tourism each year.
More than 2 million
visitors in 2013.

1. **At its height, what was the spatial extent of the Angkor Empire?**

2. **What is the relationship between the expanse of this empire and the management of both water systems and cultural and religious practices?**

▼ FIGURE 10.3.4 **Tourism at Phnom Makheng.**
Tourists gather to watch the sunset on Phnom Bakheng temple in Angkor Wat.

European Decolonization European colonial power in Southeast Asia was diminished by the Japanese invasion in World War II (**FIGURE 10.18**). The Japanese granted independence to the people of Burma in 1943 and promised independence to Indonesia on their retreat, fueling local desires for autonomy from their former European colonizers. The Philippines was granted independence from the United States directly after the war, in 1946. The British formally granted independence to Burma in 1948 and created the Federation of Malaysia in 1963 following a 12-year counterinsurgency. Singapore left the federation to become an independent country in 1965. Brunei converted from a British protectorate to an independent nation in 1983. Indonesia was granted independence by the Dutch in 1949 only after a violent struggle following a declaration of independence in 1945. Western New Guinea, formerly Dutch New Guinea, became part of Indonesia as Irian Jaya in 1963.

Apply Your Knowledge

10.7 Compare and contrast the different colonial experiences of two nations in Southeast Asia.

10.8 Using your understanding of historical geography, list the legacies the colonial experience has left in these two places.

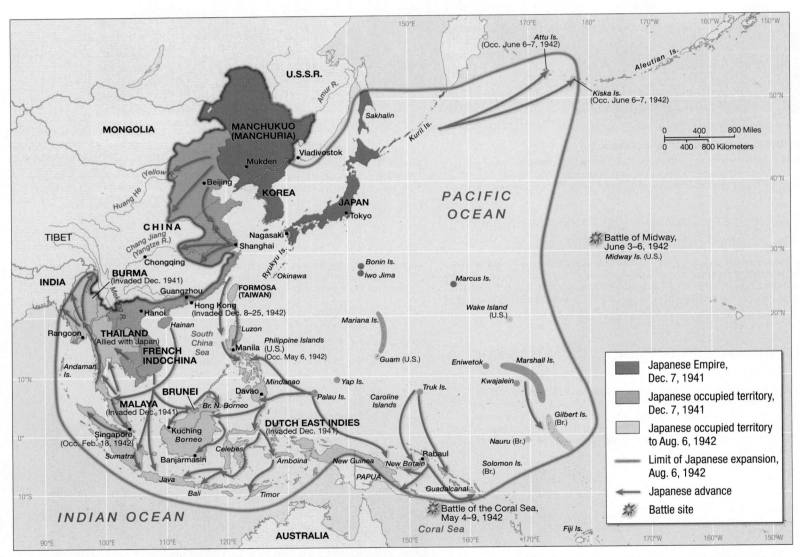

▲ **FIGURE 10.18 Japan's Occupation of Southeast Asia** The Japanese occupation of Southeast Asia during World War II had severe effects. Japan sought access to the natural resources of the region, claiming legitimacy from a shared set of cultural values with the regions that they captured under the slogan "Asia for the Asiatics." But the Japanese occupation cut off trade revenues, used forced labor, and diverted resources to Japan at the expense of local economies and food security. The destruction of bridges and roads during the conflict left Southeast Asia's infrastructure in ruins. In Vietnam, the Japanese requisitioned rice and forced farmers to grow jute fiber, resulting in a famine that killed more than 2 million people in 1944 and 1945. The Japanese were particularly harsh on the Chinese population of Southeast Asia.

Economy, Accumulation, and the Production of Inequality

The diversity of economic and political pathways for the countries of this region in the postcolonial period has led to a wide range of economic difference. In some cases, such as in Laos, a noncapitalist economy intentionally interfered with global market forces by focusing attention on social equality rather than economic development. Other countries in Southeast Asia, such as Malaysia, Singapore, Thailand, Indonesia, and Philippines, reoriented their capitalist economies in efforts to take advantage of the postwar global economy.

Postcolonial Agricultural Development Although agriculture has decreased across most of the region in the last 30 years, it still employs more than 40% of the economically active population in Indonesia, Thailand, and Vietnam and almost 70% in Cambodia and Laos. Rice production continues to dominate the land area of Southeast Asia, but the export of spices and plantation crops, such as rubber, tea, coffee, and sugar, are important as well (**FIGURE 10.19**). Because of the demand, the area planted by multinational corporations such as Del Monte and Dole with oil palm and pineapples has increased over the past 20 years, especially in Malaysia, Philippines, and Thailand.

Green Revolution Technologies A major factor that spurred on postcolonial agricultural development was the **Green Revolution**. This revolution involved the development of higher yielding seeds, especially wheat, rice, and corn, which when introduced in combination with irrigation, fertilizers, pesticides, and farm machinery, were able to increase crop production.

Green Revolution technologies contributed to dramatic increases in rice and other cereal production in Indonesia, the Philippines, and Thailand.

As a result of the Green Revolution some communal land and rights were abandoned as agriculture shifted its orientation to exports and wage labor. The introduction of mechanical rice harvesting increased rural unemployment and landlessness in many places. The rate of success was also higher among those with access to irrigation. When the new seed varieties were planted uniformly across large areas, crops became vulnerable to diseases and pests, mandating the use of toxic pesticides. Rather than encourage pesticide use, some governments, such as the one in Indonesia, sponsored the use of integrated pest management by small-scale farmers, which uses less chemically intense techniques.

Postcolonial Manufacturing and Economic Development From the 1950s through the 1970s, several Southeast Asian governments—Malaysia, Indonesia, Philippines, Thailand—developed their own local manufacturing industries in an attempt to diversify their economies away from agriculture and limit their reliance on imported manufactured goods. In Malaysia, **tariffs**—taxes placed on goods from outside the region—on imports such as clothing and plastics were increased to protect locally produced goods from competition. Local manufacturing was dominated by non-Malay (especially Chinese) investment and produced low-value goods that quickly saturated domestic markets or faced, in the case of steel, a glutted global market. These policies were largely abandoned in the region by the 1980s.

In the 1980s, Malaysia, Singapore, and Thailand shifted to export-oriented production—where countries rely heavily on producing goods for the global market—and capitalized on their global economic advantages, including a relatively well-educated workforce that has been willing to work for lower wages. Countries that were at one time fairly isolated from the global economy—such as Vietnam—have now also taken advantage of similar competitive advantages (**FIGURE 10.20**). Export-oriented development in the

▲ **FIGURE 10.19 Coffee Plantation in Indonesia** Coffee was first planted in this area in the early 1700s by the Dutch, who began exporting coffee from Java to Europe in 1711. Today, coffee remains an important export crop for Indonesia.

▼ **FIGURE 10.20 Silk Textile Factory in Vietnam** Silk production has a long history in Vietnam. Today, silk production has been industrialized and depends on a relatively inexpensive female labor force, giving the product a comparative advantage in the world silk market.

▲ **FIGURE 10.21 Singapore's Financial District** This view of the city of Singapore illustrates its modern high-rise core.

region has also benefited from very strong state involvement and high levels of **foreign direct investment (FDI)**. Manufacturing sectors that developed or relocated to Southeast Asia include automobile assembly, chemicals, and electronics. Japan has been a major source of foreign investment; many Japanese-owned firms have relocated seeking cheaper labor and land. Southeast Asian government investment in training skilled labor and generous tax incentives have attracted multinationals such as Intel, Sony, Philips, Motorola, and Hitachi to Malaysia. Global integration does make Southeast Asia vulnerable to changes in the global economy, as economic downturns in the investing countries—for example, Japan, Germany, or the United States—limit investment in Southeast Asia.

Singapore and the Asian Economic "Miracle"

Singapore is known as one of the original "Asian Tigers," which includes Hong Kong, South Korea, and Taiwan in East Asia (see Chapter 8). These countries have built large global economies. Singapore is one of the 25 wealthiest countries in the world in terms of per capita gross domestic product (GDP) and has the highest standard of living in Asia (**FIGURE 10.21**). Its infrastructure—especially the excellent port and international airport—has made it the import and shipping center for the region and the busiest port in the world. The government of Singapore made the astute decision to move to activities that required a skilled workforce, such as precision engineering, aerospace, medical instruments, specialized chemicals, and most notably, the emerging computer industry. To support this policy, the government invested heavily in education, especially engineering and computing, in high-technology industrial parks, and in state-owned pilot companies, such as Singapore Airlines. Singapore is now a **world city**,

a center for information technology and aerospace, and regional leader in **SIJORI**, which includes Singapore (SI), Johor Baharu (JO) in Malaysia, and the Riau Islands (RI) of Indonesia (**FIGURE 10.22**).

The Little Tigers In addition to Singapore, a number of other Southeast Asian economies—Thailand, Malaysia, and Indonesia—joined the "Asian tigers" in terms of economic growth. The "**little tigers**" were termed **newly industrializing economies (NIEs)**, rather than "developing" or "underdeveloped" countries. High rates of savings, balanced budgets, and low rates of inflation were indicators of economic development. Foreign investment flowed into Southeast Asia to take advantage of domestic consumer goods markets and valuable natural resources. In some cases, new industries developed locally to meet national demand, as is the case with the Proton, Malaysia's "national car."

Southeast Asia in the Global Economy The economies of Southeast Asia remain intimately linked to global economy through the production and export of inexpensive goods. This integration has facilitated dramatic growth over the last several years, despite the widespread global downtowns that took place in 2008–2009 and again in 2011–2012. Indeed, the countries of this region are expected to maintain economic growth levels on par with their larger East Asian and South Asian neighbors, China and India, over the period of 2015–2019. The growth will be more dramatic in countries such as Cambodia, Laos, and Burma, which have been more marginal to the global economy

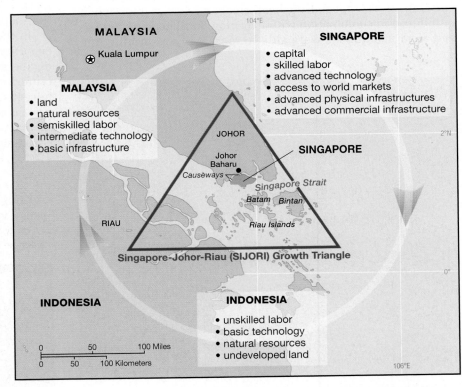

▲ **FIGURE 10.22 Malay Peninsula and Singapore** The Malay Peninsula and Singapore are at the center of economic development in Southeast Asia, especially the SIJORI growth triangle. This figure illustrates the economic complementaries of Singapore; Riau, Indonesia; and Johor, Malaysia. **From an economic standpoint, what does Singapore lack that Indonesia and Malaysia have?**

Real GDP Growth of Southeast Asia

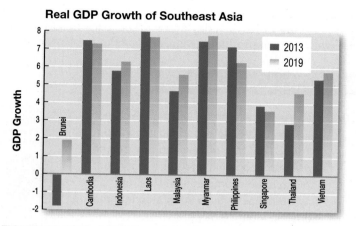

▲ **FIGURE 10.23 Real GDP Growth Southeast Asia** This world region continues to show projections of relatively strong GDP growth through 2019.

and poor (**FIGURE 10.24**). The distribution of income tends to be most unequal in the strongest economies in Southeast Asia: The wealthiest 20% controls about half the wealth in Malaysia, the Philippines, Singapore, and Thailand. Wealth concentration in Southeast Asia has sometimes also been associated with **crony capitalism**, in which leaders allow friends and family to control the economy, and **kleptocracy**, in which leaders divert national resources for their personal gain (**FIGURE 10.25**).

There is also a division of poverty rates between urban and rural Southeast Asia. In Malaysia and Thailand, about a quarter of rural people live in poverty, compared to fewer than 10% of urban residents. In terms of economic and social indicators, women are more equal to men in Southeast Asia than in many other regions. Female literacy averages more than 85%, only 10% less than the male rate. Despite these relatively positive measures, many workers in Southeast Asia are exploited in

in the past, while countries, such as Brunei, which rely heavily on oil production, will not see much growth (**FIGURE 10.23**). Many of the economies of this region also benefit economically from a mobile workforce, which sends **remittances**—monies sent back to families in home countries from working migrants—to families in countries, such as the Philippines. Domestic demand for consumer goods from a growing middle class in the region will also drive economic growth through 2019.

As mentioned earlier, Southeast Asia has long been a target of FDI. Recently, FDI into Southeast Asia collectively has outstripped FDI in China. Increases in wages in China, which has depressed FDI in the country, and the growth of the consumer market in Southeast Asia have both made Southeast Asia a target of investment from multinational corporations. Despite the dramatic increases in FDI into the region, not all countries have shared equally in that investment. In 2014, Singapore tends to receive the greatest proportion of that investment, followed by Thailand, Malaysia, Indonesia, Philippines, and Vietnam. Cambodia, Laos, and Burma receive by far the least amount of FDI.

Social and Economic Inequality

Many of the countries in Southeast Asia have seen dramatic decreases in poverty rates. This aligns with the Millennium Development Goals (MDGs) set by the United Nations (see Chapter 1). Overall, the rate of poverty in Southeast Asia has dropped from 25% in 2002 to under 8% in recent years. Despite these achievements there is still dramatic gaps between rich

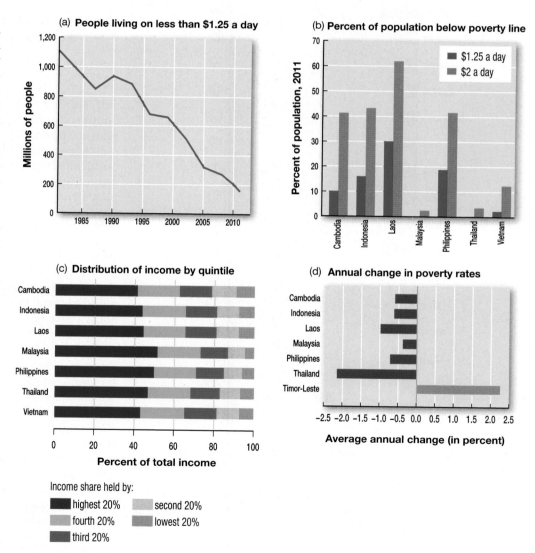

▲ **FIGURE 10.24 (a) Economic Development, (b) Poverty, (c) Income Distribution, and (d) Changing Poverty Rates in Southeast Asia** As these graphs demonstrate, the percentage of the population in poverty in many Southeast Asian countries has declines over the last two decades. Even so, the percent of the population living under the poverty line in some countries remains quite high and in many countries the wealthiest 20% of the population controls over 40% of the income. Based on the graph "Distribution of income by quintile," which country has the greatest inequality of income distribution? Explain.

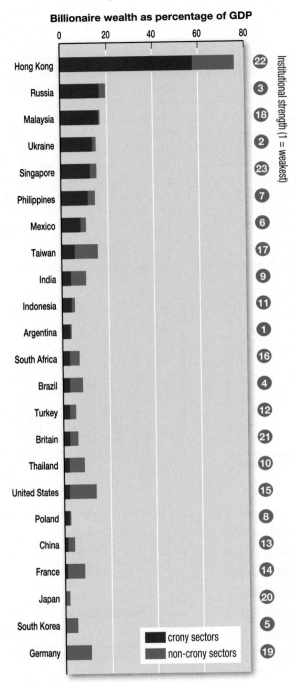

The Crony-Capitalism Index

Billionaire wealth as percentage of GDP

▲ **FIGURE 10.25 The Crony Capitalism Index, 2014** Crony capitalism represents the consolidation of wealth in the hands of a few elites who have direct access to their wealth through their connections to a country's political leadership. While Southeast Asia certainly does not have the worst rates of crony capitalism in the world, several countries of the region rank highly as countries with wealth concentrated in the hands of the very few.

terms of wages and labor conditions when compared to workers producing similar goods in North America and Europe. In the absence of unions, many work 12-hour days and seven days a week without benefits, making products such as clothing and electronics for global markets. International companies such as Nike have come under criticism for sweatshop labor practices in

countries such as Vietnam and Cambodia, where wages are just slight more than U.S. $2 a day and workers earn less than 5% of what workers earn producing similar goods in the United States.

Apply Your Knowledge

10.9 Visit the website of the Asian Development Bank to investigate economic integration in the region: http://www.adb.org, search for "Asian Economic Integration Monitor." Provide three examples that illustrate how Southeast Asia is integrated into the global economy today.

Southeast Asia Economic Integration

https://goo.gl/bfgceg

10.10 Briefly discuss who in the region has benefited the most from global integration and who has benefited the least.

10.11 What are some of the explanations for why some places are successful while others are less so?

Territory and Politics

Southeast Asia is a complex geopolitical region because of colonialism, territorial expansionism by some countries after independence, and Cold War conflicts. There have been a number of political advances in recent years, but to understand the overall geopolitics of this region it is important to appreciate the complex history of nation-building and conflict as well.

The Indochina Wars Led by **Ho Chi Minh**, the communists established a Vietnamese government in northern Vietnam with their capital in Hanoi after the French reoccupied Southern Vietnam at the end of World War II. The Vietnamese government in the north always laid claim to a unified Vietnam. When the French withdrew from the region after a devastating loss to the Vietnamese at Dien Bien Phu in 1954, Laos and Cambodia became permanently independent and North Vietnam and South Vietnam became temporarily divided. Although the split was codified in the Geneva Accords, North Vietnam quickly moved toward independence and thousands of refugees, especially Catholics, fled to South Vietnam, which established an independent capital in Saigon.

While this was taking place, communist rebels (called the **Vietcong**) fought for control of several regions of South Vietnam. Subscribing to the so-called **domino theory**, which held that the communist takeover of South Vietnam would lead to the spread of communism, the United States sent military advisors to South Vietnam in 1955. This was followed by U.S. bombing of North Vietnam in 1964 and escalation to a full-scale land war—the **Vietnam War**—with more than half a million U.S. troops, in 1967 (**FIGURE 10.26**).

The Vietnam War was probably the most serious global manifestation of the Cold War. More than a million Vietnamese people and 58,000 Americans died. U.S. forces sprayed millions of acres of Vietnam, Cambodia, and Laos with **Agent Orange** (a defoliant that removes foliage from plants, thus making it more difficult for ground forces to find cover). Agent Orange poisoned ecosystems and caused irreparable damage to human health.

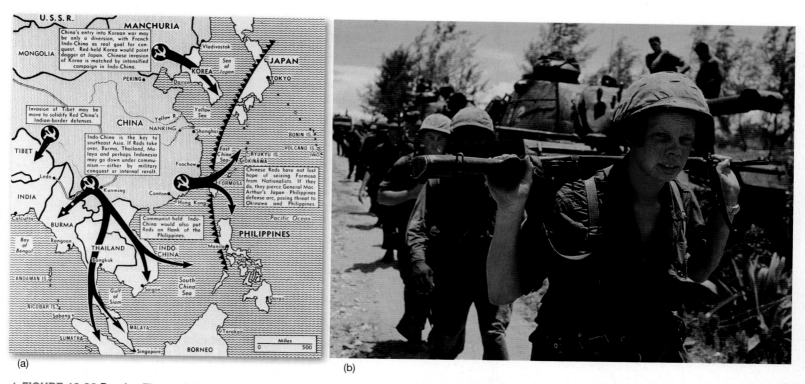

(a)   (b)

▲ **FIGURE 10.26 Domino Theory** (a) Fear of the expansion of communism in East and Southeast Asia promoted the use of map images to depict the communist threat. This map, which was produced in the 1950s, demonstrates the threat that a communist China and by association a communist Vietnam posed to U.S. interests in East Asia. Maps such as this were used later to justify the extensive, costly, and deadly Vietnam War. (b) American soldiers of 7th Marine Regiment as they march along a dirt road, Cape Batangan, Vietnam, 1965.

The war ended with the reunification of Vietnam led by the government of the North. But, 2 million Vietnamese "boat people" left South Vietnam in small, fragile boats, forming a global Vietnamese diaspora in Southeast Asia, Europe, and the Americas. The new Vietnamese government confiscated farms and factories to create state-owned enterprises, resettled ethnic minorities into intensive agricultural zones, and moved 1 million people into new economic development regions.

U.S.-led economic sanctions from 1975 to 1993 limited exports and restricted some imports such as medicines. The pockmarked landscape created by the U.S. bombing campaign promoted the spread of disease such as malaria, while these same craters provided new economic opportunities for some who have converted them to fisheries (**FIGURE 10.27**). Ethnic minorities in Laos also suffered during the Indochina Wars. Hmong (a minority ethnic group living in Laos, Thailand, Burma, and southern China) became refugees and fled Laos after fighting for the United States during the Central Intelligence Agency–led covert war in that country. Today, Hmong populations can be found in places as diverse as Merced, California; Paris, France; and Guyana, South America.

Pol Pot and the Khmer Rouge During this same period, the **Khmer Rouge** overcame the U.S.-backed military government in Cambodia. In 1975, they instituted a cruel regime under the leadership of **Pol Pot**, who suspended formal education, emptied cities, attacked the rich and educated, and isolated his country from the world. The new government renamed the country Kampuchea and engaged in mass murder of people it believed were not loyal. The places where mass murders of Cambodians

took place were later described as **killing fields**, a term that powerfully describes the farmlands that were turned into mass grave sites for the millions of people murdered by the regime (**FIGURE 10.28**). A quarter of Cambodia's population—including most of the intellectuals and professionals—died between 1975 and 1979, when Vietnam invaded and installed a new government. Many Cambodians who escaped migrated to U.S. cities such as Long Beach, California, and Boston, Massachusetts.

▲ **FIGURE 10.27 The Legacy of U.S. Bombing** The massive "carpet bombing" of many parts of Southeast Asia has left the landscape marked with deep holes that remain to this day.

▲ **FIGURE 10.28 The Cambodian Killing Fields Memorial** People bear witness to the hundreds of skulls recovered from the killing fields outside of Cambodia's capital, Phnom Penh.

Postcolonial Conflict and Ethnic Tension In addition to the legacies of the Cold War, some of the longest-lasting conflicts in the region involve ethnic minorities seeking political recognition or independence (**FIGURE 10.29**).

- **Burma** Ever since gaining its own independence in 1948 from Great Britain, the Burmese government has faced protest against an often-repressive government that has been distinguished mostly by military rule. Popular protests in 1988 resulted in martial law and the establishment of the State Law and Order Restoration

Council (SLORC). Elections were held in 1990 in which an opposition party, the **National League for Democracy (NLD)**, won 60% of the vote, compared to 21% for the existing government party. SLORC canceled the election results and remained in power. One of the leaders of the NLD, **Aung San Suu Kyi**, won the Nobel Peace Prize in 1991 for her work in bringing democratic reform to Burma, although she was placed under house arrest by the government in 1989.

After her release in 2011 she successfully gained one of the open parliamentary seats in 2012. While such a victory is hard won, the NLD's seats placed the party in a minority by law. In 2015, a new election was held and the NLD won power. But, the complex constitution, which bars people with foreign-born children to be president, disallows Aung San Suu Kyi from rising to the top post. This means that the new president will come from the NLD and will have to work closely with members of the military (**FIGURE 10.30**).

- **Indonesia** In the postcolonial period, Indonesia promoted the concept of **pancasila**—unity in diversity through belief in one God, nationalism, humanitarianism, democracy, and social justice—and an official Indonesian language. Fractures soon developed around religious differences and the Indonesian government's repression of its critics. Facing political opposition and a brief attempt by some to protest against his leadership, Sukarno cracked down on opposition in 1965. At least 500,000 people were massacred in the aftermath. Today, Indonesia operates under a representative democratic system, although the country has witnessed a number of conflicts within its borders. For example, there were tensions in areas, such as Aceh, Sumatra, where conflict was resolved when the region was given greater political and economic autonomy from the central government in 2005.

- **Thailand** Although Thailand is often described as the "land of smiles," the country has witnessed over 20 military coups since the absolute monarchy was overthrown in 1934. The most recent coup took place in 2014 and resulted in a military-appointed government. The 2015 draft constitution leaves little room for full democracy—only one-third of government representatives can be directly elected. These national concerns are coupled with subregional conflicts, such as the ones of Southern Thailand. This conflict is reflective of the tensions that exist in the south between some separatist Islamic minority groups and the majority Buddhist population.

- **Philippines** The Philippines has struggled since independence to develop a unified national identity. On the majority Muslim island of Mindanao, the Moro National Liberal Front (MILF) has struggled violently against the centralized Catholic-dominated Philippine state. The most recent conflict includes the death of 44 police commandos during a visit of Pope Francis in 2015. There is some hope for peace. Under a current

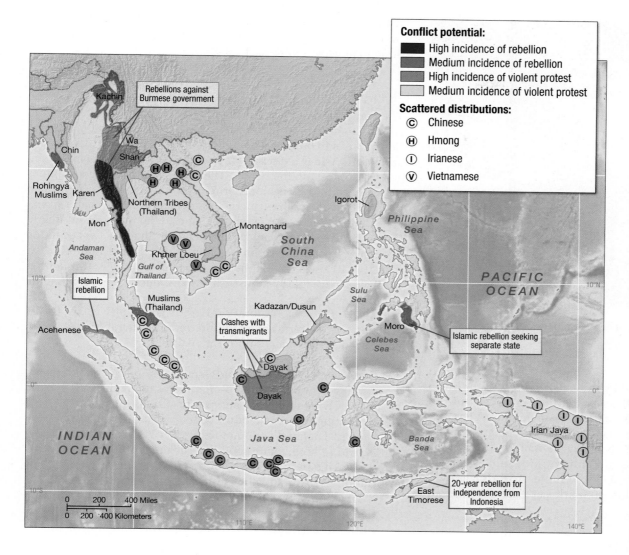

Conflict potential:
- High incidence of rebellion
- Medium incidence of rebellion
- High incidence of violent protest
- Medium incidence of violent protest

Scattered distributions:
- Ⓒ Chinese
- Ⓗ Hmong
- Ⓘ Irianese
- Ⓥ Vietnamese

◀ **FIGURE 10.29 Conflict Zones of Southeast Asia** This map shows some of the major zones of conflict in Southeast Asia, including separatist movements in the outer islands of Indonesia and the Philippines. **Which part of the region has the highest concentration of ethnic groups involved in rebellions and protests? Explain.**

agreement former rebels have registered to vote in upcoming regional and national elections.

- **Timor-Leste** Timor-Leste is a mostly Christian, former Portuguese colony that occupies the eastern portion of the island of Timor. Indonesia occupied the country after Portugal gave up control in 1976, dashing initial hopes for independence. Twenty years of resistance and more than 200,000 deaths resulted in a referendum and Timor-Leste was granted status as a newly independent country in 2002. Timor-Leste has grown from a war torn country to one of the fastest-growing economies in the region with a multiparty political system, a thriving oil industry, and a health-care complex run by over 1,000 physicians who were trained in Cuba.

Regional Cooperation Despite political conflicts, Southeast Asia provides a model for economic and, to a lesser extent, political cooperation in **ASEAN**—the Association of Southeast Asian Nations (**FIGURE 10.31**). The Asian Free Trade Association (AFTA) was created as part of ASEAN in 1993 to reduce national

▲ **FIGURE 10.30 High Level Meeting, Burma, 2014** In 2014, then president, Thein Sein, called a meeting between the opposition, government officials, and military leaders to discuss the future of democracy in the country. This was the first such meeting of its kind in a country that has seen much violence.

▲ **FIGURE 10.31 Flag of the Association of Southeast Asian Nations** This flag symbolizes the unification of the region through its supranational organization.

tariffs within the region. A policy of engagement with military and socialist governments led to invitations to the rest of the Southeast Asian region to join ASEAN (Brunei, 1994; Vietnam, 1995; Burma and Laos, 1997; Cambodia, 1999). Despite some resistance from Malaysia, ASEAN has been open to discussions with other nations and groups, including China and Australia and more recently the United States and India. ASEAN's growth shows how important the region of Southeast Asia has become over time.

Apply Your Knowledge

10.12 Briefly list some of the consequences of the struggles that took place in Vietnam and Cambodia during the Cold War.

10.13 Compare and contrast some of the more recent tensions in the region.

10.14 Visit the ASEAN website www.asean.org for information on the organization's recent initiatives. What opportunities lie in regional cooperation?

Southeast Asia Tariffs and Trade

https://goo.gl/lTwnKq

Cultures and Populations

In studying the culture of this diverse region, it is important to consider the region's interactions with other world regions. This section traces the religious and linguistic geographies of the region in relation to broader global geographies. The section expands the discussion of culture to examine local gender, sexual, drug, and ethnic practices in the region. The section concludes with a discussion of the changing demographic and settlement geographies of the region.

Religion and Language

The complex religious and linguistic geographies of this region are tied to migration patterns tied to long-term global processes. Connections to other world regions, such as East Asia (Chapter 8) and South Asia (Chapter 9) brought Buddhism, Hinduism, Islam, and Catholicism to the region. The region's linguistic geographies are also tied to the very long-term migration of Malayo-Polynesian peoples (Malay, Indonesian, or Tagalog) and Austro-Asiatic speakers (Khmer, Mon, and Vietnamese) and to the more recent migration of Tai-speaking peoples (Thai, Lao, or Shan) as well as to the even more recent migration of Chinese (e.g., Hakka) and Indian (e.g., Tamil) speakers. English is also spoken widely throughout the region, particularly in urban areas and among the region's elite.

Religion The contemporary religious geography of Southeast Asia reflects centuries of evolution under Indian, Chinese, Arab, and European influences (**FIGURE 10.32**).

- **Buddhism** Theravada Buddhism—an import from South Asia over 2,000 years ago—dominates the mainland region, where religious and political traditions intersect in the reverence for the monarchies in countries such as Thailand and Cambodia. The Thai royal family is held in high, almost godlike esteem, with pictures of the king in the majority of homes as an icon of Buddhist-inspired belief systems. Mahayana Buddhism, a Salvationist form of Buddhism that is not directly tied to local politics, is practiced throughout Vietnam, a result of that country's relationship with China.

- **Islam** The practice of Islam can be found from southern Thailand, the Malay peninsula, and through insular Southeast Asia to the southern island of Mindanao in the Philippines. Indonesia is the world's most populous Islamic country. Islamic practices in Southeast Asia are not tied directly to any particular form of government. There has been conflict in some parts of the region between minority Islamic populations and Buddhists or Catholics. Islam is widespread in Malaysia but one can also find a wide range of religions practiced openly throughout the country (**FIGURE 10.33**).

  In some cases, Islamic belief in the region has resulted in the seclusion and veiling of women and in political conflict over the enforcement of Islamic law, especially in diverse populations. Even as Indonesia is pushing to formalize laws supporting religious tolerance, there remains tension in the world's most populous Islamic country. In 2002, the bombing of a nightclub in Bali killed 202 people. Further attacks in Indonesia's tourist regions have been linked to radical Islamicist groups, including attacks on hotels in Jakarta's tourist district in July 2009. In 2015, Indonesia established a counterterrorism office in Sulawesi, eastern Indonesia, to address ongoing

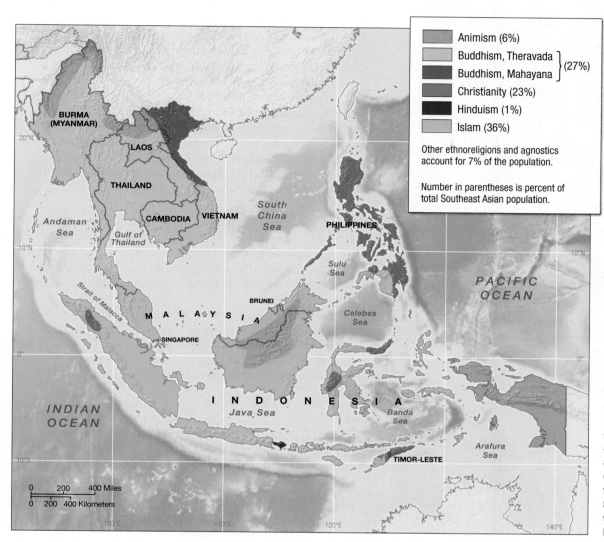

Animism (6%)

Buddhism, Theravada
Buddhism, Mahayana } (27%)

Christianity (23%)

Hinduism (1%)

Islam (36%)

Other ethnoreligions and agnostics account for 7% of the population.

Number in parentheses is percent of total Southeast Asian population.

◄ **FIGURE 10.32 Dominant Religions in Southeast Asia** Religious belief in Southeast Asia is dominated by versions of Buddhism on the mainland and Islam on the islands. Buddhism includes both Theravada and Mahayana, the former more conservative and found mostly in Burma, Thailand, and Cambodia, and the latter associated with Vietnam. Catholicism is important in the Philippines. Some ethnic minority groups in remote mountain and island areas maintain animist beliefs. **Develop a hypothesis to explain the present-day distribution of animist religions, found mainly in the mountains of the Southeast Asian mainland and on the large islands of Borneo and New Guinea.**

tensions and potential terrorist cells linked to emergent global networks.

- **Hinduism and Christianity** Although Buddhism and Islam dominate the region, other religious practices also thrive in Southeast Asia. Hinduism is common in Bali, a legacy of the early introduction of that religion into the region. Christianity, a direct result of Spanish colonialism, is the religion of 85% of the Philippine population. Christianity is also found among some of the hill tribes of Burma and in Vietnam, who were converted by French and British missionaries. Minority ethnic groups have also maintained animistic beliefs, especially in the mountains of Burma, Laos, and Vietnam and in Borneo and Irian Jaya.

▶ **FIGURE 10.33 Tibetan Buddhism Practice in Malaysia** Conflict in East Asia has long forced the migration of Tibetans out of their homeland. Many have settled in India, but Tibetans can be found throughout the world, including in Malaysia where they practice their religion and culture openly.

▲ **FIGURE 10.34 Spirit House in Laos** Spirit houses are a common feature throughout Southeast Asia. They provide shelter to local animistic spirits, who are then less likely to interfere with the everyday life of people on that land. It is common to find spirit houses on the land of practicing Buddhists, as they represent the blending of animistic traditions with Buddhism.

- **Animism** Throughout Southeast Asia, people still practice **animism**—a belief system that focuses on the souls and spirits of nonhuman objects in the natural world (**FIGURE 10.34**). Animism is strong among ethnic minority populations who live in more remote parts of the region. But, it is inappropriate to identify animistic beliefs as only present in smaller, minority communities. Animistic beliefs and practices are also part of the daily lives of people who also identify with the dominant religions of the region. The syncretism, or blending, of animism and Buddhism is commonly found throughout mainland Southeast Asia, where people worship local spirits alongside their Buddhist rituals.

Language Southeast Asian cultural diversity is reflected in more than 500 ethnic and language groups, 300 of which are located in Indonesia alone. No language has unified the region, although the various dialects of Malay share enough in common to form a broad *lingua franca* (trade language) in Malaysia, Indonesia, and Brunei.

Some national boundaries of Southeast Asia do enclose a dominant cultural and language group. Thailand (Thai), Burma (Burmese), Cambodia (Khmer), Vietnam (Vietnamese), Laos (Lao), and Malaysia (Malay) each have large majority populations that speak the same language (**FIGURE 10.35**). Indonesia has adopted a common

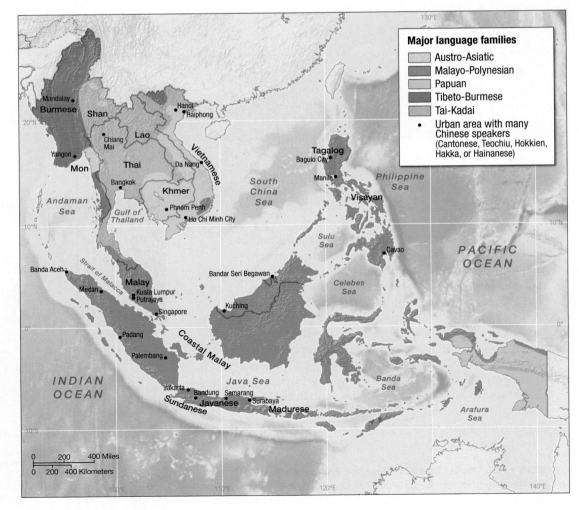

► **FIGURE 10.35 Dominant Languages in Southeast Asia** Southeast Asia has hundreds of distinct languages, which fall into five major language families: Malayo-Polynesian, spoken in insular Southeast Asia and the Malay Peninsula; Tibeto-Burmese in Burma; Tai-Kadai in Thailand Burma, Laos, and parts of Vietnam; Papuan in New Guinea; and the Austro-Asiatic languages of Vietnam and Cambodia.

Major language families

- Austro-Asiatic
- Malayo-Polynesian
- Papuan
- Tibeto-Burmese
- Tai-Kadai
- Urban area with many Chinese speakers (Cantonese, Teochiu, Hokkien, Hakka, or Hainanese)

language in an attempt to integrate diverse cultural groups. There are significant minority populations in many countries, including the Karen and Shan in Burma and the Hmong in Laos, as well as large populations of Chinese and Indians, whose first language is not the dominant national language but the language of their family's origin. So, one will find Tamil speakers in Little India in Singapore or Cantonese or Hakka speakers in Bangkok's Chinatown.

Apply Your Knowledge

10.15 Briefly describe the regional map of religion in Southeast Asia.

10.16 Explain why the regional map does not show the entire picture of how complex religious practice is in Southeast Asia.

Cultural Practices, Social Differences, and Identity

Just as there is linguistic and religious diversity in the region, there is also cultural and social diversity. One can find both **matrilocality** (where married couples move into the wife's family home) and **patrilocality** (where married couples move into the husband's family home). While in matrilocal families authority passes from father to son-in-law, not to daughters, the common practice of matrilocality gives women *some* power because they live with their own families. In general, women have more authority within Southeast Asian families and societies than in many other world regions. In many parts of Indonesia, women manage the family money. The value of daughters is reflected in a more equal preference for female and male children than is seen in either East or South Asia. Women also have employment opportunities in services and manufacturing and are preferred by many high-technology companies because they accept lower wages and are perceived as more careful and docile. This has created some conflict in home communities, where men are "left behind."

Nevertheless, the broader cultures of Southeast Asia remain **patriarchal**: Men have authority over social and political systems and socioeconomic conditions are generally better for men than for women. Men tend to receive higher wages and more education. In poorer Theravada Buddhist communities on the mainland, young boys have access to free education and housing through local monasteries, whereas girls do not (**FIGURE 10.36**). This provides boys with a double benefit, as they are granted the resources of temple life and are able to increase their merit as well as that of their parents by training as a Buddhist novice.

In recent years, **nuclear family** structures have become more common in Southeast Asia. Nuclear households contain just two generations, a pair of adults and their children. Households have also become more suburban. Agricultural land is being converted into single-family home communities, many of them gated, throughout the region. The family has shifted from parents and older people to children and from farming or industrial labor to education and service-sector work. New commercial spaces such as malls, restaurants, and vacation resorts serve a growing regional middle class. New social identities are emerging

▲ **FIGURE 10.36 Novice Monks in Class at a Temple in Thailand** In Thailand, boys, particularly from poorer families, can become novices and obtain an education at a temple. This same privilege is less often extended to girls.

in the region centered not around production—agricultural or manufacturing—but around consumption of high-end consumer goods and luxury items, such as Starbucks coffee.

Apply Your Knowledge

10.17 Explain the difference between matrilocality and patrilocality.

10.18 In what ways does the geographic organization of marriage impact gender relations in Southeast Asia?

Sexual Politics Southeast Asia, particularly Thailand, is well known for its commercial sex work industry. In Thailand, **polygyny**—the practice of having more than one wife at a time—was legal until the 1930s, when the new constitutional government formally outlawed the practice to appear more "Western." Although polygyny is illegal, the practice of *mia noi* (minor wife) remains common among wealthier men. There is also an economy of commercial sex in Thailand. Thailand's role in the Vietnam War as the rest and relaxation (R&R) capital of the region for U.S. soldiers helped foster the image of Thailand as a global commercial sex capital, and by the 1990s, tours were being advertised in places such as Japan and Germany.

Today, commercial sex work venues exist throughout Thailand, in both rural and urban areas. Women are trafficked into the country from neighboring countries, including Burma, Laos, Cambodia, and southern China. Some of the sex workers are very young women, sold into bondage by rural families or smuggled in to live in slave-like conditions. Studies suggest that for women from poor rural families, sex work may appear as a rational choice for making a living, providing opportunities for them to send money back to their villages and families.

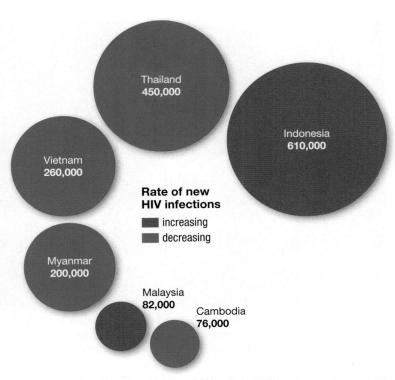

Rate of new HIV infections

■ increasing
■ decreasing

Thailand
450,000

Indonesia
610,000

Vietnam
260,000

Myanmar
200,000

Malaysia
82,000

Cambodia
76,000

▲ **FIGURE 10.37 HIV Cases and AIDS-Related Deaths in 2013** This graph represents the total number of people living with HIV in 2013. Indonesia has the highest number of HIV cases in the region. In recent years, many countries in the region, although not all, have seen declining overall rates of HIV. Thailand still has the highest reported rates of new HIV cases per capita in the region in 2013.

▲ **FIGURE 10.38 Condom Promotion Campaign, Thailand** Beginning in the 1990s, Thailand embarked on a large-scale education campaign to bring awareness to safer sex practices including the use of condoms to protect against the spread of HIV and other sexually transmitted infections (STIs).

HIV/AIDS Politics Despite successful prevention efforts, there are hundreds of thousands of human immunodeficiency virus (HIV) positive people living in Southeast Asia today (see **FIGURE 10.37**). These numbers are the result of the prevalence of the sex industry and intravenous drug use (IVDU)—heroin and methamphetamines (see "Geographies of Indulgence, Desire, and Addiction: Opium and Methamphetamine" on pp. *416–417*). Thailand, where HIV was detected the earliest, has been able to mitigate the crisis through outreach and prevention. A campaign to promote 100% condom use in commercial sex work in the late 1990s helped curb the epidemic (**FIGURE 10.38**).

The rising number of reported HIV positive cases in places, such as Indonesia and Malaysia, is creating new challenges for health-care systems in the region. The stigma that remains with HIV also means that underreporting of HIV is common. Country- and regional-based efforts have developed social support and advocate networks for people living with HIV and acquired immunodeficiency syndrome (AIDS). Despite outreach efforts in prevention and support, countries face a crisis of families in which one or more of the parents have died from HIV, leaving a legacy of "**AIDS orphans**" throughout the region. The rising number of orphans is placing a strain on the social service sector and extended family networks, creating new layers of complexity for an already extensive social, political, and economic crisis of prevention and care.

Minority Politics Southeast Asia is home to ethnic groups that historically traversed national boundaries and who are not considered citizens of any particular country. In mainland Southeast Asia these ethnic groups are called **hill tribes**, or *chao khao* in Thai. The hill tribes dwell in the sparsely populated border regions of Thailand, Burma, Laos, and Vietnam and have historically participated in dry farming and slash-and-burn agriculture, supplementing these practices with hunting-and-gathering techniques (**FIGURE 10.39**). In insular Southeast Asia, ethnic groups referred to as **sea gypsies** rely on fishing and boat travel between islands to sustain their daily caloric intake. These populations' ancestors probably arrived in the region over 5,000 years ago. The Moken, one of these groups, are under increasing pressure to settle. In the meantime, many of these people live in limbo: They are not citizens of the states of Southeast Asia and do not have the power to fully live their lives as they traditionally have.

Apply Your Knowledge

10.19 List three possible connections that might exist among the spread of HIV, drug use, and minority and ethnic politics in Southeast Asia.

10.20 Research current efforts to combat HIV in Southeast Asia at http://www.unaids.org/. What are the challenges to addressing the spread of HIV in the region?

HIV in Southeast Asia

https://goo.gl/h12kJe

▲ **FIGURE 10.39 Hmong Farmers in Laos** In villages throughout Southeast Asia, ethnic minorities practice local production techniques to cultivate, harvest, and process crops that came to the region through global trade. In this case, corn is milled using stone tools and human labor.

Demography and Urbanization

Population distributions in Southeast Asia are uneven. Rural areas remain less populated than urban ones, while regional economic processes are creating new networks of migration in and out of the region.

Population and Demographic Change The population of Southeast Asia was estimated to be about 621 million in 2015. Population growth and life expectancy in Southeast Asia surged with the dramatic reduction of malaria and improved medical care in the 1950s. After several decades of growth at more than 2% per year, overall population growth has slowed to 1.3% per year, as a result of fertility declines in Indonesia, Thailand, Singapore, and Vietnam. Indonesia's noncoercive family planning policy promotes two-child families through advertising, grassroots leadership training, and free distribution of birth-control pills and condoms. Thailand has also had an extensive family planning campaign that began in the 1970s, which has successfully reduced overall fertility in the country.

Fertility rates have remained higher in the Philippines, partly as a result of opposition to birth control by the Catholic Church, and in Malaysia, where the government encourages the Malay population to have more children. Southeast Asian population data support the theory that fertility decline is associated with higher income, lower infant mortality, and higher status of women. In wealthy Singapore, where population growth is negative, the government is promoting marriage and childbearing, especially among the highly educated. Cambodia, Laos, and Timor-Leste have lower GDP per capita, higher infant mortality, and lower levels of female literacy and schooling than do other countries in the region. Until recently, however, population growth in Cambodia, Laos, and Vietnam was reduced by high death rates from war and famine (**FIGURE 10.40**).

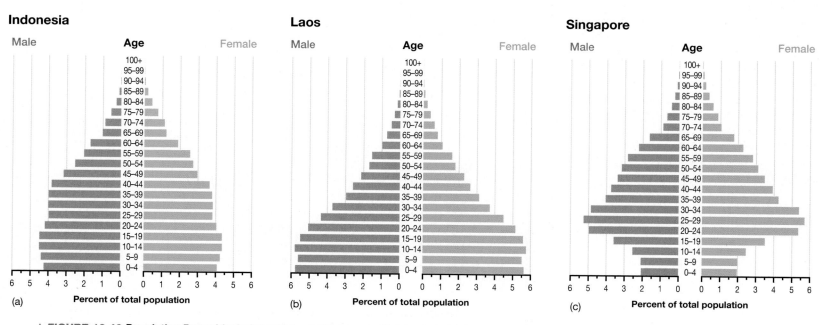

▲ **FIGURE 10.40 Population Pyramids, Indonesia, Laos, Singapore** The demographic picture across the region is quite different depending on which country one examines. (a) Indonesia's overall demographic picture demonstrates a consistent rate of new children, whereas (b) Laos still supports a very young population with higher birth rates. (c) Singapore's population pyramid represents a picture whereby the aging population is quickly outstripping younger populations. **Based on the population profiles, which of these countries has the lowest birth rate and the longest life expectancy? Explain.**

Opium and Methamphetamine

The mist-shrouded highlands of Burma, Laos, and Thailand contain much of mainland Southeast Asia's remaining forest areas; they are also known for the fields of opium poppies that historically have provided as much as half the global supply of heroin in an area now known as the Golden Triangle (**FIGURE 10.4.1**).

Opium and Heroin The allure of the opium poppy is centuries old. The narcotic was a component of trade and conflict between China and Europe (see Chapter 8, p. 320–321). In 1905, with heroin addiction on the rise, the U.S. Congress banned opium and in Thailand an active campaign to eradicate opium production was successful. Similar campaigns largely failed in Burma and Laos, which has continued to feed large markets for heroin in the United States, Europe, and Australia. The United Nations estimates that there are 11 million heroin and opium addicts worldwide. Addiction can result in overdoses and death from respiratory failure, and the use of shared needles has significantly contributed to the transmission of HIV/AIDS. Production of opium in Southeast Asia was in decline in the 1990s and early 2000s (**FIGURE 10.4.2**). Trends show that opium cultivation is once again on the rise, having tripled since 2006.

Methamphetamines More recently, this region has also become an important global and regional center for the production of methamphetamines (called *meth* in English; *yaa baa* in Thai). The drug economy has helped local ethnic groups fight off Burmese oppression, funding rebellion against the Burmese state. Although tourists still visit the region, a recent Thai government-led "war against drugs" has led to an escalation of violence in this area, particularly in rural areas that are cross-border sites for drug trafficking (**FIGURE 10.4.3**).

"Production of opium in Southeast Asia was in decline in the 1990s and early 2000s. Trends show that opium cultivation is once again on the rise, having tripled since 2006."

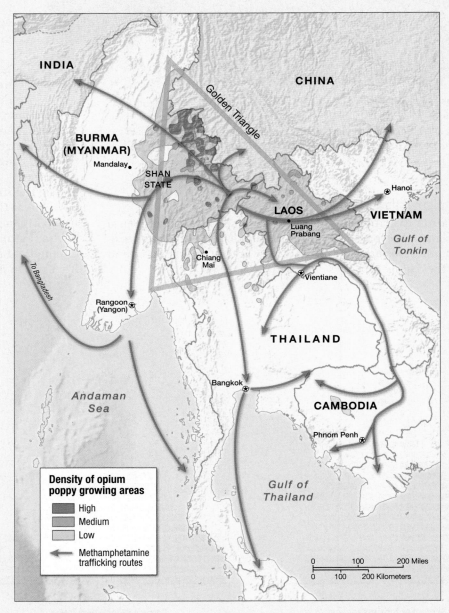

▲ **FIGURE 10.4.1 The Golden Triangle** This map highlights the density of poppy production in the region known as the Golden Triangle. The map of Thailand, which engaged in an active campaign to reduce poppy production, clearly shows the success of that program. Methamphetamine production takes place across Burma and parts of Laos, while Thailand serves as a key transshipment site for the drug.

Current Outlook for the Drug Economy The resurgence of opium and heroin production in the region suggests that Southeast Asia's historical place as a key drug-producing and consumption region remains disturbingly strong. And the increase in the number of people injecting methamphetamines suggests that Southeast Asia must continue to examine how and why such drug-use production and consumption systems are sustained in the region.

1. **Why does this region produce so many illegal drugs?**

2. **What are the consequences of drug production and use in the region?**

◀ **FIGURE 10.4.2 Opium Harvest** Members of the Wa ethnic group in northern Burma harvest opium poppies. Ethnic minorities have long relied on growing poppies and harvesting opium on small plots in the highlands to bring in cash; historically, they used the drug medicinally. More recently, drug cartels control its production and distribution, using the opium "tar" to produce heroin.

◀ **FIGURE 10.4.3 Drugs Seized in the Golden Triangle** The opium packages pictured here were seized as part of a crackdown on drug trafficking in Thailand. Although heroin is still produced in the region, new drugs, such as *yaa baa*, are now taking its place.

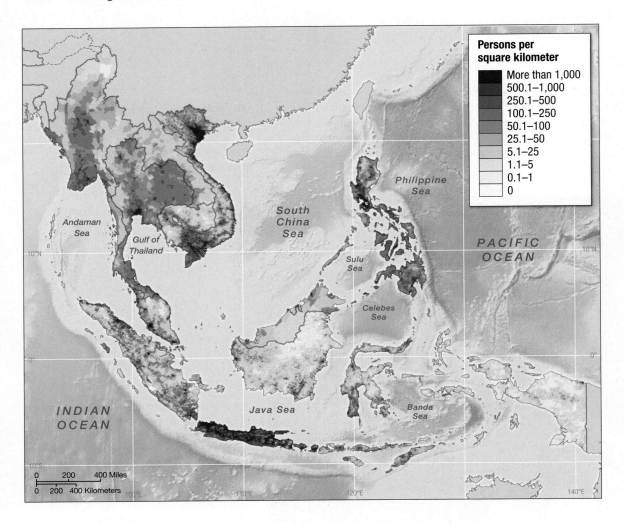

◀ **FIGURE 10.41 Population Density in Southeast Asia, 2015** The majority of people of Southeast Asia live in or near river valleys and deltas on the islands of Java, the Philippines, and Singapore. The highest population densities are in major cities, such as Singapore, although the island of Java is the most densely populated area in the entire region. More than half the population of Indonesia lives on the island of Java.

Population Distributions The map of population distribution in Southeast Asia shows people concentrated in the river valleys and deltas and on the island of Java, which is home to almost 60% of Indonesia's total population (**FIGURE 10.41**). The highest population densities (people per hectare) are in Singapore, Philippines, and Vietnam. Levels of urbanization range from 100% in the city–state of Singapore, to more than 60% in Malaysia and the Philippines, to less than 30% in Cambodia, Laos, Thailand, and Vietnam.

Migration Within Southeast Asia Three major types of migration exist within Southeast Asia.

- **Migration to Cities** People are pushed from rural areas to urban ones by landlessness and agricultural stagnation. Simultaneously, the attractions of jobs, education, health services, and access to consumer goods and popular culture pull people to rapidly growing urban areas. This phenomenon has resulted in megacities, such as Bangkok, Thailand, Jakarta, Indonesia, and Manila, Philippines.

- **Migration from Conflict Zones and by Stateless Peoples** Migration within the region stems from war and civil

unrest, such as the mass evacuations from cities in Cambodia under the Khmer Rouge and from war zones in Vietnam and Laos. More recently, refugees have left countries, such as Burma and Vietnam for neighboring countries. The UN High Commission for Refugees reported that in 2015 there were 120,000 Burmese in refugee camps in Thailand and an additional 587,000 people internally displaced within Burma. There are estimated to be an additional 506,000 stateless people who live within the boundaries of Thailand without citizenship. Their status reflects either a legacy of conflict in another country or is a result of their being a migrant ethnic minority. In recent years, the Rohingya people of Burma, an ethnic, Muslim minority, have fled to other parts of the region. This has created a humanitarian crisis, as thousands of Rohingya find themselves traveling on dangerous boats to countries that do not want to admit them (**FIGURE 10.42**).

- **Government-Mandated Resettlement Programs** A distinctively Southeast Asian form of migration is the active resettlement of populations. In 1950, the Indonesian government initiated a **transmigration** program, which redistributed populations from densely settled

▲ **FIGURE 10.42 Rohingya Refugees, Kuala Langsa, Aceh, Indonesia** A group of more than 700 rescued migrants, mostly Rohingya from Myanmar and Bangladesh, gather on arrival at the new confinement area in the fishing town of Kuala Langsa in Aceh province on May 15, 2015 after their boat sank off the coast of Indoenesia.

plans to move more Javans to Papua (**FIGURE 10.44**). Problems with the transmigration program included the lack of infrastructure in the new settlements, conflict between the settlers and indigenous groups, and the deforestation that occurred when relocated farmers accustomed to the fertile soils of Java struggled to make a living. Malaria, pests, and weed invasion also hindered the success of the program.

International Migration
International migration to and from the region also has a long history.

- **Overseas Chinese** Driven by civil wars, famine, and revolution in China, more than 20 million ethnic Chinese moved to Southeast Asia during the colonial period. Imported labor cleared forests and built irrigation and drainage systems in these vast deltas, which became the **rice bowls** of Southeast Asia. Chinese were also brought to mine tin and other minerals, to manage services in major ports, and to develop small businesses to serve demands for consumer goods. These so-called **overseas Chinese** became essential to the success of the colonial economy, and upon independence they became the entrepreneurs who ran banks, insurance companies, and shipping and agricultural businesses. As a result of this massive in-migration, ethnic Chinese make up a large percentage of the overall population in Singapore (77%). There are also large Chinese populations in Malaysia (26%) and Thailand (10%). In other countries, such as Indonesia, ethnic Chinese make up a large percentage of the economic elite. The Chinese generally live in separate urban neighborhoods often known as Chinatowns, with their own social clubs and schools.

Java and the city of Jakarta to the Moluccas, Sulawesi, Sumatra, and Kalimantan. The goals of the program were to reduce civil unrest, increase food production in peripheral regions, and further goals of regional development, national integration, and the spread of the official Indonesian language (**FIGURE 10.43**). More than 4 million people have moved with 1.7 million of the migrants receiving official government sponsorship in the form of transport, land grants, and social services. Although the program was officially terminated in 2000, the Indonesian government recently announced

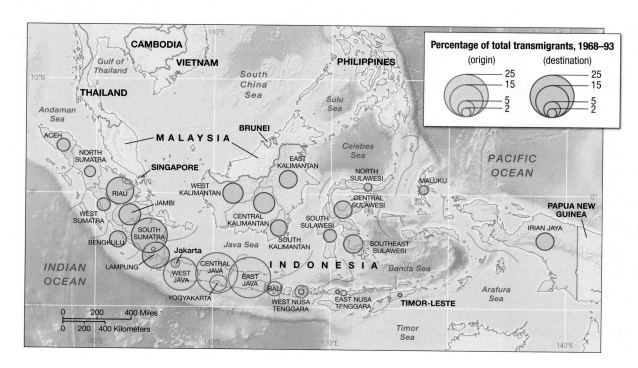

◄ **FIGURE 10.43**

Transmigration Flows in Indonesia The Indonesian government has relocated thousands of people from urban areas on Java to rural areas on the islands of Moluccas, Sumatra, Sulawesi, and Borneo. The transmigration program was formally halted in 2000, but the Indonesian government still maintains programs to support migrants moving from Java to other parts of the country.

▲ **FIGURE 10.44 A Papua Student Alliance Protest, Yogyakrata, Indonesia** Indigenous Papuan students stage an action in 2014 in stance against an agreement signed between the Dutch and the Indonesians 52 years earlier in New York that gave the Indonesian government political control of the western half of New Guinea.

- **Migration from Vietnam and Laos** After 1974, 1.5 million people left Vietnam as refugees (**FIGURE 10.45a**). About half of the Vietnamese refugees went to the United States; France, Canada, and Australia accepted others. U.S. cities such as San Francisco have large Vietnamese populations living in distinctive neighborhoods. Some Vietnamese refugees remained in refugee camps in Hong Kong for several years, while others were given permanent residency. The conflict in the region also extended into Laos, where more than 300,000 Hmong fled from Laos to Thailand, fearing persecution because they supported the United States; they were resettled in the United States among other places. Recently, with peace and reform, some overseas Vietnamese are returning to Vietnam, bringing capital earned abroad for investment in the country.

- **Migration to Middle East and Other Parts of Asia** The largest numbers of labor migrants from Southeast Asia work in the Middle East, especially in Saudi Arabia, Kuwait, and Oman, and East Asia, particularly Hong Kong and Japan. Many Southeast Asians in the Middle East are Muslim women working in the service sector as nurses and maids. Thousands of Thais and Filipinos work in Hong Kong, where their wages are typically much lower than those of local laborers. Many Philippine women work as maids as part of the *Maid Trade* in North America, Europe, and Singapore (see "Faces of the Region: The 'Maid Trade' Is Not the Fair Trade" on p. *421*).

- **The Remittance Economy** The money sent back by labor migrants (*remittances*) is very important to local economies. For example, it is estimated that over $47 billion was remitted back to the countries of Southeast Asia in 2013. That money was not evenly distributed, with $24 billion being sent back to the Philippines in 2013 and $10 billion to Vietnam (**FIGURE 10.45b**). In contrast, Singapore reported almost no remittances and Laos and Cambodia each remitted just over $1 million and $2 million, respectively.

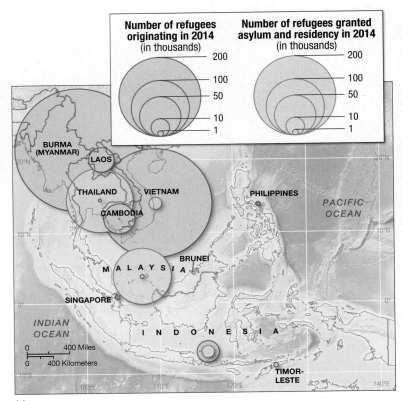

(a)

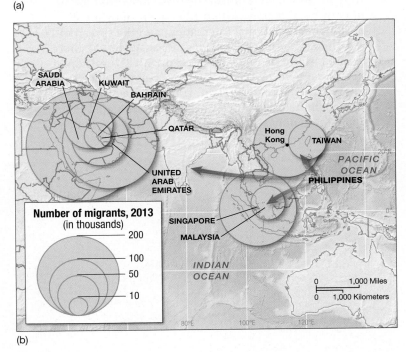

(b)

▲ **FIGURE 10.45 Migrations in Southeast Asia** (a) Southeast Asia has experienced enormous flows within and from the region, including migrants from Vietnam, Burma, Laos, and Cambodia to Thailand and beyond the region; from the Philippines to Malaysia; and from Timor-Leste to Indonesia. Between 1999 and 2007, the total refugee population was about 800,000. (b) This map shows where Philippine labor migrants were moving in 2010 and illustrates the significance of remittances from workers in Japan, Hong Kong, and the Middle East to the Philippine economy. **What are the top destinations outside Southeast Asia for Philippine economic migrants?**

The "Maid Trade" Is Not Fair Trade

On Sunday morning in downtown Hong Kong the demographics of the bustling streets of this global city shift from white collars workers in suits to thousands of Indonesian and Filipino domestic workers. This small army of laborers, who have come to Hong Kong to earn a living and send home remittances, crowds the streets of Hong Kong, laying out mats, sharing food common to their home towns, and conversing in their first languages. Many are women who have flocked to jobs as domestic workers in wealthy cities far from their home countries as part of the "Maid Trade." The numbers are staggering—over 300,000 domestic workers constitute a massive labor force that helps makes Hong Kong tick.

While many workers find a decent wage and reasonable working conditions, the story of Erwiana Sulistyaningsih is more common than most think (**FIGURE 10.5.1**). In 2014, it was reported that she had been kept as a domestic worker in largely slave-like conditions, working over 20 hours a day, seven days a week. When Erwiana

> "While not all cases are as extreme as Erwiana's, labor laws in Hong Kong create highly exploitative conditions for domestic workers, who have little to no rights under local law."

became too sick to work anymore, her employer dumped her at the Hong Kong airport. She had just U.S. $10 in her pocket. While not all cases are as extreme as Erwiana's, labor laws in Hong Kong create highly exploitative conditions for domestic workers, who have little to no rights under local law. They are paid a special wage, which is lower than the minimum wage of Hong Kong, because of their work status. And, still, thousands come each day from remote places and urban slums in Indonesia and the Philippines to make a better living for their families.

1. What are some of the reasons that women migrate as part of the Maid Trade to places, such as Hong Kong?

2. Research the working conditions of women in the Maid Trade by reading the article "Hong Kong woman found guilty of abusing Indonesian maid" at http://www.bbc.com/. What are some of the conditions under which women working as domestic laborers in Hong Kong have to suffer?

The Maid Trade

https://goo.gl/ymeQUm

◀ **FIGURE** 10.5.1

Press Conference in Support of Erwiana Sulistyaningsih Former maid Erwiana Sulistyaningsih originally from Indonesia attends a press conference in Hong Kong after her employer was convicted of beating her in a "torture" case. The case sparked international outrage and highlighted the plight of migrant domestic workers. The verdict, read out to a packed courtroom, was met with cheers by activists and supporters of Sulistyaningsih, who has become the face of a campaign for improved workers' rights in the financial hub.

Urbanization   About half of Southeast Asia's people live in urban areas. The region is also home to several enormous cities with large metropolitan populations, among them two of the world's top 30 urban areas: Manila and Jakarta. Bangkok and Ho Chi Minh City also both rank in the top 35 worldwide. The unification of North and South Vietnam in 1976 left the new country with two major urban centers—Hanoi and Ho Chi Minh City (formerly called Saigon). These cities have grown so rapidly that they have serious problems of overurbanization, such as insufficient employment opportunities, an inadequate water supply, sewerage problems, and deficient housing. With a wider metropolitan population of more than 9 million people, Bangkok is the current hub of mainland Southeast Asia and dominates Thailand with 32% of the country's urban population (more than 16 times the size of the next largest city), 90% of the trade and industrial jobs, and 50% of the GDP. There are more than 25,000 factories in the Bangkok metropolitan region, many of them labor-intensive textile producers.

Key secondary cities include Palambang, Medan, Bandung, Ujung Pandang, and Surabaya in Indonesia and Cebu and Davao in the Philippines. Many cities in Southeast Asia mix local design with that of colonial and modern planning, but they are also surrounded by unplanned growth, including desperately poor squatter settlements (**FIGURE 10.46**). Cities have extended into intensively farmed agriculture, especially rice paddies; town, industry, and agricultural villages have become intermixed. In Southeast Asia, boundaries between the city and the countryside are often blurred. There are numerous small industries, such as textiles, in rural areas and considerable agricultural production in the cities; new suburban developments are taking the place of rural farmland.

▼ **FIGURE 10.46 Shantytown in Taguig City, Manila, Philippines**   Many people live in makeshift households in tight quarters in areas of Manila, and other large cities in Southeast Asia, where it is otherwise difficult to build permanent housing. These people lack land rights and often find themselves without services, such as clean water and access to sanitation services.

Apply Your Knowledge

10.21   Identify three places in the world today that are net recipients of Southeast Asian migrant populations.

10.22   Why do Southeast Asians migrate overseas, and what are the benefits of those migrations to their home communities?

10.23   What are some of the consequences of emigration to the countries in Southeast Asia?

Future Geographies

The region, dominated by the ASEAN, has recently experienced economic growth thanks to coordinated polices in trade and development for the region. There remain questions, though, as to the regional viability of Southeast Asia and ASEAN more generally.

Regional Economic Cooperation

ASEAN is now negotiating free trade agreements with other countries and regional entities (**FIGURE 10.47**). Some activists argue that reducing trade barriers within Southeast Asia and between ASEAN and larger economies, such as India or the United States, places local economies at a distinct competitive disadvantage. How ASEAN manages its collective economic future will impact the depth of regional integration and broader global relationships between ASEAN and other economic blocs.

▲ **FIGURE 10.47 Asean Free Trade**   In December 2014, ASEAN leaders signed a new trade agreement with the South Korea government at a ceremony held in the South Korean city of Busan. ASEAN's trading presence has continued to grow over the last twenty years with the inclusion of countries, such as Vietnam, which was historically outside the supranational organization.

Intensifying Regional Integration

Activists criticize ASEAN for doing little to intervene in member countries that oppress their populations. ASEAN's reluctance to involve itself in the local affairs of Burma, Vietnam, or Philippines is causing the organization to lose credibility both within and outside the region (**FIGURE 10.48**). These debates make it clear that although regional economic cooperation has produced a more tightly connected world region, there are still many challenges to the stability of Southeast Asia as a region.

Environmental Issues and Sustainability

Future sustainable development must confront urban growth as well as air pollution, slums, and deforestation. Managing the environment and land use more equitably and ecologically may be the only way forward for improving social and economic conditions (**FIGURE 10.49**). There are questions over the management of international rivers, such as the Mekong as well as smaller river systems. Can ASEAN manage the region's resources in relation to the longer-term environmental and cultural needs of all the people in the region?

Social and Health Issues

The spread of HIV/AIDS may limit the social and economic development of the region moving forward, as countries that are poised to see economic gains, such as Vietnam and Burma, are faced with this health crisis (**FIGURE 10.50**). For countries such as Cambodia that have long struggled with economic development and political conflict, the HIV epidemic may prove to be an even more serious concern. There will be increasing pressure on governments to provide both public health care and more expensive treatments for the growing HIV-positive population.

▲ **FIGURE 10.49 Reforestation Project in the Philippines** The Philippines has the highest level of deforestation in Southeast Asia, with over one-third of its forest cover lost between the early 1990s and mid-2000s. Reforestation projects, such as this one, are an attempt to stabilize forest ecosystems in the country and mitigate the effects of long-term deforestation in the country.

▲ **FIGURE 10.48 Ang San Suu Kyi and Barack Obama** The two leaders are pictured here in 2014 at a press conference, which was held in Yangon, Burma. The visit marked the first of a US President to the country in its history.

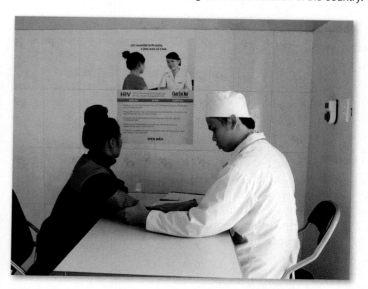

▲ **FIGURE 10.50 Rising HIV Rates for the Region** In 2008, this, then, 22 year old woman contracted HIV. She is pictured here in 2013 at a local clinic receiving care. Like many, she had children, who now must be cared for by others now that she has passed away. Despite advances in medicine to treat HIV and the complications that leads to AIDS, many lives are shortened by this deadly infection.

Learning Outcomes Revisited

▶ *Describe* the region's climate pattern, geology, and physical landscapes and explain how people have adapted to the region's physical geographies.

Southeast Asia is distinguished by its warm climates and high levels of rainfall, a product of its relationship with the ITCZ and global wind patterns. The region is also subject to high levels of volcanic activity, particularly in the island areas. Unique physical geographies and ecologies result from the relationship that the region has with these global patterns. These patterns result in a varied ecological landscape. Over time, people have adapted to the region's unique climate and physical landscape, producing large-scale agricultural systems and societies.

1. *In* what parts of the region does one find year-round rainfall? What parts are dry for at least half the year? Explain what causes these precipitation patterns.

▶ *Describe* the historical and current relationship among local, Southeast Asian regional economic cooperation and competition, economic development, and global economic issues and trends.

Southeast Asia is well integrated into the global economy, and places in this region are impacted directly by global economic changes. Not every country in Southeast Asia has been able to or has chosen to participate in the global economy at the same level. Throughout the 1990s, many countries in the region grew rapidly, as worldwide markets aggressively purchased Southeast Asian products. Growth stalled with the global economic crisis of 2008, when many countries in the region (even those not fully integrated into the global economy) were subject to high levels of inflation.

2. *Describe* the overall pattern of economic integration across the region. What tells us that some countries are more integrated than others?

▶ *Assess* the broader history of political tension in the region, noting the role that local understandings of the region's political geography and global conflicts have played.

Since World War II, when Japan thrust the region into even wider global conflict, Southeast Asia has been the site of both international proxy conflict, such as the Vietnam War, which directly encompassed not just Vietnam but Cambodia and Laos as well, and internal conflicts in countries including Cambodia and Burma. Ethnic tensions are present in the region, particularly in areas where minority populations seek autonomy, such as southern Thailand and in the Philippines.

3. *Briefly* compare and contrast three different conflicts within the region.

▶ *Trace* the unique historical development of social life in the region, explaining how and why differences and similarities between places emerge over time.

The history of this region is tied to both local developments and global flows of people and ideas. Migration into and through the region has brought important cultural and political systems from South Asia and East Asia, including forms of Buddhism and Hinduism. Global connections also introduced Islam to the region as well as Christianity, along with a diversity of economic and political systems. The differences within the region are a result of the relationship various places have to different global processes and local adaptations.

4. *What* are the different patterns of migration within the region as well as beyond the region?

▶ *Identify* the major social issues in the region today and describe how they are the result of historical, geographical, and social patterns.

Numerous social issues are key to understanding Southeast Asia. The development of nation–states in the region beginning in the 1800s has precipitated conflict over stateless groups, such as hill tribe peoples, who find themselves without a country. Drug production and use is a problem in the region and related health issues, such as the HIV epidemic, also pose a serious concern.

5. *Identify* the factors that have led to ethnic conflict in the region.

▶ *Explain* the patterns of demographic change and migration in the region and their effects.

Southeast Asia is a region of dense urban populations and much less dense rural places. In the island region, particularly the Philippines, fertility rates have remained historically high, resulting in the large populations. Since the end of the Vietnam War, Vietnam has also witnessed extensive population growth, whereas countries such as Thailand and Indonesia, which have engaged in extensive family planning, have seen decreasing population growth rates. Migration in and out of the region is tied to a number of important factors, including the availability of jobs and the role that governments play in moving populations around the region.

6. *Describe* the future map of population growth in the region. Where will populations grow most rapidly and where will there be potential population declines? What drives these differences?

Key Terms

Agent Orange (p. *406*)
AIDS orphans (p. *414*)
Angkor Wat (p. *395*)
animism (p. *412*)

archipelago (p. *389*)
ASEAN (p. *409*)
Aung San Suu Kyi (p. *408*)
biofuel (p. *390*)

chao khao (hill tribes) (p. *414*)
crony capitalism (p. *405*)
Culture System (p. *399*)
cyclone (p. *386*)

domino theory (p. *406*)
entrepôt (p. *394*)
foreign direct investment (FDI) (p. *404*)

Golden Triangle (p. *396*)
Green Revolution (p. *403*)
Ho Chi Minh (p. *406*)
Khmer (p. *395*)
Khmer Rouge (p. *407*)
killing fields (p. *407*)
kleptocracy (p. *405*)
land bridge (p. *392*)
little tigers (p. *404*)
mangrove forests (p. *392*)
matrilocality (p. *413*)
megafauna (p. *391*)

Mekong River Commission
   (MRC) (p. *396*)
monsoon (p. *385*)
National League for Democracy
   (NLD) (p. *408*)
newly industrializing economies (NIEs) (p. *404*)
nuclear family (p. *413*)
overseas Chinese (p. *419*)
paddy farming (p. *387*)
pancasila (p. *408*)
patriarchal (p. *413*)

patrilocality (p. *413*)
Pol Pot (p. *407*)
polygyny (p. *413*)
remittances (p. *405*)
rice bowls (p. *419*)
sea gypsies (p. *414*)
SIJORI (p. *404*)
Spice Islands (p. *386*)
Srivijaya (p. *395*)
sultanates (p. *398*)
swidden farming (p. *387*)
tariff (p. *403*)

Theravada Buddhism (p. *398*)
transmigration (p. *418*)
tsunami (p. *384*)
typhoon (p. *386*)
United Nations Educational,
   Scientific, and Cultural
   Organization (UNESCO)
   (p. *401*)
Vietcong (p. *406*)
Vietnam War (p. *406*)
Wallace's Line (p. *392*)
world city (p. *404*)

Thinking Geographically

1. What is the main crop in Southeast Asia and the main systems by which it is produced? How and where did the Green Revolution affect this crop?

2. What roles did India and China play in Southeast Asia prior to the colonial period, and how did these roles influence culture and religion? What political and economic roles did Japan and the United States play in Southeast Asia during the 20th century?

3. What is the legacy of European and Japanese imperialism in Southeast Asia? How did European influence extend beyond the formal colonial period and into the Cold War in the region?

4. What is ASEAN? What role does it play in the region both politically and economically?

5. What are the most pressing social issues in Southeast Asia today? What efforts are being made by local governments to address these issues?

DATA Analysis

In this chapter, you were introduced to the ASEAN, a group focused on economic, political, and sociocultural cooperation in the region. Go to the ASEAN website to further analyze the impacts of the alliance, http://www.asean.org

1. Using the "ASEAN" tab on the home page, list the current member states in ASEAN.

2. Using the "Communities" tab, go to the "ASEAN Political-Security Community" page, and download the ASEAN Charter.

3. What are three purposes of ASEAN, according to Article 1?

Southeast Asia Cooperation

https://goo.gl/vj5TCl

4. Using the "Resources" tab, download the "ASEAN Roadmap for the Attainment of the Millennium Development Goals."

   a. Read the "Introduction" and define the relationship of ASEAN to the region.
   b. In the "Objectives" find the eight thematic areas that comprise the MDGs.
   c. Go the table, "Summary of the Roadmap," and list three action items to support the MDGs.

5. Finally, do an Internet search on the recent news and activities of ASEAN. Compare and contrast the MDGs of ASEAN to these activities.

6. Based on your analysis, state how you think ASEAN contributes to the region's economic, political, and social stability.

MasteringGeography™

Looking for additional review and test prep materials? Visit the Study Area in MasteringGeography™ to enhance your geographic literacy, spatial reasoning skills, and understanding of this chapter's content by accessing a variety of resources, including MapMaster interactive maps, videos, *In the News* RSS feeds, flashcards, web links, self-study quizzes, and an eText version of *World Regions in Global Context*.

Sea-Level Rise and the Pacific Islands. Low lying islands in the Pacific are at risk from rising seas as the oceans warm and polar ice melts as a result of climate change. In the Kiribati islands, such as on the island of Abaiang, some of the freshwater lagoons have already been flooded by sea water.

Inset: Representatives from Pacific Islands attend international climate negotiations to protest the impact of climate change on their homes and culture.

Oceania

"If you were faced with the threat of the disappearance of your nation, what would you do?" asked the president of Tuvalu at a United Nations (UN) meeting in 2014. Tuvalu is a small country consisting of 9 islands in the central Pacific with a total area of less than 10 square miles (26 km²) and a population around 11,000 people. The economy is based on fishing, government employment, remittances from overseas workers and seamen, and revenues from the ".tv" domain name. Most settlements are near sea level. Several hundred miles to the northeast, Kiribati (pronounced "kiribas") is a nation of about 100,000 people and 33 low-lying atoll islands in the central Pacific, straddling the equator. The country has few natural resources and incomes are low. Copra (from coconuts) and fish are the main products and tourism is becoming more important. Most of the land is less than 6 feet (2 meters) above sea level. Water supplies are fragile, often polluted by high tides.

The warming of Earth as a result of increasing greenhouse gases is causing sea-level rise as warmer ocean water expands and ice sheets melt. If emissions continue, sea levels may rise more than a meter (3 feet) by 2100. This could be devastating for these low-lying islands in the Pacific, causing erosion, flooding fields with saltwater and damaging crops, contaminating water supplies, and destroying infrastructure.

Tuvalu and Kiribati are members of the **Alliance of Small Island States (AOSIS)**, which was established in 1990 to represent islands in international negotiations, especially those relating to climate change. Oceania is represented in AOSIS by 16 members, ranging from Papua New Guinea to tiny Niue and Nauru. Small islands have lacked the power to influence international discussions as individual countries and joined together seeking a more powerful voice. These small islands produce almost no greenhouse gas emissions but are tremendously vulnerable to climate change. Millions in development funds are being used to plan and implement adaptations to sea-level rise including new sea walls and vegetation barriers, rainwater harvesting, and desalination plants. In 2012, the government of Kiribati purchased a large estate in Fiji for possible migration and relocation.

Learning Outcomes

► *Explain* the factors that influence the climate of Oceania (including Antarctica and the Pacific Ocean) and the impacts of climate change and other environmental threats.

► *Describe* the unique ecosystems and the impacts that the introduction of nonnative species have had in the region.

► *Analyze* how colonialism and geopolitical struggles, such as World War II, the Cold War, and nuclear policy, have shaped the region, and describe the current international and regional political and economic alignments of major countries in the region.

► *Explain* how exports connect Australia and New Zealand to the rest of the world and describe changes resulting from globalization, free trade, and a shift in demand from Europe to Asia.

► *Explain* the importance of tourism in the region and the economic, social, and environmental impacts of the tourist industry.

► *Summarize* the main shifts in Australian and New Zealand policies toward indigenous peoples and on immigration issues and compare and contrast the differences in culture and gender relations of different groups in Oceania.

Environment, Society, and Sustainability

Oceania includes the continental landmass of Australia, the large islands of New Zealand (Maori name "Aotearoa") and eastern New Guinea, and smaller nations and territories that are often referred to collectively as the Pacific Islands (**FIGURE 11.1**). One of the most isolated world regions in terms of its physical geography, Oceania occupies one-third of Earth's surface and includes more than 20,000 islands. Island groups include Fiji, the Solomon Islands, Vanuatu, and New Caledonia, which with Papua New Guinea are known as **Melanesia**, meaning black islands;

Nauru, Kiribati, Palau, the Marshall Islands, the Federated States of Micronesia, Guam, and the Northern Mariana Islands, which are known as **Micronesia**, meaning small islands; and Samoa, Tonga, the Cook Islands, Wallis and Futuna, Easter Island, Tuvalu, Niue, Tokelau, French Polynesia/Tahiti and sometimes Hawaii, which are known as **Polynesia**, meaning many islands.

This chapter also includes Antarctica, the continent that lies to the south of the Pacific Ocean around the South Pole. It is always a challenge to place Antarctica within the groupings of world regions. However, it fits well within Oceania, because of the importance of the marine environment in Antarctica, its relationship with New Zealand and Australia, and its experience as a simultaneously globally interconnected and isolated regional space (see "Emerging Region: Antarctica" on pp. 430–431).

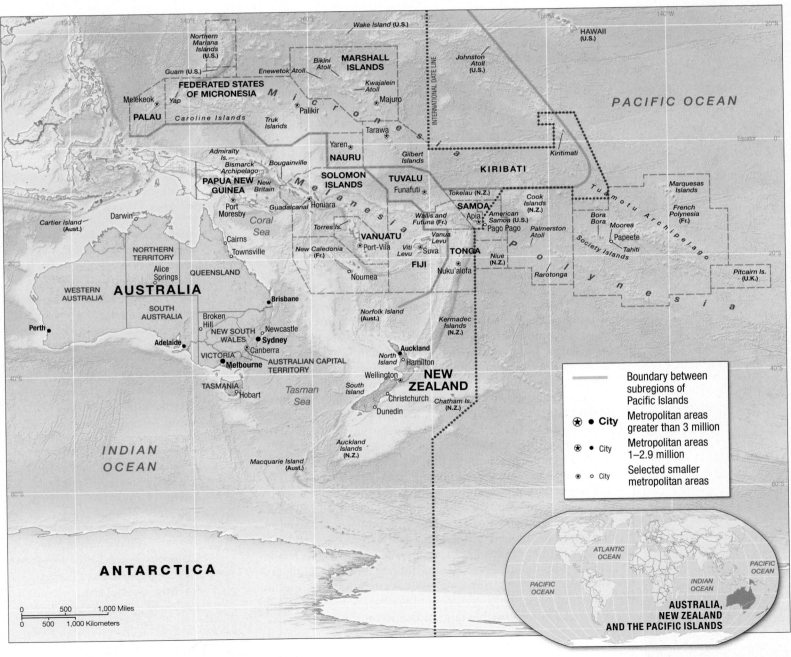

▲ **FIGURE 11.1 Oceania: Countries, major cities, and physical features** Identify those Pacific islands under the dominion of New Zealand (N.Z.), United Kingdom (U.K.), United States (U.S.), Australia (Aust.) and France (Fr.).

The countries in this region share an orientation to the ocean, relatively low populations, and dramatic coastal and mountain landscapes. This region is geologically rich in minerals, bringing it into the wider global economy of resource extraction. Oceania's diverse ecosystems and rich biodiversity are highly susceptible to human-induced change such as invasive species, ocean pollution and climate change.

Climate and Climate Change

Most of Oceania lies within the tropics and enjoys the warm seas and moisture-bearing winds of tropical latitudes. Australia and New Zealand reach farther south toward the South Pole and thus include cooler temperate climates of the southern westerly wind belts, with annual average temperatures declining from north to south (**FIGURE 11.2**).

Australian Climate The interiors of large continents in the tropics have very hot, dry conditions. Australia is no exception. It is one of the most arid areas on Earth, with two-thirds of the country receiving less than 50 cm (20 inches) of rainfall a year. This harsh desert climate limits human activity and has required complex adaptations from both people and ecosystems in the continent's interior. Droughts in Australia are associated with severe wildfires that can race across the tinder-dry bush vegetation, especially where oily eucalyptus fuels the fire (**FIGURE 11.3**). The Black Saturday bushfires in southern Australia in 2009 killed 173 people and burned more than a million acres (450,000 hectares) after a heat wave in which temperatures reached 46°C (115°F).

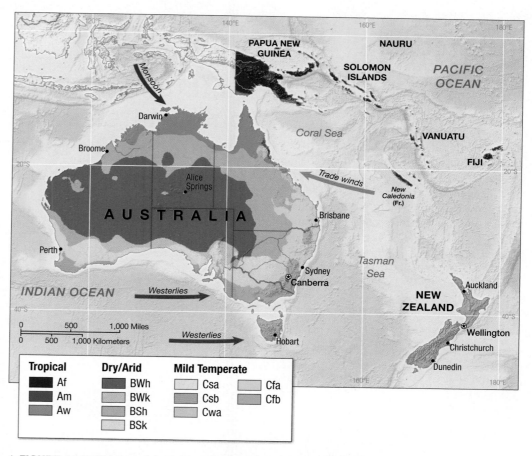

▲ **FIGURE 11.2 Climates of Australia and New Zealand** Australia's interior is dominated by dry, arid desert climates (BWh and BWk) and semi-arid steppe (BSh and BSk). The coastal climates vary from warm, seasonally wet, tropical climate in the far north (Aw), to moderate temperate climates with summer rain (Cfa) in the southeast, and Meditteranean in the south (Csa, Csb). Tasmania and New Zealand have a cooler temperate maritime climate with rain most of the year. The Pacific Islands have hot, wet tropical climates with rain for most of the year (Af). For more detail on the legend see Chapter 1, Figure 1.9. **What are the name of the winds that bring rain to Tasmania and to northern Australia?**

"Oceania's diverse ecosystems and rich biodiversity are highly susceptible to human-induced change such as invasive species, ocean pollution, and climate change."

▶ **FIGURE 11.3 Bushfire in Australia** Drought, high temperatures, and combustible vegetation have set the stage for intense and devastating fires across much of southern Australia, destroying homes, wildlife, and trees, such as here in 2013 near Sydney.

Antarctica

Antarctica is an emerging region of global concern as an enormous yet remote continent claimed by many countries, yet owned and occupied permanently by none, valued by many environmentalists, but threatened by climate change and mining interests. Antarctica has a severe climate. Temperatures average –51°C (–60°F) during its six-month winter, when the sun does not rise above the horizon and the continent is in perpetual twilight surrounded by vast areas of frozen pack ice. Antarctica inspires scientific and public

"Antarctica inspires scientific and public interest and each summer about 35,000 tourists visit, traveling mostly by ship to see the landscape and wildlife."

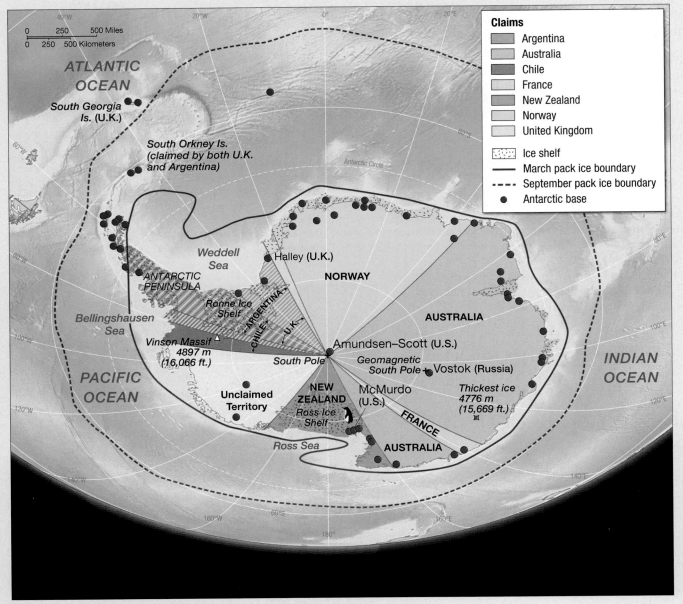

◄ **FIGURE 11.1.1**
Territorial Claims in Antarctica Seven countries have claims in Antarctica, but more than 50 have signed the Antarctic Treaty. More than twenty countries maintain permanent scientific research bases on the continent.

▲ FIGURE 11.1.2 Tourists Visit a King Penguin Colony on an Antarctic Cruise The photo cannot convey the intense noise or smell of the colony.

interest and each summer about 35,000 tourists visit, traveling mostly by ship to see the landscape and wildlife, which includes whales and millions of penguins (**FIGURE 11.1.2**). About 5,000 scientists live in Antarctica during the polar summer at research stations where they study ecology and climate.

The **Antarctic Treaty** governs international relations on the continent. The treaty covers all the area south of 60°S and is one of the most successful international environmental and political agreements. Entered into force in 1961 and now signed by 53 countries, the treaty bans nuclear tests and the disposal of radioactive waste, and ensures that the continent can be used only for peaceful purposes and is set aside for scientific research. In 1991, the treaty added a 50-year ban on mineral and oil exploration. Nevertheless, several countries—Australia, Argentina, Chile, France, New Zealand, Norway, and Great Britain—still claim specific slices of the Antarctic pie, hoping perhaps to be able to assert rights to offshore fisheries and onshore mineral exploitation (**FIGURE 11.1.1**).

Global changes are having an impact on Antarctica. The ozone hole continues despite international efforts to control ozone-depleting gases. Climate change is causing temperatures to warm and ice cover is declining. The ecology of birds and marine life is changing. If Antarctic ice continues to melt as a result of climate change, it may contribute to a worldwide rise in sea levels (**FIGURE 11.1.3**).

Ice loss in Antarctica

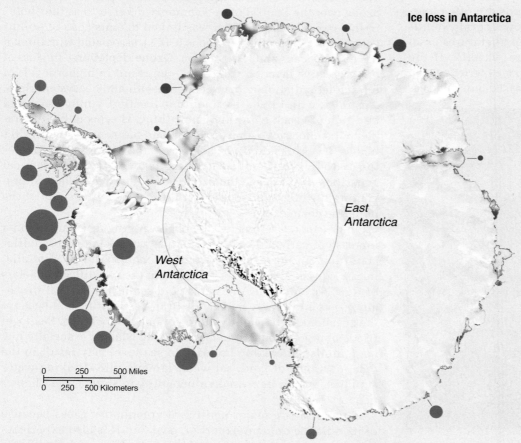

East Antarctica

West Antarctica

0   250   500 Miles
0   250   500 Kilometers

Rate of thickness change (m/decade)

−25   −10   0   10

Percent of thickness change, 1994–2012

 5% Loss   5% Gain

1. **Consult the website for the Antarctic Treaty www.ats.aq/e/ats.htm and read about at least one of the aspects of the Treaty (e.g., science, tourism) and their guidelines.**

Antarctic Treaty

https://goo.gl/u1m00z

2. **Use the Internet to find a company offering tours to Antarctica. What do they describe as the highlights of the trip?**

◄ FIGURE 11.1.3 Ice Retreat Around the Antarctic Peninsula Using maps and satellites, scientists have mapped the retreat of ice caps and glaciers around Antarctica. This map shows areas where the ice sheets are thinning (red colors, larger circles have greater loss) or increasing (blue colors).

Most of coastal Australia has higher precipitation than its interior. Queensland in the northeast has heavy rainfall from the humid southeasterly trade winds that rise over the highlands near the coast. Northern Australia receives most of its rain from monsoon winds. The rainy season sometimes brings tropical cyclones and severe flooding to northern Australia. In 2011, 75% of Queensland was declared a flood disaster zone. Southern Australia receives rainfall from storms associated with the westerly wind zone, mostly during winter (June to August), when storm tracks shift northward. The southern and southwestern coasts around the cities of Adelaide and Perth have mild temperatures and winter rainfall similar to the Mediterranean climate at this latitude in Chile and South Africa.

New Zealand Climate Two major islands spanning 1,600 kilometers (976 miles) from north to south—called the North Island and the South Island— comprise New Zealand, along with a number of smaller islands. The two islands sit in the middle of the westerly wind belt, where frequent storms release heavy rain on the western coasts as they rise over the high mountain ranges. The eastern coasts of New Zealand are drier and sunnier because they lie in the rain shadow east of the mountains. The North Island is generally much warmer than the South Island, but mountain climates are cooler and wetter throughout the country, often with heavy snow that favors ski tourism on the South Island in the southern hemisphere winter (June to October).

Pacific Island Climate The Pacific Islands are often classified into the "high" islands and the "low" coral islands (called **atolls**). All the islands have warm temperatures associated with year-round high sun and the warmth of the tropical ocean. Higher islands with mountains experience substantial orographic precipitation (**FIGURE 11.4**). Islands and seas

▲ **FIGURE 11.4 Pacific Island Climate** The island of Moorea in French Polynesia has mountains rising from the coast. As moist winds blow over the island and rise over the mountains, clouds form, and may produce orographic precipitation.

throughout the region also receive some rainfall from the towering cumulus clouds that form as convection heats the air at the Equator. The lower islands are much drier because they do not benefit from orographic precipitation and are small enough to escape the convective downpours. Many low-lying islands experience near-desert conditions and shortages of fresh water. When there is an El Niño (see Chapter 7) and ocean currents shift direction and change sea-surface temperatures over the Pacific, changes in wind patterns can cause severe drought in Papua New Guinea, Australia, and Micronesia, resulting in crop failures, food shortages, and costly shipments of drinking water to smaller islands.

Antarctica (see "Emerging Regions: Antarctica" on pp. 430–431) has the coldest climate on Earth because polar latitudes receive little sun and the continent has elevations averaging 2,286 meters (7,500 feet). It is mostly covered by ice that is more than 3 km (2 miles) thick in some places. Temperatures can drop to −73°C (−100°F) during the dark winter. It is dry—a polar desert—despite the water stored in the thick ice sheets.

Climate Change and Ozone Depletion Oceania is especially vulnerable to global environmental changes. Ozone is found high in the atmosphere, especially over the polar regions. It absorbs ultraviolet radiation that if it reaches Earth can cause skin cancer, cataracts, and damage to marine organisms. In the 1980s, scientists first noticed a dramatic drop in the amount of ozone over the South Pole—commonly referred to as the "hole" in the ozone layer. The loss was linked to emissions of pollutants, such as chlorofluorocarbons (CFCs), associated with human use of spray cans and refrigeration. **Ozone depletion**—the loss of the protective layer of ozone gas—can result in higher levels of ultraviolet radiation and associated health and ecosystem risks. Australia, with its southern latitude location, sunny days, and tradition of beach going and sunbathing, is especially vulnerable to the effects of ozone depletion and has some of the world's highest levels of skin cancer. However, the **Montreal Protocol**— the 1987 international treaty established to reduce chemical emissions that damage the ozone layer—has succeeded in preventing further ozone depletion through banning some of the polluting chemicals.

In Oceania, the impacts of global warming resulting from an increase in carbon dioxide emissions worldwide include drier conditions in the already drought-prone interior of Australia, increased risk of forest fires, melting of the magnificent glaciers of New Zealand and Antarctica, and sea-level rise that threatens coasts and low-lying islands (from ice melting and because the ocean takes up slightly more volume as it warms). Sea-level rise is of urgent concern to many Pacific Islands, especially low coral atolls, where any increase in sea level may result in the disappearance of land, saltwater moving into drinking water supplies, and an increased vulnerability to storms (see Chapter opener).

Australia is a major emitter of greenhouse gases because of its reliance on, and export of, coal, but it is also experiencing observable climate change. Elections have included fierce debates about whether the country should reduce emissions, and as of 2015 a tax on carbon had been canceled. Meanwhile Australians are adapting to warmer temperatures by relocating agricultural zones, moving vulnerable species from warmer to cooler regions, building desalination plants to ensure urban drinking water supplies, and reviewing wildfire and flood emergency plans.

Apply Your Knowledge

11.1 Find a government or other website of the smaller Pacific Islands, such as Tuvalu, Kiribati, or Samoa. How does the website describe local vulnerability to climate change? What governmental policies are being proposed to reduce the risk of climate change?

Ozone and Montreal Protocol

https://goo.gl/MuYW02

11.2 Consult information sources about the ozone layer and Montreal Protocol (e.g., http://www.epa.gov/ or http://ozone.unep.org) to assess the success of policies to stop ozone depletion. What is the role of older motor vehicles in the ozone problem?

Geological Resources, Risks, and Water

Geological history, especially tectonic activity, influences the geography of many of the major subregions of Oceania (**FIGURE 11.5**). Australia is a very old and stable landmass that sits on the Australian tectonic plate (see Chapter 1, Figure 1.12), whereas New Zealand sits at the boundary where major tectonic plates produce high levels of volcanic and earthquake risk (see "Visualizing Geography: New Zealand's Physical Landscape" on pp. 434–435).

Australian Physical Landscape The vast Australian continent—almost as large as the United States, minus Alaska— has three distinct physiographic regions.

- The *eastern highlands* of Australia are the remnants of an old folded mountain range with a steep escarpment on the eastern flanks. The highland crest is called the Great Dividing Range because it separates the rivers that flow to the east coast from those flowing inland or to the south.

- The middle of the country—*the interior lowlands*—has a series of dry basins and then rise to the third physiographic region, the western Australian Plateau. A shallow ocean once flooded the lowlands. However, only Lake Eyre now remains and it is often dry, filled only occasionally by inland-draining rivers. The **Great Artesian Basin** occupies a large part of the lowlands. The basin holds the world's largest groundwater aquifer, a massive reservoir of underground water in porous rocks. The basin is described as artesian because overlying rocks have placed pressure on the underground water so that when a well is drilled, the water rises rapidly to the surface and discharges as if from a pressurized tap.

- The *western plateau* of ancient shield rocks, with a few low mountains and large areas of flatter desert plains and plateaus, occupies two-thirds of Australia. This region has numerous mineral deposits—the basis for Australia's mining industry—and old, weathered soils that are too nutrient poor or salty for agriculture.

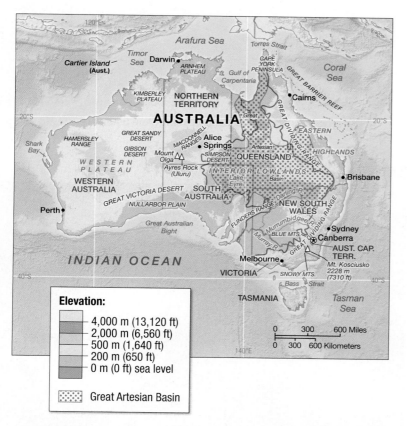

▲ **FIGURE 11.5 Physical geography of Australia** In Australia, key features include the Great Dividing Range of the eastern highlands, central deserts, Great Artesian Basin, and Great Barrier Reef.

The Great Artesian Basin and Outback The wells that tap the Great Artesian Basin are critical to the human settlements and livestock of arid east-central Australia, although the cost of drilling is increasing and the water is often warm and salty. Although inland Australia is very dry, the exploitation of underground water from the Great Artesian Basin supports scattered homesteads where ranchers raise livestock on sheep and cattle stations (the name used for a farm or ranch).

The remote and drier inland areas of Australia are often called the **Outback**—a long way from schools, shops, and hospitals. Adaptations to this isolation include distance education, with children taught by radio or Internet, and a flying doctor service for emergency medical care.

New Zealand's Physical Geography

The Lord of the Rings and Narnia movies—partly filmed (and then enhanced digitally) among the glaciated mountains of the South Island—have popularized the dramatic landscape of New Zealand. Tourists flock to spectacular fjords such as Milford Sound, the mountain ranges around Mount Cook (Aoraki in Maori) and glaciers reaching almost to the coast. What processes created these landscapes and how and why do they vary across the South Island?

1 Tectonic activity

Because New Zealand straddles a plate boundary, it is tectonically active with earthquakes and, in the north island, volcanic activity. The Pacific Plate is colliding with the Australian Plate at a rate of about 40 mm/yr.

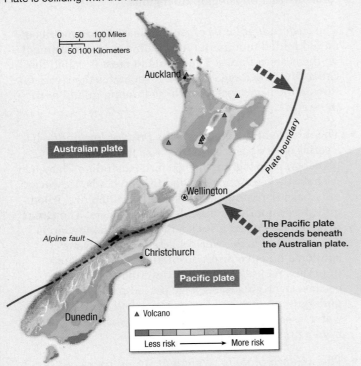

▲ **FIGURE 11.2.1a** Seismic and volcanic activity in New Zealand. This map shows the regions with risk of ground shaking.

> **"New Zealand has protected many of the landscapes shaped by ice and tectonic activity as national parks and conservation areas."**

▲ **FIGURE 11.2.1b** Subduction at the plate boundary.

▼ **FIGURE 11.2.2** 2010 earthquake in Christchurch.

Christchurch earthquake

https://goo.gl/eHk7On

The 2010 and 2011 Canterbury earthquakes (also known as the Christchurch earthquakes) struck the South Island of New Zealand causing considerable damage. The photo shows dust rising immediately after the 2010 earthquake that killed 185 people and damaged many buildings.

2 Glaciation

New Zealand has dozens of impressive glaciers, mostly in the South Island where the Frans Josef and Fox glaciers flow from mountains to the coast (fig 11.2.3b). These glaciers are remnants from the last major glaciation 18,000 to 20,000 years ago, when sea level was lower and ice covered the southern Alps (fig 11.2.3a). During the glaciations, rivers of ice dug deep valleys near the coasts, several of which were flooded by the ocean as the ice retreated (the process of glaciation is described in chapter 1). The deep coastal inlets include Milford Sound, one of the wonders of the natural world.

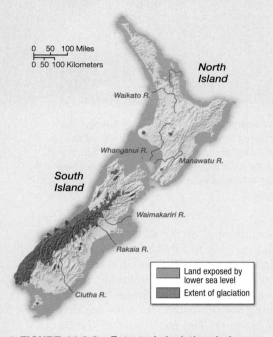

▲ FIGURE 11.2.3a **Extent of glaciation during the ice age.**

▲ FIGURE 11.2.3b **Extent of current glaciers.**

Each glacier in the New Zealand Glacier Inventory is plotted here as a red dot. Most are found in the central (and highest) part of the Southern Alps. Apart from several small glaciers near the crest of Mt Ruapehu, glaciers are confined to the South Island.

1. **Find a web site promoting tourism to New Zealand's South Island. What aspects and photos of the physical landscape are used to encourage people to visit?**

2. **Which other world regions feature glaciated landscapes and what are some of the typical landforms left by ice?**

3 Conservation

New Zealand has protected many of the landscapes shaped by ice and tectonic activity as national parks and conservation areas.

▲ FIGURE 11.2.4a **Milford Sound.**

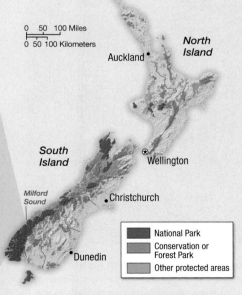

▲ FIGURE 11.2.4b **Map of protected areas.**

Causes of New Zealand earthquakes

https://goo.gl/7Qr9bT

Desert Landforms Australia has impressive examples of desert landforms, including wind-shaped undulating ridges of sand dunes, stony plains with varnished rock fragments called desert pavement, and dry interior drainage basins called *playas.* Centuries of erosion by wind and water have left erosion-resistant domes of rocks standing above the surrounding landscape. Thousands of tourists visit the most famous of these isolated rock domes, which are also sacred sites for Australia's **Aborigines**— the indigenous peoples who have lived in Australia for thousands of years. Famous rock domes include *Uluru* (also known as Ayres Rock), and the Olgas, called *Kata Tjuta* (**FIGURE 11.6**).

New Zealand Physical Landscape In contrast to Australia, New Zealand is much younger geologically and more tectonically active because it is located where the Pacific Plate is moving under the Australian Plate (see "Visualizing Geography:

▲ FIGURE 11.7 **New Zealand Volcanism and Geothermal Energy** The North Island of New Zealand has hot springs, geysers, and steam vents associated with volcanic activity. Geothermal energy sources, such as the 140 MW Kawerau power plant shown in this photo, contribute 13% of New Zealand's electricity production. **Which other countries generate a lot of energy from geothermal sources?**

New Zealand's Physical Landscape" on pp. 434–435). The South Island has rugged mountains rising to 3,754 meters (12,316 feet) on Mount Cook in the Southern Alps. This is far enough south (and thus cold enough) to have extensive permanent snowfields and more than 300 glaciers, some flowing almost to sea level. The west coast has stunning fjords created when the sea flooded the deep valleys cut by glaciers. The east coast of the South Island has much gentler relief, with rolling foothills, long valleys with braided rivers and freshwater lakes, and plains formed from stream deposits that are used for agriculture. The North Island has much more volcanic activity than the South Island. On the North Island, volcanically heated water that emerges from hot springs, geysers, and steam vents is captured in geothermal facilities and used as a nonpolluting energy resource (**FIGURE 11.7**). The North Island also has extensive areas of rolling hills and valleys, where the warmer climate and rich volcanic and river-deposited soils nourish a productive agriculture.

Pacific Island Physical Landscapes There are two main types of Pacific Islands: high and low.

The *high Pacific Islands,* which are mostly volcanic in origin, rise steeply from the sea and have very narrow coastal plains and deep narrow valleys (see Figure 11.4). Many high islands, such as those of Samoa (and Hawaii), are created in linear chains as tectonic plates move over hot spots where molten rock reaches the surface. Others, such as the Marianas Islands and Vanuatu islands, form along the edge of tectonic plates.

The high island of New Guinea is the second-largest island in the world (after Greenland; because Australia is a continent). The mountain spine of the island of New Guinea rises to more than 4,000 meters (13,000 feet) and features many extinct volcanoes and high isolated basins.

The *low Pacific Islands* are mostly atolls created from the buildup of skeletons of coral organisms that grow in shallow

▲ FIGURE 11.6 **Uluru-Ayres Rock** The Ayres Rock tourist resort serves visitors who come to see the dramatic landscapes of central Australia and generates jobs for local Aboriginal peoples. Ayres Rock can be seen in the distance with the Uluru-Kata Tjuta cultural center in the foreground.

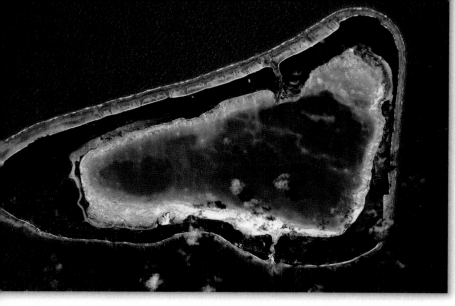

▲ **FIGURE 11.8 Pacific Islands and Atolls** Low islands and reefs surround a lagoon at Marakei Atoll in Kiribati. These low islands are vulnerable to sea-level rise.

tropical waters. Atolls are usually circular, with a series of coral reefs or small islands ringing and sheltering an interior lagoon (**FIGURE 11.8**). These lagoons may contain the remnants of earlier islands or a volcanic island that has sunk below the surface. Many of these islands are at sea level making them vulnerable to storms, tidal waves, and rising seas (see Chapter Opener).

Mining

The Pacific mining boom began in the 1840s when valuable mineral deposits were identified in southern Australia, including silver, copper, and gold. By the 1850s, 40% of the world's gold was coming from Australia and many European immigrants were arriving to work in the mines. Other globally significant mineral deposits in the ancient rocks of Australia include nickel, opals, uranium, and coal, and major mining centers are now found in southern, western, and northern Australia (**FIGURE 11.9**).

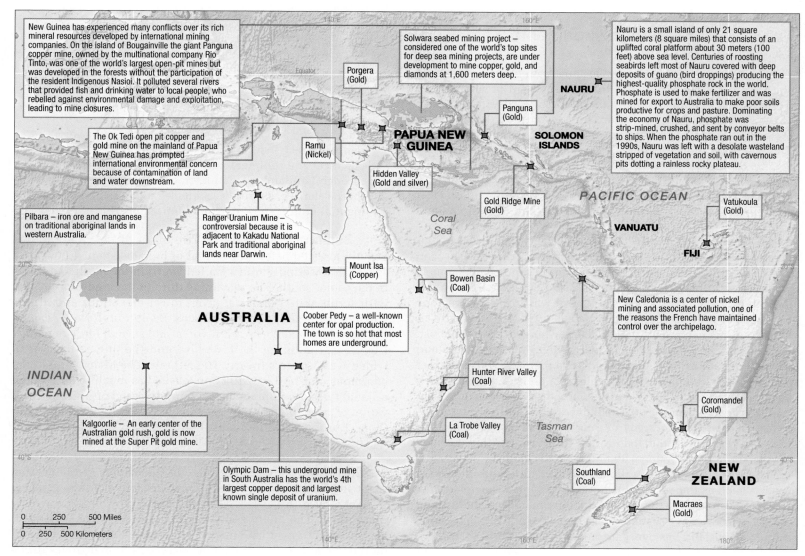

▲ **FIGURE 11.9 Mining in Oceania** Mining is a key economic activity across the region. This map highlights important and sometimes controversial mines in Australia, New Zealand, New Guinea, New Caledonia, Fiji, and the Solomon Islands, as well as the Solwara project to mine the ocean northeast of New Guinea in the Bismarck Sea. **What are the key minerals mined across Oceania?**

New Zealand also has abundant mineral resources, including iron, coal, and gold. Minerals are also significant to the economies of a distinctive set of islands across Oceania, where mining has destroyed landscapes and created social tensions over the wealth that flows from mineral exports.

11.3 Conduct Internet research to understand a conflict over mining on a Pacific Island—what are some of the complaints of local people and how, if at all, have they been resolved?

11.4 Read about the Gold Rush in Australia (there are many websites) and consider parallels with the mining boom in the United States or other world region.

Ecology, Land, and Environmental Management

The relatively long isolation of Oceania from other regions and continents has contributed to the development of some of the world's most unique, diverse, and vulnerable ecosystems. Many of the species that evolved from the isolated populations have remarkable adaptations to the physical environment and are found only in that locality. Contact of the region with other regions through exploration, trade, and migration often changed these ecosystems significantly. As such, the region bears the ecological marks of both its isolation and global connectivity. Oceania's marine ecosystems are highly diverse and of great value.

Australian Ecosystems Australia has very high levels of biodiversity, including more than 20,000 different plant species, 650 species of birds, and 380 different species of reptiles. Its fauna is distinctively different from that of adjacent Southeast Asia because even during glaciations, when sea level was lower, the regions were separated by deep ocean. This faunal boundary is called Wallace's line after naturalist Alfred Wallace, who noticed the contrast.

Australia has several remarkable species of **marsupials** and **monotremes**. A marsupial gives birth to a premature offspring that then develops and feeds from nipples in a pouch on the mother's body. Australian marsupials range in size, from the large gray kangaroos to koalas, wombats, to small mice and voles. Monotremes—such as the platypus and spiny anteater—are very unusual mammals in that they lay eggs rather than gestate their young within the body but then nurture the young with milk from the mother.

The vegetation of Australian ecosystems includes dominant species such as the eucalyptus or gum tree and acacia or *wattle.* The northern interior has extensive but sparse grasslands, and the driest zones have scattered desert grasses and shrubs (but no cacti). Queensland has dense tropical forests (**FIGURE 11.10**).

- **Aborigines and Australian Ecosystems** As Aborigine populations grew, they may have reduced local populations of major game species such as kangaroos, but the most significant environmental change they implemented was the transformation of vegetation through the use of fire. Aborigines used fire both to improve grazing for game and to drive animals to hunters. Ecologists believe that over thousands of years, some Australian vegetation became more resistant to these fires.

- **The Great Barrier Reef** Australia has some of the world's most iconic marine and coastal ecology such as the **Great Barrier Reef**, which is a United Nations Educational, Scientific, and Cultural Organization (UNESCO) World Heritage Site. Fringing the northeast coast of Australia, the reef is more than 2,000 km (1,250 miles) long and easily visible from space. The reef's biodiversity is remarkable: Diving below the surface, the visitor encounters intricate and colorful corals and waving sea grasses that are home to millions of brilliant fish and marine animals such as turtles, whales, and dugong (large seal-like mammals) (**FIGURE 11.11**). The reefs were formed over millions of years from the skeletons of marine coral organisms in the warm tropical waters of the Coral Sea. More than 2 million tourists a year visit the reef and participate in activities that include fishing, diving, snorkeling, and reef walking. The reef is at risk from human activity that includes climate change (which causes ocean acidification and coral bleaching from higher ocean temperatures); fishing with large nets that damage reefs; chemical and sediment pollution from rivers that run through farmland; predation on coral by the crown-of-thorns starfish; shipping accidents; and coastal development, including new ports and tourist facilities.

New Zealand Ecosystems Two-thirds of New Zealand was heavily forested before the arrival of humans about 1,000 years ago, with towering conifers (called kauri) in the north and beech in the cooler south. The remaining third of the land was covered with scrub, with grasses at drier, lower elevations and alpine grasslands (tundra) at high altitudes. Evolution in New Zealand produced no predators or carnivores, and several birds remained flightless, including the kiwi bird and the now extinct moa, which was 3 meters (nearly 10 feet) tall. The Maori hunted the enormous moa bird to extinction, cleared as much as 40% of the original forests, and practiced agriculture based on shifting cultivation of sweet potatoes—a knowledge system they brought with them. European settlers and their animals further decimated local ecosystems.

Pacific Island Ecosystems On the islands of the Pacific, plant species that can be easily transported by ocean (e.g., coconuts) or air (e.g., fruit seeds eaten and excreted by birds) are widely distributed. The biodiversity declines as one moves eastward, away from the larger landmasses. Luxuriant rain forests are found on the wetter and higher islands; marshes and mangroves thrive along the coastal margins. The larger islands, such as New Guinea and Hawaii, also have extensive middle-elevation grasslands.

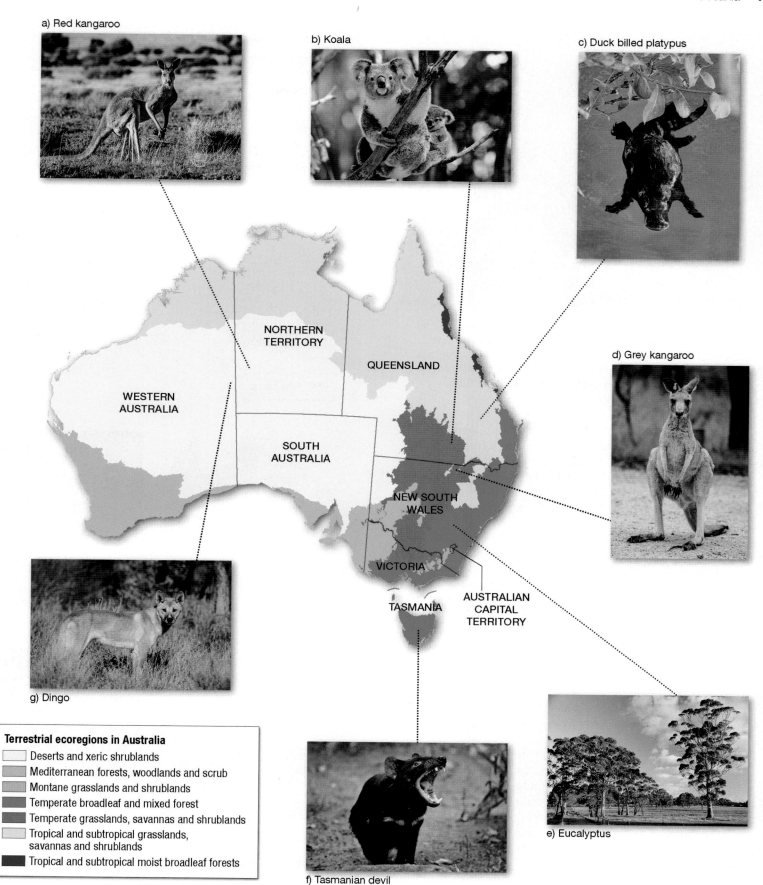

a) Red kangaroo

b) Koala

c) Duck billed platypus

d) Grey kangaroo

g) Dingo

f) Tasmanian devil

e) Eucalyptus

NORTHERN
TERRITORY

QUEENSLAND

WESTERN
AUSTRALIA

SOUTH
AUSTRALIA

NEW SOUTH
WALES

VICTORIA

TASMANIA

AUSTRALIAN
CAPITAL
TERRITORY

Terrestrial ecoregions in Australia

Deserts and xeric shrublands

Mediterranean forests, woodlands and scrub

Montane grasslands and shrublands

Temperate broadleaf and mixed forest

Temperate grasslands, savannas and shrublands

Tropical and subtropical grasslands,
savannas and shrublands

Tropical and subtropical moist broadleaf forests

▲ FIGURE 11.10 **Ecosystems in Australia** Australia has large areas of desert and shrub ecology and tropical and temperate forests along the east coast and in the southwest. Several unique species occupy the landscape including a) red kangaroos, b) koalas, c) duck billed platypus, d) grey kangaroos, e) eucalyptus trees, f) the Tasmanian devil, and g) dingos.

▲ **FIGURE 11.11 The Great Barrier Reef** The undersea landscape of the Great Barrier Reef has become a major destination for tourists who swim among the corals and grasses and view the colorful tropical fish. It consists of 3,400 coral reefs, incorporating more than 300 species of coral and hosting more than 1,500 species of fish and 4,000 different types of mollusks. **What are some of the threats to the reef?**

The smaller and drier coral islands have much sparser vegetation, but coconut palms, which are a basis of human subsistence and grow along many beaches, are ubiquitous. There are few native mammals on the Pacific Islands (with the exception of New Guinea). The richest fauna include the birds and marine organisms, especially those of reefs and lagoons, including turtles, shellfish, tuna, sharks, and octopus.

Introduced Exotic and Feral Species Beginning with the introduction of the dingo from Southeast Asia by Australian Aborigines about 3,500 years ago, introduced species have devastated the native species of Oceania. (Introduced species are also called **exotics** because they come from elsewhere.) The dingo, a canine similar to a coyote, is believed to have outcompeted and outhunted the marsupial predators, such as the now-extinct Tasmanian wolf. **Ecological imperialism**—the process of European organisms taking over the ecosystems of other regions of the world—led to the endangerment and extinction of numerous other native species in this region as introduced species came to dominate and local species lost out due to pressures from hunting, competition, and habitat destruction.

The flightless birds of Oceania were the most vulnerable to introduced predators such as rats, cats, dogs, and snakes. Conservation efforts to protect birds today include the establishment of reserves and the careful monitoring and elimination of predators. On the Pacific Islands, there are great efforts to contain the spread of the introduced brown tree snake, the mongoose (a very aggressive small mammal), and a carnivorous snail, all of which prey on local species.

Feral animals of Australia—domesticated species that end up in the wild—include horses, cattle, sheep, goats, pigs, and camels (**FIGURE 11.12**). Escaped populations are the

TABLE 11.12 Feral Animals in Australia		
Species	Reason for introduction	Number
Camel	Transport—feral	300,000
Cane toad	Beetle control	200 million
Cat	Pet/rodent control—feral	15 million
Donkey	Transport—feral	5 million
European rabbit	Hunting	200 million+
Horse (brumby)	Farmwork—feral	300,000+
Pig	Livestock—feral	23 million
Red fox	Hunting	7.2 million+

(a)

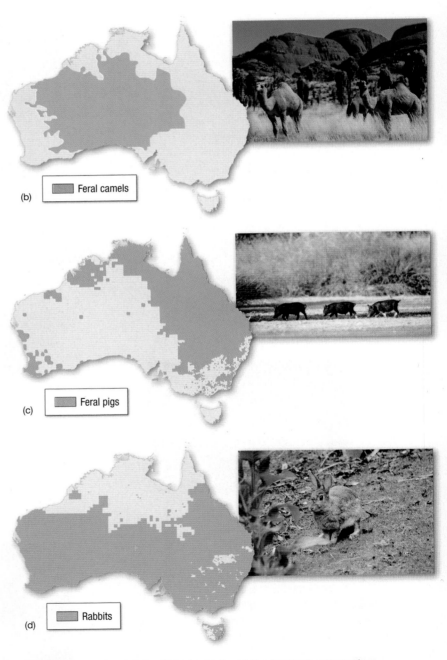

(b) Feral camels

(c) Feral pigs

(d) Rabbits

▲ **FIGURE 11.12 Feral Animals in Australia** (a) Feral animals, escaped from farms and captivity are common in the Australian landscape and include millions of toads, cats, donkeys, camels (b), pigs (c) and rabbits (d).

Ocean Plastic Pollution

The world's oceans are becoming increasingly polluted as a result of human activity. Fertilizer pollution creates "dead zones" along coasts where marine life lacks oxygen. Carbon dioxide increases contribute to the acidification of the ocean and destroy reefs. But much of the recent concern about ocean pollution focuses on plastic pollution. Plastic breaks down into smaller particles but stays in the ocean for hundreds of years, contaminating and killing marine life. Plastic waste is dumped into the ocean in the form of fishing nets and port garbage but also from untreated storm and wastewater, winds and tsunami waves that carry plastic bags and bottles, and microbeads from cleansers into the sea. A lot of the plastic is carried by ocean currents into "gyres" where currents circulate and trap waste creating vast mats of floating trash such as the **Pacific Garbage Patch** (**FIGURE 11.3.1**). The solutions to ocean plastic pollution include reducing the use of plastic (such as through bans on use of plastic bags or coastal dumping) and increasing recycling.

"The solutions to ocean plastic pollution include reducing the use of plastic (such as through bans on the use of plastic bags or coastal dumping) and increasing recycling."

Proposals have been made to clean up the ocean garbage patches through collecting the waste with large booms such as those used to manage oil spills.

1. National Geographic's Ocean website has a number of articles on marine pollution http://ocean.nationalgeographic.com/. Read one of these articles and take note of three key things you learn from the article.

2. Use a news site (such as Google News or guardian.co.uk) to find a news story on ocean plastic pollution. What is proposed as a solution? What can you do yourself?

Ocean Plastic Pollution

https://goo.gl/gI0MQx

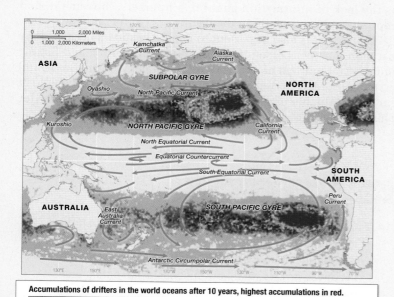

Accumulations of drifters in the world oceans after 10 years, highest accumulations in red.

0.01   2   4   6   8   10

(a)

TOP 10 ITEMS FOUND

① 2,117,931 cigarettes / cigarette filters
② 1,140,222 food wrappers / containers
③ 1,065,171 beverage bottles (plastic)
④ 1,019,902 bags (plastic)
⑤ 958,893 caps, lids
⑥ 692,767 cups, plates, forks, knives, spoons
⑦ 611,048 straws, stirrers
⑧ 521,730 beverage bottles (glass)
⑨ 339,875 beverage cans
⑩ 298,332 bags (paper)

(b)

(c)

▲ **FIGURE 11.3.1 The Pacific Garbage Patch** (a) Global ocean currents include several regions of circular patterns, called gyres, that trap plastic and other waste into accumulations of rubbish that damage marine ecosystems. This map shows where "drifters" used by scientists accumulate in the same regions as waste. (b) Top 10 items found during the 2012 International Coastal Cleanup. (c) Plastic bottles, shoes, and cans wash up on beaches along with coconuts in Fiji.

largest population in the world and include 23 million pigs, 2.6 million goats, 300,000 horses (called brumbies), and 300,000 camels.

Another ecological disaster was the introduction of the European rabbit to Australia in 1859, whose populations exploded and devastated pasturelands. The rabbit population was partially eradicated in the 1950s by the deliberate introduction of a disease called myxomatosis.

Apply Your Knowledge

11.5 Use the Internet to research the current conservation and management status of one of Oceania's unique species (e.g., platypus, Tasmanian devil, etc.) or an invasive species in a specific country of Oceania (e.g., prickly pear cactus, cane toad, etc.). What policies have been put in place, and have they been successful?

11.6 Consult the Encyclopedia of the Earth and read about the Biodiversity of New Zealand http://www.eoearth.org/. What are some of the threats to biodiversity and efforts to conserve it? Or consult the Australian government's Department of Environment website and read about one of the National Biodiversity Hot Spots: http://www.environment.gov.au/. Why is it a hotspot?

Australia and New Zealand Biodiversity
https://goo.gl/fbWJ1z

Sustainability of Marine Resources and Agricultural Land

Overfishing and mining threaten marine environments, and land is at risk from drought, overgrazing, and erosion.

Agricultural Sustainability Farming throughout the region faces environmental challenges. In Australia, drought and poor irrigation practices have led to the buildup of salts and toxic minerals in soils, and the overuse of land and fertilizer has resulted in pollution of waterways and coastal marine systems. The Landcare grassroots network links thousands of farmers who practice more sustainable farming by minimizing chemical use, maintaining trees and irrigation systems, sharing ideas, and protecting wildlife habitat. Traditional hunting and gathering by Aboriginal groups includes the collection of roots, seeds, grubs, insects, and lizards gathered for essential calories and proteins and fishing in coastal areas.

In New Zealand, farmers have responded to consumer demand by increasing production of organic fruit (e.g., kiwi and apples), dairy, and vegetables. While farming is difficult on many of the Pacific Islands, their isolation means agriculture uses few chemicals and some producers are now serving international organic markets with coconut, vanilla, and spices.

Fisheries Oceania provides many interesting examples of how communities manage renewable marine resources such as fisheries, which are often **common property resources** (managed collectively by a community that has rights to the resource, rather than owned by individuals). Strategies for managing marine resources in the Pacific include moratoriums—periods or places where fishing is not permitted by local communities—called *tabu* in the Pacific. Family or group access based on customary rights to harvest a resource is also recognized. Recently, some countries have brought fisheries under government control or have regulated harvesting more formally through permits and quotas. The establishment of the international 200 nautical mile (370 km) **exclusive economic zone (EEZ)**, which was formalized in 1982 in the **United Nations Convention on the Law of the Sea (UNCLOS)**, was of tremendous significance to Oceania because it allowed countries with a small land area but many scattered islands, such as Tonga and the Cook Islands, to lay claim to immense areas of ocean (**FIGURE 11.13**). These island nations can now demand licensing fees from the international fleet that seeks to catch tuna, for example, within their zones.

Outside of the EEZs, fisheries are open to all and are vulnerable to the so-called **tragedy of the commons**, in which an open-access common resource is overexploited by individuals who do not recognize how their own use of the resource can add to that of many others to degrade the environment—for example, overfishing a given species to the point of extinction. One of the greatest challenges in the sustainable management of fisheries is the lack of information about fish numbers, movement, and reproduction, especially in the Pacific.

Ocean law is also important to Pacific Islanders because they eat more fish per person than any other population, and fishing and coastal tourism are critical to the majority of smaller island economies. Marine resources include not only fish and shellfish but also valuable exports such as pearls and shells (mostly for shirt buttons).

Apply Your Knowledge

11.7 Conduct some Internet research to learn about efforts to reduce the pollution in the world's oceans. Who is undertaking these efforts, what do cleanup initiatives involve, and how successful have they been to date?

11.8 The UN Food and Agricultural Organization monitors the state of fisheries around the world. Consult a recent report on the Pacific Islands (e.g., at http://www.fao.org) and use the example of one island to understand what the main trends, issues, and management approach are with their fisheries.

Pacific Islands Fisheries
https://goo.gl/FynrfN

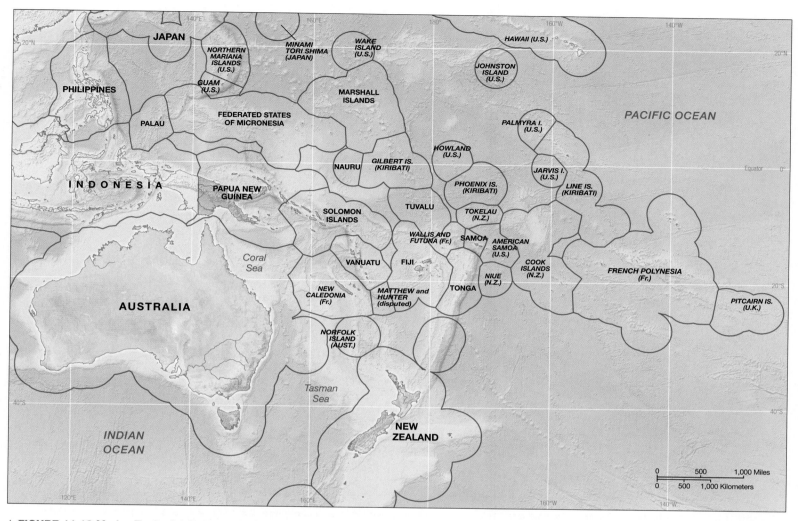

▲ **FIGURE 11.13 Marine Territorial Claims in the Pacific** The EEZs of Pacific nations and territories leave very little unclaimed ocean in the region. UNCLOS, the international treaty that established these zones, was ratified by 119 member states of the UN in 1982 and now consists of more than 130 signatory members. One result of the treaty is that 90% of the world's fisheries are now claimed. **Which Pacific islands seem to claim the largest areas of ocean?**

History, Economy, and Territory

Oceania's history begins with early human occupation and the spread of populations across the region, then European contact and colonization, independence, the impact of World War II, and integration into a global economy. The historical geography in the region has produced economies that are globally and regionally significant and others that are dependent or underdeveloped, such as French Polynesia and Papua New Guinea. Some of the Pacific Islands remain colonies. Australia has strong connections with Southeast Asia and New Zealand, and exports and goods around the world.

Historical Legacies and Landscapes

The early human history of Oceania begins with the migration of humans from Southeast Asia into Australia, New Guinea, and nearby islands about 40,000 years ago. A second dispersal to the Pacific Islands, such as Fiji, Tonga, and Samoa, followed about 3,500 years ago, along with dispersal from India to Australia (**FIGURE 11.14**). New skills in tracking human DNA and bacteria, as well as linguistic connections, have enhanced the understanding of these movements.

The early inhabitants of Australia were the ancestors of the Aboriginal and Torres Strait people, who still live in Australia today. Early Aborigines left a legacy of rock paintings and traditions that include respect for the land and ancestors. In New Guinea, indigenous communities developed agriculture around 7000 B.C.E., domesticating plants such as taro and banana and clearing forests and grasslands with fire. The **Maori**, the indigenous people of New Zealand, arrived in New Zealand from eastern Polynesia sometime before 1300 C.E.

Early European Exploration and Colonization

The Portuguese explorer Ferdinand Magellan crossed the Pacific in 1520, encountering the Mariana Islands on the way to the Philippines.

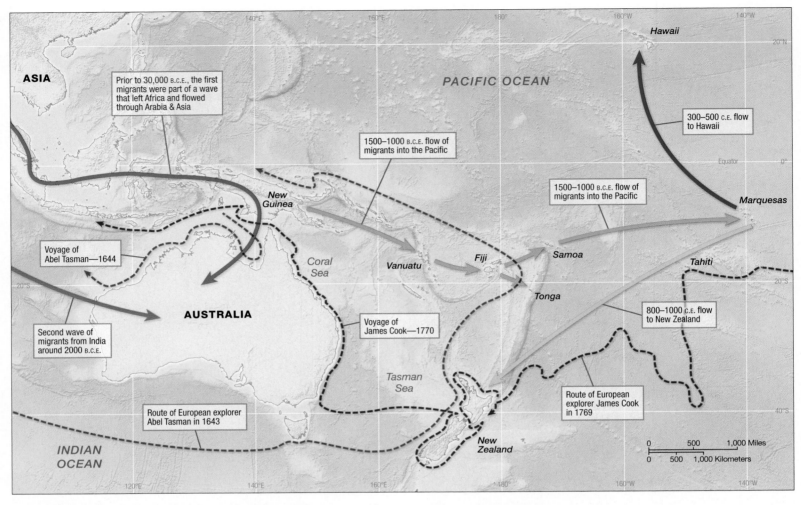

▲ **FIGURE 11.14** **The Peopling of Oceania** According to DNA research, people arrived in Oceania from Asia, first to Australia and New Guinea, then eastward across the Pacific. New Zealand was populated from the eastern Pacific.

European contact with Oceania began in earnest in the 1600s with Dutch exploration of the west coast and south coast of Australia, claiming it as Van Diemen's Land in 1642. The Spanish settled Guam in 1668 as a base to control sailing routes. The Dutch explorer Abel Tasman also approached New Zealand and made contact with the Maori in 1642, but the encounter was violent and the Dutch did not land. They did, however, begin calling the islands *Nieuw Zeeland* (after a region of the Netherlands). It took another 100 years for explorer Captain James Cook, who sailed via the Pacific Island of Tahiti to land in New Zealand and then at Botany Bay, Australia, to claim the land for Britain in 1770 (Figure 11.14).

Australia Based on Cook's reports, the British government decided to populate New South Wales by sending more than 160,000 convicts (mostly petty thieves and political activists) to the region to relieve pressure on their prisons, reinforce territorial claims, and provide cheap labor for economic development. Many eventually gained freedom through pardons and decided to stay in Australia, often as farmers. Other free settlers were offered cash rewards for emigrating and were assigned convicts as laborers. Farming was often difficult because of poor soils, plant disease, and a variable climate, and many of the new communities depended on imports of food and other goods from Britain. The most successful agricultural areas produced wheat in coastal valleys, such as the Derwent Valley in Tasmania, and around Adelaide. The whaling trade, based in Hobart, Tasmania, generated the main exports from the initial settlements.

A momentous shift took place with the import of the first livestock, especially the Merino sheep, which thrived in central New South Wales and Tasmania. High global prices for wool exports and the success of the British textile industry encouraged sheep raising. By 1860, investment was increasing, exports were booming, and there were 21 million sheep in Australia. Stockmen began to move inland, especially to the grasslands on the western slopes of the Great Dividing Range. European frontier settlers came into conflict with the Aborigines, who defended their traditional lands and lost their lives in the process.

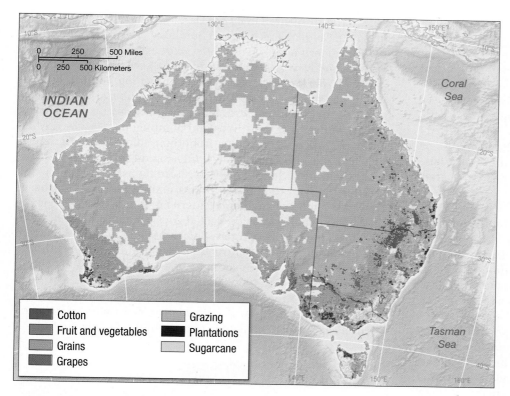

▲ FIGURE 11.15 Australian Agriculture Although much of Australia is dry and used primarily for grazing, there are important agricultural areas across the country where grains, such as wheat, are grown. There are also areas along the tropical coast of Queensland that grow sugar cane. The Mediterranean climate of southern Australia supports cotton, rice, and a thriving wine industry. **Where are most crops grown in Australia?**

Legend:
- Cotton
- Fruit and vegetables
- Grains
- Grapes
- Grazing
- Plantations
- Sugarcane

established until the 1840s. The British initially treated New Zealand as part of Australia. Small whaling settlements established trading relationships with the Maori. This contact gave the Maori access to firearms, exposed them to disease and missionaries, and ushered them into the capitalist economy.

The British annexed New Zealand in 1840 through the **Treaty of Waitangi**, a pact with 40 Maori chiefs on the North Island (**FIGURE 11.16**). This treaty purported to protect Maori rights and land ownership if the Maori accepted the British monarch as their sovereign, granted the British crown monopoly on land purchases, and became British subjects. Alarmed by European settlement, land purchases, and the expansion of sheep pastures, some Maori resisted the British and waged warfare through the mid-1800s.

After 1882, the technological innovation of refrigerated shipping allowed the economies of New Zealand and Australia to expand and shift from producing nonperishables such as wool, metals, and wheat to exporting the more profitable fresh and frozen meat and dairy products. The opening of the Suez Canal in 1867 and the Panama Canal in 1914 facilitated trade by reducing the time and cost of ocean transport to Europe. Australia and New Zealand became

Commercial wheat production centered in southeastern Australia (**FIGURE 11.15**). The frost-free climate of the central coast of eastern Australia also allowed for sugar plantations around Brisbane beginning in the 1860s. These plantations depended on indentured laborers brought in from the Pacific Islands of Vanuatu and the Solomon Islands. Cattle were also introduced into the warmer and drier regions of central and northern Australia after wells were drilled into the Great Artesian Basin in the 1880s. Cattle survived more easily than sheep on the sparse vegetation. The construction of railroads, which radiated from ports to livestock yards, grain elevators, and mines, further facilitated development in the region.

Another major transformation of the Australian economy occurred with the discovery of gold in 1851. By this time the country was moving toward independence with two-thirds of the legislatures elected by popular vote (and the rest appointed by the British).

New Zealand European settlement and economic development in New Zealand began with the establishment of a sealing station on the South Island in 1792, but the first official settlers' colonies were not

▲ FIGURE 11.16 Treaty of Waitangi Waitangi Day commemorates the treaty signed by Britain's Queen Victoria and the indigenous Maori peoples. Maori often protest on this annual holiday, claiming that many promises remain unfulfilled, including those addressing land and resource rights. Here, hundreds of demonstrators in a *hikoi* (march) head for the Waitangi Treaty Grounds in Waitangi, Bay of Islands, New Zealand.

staple economies that depended on the export of a few natural resources.

Pacific Islands Initially the Pacific Islands were of little interest to Europe. However, passing European sailors brought fatal diseases to local populations who had no resistance to them.

As Britain and France rose to power in Europe in the 18th century, their adventurers set out for the Pacific Islands. British explorer Samuel Wallis and Luis Antoine de Bougainville of France were made welcome by the people of Tahiti in the 1760s, and their reports of friendly people and abundance cultivated the myth of a tropical paradise.

By the beginning of the 19th century, hundreds of whaling ships called regularly at islands such as Tahiti, Fiji, and Samoa for supplies. Missionaries sought conversions in Tahiti, Samoa, Tuamotu, Tuvalu, the Cook Islands, Tonga, and Fiji. The missionaries often advised native chiefs to alter local laws and traditions to conform to European principles, disrupting traditional social ties, beliefs, and political structures and provoking some rebellions.

Coconut, the staple product of the Pacific, became part of European trade from about 1840 in the form of copra, dried coconut meat, used to make coconut oil for soaps and food. From the 1840s to 1904, people from the Pacific were kidnapped and enslaved in a process called **blackbirding** that brought thousands of laborers, collectively called *kanakas* (because many of them were of Kanak origin from the islands now known as Vanuatu), to Australian cotton and sugar plantations.

As in sub-Saharan Africa, European nations governed the islands according to their different colonial styles, modified to local conditions. Britain ruled through governors who incorporated native leadership into their administrations in a form of indirect rule. The French practiced direct rule and assimilation into French culture and institutions. The colonial powers restructured the islands' economies and societies in ways that left enduring legacies. For example, the British brought numerous contract workers from India to work on plantations in Fiji, creating a divided society of ethnic Asians and Pacific Islanders that produces significant political tensions even today.

Independence, World War II, and Global Reorientations
Britain granted independence to Australia in 1901, creating the Commonwealth of Australia with a federal structure composed of six states, plus eventually the Northern Territory and the Australian Capital Territory around Canberra. New Zealand declined to join this new nation, instead choosing dominion status as a self-governing colony of Britain in 1907. Both countries gained their own colonies in the Pacific in 1901, when Britain turned over Papua New Guinea to Australia and the Cook Islands to New Zealand.

Australia and New Zealand joined the Allies in World War I and more than 33,000 of their soldiers were killed or wounded at Gallipoli, Turkey, in 1915. World War II further influenced Australia's geopolitical orientation. Although Australians fought with Britain in Europe and North Africa, Australia linked more closely with the United States to fight the rapid advance of the Japanese in Asia and the Pacific, which saw the bombing of Pearl Harbor in Hawaii, the capture of 15,000 Australians in Singapore, and the bombing of the northern Australian city of Darwin. This alliance meant that during the Cold War, as East and Southeast Asia became the focus of U.S. concern about communist expansion from China, both Australia and New Zealand sent troops to Korea and Vietnam.

World War II also marked a critical turning point in the history of the Pacific Islands. During the war, thousands of foreign soldiers fought and constructed military bases on the islands. Many islanders lost their lives and bombing damage was extensive as the Japanese advanced from their colonies in Micronesia (such as Palau) to Guam, New Guinea, and the Solomon Islands and as the United States fought to retake the islands from Japan.

After the war, the United States maintained military bases on islands such as Guam and American Samoa (**FIGURE 11.17**). Self-government began with elected governments and small independence movements. Western Samoa became independent from New Zealand in 1962; Nauru from Australia in 1968; Fiji and Tonga from the United Kingdom in 1970; Papua New Guinea from Australia in 1975; and the Solomon Islands, Kiribati, and Tuvalu from the United Kingdom and Vanuatu from the United Kingdom and France by 1980. Some islands, such as New Caledonia. French Polynesia and American Samoa, remain under foreign control.

Apply Your Knowledge

11.9 Use the Internet to find government or museum websites from Oceania that describe the early history of a country. What evidence do they use to tell the early history?

11.10 Find a film or book about World War II set in Oceania, such as the classic 1949 Broadway musical *South Pacific,* which was made into a film released in 1958. How do the stories portray local residents? Do you think these portrayals are accurate?

▼ **FIGURE 11.17 Andersen Air Force Base on Guam** F-15 and B-2 planes fly over the U.S. base on the island of Guam. The U.S. established military bases on islands seized from the Japanese during World War II and has maintained a presence across the Pacific since.

Economy, Accumulation, and the Production of Inequality

While some countries, such as Australia and New Zealand, have diversified their economies and exports and built strong domestic demand, other countries have economies that are based in part subsistence, dependent on tourism or a few exports, or have low levels of consumption.

Agriculture and Economy As noted earlier in the chapter, the Australian economy developed around production and export of agricultural commodities and minerals from key parts of the country. For example, southeastern Australia has rich agricultural lands from southern Queensland along the coast to the cities of Sydney, Melbourne, and Adelaide, and inland to Australia's capital city of Canberra and the agricultural regions of the Darling Downs, Murrumbidgee Irrigation Area, and the Barossa Valley (see Figure 11.15). These rich agricultural lands host a livestock industry of milk, beef, and lamb production, with animals that graze on pastures improved by fertilizer (especially phosphate) and introduced grasses. Wheat and other grains are often grown in rotation with sheep raising on larger farms. Wheat from this area and from Western Australia combines to support an export wheat industry that ranks with that of Canada and France (only the United States is ahead of this group). Cotton from New South Wales contributes to Australia's dominance in world cotton exports. Southeastern Australia also includes several areas of intensive horticulture (fruits and vegetables) and vineyards. Australian wine exports are now the fourth largest in the world (behind France, Italy, and Spain) and Australia's wineries support a thriving tourist industry.

New Zealand also developed an economy based on agriculture, dominated by livestock, particularly sheep, lamb, and dairy cattle farming, as well as wheat and barley production. New Zealand has put some of its land into wine vineyards and forest exports. The future of agriculture in Australia and New Zealand amid global economic change is unclear, as competition from Latin America and Asia, the high cost of inputs, the loss of subsidies, and the changing structure of demand create a new geography of international agricultural trade.

From Import Substitution to Free Trade in Australia and New Zealand

From the 1920s, regional policies to substitute expensive imports with domestic production resulted in heavy subsidies of manufacturing and high tariffs on imported goods, especially in Australia. The goal was to create new jobs, diversify the economy, and reduce the sensitivity to global demand for wool and minerals. As in other regions, these import substitution policies had mixed success. Although Australia and New Zealand had middle-class populations that generated a demand for manufactured goods, the overall market was small, and labor costs were high as a result of labor union activism. The new industries were often inefficient because they were protected from competition in the world market. New Zealand had a particularly high level of government involvement in the economy, with state-run marketing boards controlling exports such as wool, meat, dairy products, and fruit, and government ownership of banking, telecommunications, energy, rail, steel, and forest enterprises.

Although Australia protected manufacturing and subsidized agriculture, the government set up few barriers to foreign investment. Many sectors, including minerals and land, had high levels of foreign ownership, especially by British firms. During the 1970s, investment patterns began to change, when Asian, especially Japanese, capital started to flow into Australia. Another major change occurred when the United Kingdom joined the European Community in 1973 and ended preferential trade relationships with British Commonwealth nations, including Australia and New Zealand. Trade patterns in both Australia and New Zealand have shifted from exports to Europe to exports to Asia (**FIGURE 11.18**).

Both Australia and New Zealand decided to reduce government intervention in and regulation of the economy in the 1980s. The Australian government deregulated banking, reduced subsidies to industry and agriculture, and sold off public-sector energy industries. New Zealand eliminated agricultural subsidies, removed trade tariffs, reduced welfare spending, and privatized government-owned enterprises, including airlines, postal services, and forests. Although these policies succeeded in reducing the national debt, they also increased economic inequality and unemployment.

The most rapidly growing sector of major regional economies has been services. Employment in this sector has increased from 58% in 1980 to 77% in 2015 in Australia, and from 58% to 75% in New Zealand during the same time period. Finance, tourism, media, and business services expanded the most and were associated with an increase in female employment and considerable foreign investment.

The most important economic regions in Australia for manufacturing are around the cities of Sydney, Melbourne, Brisbane, and Adelaide, and include food processing, electronics, petrochemicals, metals, and auto manufacturing. New Zealand has a significant food industry (linked to the agricultural economy), as well as chemical, wood and paper, and machinery manufacturing.

Economic Development in the Pacific Islands Many Pacific Islands are now integrated into the world system through their dependence on imported goods, people working abroad, aid and transfer payments, air transportation, and the emergence of international tourism. They are sometimes referred to as **MIRAB (migration, remittances, aid, and bureaucracy) economies** because of their dependence on labor migration, money sent back from overseas workers, foreign aid, and jobs in government.

Most of the nations in Polynesia spend twice as much on imports as they gain from exports, but compared to countries in Latin America or Southeast Asia, total international debt is relatively low in Oceania. This low level of debt has been maintained despite the trade deficits because of the large flows of official aid, remittances from overseas workers, and interest on savings held outside the country.

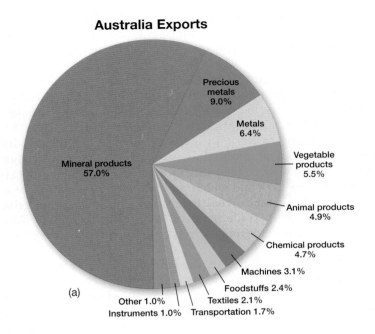

Australia Exports

- Mineral products 57.0%
- Precious metals 9.0%
- Metals 6.4%
- Vegetable products 5.5%
- Animal products 4.9%
- Chemical products 4.7%
- Machines 3.1%
- Foodstuffs 2.4%
- Textiles 2.1%
- Transportation 1.7%
- Instruments 1.0%
- Other 1.0%

(a)

Australia Export Destinations

- China 35.0%
- Others 19.6%
- Japan 17.0%
- South Korea 7.3%
- India 4.2%
- Hong Kong 3.9%
- United States 3.2%
- Malaysia 2.3%
- Thailand 2.0%
- New Zealand 2.0%
- Indonesia 1.8%
- Singapore 1.7%

(c)

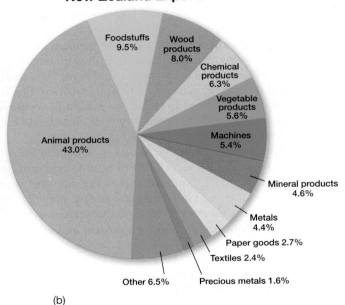

New Zealand Exports

- Animal products 43.0%
- Foodstuffs 9.5%
- Wood products 8.0%
- Chemical products 6.3%
- Vegetable products 5.6%
- Machines 5.4%
- Mineral products 4.6%
- Metals 4.4%
- Paper goods 2.7%
- Textiles 2.4%
- Precious metals 1.6%
- Other 6.5%

(b)

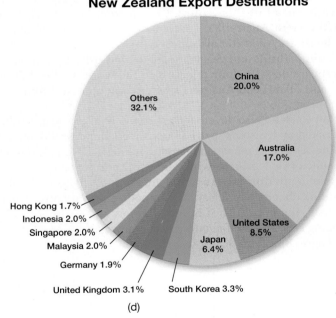

New Zealand Export Destinations

- Others 32.1%
- China 20.0%
- Australia 17.0%
- United States 8.5%
- Japan 6.4%
- South Korea 3.3%
- United Kingdom 3.1%
- Germany 1.9%
- Malaysia 2.0%
- Singapore 2.0%
- Indonesia 2.0%
- Hong Kong 1.7%

(d)

▲ **FIGURE 11.18 Australia and New Zealand Exports** These diagrams show the structure of Australian and New Zealand exports by value in 2012. (a) Australia's exports are dominated by minerals, metals and foodstuffs, and (b) New Zealand's by livestock, food, and wood. (c) The focus of Australian exports is now on Asia (d) New Zealand exports go to China, Australia and a diversity of other destinations. **What is the overall percent of each countries exports that are foodstuffs, vegetable or animal?**

Tourism The economies and employment in the Pacific Islands are extremely dependent on tourism with more than 3 million visitors a year (**FIGURE 11.19a**). The widespread image of the islands as a vacation paradise originated in the accounts of the first European visitors, who described tropical abundance and peaceful locals. The appeal of the region as a tourist destination grew as international air and cruise routes included stops at island groups such as Hawaii and Fiji, often en route to Australia or Asia. But the big boom in Pacific tourism occurred after 1980

as air travel became more accessible and increasing numbers of North Americans and Asians (especially Japanese) sought luxury tropical vacations.

Tourism is the major source of foreign exchange for the Cook Islands, Fiji, French Polynesia, Samoa, Tonga, Tuvalu, and Vanuatu (**FIGURE 11.19b**). Challenges faced by these tourism-dependent economies include vulnerability to international trends in tourism, political unrest that dissuades tourists, ensuring that the benefits of tourism reach everyone, and minimizing

(a)

Receipts from International Tourism, 2012

(b)

▲ **FIGURE 11.19 Tourism in the Pacific** (a) Luxury resorts like Bora Bora in French Polynesia attract tourists with descriptions such as the "islands of dreams" and an "emerald in a setting of turquoise, encircled by a necklace of pearls." Visitors bring valuable foreign exchange to the economy, but resorts put pressure on the local resource base and encourage local people to modify their livelihoods and rituals to cater to foreign visitors. (b) Tourism is an important source of revenue across Oceania. **Which three islands seem to be most dependent on tourism for revenue?**

negative effects on local cultures and environments. These problems can include overuse of resources, damage to ecosystems, disrespect for local culture, sex tourism, and revenue that is accrued mostly by companies outside the region.

Apply Your Knowledge

11.11 Use the Massachusetts Institute of Technology (M.I.T) Observatory of Economic Complexity (https://atlas.media.mit.edu) to analyze the imports and exports of the Cook Islands, Fiji, French Polynesia, Papua New Guinea, Samoa, the Solomon Islands, or Tonga. What do they export and import and what does this tell you about the potential vulnerabilities of their economy?

Pacific Islands Imports and Exports

https://goo.gl/eyo1d4

11.12 What are the risks for countries whose economies depend on tourism? Pick a country in the region and research economic data on how its tourism industry fared over the last decade.

Poverty and Inequality Although Oceania has generally higher incomes and better living conditions than many other world regions, there are significant differences between and within countries of the region. Australia and New Zealand are distinctive for their high average incomes. The per capita annual gross national income (GNI) was U.S. $42,120 (in purchasing power parity [PPP]) for Australia and U.S. $33,760 (in PPP) for New Zealand in 2013. Some Pacific Islands have an average GNI in PPP near or above U.S. $10,000 per person as a result of their associations and jobs with the United States (Guam) or France (French Polynesia, New Caledonia). Others, such as Papua New Guinea and the Solomon Islands, registered less than U.S. $3,000 in GNI per capita in 2013. Although inequality is apparently less than in many other world regions, poverty persists throughout the region, including in Australia and New Zealand, where the Aborigine and Maori populations are particularly disadvantaged.

Australia and New Zealand were for many years reputed to have strong welfare systems and equitable societies, at least for nonindigenous populations. However, income inequality has increased in the last two decades, as governments have reduced or privatized social services. Larger populations of

The Stolen Generations

In February 2008, Australian Prime Minister Kevin Rudd formally apologized for the removal of Aboriginal and Torres Strait Islander children from their families and for the associated indignity and degradation.

Between the late 1800s and the 1960s tens of thousands of Australian Aboriginal children were removed from their families by government and church authorities using the excuse of child protection, including the supposed neglect of mixed-race children, and the need to assimilate them into modern life. Some were psychologically, physically, and sexually abused while living in state care or with their adoptive families and efforts to make these children reject their culture often caused them to feel ashamed of their indigenous heritage. Children were wrongly told that their parents had died or abandoned them, and many never knew where they had been taken from or who their biological families were. In 1997, a national inquiry into the separation of Aboriginal and Torres Strait Islander children from their families produced the Bringing Them Home report that documented the story of these children, partly based on the oral testimony of those who had been taken. The report concluded that "indigenous families and communities have endured gross violations of their human rights. These violations continue to affect indigenous people's daily lives. They were an act of genocide, aimed at wiping out indigenous families, communities, and cultures, vital to the precious and inalienable heritage of Australia." The story of the children is told in the award winning film *Rabbit-Proof Fence* as well as several

> "Almost every Aboriginal family has been affected by the forcible removal of one or more children across generations. Many people, families, and communities are still coming to terms with the trauma that this has caused."

books and plays. Many ordinary Australians were shocked by the report and in 2000, more than 300,000 people marched across the Sydney Harbor Bridge in a walk of reconciliation, and more than 1 million Australians signed "Sorry Books" that stated, "We stole your land, stole your children, stole your lives. Sorry," as a way of apologizing for the treatment of Aboriginal peoples (FIGURE 11.4.1).

Although many of Australia's state governments apologized soon after the report, it was not until 2008 that the national government issued a formal apology to the children and their families. Lawsuits have not resulted in much compensation, although people have received support to document their experience and find their families. Almost every Aboriginal family has been affected by the forcible removal of one or more children across generations. Many people, families, and communities are still coming to terms with the trauma that this has caused.

1. Read Chapters 10, 11, or 12 of the Bringing Them Home report at www.humanrights.gov.au to understand the personal experience of the children or families or watch testimonies at www.stolengenerationstestimonies.com/. What are the main impacts on them?

2. Investigate how this reflects similar events in another region such as the history of American Indian Boarding Schools in the United States or the children of the disappeared in Argentina.

Stolen Generations

https://goo.gl/vs1DhB

▼ FIGURE 11.4.1 Many Australians were ashamed when the Bringing Them Home inquiry revealed the details about the removal of Aboriginal children from their families. (a) People recorded their apologies by displaying "sorry" and (b) eventually the government made a formal apology to Aboginal People.

(a)

(b)

single parents, the elderly, refugees, and workers in low-paid service-sector jobs are also diminishing the overall ranking of Australia and New Zealand as places where everyone can make a good living.

Monetary measures such as gross domestic product (GDP) and PPP are of limited use where many people are living in economies based on exchanges and barter or on subsistence. The concept of **subsistence affluence** has been used to describe Pacific Island societies. In these societies, monetary incomes may be low, but local resources such as coconut and fish provide a reasonable diet, and extended family and reciprocal support prevent serious deprivation. Adequate diets and relatively effective health and education systems contribute to comparatively high life expectancies and literacy and low infant mortality throughout the Pacific Islands. Literacy is above 90% for both men and women in much of the region. Notably, Papua New Guinea, Kiribati, and Vanuatu have higher infant mortality and lower literacy than other parts of Oceania.

Territory and Politics

Oceania has seen substantial political changes in recent years, including the shift in alignment from Europe to North America and Asia and the challenges of coping with its relative geographic isolation within a global economy. The stability of some independent democracies and dependencies has been threatened by internal tensions. Political and economic integration has been sought through regional cooperation agreements. And inequalities within Australia and New Zealand have highlighted the conditions of indigenous groups, while at the same time, these countries have embraced multicultural identities.

Regional Cooperation Regional agreements include the South Pacific Commission founded in 1947, which focuses on social and economic development and includes 21 island nations and territories, Australia, New Zealand, the United States, France, and the United Kingdom. The **Pacific Islands Forum**, established in 1971, promotes discussion and cooperation on trade, fisheries, and tourism among all the independent and self-governing states of Oceania. It has supported maritime territorial rights and a nuclear-free Pacific as well as the independence goals of French Polynesia and New Caledonia. In 2000, the Forum legitimized peacekeeping military operations led by New Zealand and Australia in the Solomon Islands and Nauru, and in 2009 Fiji was suspended until it held democratic elections in 2014. The Pacific Island Countries Trade Agreements remove trade barriers within the region.

Australia and New Zealand are members of larger economic and political alliances such as the 21 member Asia-Pacific Economic Cooperation group (APEC), which also includes Papua New Guinea and focuses on improving transportation links and liberalizing regional trade around the Pacific Rim. Both Australia and New Zealand have been able to take advantage of APEC to increase exports to Asia, especially to Japan.

Independence and Secessionist Movements Oceania is relatively politically stable compared to many other regions. A number of smaller or resource-poor islands in the region have maintained close associations with, or are still under the control of, the United States or New Zealand. U.S. dependencies receive financial subsidies, called transfer payments, in return for military base sites and security control. France has also maintained its control over several islands, insisting that they are integral parts of the French state.

The most serious recent political conflicts in Oceania have involved encounters between ethnic groups in Fiji and demonstrations by independence or secessionist movements in New Caledonia and Papua New Guinea. Although independence movements and ethnic rivalries endanger regional peace, one of the greatest threats to stability may be the lack of jobs for young people on Pacific Islands.

- **Fiji** The conflict in Fiji is a legacy of the British colonial policies that brought Asian Indians as indentured workers to local sugar plantations from 1879 to 1920. By the 1960s, Indo-Fijians almost outnumbered the ethnic Fijian population, dominating commerce and urban life and maintaining a separate existence with little intermarriage and continued cultural and religious segregation. The indigenous Fijians took over the government at the time of its independence in 1970, but subsequent elections have produced contested wins for Indo-Fijian parties, and military takeovers occurred in 1987, 2000, and 2006. Only in 2014 did observers certify a democratic election.

- **New Caledonia** In the nickel-rich islands of New Caledonia, the indigenous Melanesian population, known as Kanaks, has been militantly pressing for independence for years, but has been outvoted by those of French descent (called *demis*), who prefer to remain part of France. The Nouméa Accord of 1998 promises a referendum on independence by 2018.

- **New Guinea** Residents of Bougainville Island are trying to secede from Papua New Guinea, claiming ethnic affiliation with the other Solomon Islands that are independent, and complaining that they do not receive a fair share of the profits from local mines. There are also conflicts over mining and between different tribal groups in New Guinea.

Multiculturalism and Indigenous Social Movements The rights of indigenous peoples and the creation of a multicultural society and national identity are high-profile issues in Australia and New Zealand. The countries share a history of British colonialism and dispossession of indigenous lands and cultures but have distinctly different contemporary approaches to intercultural politics and relations.

In New Zealand, the 1840 Treaty of Waitangi frames Maori rights (see Figure 11.16). Although the Maori interpreted the treaty as guaranteeing their land and rights, the century that followed saw large-scale dispossession of Maori land and disrespect for Maori culture. Maori landholdings—over most of New Zealand—were reduced to just 3% of the total area of New Zealand. A series of protests, court cases, and reawakening to Maori tradition led to the establishment of the Waitangi tribunal in 1975, which eventually reinterpreted the Treaty of Waitangi as more favorable to the Maori and investigated a series of Maori land and fishery claims. The Maori were established as *tangata whenua* (the "people of the land"), and Maori was recognized as an official language of New Zealand with the country also known by its Maori name of Aotearoa. Some land claims were settled or compensated through money or grants of government land. Others are too large or threatening to private interests to be easily recognized.

A bicultural Maori and *Pakeha* (a Maori term for whites) society has been adopted rather than a multicultural policy that would encompass other immigrant groups, such as Pacific Islanders and Asians, or recognize the differences within Maori cultures. New Zealand's recognition of Maori rights, language, and culture as part of a national identity has not solved some of the deeper problems of racism toward the Maori or of their poverty and alienation. Maori unemployment is twice that of white residents; average incomes, home ownership, and educational levels are less than half; and welfare dependence is much higher.

Australian Aborigines, in contrast, have had no recourse to a treaty to assert their rights. The European colonists saw the indigenous peoples as primitive and their land as *terra nullius,* owned by no one, and therefore freely available to settlers. Only in the 1930s were reserves set aside for Aboriginal populations, mostly in very marginal environments with little autonomy or access to services. In many ways, the Aboriginal population had been made "invisible"; they were not counted in the census or allowed to vote until the 1960s. It was also stereotyped as a primitive and homogeneous nomadic culture, when in fact the Aboriginal population encompassed many different cultures.

One of the most misguided programs set out to assimilate the Aboriginal population from the 1800s to 1964 by forcibly removing children from their families and communities and placing them in white foster homes and institutions. These **stolen generations** of as many as 100,000 Aboriginal children was given voice and officially acknowledged by the Australian government in a national inquiry in the 1990s (see "Faces of the Region: The Stolen Generations" on p. 450).

The Aboriginal Land Rights Act of 1976 gave Aborigines title to almost 20% of the Northern Territory and opened government land to claims through regional land councils. The states of South Australia and Western Australia have also handed over land to Aboriginal ownership or leases. In 1992, the Australian High Court effectively overruled the doctrine of *terra nullius,* encouraging Aboriginal claims for land and compensation. Aboriginal control now extends over about 15% of Australia, and claims have been made to at least another 20%. The more contentious claims involve land with valuable mineral resources, especially uranium (see "Geographies of Indulgence, Desire, and Addiction: Uranium" on pp. 456–457), or where development threatens sites that are considered sacred or spiritual by the Aborigines.

Despite the apologies and recognition of land claims, indigenous Australians remain disadvantaged on almost all economic and social indicators. Their unemployment rate is at four times the national average and they have much lower average incomes, housing quality, and educational levels, and higher levels of suicide, substance abuse, disease, and violence. Australian Aborigines still have much less power and recognition than the Maori of New Zealand; this is reflected in Australia's adoption of a multicultural rather than bicultural policy of national integration. Multiculturalism emerged in the 1970s and in an official effort to embrace the distinctive cultures of many different ethnic and immigrant groups. The National Agenda for a Multicultural Australia set out to promote tolerance and cultural rights and to reduce discrimination, while still maintaining English as the official language and avoiding special treatment for any one group such that Aborigines are just one of many groups in a multicultural society.

There has been considerable right wing opposition to immigration, Aboriginal rights, and multiculturalism in the last decade, much of this exacerbated by recent global economic conditions. It is clear that land rights and reconciliation for Aboriginal peoples will remain contested as the Australian government strives to balance competing interests and needs.

Apply Your Knowledge

11.13 Consult the website of the Pacific Islands Forum (http://www.forumsec.org/) and read one of the recent Forum Communiques. Summarize the key concerns expressed.

Pacific Islands Cooperation

https://goo.gl/eUllX1

11.14 Summarize the different ways that Australia and New Zealand have treated their indigenous populations. Can you see differences or similarities to the way indigenous peoples have been treated in other countries such as the United States, Canada, and China?

Culture and Populations

The culture and politics of this region are intimately interconnected to the wider global processes of migration, colonialism, conflict, globalization, and environmental change. The linguistic and religious practices of many people in this region, for example, are partially related to the region's colonial experience. And the culture of migration in the region is tied to both the long-term experience of movement and recent environmental and economic changes that are creating push and pull factors for migrants.

Language

Many of the estimated 260 interrelated languages spoken by Australian Aborigines that were unique to the continent have become extinct or have only a few surviving speakers. Many Aborigines no longer speak anything but English; of the indigenous languages, only Mabuiag (the language of the Western

Torres Strait islands) and the Australian Western Desert languages are spoken by more than a few hundred people. British settlement in Australia and New Zealand resulted in English being the most widely spoken, as well as the official, language.

The Pacific Islands have an enormous number and variety of languages, divided into two general families—Austranesian and Papuan. It is believed that almost 20% of all living languages are spoken on the island of New Guinea (**FIGURE 11.20**). Most languages on the island are not understood by even the people in the next valley. To compensate for Papua New Guinea's 820 distinct living languages, a widespread trade language (or *lingua franca*) called *Tok Pisin* is used, that combines local and English words. Other island groups also have a large number of languages. For example, Vanuatu has 110 living languages, many spoken by at least 500 people, and the Solomon Islands have 70 different languages, most with at least 500 speakers. This diversity of languages has been explained by the fragmentation and isolation of New Guinea's physical geography and Vanuatu's many islands.

In New Zealand, both English and Maori are official languages. Almost all Maori speak English, and about 100,000 also speak or understand Maori, a language related to those of Polynesia and especially to Hawaiian and Samoan.

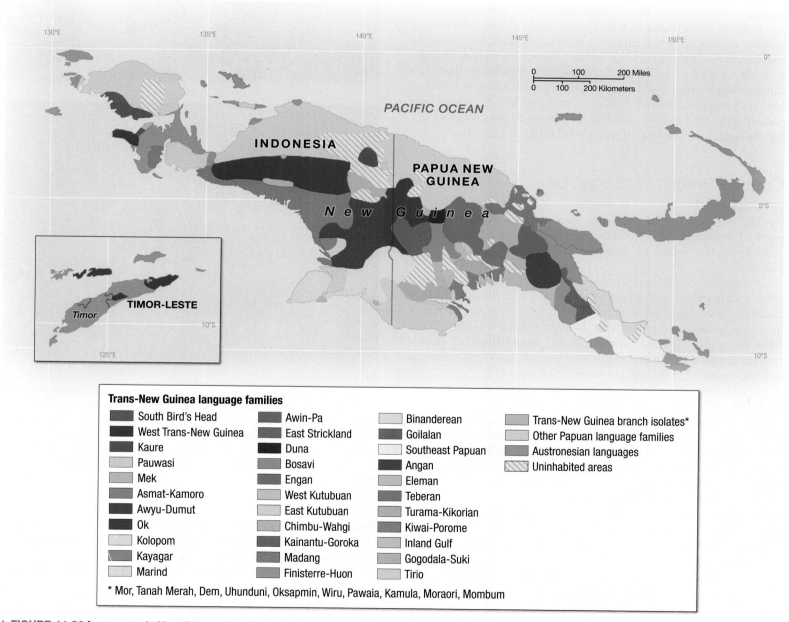

Trans-New Guinea language families

- South Bird's Head
- West Trans-New Guinea
- Kaure
- Pauwasi
- Mek
- Asmat-Kamoro
- Awyu-Dumut
- Ok
- Kolopom
- Kayagar
- Marind
- Awin-Pa
- East Strickland
- Duna
- Bosavi
- Engan
- West Kutubuan
- East Kutubuan
- Chimbu-Wahgi
- Kainantu-Goroka
- Madang
- Finisterre-Huon
- Binanderean
- Goilalan
- Southeast Papuan
- Angan
- Eleman
- Teberan
- Turama-Kikorian
- Kiwai-Porome
- Inland Gulf
- Gogodala-Suki
- Tirio
- Trans-New Guinea branch isolates*
- Other Papuan language families
- Austronesian languages
- Uninhabited areas

* Mor, Tanah Merah, Dem, Uhunduni, Oksapmin, Wiru, Pawaia, Kamula, Moraori, Mombum

▲ **FIGURE 11.20 Languages in New Guinea** New Guinea has more than 800 languages developed by tribes isolated by mountains and conflict. Many Papuan languages have only a few hundred speakers but some—such as Enga and Huli—are spoken by more than 100,000. The map above shows some of the main languages spoken across the island in both the Indonesian and Papua New Guinean regions. There are also areas of Austronesian languages, such as Motu, associated with migration across the region 3,500 years ago. Several million people speak the English-based pidgin language *Tok Pisin* as a second language. **List the factors that produced New Guinea's extreme linguistic diversity and explain what led to the development of Tok Pisin.**

Religion

The British colonial heritage of Australia and New Zealand is reflected in the dominance of Christianity in these countries—mostly Protestant in New Zealand and about half Protestant and half Catholic in Australia. Irish and southern European immigrants (particularly those from Italy) brought Catholicism to the region. Missionaries successfully converted most Pacific Islanders to Christianity; a range of Protestant denominations and Catholicism are prevalent in French Polynesia and evangelical Christianity has spread to many islands. However, today, many people report having no strong religious beliefs.

Non-Christian, local beliefs are still important in the Pacific Islands, especially in Papua New Guinea, and some communities have fused local and Christian practices by combining harvest rituals with the celebration of Christmas, for example. Hinduism is important in Fiji among the Asian Indian population. As a result of Asian immigration, both Buddhism and Hinduism are growing in significance in Australia. In New Zealand, Maori communities abandoned many local beliefs when they converted to Christianity and lost their land in the 19th century, but there is now a revival of traditional religion and ritual.

Cultural Practices, Social Differences, and Identity

Contact with the rest of the world disrupted the indigenous cultures of Australia, New Zealand, and the Pacific Islands. At the same time, the outsiders romanticized and commodified those cultures.

Aboriginal culture in Australia includes a complex spiritual relationship to the land, a nomadic lifestyle, and the use of fire in hunting. The Aborigine worldview links people to each other, to ancestral beings, and to the land through rituals, art, and taboos. This worldview is associated with the **Dreamtime**, a concept that joins past and future, people and places, in a continuity that ensures respect for the natural world.

Aboriginal cultures, based on strong spiritual ties to land, were marginalized and homogenized when Aborigines were removed from their ancestral lands and resettled on reserves. Today, however, Aboriginal art, often based on rock art designs in dotted forms, silhouettes, and so-called X-ray styles, has become very popular in contemporary markets (**FIGURE 11.21a**), and native dances and songs, such as those that are part of the social gathering known as *corroboree,* are performed for tourists. These representations fail to fully convey the complexity of Aboriginal life in Australia; however, a variety of grassroots movements and educational efforts have emerged to promote Aboriginal culture and traditions.

Maori tradition—such as the welcome ceremonies that include the *hongi* (pressing noses together) and the *haka* war dance (**FIGURE 11.21b**) now performed at international sporting events—is celebrated as integral to New Zealand's official bicultural identity. Maori architecture includes distinctive carved and decorated meetinghouses called *wharenui* (big

(a)

(b)

▲ **FIGURE 11.21 Aborigine and Maori Culture** (a) Contemporary Aboriginal artists based some of their work on traditional Aboriginal "X-ray" painting of animals on rocks. (b) The New Zealand national rugby team—the All Blacks—perform a traditional Maori *haka* dance before an international match against Wales in Cardiff, U.K. **Use the internet to research the life and achievements of a famous Aboriginal painter or Maori rugby player.**

houses). Artistic expressions include intricate carvings such as those found on war canoes and decorative masks and tattoos (*moko*), also found on other Pacific Islands.

British culture still influences Australia and New Zealand, but their cultures are seen as less hierarchical and more informal. Connections with Britain provoke considerable ambivalence, and there are efforts to establish distinct identities by embracing indigenous traditions and new immigrant cultures, such as those from Asia. However, a British legacy endures in the popularity of sports such as cricket and rugby, where the national teams, such as the New Zealand All Blacks rugby team (see Figure 11.21b), have gained international renown and enjoy enthusiastic local support. Surfing is often seen as the sport that characterizes Australia.

The cultures of the different Pacific Islands have been a popular topic of study for anthropologists—such as Margaret Mead, who worked in Samoa and New Guinea—but early studies have been criticized for overdramatizing or over-romanticizing aspects of island culture and for underestimating the diversity and politics of traditions.

To most of the world, the Pacific Islands form a distinctive stereotype—tropical paradises where local people fish, collect coconuts, and make crafts while tourists relax on beautiful beaches and swim in peaceful lagoons fringed with coral reefs. But island culture is highly varied. For example, in Samoa, traditions include communal living and eating arrangements, extensive symbolic tattoos, and a traditional system of behavior—the Samoan Way—that includes obligations to community and church and respect for elders. Other distinctive customs include the strict separation of the male and female in much of New Guinea and Melanesia; the traditions of ritual warfare and reciprocal gift exchange; the importance of local leaders, or "big men"; and close links with kin and extended families. Pigs are still considered a measure of wealth on many islands, including New Guinea, yams are the most important food source (**FIGURE 11.22**), and a traditional plant, kava, is consumed as a recreational and ritual relaxant on many islands. Globalization is changing culture rapidly on some islands with the spread of TV, processed foods, tourism, and the Internet.

Gender and Sexual Identity
Gender roles in Oceania are rapidly changing and complex. In Australia, the image of the frontier rancher or miner is associated with heavy drinking, gambling, male camaraderie, and a tough, laconic attitude. The country has a persistent pay gap with women earning only 80% of men's incomes on average and being underrepresented in management.

In recent years, Australian cities, such as Sydney, have celebrated gay and transgendered identities through events such as annual Mardi Gras parades. Same-sex marriages are not recognized (although partners are recognized in many cases).

Across the Pacific Islands, traditional gender roles vary from cultures controlled by women (such as the matriarchal traditions of the Marshall and Solomon Islands), to those where males and females lead mostly separate lives (such as in parts of New Guinea), to others controlled by men.

Although it shares the image of the strong frontier male with Australia, New Zealand has a long history of promoting women's equality and was the first country to extend the right to all adult women to vote (1893). New Zealand legalized same-sex marriage in 2013.

Many women in Australia and New Zealand have shifted from roles as traditional housewives into a multitude of careers and to senior political positions (such as former prime ministers

▲ FIGURE 11.22 **Traditional Village in New Guinea** The yam house of a chief in Tukwaukwa village on Kiriwina in the Trobriand Islands of New Guinea is shown here. The islanders traditionally use yams as currency and accumulate them as a sign of wealth and power.

Helen Clark of New Zealand and Julia Gillard of Australia), and their efforts are supported by a strong feminist movement.

Social Problems Serious social problems in Oceania include alcohol abuse and high levels of domestic violence. Papua New Guinea has some of the world's highest levels of violence against women. Violence and alcoholism, associated with poverty and discrimination, are also a problem in some Aboriginal and Maori communities. With shelters and support groups, women in Papua New Guinea are building "safe houses" for women fleeing violence.

Arts and Film Oceania has an international reputation in film, literature, art, and music. The film industry in Australia and New Zealand is especially renowned. The most famous movies associated with New Zealand are probably the *Lord of the Rings* series directed by New Zealander Peter Jackson. These Oscar-winning spectacles feature the country's dramatic landscapes, and their popularity has attracted many tourists to the region (**FIGURE 11.23**). Australia is known for its actors, many of whom have become international stars, such as Cate Blanchett, Nicole Kidman, and Hugh Jackman, and for cult films such as *Mad Max* and *Priscilla, Queen of the Desert*.

In addition to traditional arts and music such as the Australian Aboriginal dot paintings, there is a vibrant arts culture in cities such as Sydney, Melbourne, and Auckland, with galleries, concert halls, and popular music venues. Musicians from Australia and New Zealand include famous opera singers such as Joan Sutherland and Kiri Te Kanawa, and popular singers Kylie Minogue, Lorde, and Iggy Azalea.

Uranium

Much like oil in the Middle East, uranium links Oceania to the global hunger for cheap energy and to geopolitical conflicts. Uranium is a radioactive element mined for its value as the fuel for nuclear power plants that generate electricity—and as an ingredient in nuclear bombs.

After World War II, the United States, Britain, and France joined the Cold War arms race to develop powerful nuclear weapons containing uranium and tested them near or on Pacific Islands, with devastating implications for local residents and environments. Radiation can have serious short- and long-term effects on people, including radiation poisoning, leukemia, and birth defects. The United States conducted nuclear tests between 1946 and 1958 in the Marshall Islands, relocating the residents of Bikini and Eniwetok Atolls (**FIGURE 11.5.1**). Although the prevailing winds were supposed to carry the bombs radioactive fallout away from inhabited

> "The United States, Britain, and France joined the Cold War arms race to develop powerful nuclear weapons containing uranium and tested them near or on Pacific Islands, with devastating implications for local residents and environments."

islands, radioactive ash dusted the islands. The U.S. government evacuated some residents at short notice without warning them that they might not be able to return or sharing information about health risks. It was many years before they were

allowed to return, although there is still some risk from foods grown on the islands. Although the United States has monitored the health of the islanders and established a U.S. $700-million trust fund, many residents of the islands are angry and

◀ **FIGURE 11.5.1**
Nuclear explosion on Eniwetok Atoll The United States tested atomic bombs on atolls in the Marshall Islands from the 1950s, causing serious health risks to islanders and resulting in the evacuation of some atolls.

resentful about the experiments that disrupted their lives. In 2014, the Marshall Islands sought attention for their concerns when they sued the United States and eight other countries for failing to control the spread of nuclear weapons.

France conducted more than 150 bomb tests on the tiny atolls of Moruroa and Fangataufa in French Polynesia beginning in 1966. The bombs showered the surrounding regions with radioactive plutonium far above acceptable levels and reaching as far as Tahiti, Samoa, and Tonga, hundreds of miles to the west. Opposition from other Pacific Islands and New Zealand and Australia culminated in boycotts of French products, including wine and cheese, during the 1970s. Locals in Polynesia used the bomb tests as a reason to seek independence from France, and international activists tried to halt testing. The 1985 Treaty of Raratonga declared a South Pacific Nuclear-Free Zone, which bans the use, testing, and possession of nuclear weapons.

The environmental group Greenpeace protested bomb tests by sailing its ship *Rainbow* *Warrior* toward the testing zones, and French intelligence agents eventually bombed the ship while it was moored in the harbor of Auckland, New Zealand. The resulting international scandal prompted New Zealand to take a strong stand against nuclear proliferation, banning all nuclear-powered and nuclear-armed vessels from its harbors (against the objections of the United States), breaking off diplomatic relations with France, and taking a leadership role in the antinuclear movement in the Pacific. New Zealand's actions contributed to the announcement by France in 1996, after riots in Tahiti, that it would end nuclear testing.

Australia has more than 30% of the world's uranium reserves but has not developed nuclear power partly because of ample coal. Mining uranium can release radioactivity into the landscape and creates risks for mine workers. Some of the most important Australian resources of uranium are found on or near Aboriginal lands and in conservation areas. For example, the Ranger Uranium mine began operations in the Northern Territory in 1980 within the boundary of Kakadu National Park, which has great natural beauty and cultural value. The mine has produced more than 16 million tons (35 billion pounds) of radioactive waste and has created serious water-pollution problems in the area (**FIGURE 11.5.2**).

---

1. **UNESCO has declared the Bikini Atoll Nuclear Test Site a world heritage site. Consult their website to understand why it was made a UNESCO site http://whc.unesco.org.**

Uranium in Oceania

https://goo.gl/sl8fex

2. **Consult the website of the World Nuclear Association to research where Australia exports uranium and look at the section on operating mines to identify how one mine manages its environmental impacts www.world-nuclear.org.**

◄ **FIGURE 11.5.2 Uranium Mining in Australia** The open pit Ranger Uranium mine in northern Australia is surrounded by Kakadu National Park. Falling uranium prices (especially after Japan cancelled nuclear power expansion after the Fukushima accident) and lack of support from the local Aboriginal authorities means that mining is unlikely to continue beyond 2020.

▲ **FIGURE 11.23 Lord of the Rings** Peter Jackson chose his native country, New Zealand, to film Tolkien's epic *Lord of the Rings* and *The Hobbit.* The films earned box office revenues of more than U.S. $5 billion, exposing millions of people to New Zealand landscapes that were often modified with special effects in the films. Tourists flocked to tours featuring the film locations.

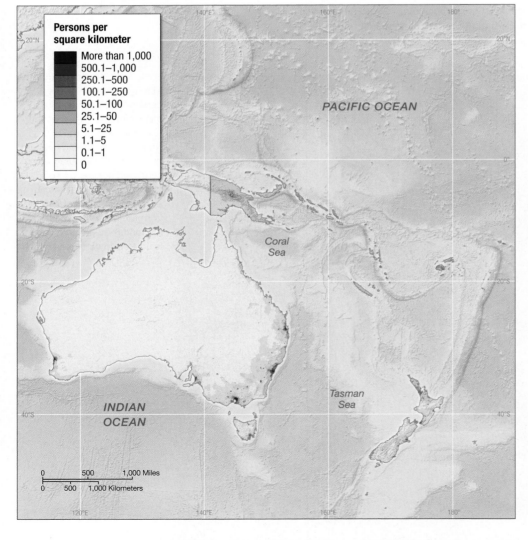

◀ **FIGURE 11.24 Population Density Map of Oceania, 2010** Although most of Oceania appears sparsely populated on this map, several small islands have high population density. Australia's population is concentrated in the cities and in the southeastern region of the country. **Compare Australia's population density with the climate zones in Figure 11.2, then write a sentence explaining the distribution of population in Australia.**

Apply Your Knowledge

11.15 See or read a recent film or novel about Australia or New Zealand and assess how it portrays culture, gender, and ethnic relations in the region.

11.16 Research the traditional culture of one of the Pacific Islands using the Internet and compare the island's traditional culture with its present-day way of life.

Demography and Urbanization

Oceania is one of the least-populated regions of the world. Of its 2015 population of 40 million people, 24 million live in Australia, 7.7 million live in Papua New Guinea, and 4.6 million live in New Zealand. Most people live near water on the larger land areas in the region, especially along the coasts of southeast Australia and coastal Papua New Guinea. Overall population densities of the larger countries are very low: 3 people per square kilometer in Australia (7.8 people per square mile), 16 per square kilometer in Papua New Guinea (39 people per square mile), and 17 per square kilometer (44 people per square mile) in New Zealand in 2014. The smaller islands, in contrast, can have fairly high population densities, reaching more than 150 per square kilometer (388 people per square mile) in Guam, the Marshall Islands, Nauru, and Tonga (**FIGURE 11.24**).

Demographic Change The overall population of the region is growing quite slowly (at 1% per year). This slow growth is in large part due to the low growth and fertility rates (less than 2 children per woman) in Australia and New Zealand. The low birthrate, combined with long life expectancy, is creating a growing proportion of older residents and raising concerns about the provision of services such as social security, health care, and pensions for the aging population. Fertility rates are higher in the Pacific Islands, up to more than 4 children per woman in Papua New Guinea and Samoa.

The contrasts in age structure can be seen in population pyramids for New Zealand and Papua New Guinea reflecting several decades of low birthrates in New Zealand and longer

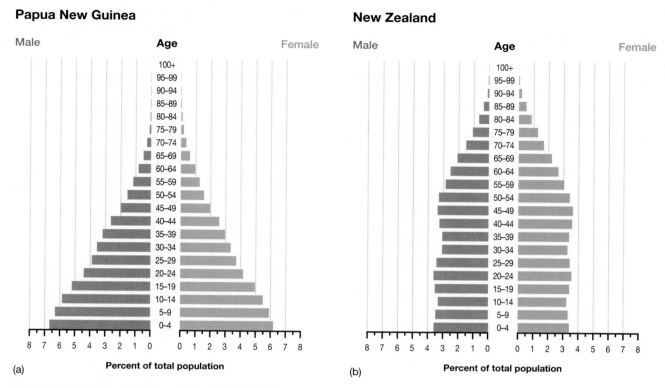

▲ FIGURE 11.25 Age Pyramids for New Guinea and New Zealand (a) Papua New Guinea has a higher birthrate and shorter life expectancy with more than 50% of the population under 20 years old. (b) Fertility and birthrates have fallen in New Zealand and people are living longer so that a significant portion of the population is over 65 years old and each age cohort is about the same size under 50.

life expectancies, high birthrates, and shorter lives for the elderly in Papua New Guinea (**FIGURE 11.25**).

Immigration and Ethnicity There were at least half a million Australian Aborigines and perhaps a quarter of a million New Zealand Maori in Oceania when the Europeans arrived. The introduction of European diseases and the violence of some colonial encounters reduced native populations significantly and resulted in majority European populations in Australia and New Zealand within a century or so of initial settlement. The contemporary ethnic composition of Australia and New Zealand is strongly influenced by the larger history of immigration. Almost a quarter of Australia's current residents were born elsewhere. Some Chinese arrived in Australia during the gold rush of the 19th century along with as migrants of Irish, Scottish, and German origins.

Australia sought to maintain a European "look" through the adoption of the **White Australia policy** after its independence in 1901, which restricted immigration to people from northern Europe through a ranking that placed British and Scandinavians as the highest priority, followed by southern Europeans. After World War II, the government made extra efforts to attract immigrants as labor for manufacturing, as a source of military volunteers, and in response to humanitarian pressure to resettle refugees from Eastern Europe.

In 1973, the racist restrictions on immigration were removed and replaced with skills criteria, and a new wave of immigration from Asia began. Large numbers of migrants from Vietnam,

Hong Kong, and the Philippines arrived as a result. By the end of the 20th century, Australia's population was much more diverse, with about 73% of the population of British or Irish heritage, 20% from elsewhere in Europe, 5% from Asia, and only 2% Aborigine. The economic crisis of 2008–2009 slowed Australian immigration.

New Zealand also adopted an immigration policy designed to attract white immigrants. Before World War II, this included a bias not just against nonwhites but also against Irish Catholics. After the war, labor shortages were met by temporary labor migration from the Pacific Islands. More recent changes in immigration policy have opened New Zealand to more permanent settlers from the Pacific as well as to Asian immigrants who can bring capital with them. According to the 2013 census, New Zealand was still dominated by people of English and Scottish ancestry, but with a growing percentage of people identifying with Pacific Island (about 7.5%) or Asian (12%) heritage as well as those who are Maori (15%).

The Pacific Islands have much higher proportions of indigenous people than Australia or New Zealand, averaging more than 80% indigenous, 14% Asian, and 6% Europeans. The exceptions are Fiji, which has a very large Asian Indian population associated with the importation of contract labor during colonial times; New Caledonia, where about 40% of the population is indigenous, 40% French, and 20% descendants of Southeast Asian contract workers; and Guam, which has significant American and Filipino populations.

Oceanic Migration and Diasporas One of the largest diasporas within Oceania has been the emigration of Pacific Islanders to work and live in New Zealand and Australia. This emigration has been encouraged in the last 50 years by labor needs, refugee movements, and lenient rules allowing migration from current and former colonies and, more recently, by global climate change. More people from Niue, Tokelau, and the Cook Islands now live in Auckland, the capital of New Zealand, than remain on those islands. About one-eighth of all Samoans and migrants from Tonga and Fiji also live in Auckland. Australia has accepted thousands of refugees from Timor-Leste in Southeast Asia and from Fiji.

A second diaspora is from the Pacific Islands to North America, especially to Hawaii and California. Many Samoans and Tongans have moved to the United States, and unrest in Fiji sent a wave of several thousand Indo-Fijians to the United States and Canada. There is also a small emigration flow of Australians and New Zealanders to Europe and North America, including young people seeking educational and work opportunities in England, Canada, or the United States. Within Australia, there is an internal migration flow from southeastern Australia northward along the east coast and to western Australia as economic opportunities open up, air travel and the Internet expand, and life becomes more expensive in the southeast.

Urban Areas Australia is about 89% urban, and New Zealand is 86% urban. Three-quarters of future population growth in these two countries is forecast to be in cities. The overall high levels of urbanization in Australia and New Zealand are due in part to the settlement patterns of European colonists and migrants and the harsh climatic conditions of much of Australia, which have limited large-scale rural development. This contrasts with other parts of the region. The average urban population for Pacific Island states is only 34% and Papua New Guinea is only 12.5% urban.

- **Sydney** Despite popular media images of the Australian Outback, most Australians live close to the southeastern coast. Within the coastal region, Sydney is the largest metropolitan area. It has a population of almost 4.8 million people; one in five Australians lives in and around Sydney. High levels of car ownership characterize the city, and the population sprawls into surrounding suburbs. The wealthier areas of the city are to the east, and poorer residents concentrate in the western suburbs. As immigrants from different regions settled in groups based on their origin, Sydney developed several ethnic neighborhoods, including Greek, Italian, and Vietnamese. Many Aborigines who migrated to the city settled in one suburb, and poverty and discrimination have made their neighborhood a focus of indigenous action and social programs. A number of older downtown neighborhoods, such as Paddington, have been renovated through **gentrification**, the process in which older working-class neighborhoods are converted to serve higher-income households. Some of the older warehouse and manufacturing districts near the harbor have been redeveloped into shopping, museum, convention, and entertainment areas such as the Darling Harbor district. Sydney is also surrounded by parks and protected areas, and by beautiful beaches that encourage swimming, surfing, and sailing (**FIGURE 11.26**).

- **Melbourne** Australia's second-largest city, Melbourne was the manufacturing center of southeastern Australia (**FIGURE 11.27**). Located on a sheltered harbor on Port Phillip Bay, Melbourne has a local population of 4 million people. The city first developed as the transport hub for the 19th-century gold rush; it grew further when thousands of refugees and migrants were sponsored to come to Australia after World War II and sent to work in Melbourne's industrial sector, which manufactured textiles and clothing and processed metal. Contemporary industry includes chemicals, food processing, automobiles, and computers. The city has developed a reputation as a cultural center.

- **Canberra** The capital of Australia, Canberra, is a planned city. Canberra's inland site was chosen because it lies between the country's two largest cities, Melbourne and Sydney.

▲ **FIGURE 11.26 Sydney Harbor** This panoramic view of Sydney shows the downtown business core surrounded by the sea. Although Sydney has some manufacturing, the city economy is overwhelmingly service oriented, focusing on trade, banking, and tourism.

▲ **FIGURE 11.27 Melbourne** Australia's second-largest city, Melbourne, has a downtown core surrounded by industrial activities and neighborhoods and cafes that attract many young people.

as well as Sydney, Melbourne, and Adelaide in Australia, are considered some of the world's most livable based on education, health, low crime, environmental quality, and access to recreation.

- **Port Moresby** The capital of Papua New Guinea, Port Moresby, is a trading port and has been called one of the world's least livable cities, with high levels of crime, gender violence, and growing squatter settlements (**FIGURE 11.28**).

Apply Your Knowledge

11.17 Visit the website of a major city government in Australia or New Zealand. How does the city promote itself online? How does it discuss its problems and plans for the future?

11.18 Choose a Pacific Island and research its demographic profile (population, density, fertility, age structure). What does the age structure indicate about future society?

- **Auckland** With 1.38 million residents, the city of Auckland remains the largest in New Zealand. Other cities, such as Wellington and Christchurch, have fewer than 400,000 residents. Cities in New Zealand,

▼ **FIGURE 11.28 Port Moresby** The capital of New Guinea, Port Moresby has slums adjacent to more modern buildings in this crowded coastal city.

Australian Total Greenhouse Gas Emissions by Gas

Total F-Gas*
(MtCO₂e)

Total N₂0
(MtCO₂e)

Total CH₄
(MtCO₂e)

Total CO₂
(excluding Land-Use
Change and Forestry)
(MtCO₂e)

*Fluorinated greenhouse gases such
as HFCs used as refrigerants

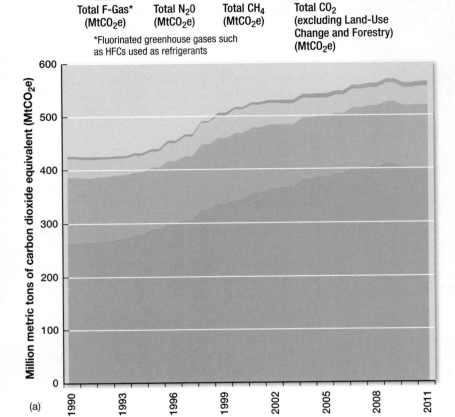

(a)

New Zealand Total Greenhouse Gas Emissions by Gas

Total F-Gas*
(MtCO₂e)

Total N₂0
(MtCO₂e)

Total CH₄
(MtCO₂e)

Total CO₂
(excluding Land-Use
Change and Forestry)
(MtCO₂e)

*Fluorinated greenhouse gases such
as HFCs used as refrigerants

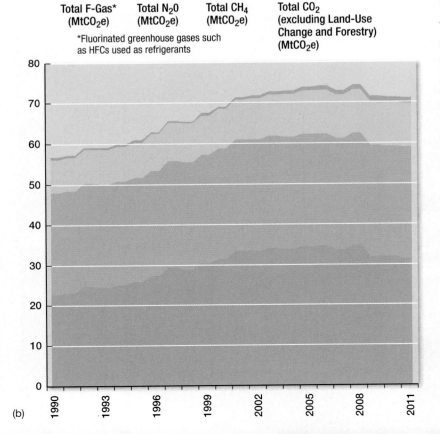

(b)

Future Geographies

Often viewed as a relatively isolated world region, Oceania, which stretches across a vast waterscape, is increasingly interconnected to the rest of the world in terms of its physical and human geographies. The future of this region depends on the environmental, economic, and sociocultural scenarios that unfold over the coming decades, especially those associated with the climate and oceans, the global economy, and the protection of local cultures.

The Challenges of Climate Change and Ocean Sustainability For some countries in Oceania, geography will literally change as sea-level rise drowns their coasts and reshapes their maps. Australia and New Zealand, with coal exports and high livestock numbers, will need to consider how to reduce greenhouse gas emissions while adapting their agriculture and settlements to warmer and drier climates (**FIGURE 11.29**). The islands of the Pacific are likely to make climate adaptation a priority and hope for assistance from the international community. The populations of some islands may need to migrate to other countries as seas rise. Another shared environmental challenge for the future is the protection of ocean and marine environments from pollution, overexploitation, and acidification.

Connections to the Global Economy The shift in trade to Asia is an important element of Australia and New Zealand's economic futures. Across the Pacific Islands, challenges include the stability and diversification of economies, especially those that depend on foreign military bases, tourism, imported food, and migrant remittances.

Traditional Cultures The future of traditional cultures in Oceania, as is the case in many other world regions, is at risk from globalized culture, economic development, and political conflicts. While efforts in Australia and New Zealand have fostered growing respect for Aboriginal and Maori cultures, poverty, inequality, and racism continue. In the Pacific, local cultures are disappearing as young people are attracted by global popular culture or choosing to migrate because of unemployment and the attraction of urban life.

◄ **FIGURE 11.29 Greenhouse Gas Emission Trends for Australia and New Zealand** Australia's greenhouse gas emissions have increased by more than 30% from 1990 to 2011 (mostly carbon dioxide (CO₂) from fossil fuels), although at one point it signed an international agreement to stabilize emissions at no more than 8% above 1990 levels (the Kyoto protocol that they then withdrew from). New Zealand's emissions are much lower as a result of smaller population and lower per capita consumption. But they have increased by almost 25%, with a large share of emissions from the methane emitted by livestock. Although both countries are considering shifts from fossil fuels to renewables, reducing their contribution to global climate change—which will have devastating impacts on some of the small island countries in Oceania—is a challenge for both countries. The figures show trends in emissions in carbon equivalent (MtCO₂e) major greenhouse gases including industrial fluorinated gases (F-gas), nitrous oxide (N₂O), methane (CH₄), and carbon dioxide (CO₂)

Learning Outcomes Revisited

▶ *Explain* the factors that influence the climate of Oceania (including Antarctica and the Pacific Ocean) and the impacts of climate change and other environmental threats.

Most of Oceania has a tropical climate where rains are associated with major wind belts such as the trades, monsoons, and Westerlies. The interior of Australia is extremely dry because of its distance from the sea and the subtropical high-pressure zone. Climate change is already increasing the risks of heat waves and droughts in Oceania, especially in Australia, and sea-level rise, ocean acidification, and extreme weather events associated with climate change are a particular risk for islands in the Pacific. Adaptation may moderate these impacts. In addition to climate change, the Pacific and Antarctica are threatened by pollution and the disturbance of fragile ecosystems, by the overharvesting of fish and whales, and by mineral exploitation. The Law of the Sea (UNCLOS) and the Antarctic Treaty provide some protection.

1. *What* are the major environmental changes that have affected the climate and oceans of the region?

▶ *Describe* the unique ecosystems and the impacts that the introduction of nonnative species have had in the region.

Oceania is home to many species that evolved on isolated islands that lacked predators. It is also home to species with unique characteristics or adaptations such as marsupials and monotremes. Many of these are vulnerable to competition and predation from exotic nonnative plants and animals introduced during colonization and that may have become feral, such as the rabbit and pig in Australia or mongoose on the Pacific Islands.

2. *What* are some of the more unusual characteristics of ecosystems in Australia and New Zealand and how have they been affected by introduced species?

▶ *Analyze* how colonialism and geopolitical struggles, such as World War II, the Cold War, and nuclear policy, have shaped the region, and describe the current international and regional political and economic alignments of major countries in the region.

In addition to ecological changes, Europeans introduced grains and livestock to Australia and New Zealand that altered land use, expropriated land from native peoples and changed their cultures, and ruthlessly pursued minerals to export. Colonists settled in Australia and New Zealand and people of European ancestry have dominated population, economy, culture, and politics. Colonization of the Pacific was more variable but has included whaling, mining, and military bases.

The region was mostly allied with Europe and North America in World War II, and Australia also sent troops to support the United States in Vietnam. The Japanese took over many of the Pacific Islands during World War II. The United States and its allies eventually reclaimed most of these and established lasting key military bases, territories, and cultural influences across the Pacific. France and other countries used the Pacific Islands for testing nuclear weapons; these tests were met by opposition from antinuclear activists and the New Zealand government. The region still has strong links to Europe and the United States but is creating new alignments with Asia. Small islands in the region have joined with other islands as a lobbying group within the UN; their efforts are especially focused on issues relating to climate change.

3. *Describe* the major political alliances and conflicts that have affected the region.

▶ *Explain* how exports connect Australia and New Zealand to the rest of the world and describe changes resulting from globalization, free trade, and a shift in demand from Europe to Asia.

The most important exports from the region include minerals, such as gold, copper, and coal; agricultural products, such as meat, dairy, sugar, and grain; and manufactured goods. The end of preferential trade links to Europe and the adoption of free-trade policies reoriented economies to global markets, with their vulnerabilities and opportunities, and to Asia, with its growing demand for food, minerals, and other goods.

4. *How* has trade in Australia and New Zealand changed over time?

▶ *Explain* the importance of tourism in the region and the economic, social, and environmental impacts of the tourist industry.

Tourism is critical to the economies of most of the Pacific Islands and is also important to Australia and New Zealand. Tourism can have negative environmental and social impacts in the region if not managed carefully; for example, tourism can damage reefs. Visitors contribute money to the economy, but the resorts drain local resource bases and encourage local people to modify their livelihoods and rituals to appeal to foreign tourists' idealized images of the region.

5. *Contrast* the benefits and costs of tourism to the region.

▶ *Summarize* the main shifts in Australian and New Zealand policies toward indigenous peoples and on immigration issues. and compare and contrast the differences in culture and gender relations of different groups in Oceania.

European colonists who came to Australia did not recognize the land and cultural rights of Aborigines and claimed much of the best land. The Australian government, dominated by whites of European origin, created reserves in marginal areas for indigenous people and removed children from their families to assimilate them. Some efforts have been made to remedy these policies by giving Aborigines title to about 15% of the land in Australia, and apologizing for the stolen generations. The Maori of New Zealand were granted rights under the Treaty of Waitangi and are recognized in the country's bicultural policy. Australia adopted the White Australia policy in 1901 that prioritized northern European immigrants. This ended in 1973, opening the door for immigrants to arrive from Asia.

Cultures of Australia and New Zealand include the Aborigines and Maori, groups that have traditionally held strong spiritual beliefs and fostered connections to the land. Aboriginal languages are in decline in Australia, but Maori is sustained as a part of New Zealand's bilingual policy. The culture, religion, sport, and language of both Australia and New Zealand were influenced by the British and by the image of the rugged Outback male. Australia has shifted to a more multicultural society that respects the religion, language, and traditions of newer immigrants as well as Aboriginal practices. The Pacific Islands are home to an enormous number of languages and cultures, with varied gender relations, especially in New Guinea. Many islands have cultural traditions tied to the sea. Traditional island dress, dances, and foods have often been appropriated to encourage tourism.

6. *Briefly* describe how the treatment of indigenous peoples and immigrants has changed in Australia and New Zealand and some of the current problems.

Key Terms

Aborigines (p. *436*)
Antarctic Treaty (p. *431*)
AOSIS (Alliance of Small Island States) (p. *427*)
atoll (p. *432*)
blackbirding (p. *446*)
common property resources (p. *442*)
Dreamtime (p. *454*)
ecological imperialism (p. *440*)

exclusive economic zone (EEZ) (p. *442*)
exotics (p. *440*)
feral (p. *440*)
gentrification (p. *460*)
Great Artesian Basin (p. *433*)
Great Barrier Reef (p. *438*)
Maori (p. *443*)
marsupial (p. *438*)

Melanesia (p. *428*)
Micronesia (p. *428*)
MIRAB (migration, remittances, aid, and bureaucracy) economies (p. *447*)
monotreme (p. *438*)
Montreal Protocol (p. *432*)
Outback (p. *433*)
ozone depletion (p. *432*)
Pacific Garbage Patch (p. *441*)

Pacific Islands Forum (p. *451*)
Polynesia (p. *428*)
stolen generations (p. *452*)
subsistence affluence (p. *451*)
tragedy of the commons (p. *442*)
Treaty of Waitangi (p. *445*)
United Nations Convention on the Law of the Sea (UNCLOS) (p. *442*)
White Australia policy (p. *459*)

Thinking Geographically

1. How have the large expanse of ocean and the general isolation of the region defined the cultures and the economies in Oceania?

2. Describe some of the physical, climate-related, and ecological differences between high volcanic islands and low coral atolls.

3. Compare and contrast the different ways in which European colonial powers changed land use across the Pacific.

4. How have changes in global trade in the past 75 years altered the economies and landscapes of Oceania?

5. How did World War II and the Cold War affect the Pacific Islands politically, economically, and demographically?

6. Compare and contrast the ways Australia and New Zealand have dealt with indigenous peoples, immigration, and multicultural movements.

DATA Analysis

Download the latest Statistical Yearbook for Asia and the Pacific (e.g., 2014) from http://www.unescap.org from the Oceania region as defined in this chapter and compare two countries on the following issues:

1. What are the current fertility rates and life expectancies in the two countries and how have they changed since 1990?

2. What are the latest infant/child mortality rates and how have they changed since 1990?

3. What are the latest per capita CO_2 emissions and how have they changed in 1990?

4. What are the latest per capita GDP (PPP per capita) and how have they changed since 1990?

5. What are the latest net external debts as percentage of GDP and how have they changed since 1990?

6. Based on this analysis, which of the two countries has the best social and economic indicators currently and which has shown the most improvement since 1990? How can you explain the differences?

Pacific Islands Statistics

https://goo.gl/vW6Rrz

MasteringGeography™

Looking for additional review and test prep materials? Visit the Study Area to enhance your geographic literacy, spatial reasoning skills, and understanding of this chapter's content by accessing a variety of resources, including MapMaster interactive maps, videos, *In the News* RSS feeds, flashcards, web links, self-study quizzes, and an eText version of *World Regions in Global Context.*

APPENDIX   Maps and Geospatial Technologies

Maps are representations of the world. They are usually two-dimensional, graphic representations that use lines and symbols to convey information or ideas about spatial relationships. Maps express particular interpretations of the world, and they affect how we understand the world and see ourselves in relation to others. As such, all maps are social products. In general, maps reflect the power of the people who draw them. Just including things on a map—literally "putting a place on the map"—can be empowering. The design of maps—what they include, what they omit, and how their content is portrayed—inevitably reflects the experiences, priorities, interpretations, and intentions of their authors. The most widely understood and accepted maps reflect the view of the world that is dominant in universities and government agencies.

Many maps are designed to represent the *form* of Earth's surface and to show permanent (or at least long-standing) features such as buildings, highways, field boundaries, and political boundaries. *Topographic maps* use a *contour,* a line that connects points of equal vertical distance above or below a zero data point, usually sea level to represent the form of Earth's surface (**FIGURE A.1**).

Maps that are designed to represent the spatial dimensions of particular conditions, processes, or events are called *thematic maps.* These can be based on any of a number of devices that allow cartographers or mapmakers to portray spatial variations or spatial relationships. One device is the *isoline,* a line (similar to a contour) that connects places of equal data value (e.g., air pollution, as in **FIGURE A.2**, p. A-2). Maps based on isolines are known as *isopleth maps.* Another common device used in thematic maps is the *proportional symbol.* For example, circles, squares, spheres, cubes, or some other shape can be drawn in proportion to the frequency of occurrence of some phenomenon

▶ **FIGURE A.1** **Topographic Maps**
Topographic maps represent the form of Earth's surface in both horizontal and vertical dimensions. This extract is from the Lewis Lake region of Yellowstone National Park in Wyoming, U.S. The height of landforms is represented by contours (lines that connect points of equal vertical distance above sea level), which on this map are drawn every 100 feet (30.5 meters). Features such as rivers and land cover, are also shown on the map. Note how the closely spaced contours of the hill slopes represent the shape and form of the land.

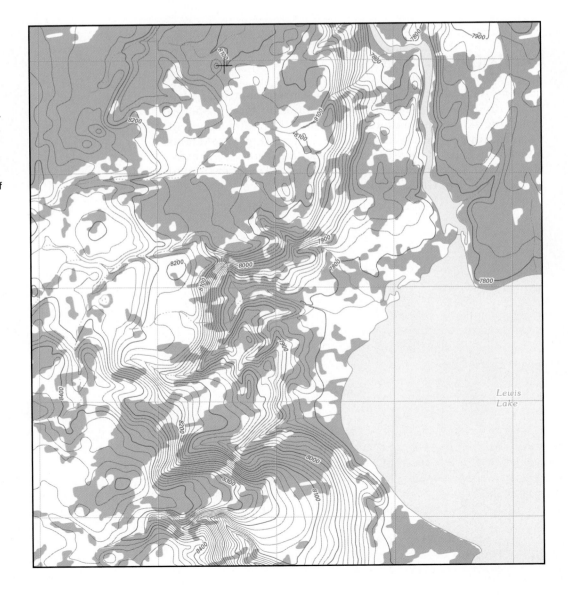

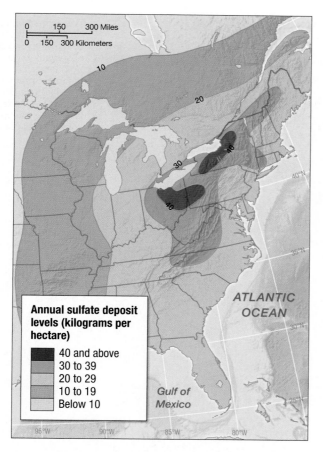

▲ **FIGURE A.2** **Isoline Maps** Isoline maps portray spatial information by connecting points of equal data value. Contours on topographic maps (see Figure A.1) are a type of isoline. This map shows one type of air pollution in the eastern United States.

or event at a given location. Figure 1.46, p. 39, shows an example using proportional circles. Symbols such as arrows or lines can also be drawn proportionally to portray flows between particular places. Simple distributions can be effectively portrayed through *dot maps,* in which a single dot represents a specified number of occurrences of some phenomenon or event. Yet another type of map is the *choropleth map,* in which tonal shadings are graduated to reflect area variations in numbers, frequencies, or densities (see, e.g., Figure 1.27, p. 28).

Map Scales

A *map scale* is simply the ratio between linear distance on a map and linear distance on Earth's surface. It is usually expressed in terms of corresponding lengths, as in "one centimeter equals one kilometer," or as a *representative fraction* (in this case, 1/100,000) or ratio (1:100,000). *Small-scale* maps are based on small representative fractions (e.g., 1/1,000,000 or 1/10,000,000). They cover a large part of Earth's surface on the printed page. A map drawn on this page to the scale of 1:10,000,000 would cover about half of the United States; a map drawn to the scale of 1:16,000,000 would easily cover the whole of Europe. *Large-scale* maps are based on larger

representative fractions (e.g., 1/25,000 or 1/10,000). A map drawn on this page to the scale 1:10,000 would cover a typical suburban subdivision; a map drawn to the scale of 1:1,000 would cover just a block or two.

Map Projections

A *map projection* is a systematic rendering on a flat surface of the geographic coordinates of the features found on Earth's surface. Because Earth's surface is curved and not a perfect sphere, it is impossible to represent on a flat plane, sheet of paper, or monitor screen without some distortion. Cartographers have devised a number of techniques for projecting *latitude* and *longitude* (**FIGURE A.3**, p. A-3) onto a flat surface, and the resulting representations each have advantages and disadvantages. None can represent distance correctly in all directions, though many can represent compass bearings or area without distortion. The choice of map projection depends largely on the purpose of the map.

Projections that allow distance to be represented as accurately as possible are called *equidistant projections*. These can represent distance accurately in only one direction (usually north–south), although they usually provide accurate scale in the perpendicular direction (which in most cases is the Equator). Equidistant projections are often more aesthetically pleasing for representing Earth as a whole, or large portions of it. An example is the Polyconic projection (**FIGURE A4A**, p. A-4).

Projections on which compass directions are rendered accurately are known as *conformal projections*. On the Mercator projection (see **FIGURE A4B**), for example, a compass bearing between any two points is plotted as a straight line. As a result, the Mercator projection has been widely used in navigation for hundreds of years. The Mercator projection was also widely used as the standard classroom wall map of the world for many years, and its image of the world has entered deeply into general consciousness. As a result, many Europeans and North Americans have an exaggerated sense of the size of the northern continents and underestimate the size of Africa.

Some projections are designed such that compass directions are correct from only one central point. These are known as *azimuthal projections*. They can be equidistant, as in the Azimuthal Equidistant projection (see **FIGURE A4C**), which is sometimes used to show air-route distances from a specific location.

Projections that portray areas on Earth's surface in their true proportions are known as *equal-area projections*. Such projections are used when the cartographer wishes to compare and contrast distributions on Earth's surface—the relative area of different types of land use, for example. Examples of equal-area projections include the Eckert IV projection, Bartholomew's Nordic projection (used in Figure 1.22, p. 22), and the Mollweide projection (see **FIGURE A4D**). Equal-area projections such as the Mollweide are especially useful for thematic maps showing economic, demographic, or cultural data. Unfortunately, preserving accuracy in terms of area tends to result in world maps on which many locations appear squashed.

For some applications, aesthetic appearance is more important than conformality, equivalence, or equidistance, so cartographers

▶ **FIGURE A.3 Latitude and Longitude** Lines of latitude and longitude provide a grid that covers Earth, allowing any point on Earth's surface to be accurately referenced. Latitude is measured in terms of angular distance (i.e., degrees and minutes) north or south of the Equator, as shown in (a). Longitude is measured in the same way, but east and west from the Prime Meridian, a line around Earth's surface that passes through both poles (North and South) and the Royal Observatory in Greenwich, just to the east of Central London, in England. Locations are always stated with latitudinal measurements first. The location of Paris, France, for example, is 48°51' N and 2°20' E, as shown in (b).

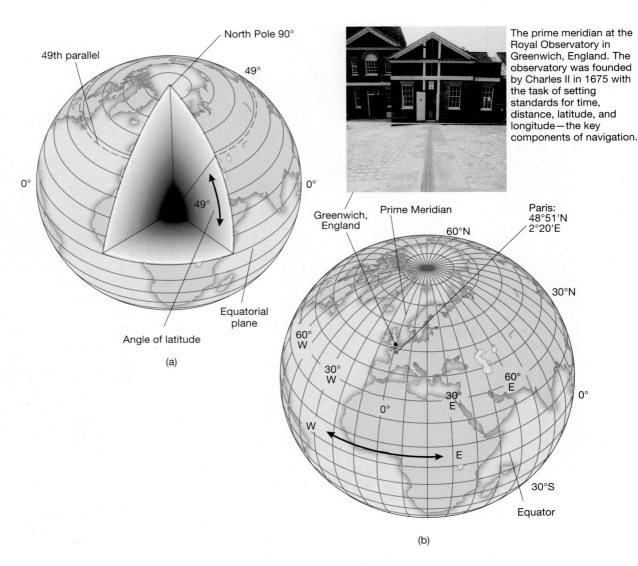

The prime meridian at the Royal Observatory in Greenwich, England. The observatory was founded by Charles II in 1675 with the task of setting standards for time, distance, latitude, and longitude—the key components of navigation.

(a)

(b)

have devised a number of other projections. Examples include the Times projection, which is used in many world atlases, and the Robinson projection, which is used by the National Geographic Society in many of its publications. The *Robinson projection* (**FIGURE A.4E**, p. A-4) is a compromise projection that distorts both area and directional relationships but provides a general-purpose world map.

There are also political considerations. Countries may appear larger and thus more important on one projection than on another. The *Peters projection*, for example (**FIGURE A.4F**, p. A-4), is a deliberate attempt to give prominence to the underdeveloped countries of the Equatorial regions and the southern hemisphere. As such, it has been officially adopted by the World Council of Churches and numerous agencies of the United Nations and other international institutions. Its unusual shapes give it a shock value that gets people's attention. For some, however, those shapes are ugly and many have rejected the Peters projection. This emphasizes both the aesthetic and political nature of map projections and decisions surrounding their use.

One kind of projection that is sometimes used in small-scale thematic maps is the *cartogram*. In this projection, space is transformed according to statistical factors, with the largest mapping units representing the greatest statistical values. **FIGURE A.5A**, p. A-5, shows a cartogram of the world in which countries are represented in proportion to the threatened amphibian species that are present there. This sort of projection is particularly effective in helping us visualize relative inequalities in the global distribution of biodiversity and endangered species. **FIGURE A.5B**, p. A-5, shows a cartogram of the world in which the cost of telephone calls made from the United States has been substituted for linear distance as the basis of the map. The deliberate distortion of the shapes of the continents in this sort of projection dramatically emphasizes spatial variations.

Finally, advancements in computer graphics has made it possible for cartographers to create more dynamic two-dimensional representations of Earth's surface and expand the types of maps that are designed in three-dimensional space. Computer software that renders three-dimensional statistical data onto the flat surface of a monitor screen or a piece of paper facilitates the visualization of many aspects of human geography in innovative and provocative ways (**FIGURE A.6**, p. A-6).

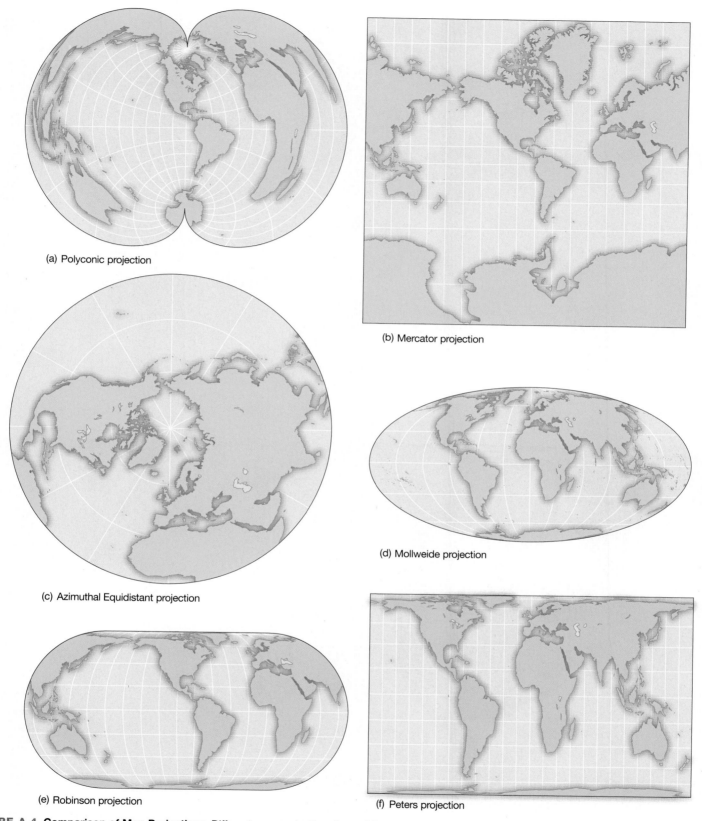

(a) Polyconic projection

(b) Mercator projection

(c) Azimuthal Equidistant projection

(d) Mollweide projection

(e) Robinson projection

(f) Peters projection

▲ **FIGURE A.4 Comparison of Map Projections** Different map projections have different properties. (a) The Polyconic projection is true to scale along each east–west parallel and along the central north–south meridian. It is free of distortion only along the central meridian. (b) On the Mercator projection, compass directions between any two points are true, and the shapes of landmasses are true, but their relative size is distorted. (c) On the Azimuthal Equidistant projection, distances measured from the center of the map are true, but direction, area, and shape are increasingly distorted with distance from the center point. (d) On the Mollweide projection, relative sizes are true, but shapes are distorted. (e) On the Robinson projection, distance, direction, area, and shape are all distorted in an attempt to balance the properties of the map. It is designed purely for appearance. (f) This equal-area projection offers an alternative to traditional projections, which, Arno Peters argued, exaggerate the size and apparent importance of the higher latitudes. It has been criticized by cartographers in the United States on the grounds of aesthetics: one consequence of equal-area projections is that they distort the shape of landmasses.

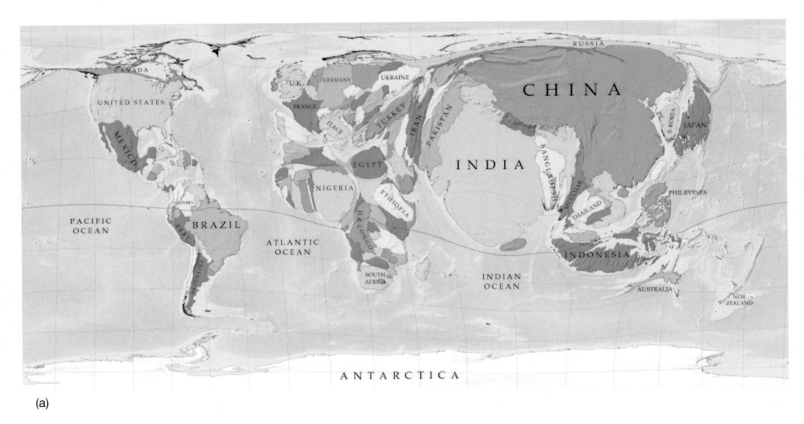

(a)

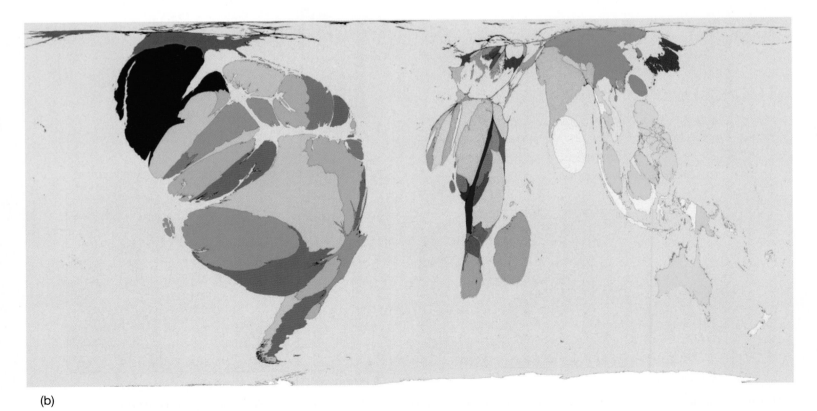

(b)

▲ **FIGURE A.5 Cartograms** In a cartogram, space is distorted to emphasize a particular attribute of places or regions. (a) This example shows the relative size of countries based on their population rather than their area; the cartographers have maintained the shape of each country as closely as possible to make the map easier to read. As you can see, population-based cartograms are very effective in demonstrating spatial inequality. (b) In this example, countries are sized based on the number of threatened amphibian species that are present there. In this case, the countries of the northern latitudes disappear while the countries of Central Word South America stand out. These are centers of some of the world's most important biodiversity.

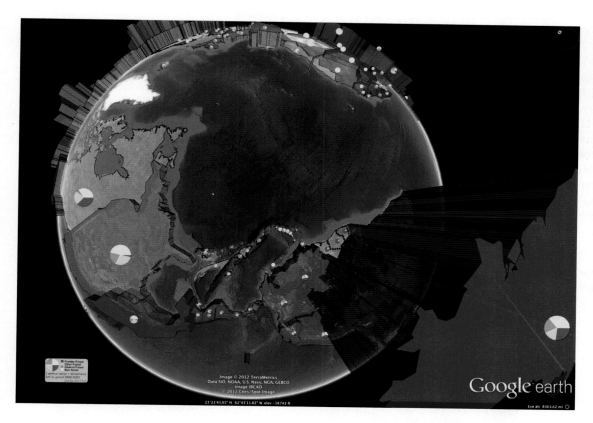

◄ **FIGURE A.6 Geovisualization**
A three-dimensional visualization of global deforestation using Google Earth, developed by David Tryse. The red/green 3D visualization show the levels of deforestation, while the circle graphs depict how much is frontier (dark green), other forest (light green), cleared (orange), or non-forest (clear).

Geospatial Technologies: The GIS Revolution

Geographic information systems (GIS)—organized collections of computer hardware, software, and geographic data that are designed to capture, store, update, manipulate, and display geographically referenced information—have rapidly grown to become a predominant method of geographic analysis, particularly in the military and commercial worlds. The software in GIS incorporates programs to store and access spatial data, to manipulate those data, and to draw maps.

As an industry, GIS is enormous, as GIS systems and analysis underpin many decisions in business and government. In 2013, the global geospatial industry generated U.S. $270 billion in revenue, and is forecast to grow 10–15% annually over 2013–2018. Jobs in GIS are numerous and growing and include cartographic design, data analysis, programming, and the management of geographic analysis projects and databases. The number of jobs in geospatial industries is growing at approximately 35% annually, with much of that growth coming from the commercial subsection of the market.

The primary requirement for data to be used in GIS is that the locations for the variables are known. Location may be annotated by x, y, and z coordinates of longitude, latitude, and elevation or by such systems as zip codes or highway mile markers. Any variable that can be located spatially can be fed into a GIS. Data capture—putting the information into the system—is the most time-consuming component of GIS work. Different sources of data, using different systems of measurement, scales, and systems of representation, must be integrated with one another; changes must be tracked and updated. Many GIS operations in the United States, Europe, Japan, and Australia have begun to contract out such work to firms in countries where labor is cheaper; India has emerged as a major data conversion center for GIS.

Applications of GIS

GIS is now common in most people's daily lives, through the use of online mapping and navigation technology in smartphones and automobiles. Beyond these daily applications, however, GIS can be used to answer complex questions, by merging data from different sources, on different topics, and at different scales. A GIS makes it possible to link, or integrate, information that is difficult to associate through any other means. For example, using data on levels of income, reported health problems, and the distribution of infrastructure and roads in a rural area, GIS analysis can reveal the best and most effective locations for new hospitals or clinics.

GIS can also be used to support complex decision making and prioritizing among diverse stakeholders. For example, in trying to determine locations for economic development near a sensitive environment or national park, diverse stakeholders (like business owners, local residents, and environmentalists) can be brought together to examine different possible outcomes and scenarios using computer-generated maps. As each group stresses different priorities, new maps can be generated allowing for compromise and further discussion. In this way, GIS can be a tool for democratic decision making.

GIS technology can render visible many aspects of geography that were previously unseen. For example, it can produce incredibly detailed maps based on millions of pieces of information—maps that could never have been drawn by human hands. At the other extreme of spatial scale, GIS can put places under the microscope, creating detailed new insights using huge databases and effortlessly browsable media (**FIGURE A.7**).

How GIS data are shared and stored determines access to spatial information. Access to spatial data has grown from government and corporate entities to the global public over the past few decades. Web GIS is an example of a data source from which

anyone can download shape files—a standardized vector data format that uses points, lines, and polygons to represent key geographic features—and other forms of data to create maps. Cloud-based tools like Cloud GIS, which can be accessed through apps, allow different levels of sharing data between private, public, community, and hybrid clouds.

As GIS has expanded access to spatial data, cartographic methods have also developed. *Geovisualization* is the set of tools and techniques used in geospatial data analysis to display space, time, and any number of thematic attributes. Maps are representations of Earth's surface, but can also communicate many different attributes of surface features, including change over time and the processes that cause change. Spatial data are

▶ **FIGURE A.7 GIS-Derived Map of Sea-Level Rise** This detailed map shows the predicted land cover of the May River Estuary, assuming a sea-level rise of 1 foot, using a composite digital terrain model obtained through remote sensing of multiple types of terrain.

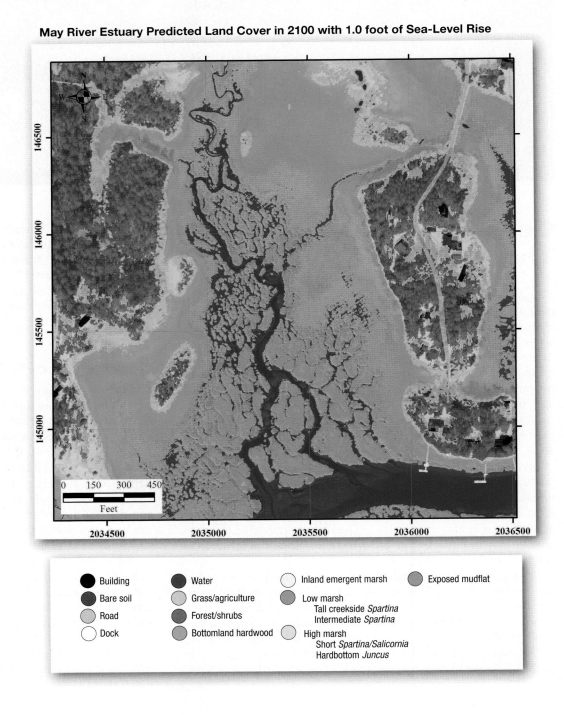

May River Estuary Predicted Land Cover in 2100 with 1.0 foot of Sea-Level Rise

often linked to a temporal component such as climatic, environmental, and biological changes. A geovisual tool or program can utilize GIS data and then display those as an animation or sequence of steps. Geovisualization can generate three-dimensional displays, interactive maps, time lapse maps, and other map animations.

Many advances in GIS have come from military applications and this has been a point of debate within the discipline of geography. GIS, for example, allows infantry commanders to calculate line of sight from tanks and defensive emplacements, allows cruise missiles to fly below enemy radar, and provides a comprehensive basis for military intelligence. Beyond the military, GIS technology allows an enormous range of problems to be addressed. It can be used to decide how to manage farmland, to monitor the spread of infectious diseases, to monitor tree cover in metropolitan areas, to assess changes in ecosystems, to analyze the impact of proposed changes in the boundaries of legislative districts, to identify the location of potential business customers, to map the spatial distributions and patterns related to crime, and to provide a basis for urban and regional planning.

Data Acquisition in GIS

Spatial data are acquired through many sources. Existing maps are scanned or digitized in the form of *vector* or *raster data* (**FIGURE A.8**). Vector data are spatial representations constructed using points, lines, and polygons, whereas raster data are spatial representations that use a grid cell approach

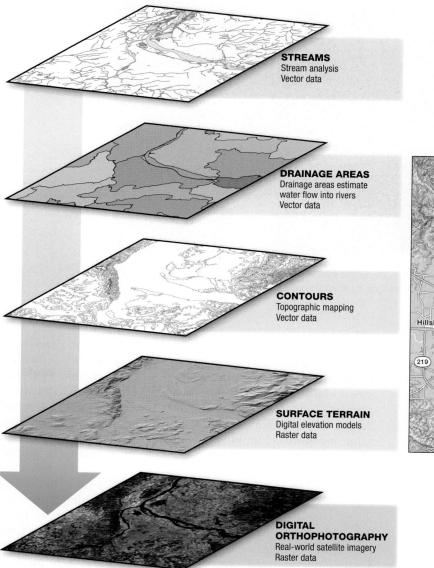

STREAMS
Stream analysis
Vector data

DRAINAGE AREAS
Drainage areas estimate water flow into rivers
Vector data

CONTOURS
Topographic mapping
Vector data

SURFACE TERRAIN
Digital elevation models
Raster data

DIGITAL ORTHOPHOTOGRAPHY
Real-world satellite imagery
Raster data

COMPILATION MAP

▲ **FIGURE A.8** **An example of map layers used together in GIS** A Geographic Information System can bring together different types of data and layer those data onto one map, showing relationships between different types of data and processes.

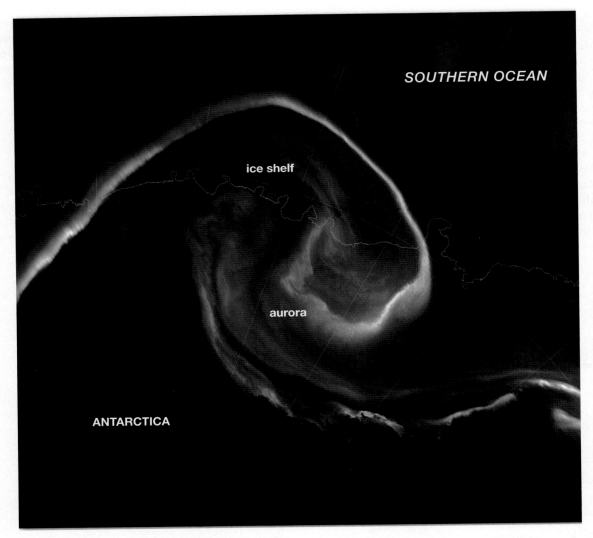

▲ **FIGURE A.9 Antarctic Ice Sheet at Night** Light from the *aurora australis*, or "southern lights," is bright enough to reveal the ice edge in Antarctica's Queen Maud Land. Images such as this one can be digitized and used to analyze changes in Earth's surface.

to managing data. Vector data are often used to show roads, borders, and physical systems, such as rivers, while raster data are often used when mapping data from aerial photography or satellite imagery. *Survey data* are entered into GIS from global navigational satellite positions, like the Global Positional System (GPS). Field coordinates can be directly uploaded into GIS from GPS units on the ground. *Remote sensing*, a subfield in geography, is a form of data acquisition that uses sensors to capture data at a distance. Aerial photographs and satellite sensors are two forms of remote sensing data utilized in GIS. Different categories of data comprise GIS *layers*, which can be stacked in mapping applications to display the distribution of the various types of data in relation to each other (**FIGURE A.9**).

Local or community knowledge is another valuable source of spatial data and is derived from and produced for inhabitants of specific geographies. Often developers, politicians, and academics have created land-use plans without input from the

communities directly impacted. *Public Participation GIS (PGIS)* began as a way to provide community groups with information for decision making (**FIGURE A.10**, p. A-10). Over time, PGIS has grown to promote local knowledge, build geographic education, stimulate innovative cartography, and encourage positive social change in marginalized communities. This is done by collaborating between community members and geographers through interviews, community maps, place narratives and focus groups on land-use patterns.

Collaborative mapping exists in other forms including *Voluntary Geographic Information (VGI)* in which individuals in specific areas voluntarily contribute geospatial data. VGI can incorporate traditional geographic information as well as subjective, emotional, and other forms of data, including photographs and observations. Sources of these data include platforms such as social media (Twitter, Flickr, Instagram), travel sites (TripAdvisor), and public mapping forums (Panoramio, Wikimapia, OpenStreetMap). Collaborators and

▲ **FIGURE A.10 Public Participatory GIS in Laramie, Wyoming, USA**
The Wyoming Geographic Information Science Center (WyGISC) organized a community mapping project in Laramie, WY in 2011. Local residents were trained to use geographic positioning systems (GPS) devices so that they could collect data about their city. The project's goal was not only to raise geographic awareness but to collect geographic data that could be used by the community in the future.

volunteers have contributed a substantial amount of environmental data, producing "citizen science." *Citizen Science* refers to communities or networks of citizens who act as observers in a domain of science. Programs like Colorado River Watch, Project GLOBE, and Christmas Bird Count are regional and international networks of trained citizens who monitor different processes and phenomena in the planet, and share this information for analysis.

Participatory methods in GIS are critical tools for social change. Using collaborative spatial data, geographers codevelop projects that are relevant to a community and provide a platform for underrepresented voices. Social scientists and planners have deployed PGIS to build more sustainable communities. A recent study in Tinto, Cameroon, utilized PGIS to discern how different stakeholders manage the shared resource of a forest. Collaborators found that by using PGIS to monitor use, management, and governance, they were able to analyze community input and recommend sustainable land-use plans. Tinto community members used that data to influence their resource management decisions. In addition to environmental projects, PGIS has been used in geographies of conflict to rebuild communities and heal trauma. After the civil war in Lebanon (1975–1990), city planners, social scientists, and community members created mental maps of Beirut using their memories and stories about places. From those mental maps, a new vision was created to rebuild the war-torn city.

Critiques of GIS

In just the past few years, GIS has resulted in the creation of more maps than were created in all previous human history. One result is that as maps have become more commonplace, more people and more businesses have become more spatially aware. Nevertheless, some critics have argued that GIS has a number of drawbacks. First, the quality and utility of maps are only as good as the data and decisions that lie behind them. Since all maps are only partial representations of the world, greater ease of mapmaking provides new opportunities for propaganda, dissimulation, and deceit. Though GIS is available to many groups in society, moreover, the expertise and expense of GIS often mean that the power of mapmaking has remained among elites, companies, and governments.

GIS is also an unquestionable part of the increasing surveillance of populations, made further possible by high-resolution satellite imagery, surveillance cameras, and other technology. Knowing and tracking where people are and where they are going is a powerful tool for corporations, governments, and other powerful institutions. A further fear is that GIS may be helping create a world in which people are not treated and judged by who they are and what they do, but more by where they live. People's credit ratings, ability to buy insurance, and ability to secure a mortgage, for example, are all routinely judged, in part, by GIS-based analyses that take into account the attributes and characteristics of their neighbors and neighborhoods. Like many new technologies, therefore, GIS has the power to support decision making and community empowerment, even while it can be a dangerous tool in the hands of the powerful.

Glossary

A

Aborigines: indigenous peoples of Australia.

acid rain: precipitation that has mixed with air pollution to produce rain that contains levels of acidity—often in the form of sulfuric acid—that are harmful to vegetation and aquatic life.

acquired immunodeficiency syndrome (AIDS): the final stage of HIV disease when the body has lost most of its immunity to defend against infections.

adivasi: an umbrella term for a heterogeneous set of ethnic and tribal people in India who represent a significant part of the manual labor force in the regions where they predominate.

afforestation: programs that convert previously non-forested land to forest by planting seeds or trees.

Afrikaners: people living in South Africa, who are descended from predominantly Dutch, French, and German settlers and speak the Afrikaans language.

Agent Orange: a chemical agent used by the U.S. military in Vietnam to remove the foliage from plants, making it less difficult for military ground forces to find cover.

AIDS orphan: a child who has lost one or more parents to an AIDS-related death.

ALBA: Bolivarian Alliance for the Peoples of Our America is an intergovernmental organization that includes Bolivia, Cuba, Ecuador, Nicaragua, Venezuela, and several other Caribbean countries. The group tends to be against U.S. influence in the Latin America and includes mostly socialist governments.

al-Qaeda: a global militant organization that operates as a network composed of a multinational, stateless army and a radical Islamist movement.

alternative development: a development approach that calls into question the basic tenets of neoliberalism such as a universal faith in markets, a focus on reducing government programs and roles, and the assumption that development is a purely economic process.

altiplano: high-elevation plateaus and basins that lie within even higher mountains, especially in Bolivia and Peru, at more than 3,000 meters (9,500 feet) in the Andes of Latin America.

altitudinal zonation: a vertical classification of environment and land use according to elevation based mainly on variations in climate and vegetation.

American Revolution: the revolution launched by the residents of the original 13 U.S. colonies became disillusioned with the taxes they were forced to pay to help Britain recoup the high cost of the war.

Americanization: a process by which a generation of individuals born elsewhere felt less loyalty and fewer cultural ties to their countries of origin and developed a new ethos with ties to the United States.

ancestor worship: a belief that the living can communicate with the dead and that dead spirits, to whom offerings are ritually made, have the ability to influence people's lives.

Angkor Wat: a 12th-century Hindu (and Buddhist) temple compound that is one of the largest religious structures ever built and a focus of tourism in contemporary Cambodia.

anime: a distinctive stylized form of Japanese animated film.

animism: a belief system that focuses on the souls and spirits of nonhuman objects in the natural world.

Antarctic Treaty: an agreement that governs international relations on the Antarctic continent. Fifty-three nations have agreed to ban nuclear testing and waste disposal and mineral and oil exploration, setting aside the area south of 60°S latitude for peaceful purposes and scientific research.

Anthropocene (the "human era"): a term used to describe the current geological epoch in which human activity dominates the planet.

anthropogenic: created or retained by human beings.

AOSIS (Alliance of Small Island States): an association of 44 low lying, mostly island, countries that have formed an alliance to combat global warming, which threatens their existence through sea-level rise.

apartheid: a South African policy of racial separation that, prior to 1994, structured space and society to separate black, white, and colored populations.

apostasy: the abandonment or renunciation of religious faith or beliefs.

apparatchik: a state bureaucrat in the Russian Communist Party during the Soviet era.

Arab League: a regional alliance in the Middle East founded in 1945 to strengthen ties among member states, set policies, and promote common interests. Member states include Egypt, Iraq, Lebanon, Saudi Arabia, Syria, Jordan, Yemen, Algeria, Bahrain, Kuwait, Morocco, Oman, Qatar, Sudan, Tunisia, the United Arab Emirates, and the Palestinian Authority.

Arab Spring: a term that refers to the numerous recent protests against dictatorships across the Middle East region in which people motivated by economic inequalities and lack of access to basic resources have taken to the streets to voice their concerns.

archipelago: a group of islands or expanse of water with many islands.

aridity: a characteristic of a climate with insufficient moisture to support trees or woody plants.

Aryan: a group of livestock-keeping peoples from the northwest who populated the plains of the Ganga (Ganges) River in India from 1500–500 b.c.e.

ASEAN: an international organization of the nations of Southeast Asia established to promote economic growth and regional security.

Asian Brown Cloud: a blanket of air pollution 3 kilometers (nearly 2 miles) thick that hovers over most of the tropical Indian Ocean and South, Southeast, and East Asia, stretching from the Arabian Peninsula across India, Southeast Asia, and China almost to Korea.

Asian Tigers: newly industrialized territories of Hong Kong, Taiwan, South Korea, and Singapore that have experienced rapid economic growth.

aspect: the direction in which a sloping piece of land faces.

assimilation: the process by which peoples of different cultural backgrounds who occupy a common territory achieve sufficient cultural solidarity to sustain a national existence.

atheism: the denial and/or lack of belief in God or gods.

atmospheric circulation: the global movement of air that transports heat and moisture around the earth.

atoll: a low-lying island landform consisting of a circle of coral reefs around a lagoon, often associated with the rim of a submerged volcano or mountain.

Aung San Suu Kyi: one of the leaders of the National League for Democracy in Burma and a recipient of the Nobel Peace Prize in 1991 for her work on bringing democratic reform to Burma.

B

Balfour Declaration: a 1917 British mandate that required the establishment of a Jewish national homeland.

balkanization: the division of a territory into smaller and often mutually hostile units.

Berlin Conference: a meeting convened in 1884–1885 by German Chancellor Otto von Bismarck to divide Africa among European colonial powers.

Bible Belt: a region of the United States that stretches from Texas to Missouri and is dominated by Protestant denominations, many of them fundamentalist and evangelical.

biodiversity: the variety in the types and numbers of species in particular regions of the world.

biofuel: fuels produced from raw, often renewable, biological resources, such as ethanol made from corn or sugar.

biomes: regions of similar climate, physical conditions, and plants and animals.

bioprospecting: a search for plants, animals, and other organisms that may be of medicinal value or may have other commercial use.

birthrates: the measure of the number of births in a population, usually expressed as births per 1,000 people per year or as a percentage.

blackbirding: a process where thousands of laborers from the Pacific, collectively called kanakas, were kidnapped and enslaved to work on Australian cotton and sugar plantations.

Bollywood: the popular name given to the regional site of film production in Mumbai, India.

Bolshevik: the majority faction of the Communist Party that led a successful revolution against the tsar in Russia in 1917.

braceros: a guest worker from Mexico given a temporary permit to work as a farm laborer in the United States between 1942 and 1964.

brain drain: a situation the occurs when the most talented and well-educated young people emigrate out of their home nation in search of opportunities elsewhere.

BRICS: the acronym for the emerging world region that includes Brazil, India, China, and South Africa.

Buddhism: a belief system that stresses nonviolence, moderation, and the cessation of suffering and originated in South Asia and is traced to a religious leader who came to be known as Buddha (the Enlightened One).

buffer zone: a group of smaller or less powerful countries situated between larger or more powerful countries that are geopolitical rivals.

bureaucratic class: nonelected government officials.

bush fallow: the modification of shifting cultivation, where crops are rotated and land is left fallow (uncultivated) for short periods.

bushmeat: the flesh of wild animals hunted for human consumption. The sale of bushmeat has created controversy in regions where endangered species are eaten for food.

C

call centers: an industry in which workers respond to sales and customer service phone calls from around the world and work long hours for modest pay.

cap-and-trade programs: regulatory systems in which (1) the "cap" is a government-imposed limit on carbon (greenhouse gas) emissions, (2) emission permits or quotas are given or sold to states (or firms), and (3) the "trade" occurs when unused permits are sold by those who have been able to reduce their emissions beyond their quota to those who are unable to meet their quota.

capitalism: a form of economic and social organization characterized by the profit motive and the control of the means of production, distribution, and exchange of goods by private ownership.

carbon footprint: a method of estimating how much an individual, a region, or a country is contributing to global climate change through its own greenhouse gas emissions.

carbon market: a system in which countries and companies are required to reduce emissions and can earn so-called carbon credits toward meeting their reduction obligations by investing in reducing emissions in developing countries; this system allows developing countries that have ratified the Kyoto Protocol agreement to receive payments as incentives for investments in climate-friendly projects that reduce greenhouse gas emissions.

cartels: a collection of independent businesses formed to regulate production, prices, and marketing.

caste: a system of hereditary, endogamous groupings that is reinforced by language, religion, and occupation.

chador: a loose robe worn by Muslim women that covers most of the body.

chaokhao (hill tribes): ethnic minorities living in mainland Southeast Asia, particularly, but not exclusively, in the northern more mountainous parts of the region.

child soldier: a child under 18 years of age who is forced or recruited to join armed groups.

cirque: a deep, bowl-shaped basin on a mountainside, shaped by ice action.

civil society: a network of social groups and cultural traditions that operate independently of the state and its political institutions.

class: a social group distinction, typically assigned based on wealth and access to resources that may be internalized culturally by communities and coalesced into a self-conscious identity.

climate: typical conditions of the weather that is expected at a location, often measured by long-term averages of temperature and precipitation.

climate change: long-term shifts in temperature and precipitation at a regional or global scale, resulting from systemic alterations in the Earth system, changes incoming solar radiation, and/or from the loading of greenhouse gases in the atmosphere including those produced by human activity.

climate system: interactions of air, water, and the Sun's energy circulating around the globe; weather and climate are the product of the climate system. A climate system consists of five major components: the atmosphere, hydrosphere, cryosphere, a land surface, and biosphere.

climate vulnerability: a condition describing the relative sensitivity of geophysical, biological, and socio-economic systems to harm from climate variability and change.

Cold War: the period between World War II and 1990 in which the United States and the Soviet Union established their role as global powers; their struggles took place through a variety of proxy conflicts.

collectivization: a process by which peasants are relocated onto collective farms by the state where their labor is expected to produce bigger yields.

colonialism: a political and economic system in which regions and societies are legally, economically, and politically dominated by an external society.

Columbian Exchange: the interchange of crops, animals, people, and diseases between the Old World of Europe and Africa and the New World of the Americas that began with the voyages of Christopher Columbus in 1492.

command economy: a national economy in which all aspects of production and distribution are centrally controlled by government agencies.

commodity chain: a network of labor and production processes that originates in the extraction or production of raw materials and whose end result is the delivery and consumption of a finished commodity.

Common Agricultural Policy (CAP): the system of European Union support of wholesale prices for agricultural produce created to bolster the agricultural sector of the EU.

common property resources: resources, such as fish or forests, that are managed collectively by a community that has rights to the resource, rather than their being owned by individuals.

Commonwealth of Independent States (CIS): a forum for discussing the management of economic and political problems, including defense issues, transport and communications, regional trade agreements, and environmental protection; members are the Russian Federation, Belarus, Ukraine, the Central Asian states, and some Transcaucasus states.

communism: a form of economic and social organization characterized by the common ownership of the means of production, distribution, and exchange.

Communist Council for Mutual Economic Assistance (COMECON): an organization established in 1949 to reorganize Eastern European economies in the Soviet mold; member countries were to pursue independent, centralized plans designed to produce economic self-sufficiency.

convergent plate boundary: an area where tectonic plates of Earth's crust meet and create subduction zones.

Confederate States of America: the government founded by the southern states of the United States in 1861 with the goal of seceding from the Union; the division from the northern states was rooted in slavery and economic issues.

conflict diamond: a diamond mined in a war zone and/or used to finance conflicts. Also known as "blood" diamonds.

Confucianism: a spiritual and philosophical tradition native to China expressed in a philosophy that emphasizes the importance of ethics, good governance, education, family, and hard work.

Convention on International Trade in Endangered Species (CITES): a global treaty first signed in 1990 that restricts trade in rare or endangered animals and their parts between member countries.

creative destruction: the withdrawal of investments from activities (and regions) that yield low rates of profit in order to reinvest in new activities (and new places).

crony capitalism: a system of capitalism in which political leaders and the friends of those leaders are privileged.

Cultural Revolution: a sustained attack on Chinese traditions and cultural practices in which millions of people were displaced, tens of thousands lost their lives, and much of urban China was plunged into a terrifying climate of suspicion and recrimination; launched in 1966 by Mao Zedong, who called it a "Great Proletarian Cultural Revolution."

culture: a shared set of attitudes and behavior characteristic of a particular social group that also includes shared symbols and everyday practice and activities.

culture system: the Dutch colonial policy from 1830 to 1870 that required farmers in Java to devote one-fifth of their land and their labor to production of an export crop.

cyberwarfare: activities by nations or firms/corporations to gain access to, hack, or sabotage the computers, networks, and information infrastructure of another country or company.

cyclone: a large rotating storm produced by warm ocean temperatures and eddies in the trade winds from August to October.

D

Dalai Lama: the spiritual leader of Tibet; the latest reincarnation of a series of spiritual leaders who, according to Tibetan Buddhism, have chosen to be reborn in order to enlighten others.

death rates: the measure of deaths in a population expressed as either 1,000 deaths per population or as a percentage.

deforestation: the clearing, thinning, or elimination of tree cover in historically forested areas, most typically referring to human-caused treecover loss.

deindustrialization: the decline in industrial employment in core regions as firms scale back activities in response to lower levels of profitability.

demilitarized zone (DMZ): a designated area between two or more countries where military activity is prohibited.

democracy: an egalitarian form of government in which all citizens determine public policy and the laws and the actions of their state and all citizens have a right to express their opinions.

demographic collapse: the rapid die-off of indigenous populations of the Americas that began occurring in about 1500 as a result of diseases such as smallpox introduced by the Europeans to which indigenous peoples had little or no resistance.

demographic transition: the replacement of high birthrates and death rates by low birthrates and death rates.

demographics: the population densities, movements, and changes of people over time and space.

dependency: an economic system in which resources flow from poorer to richer nations as a result of unfair trade, colonial control, and other unequal power structures.

desalinization: a process that converts saltwater into drinking water.

desertification: the process by which arid and semiarid lands become degraded and less productive, leading to desertlike conditions.

development: the improvement in the economic well-being and standard of living of people.

dharma: a core aspect of Hinduism and Buddhism; the principle of cosmic order, in which actions are aligned with a right way of being and doing in the universe (virtue, righteousness, social duty).

dialects: regional variations within a language.

diaspora: the spatial dispersion of a previously homogeneous group.

divergent plate boundary: an area where tectonic plates of Earth's crust are pulling apart.

domestication: the adaptation of wild plants and animals through selective breeding for preferred characteristics into cultivated or tamed forms.

domino theory: the view that political unrest in one country can destabilize neighboring countries and start a chain of events.

dowry: a cultural system in which significant payments are made during a marriage by the bride's family to the groom's family.

Dreamtime: the aboriginal worldview that links past and future and people and places in a continuity that ensures respect for the natural world.

drone: an unmanned, remote-controlled aerial military vehicle used to spy on and strike enemy targets.

dry farming: arable farming techniques that allow the cultivation of crops without irrigation in regions of limited moisture.

E

Ebola: A highly infectious viral disease spread by contact with bodily fluids that emerged in Sub-Saharan Africa in the 1970s—probably from wild animal populations—and has killed thousands of people.

ecological imperialism: a concept developed by historian Alfred Crosby to describe the way European organisms, including diseases, pests, and domestic animals, took over the ecosystems of other regions of the world, often with devastating impacts on local peoples, flora, and fauna.

ecosystem: a complex of living organisms, their physical environment, and all their relationships in a particular place.

ecosystem services: resources and processes that are supplied by natural ecosystems that benefit people, regions, and the planet as a whole, such as water and soil protection provided by wetlands or the removal of greenhouse gases from the atmosphere provided by forests.

ecotourism: tourism designed to be environmentally conscious and sensitive as well as provide economic opportunities for local people.

ECOWAS: an organization established in 1975 to promote trade and cooperation in West Africa; includes the countries of Benin, Burkina Faso, Cape Verde, Côte d'Ivoire, Gambia, Ghana, Guinea, Guinea-Bissau, Liberia, Mali, Mauritania, Niger, Nigeria, Senegal, Sierra Leone, and Togo.

El Niño: the periodic warming of sea-surface temperatures in the tropical Pacific off the coast of Peru that results in worldwide changes in climate, including droughts and floods.

emerging region: an area where loosely connected locations are developing greater internal coherence.

emissions trading scheme (ETS): a system that allows energy-intensive facilities (e.g., power generation plants and iron and steel, glass, and cement factories) to buy and sell permits that allow them to emit carbon dioxide (CO_2) or other greenhouse gases into the atmosphere.

enclave: a culturally distinct territory that is encompassed by a different cultural group or groups.

endogamous: a description of culture where families are expected to find marriage partners for their children among other members of the kinship group.

Enlightenment: an 18th-century movement in Europe and North America marked by a belief in the sovereignty of reason and empirical research in the sciences.

entrepôt: a seaport that is an intermediary center of trade and shipment.

environmental geography: the study of the relationship between humans and the natural and built environments in which they live.

ethnic cleansing: the systematic and forced removal of members of an ethnic group in order to change the ethnic composition of a region.

ethnic group: a group of people whose members share cultural characteristics and commonalities, such as language or religion.

ethnicity: state of belonging to a social group that has a common national or cultural tradition; socially created system of rules about who belongs to a particular cultural group.

ethnonationalism: nationalism based on ethnic identity.

European Union (EU): a supranational economic and political organization made up of 27 European countries founded to recapture prosperity and power through economic and political integration; it is formerly known as the European Economic Community.

Europeanization: a process by which a non-European subject gradually adopts the cultural norms of European countries.

exclave: a portion of a country or a cultural group's territory that lies outside its contiguous land area.

exclusive economic zone (EEZ): an international 200-nautical-mile (370-kilometer) area that was formalized in 1982 in the United Nations Convention on the Law of the Sea (UNCLOS); of tremendous significance to Oceania because it allows countries with a small land area but many scattered islands, such as Tonga and the Cook Islands, to lay claim to immense areas of ocean.

exotics: species from places/regions outside their current location. These species may or may not be invasive in character.

extractive reserves: areas of land, often forest, that are protected for extractive uses by local groups; one of the best known is in the Amazon Rainforest and named for Chico Mendes, a rubber tapper who organized resistance to deforestation by ranchers and was murdered in 1988.

F

failed state: a nation in which the central government is so weak that it has little control over its territory.

fair trade movement: a movement concerned with ensuring that producers and workers are paid a reasonable wage and may also imply that crops, including flowers and coffee, are produced with more sustainable methods.

favelas: the Brazilian term for informal settlements in cities that lack good housing and services.

federal system: political authority divided between autonomous sets of governments, one national and the others at lower levels, such as state/province, county, city, and town; in a federal system, many decisions are made at the local level.

fengshui: the belief that the physical attributes of places can be analyzed and manipulated to improve the flow of cosmic energy, or *qi*, that binds all living things. Feng shui involves strategies of siting, landscaping, architectural design, and furniture placement to direct energy flows.

feral: a domesticated species that ends up in the wild.

Fertile Crescent: a region arching across the northern part of the Syrian Desert and extending from the Nile Valley to the Mesopotamian Basin in the depression between the Tigris and Euphrates Rivers.

feudal system: regional hierarchies composed, at the bottom, of local nobles and, at the top, of lords or monarchs owning immense stretches of land. In this system, landowners delegated smaller parcels of land in return for political allegiance and economic obligations in the form of money dues or labor.

First Arab–Israeli War: a conflict that took place in 1948 when British forces withdrew from Palestine. The war pitted Arab forces against the recently established Jewish state with the goal of eradicating Israel.

Five-Year Plan: social and economic development initiatives typically in centrally planned economies, especially including communist China and the Soviet Union.

fjord: a steep-sided, narrow inlet of the sea formed when deeply glaciated valleys are flooded by the sea.

foreign direct investment (FDI): an active financial stake in production in a country by a company or an individual located in another country.

fossil fuels: deposits of hydrocarbon that have developed over millions of years from the remains of plants and animals, which have been converted to potential energy sources under extreme pressure below Earth's surface; includes coal, oil, and gas.

French and Indian War: a conflict in North America (1754–1763) among European colonial groups who were vying for control over indigenous lands; also called the **Seven Years' War**.

Fundamentalism: the belief in or strict adherence to a set of basic principles; often arises in reaction to perceived doctrinal compromises with modern social and political life.

G

G8: Group of Eight countries (Canada, France, Germany, Italy, Japan, Russia, the United Kingdom, and the United States) whose heads of state meet each year to discuss issues of mutual and global concern.

Gaza Strip: the part of the Occupied Territories in which Palestinians live, located on the Mediterranean coast.

gender: social differences between men and women rather than the anatomical differences related to sex.

genetically modified organisms (GMOs): an organism that has had its DNA modified in a laboratory rather than through cross-pollination or other forms of evolution.

genocide: deliberate effort to destroy an ethnic, tribal, racial, religious, or national group.

gentrification: the renovation of older, centrally located, working-class neighborhoods by higher-income households.

geoengineering: deliberate modification of the earth system, most commonly to combat climate change by managing solar radiation (e.g., through adding particles to block sunlight and cool the atmosphere) or removing carbon dioxide from the atmosphere (e.g., through increasing ocean take up or large scale tree planting).

geographical imagination: a term referring to the images that people have of the world around them.

geography: the study of the physical features of Earth and its atmosphere, the spatial organization and distribution of human activities, and the complex interrelationships between people and the natural and human-made environments in which they live.

geomorphology: the study of landforms.

geosyncline: a geological depression of sedimentary rocks.

glasnost: an official policy change by Mikhail Gorbachev initiated in 1985 in the Soviet Union that stressed open government and increased access to information as well as a more honest discussion about the country's social issues and concerns.

global financial center: a city or city district that has a concentration of financial institutions that may serve the interests of small groups of elites in that host country and provide opportunities for multinational corporations to move their finances out of their home country to protect against taxes.

global north: those parts of the globe experiencing the highest levels of economic development, which are historically—though by no means exclusively— found in northern latitudes.

global south: those parts of the globe experiencing the lowest levels of economic development, which are historically—though by no means exclusively— found in southern latitudes.

global warming: the increase in world temperatures and change in climate associated with increasing levels of carbon dioxide and other gases including those from human activities, such as deforestation, livestock production, and fossil-fuel burning.

globalization: the increasing interconnectedness of different parts of the world through common processes of economic, environmental, political, and cultural change.

Golden Triangle: an area, which includes the highlands of Burma, Laos, and Thailand, that contains most of the region's forest areas and fields of opium poppies—producing as much as half of the world's supply of heroin—and more recently, a global and regional center for methamphetamine production.

Gondwanaland: a vast continental area, also known as Gondwana, believed to have formed 200 million years ago when the southern part of the ancient supercontinent Pangaea broke off and which resulted in the continents now mostly in the Southern Hemisphere.

government: an entity that has the power to make and enforce laws.

Grameen Bank: a nongovernmental grassroots organization formed to provide small loans—referred to as microfinance—to the rural poor in Bangladesh, which has grown into an international institution that lends worldwide to poor borrowers who have no credit.

gravity system: a type of water mining where holes are drilled into the ground at the foot of a mountain or hill to tap local groundwater; known as *qanat* in Iran, *flaj* on the Arabian Peninsula, and *foggara* in North Africa.

Great Artesian Basin: the world's largest reserve of underground water (groundwater); it is under pressure so that water rises to the surface when wells are bored; located in central Australia.

Great Barrier Reef: a giant coral reef in the western Pacific Ocean, off the coast of Queensland, Australia, that is more than 2,000 kilometers (1,250 miles) long. The reefs were formed over millions of years from the skeletons of marine coral organisms in the warm tropical waters of the Coral Sea.

Great Depression: a severe decline in the world economy that lasted from 1929 until the mid 1930s.

Great Game: a conflict between Great Britain and Russia over the border region of Afghanistan and China in the 1800s and 1900s.

Great Leap Forward: a Chinese policy scheme launched in 1958 to accelerate the pace of economic growth by merging land into huge communes and seeking to industrialize the countryside; largely viewed as a massive failure and partly responsible for starvation in the period that followed.

Green Belt Movement: a nongovernmental organization created to protect trees in and around Nairobi, Kenya, led by Nobel Prize–winning environmental and political activist Wangari Maathai.

Green Revolution: a technological package of higher-yielding seeds, especially wheat, rice, and corn, that in combination with irrigation, fertilizers, pesticides, and farm machinery was able to increase crop production in the developing world after about 1950.

greenhouse gases (GHGs): gases in the atmosphere, such as carbon dioxide, methane, and nitrous oxide that trap energy, leading to planetary warming. Since the industrial revolution human activity such as fossil fuel burning has increased emissions resulting in global climate changes.

griot: a respected storyteller and singer in West Africa.

gross domestic product (GDP): an estimate of the total value of all materials, foodstuffs, goods, and services that are produced in a country in a particular year.

gross national happiness: a term coined by the king of Bhutan to encourage the measurement of the spiritual development and quality of life of the population as opposed to per capita economic activity.

gross national income (GNI): an estimate similar to GDP but including the net value of income from abroad—flows of profits or losses from overseas investments.

guest workers: a foreigner who is permitted to work in another country on a temporary basis.

Gulf Cooperation Council (GCC): an organization that coordinates political, economic, and cultural issues of concern to its six member states— Saudi Arabia, Kuwait, Bahrain, Qatar, the United Arab Emirates, and Oman.

Gupta Empire: the empire that united northern India from 320–480 c.e. and produced the decimal system of notation, the golden age of Sanskrit and Hindu art, and contributions to science, medicine, and trade.

H

haciendas: a large agricultural estate in colonial Latin America and Spain that grows crops for domestic consumption (e.g., for mines, missions, and cities) and sometimes for export.

hajj: a pilgrimage to Mecca required of all Muslims.

Hamas: the Islamic Palestinian party founded in 1987 that largely governs the Gaza Strip and whom the Israeli government refuses to recognize or negotiate with.

Harappan Civilization: considered the first urban civilization in the Indus River Valley; employed sophisticated agricultural techniques to produce surpluses, primarily in cotton and grains.

harmattan: a hot, dry wind that blows out of inland Africa.

Hasidism: a mystical offshoot of Judaism, practiced by a very small percentage of Jews.

hate crime: an act of violence committed because of prejudice against women; homosexuals; or ethnic, racial, and religious minorities.

heathland: an open, uncultivated land with poor soil and scrub vegetation.

Hezbollah: a political and military organization based in Lebanon, in which some members are part of militias with a main goal of destroying the state of Israel.

Hinduism: a major religious and cultural tradition, most predominantly practiced by roughly a billion people in South Asia. Developed from ancient Vedic traditions, Hinduism has no unifying organizational structure and exists in different forms, communities, and practices.

Hindutva: a social and political movement that calls on India to unite as an explicitly Hindu nation; it has given birth to a range of political parties over the years, including the Bharatiya Janata Party (BJP).

Ho Chi Minh: the Vietnamese leader who led a revolutionary war of independence against Japanese, French, and then U.S. forces; he was also one of the founders of the Vietnamese Communist Party.

Holocaust: Nazi Germany's systematic genocide of various ethnic, religious, and national minority groups—including 2 out of every 3 European Jews—during the 1930s and 1940s.

Hukou: a record system in China for the registration of households, which differentiates rural from urban populations and has been used in attempts to control and rationalize the rate of population migration.

Human Development Index (HDI): a United Nations metric based on measures of life expectancy, educational attainment, and personal income.

human geography: the study of the spatial organization of human activity and how humans make Earth into a home.

human immunodeficiency virus (HIV): a virus usually transmitted through sexual contact, blood transfusions, and needle sharing and from mother to child, causing severe damage to the immune system and making it difficult for the body to fight infections.

hutongs: narrow lanes in areas of Chinese cities that characterize traditional, tightly packed residential neighborhoods.

hydraulic civilizations: large state societies hypothesized to have arisen from the needs of organizing massive irrigation systems; civilizations that survive, thrive, and expand based on their capacity for controlling water.

I

ideograph: a linguistic symbol representing a single idea or object rather than a sound.

imperialism: the extension of the power of a nation through direct or indirect control of the economic and political life of other territories.

import substitution: a process by which domestic producers provide goods or services that were formerly bought from foreign producers.

income inequality: the extent to which **income** is distributed in an uneven manner among a population.

indentured servant: an individual bound by contract to the service of another for a specified term.

Indian National Congress Party: a party established in India to promote democracy and freedom in 1887 led by Mohandas Gandhi.

Industrial Revolution: the rapid development of mechanized manufacturing that gathered momentum in the early 19th century.

informal economy: economic activities that take place beyond the official record and not subject to formalized systems of regulation or remuneration (e.g., street selling and petty crime).

information technology (IT): the use of computer systems for storing, retrieving, and sending information.

intercropping: the mixing of different crop species, which typically results in higher productivity, nitrogen fixation, and/or pest or drought tolerance.

intermontane: a set of basins, plateaus, and smaller ranges that lie between mountains.

internal migration: the movement of populations within a national territory.

Internally Displaced Person (IDP): an individual who is uprooted within his or her own country due to civil conflict or human rights violations.

International Monetary Fund (IMF): an international organization that monitors the international financial system and provides loans to governments throughout the world.

internationalist: a person who believes in equal rights for all nations and wants to break down national barriers and end ethnic rivalries.

intertropical convergence zone (ITCZ): a region where air flows together and rises vertically as a result of intense solar heating at the Equator, often with heavy rainfall, and shifting north and south with the seasons.

intifada: an uprising of Palestinians in the 1980s against the rule of Israel in the Occupied Territories.

invasive species: plants or animals from one place that "hitchhike" to new locations on human transport, or are deliberately introduced and thrive at their destination, interacting with native species and habitats to create new ecological mixes, land covers, and novel ecosystems sometimes through destruction of native species.

Iranian Green Revolution: the 2009 uprising in Iran based on allegations of election fraud against President Mahmoud Ahmadinejad.

Iron Curtain: the militarized frontier zone across which Soviet and East European authorities allowed the absolute minimum movement of people, goods, and information during the Cold War.

irredentism: the assertion by the government of a country that a minority living outside its borders belongs to it historically and culturally.

Islam: the religion of the Muslims that is based on submission to God's will according to the Qur'an.

Islamism: political movement or political identity that promotes Islamic law, pan-Islamic unity, and rejection of Western influences in the Muslim world.

J

Janjaweed: the militia of African Arabs supported by the authoritarian Islamic Sudanese government to fight the rebellion of native Africans in Darfur (beginning in 2003); this resulted in one of the most brutal campaigns of ethnic cleansing in African history.

Jasmine Revolution: the protest against dictatorial government policies in Tunisia in 2011 that was part of the Arab Spring movement around the Middle East and North Africa.

jati: endogamous kinship groups associated with Hinduism and therefore most common in India and Nepal.

jihad: a sacred struggle or striving to carry out God's will according to the tenets of Islam; the term connotes both an inward spiritual struggle to attain perfect faith and an outward material struggle to promote justice and the Islamic social system.

jihadist: a member of the global movement that seeks war on behalf of Islam against those who oppose the religion.

Joseph Stalin: the leader of the Soviet Union who developed a command economy, employed police terror for state compliance, and led the U.S.S.R. during World War II.

K

karma: the idea (in Hinduism and Buddhism) of cosmic responsibilities for actions and deeds visited upon eternal souls throughout an endless cycle of reincarnated lifetimes.

keiretsu: Japanese business networks facilitated after World War II by the Japanese government in order to promote national recovery.

khan: the ruler or leader in the Central Asian Muslim kingdoms (khanates).

Khmer: the Cambodian language; the Khmer Empire emerged in the 9th century and later constructed the Angkor Wat.

Khmer Rouge: the U.S.-backed military government in Cambodia (1975– 1979) led by Pol Pot; he isolated the country, suspended education, and committed the mass murders of millions.

killing fields: fields found throughout Cambodia where millions of people, murdered by the Khmer Rouge regime, were buried.

Kim Il Sung: an anti-Japanese Marxist–Leninist nationalist leader who came to power in 1949 in North Korea and imposed an austere regime and regimented way of life.

Kim Jong Il: a son of Kim Il Sung who succeeded his father in North Korea and was known as the "Dear Leader"; during his regime, food shortages increased between 1995 and 2005.

Kim Jong-un: a son of Kim Jong Il who succeeded him in 2012; the youngest head of state in the world, he governs North Korea as one of the most highly militarized countries in the world.

kinship: the shared notion of relationship among members of a group often, but not necessarily, based on blood, marriage, or adoption.

kleptocracy: a form of government in which leaders divert national resources for their personal gain.

Kurds: an ethnic group and non-Arabic people who are mostly Sunni Muslims and who have been struggling with both the Turkish and Iranian governments for autonomy; twenty million Kurds live in the mountainous region along the borders of Iraq, Iran, Syria, Turkey, and a small area in Armenia called Kurdistan by the Kurds.

L

La Niña: the periodic abnormal cooling of sea-surface temperatures in the tropical Pacific off the coast of Peru that results in worldwide changes in climate, including droughts and floods that contrast with those produced by El Niño.

land bridge: a dry land connection between two continents or islands, exposed, for example, when sea level falls during an ice age.

land grabs: large-scale land purchases made by foreign investors in developing countries.

landscape: a portion of Earth's surface that has been shaped and transformed by human activity over time.

language: a method of human communication, either spoken or written, consisting of the use of words in a structured and conventional way. Language is a central aspect of cultural identity, reflecting the ways that different groups understand and interpret the world around them.

liberation theology: a Catholic movement originating in Latin America focused on social justice and on helping the poor and oppressed.

lingua franca: a common language used to communicate among people of different backgrounds and languages, often for trading purposes.

Little Ice Age: the period of cooler climate that significantly reduced the growing season for crops that occurred in about 1300 c.e.

Little Tigers: the descriptive nickname for the economies of Thailand, Malaysia, and Indonesia, which have had relatively high rates of economic growth since the early 1980s.

local food: usually organically grown food that it is produced within a fairly limited distance from where it is consumed.

loess: a surface cover of fine-grained silt and clay deposited by wind action and usually resulting in deep layers of yellowish, loamy soils.

M

Maastricht Treaty: an agreement signed by members of the European Community in 1992, committing them to economic and monetary union and the creation of the European Union.

machismo: a Spanish word that constructs the ideal Latin American man as fathering many children, dominant within the family, proud, and fearless.

Mafia: an organized crime group.

majority world: a term that is sometimes used to refer to the global south, drawing attention to the fact that the majority of the world's population lives in this part of the world today.

malaria: a disease transmitted to humans by mosquitoes that causes fever, anemia, and often-fatal complications.

mandates: a delegation of political power over a region, province, or state.

manga: Japanese print cartoons books (or *komikku*), which date at least to the last century.

mangrove forests: groups of evergreen trees that form dense, tangled thickets in marshes and along muddy tidal shores; found along regional coastlines.

Manufacturing Belt: the region of the Northeast and Midwest of the United States where manufacturing was concentrated through the mid- to late-20th century.

Mao Zedong: a communist leader and founder of the People's Republic of China.

Maori: indigenous peoples of New Zealand.

map: a visual and symbolic representation of geography, depicting spatial relationships, themes, distances, places, and features.

maquiladora: industrial plant in Mexico, originally within the border zone with the United States and often owned or built with foreign capital, that assembles components for export as finished products free from customs duties.

marianismo: a Latin American version of the ideal woman in the image of the Virgin Mary; she is chaste, submissive, maternal, dependent on men, and closeted within the family.

maroon communities: settlements in the Caribbean and Latin America in the 1700s and 1800s created by escaped and liberated African and sometimes indigenous slaves.

Marshall Plan: the strategy (named after U.S. Secretary of State George Marshall) of quickly rebuilding the West German economy after World War II with U.S. funds in order to prevent the spread of socialism or a recurrence of fascism.

marsupial: an Australian mammal such as the kangaroo, koala, and wombat that gives birth to premature infants that develop and drink milk from nipples in a pouch on the mother's body.

massif: a mountainous block of Earth's crust bounded by faults or folds and displaced as a unit.

matrilocality: the cultural practice in which a married male–female couple lives with the family of the woman.

Mauryan Empire: the first empire to establish rule across South Asia; promoted a policy of "conquest by dharma."

Mecca: the city in present day Saudi Arabia where Muhammad was born in 570 c.e.; Muslim religious practice includes praying five times a day facing in the direction of Mecca.

medical tourism: the travel of international visitors to a destination, seeking medical treatments that are either unavailable or overly expensive in their home country.

megacities: one of the world's largest metropolitan areas.

megafauna: large or giant animals such as elephants; the term is often used to describe animals that are now extinct such as the mammoths of the Pleistocene era.

Mekong River Commission (MRC): the intergovernmental organization that coordinates the management of the Mekong River Basin; its members include Cambodia, Laos, Thailand, and Vietnam, as well as "dialogue partners" Burma and the People's Republic of China.

Melanesia: the region of the western Pacific that includes the westerly and largest islands of Papua New Guinea, the Solomon Islands, Fiji, Vanuatu, and New Caledonia.

merchant capitalism: a form of capitalism characterized by trade in commodities and a highly organized system of banking, credit, stock, and insurance services.

mestizo: a term used in Latin America to identify a person of mixed white (European) and American Indian ancestry.

microfinance: programs that provide small-scale credit and savings to the self-employed poor, including those in the informal sector who cannot borrow money from commercial banks.

Micronesia: the region of island states in the South Pacific that includes Guam, Kiribati, the Marshall Islands, the federate states of Micronesia, Nauru, Northern Mariana Islands, and Palau.

Millennium Development Goals (MDGs): eight goals, agreed to by members of the United Nations, that include the eradication of poverty, universal primary education, gender equality, the reduction of child mortality, the improvement of maternal health, the combating of disease, environmental sustainability, and the creation of global partnerships.

minority world: a term sometimes used to refer to the global north, drawing attention to the fact that only a minority of the world's population lives in this part of the world today.

MIRAB (migration, remittances, aid, and bureaucracy) economies: economies, such as those of many Pacific islands, that depend on labor migration, money sent back from overseas workers, foreign aid, and jobs in government.

modernity: a way of thinking that emphasizes innovation over tradition, rationality over mysticism, and utopianism over fatalism emphasizes reason, scientific rationality, creativity, novelty, and progress.

Mohandas Gandhi: inspirational leader who made a case of India's independence from Great Britain through nonviolence.

monotreme: an egg-laying mammal, such as the platypus, most often found in Australia and New Guinea.

monsoon: a seasonal reversal of wind flows in parts of the lower to middle latitudes, driven by atmospheric pressure gradients, which governs precipitation patterns in many regions, including especially South and Southeast Asia.

Montreal Protocol: a 1987 international treaty established to reduce chemical emissions that damage the ozone layer.

moraine: the accumulation of rock and soil carried forward by a glacier and eventually deposited at the glacier's frontal edge or along its sides.

Mughals: the 15th-century clan of Turks from Persia who conquered Afghanistan and most of India from 1526 to 1707; their rule promoted Islam in India, and they were famous for spectacular architecture such as the Taj Mahal.

mujahideen: a zealous group of fundamentalist Islamic tribal leaders in Afghanistan.

multiculturalism: the process of immigrant incorporation in which each ethnic group has the right to enjoy and protect its officially recognized "native" culture.

multinational corporations (MNCs): companies with locations throughout the world; These companies may use their various sites to outsource unskilled jobs from places with expensive and highly regulated labor laws to places with cheaper labor costs and less regulation of working conditions; they also use their networks to establish and sustain not only places of production but international consumer markets as well.

Muslims: a member of the Islamic religion.

N

nation: a group of people sharing common elements of culture, such as religion, language, a history, or political identity.

National League for Democracy (NLD): an opposition party to the military regime in Burma that won parliamentary seats in 2012, including one for leader and Nobel Peace Prize–winner Aung San Suu Kyi.

nationalism: the feeling of belonging to a nation as well as the belief that a nation has a natural right to determine its own affairs.

nationalist movements: organized groups of people sharing common elements of culture, such as language, religion, or history, who wish to determine their own political affairs.

nationalization: the process of converting key industries from private to governmental organization and control.

nation–state: an idealized form of a state, consisting of a homogeneous group of people living in the same territory.

Naxalite: a member of an armed revolutionary group in the Indian subcontinent advocating communism.

Near Abroad: independent states that were formerly republics of the Soviet Union.

Nelson Mandela: the anti-apartheid movement leader in South Africa; imprisoned for 27 years under apartheid, he became the first black president of South Africa in 1994 and leader of the African National Congress. He died in 2013.

neoliberal development: a philosophy that suggests that development is best achieved when government budgets are reduced, public ownership of industries or utilities is turned over to private parties, trade barriers are removed, and government regulation of working conditions and the environment are minimized.

neoliberalism: an economic doctrine based on a belief in a minimalist role for the state, which assumes the desirability of free markets and private ownership as the ideal condition not only for economic organization but also for social and political life.

newly industrializing economies (NIEs): countries whose economies feature high rates of savings, balanced budgets, and low rates of inflation, which are indicators of a successful transition to industrial economy; a term used to describe "little tigers" in Southeast Asia.

Nollywood: the popular for Nigeria's film industry.

nomadic pastoralist: a person who herds animals by moving from place to place and carefully and deliberately following rainfall and plant growth to maintain their flocks.

nongovernmental organization (NGO): a formally constituted organization that is not a part of the government and are not conventional for-profit business (e.g., environmental and humanitarian groups).

nontraditional agricultural exports (NTAEs): newer export crops, such as vegetables and flowers, that contrast with the traditional exports such as sugar and coffee, and often require fast, refrigerated transport to market.

North America Act of 1867: a law that created the Dominion of Canada, dissolving its colonial status and effectively establishing it as an autonomous state with its own constitution and parliament.

North American Free Trade Agreement (NAFTA): a 1994 agreement among the United States, Canada, and Mexico to reduce barriers to trade among the three countries, through, for example, reducing customs tariffs and quotas.

Northwest Passage: the ice-choked waterway spanning the Arctic Sea between the Atlantic and Pacific Oceans, north of the Canadian and Russian mainlands.

nuclear family: a household that contains just two generations, a set of parents and their children.

O

oases: places in arid and desert environments where underground water percolates to the surface producing fertile soils that can support plant and animal life.

Occupied Territories: a region under Israeli occupation that includes the West Bank, Gaza Strip, and the Golan Heights where many Palestinians live as refugees.

Occupy Wall Street: a social movement founded in Canada to express dissatisfaction with the growing wealth gap between a small number of superrich individuals and the rest of the population.

offshoring: a practice whereby multinational corporations (MNCs) can take advantage of tax benefits in host banking countries.

old-age dependency ratio: the number of people age 65 and older compared with the number of working-age people (age 15 to 64).

oligarch: a business leader who wields significant political and economic power.

one-child policy: a Chinese policy introduced in 1978 involving rewards for families that give birth to only one child, including work bonuses and priority in housing.

Opium Wars: conflicts between China and Great Britain in the middle of the 1800s that resulted in China's defeat and the signing of the Treaty of Nanking.

organic farming: farming or animal husbandry that occurs without commercial fertilizers, synthetic pesticides, or growth hormones.

Organization of Petroleum Exporting Countries (OPEC): a specialist economic organization with the central purpose of fixing crude oil prices among its member states. OPEC has 12 member states and is dominated by Middle Eastern Arab states.

orographic effect: the influence of hills and mountains in lifting air and winds, cooling the air, and inducing precipitation.

orographic precipitation: rain or snow that falls when moisture-laden air, which has blown over warm oceans, encounters a landmass, especially coastal mountains; results in the formation of a dry rain shadow region on the inland, or lee, side of the mountains, where sinking air that has lost its moisture becomes even drier.

Orthodox Judaism: a group that represents a small percentage of practicing Jewish people in the Middle East and North Africa who live according to strict adherence to the religious texts of the Old Testament.

Outback: the remote, drier, and thinly populated interior of Australia.

outsource: an arrangement where jobs from places with expensive and highly regulated labor laws are moved to places with less expensive labor costs and less regulated work environments.

overseas Chinese: migrants from China who settled outside of China as early as the 14th century but mainly during the period of European colonialism, who worked migrated out of China as contract plantation, mine, and rail workers and then moved into clerical and business roles.

ozone depletion: the loss of the protective layer of ozone gas that prevents harmful ultraviolet radiation from reaching Earth's surface. Ozone loss causes increases in skin cancer and other ecological damage.

P

Pacific Garbage Patch: trash accumulation in both the western and eastern Pacific Ocean caused by currents, called gyres; these currents trap plastic and other waste into accumulations of rubbish that damage marine ecosystems.

Pacific Islands Forum: a forum that promotes discussion and cooperation on trade, fisheries, and tourism among all the independent and self-governing states of Oceania.

Pacific Rim: a loosely defined region of countries that border the Pacific Ocean.

paddy farming: a system of farming in which terraces are cut into steep hillsides to provide level surfaces for water control; the construction of dikes (ridges) allows fields to be flooded, plowed, planted, and drained before harvest.

Pan-Arabism: a movement across the Middle East and North Africa to ally Arab peoples against the Ottomans prior to World War I and later Europeans during the colonial and postcolonial periods.

pancasila: the Indonesian, postcolonial nation-building ideology whereby all Indonesians are connected through unity in diversity through belief in one god, nationalism, humanitarianism, democracy, and social justice.

Pangaea: an ancient supercontinent, comprising all the continental crust of Earth, of which Africa is the heart.

Paris Agreement: The United Nations Framework Convention on Climate Change conference held in Paris in December 2015 where 196 nations (the Conference of Parties or COP21) agreed to steep cuts in greenhouse gas emissions and support for climate change adaptation in efforts to keep global warming to less than 2 degrees.

partition: the British division of India and Pakistan in 1947 along ethnic lines, particularly language and religion.

pastoralism: a system of farming and way of life based on keeping herds of grazing animals (e.g., cattle, sheep, goats, horses, camels, yaks).

patriarchal: a description of society where men have authority over family and society in social and political systems and socioeconomic conditions are generally better for men than women.

patrilocality: the cultural practice by which a married male–female couple lives with the family of the man.

payments for environmental services (PES): incentives provided to farmers or other groups in return for protecting the environment and ecosystem services such as forests, carbon, watersheds, or biodiversity.

permafrost: a permanently frozen subsoil, which may extend for several meters below the surface layer and may defrost up to a depth of a meter (3 feet) or so during summer months.

petrodollars: revenue generated by the sale of oil.

petroleum: a liquid compound that can be converted into fuel and developed into energy sources, lubricants, waxes, asphalt, and medicine.

physical geography: a branch of geography dealing with natural features and processes (*see also* physiography).

physiographic region: a broad region within which there is a coherence of geology, relief, landforms, soils, and vegetation.

pinyin: a system of writing Chinese language using the Roman alphabet.

place: a specific geographic setting with distinctive physical, social, and cultural attributes.

plantation economy: an economic system typical of colonial trade made up of extensive, European-owned, operated, and financed enterprises where single crops were produced by local or imported labor for a world market.

plantation: a large agricultural estate that is usually tropical or semitropical, monocultural (one crop), and commercial or export oriented; most plantations were established in the colonial period.

plate tectonics: the theory that Earth's crust is divided into large solid plates that move relative to each other and cause mountain building and volcanic and earthquake activity when they separate or meet.

pluralist democracy: a society in which members of a diverse group continue to participate in their traditional cultures and special interests.

Pol Pot: the leader of the brutal military regime in Cambodia, starting in 1975.

polder: an area of low land reclaimed from a body of water by building dikes and draining the water.

polygyny: the practice of having more than one wife.

Polynesia: central and southern Pacific islands that include the independent countries of Samoa, Tonga, the Cook Islands, Niue, and Tuvalu; the U.S. territory of American Samoa; the French territories of Wallis, Fortuna, and French Polynesia (including the island of Tahiti, the Society Islands, the Tuamotu archipelago, and the Marquesas Islands); the New Zealand territory of Tokelau; and the British territory of the Pitcairn Islands.

population pyramid: a graphical illustration that shows the distribution of various age groups relative to the total of male and females in a specific population.

primary sector: refers to economic activity that is concerned directly with natural resources of any kind.

privatization: the turnover or sale of state-owned industries and enterprises to private interests.

purchasing power parity (PPP) per capita: a measure of how much of a common "market basket" of goods and services a currency can purchase locally, including goods and services that are not traded internationally; PPP makes it possible to compare levels of economic prosperity between countries where the price of goods might be relatively much higher or lower.

purdah: a traditional social and religious practice in which women in communities and households are segregated from men entirely.

Q

qital: the Arabic word for fighting or warfare; refers to a form of jihad in terms of conquest or conversion against nonbelievers.

Qu'ran: the Islamic sacred book; Muslims believe the contents are the direct spoken words of God to Muhammad.

quaternary sector: the term refers to economic activity that deals with the handling and processing of knowledge and information.

R

race: a problematic and illusory classification of human beings based on skin color and other physical characteristics; biologically speaking, no such thing as race exists within the human species.

racialization: the practice of creating unequal castes where whiteness is considered the norm, despite the biological reality that no such thing as race exists within the human species.

rain shadow: a phenomenon that occurs when mountains cause most of the moisture contained in the air masses passing over them to condense and fall as rain on the mountains, dry air then descends and warms creating dry conditions and deserts on the landward side of the mountains.

Raj: the rule of the British in India, typified by indirect control over local populations through local rulers, bolstered by colonial bureaucracies and armies.

REDD: programs that allow countries and companies with high greenhouse gas emissions to get credit for emission reductions by providing financial and other incentives for forest planting and protection in the developing world.

region: a large territory that encompasses many places, all or most of which share similar attributes in comparison with the attributes of places elsewhere.

regionalism: strong feelings of collective identity shared by religious or ethnic groups that are concentrated within a particular region.

regionalization: the process through which distinctive areas come into being.

religion: belief systems and practices that recognize the existence of powers higher than humankind.

remittances: money sent home to family or friends by people working temporarily or permanently in other countries.

reservation/reserve: an area of land managed by an indigenous tribe under the U.S. Department of the Interior's Bureau of Indian Affairs, or in Canada, under the Minister of Indian Affairs.

rewilding: a conservation practice where ecological functions and evolutionary processes, which are thought to have existed in past ecosystems or before human influence, are deliberately restored or created; rewilding often requires the reintroduction of large predators to ecosystems where they have been long extinct.

rice bowls: commonly refers to regions in Asia where large-scale, wet-rice agriculture production takes place, providing a continuous source of this staple crop.

rift valley: a block of land that drops between two others, forming a steepsided trough, often at faults on a divergent plate boundary.

Ring of Fire: a chain of seismic instability and volcanic activity that stretches from Southeast Asia through the Philippines, the Japanese archipelago, the Kamchatka Peninsula, and down the Pacific coast of the Americas to the southern Andes in Chile. It is caused by the tension built up by moving tectonic plates.

Rustbelt: a core of North American industrialization in the northeast and Midwest; also called the Snow Belt.

S

Sahel: an area at the southern border of the Sahara Desert in Africa that is arid, with highly variable rainfall and a sparse population dependent on pastoralism.

satellite state: a state that is economically dependent and politically and militarily subservient to another state.

savanna: grassland vegetation found in tropical areas with a pronounced dry season and periodic fires.

scale: the term that defines the relationship between the distance on the map and the distance on the surface of Earth the map represents in the context of a map. Scale is also used to differentiate levels of geographic area such as local, regional, national, and global.

sea gypsies: nomadic fisherpeoples and ethnic minorities who commonly make their living and homes in the coastal waters of South and Southeast Asia.

secondary sector: an economic activity involving the processing, transformation, fabrication, or assembly of raw materials or the reassembly, refinishing, or packaging of manufactured goods.

secularism: nonreligious.

Semitic: the term that refers to a language family including Arabic, Hebrew, and Aramaic.

sense of place: the feelings evoked among people as a result of the experiences and memories they associate with a location and to the symbolism they attach to that space and context.

serfdom: a social practice whereby members of the lowest class were attached to a lord and his land.

Seven Years' War: *see* French and Indian War.

sexuality: a set of practices and identities related to sexual acts and desires.

Shari'a: Islamic canonical law and the foundation of political institutions in Saudi Arabia, Iran, Oman, and Yemen.

Shi'a: a sect of Islam whose beliefs are based on an interpretation of Islam in 7th century C.E.; adherents are mostly located in Iran and Iraq. The Shi'a argue that political leadership is divine and leaders must be descendants of the Prophet.

shifting cultivation: an agricultural system that preserves soil fertility by moving crops from one plot to another.

Shinto: a Japanese indigenous religious culture, which stresses a belief in the nature of sacred powers that can be recognized in existing things, and which include practices entailing ritual purification, the offering of food to sacred powers, sacred music and dance, solemn worship, and joyous celebration.

shogun: a local noble lord in dynastic Japan.

SIJORI: an economically integrated growth region that includes Singapore, Johor Baharu (Malaysia), and the Riau Islands (Indonesia).

Silk Road: an ancient east–west trade route between Europe and China.

slash-and-burn: an agricultural system often used in tropical forests that involves cutting trees and brush and burning them so that crops can benefit from cleared ground and nutrients in the ash.

slow food: a movement to resist fast food by preserving the cultural cuisine and the associated food and farming of an area.

social housing: rental housing that is owned and managed by a public institution or nonprofit organization.

social movements: organized political activism by groups or individuals.

Southern Crescent: a secondary, prosperous, emergent region that straddles the Alps, running from Frankfurt through Stuttgart, Zürich, and Munich, and finally to Milan, Turin, and northern Italy.

sovereign state: a political unit that exercises power over a territory and its people and is recognized by other states; a sovereign state's independent power is codified in international law.

soviets: a network of grassroots councils of workers that emerged in Russia at the turn of the 20th century.

Special Administrative Regions: territories and geographical areas within China that fall within the sovereignty of the People's Republic of China, but exercise relatively high levels of autonomy and self-governance.

Spice Islands: an archipelago in Southeast Asia (modern-day eastern Indonesia in particular) where items, such as nutmeg, pepper, and mustard, were domesticated and globally traded to other world regions, particularly beginning with the European colonial period, when the Portuguese and Dutch occupied the region.

Srivijaya: the island culture of Sumatra, an urban-based kingdom that emerged in Southeast Asia through maritime trade.

staples economy: a financial system based on natural resources that are unprocessed or minimally processed before they are exported to other areas where they are manufactured into end products.

state: an independent political unit with territorial boundaries that are internationally recognized by other states.

state capitalist: a market-based economy with private ownership and investment in which the state continues to own some firms, to seek and obtain technology, and to carefully control the value of its currency.

state socialism: a form of economy based on principles of collective ownership and administration of the means of production and distribution of goods dominated and directed by state bureaucracies.

steppe: semiarid, cooler, treeless, grassland plains.

stolen generation: Aboriginal children who were forcibly removed from their homes in Australia and placed in white foster homes or institutions.

structural adjustment programs (SAPs): economic reforms in the 1980s and 1990s that involved the removal of subsidies and trade barriers, the privatization of government-owned enterprises such as telephone and oil companies, reductions in the power of unions, and an overall focus on export expansion. These policies, while reducing inflation and debt, had negative effects on the poor.

subsistence affluence: a decent standard of living achieved with little cash income through reliance on local foods and community resources.

subtropical high: a zone of descending air, which results in dry, stable desert conditions over tropical deserts such as the Sahara.

suburbanization: the growth of population along the fringes of large metropolitan areas.

Sudd: the vast wetland in South Sudan fed by the Nile river.

sultanate: a Muslim state ruled by supreme leaders or sultans.

Sunbelt: the region of the United States that experienced substantial growth (due to the growth of the computer and information technology economy) during the decline of industrialization in the Rust Belt during the 1960s and the 1970s.

Sunna: a set of practical guidelines for Islamic behavior—the body of traditions derived from the words and actions of the prophet Muhammad; not a written document.

Sunni: the sect of Islam practiced by the majority of Muslims in the Middle East and North Africa.

supranational organization: a collection of states with a common economic and/or political goal.

sustainability: the ability to meet current human needs, while preserving the environment and resources for the future.

swidden farming: the form of agriculture in which land is cleared for cultivation by cutting and burning shrubs or trees, allowing multiple years of planting until forest regrowth occurs.

syncretic: the term that refers to religious practices that have co-evolved and merged with one another over the centuries.

system: a set of elements linked together so that changes in one element often result in changes in another.

T

taiga: an ecological zone of boreal coniferous forest.

Taliban: a fundamentalist Muslim group that ruled much of Afghanistan between 1996 and 2001.

tariff: a tax placed on goods from outside the region to be paid on a particular class of imports or exports.

tar sands: is the oil that occurs in deposits of sand saturated with bitumen (tar), whose extraction is toxic.

technology systems: a cluster of interrelated energy, transportation, and production technologies that dominates economic activity for several decades; since the beginning of the Industrial Revolution, we can identify four of them.

terms of trade: the relationship between the prices a country pays for imports compared to the price it receives for its exports; poor terms of trade are when import prices are much higher than export prices.

terracing: the creation of a distinctive landscape of stepped and reinforced flat agricultural fields cut into steep slopes in order to stabilize the land for cropping by reducing soil erosion and frost risks and creating flat land that can be irrigated.

territorial production complexes: regional groupings of production facilities in Soviet state socialism; complexes were based on local resources that were suited to clusters of interdependent industries.

tertiary sector: an activity involving the sale and exchange of goods and services.

Theravada Buddhism: a conservative form of Buddhism, originating in South Asia, and currently practiced in Burma, Thailand, and Cambodia.

trade winds: prevailing westerly winds in the tropics that blow toward the Equator from the northeast in the Northern Hemisphere or the southeast in the Southern Hemisphere.

trading empire: large-scale political economies that emerged as the industrial nations of Europe pursued overseas expansion in the early 19th century.

tragedy of the commons: when an open-access common resource is over exploited by individuals who do not recognize how their own use of the resource can add to that of many others to degrade the environment, for example, overfishing a given species to the point of extinction or overgrazing.

transhumance: the action of moving herds according to seasonal rhythms: in Europe to warmer, lowland areas in the winter and cooler, highland areas in the summer, and in Africa from wetter to drier areas.

transmigration: a policy of resettling people from densely populated areas to less populated, often frontier regions.

Treaty of Nanking: the 1842 treaty that ended the first Opium War between China and Great Britain and ceded the island of Hong Kong to the British; the treaty allowed European and American traders access to Chinese markets through a series of treaty ports (ports that were opened to foreign trade as a result of pressure from the major powers).

Treaty of Tordesillas: an agreement to divide the world between Spain and Portugal along a north–south line 370 leagues (about 1800 kilometers, or 1100 miles) west of the Cape Verde Islands. Approved by the pope in 1494, Portugal received the area east of the line, including much of Brazil and parts of Africa, and Spain received the area to the west.

Treaty of Waitangi: the 1840 agreement in which 40 Maori chiefs gave the queen of England governance over their land and the right to purchase it in exchange for protection and citizenship; reinterpreted by the Waitangi tribunal in the 1990s, this treaty provides the basis for Maori land rights and New Zealand's bicultural society.

tribe: a group that shares a common set of ideas about collective loyalty and political action; in tribes, group affiliation is often based on shared kinship, language, and territory.

tsar: a ruler of the Russian Empire.

tsetse fly: a blood-sucking flying insect that lives in African woodland and scrub regions; associated with both human and livestock diseases such as sleeping sickness (tryptosomiasis) and nagana.

tsunami: a sometimes-catastrophic coastal wave created by offshore seismic (earthquake) activity.

tube well: a pipe drilled to pump groundwater that is intended to provide irrigation or drinking water free of contamination. In Bangladesh the drilling of such wells resulted in high levels of arsenic contamination.

tundra: an arctic wilderness where the climate precludes any agriculture or forestry; permafrost and very short summers mean that the natural vegetation consists of mosses, lichens, and certain hardy grasses.

typhoon: a large rotating storm in Pacific Asia, which is produced by warm ocean temperatures and eddies in the trade winds from August to October (see cyclone above).

U

U.S. Civil War: the war between the North and the South in the United States from 1861 to 1865; precipitated the official end of slavery in the United States.

multinational corporations (MNCs): companies with locations and facilities throughout the world, and controlling interests beyond their home country.

Union for the Mediterranean: a multilateral partnership that encompasses 43 countries from Europe and the Mediterranean Basin established to form links across the Mediterranean with the European Union; formerly the Euro-Mediterranean Partnership, relaunched in 2008 as the Union for the Mediterranean.

Union of Soviet Socialist Republics (U.S.S.R.): a federal system created from the Russian empire in the aftermath of the 1917 Russian Revolution and formally dissolved in 1991; initiated by Lenin in 1922 to recognize the regional nationalities, which were to unite as a single Soviet people.

United Nations (UN): a supranational/international organization founded in 1945 aimed at facilitating cooperation in international law, security, economic development, human rights, and world peace; there are currently 193 member states in the U.N.

United Nations Convention on the Law of the Sea (UNCLOS): a U.N. convention, formalized in 1982, established a 200-mile exclusive economic zone (EEZ) for all nations, while also providing for the protection of international waters and the creation of rights of passage through key strategic waterways throughout the world.

United Nations Educational, Scientific, and Cultural Organization (UNESCO): the arm of the U.N. whose mission is to contribute to the building of peace, eradication of poverty, sustainable development, and intercultural dialogue through education, the sciences, culture, communication, and information.

United Nations Framework Convention on Climate Change (UNFCCC): The international treaty, signed at the Rio Earth Summit in 1992, with a goal of reducing the risks of dangerous human-causes climate change through stabilizing greenhouse gas concentrations in the atmosphere through protocols (such as Kyoto) and agreements (such as Paris).

United Nations Kyoto Protocol: a 1997 legally binding global agreement in which industrialized countries had committed to reduce greenhouse gas emissions.

uplands: high, hilly land.

urban bias: the tendency to concentrate investment and attention in urban rather than rural areas.

urban primacy: a condition in which the population of the largest city in an urban system is disproportionately large in relation to the second- and third-largest cities in that system.

V

Vietcong: communist guerilla rebels who took control of portions of South Vietnam during the Indochina wars and fought the South Vietnamese government forces from 1954 to 1975 and U.S. forces during the Vietnam War with the support of the North Vietnamese army.

Vietnam War: a conflict between communist North Vietnam and U.S.-backed South Vietnam that began in 1964 after the United States bombed North Vietnam; the war resulted in over a million Vietnamese and 58,000 American deaths and ended with the unification of North and South Vietnam.

Vladimir Ilyich Lenin: a revolutionary who led the Bolshevik takeover of power in Russia in 1917, and was the architect and first head of the U.S.S.R.

W

Wallace's Line: an imaginary line drawn in 1859 that serves as a division between species; the line is associated with the deep-ocean trench between the islands of Bali and Lombok in Indonesia that could not be crossed by animals and plants even during the low sea levels of the ice ages.

War on Terror: the U.S. response to the terrorist attacks of September 11, 2001; a global war against terrorism in which first Afghanistan and then Iraq and Pakistan were identified as the greatest threats to U.S. security.

watershed: the drainage area of a particular river or river system.

wealth inequality: an inequality that measures the difference in how much money and other assets combined, individuals have accumulated.

weapons of mass destruction (WMD): nuclear, radiological, chemical, or biological devices that can kill large number of people, and/or cause widespread damage to the built environment or the biosphere.

weather: the instantaneous or immediate state of the atmosphere (e.g., it is raining).

welfare state: a system in which the government undertakes to protect the health and well-being of its citizens with the aim of distributing income and resources to the poorer members of society.

westerlies: air in the midlatitudes flowing poleward from the tropics from west to east.

wet farming: agriculture that involves irrigation.

White Australia policy: a government policy in effect in Australia until 1975 that restricted immigration to people from northern Europe through a ranking; British and Scandinavian immigrant candidates were given the highest priority, followed by southern Europeans.

world cities: a city in which a disproportionate share of the world's most important business—economic, political, and cultural—is transacted.

world region: a large-scale geographic division based on continental and physiographic settings that contain major clusters of humankind with broadly similar cultural attributes.

world regional geography: a term referring to global geographic processes, while at the same time explaining why and how certain patterns emerge on Earth.

X

xenophobia: a hatred and/or fear of foreigners.

Z

zero population growth: a demographic state where the number of births match the number of deaths in a population in such a way that no natural population growth occurs.

Zionism: a movement whose chief objective has been the establishment of a legally recognized home in Palestine for the Jewish people.

zone of alienation: the area surrounding the Chernobyl reactor in Ukraine where radiation levels remain high as a result of the 1986 nuclear reactor accident; only a small number of residents and scientific teams reside in the zone of alienation.

Photo, Illustration, and Text Credits

Pearson Education, Inc 4.27: Iain Masterton/Alamy 4.28: United Nations Statistics Division 4.29: Neyya/Getty Images 4.30: Parker Photography/Alamy 4.31: International Mapping/Pearson Education, Inc 4.32: International Mapping/Pearson Education, Inc 4.33: International Mapping/Pearson Education, Inc 4.34: International Mapping/Pearson Education, Inc 4.35: International Mapping/Pearson Education, Inc 4.35: International Mapping/Pearson Education, Inc. 4.36: Issam Rimawi/Anadolu Agency/Getty Images 4.37: Michael Best/UPPA/PhotoShot 4.38: International Mapping/Pearson Education, Inc. 4.39: Handout/Alamy 4.40: International Mapping/Pearson Education, Inc 4.41: Muhammad Hamed/Reuters 4.42: P. Svarc/ARCO/Glow Images 4.43: International Mapping/Pearson Education, Inc 4.44: Westend61 GmbH/Alamy 4.45: Everett Collection 4.46: Youssef Boudlal/Reuters 4.47: Bart Pro/Alamy 4.48: International Mapping/Pearson Education, Inc. 4.48a: Juanmonino/Getty Images 4.48b: Image Source/Getty Images 4.48c: ADB Travel/dbimages/Alamy 4.48d: Moodboard/Alamy 4.48e: Tempura/Getty Images 4.48f: Itani/Alamy 4.49: International Mapping/Pearson Education, Inc 4.50: International Mapping/Pearson Education, Inc 4.51: CIESIN 4.52: International Mapping/Pearson Education, Inc. 4.53: Thomas Koehler/Getty Images 4.54: International Mapping/Pearson Education, Inc 4.55: International Mapping/Pearson Education, Inc 4.56: Fedal Senna/AFP/Getty Images 4.57: International Mapping/Pearson Education, Inc 4.58: Valery Voennyy/Alamy 4.59: Balkis Press/Sipa USA/AP Images

CHAPTER 5

CO.5: Karen Kasmauski/Latitude/Corbis 5.1: International Mapping/Pearson Education, Inc. 5.1.1a: Finbarr O'Reilly/Reuters 5.2: International Mapping/Pearson Education, Inc. 5.2.1a: ImageBroker/Alamy 5.3: International Mapping/Pearson Education, Inc. 5.3.1: International Mapping/Pearson Education, Inc. 5.3.2: International Mapping/Pearson Education, Inc. 5.3.3: International Mapping/Pearson Education, Inc. 5.3.4: International Mapping/Pearson Education, Inc. 5.3.5: International Mapping/Pearson Education, Inc. 5.4.1: Siphiwe Sibeko/Reuters 5.40: International Mapping/Pearson Education, Inc. 5.41: International Mapping/Pearson Education, Inc. 5.4a: Sean Sprague/Alamy 5.4b: Gilles Paire/Fotolia 5.5: UNEP/GRID-Arendal, http://www.grida.no/graphicslib/detail/climate-change-vulnerability-in-africa_7239. Designed by Delphine Digout, Revised by Hugo Ahlenius, UNEP/GRID-Arendal. Used with permission. 5.6: International Mapping/Pearson Education, Inc. 5.7: International Mapping/Pearson Education, Inc. 5.7b: Panoramic Images/Getty Images 5.7c: Photoromano/Fotolia 5.8: Nic Bothma/EPA/Newscom 5.9: International Mapping/Pearson Education, Inc. 5.10: International Mapping/Pearson Education, Inc. 5.11a: International Mapping/Pearson Education, Inc. 5.11b: Anton Ivanov/Shutterstock 5.12: Jason Edwards/Getty Images 5.13: International Mapping/Pearson Education, Inc. 5.14a: AfriPics.com/Alamy 5.14b: Didier Ruef/Cosmos/Redux Pictures 5.14c: John Shaw/NHPA/Photoshot/Newscom 5.15a: Martin Harvey/Getty Images 5.15b: David Cayless/Getty Images 5.16a: John Reader/Science Source 5.16b: Dmitry Chulov/Fotolia 5.16c: Sandro Vannini/Corbis 5.16d: Nick Greaves/Images of Africa Photobank/Alamy 5.16e: Nik Wheeler/ Encyclopedia/Corbis 5.16f: Seth Lazar/Alamy 5.17: International Mapping/Pearson Education, Inc. 5.18: Adapted from A. Thomas and B. Crow (eds.), Third World Atlas. Buckingham, UK: Open University Press, 1994, p. 35. 5.19a: Michael Lewis/Getty Images 5.19b: Nomad/SuperStock 5.20: International Mapping/Pearson Education, Inc. 5.21a: Jon Hrusa/EPA/Newscom 5.21b: Giulio Napolitano/FAO/Food and Agriculture Organization of the United Nations 5.22: International Mapping/Pearson Education, Inc. 5.22b: Ton Koene/ VWPics/Newscom 5.23: International Mapping/Pearson Education, Inc. 5.24: International Mapping/Pearson Education, Inc. 5.25: International Mapping/Pearson Education, Inc. 5.26: AP Images 5.27: Lynn Hilton/Alamy 5.28: Akintunde Akinleye/Reuters 5.29: Alison Shelley Photography 5.30: Redrawn from S. Aryeetey-Attoh (ed.), The Geograpy of Sub0Saharan Africa. Upper Saddle River, NJ: Prentice hall, 1997, Fig. 4.6. 5.31: Jenny/Newscom 5.31: Linda Hughes/Shutterstock 5.31: Valentin Flauraud/Reuters 5.32: Jed Stone/Alamy 5.33: The Image Works 5.34: CIESIN 5.35: International Mapping/Pearson Education, Inc. 5.36: International Mapping/Pearson Education, Inc. 5.37: International Mapping/Pearson Education, Inc. 5.38b: Juda Ngwenya/Reuters 5.39: Dai Kurokawa/EPA/newscom

CHAPTER 6

CO.6: Mladen Antonov/AFP/Getty Images 6.1: International Mapping/Pearson Education, Inc 6.1.1: International Mapping/Pearson Education, Inc. 6.1.2: Based on infographic from http://wafoodcoalition.blogspot.com/2013/10/the-average-american-dietits 6.2: International Mapping/Pearson Education, Inc 6.2.1: Data from http://www.cantechletter.com/wp-content/uploads/2014/03/marijuana-use-byprovince. 6.2.2: Based on http://hempedification.blogspot.com/2015/02/facts-about-cannabis.html and NORML, Drug Policy Alliance and the Marijuana Policy Project 6.3: "How Will Climate Change Impact the EPA Region 6 Area?" U.S. Environmental Protection Agency, available at http://www.epa.gov/region6/climatechange/maps.htm 6.3.1: International Mapping/Pearson Education, Inc. 6.3.2: International Mapping/Pearson Education, Inc. 6.4: International Mapping/Pearson Education, Inc 6.4.1: Katherine Frey/The Washington Post/Getty Images 6.5a: David Cobb/Alamy 6.5b: Mike Grandmaison/All Canada Photos/Alamy 6.5c: Christopher Price/Alamy 6.6: Department of the Army, Mississippi River Commission Corps of Engineers, Vicksburg Mississippi, 1986. 6.7: International Mapping/Pearson Education, Inc 6.8: B. Brown/Shutterstock 6.9: Rob Crandall 6.11: Teschner/Agencja Fotograficzna Caro/Alamy 6.12: International Mapping/Pearson Education, Inc 6.12: Vicki Beaver/Alamy 6.13: International Mapping/Pearson Education, Inc. 6.14: Knox, Paul L.; McCarthy, Linda M., Urbanization: An Introduction To Urban Geography, 3rd, © 2012. Printed and Electronically reproduced by permission of Pearson Education, Inc., Upper Saddle River, New Jersey. 6.15: Based on http://www.wired.com/2011/05/ff_jobsclustermap/ 6.16: David Leadbitter/Alamy 6.17: Data from http://w.baike.com/d6301443985f4a879acd2a5923feb32f.html 6.18: International Mapping/Pearson Education, Inc. 6.19: Data from http://www.infoplease.com/ipa/A0004615.html 6.20: CBP Photo/Alamy 6.21: Jonathan Ernst/Reuters 6.22: MediaPunch/REX/AP Images 6.23: Data from http://www.everyoneelseisdoingit.com/maps/MLB_2012.html 6.24: Michael Tullberg/Getty Images 6.25: Josh Edelson/AFP/Getty Images 6.26: Based on Infographic "U.S. Immigration Has Occurred in Waves, With Peaks Followed by Troughs" by Philip Martin, "Trends in Migration to the U.S." Population Reference Bureau. May 2014. 6.27: International Mapping/Pearson Education, Inc. 6.28: International Mapping/Pearson Education, Inc 6.29: International Mapping/Pearson Education, Inc 6.30: James Quine/Alamy 6.31: Michael Wheatley/Alamy 6.32: Infographic from Bishop, Jason and Matt Owens. "The Three-Trillion-Dollar War: Its Cost In Ten Steps" GOOD Magazine. 6 November 2008 6.33: International Mapping/Pearson Education, Inc.

CHAPTER 7

CO.7: Desmond Boylan/Reuters 7.1: International Mapping/Pearson Education, Inc 7.1.1a: Jorge Uzon/AFP/Getty Images 7.1.1b: International Mapping/Pearson Education, Inc. 7.1.1c: International Mapping/Pearson Education, Inc. 7.1.2: International Mapping/Pearson Education, Inc. 7.1.3: International Mapping/Pearson Education, Inc. 7.1.4: Google Earth/International Mapping/Pearson Education, Inc. 7.1.5: International Mapping/Pearson Education, Inc. 7.2: Kristian Peetz/Westend61/Corbis 7.2.1a: International Mapping/Pearson Education, Inc. 7.2.1b: International Mapping/Pearson Education, Inc. 7.3: International Mapping/Pearson Education, Inc 7.3.1: John Coletti/Getty Images 7.3.2a: International Mapping/Pearson Education, Inc 7.3.2b: International Mapping/Pearson Education, Inc. 7.3.2a: International Mapping/Pearson Education, Inc 7.3.2b: International Mapping/Pearson Education, Inc. 7.4: International Mapping/Pearson Education, Inc. 7.4.1: International Mapping/Pearson Education, Inc 7.5.1: Enrique Marcarian/Reuters 7.5a: Swoan Parker/Reuters 7.5b: NASA 7.6: International Mapping/Pearson Education, Inc 7.7a: International Mapping/Pearson Education, Inc. 7.7b: Lonnie Thompson 7.7c: Gary Braasch/ZUMA Press/Newscom 7.7d: Lonnie Thompson 7.8a: International Mapping/Pearson Education, Inc. 7.8b: Diana Liverman 7.8c: Eduardo Verdugo/Ap Images 7.9: International Mapping/Pearson Education, Inc 7.10a: International Mapping/Pearson Education, Inc 7.10b: Daniel Aguilar/Reuters 7.11a: Reuters/Corbis 7.11b: Ann Johansson/Corbis 7.12: Joseph Sohm/Spirit/Corbis 7.13a: Jim Zuckerman/Comet/Corbis 7.13b: Alamy 7.13c: Think4photop/Shutterstock 7.14a: International Mapping/Pearson Education, Inc 7.14b: Ted Spiegel/Encyclopedia/Corbis 7.14c: Steve Percival/Science Photo Library/Alamy 7.15: Witold Skrypczak/Getty Images 7.16a: International Mapping/Pearson Education, Inc. 7.16b: P. Henry/ARCO/Glow Images 7.16c: Jorge Adorno/Reuters 7.17: Devon Stephens/Alamy 7.18: International Mapping/Pearson Education, Inc 7.19: International Mapping/Pearson Education, Inc 7.20: History of Cuernavaca and Morelos: Sugar Plantation, Tealtenango Morelos (Detail) (1931), Diego Rivera. Fresco, 435 x 282 cm, west wall. Museo Quaunahuac, Instituti Nacional de Antropologia e Historia, Cuernavaca, Mexico. Gianni Dagli Orti/The Art Archive/Art Resource, New York. © ARS, New York 7.21: Gabriel M. Covian/The Image Bank/Getty Images 7.22: ZUMA Press/Alamy 7.23: International Mapping/Pearson Education, Inc. 7.24: International Mapping/Pearson Education, Inc. 7.25: Alan Dykes/Alamy 7.26: Patrick Frilet/REX/Newscom 7.27: International Mapping/Pearson Education, Inc 7.27a: Rafael Perez/Reuters 7.27b: Bettmann/Corbis 7.27c: Transcendental Graphics/Getty Images 7.27d: Bettmann/Corbis 7.28: Sergio Pitamitz/Encyclopedia/Corbis 7.29: Jorge Dan Lopez/Reuters 7.30: Guillermo Granja/Reuters 7.31: International Mapping/Pearson Education, Inc 7.31b: Jorge Dan Lopez/Reuters 7.32: International Mapping/Pearson Education, Inc 7.33: International Mapping/Pearson Education, Inc. 7.34a: EPA/Alamy 7.34b: Von Der Laage/Gladys Chai/ZUMA Press/Newscom 7.35: International Mapping/Pearson Education, Inc. 7.36: International Mapping/Pearson Education, Inc 7.37a: International Mapping/Pearson Education, Inc. 7.37b: International Mapping/Pearson Education, Inc. 7.38: International Mapping/Pearson Education, Inc 7.39: Filipe Frazao/Shutterstock 7.40: Dabldy/Getty Images 7.41: International Mapping/Pearson Education, Inc. 7.42: Enrique Cuneo/AP/Corbis 7.43: International Mapping/Pearson Education, Inc.

CHAPTER 8

CO.8: DigitalGlobe/ScapeWare3d/Getty Images 8.1: International Mapping/Pearson Education, Inc 8.1.1: Stocktrek Images/Getty Images 8.1.2: International Mapping/Pearson Education, Inc 8.1.4: Based on I. Barnes and R. Hudson, History Atlas Of Asia. New York: Macmillan, 1998, p. 129. 8.2: International Mapping/Pearson Education, Inc 8.2.1: Jiji Press/AFP Photo/Getty Images 8.2.2: Reynold Sumayku/Alamy 8.3: International Mapping/Pearson Education, Inc 8.3.1a: International Mapping/Pearson Education, Inc. 8.3.1b: "Chinese Opium Smokers," drawn by Thomas Allom, engraved by G. Paterson. From The Chinese Empire, Illustrated: Being a Series of Views from Original Sketches, Displaying the Scenery, Architecture, Social Habits, &c., of that Ancient and Exclusive Nation by Thomas Allon and George Newenham. London Printing and Publishing, 1858. 8.3.2: International Mapping/Pearson Education, Inc. 8.3.3a: International Mapping/Pearson Education, Inc. 8.3.3b: International Mapping/Pearson Education, Inc. 8.3.4: International Mapping/Pearson Education, Inc. 8.3.4: International Mapping/Pearson Education, Inc. 8.4: Ingram Publishing/Newscom 8.4.1: EvrenKalinbacak/Shutterstock 8.4.2: http://www.economist.com/news/special-report/21631795-under-american-leadership-pacific-hasbecome-engine-room-world-trade 8.5: Flocu/Shutterstock 8.5.1: ChinaFotoPress/Getty Images 8.6: John Bill/Shutterstock 8.7: Reference source: http://qz.com/159105/2013-will-be-remembered-as-the-year-that-deadly-suffocating-smog-consumedchina/ 8.8: International Mapping/Pearson Education, Inc 8.9: Kyodo/Reuters 8.10: Prill/Shutterstock 8.11: Kevin Frayer/Getty Images 8.12: Bruno Morandi/

Robert Harding World Imagery 8.13: Nobythai/Getty Images 8.14: Keren Su/Getty Images 8.15: Javarman/Fotolia 8.16: Alex Hofford/EPA/Newscom 8.17: Jeremy Sutton Hibbert/Alamy 8.18: Jung Yeon Je/AFP/Getty Images 8.19: International Mapping/Pearson Education, Inc 8.19a: Izmael/Shutterstock 8.20: Based on I. Barnes and R. Hudson, History Atlas of Asia. New York: Macmillan, 1998, pp. 45 and 46-47. 8.21: International Mapping/Pearson Education, Inc 8.22: Tito Wong/Shutterstock 8.23: Keiretsu Diagram reprinted courtesy of Allegis Capital, http://www.allegiscapital.com/partners-keiretsu. html 8.24: International Mapping/Pearson Education, Inc. 8.25: TPG Top Photo Group/Newscom 8.26: Rob Crandall/Stock Connection Worldwide/Newscom 8.27: International Mapping/Pearson Education, Inc. 8.28: Pedro Ugarte/Newscom 8.29: EPA/Alamy 8.30: Chuong Vu/Shutterstock 8.31: Craig Hanson/Shutterstock 8.32: Colin Sinclair/Dorling Kindersley, Ltd. 8.33: International Mapping/Pearson Education, Inc 8.34: Christian Kober/Robert Harding Picture Library/Alamy 8.35: Martin A. Doe/Alamy 8.36: Alamy 8.37: Charles Platiau/Reuters 8.38: From Center for International Earth Science Information Network [CIESIN], Columbia University; International Food Policy Research Institute [IFPRI]; and World Resources Institute [WRI], 2000. Gridded Population Of The World [GPW], Version 3, Palisades, NY: CIESIN, Columbia University, 2005. Available at http://www.sedac.ciesin.columbia.edu/gpw/. 8.39: Edmund Sumner/VIEW Pictures, Ltd./Alamy 8.40: International Mapping/Pearson Education, Inc. 8.41: International Mapping/Pearson Education, Inc 8.42: David Sacks/Getty Images 8.43: Haver Analytics; The Economist estimates http://www.economist.com/news/special-report/21600797-2030-chinese-cities-will-be-home-about-1-billion-people-getting-urban-china-work 8.44: International Mapping/Pearson Education, Inc 8.45: Lou Linwei/Alamy 8.46: International Mapping/Pearson Education, Inc. 8.47: International Mapping/Pearson Education, Inc. 8.48: Tong Jiang/Imaginechina/AP Images

Images 10.5.1: Philippe Lopez/AFP/Getty Images 10.6: : Based on information from T.R. Leinbach and R. Ulack, Southeast Asia: Diversity And Development. Upper Saddle River, NJ: Prentice Hall, 2000, Map 2.1 and H.C. Brookfield and Y. Byron, South-East Asia's Environmental Future: The Search For Sustainability. New York: United Nations University Press, 1993, 13.1. Bradshaw, p. 251, Fig. 6.6 10.7a: International Mapping/Pearson Education, Inc 10.7b: Chris Stowers/Dorling Kindersley, Ltd. 10.8: BlueOrange Studio/Shutterstock 10.9: Supri/Reuters 10.11: Jiang Liu/Shutterstock 10.13: The Natural History Museum, London/Alamy 10.14a: International Mapping/Pearson Education, Inc. 10.14c: Tang Chhin Sothy/AFP/GettyImages 10.15: Pigprox/Shutterstock 10.16: Third World Atlas, Open University Press 10.17: Dallas/John Heaton/Alamy 10.18: Atlas of 20th Century World History, HarperCollins Publisher (Canada) 10.19: YT Haryono/Reuters 10.20: Gerhard Zwerger-Schoner/ImageBroker/Alamy 10.21: Ekkachai/Shutterstock 10.23: International Mapping/Pearson Education, Inc. 10.24a: International Mapping/Pearson Education, Inc. 10.26b: Paul Schutzer/The LIFE Picture Collection/Getty Images 10.27: David Longstreath/AP Images 10.28: Tang Chhin Sothy/AFP/Getty Images 10.29: Leinbach, Thomas R.; Ulack, Richard, Southeast Asia: Diversity And Development, 1st ed., c.2000. Reprinted and Electronically reproduced by permission of Pearson Education, Inc. Upper Saddle River, New Jersey. 10.30: Lynn Bo Bo/EPA/Alamy 10.31: Romeo Gacad/AFP/Getty Images 10.32: International Mapping/Pearson Education, Inc. 10.33: Tengku Bahar/AFP/Getty Images 10.34: Luca Tettoni/Robert Harding Picture Library/Alamy 10.36: Ace Stock Limited/Alamy 10.37: International Mapping/Pearson Education, Inc. 10.38: AFP/Getty Images 10.39: John Elk III/Alamy 10.41: International Mapping/Pearson Education, Inc 10.42: Januar/AFP/Getty Images 10.43: Redrawn from T.R. Leinbach and R. Ulack, Southeast Asia: Diversity And Development, Upper Saddle River, NJ: Prentice Hall, 2000, Map 12.7. 10.44: Slamet Riyadi/AP Images 10.45a: International Mapping/Pearson Education, Inc. 10.45b: International Mapping/Pearson Education, Inc. 10.46: Cheryl Ravelo/Reuters 10.47: Jung Yeon-Je/AFP/Getty Images 10.48: Wai Yan/Xinhua/Alamy 10.49: International Mapping/Pearson Education, Inc.

CHAPTER 9

CO.9: Indranil Mukherjee/AFP/Getty Images 9.1: International Mapping/Pearson Education, Inc. 9.1.1: Aditya "Dicky" Singh/Alamy 9.1.1a: International Mapping/Pearson Education, Inc. 9.1.2: ErickN/Shutterstock 9.2: International Mapping/Pearson Education, Inc. 9.2.1: Blickwinkel/Alamy 9.2.1: International Mapping/Pearson Education, Inc. 9.3: Frank Bienewald/ImageBroker/Alamy 9.3.1: Haroon Sabawoon/Anadolu Agency/Getty Images 9.4: International Mapping/Pearson Education, Inc. 9.4.1: International Mapping/Pearson Education, Inc. 9.4.2: International Mapping/Pearson Education, Inc. 9.4.3: International Mapping/Pearson Education, Inc. 9.4.4: International Mapping/Pearson Education, Inc. 9.5: Bruno Morandi/Robert Harding Picture Library/Alamy 9.5: : http://www.wsj.com/articles/SB10001424052748704905604575026800522011226 9.6: International Mapping/Pearson Education, Inc. 9.7: Mohammed Seeneen/AP Images 9.8: Biswaranjan Rout/AP Images 9.9: http://geology.about.com/od/seishazardmaps/ss/World-Seismic-Hazard-Maps_15.htm 9.9c: Adnan Abidi/Reuters 9.10: Adapted from B. L. C. Johnson, South Asia, 2nd ed. London: Heinemann, 1982, p. 9. 9.11: Ashok Captain/Ephotocorp/Alamy 9.12: Daniel Prudek/Shutterstock 9.13: Russell Kord/Alamy 9.14: Finnbarr Webster/Alamy 9.15: Danish Siddiqui/Reuters 9.16: Jayanta Shaw/Reuters 9.17a: BGS and DPHE. 2001. Arsenic contamination of groundwater in Bangladesh. Kinniburgh, D G and Smedley, P L (Editors). British Geological Survey Technical Report WC/00/19. British Geological Survey: Keyworth. http://www.bgs.ac.uk/arsenic/bangladesh/ 9.18: State of Forest Report, 2009. 9.19: Nicolas Marino/Novarc Images/Alamy 9.20: Mustafa Quraishi/AP Images 9.21: Robert Harding/Getty Images 9.22: Redrawn from India: A History, Grove-Atlantic Monthly Press 9.23: International Mapping/Pearson Education, Inc. 9.24: Alamy 9.25: International Mapping/Pearson Education, Inc. 9.26: International Mapping/Pearson Education, Inc. 9.27: http://www.theguardian.com/news/datablog/2013/feb/22/cameron-india-trade-exports-importspartners 9.28: David Pearson/Alamy 9.29: Fredrik Renander/Alamy 9.30: Indranil Mukherjee/AFP/Getty Images 9.31: Allgoewer/Blickwinkel/Alamy 9.32: Flowerphotos/Alamy 9.33: Randy Olson/National Geographic Image Collection/Alamy 9.34: Philippe Lissac/Godong/Newscom 9.35: International Mapping/Pearson Education, Inc 9.36: Harish Tyagi/EPA/Newscom 9.37: International Mapping/Pearson Education, Inc 9.38: Roberto Fumagalli/Alamy 9.39: Mariia Pazhyna/Fotolia 9.40: Diwakar Prasad/Hindustan Times/Newscom 9.41: http://hindiurduflagship.org/about/two-languages-or-one/9.40 9.42: Adnan Abidi/Reuters 9.43: International Mapping/Pearson Education, Inc. 9.44: Ganesh Patil/Aanna Films/Everett Collection 9.45: Boisvieux Christophe/Hemis.Fr 9.46: Joe Gough/Shutterstock 9.47: Gridded Population of the World [GPW]. Version 2. Palisades, NY: CIESIN, Columbia University. Available at http://sedac.ciesin.org/plue/gpw. 9.48: International Mapping/Pearson Education, Inc. 9.49: International Mapping/Pearson Education, Inc. 9.49: National Family Health Survey, India http://www.nfhsindia.org/ 9.51: Zohra Bensemra/Reuters 9.52: International Mapping/Pearson Education, Inc. 9.53: Ajay Verma/Reuters 9.54: http://qz.com/342923/no-indias-economy-is-not-about-to-catch-up-with-chinasanytime- soon/

CHAPTER 10

CO.10: Erik De Castro/Reuters 10.1.1: Nicolas Asfouri/AFP/Getty Images 10.1.2: Jay Directo/AFP/Getty Images 10.1.3: International Mapping/Pearson Education, Inc. 10.2.2: Ian Cruickshank/Alamy 10.2.3: Roland Seitre/Nature Picture Library 10.3: Taylor Weidman/Getty Images 10.3.1: International Mapping/Pearson Education, Inc. 10.3.2: NASA 10.3.3: Tom Roche/Shutterstock 10.3.4: A. Howden/Cambodia Stock Photography/Alamy 10.4: Supri/Reuters 10.4.1: International Mapping/Pearson Education, Inc. 10.4.1: A. Howden/Cambodia Stock Photography/Alamy 10.4.2: Flowerphotos/Alamy 10.4.3: Khin Maung Win/AFP/Getty Images 10.5: AGF Srl/Alamy 10.5: Chau Doan/Light Rocket/Getty

CHAPTER 11

CO.11: Justin Mcmanus/The AGE/Fairfax Media/Getty Images CO.11 Inset: Nicky Park/EPA/Corbis 11.1: International Mapping/Pearson Education, Inc 11.1.1: Based on G. Lean and D. Hinrichsen (eds.), Atlas of the Environment and Terraquest, Virtual Atarctica Expedition. Available at http://www.terraquest.com/va/expedition/maps/cont.map.html. 11.1.2: International Mapping/Pearson Education, Inc 11.1.3: Niebrugge Images/Alamy 11.2: International Mapping/Pearson Education, Inc 11.2.1a: International Mapping/Pearson Education, Inc 11.2.1b: International Mapping/Pearson Education, Inc 11.2.2: Greg Bowker/New Zealand Herald/AP Images 11.2.3a: International Mapping/Pearson Education, Inc 11.2.3b: International Mapping/Pearson Education, Inc 11.2.4a: Rawpixel/Fotolia 11.2.4b: International Mapping/Pearson Education, Inc 11.3: EPA/Alamy 11.3.1a: International Mapping/Pearson Education, Inc 11.3.1b: Ocean Conservancy 11.3.1c: Ashley Cooper/Alamy 11.4: Manfred Gottschalk/Alamy 11.4.1a: Alamy 11.4.1b: Sheldon Levis/Getty Images 11.5: International Mapping/Pearson Education, Inc 11.5.1: Bettmann/Corbis 11.5.2: Blphoto/Alamy 11.6: John Gollings/Arcaid Images/Alamy 11.7: Tim Graham/Robert Harding/Newscom 11.8: NASA 11.9: International Mapping/Pearson Education, Inc 11.10: Map of Terrestrial Ecoregions in Australia, Commonwealth of Australia. Available at http://www.environment.gov.au/land/nrs/science/ibra/australias-ecoregions 11.10a: Manfred Gottschalk/Getty Images 11.10b: Shane Partridge/Alamy 11.10c: Dave Watts/Alamy 11.10d: Greg Brave/Alamy 11.10e: Metriognome/Shutterstock 11.10f: EPA/Alamy 11.10g: John Carnemolla/Shutterstock 11.11: Auscape/UIG/Getty Images 11.12b: Stanislav Fosenbauer/Shutterstock 11.12c: Auscape/UIG/Getty Images 11.12d: Rob Walls/Alamy 11.13: International Mapping/Pearson Education, Inc 11.14: International Mapping/Pearson Education, Inc 11.15: International Mapping/Pearson Education, Inc 11.16: Malcolm Pullman/AP Images 11.17: US Air Force 11.18: International Mapping/Pearson Education, Inc. 11.19: Olga V/Fotolia 11.19: International Mapping/Pearson Education, Inc. 11.20: International Mapping/Pearson Education, Inc 11.21a: Travelscape Images/Alamy 11.21b: Newscom 11.22: Eric Lafforgue/Alamy 11.23: New Line/Everett Collection 11.24: Columbia University, Center for International Earth Science Information Network [CIESIN]; International Center for Tropical Agriculture [CIAT]. 2005 Gridded Population of the World. Available at http://sedac.ciesin.org/plue/gpw. 11.25a: International Mapping/Pearson Education, Inc. 11.25b: International Mapping/Pearson Education, Inc. 11.26: Shutterstock 11.27: Juergen Hasenkopf/Alamy 11.28: Marc Anderson/Alamy 11.29: International Mapping/Pearson Education, Inc.

APPENDIX

A.1: International Mapping/Pearson Education, Inc. A.2: International Mapping/Pearson Education, Inc. A.3a: International Mapping/Pearson Education, Inc. A.3b: International Mapping/Pearson Education, Inc. A.3c: Tom McKnight A.4: International Mapping/Pearson Education, Inc. A.5: International Mapping/Pearson Education, Inc. A.6: International Mapping/Pearson Education, Inc. A.7a: International Mapping/Pearson Education, Inc. A.7b: International Mapping/Pearson Education, Inc. A.8: Google Earth A.9: International Mapping/Pearson Education, Inc. A.10: International Mapping/Pearson Education, Inc. A.11: NASA Earth Observatory and NOAA National Geophysical Data Center A.12: Katie Glennemeier/Laramie Boomerang/AP Images Cover: WSBoon Images/Getty Images

Index

World – Physical

Great Basin	Land features
Caribbean Sea	Water bodies
Aleutian Trench	Underwater features

ARCTIC OCEAN

80°N 60°N 40°N 20°N Tropic of Cancer Equator Tropic of Capricorn 20°S 40°S 60°S 80°S Antarctic Circle

160°W 140°W 120°W 100°W 80°W

QUEEN ELIZABETH ISLANDS
Ellesmere Island
GREEN
Beaufort Sea
Victoria Island
Baffin Island
Baffin Bay
Great Bear Lake
Great Slave Lake
Hudson Bay
Davis Strait
Yukon R.
Bering Strait
Bering Sea
MACKENZIE MTS.
Mackenzie R.
Denali 20,310 ft (6,190 m)
NORTH AMERICA
Canadian Shield
Labrador
Labrador Sea
Aleutian Islands
Aleutian Trench
Gulf of Alaska
ROCKY MOUNTAINS
Saskatchewan R.
Lake Winnipeg
Great Lakes
Island of Newfound
Vancouver I.
CASCADE RANGE
Columbia R.
GREAT PLAINS
Missouri R.
Cape Cod
Sohm Plain
Northeast
Mendocino Fracture Zone
SIERRA NEVADA
Great Basin
Colorado R.
Ohio R.
Mississippi R.
APPALACHIAN MTS.
Cape Hatteras
ATLANTIC OCEAN
Hatteras Plain
Murray Fracture Zone
Rio Grande
SIERRA MADRE
Bahama Is.
Bermuda Rise
Hawaiian Ridge
Tropic of Cancer
Molokai Fracture Zone
Pacific
Baja California
Mexican Plateau
Gulf of Mexico
Cuba
Puerto Rico Trench
Mid-Atlantic Rid
Hawaiian Is.
Johnston Atoll
Clarion Fracture Zone
CENTRAL AMERICA
Greater Antilles
Caribbean Sea
West Indies
Central Pacific Basin
Line Islands
PACIFIC OCEAN
Middle America Trench
ANDES
Orinoco R.
Guiana Highlands
Demerara Plain
Clipperton Fracture Zone
Galápagos Is.
AMAZON
Amazon R.
Equator
BASIN SOUTH AMERICA
Phoenix Is.
POLYNESIA
Basin
Marquesas Is.
East Pacific Rise
Mato Grosso Plateau
Brazilian Shield
Samoa Is.
Tonga Is.
Cook Is.
Society Is.
Tahiti
Tuamotu Archipelago
Peru-Chile
Atacama Desert
ANDES
Gran Chaco
Tonga Trench
Austral Islands
Pitcairn I.
Sala y Gómez
Easter I.
Ridge
Nazca Ridge
Mt. Aconcagua 22,834 ft (6,960 m)
Paraná R.
Kermadec Tr.
Southwest Pacific Basin
Louisville Ridge
Tropic of Capricorn
Challenger Fracture Zone
Juan Fernández Is.
Pampas
Rio de la Plata
Patagonia
Argentine Plain
0 1,000 2,000 Miles
0 1,000 2,000 Kilometers
Southeast Pacific Basin
Humboldt Plain
Strait of Magellan
Falkland Is.
South Georgia Rid
Eltanin Fracture Zone
Cape Horn
Udintsev Fracture Zone
Drake Passage
Pacific-Antarctic Ridge
Antarctic Circle
SOUTHERN OCEAN

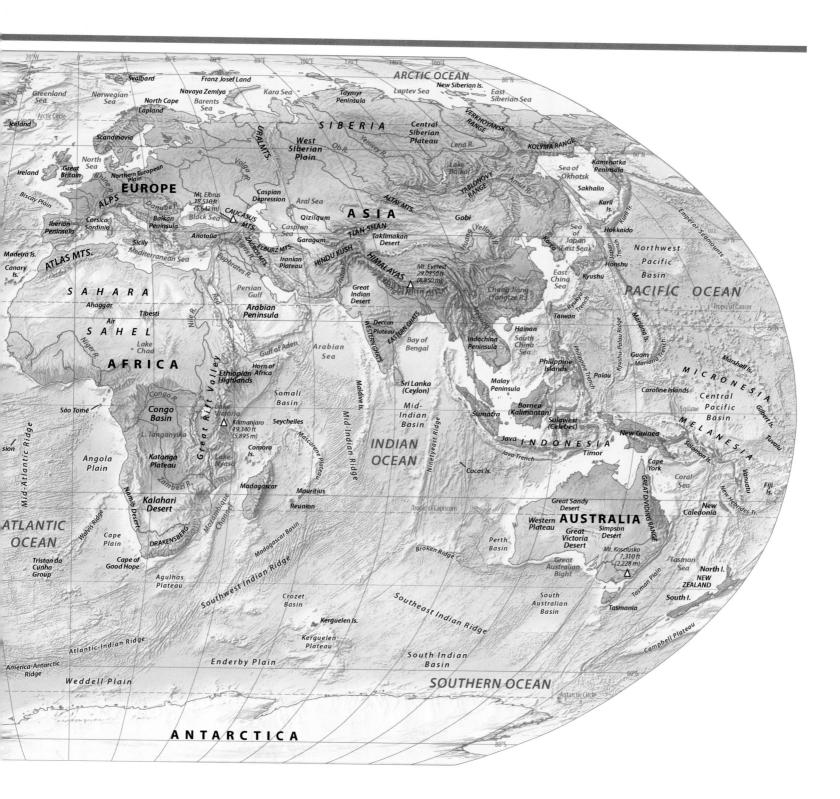

ARCTIC OCEAN

Greenland
Sea

Norwegian
Sea

Svalbard

Franz Josef Land

Novaya Zemlya

Kara Sea

Taymyr
Peninsula

Laptev Sea

New Siberian Is.

East
Siberian Sea

Iceland

Arctic Circle

North Cape
Lapland

Barents
Sea

SIBERIA

Central
Siberian
Plateau

VERKHOYANSK
RANGE

KOLYMA RANGE

Scandinavia

North
Sea

Ireland

Great
Britain

Northern European
Plain

URAL MTS.

West
Siberian
Plain

Ob R.

Yenisey R.

Lena R.

Kamchatka
Peninsula

Sea of
Okhotsk

Sakhalin

EUROPE

Biscay Plain

ALPS

Danube R.

Volga R.

Caspian
Depression

Aral Sea

ASIA

Lake
Baikal

YABLONOVY
RANGE

Amur R.

Kuril
Is.

Kuril Tr.

Emperor Seamounts

Iberian
Peninsula

Corsica
Sardinia

Balkan
Peninsula

Black Sea

Mt. Elbrus
18,510 ft
(5,642 m)

CAUCASUS
MTS.

Caspian
Sea

Qizilqum

Garagum

TIAN SHAN

Taklimakan
Desert

Gobi

Sea
of
Japan
(East Sea)

Korea

Hokkaido

Honshu

Kuril Tr.

Northwest
Pacific
Basin

Madeira Is.

ATLAS MTS.

Sicily

Mediterranean Sea

Anatolia

ELBURZ MTS.

ZAGROS MTS.

Euphrates R.

Tigris R.

Iranian
Plateau

HINDU KUSH

HIMALAYAS

Mt. Everest
29,035 ft
(8,850 m)

Chang Jiang
(Yangtze R.)

East
China
Sea

Kyushu

Ryukyu Is.

Japan Trench

PACIFIC OCEAN

Canary
Is.

SAHARA

Ahaggar

Tibesti

Red R.

Nile R.

Persian
Gulf

Arabian
Peninsula

Great
Indian
Desert

Deccan
Plateau

WESTERN GHATS

EASTERN GHATS

Bay of
Bengal

Indochina
Peninsula

Hainan

South
China
Sea

Taiwan

Mariana Is.

Guam

Kyushu-Palau Ridge

Mariana Trench

Tropic of Cancer

Marshall Is.

MICRONESIA

Air

SAHEL

Lake
Chad

Gulf of Aden

Arabian
Sea

Maldive Is.

Sri Lanka
(Ceylon)

Malay
Peninsula

Philippine
Islands

Palau

Caroline Islands

Central
Pacific
Basin

AFRICA

Niger R.

Ethiopian
Highlands

Horn of
Africa

Somali
Basin

Mid-
Indian
Basin

Philippine Trench

Equator

São Tomé

Congo R.

Congo
Basin

Lake
Victoria

Great Rift Valley

Kilimanjaro
19,340 ft
(5,895 m)

Seychelles

Mascarene Plateau

Ninetyeast Ridge

INDIAN
OCEAN

Sumatra

Borneo
(Kalimantan)

Sulawesi
(Celebes)

New Guinea

MELANESIA

Tuvalu

sion

Mid-Atlantic Ridge

Angola
Plain

Katanga
Plateau

L. Tanganyika

Lake
Nyasa

Comoro
Is.

Zambezi R.

INDONESIA

Java

Java Trench

Timor

Cape
York

Solomon Is.

New Hebrides Tr.

Vanuatu

Fiji
Is.

ATLANTIC
OCEAN

Tristan da
Cunha
Group

Cape
Plain

Walvis Ridge

Namib Desert

Kalahari
Desert

DRAKENSBERG

Cape of
Good Hope

Mozambique
Channel

Madagascar

Agulhas
Plateau

Madagascar Basin

Mauritius

Reunion

Broken Ridge

Tropic of Capricorn

Perth
Basin

Great Sandy
Desert

Western
Plateau

AUSTRALIA

Great
Victoria
Desert

Simpson
Desert

GREAT DIVIDING RANGE

Coral
Sea

New
Caledonia

Crozet
Basin

Mt. Kosciusko
7,310 ft
(2,228 m)

Tasman
Sea

North I.

NEW
ZEALAND

Southwest Indian Ridge

Southeast Indian Ridge

South
Australian
Basin

Great
Australian
Bight

Tasmania

Tasman Plain

South I.

Atlantic-Indian Ridge

Kerguelen Is.

Kerguelen
Plateau

South Indian
Basin

Campbell Plateau

America-Antarctic
Ridge

Enderby Plain

SOUTHERN OCEAN

Antarctic Circle

Weddell Plain

ANTARCTICA